Richard Brauer: Collected Papers
Volume II
Finite Groups

Mathematicians of Our Time

Gian-Carlo Rota, series editor

Richard Brauer: Collected Papers
Volume I
Theory of Algebras, and Finite Groups
edited by Paul Fong and Warren J. Wong [17]

Richard Brauer: Collected Papers
Volume II
Finite Groups
edited by Paul Fong and Warren J. Wong [18]

Richard Brauer: Collected Papers
Volume III
Finite Groups, Lie Groups, Number Theory, Polynomials and Equations; Geometry, and Biography
edited by Paul Fong and Warren J. Wong [19]

Paul Erdös: The Art of Counting
edited by Joel Spencer [5]

Einar Hille: Classical Analysis and Functional Analysis
Selected Papers of Einar Hille
edited by Robert R. Kallman [11]

Mark Kac: Probability, Number Theory, and Statistical Physics
Selected Papers
edited by K. Baclawski and M. D. Donsker [14]

Charles Loewner: Theory of Continuous Groups
notes by Harley Flanders and Murray H. Protter [1]

Percy Alexander MacMahon: Collected Papers
Volume I
Combinatorics
edited by George E. Andrews [13]

George Pólya: Collected Papers
Volume I
Singularities of Analytic Functions
edited by R. P. Boas [7]

Goerge Pólya: Collected Papers
Volume II
Location of Zeros
edited by R. P. Boas [8]

Collected Papers of Hans Rademacher
Volume I
edited by Emil Grosswald [3]

Collected Papers of Hans Rademacher
Volume II
edited by Emil Grosswald [4]

Stanislaw Ulam: Selected Works
Volume I
Sets, Numbers, and Universes
edited by W. A. Bayer, J. Mycielski, and G.-C. Rota [9]

Norbert Wiener: Collected Works
Volume I
Mathematical Philosophy and Foundations; Potential Theory; Brownian Movement, Wiener Integrals, Ergodic and Chaos Theories; Turbulence and Statistical Mechanics
edited by P. Masani [10]

Norbert Wiener: Collected Works
Volume II
Generalized Harmonic Analysis and Tauberian Theory; Classical Harmonic and Complex Analysis
edited by P. Masani [15]

Oscar Zariski: Collected Papers
Volume I
Foundations of Algebraic Geometry and Resolution of Singularities
edited by H. Hironaka and D. Mumford [2]

Oscar Zariski: Collected Papers
Volume II
Holomorphic Functions and Linear Systems
edited by M. Artin and D. Mumford [6]

Oscar Zariski: Collected Papers
Volume III
Topology of Curves and Surfaces, and Special Topics in the Theory of Algebraic Varieties
edited by M. Artin and B. Mazur [12]

Oscar Zariski: Collected Papers
Volume IV
Equisingularity on Algebraic Varieties
edited by J. Lipman and B. Teissier [16]

Note: Series number appears in brackets.

Richard Brauer

Collected Papers

Volume II
Finite Groups

Edited by
Paul Fong and Warren J. Wong

The MIT Press
Cambridge, Massachusetts, and London, England

Publisher's note: Richard Brauer left an annotated set of his publications. Following his wishes, these annotations have been incorporated by hand into these reprints.

Publication of this volume was made possible in part by a grant from the International Business Machines Corporation.

This book was printed and bound by The Murray Printing Company in the United States of America.

Library of Congress Cataloging in Publication Data

Brauer, Richard, 1901–
Collected papers.

(Mathematicians of our time; v. 17–19)
Text in English and German.
Bibliography: p.
CONTENTS: v. 1. Theory of algebras and finite groups.—v. 2. Finite groups.—v. 3. Finite groups, Lie groups, number theory, polynomials and equations; geometry, and biography.
1. Mathematics—Collected works. I. Fong, Paul, 1933– II. Wong, Warren J. III. Series.
QA3.B83 510 80–17622

ISBN 0-262-02135-8 (v. 1)
0-262-02148-X (v. 2)
0-262-02149-8 (v. 3)
0-262-02157-9 (complete set)

ISBN: 0-262-52383-3 (v. 2, paperback)

1901–1977

Contents

(Bracketed numbers are from the Bibliography in Volume I)

Volume II

Finite Groups (Continued)

Finite Groups (Continued)

Annals of Mathematics
Vol. 62, No. 3, November, 1955
Printed in U.S.A.

ON GROUPS OF EVEN ORDER

By Richard Brauer and K. A. Fowler

(Received September 14, 1954)

I. Introduction

1. Let $\mathfrak{G}$ be a group of finite order g. We prove first that if $g > 2$ is even, then there exists a proper subgroup of order $v > \sqrt[3]{g}$. The proof is quite elementary but the method cannot be applied if g is odd, though it seems probable that a similar statement holds in that case too. Indeed, if $\mathfrak{G}$ is soluble of an order $g > 1$, g not a prime, it is very easy to see that $\mathfrak{G}$ has a proper subgroup of order $v \geqq \sqrt{g}$. It is a well known unproved conjecture that all groups of odd order are soluble.

We shall use the term *involution* for a group element of order 2. If m is the total number of involutions of $\mathfrak{G}$ and if we set $n = g/m$, the same method shows that $\mathfrak{G}$ contains a normal subgroup $\mathfrak{L}$ distinct from $\mathfrak{G}$ such that the index of $\mathfrak{L}$ is either 2 or is less than $[n(n + 2)/2]!$ (where $[x]$ denotes the largest integer not exceeding the real number x). If J is an involution in $\mathfrak{G}$ and if $n(J)$ is the order of its normalizer $\mathfrak{N}(J)$ in $\mathfrak{G}$, then $n \leqq n(J)$. It then follows that there exist only a finite number of simple groups in which the normalizer of an involution is isomorphic to a given group.

2. The following terminology will be useful. An element G of a group $\mathfrak{G}$ will be said to be *real in* $\mathfrak{G}$ if G and G^{-1} are conjugate in $\mathfrak{G}$.[1] Real elements different from the identity 1 occur only in groups of even order. With G, every conjugate element is real. We may therefore speak of real and non-real classes of conjugate elements in $\mathfrak{G}$.

Let $\mathfrak{G}^{\#}$ denote the set of elements different from 1 in $\mathfrak{G}$. We introduce a "distance" $d(G, H)$ for any two elements G, H of $\mathfrak{G}^{\#}$. If $G = H$, set $d(G, H) = 0$. If $G \neq H$ and if there exists a chain $G_0, G_1, \cdots, G_l$ of elements of $\mathfrak{G}^{\#}$ with $G_0 = G$, $G_l = H$ such that G_{i-1} and G_i commute, let $d(G, H)$ denote the length l of the shortest such chain connecting G and H. In particular, $d(G, H) = 1$ if and only if G and H commute and $G \neq H$. If there does not exist a chain connecting G and H, set $d(G, H) = \infty$. It is clear that this distance has all the usual properties except that it can be infinite.

If $\mathfrak{M}$ is a subset of $\mathfrak{G}^{\#}$ and $G \in \mathfrak{G}^{\#}$, we define the distance $d(G, \mathfrak{M})$ of G from $\mathfrak{M}$ to be the minimum of the distances $d(G, H)$ for $H \in \mathfrak{M}$.

3. In Section III, it is shown that if the group $\mathfrak{G}$ contains more than one class of involutions, then any two involutions have distance at most 3. This implies that if the two elements G and H both have normalizers of even order

[1] An element G is real in $\mathfrak{G}$ if and only if every character of $\mathfrak{G}$ has a real value for G. Note that if G is not real in $\mathfrak{G}$, it may be real in groups containing $\mathfrak{G}$ as a subgroup.

in a group of this kind, then $d(G, H) \leqq 5$. The values 3 and 5 given here cannot be replaced by smaller values.

4. Section IV deals with properties of real elements G different from 1. If G has distance at least 4 from the set $\mathfrak{M}$ of involutions, this distance is infinite. Actually, the normalizer $\mathfrak{H} = \mathfrak{N}(G)$ of G in $\mathfrak{G}$ is an abelian group which is the normalizer of each of its elements $H \neq 1$. This implies that $d(H, L) = \infty$ if H is an element of $\mathfrak{H}$, $H \neq 1$, and L is an element of $\mathfrak{G}$ not in $\mathfrak{H}$. Subgroups $\mathfrak{H}$ of this type occur fairly frequently in groups of even order. They have a number of interesting properties. In particular, the order h of $\mathfrak{H}$ is relatively prime to its index. There exist involutions J which transform every element of $\mathfrak{H}$ into its inverse. If $n(J)$ is the order of the normalizer $\mathfrak{N}(J)$ of J in $\mathfrak{G}$, then $h \leqq n(J) + 1$, unless $\mathfrak{H}$ is a normal subgroup of $\mathfrak{G}$; in the latter case, $\mathfrak{G}$ "splits" into $\mathfrak{H}$ and a subgroup $\mathfrak{W}$ of $\mathfrak{N}(J)$. There exist infinitely many simple groups $\mathfrak{G}$ each of which contains a subgroup $\mathfrak{H}$ of the type here discussed with $h = n(J) + 1$.

Our results concerning real elements $G \neq 1$ of distance at most 3 from the set $\mathfrak{M}$ of involutions are rather fragmentary. It can happen that the distance from G to $\mathfrak{M}$ is actually equal to 3. In this connection, we show that, under certain conditions, some of the Sylow subgroups of $\mathfrak{G}$ are abelian. It is, of course, very easy to construct groups in which no Sylow subgroup is abelian. However, a large number of "interesting" groups seem to possess some abelian Sylow subgroups. Perhaps, in view of this, our result deserves consideration.

5. The last section deals with properties of the characters of groups of even order. If n has the same significance as in **1**, there exists an irreducible real character, not the 1-character, of a degree less than n. On the basis of this remark, one can study the cases where n is small. If the results of C. Jordan and H. F. Blichfeldt on linear groups of a given degree could be improved materially, this would make it possible to improve the results mentioned above in **1**.

If p is a prime dividing g with the exact exponent a, it may be that $\mathfrak{G}$ does not possess irreducible characters of defect 0 for p, that is, characters whose degrees are divisible by p^a. On the other hand, many "interesting" groups do have such characters. In Section V we give some sufficient conditions for the existence of characters of defect 0.

Finally, if $\mathfrak{G}$ contains a subgroup $\mathfrak{H}$ of the type discussed in **4**, rather detailed information concerning the values of the irreducible characters of $\mathfrak{G}$ for the elements of $\mathfrak{H}$ can be given. This is of great help in constructing the characters of $\mathfrak{G}$.

6. Notation. The normalizer of an element G of $\mathfrak{G}$ will be denoted by $\mathfrak{N}(G)$ and its order by $n(G)$. The set of elements X of $\mathfrak{G}$ which transform G into G or G^{-1}, that is, for which

$$X^{-1}GX = G \quad \text{or} \quad X^{-1}GX = G^{-1}$$

forms a subgroup $\mathfrak{N}^*(G)$. If G is non-real or if G is an involution, $\mathfrak{N}^*(G) = \mathfrak{N}(G)$. If G is real and of order greater than 2, $\mathfrak{N}^*(G)$ has order $2n(G)$.

The classes of conjugate elements of $\mathfrak{G}$ will be denoted by $\mathfrak{K}_0, \mathfrak{K}_1, \cdots, \mathfrak{K}_{k-1}$. Here, $\mathfrak{K}_0$ will be the class containing 1. Then the classes containing involutions are taken, say these are the classes $\mathfrak{K}_1, \cdots, \mathfrak{K}_r$. Next we take the other real classes and finally the non-real classes. Usually, G_i will denote a representative element for $\mathfrak{K}_i$. If $n_i = n(G_i)$, then $\mathfrak{K}_i$ consists of g/n_i elements.

The group ring of $\mathfrak{G}$ formed over the field of rational numbers will be denoted by Γ. With each $\mathfrak{K}_i$ we associate an element K_i of Γ. Here, K_i is the sum of the g/n_i elements of $\mathfrak{K}_i$. As is well known, the elements $\mathrm{K}_0, \mathrm{K}_1, \cdots, \mathrm{K}_{k-1}$ form a basis for the center Λ of Γ, and hence we have formulae

$$\mathrm{K}_i\mathrm{K}_j = \sum_{\mu=0}^{k-1} a_{ij\mu}\mathrm{K}_\mu . \tag{0}$$

Here, the $a_{ij\mu}$ are non-negative rational integers.

If $\mathfrak{H}$ is a subgroup of $\mathfrak{G}$, the index of $\mathfrak{H}$ in $\mathfrak{G}$ will be denoted by $(\mathfrak{G}:\mathfrak{H})$.

II. Existence of large subgroups

7. Let $\mathfrak{M}$ be the set of involutions of the group $\mathfrak{G}$ of even order. Then $\mathfrak{M}$ is the union of $\mathfrak{K}_1, \mathfrak{K}_2, \cdots, \mathfrak{K}_r$. Set

$$\mathrm{M}^{\cdot} = \mathrm{K}_1 + \mathrm{K}_2 + \cdots + \mathrm{K}_r . \tag{1}$$

Then M is that element of Λ which is the sum of all the involutions in $\mathfrak{G}$. It follows from (0) that we have formulae

$$\mathrm{M}^2 = \sum_{i=0}^{k-1} c_i\mathrm{K}_i . \tag{2}$$

Clearly, the coefficient c_i is equal to the number of ordered pairs (X, Y) of involutions X, Y such that

$$XY = G_i \qquad (X, Y \in \mathfrak{M}). \tag{3}$$

We show

Lemma (2A). *If $G_i^2 \neq 1$, then c_i is the number of involutions of $\mathfrak{G}$ which transform G_i into G_i^{-1}. If G_i is an involution, then $c_i = \nu_i - 1$ where ν_i is the number of involutions in $\mathfrak{N}(G_i)$. Finally, for $G_i = 1$, $c_i = m$.*

Proof. If X, Y satisfy the conditions (3), then

$$G_i^{-1} = Y^{-1}X^{-1} = YX = X^{-1}(XY)X = X^{-1}G_iX.$$

Conversely, if $X^{-1}G_iX = G_i^{-1}$ and $X \in \mathfrak{M}$, then the element $Y = XG_i$ satisfies the equation $Y^2 = 1$. Hence the conditions (3) are satisfied if $Y \neq 1$. We have $Y = 1$ if and only if $G_i = X$, and then G_i itself is an involution. All the statements of the lemma are now readily obtained.

Corollary (2B). *If G_i is non-real, $c_i = 0$. For any G_i, $c_i \leqq n(G_i)$. If G_i is an involution, $c_i \leqq n(G_i) - 2$.*

Proof. If G_i is not real, the lemma shows that $c_i = 0$. If G_i is real, there are exactly $n(G_i)$ elements which transform G_i into G_i^{-1}, and hence $c_i \leqq n(G_i) = n_i$. Finally, if G_i is an involution, $\nu_i \leqq n_i - 1$ and hence $c_i \leqq n_i - 2$.

6. Counting the number of group elements occurring on both sides of (2), we obtain

$$m^2 = \sum_i c_i g/n_i ,$$

where m is the number of involutions in $\mathfrak{G}$. We now apply (2A) and (2B). For each real G_i, $c_i g/n_i \leqq g$. If k_1 is the number of real classes, we obtain

$$m^2 \leqq m + \sum_{i=1}^{r} (\nu_i - 1)g/n_i + (k_1 - r - 1)g \tag{4}$$

where the last term originates from the $k_1 - r - 1$ real classes $\mathfrak{K}_i$ of elements of order larger than 2.

The number of involutions is given by

$$m = \sum_{i=1}^{r} g/n_i . \tag{5}$$

If we set

$$\nu = \text{Max}\ \{\nu_1, \nu_2, \cdots, \nu_r\}$$

we obtain

$$m^2 \leqq \nu m + (k_1 - r - 1)g. \tag{4*}$$

On the other hand, since the total number of real elements is at most g, we have

$$1 + m + \sum_{i=r+1}^{k_1-1} g/n_i \leqq g. \tag{6}$$

Thus if we set

$$h = \text{Max}\ \{n_i\} \qquad \text{for } i = r+1, r+2, \cdots, k_1 - 1$$

we have

$$(k_1 - r - 1)g \leqq h(g - m - 1).$$

This is still true for $k_1 = r + 1$, if we set $h = 0$. If we substitute this in (4*) we obtain

(2C). *Let $\mathfrak{G}$ be a group of even order g which contains m involutions. If ν is the maximal number of involutions which can occur in the normalizer of an involution, then*

$$m^2 \leqq hg + (\nu - h)m - h$$

where h is either the order of the normalizer $\mathfrak{N}(H)$ of a real element H of order greater than 2 or $h = 0$.

We now prove

THEOREM (2D). *If $\mathfrak{G}$ is a group of even order $g > 2$, there exists a subgroup $\mathfrak{V} \neq \mathfrak{G}$ of order $v > \sqrt[3]{g}$.*

PROOF. If we set $m = g/n$, then (2C) yields

$$g \leqq hn^2 + (\nu - h)n = hn(n-1) + \nu n.$$

One of the groups $\mathfrak{N}(G_i)$ with $1 \leqq i \leqq r$ contains ν involutions and hence $\nu \leqq n_i - 1$ for some such i. On the other hand, (5) yields

$$n^{-1} = n_1^{-1} + n_2^{-1} + \cdots + n_r^{-1} \tag{5*}$$

and hence $n \leqq n_i$. It now follows that

$$g \leqq hn_i(n_i - 1) + (n_i - 1)n_i = (h + 1)n_i(n_i - 1).$$

If $n_i \leqq h$, then $g < h^3$. Now h is the order of a subgroup $\mathfrak{N}(H)$ and, since H is conjugate to $H^{-1} \neq H$ in $\mathfrak{G}$, we have $\mathfrak{N}(H) \neq \mathfrak{G}$. Hence the theorem holds in this case.

If $h < n_i$, we have $g < n_i^3$. If $n_i \neq g$, the theorem is true with $\mathfrak{V} = \mathfrak{N}(G_i)$.

It remains to deal with the case $n_i = g$. Then $\mathfrak{G}$ contains an invariant involution G_i. We may assume that the theorem has been proved for groups of even order less than g. If $g/2$ is even and $g \neq 4$, we may apply the theorem to $\mathfrak{G}/\{G_i\}$. It follows that $\mathfrak{G}/\{G_i\}$ contains a subgroup $\mathfrak{V}^* \neq \mathfrak{G}/\{G_i\}$ of an order $v^* > \sqrt[3]{g/2}$. Since we may set $\mathfrak{V}^* = \mathfrak{V}/\{G_i\}$ where $\mathfrak{V}$ is a subgroup of order $v = 2v^*$ of $\mathfrak{G}$, we have

$$g > v > \sqrt[3]{4g} > \sqrt[3]{g}.$$

Again the theorem holds. It is trivial for $g = 4$.

Finally, if $g/2$ is odd, it is well known[2] that $\mathfrak{G}$ contains a subgroup of order $g/2$ and $g/2 > \sqrt[3]{g}$. Hence the theorem holds in all cases.

(2E). *If $\mathfrak{G}$ does not contain invariant involutions, the group $\mathfrak{V}$ in* (2D) *can be chosen as the normalizer of a suitable real element of $\mathfrak{G}$.*

9. We use a method very similar to that in **8** in order to obtain a slightly stronger result.

Some of the classes $\mathfrak{K}_1, \mathfrak{K}_2, \cdots, \mathfrak{K}_r$ may consist of invariant involutions. Assume that the notation is chosen such that these are the classes $\mathfrak{K}_i$ with $i = 1, 2, \cdots, r'$. If there is no invariant involution, set $r' = 0$. Since we have $n_i = g$, $\mathfrak{N}(G_i) = \mathfrak{G}$ for $i = 1, 2, \cdots, r'$, we have $c_i = \nu_i - 1 = m - 1$ for $1 \leqq i \leqq r'$. Now (4) becomes

$$m^2 \leqq m + r'(m - 1) + \textstyle\sum_{i=r'+1}^{r} (n_i - 2)g/n_i + (k_1 - r - 1)g.$$

The formula (5) can be written in the form

$$m = r' + \textstyle\sum_{i=r'+1}^{r} g/n_i$$

and we obtain

$$m^2 \leqq m + r'(m - 1) + (r - r')g - 2(m - r') + (k_1 - r - 1)g$$

and hence

$$m^2 \leqq (r' - 1)m + r' + (r - r')g + (k_1 - r - 1)g. \tag{7}$$

[2] For instance, this is a simple consequence of a Theorem of Burnside. See [3], p. 133.

Let u denote the minimal index of all the subgroups of $\mathfrak{G}$ which are distinct from $\mathfrak{G}$, and assume that $u > 2$. Since $n_i < g$ for $i = r' + 1, \cdots, r$, we have $u \leqq g/n_i$ for these i, and (5*) yields

$$(r - r')u \leqq m - r'.$$

For $i = r + 1, r + 2, \cdots, k_1 - 1$, let $\mathfrak{N}^*(G_i)$ denote the group of order $2n_i$ introduced in **6**. Since $\mathfrak{N}^*(G_i)$ has a subgroup $\mathfrak{N}(G_i)$ of index 2 and we assumed $u > 2$, we must have $\mathfrak{N}^*(G_i) \neq \mathfrak{G}$. It follows that u cannot exceed the index $g/2n_i$ of $\mathfrak{N}^*(G_i)$ in $\mathfrak{G}$. Now (6) implies

$$2u(k_1 - r - 1) \leqq g - 1 - m.$$

If we combine the last two inequalities with (7), we obtain

$$m^2 \leqq (r' - 1)m + r' + (m - r')g/u + g(g - 1 - m)/2u.$$

Again set $m = g/n$. An easy computation yields

$$(8) \qquad \begin{aligned} g &< (r' - 1)n + gn/2u + gn^2/2u, \\ g - gn(n + 1)/2u &< (r' - 1)n. \end{aligned}$$

In particular, if $r' \leqq 1$ we must have $n(n + 1)/2u > 1$, that is, $u < n(n + 1)/2$.

Suppose that $r' > 1$ and assume that $u \geqq n(n + 2)/2$, $u \neq 2$. Then (8) yields

$$(r' - 1)n > gn/2u$$

and hence

$$g < 2u(r' - 1).$$

The r' invariant involutions together with 1 form a subgroup $\mathfrak{C}$ of order $r' + 1$ of the center of $\mathfrak{G}$. Since any subgroup of $\mathfrak{G}/\mathfrak{C}$ different from $\mathfrak{G}/\mathfrak{C}$ has an index at least u, the order of such a subgroup is not greater than $g/u(r' + 1)$, and

$$g/u(r' + 1) < g/u(r' - 1) < 2.$$

It follows that $\mathfrak{G}/\mathfrak{C}$ is certainly cyclic. Hence $\mathfrak{G}$ is generated by $\mathfrak{C}$ and at most one additional element. Since $\mathfrak{C}$ lies in the center of $\mathfrak{G}$, it follows that $\mathfrak{G}$ is abelian. But then $\mathfrak{G}$ has a subgroup of index 2, whereas we assumed $u > 2$. Thus the assumptions $u > 2$, $u \geqq n(n + 2)/2$ lead to a contradiction. This proves the theorem:

THEOREM (2F). *Let $\mathfrak{G}$ be a group of even order g which contains exactly m involutions. If $n = g/m$, then $\mathfrak{G}$ contains a subgroup of index u such that either $u = 2$ or*

$$1 < u < n(n + 2)/2.$$

If $\mathfrak{G}$ contains at most one invariant involution, the number $n(n + 2)/2$ can be replaced by $n(n + 1)/2$.

We now are able to obtain a slight improvement of (2D). We state:

COROLLARY (2G). *If $\mathfrak{G}$ is a group of even order $g > 2$, there exists a subgroup $\mathfrak{V} \neq \mathfrak{G}$ of order $v > \sqrt[3]{2g} - 1/3$.*

PROOF. If $\mathfrak{G}$ contains a subgroup of index 2, the statement is true, since $g/2 \geqq \sqrt[3]{2g}$ for $g \geqq 4$. Assume that $\mathfrak{G}$ does not have subgroups of index 2.

Suppose first that $\mathfrak{G}$ does not contain invariant involutions. If $\mathfrak{V}$ is a subgroup of $\mathfrak{G}$ of maximal order $v < g$, Theorem (2F) shows that $g/v < n(n+1)/2$. It follows from (5*) that $n \leqq n_1$, and since n_1 is the order of a subgroup $\mathfrak{N}(G_1) \neq \mathfrak{G}$, we have $n_1 \leqq v$. Thus $2g < v^2(v+1)$. One easily sees that this implies $v > \sqrt[3]{2g} - 1/3$.

The case in which $\mathfrak{G}$ contains invariant involutions can be treated in a manner similar to that used in the proof of (2D).

THEOREM (2H). *If $\mathfrak{G}$ is a group of even order g which contains m involutions and if $n = g/m$, then there exists a normal subgroup $\mathfrak{L} \neq \mathfrak{G}$ of $\mathfrak{G}$ such that $\mathfrak{G}/\mathfrak{L}$ is isomorphic to a subgroup of the symmetric group on u letters with $u = 2$ or $u < n(n+2)/2$. In particular, $(\mathfrak{G}:\mathfrak{L}) = 2$ or $(\mathfrak{G}:\mathfrak{L}) < [n(n+2)/2]!$ If $\mathfrak{G}$ contains at most one invariant involution, the number $n(n+2)/2$ can be replaced by $n(n+1)/2$.*

PROOF. If $\mathfrak{V}$ is the subgroup mentioned in (2F), there corresponds to $\mathfrak{V}$ a permutation representation of $\mathfrak{G}$ of degree u. We obtain (2H) by taking for $\mathfrak{L}$ the kernel of this representation.

COROLLARY (2I). *If $\mathfrak{G}$ is a simple group of even order $g > 2$ which contains m involutions and if $n = g/m$, then*

$$g < [n(n+1)/2]!$$

If J is an involution of $\mathfrak{G}$, then n in the inequality can be replaced by $n(J)$. There exist only a finite number of simple groups in which the normalizer of an involution is isomorphic to a given group.

PROOF. The first statement follows at once from (2H), since a simple group of even order $g > 2$ cannot contain an invariant involution. The second statement then follows from (5*), which implies $n \leqq n(J)$.

10. For a later application, we mention still another result which is obtained by the method used above.

THEOREM (2J). *If $\mathfrak{G}$ is a group of even order g which contains exactly m involutions, then the number k_1 of real classes of $\mathfrak{G}$ satisfies the inequality*

$$k_1 - 1 \geqq m(m+1)/g.$$

PROOF. In (4), ν_i can be replaced by $n_i - 1$. Then

$$m^2 \leqq m + rg - 2\textstyle\sum_{i=1}^{r} g/n_i + (k_1 - r - 1)g.$$

This, together with (5), yields

$$m^2 \leqq -m + (k_1 - 1)g.$$

III. Groups with more than one class of involutions

11. We now prove

LEMMA (3A). *If the two involutions X and Y of $\mathfrak{G}$ are not conjugate in $\mathfrak{G}$, there exists an involution J which commutes with X and Y.*

PROOF. Set $G = XY$. As we have seen in the proof of (2A), both X and Y transform G into G^{-1}. Then G is real, and X and Y are elements of the group $\mathfrak{N}^*(G)$ introduced in Section **6**. If $G = 1$, then $X = Y$. If G has order 2, then G is an involution which commutes with X and Y. Thus we may suppose that G has order greater than 2. Then $\mathfrak{N}^*(G)$ has order $2n(G)$. If $n(G)$ were odd, both $\{X\}$ and $\{Y\}$ would be 2-Sylow groups of $\mathfrak{N}^*(G)$. Then X and Y would be conjugate in $\mathfrak{N}^*(G)$ and hence in $\mathfrak{G}$, contrary to the hypothesis. Hence $n(G)$ is even. Let $\mathfrak{P}^*$ be a 2-Sylow group of $\mathfrak{N}^*(G)$ which contains X. Since $X \notin \mathfrak{N}(G)$, the intersection $\mathfrak{P} = \mathfrak{P}^* \cap \mathfrak{N}(G)$ must have index 2 in $\mathfrak{P}^*$, and $\mathfrak{P}$ is a 2-Sylow group of $\mathfrak{N}(G)$. The order of $\mathfrak{P}$ is at least 2, and since $\mathfrak{P}$ is normal in $\mathfrak{P}^*$ we can find a normal subgroup $\{J\}$ of order 2 of $\mathfrak{P}^*$ such that $\{J\} \subseteq \mathfrak{P}$. Then J is an involution which commutes with X and G and hence with $Y = XG$.

We now prove

THEOREM (3B). *Let $\mathfrak{G}$ be a group of order g which contains $r \geqq 2$ classes of involutions $\mathfrak{K}_1, \mathfrak{K}_2, \cdots, \mathfrak{K}_r$. For $G_i \in \mathfrak{K}_i$, let n_i denote the order of the normalizer $\mathfrak{N}(G_i)$ of G_i, and let ν_i denote the number of involutions in $\mathfrak{N}(G_i)$. If ν is the maximum of the ν_i, then*

$$g \leqq (\nu - 2)(\nu_1 - 1)/(n_2^{-1} + n_3^{-1} + \cdots + n_r^{-1}).$$

PROOF. Let J range over the $\nu_1 - 1$ involutions of $\mathfrak{N}(G_1)$ distinct from G_1, and for each J denote by $\mathfrak{A}(J)$ the set of involutions of $\mathfrak{N}(J)$ which do not commute with G_1. For a fixed J, $\mathfrak{A}(J)$ contains at most $\nu - 3$ elements, since G_1, J, and G_1J are distinct involutions in $\mathfrak{N}(J)$ which commute with G_1. It follows from (3A) that each involution not in $\mathfrak{K}_1$ and not in $\mathfrak{N}(G_1)$ must lie in one of the sets $\mathfrak{A}(J)$. If $\mathfrak{U}$ denotes the union of the classes $\mathfrak{K}_2, \mathfrak{K}_3, \cdots, \mathfrak{K}_r$, then at most $(\nu_1 - 1)(\nu - 3)$ elements of $\mathfrak{U}$ do not commute with G_1. Since at most $\nu_1 - 1$ elements of $\mathfrak{U}$ do commute with G_1, and since $\mathfrak{U}$ contains exactly $g(n_2^{-1} + \cdots + n_r^{-1})$ elements, we obtain

$$g(n_2^{-1} + \cdots + n_r^{-1}) \leqq (\nu - 2)(\nu_1 - 1).$$

This proves the theorem.

COROLLARY (3C). *If the notation is the same as in* (3B), *and if the notation is chosen such that $n_1 \leqq n_2 \leqq \cdots \leqq n_r$, then*

$$g < n_1 n_2 n_r ; \qquad (r - 1)g < n_1 n_r \sqrt[r-1]{n_2 n_3 \cdots n_r} \leqq n_1 n_r^2 .$$

Indeed, since $\nu_i \leqq n_i - 1$, we have

$$(\nu - 2)(\nu_1 - 1) < n_1 n_r .$$

On the other hand

$$n_2^{-1} + \cdots + n_r^{-1} \geqq n_2^{-1}$$

and

$$n_2^{-1} + \cdots + n_r^{-1} \geqq (r-1)\sqrt[r-1]{(n_2 \cdots n_r)^{-1}} \geqq (r-1)n_r^{-1}.$$

In particular, we have $n_r > \sqrt[3]{(r-1)g}$. This shows that if $\mathfrak{G}$ does not contain invariant involutions and if $r \geqq 2$, then the group $\mathfrak{B}$ in (2D) can be chosen as the normalizer of an involution. For $r \geqq 3$, we obtain an improvement of (2G).

12. Using the terminology introduced in the introduction, we state

THEOREM (3D). *If $\mathfrak{G}$ contains more than one class of involutions, then any two involutions of $\mathfrak{G}$ have distance at most* 3.

PROOF. If the two involutions X and Z belong to different classes, it follows from (3A) that $d(X, Z) \leqq 2$. Suppose then that X, Z belong to the same class $\mathfrak{K}_i$. It follows from (3A) that some element X_1 of $\mathfrak{K}_i$ must commute with an involution Y not in $\mathfrak{K}_i$. After replacing X_1 and Y by conjugates, we may assume X equals X_1. Since Z and Y belong to different classes, we have $d(Z, Y) \leqq 2$. On the other hand, $d(X, Y) = 1$. Hence $d(X, Z) \leqq 3$.

COROLLARY (3E). *If $\mathfrak{G}$ contains more than one class of involutions, then any two elements G_1 and G_2 with even $n(G_1)$, $n(G_2)$ have distance at most* 5.

Indeed, if $n(G_i)$ is even, there exists an involution X_i which commutes with G_i. Hence

$$d(G_1, X_1) \leqq 1, \qquad d(X_1, X_2) \leqq 3, \qquad d(X_2, G_2) \leqq 1.$$

It follows that $d(G_1, G_2) \leqq 5$.

REMARK. There exist groups with more than one class of involutions in which involutions of distance 3 occur. For instance, let $\mathfrak{G}$ be the symmetric group on p letters, where p is a prime and $p \geqq 5$. If G is a cycle of length p, there exists an involution X which transforms G into G^{-1}. Then $Y = XG$ also is an involution. If an element Z commutes with both X and Y, then Z would commute with G and hence Z would be a power of G. The only power of G which commutes with an involution is 1. Hence $Z = 1$ and this shows that $d(X, Y) > 2$.

A similar argument can be used to prove

(3F). *If $\mathfrak{G}$ is a group of even order which contains a real element G such that $n(H)$ is odd for every H different from 1 in $\mathfrak{N}(G)$, then $\mathfrak{G}$ contains involutions which have distance greater than* 2.

One can also show by examples that the number 5 in (3E) cannot be replaced by a smaller value.

IV. The set of real elements

13. We prove

LEMMA (4A). *If G is a real element of the group $\mathfrak{G}$ of even order and if $n(G)$ is odd, then there exists an involution J which transforms G into G^{-1}. All involutions which transform G into G^{-1} are conjugate in $\mathfrak{G}$, and the number of such involutions is equal to the index of $\mathfrak{N}(G) \cap \mathfrak{N}(J)$ in $\mathfrak{N}(G)$.*

PROOF. Since $n(G)$ is odd, G is not an involution. Then the group $\mathfrak{N}^*(G)$ has even order $2n(G)$ and therefore it contains an involution J. Since $n(G)$ is odd, J cannot transform G into G. Hence J transforms G into G^{-1}.

If X is any involution such that $X^{-1}GX = G^{-1}$, one sees as in the proof of (3A) that X is conjugate to J in $\mathfrak{N}^*(G)$. The number of elements in the class of J in $\mathfrak{N}^*(G)$ is equal to the index of $\mathfrak{N}^*(G) \cap \mathfrak{N}(J)$ in $\mathfrak{N}^*(G)$, and every element in this class is an involution which transforms G into G^{-1}. Since J does not belong to the subgroup $\mathfrak{N}(G)$ of index 2 of $\mathfrak{N}^*(G)$, it follows that $\mathfrak{N}(G) \cap \mathfrak{N}(J)$ has index 2 in $\mathfrak{N}^*(G) \cap \mathfrak{N}(J)$. Hence the index of $\mathfrak{N}(G) \cap \mathfrak{N}(J)$ in $\mathfrak{N}(G)$ is equal to the index of $\mathfrak{N}^*(G) \cap \mathfrak{N}(J)$ in $\mathfrak{N}^*(G)$. This completes the proof.

The following theorem is essentially a restatement of a result due to Burnside, [2], pp. 229, 230.

THEOREM (4B). *Let $\mathfrak{H}$ be a subgroup of $\mathfrak{G}$. If there exists an involution J in the normalizer of $\mathfrak{H}$, and if J commutes with no element of $\mathfrak{H}$ different from* 1, *then $\mathfrak{H}$ is abelian of odd order and J transforms every element of $\mathfrak{H}$ into its inverse.*

PROOF. Every element of $\mathfrak{H}$ can be written in the form $JH^{-1}JH$ for $H \in \mathfrak{H}$. Hence J transforms every element of $\mathfrak{H}$ into its inverse. This implies that $\mathfrak{H}$ cannot contain an involution. For $H, K \in \mathfrak{H}$,

$$KH = J(H^{-1}K^{-1})J = (JH^{-1}J)(JK^{-1}J) = HK,$$

and hence $\mathfrak{H}$ is abelian.

As an immediate consequence of (4B), we have

COROLLARY (4C). *Assume that $G \neq 1$ is an element of $\mathfrak{G}$ which is transformed into its inverse by the involution J. If $d(G, J) \geqq 3$, then $\mathfrak{N}(G)$ is abelian of odd order and J transforms each element of $\mathfrak{N}(G)$ into its inverse.*

We now prove

THEOREM (4D). *Assume that $d(G, J) \geqq 4$ in (4C). Then $\mathfrak{N}(G)$ is the normalizer of each of its elements different from* 1; *and $d(H, Z) = \infty$ for $H \in \mathfrak{N}(G)$, $Z \notin \mathfrak{N}(G)$, $H \neq 1$.*

PROOF. It follows from (4C) that $\mathfrak{N}(G)$ is abelian. If $H \in \mathfrak{N}(G)$, then $\mathfrak{N}(G) \subseteq \mathfrak{N}(H)$. For $H \neq 1$, we have $d(H, J) \geqq 3$ and hence (4C) can be applied to H instead of G: Since $\mathfrak{N}(H)$ is abelian and $G \in \mathfrak{N}(H)$, we have $\mathfrak{N}(H) \subseteq \mathfrak{N}(G)$ and hence $\mathfrak{N}(G) = \mathfrak{N}(H)$.

It is now clear that any element $H \neq 1$ of $\mathfrak{N}(G)$ has distance at most 1 from every element different from 1 of $\mathfrak{N}(G)$ and distance ∞ from every element not in $\mathfrak{N}(G)$.

COROLLARY (4E). *If a real element $G \neq 1$ has distance at least 4 from the set $\mathfrak{M}$ of involutions, then $d(G, \mathfrak{M}) = \infty$ and $\mathfrak{N}(G)$ is the normalizer of each $H \neq 1$ in $\mathfrak{N}(G)$.*

Indeed, (4A) shows that there exists an involution J such that $J^{-1}GJ = G^{-1}$. Then (4C) and (4D) apply. Since $n(G)$ is odd, all involutions lie outside $\mathfrak{N}(G)$.

14. We consider a subgroup $\mathfrak{H}$ of an arbitrary group $\mathfrak{G}$ of finite order g such that $\mathfrak{H}$ is the normalizer of each of its elements different from 1. Our results will apply to the subgroup $\mathfrak{N}(G)$ in (4D).

It is clear that $\mathfrak{H}$ will be abelian. If p is a prime dividing the order h of $\mathfrak{H}$, there exists an element P of order p in $\mathfrak{H}$. Let $\mathfrak{P}$ be a p-Sylow group of $\mathfrak{G}$ which contains P, and let $P_0 \neq 1$ be an element of the center of $\mathfrak{P}$. Then $P_0 \in \mathfrak{N}(P) = \mathfrak{H}$ and hence $\mathfrak{N}(P_0) = \mathfrak{H}$. Since $\mathfrak{P} \subseteq \mathfrak{N}(P_0) = \mathfrak{H}$, it follows that p is prime to the index g/h of $\mathfrak{H}$.

Let $\mathfrak{N}$ denote the normalizer of $\mathfrak{H}$ in $\mathfrak{G}$. Since $\mathfrak{H}$ is a normal subgroup of $\mathfrak{N}$ and since h is relatively prime to $(\mathfrak{N}:\mathfrak{H})$, there exists a subgroup $\mathfrak{W}$ such that ([3], p. 125)

$$\mathfrak{N} = \mathfrak{H}\mathfrak{W}, \qquad \mathfrak{H} \cap \mathfrak{W} = \{1\}.$$

If $\mathfrak{W}$ has order w, then $\mathfrak{N}$ has order hw.

In $\mathfrak{N}$, the $h - 1$ elements $H \neq 1$ of $\mathfrak{H}$ have normalizers of order h. Hence they are distributed into $(h - 1)/w$ classes of conjugate elements each consisting of w elements. In particular, w divides $h - 1$. Actually, if p^a is the highest power of a prime p dividing h, then w divides $p^a - 1$. This is seen by considering a Sylow group of $\mathfrak{H}$.

If an element A of $\mathfrak{G}$ transforms an element $H \neq 1$ of $\mathfrak{H}$ into an element K of $\mathfrak{H}$, we have $A^{-1}\mathfrak{N}(H)A = \mathfrak{N}(K)$, that is, $A^{-1}\mathfrak{H}A = \mathfrak{H}$ and hence A lies in $\mathfrak{N}$.

No two distinct conjugates of $\mathfrak{H}$ can have an intersection different from 1, since each conjugate of $\mathfrak{H}$ is the normalizer of each of its elements different from 1. Now the arguments leading to Sylow's theorem show that the number of conjugates is congruent to 1 modulo h. If we denote this number by $1 + Nh$, where N is a rational integer, then $g/(hw) = 1 + Nh$ and hence $g = wh(1 + Nh)$.

We have proved

THEOREM (4F). *If $\mathfrak{G}$ is a group of finite order g and $\mathfrak{H}$ is a subgroup such that $\mathfrak{H}$ is the normalizer of each of its elements different from* 1, *then $\mathfrak{H}$ is abelian and its order h is relatively prime to its index g/h. We can set*

$$g = hw(1 + Nh); \qquad h - 1 = wt, \tag{9}$$

where t, w, and N are rational integers, $t \geqq 0$, $w \geqq 1$, $N \geqq 0$. The normalizer $\mathfrak{N}$ of $\mathfrak{H}$ has order hw and there exists a subgroup $\mathfrak{W}$ of order w such that

$$\mathfrak{N} = \mathfrak{H}\mathfrak{W}; \qquad \mathfrak{H} \cap \mathfrak{W} = \{1\}. \tag{10}$$

Each element of $\mathfrak{H}$ different from 1 *is conjugate in $\mathfrak{G}$ to exactly w elements of $\mathfrak{H}$ and any two of these w elements are conjugate in $\mathfrak{N}$.*

15. We now take for $\mathfrak{H}$ the group $\mathfrak{N}(G)$ in (4D). Since J maps each $H \in \mathfrak{H}$ on its inverse, we have $J \in \mathfrak{N}$. After replacing $\mathfrak{W}$ by a conjugate group in $\mathfrak{N}$, we may assume $J \in \mathfrak{W}$. Let H be a fixed element of $\mathfrak{H}$ different from 1. Since $\mathfrak{N}(H) = \mathfrak{H}$, only the elements of $\mathfrak{H}J$ will transform H into H^{-1}. On the other hand, for $W \in \mathfrak{W}$, the element $W^{-1}JW$ transforms each element of $\mathfrak{H}$ into its inverse. Hence we have $W^{-1}JW = H_0J$ with $H_0 \in \mathfrak{H}$. Since J, $W \in \mathfrak{W}$, it follows that $H_0 \in \mathfrak{W}$. Now (10) shows that $H_0 = 1$ and hence that $W^{-1}JW = J$. We now have

THEOREM (4G). *If $\mathfrak{G}$ is a group of even order g and G is a real element different from 1 which has distance at least 4 (and hence distance ∞) from the set of involutions, the results of (4F) apply to $\mathfrak{H} = \mathfrak{N}(G)$. If J is an involution which transforms G into G^{-1} and hence every element of $\mathfrak{H}$ into its inverse, we may choose $\mathfrak{W}$ in (10) such that $J \in \mathfrak{W}$ and $\mathfrak{W} \subseteq \mathfrak{N}(J)$. The number w is even.*

16. Let $H \in \mathfrak{H}$, $H \neq 1$. Our results show that there exist exactly h involutions which transform H into H^{-1}. It then follows from (2A) and its proof that there exist exactly h ordered pairs (X, Y) of involutions with the product H; moreover, if (X, Y) is any such pair, then both X and Y transform H into H^{-1}. Then, by (4A), X and Y belong to the same class of $\mathfrak{G}$ as J does. Denote this class by $\mathfrak{K}_1$. We now see that there are exactly h ordered pairs of elements of $\mathfrak{K}_1$ with the product H.

From (0) we have

$$\text{(11)} \qquad \mathrm{K}_1^2 = \sum_{i=0}^{k-1} a_{11i}\mathrm{K}_i .$$

If $H \in \mathfrak{K}_\lambda$, then $a_{11\lambda}$ is equal to the number of ordered pairs of elements of $\mathfrak{K}_1$ with the product H. Hence $a_{11\lambda} = h$.

The number of elements in $\mathfrak{K}_1$ is g/n_1. It follows easily that $a_{110} = g/n_1$. Counting the number of group elements occurring on both sides of (11), we obtain

$$\text{(12)} \qquad (g/n_1)^2 = g/n_1 + \sum_{i=1}^{k-1} a_{11i}g/n_i .$$

For the class $\mathfrak{K}_\lambda$, we have $a_{11\lambda} = h$, $n_\lambda = h$, and thus the term in (12) for $i = \lambda$ is g. There are exactly $(h - 1)/w = t$ classes $\mathfrak{K}_\lambda$ which contain elements of $\mathfrak{H}$ different from 1.

It may happen that there are several non-conjugate groups of the same type as $\mathfrak{H}$ whose elements are transformed into their inverses by the same involution J. Let us denote these groups by $\mathfrak{H}_i$ $(i = 1, 2, \cdots, s)$ and let t_i have the same significance for $\mathfrak{H}_i$ as t had for $\mathfrak{H}$. Then the elements of $\mathfrak{H}_i$ different from 1 will contribute t_i terms g to the sum in (12) and different $\mathfrak{H}_i$ must contribute different terms. Hence

$$g^2/n_1^2 \geqq g/n_1 + g\sum_{i=1}^{s} t_i , \qquad g \geqq n_1 + n_1^2\sum t_i .$$

We have shown

(4H). *Suppose that J is an involution of $\mathfrak{G}$ and that there exist elements $X_1, X_2, \cdots, X_s$ each of which is transformed into its inverse by J, each of which has distance at least 4 from J, and which are such that no element of the class of X_i commutes with X_j for $i \neq j$. If the elements different from 1 of $\mathfrak{H}_i = \mathfrak{N}(X_i)$ belong to t_i different classes in $\mathfrak{G}$, then*

$$\text{(13)} \qquad g \geqq n(J) + n(J)^2\sum_{i=1}^{s} t_i .$$

17. Again let $\mathfrak{H}$ be the group mentioned in (4G), and consider the h sets $H\mathfrak{M}$ with $H \in \mathfrak{H}$. If two such sets $H_1\mathfrak{M}$ and $H_2\mathfrak{M}$ with $H_1, H_2 \in \mathfrak{H}$, $H_1 \neq H_2$, have an element D in common, there exist involutions J_1 and J_2 such that $D = H_1J_1 = H_2J_2$, $H_1^{-1}H_2 = J_1J_2$. Then, as we have seen in the proof of (2A), J_1 transforms $H_1^{-1}H_2$ into its inverse. Hence $J_1 \in \mathfrak{H}J$, $H_1J_1 \in \mathfrak{H}J$, and hence $H_1\mathfrak{M} \cap H_2\mathfrak{M} \subseteq \mathfrak{H}J$. Conversely, if H and H_0 are any two elements of $\mathfrak{H}$, then $J^{-1}H^{-1}H_0J = H_0^{-1}H$. It follows that $(H^{-1}H_0J)^2 = 1$. Since J does not lie in $\mathfrak{H}$, $H^{-1}H_0J \in \mathfrak{M}$. This shows that $\mathfrak{H}J$ is contained in each of the sets $H\mathfrak{M}$. Thus, $H_1\mathfrak{M} \cap H_2\mathfrak{M} = \mathfrak{H}J$.

If m again denotes the number of involutions in $\mathfrak{G}$, then each $H\mathfrak{M}$ contains the h elements of $\mathfrak{H}J$ and $m - h$ elements which do not appear in any of the other sets $H_1\mathfrak{M}$. The number of elements in the union of the sets $H\mathfrak{M}$ is therefore $h + h(m - h)$. No element of $\mathfrak{H}$ can appear in a set $H\mathfrak{M}$ since $\mathfrak{H}$ has odd order and cannot contain an involution. Since we have g elements in $\mathfrak{G}$, at most $g - h$ distinct elements can lie in the union of the sets $H\mathfrak{M}$. Hence

$$h + h(m - h) \leqq g - h.$$

This yields $mh - g \leqq h(h - 2)$. If we again set $m = g/n$, we obtain

$$g(h - n) \leqq h(h - 2)n.$$

(4I). *Let h be the order of the group $\mathfrak{H}$ in* (4G), *and let m be the number of involutions in $\mathfrak{G}$. If $h > g/m = n$, then*

$$g \leqq \frac{h(h-2)n}{h-n}. \tag{14}$$

18. Set $n(J) = n_1$. Since w divides n_1 (cf. (4G)), we can set $n_1 = wz$, where z is a positive rational integer. On the other hand, n_1 divides g/h. Indeed, if this were not so, there would exist an element P in $\mathfrak{N}(J)$ of a prime order dividing h. Some conjugate P_1 of P then belongs to $\mathfrak{H}$ and hence $n(P_1) = h$. This implies that $n(P) = h$. However, since $J \in \mathfrak{N}(P)$, $n(P)$ must be even and we have a contradiction. Thus n_1 divides g/h. Since $n_1 = wz$, it follows from (9) that z divides $1 + Nh$. Set $1 + Nh = zx$, where x is a positive rational integer. Then

$$\begin{gathered} g = hwxz; \\ h - 1 = wt, \qquad 1 + Nh = xz, \qquad n_1 = wz. \end{gathered} \tag{15}$$

Suppose that $h > n_1$. It then follows from (5*) that $h > n$ and hence (14) holds. Since $n(h - n_1) \leqq n_1(h - n)$,

$$g \leqq \frac{h(h-2)n_1}{h-n_1}. \tag{16}$$

Set $v = h - n_1 > 0$. We then have

$$xv = xh - xn_1 = xh - xwz = -w + h(x - wN).$$

It follows from (15) and (16) that $x \leqq (h - 2)/v$. Thus, we must have

$$x - wN = 1; \qquad xv = -w + h. \tag{17}$$

Since $wz = n_1 = h - v$, the equation (15) becomes

(18) $$g = hn_1(h - w)/(h - n_1) = h(h - v)(h - w)/v.$$

On the other hand, (13) shows that

$$n_1 + n_1^2 t \leqq g.$$

This in conjunction with (18) yields

(19) $$(1 + n_1 t)(h - n_1) \leqq h(h - w).$$

Multiply both members of (19) by w and recall that $wt = h - 1$. An easy computation yields

(20) $$h^2(n_1 - w) - h(n_1^2 + n_1 - w^2) + (w + n_1^2) \leqq 0.$$

If $w = n_1$, then (15) shows that $z = 1$, $x = 1 + Nh$, and since (17) implies $wN = hN$, where $w \leqq h - 1$, we must have $N = 0$. Then $\mathfrak{H}$ is normal in $\mathfrak{G}$. If this case is excluded, then by (15) $n_1 > w$, $z > 1$. If we had $h \geqq (n_1^2 + n_1 - w^2)/(n_1 - w)$, the left hand side of (20) would be positive, which is impossible. Hence

$$h < (n_1^2 + n_1 - w^2)/(n_1 - w) = n_1 + w + n_1/(n_1 - w).$$

Now $n_1/(n_1 - w) = wz/(wz - w) = z/(z - 1) \leqq 2$. Thus $h \leqq n_1 + w + 1$. By (15), $h \equiv 1 \equiv 1 + n_1 \pmod{w}$, and since we assumed $h > n_1$, it follows that $h = 1 + n_1$ or $h = 1 + n_1 + w$. In the latter case, (19) becomes

$$(1 + n_1 t)(1 + w) \leqq h(1 + n_1) = (tw + 1)(1 + n_1),$$

and then $n_1 t + w \leqq wt + n_1$, $n_1(t - 1) \leqq w(t - 1)$. Since $n_1 > w$, we must have $t = 1$. Then $h - 1 = w$, and since $h = 1 + n_1 + w$, we obtain $n_1 = 0$, a contradiction. It follows that we must have $h = 1 + n_1$. This yields the result

THEOREM (4J). *Let $\mathfrak{G}$ be a group of even order g which contains a real element G different from 1 and with distance at least 4 from the set of involutions. Let J be any involution which transforms G into G^{-1}. If the group $\mathfrak{N}(G)$ is not normal in $\mathfrak{G}$, then its order h is at most $n(J) + 1$. The case $h = n(J) + 1$ occurs only if $g = h(h - 1)(h - w)$, where w is the order of the group $\mathfrak{W}$ mentioned in* (4G).

There exist infinitely many groups in which the case $h = n(J) + 1$ occurs. If $\mathfrak{G}$ is the group $LF(2, 2^a)$ of order $g = (2^a + 1)2^a(2^a - 1)$, there exists only one class of involutions, $r = 1$, $n = n_1 = 2^a$, and there exists a subgroup $\mathfrak{H}$ of the type here discussed with $h = 2^a + 1$. Here, $w = 2$, $t = (h - 1)/2 = 2^{a-1}$.

19. We conclude this section with a few simple remarks:

(4K). *If $\mathfrak{M}_0$ is a set consisting of m_0 involutions of $\mathfrak{G}$, any subgroup $\mathfrak{L}$ of order $l > g/(m_0 + 1)$ contains an element L_0 different from 1 which is transformed into its inverse by some element J of $\mathfrak{M}_0$.*

PROOF. If $\mathfrak{L}$ contains an element J of $\mathfrak{M}_0$, we may take $L_0 = J$ and we have $J^{-1}L_0J = L_0^{-1}$. Assume then that $\mathfrak{L}$ and $\mathfrak{M}_0$ are disjoint.

If two sets $L_1\mathfrak{M}_0$ and $L_2\mathfrak{M}_0$, with $L_1, L_2 \in \mathfrak{L}$, $L_1 \neq L_2$, are not disjoint, there exist involutions $J_1, J_2 \in \mathfrak{M}_0$ such that $L_1J_1 = L_2J_2$. Then $L_1^{-1}L_2 = J_1J_2$ and J_1 transforms the element $L_0 = L_1^{-1}L_2 \in \mathfrak{L}$ into its inverse.

If the l sets $L\mathfrak{M}_0$, $L \in \mathfrak{L}$, are pairwise disjoint, their union contains lm_0 distinct elements. Since no element of $\mathfrak{L}$ appears, we have $lm_0 \leqq g - l$ and hence $l \leqq g/(m_0 + 1)$.

As a special case, we note

(4L). *If $\mathfrak{G}$ contains m involutions, any subgroup $\mathfrak{L}$ of order $l > g/(m + 1)$ contains real elements different from* 1. *In particular, if $n = g/m$, any subgroup of order $l \geqq n$ contains real elements different from* 1.

The following example shows that this result cannot be improved substantially. If $\mathfrak{G} = LF(2, q)$ and q is a prime power with $q \equiv -1 \pmod 4$, the subgroups of order q do not contain real elements different from 1. On the other hand, $m = q(q-1)/2$ and $g/(m+1) < q + 1$.

As another consequence of (4K), we note that if the group $\mathfrak{G}$ in (4G) contains $r \geqq 2$ classes of involutions $\mathfrak{K}_1, \mathfrak{K}_2, \cdots, \mathfrak{K}_r$ and if $J \in \mathfrak{K}_1$, then

$$h \leqq (n_2^{-1} + n_3^{-1} + \cdots + n_r^{-1} + g^{-1})^{-1}.$$

Indeed, as we have seen above, no element of one of the classes $\mathfrak{K}_2, \mathfrak{K}_3, \cdots, \mathfrak{K}^r$ can transform an element of $\mathfrak{H}$ different from 1 into its inverse.

It is a consequence of (4G) that if subgroups $\mathfrak{H}$ of the type discussed there occur, then some of the Sylow groups of $\mathfrak{G}$ are abelian and consist entirely of real elements. We can obtain the same conclusion under slightly different assumptions.

(4M). *Let p be an odd prime. If $\mathfrak{G}$ contains real elements of an order divisible by p, and if every real element of order p has distance greater than* 2 *from the set of involutions of $\mathfrak{G}$, then the p-Sylow groups $\mathfrak{P}$ of $\mathfrak{G}$ are abelian and consist entirely of real elements. All the elements of $\mathfrak{P}$ different from* 1 *have the same normalizer, which is abelian and consists entirely of real elements.*

PROOF. Let G be a real element of an order divisible by p. After replacing G by a suitable power, we may assume that G has order p. Then (4A) and (4C) show that there exist involutions J which transform G into G^{-1}, that $\mathfrak{N}(G)$ is abelian, and that J transforms every element of $\mathfrak{N}(G)$ into its inverse. Let $\mathfrak{P}$ be a p-Sylow subgroup of $\mathfrak{G}$ which contains G, and let G_0 be an element of order p in the center of $\mathfrak{P}$. Since $G_0 \in \mathfrak{N}(G)$, we can apply the above argument to G_0 instead of G. Since $\mathfrak{P} \subseteq \mathfrak{N}(G_0)$, $\mathfrak{P}$ is abelian and consists entirely of real elements.

Now let P be an element of $\mathfrak{P}$ different from 1. A suitable power P^s of P has order p. The above argument shows that $\mathfrak{N}(P^s)$, and hence $\mathfrak{N}(P)$, is abelian and consists entirely of real elements. It now follows easily that all the elements of $\mathfrak{P}$ different from 1 have the same normalizer.

V. Results concerning the characters

20. Let $\mathfrak{G}$ be a group of even order g. Let $\chi_0, \chi_1, \cdots, \chi_{k-1}$ denote the ordinary irreducible characters of $\mathfrak{G}$, and let f_i denote the degree of χ_i. We take χ_0 to be the 1-character. It is well known that the number of real characters is equal to the number k_1 of real classes of conjugate elements. Let f be the minimal degree of a real character χ_i with $i > 0$. Since

$$\sum_{i=0}^{k-1} f_i^2 = g,$$

it follows that

$$1 + (k_1 - 1)f^2 \leqq g.$$

Now (2J) shows that $m(m + 1)f^2 \leqq g(g - 1)$. This yields

(5A). *If $\mathfrak{G}$ is a group of even order g which contains m involutions, there exists a real character, not the* 1*-character, of a degree f such that*

$$f \leqq \sqrt{g(g - 1)/(m^2 + m)}.$$

In particular, if $n = g/m$, then $f < n$.

Thus if $n \leqq 2$, that is, if at least half of the elements of $\mathfrak{G}$ are involutions, $\mathfrak{G}$ has a real character of degree 1 which is not the 1-character. This implies that $\mathfrak{G}$ has a normal subgroup of index 2. One can also show that the elements in $\mathfrak{G}$ of odd order form a normal subgroup of $\mathfrak{G}$. Using the known groups of degree 2, one can obtain

(5B). *If the group $\mathfrak{G}$ of even order g contains at least $g/3$ involutions, then $\mathfrak{G}$ has a normal subgroup $\mathfrak{G}_0$ such that $\mathfrak{G}/\mathfrak{G}_0$ either is cyclic of order* 2 *or* 3 *or is the icosahedral group of order* 60.

If (5A) is combined with Jordan's and Blichfeldt's Theorems on linear groups of given degrees, results similar to (2H) can be obtained. However, our present knowledge in this matter does not enable us to improve the results given in II.

21. We denote by $\chi_{\rho i}$ the value of the character χ_ρ for the class $\mathfrak{K}_i$. It is well known that to every character χ_ρ of $\mathfrak{G}$ there corresponds a character ω_ρ of degree 1 of the center Λ of the group algebra Γ. The values of ω_ρ for the elements of the basis $\mathrm{K}_0, \cdots, \mathrm{K}_{k-1}$ of Λ are given by

$$\omega_\rho(\mathrm{K}_i) = \frac{g\chi_{\rho i}}{n_i f_\rho}.$$

It follows from (0) that

$$\omega_\rho(\mathrm{K}_i)\omega_\rho(\mathrm{K}_j) = \sum_\mu a_{ij\mu}\omega_\rho(\mathrm{K}_\mu),$$

and hence

$$g n_i^{-1} n_j^{-1} \chi_{\rho i}\chi_{\rho j} = f_\rho \sum_\mu a_{ij\mu}\chi_{\rho\mu} n_\mu^{-1}.$$

If we multiply both members of the last equation by $\bar{\chi}_{\rho\lambda}$ for a fixed value of λ, add over all ρ, and apply the orthogonality relations for group characters, we obtain the well known formulae

$$a_{ij\lambda} = g n_i^{-1} n_j^{-1} \sum_\rho \chi_{\rho i}\chi_{\rho j}\bar{\chi}_{\rho\lambda} f_\rho^{-1}. \tag{21}$$

As above, we denote the classes containing involutions by $\mathfrak{K}_1, \mathfrak{K}_2, \cdots, \mathfrak{K}_r$. It follows from (1), (0), and (2) that

$$c_\lambda = \sum_{i,j=1}^{r} a_{ij\lambda}. \tag{22}$$

If we take into account that $\chi_{\rho i}$ is real for $i = 1, 2, \cdots, r$, we obtain

$$c_\lambda = g\sum_{i,j=1}^{r} n_i^{-1}n_j^{-1}\sum_\rho \chi_{\rho i}\chi_{\rho j}\chi_{\rho\lambda}f_\rho^{-1}. \tag{23}$$

We now show

(5C). *Suppose that G is a real element of the group $\mathfrak{G}$ of even order g for which $n(G)$ is odd and that J is an involution which transforms G into G^{-1}* (cf. 4A). *If p is a prime which divides $n(G)$ with the exact exponent ν but which does not divide $n(J)$, then $\mathfrak{G}$ possesses a p-block B of defect $d \leqq \nu$. We may choose B such that it contains characters of positive defect for 2.*

Proof. It follows from (4A) that all involutions transforming G into G^{-1} lie in one class, say in $\mathfrak{K}_1$. If G belongs say to $\mathfrak{K}_\lambda$, then (2A) and (4A) show that $c_\lambda = a_{11\lambda}$ and that c_λ is a divisor n_λ/s of n_λ, where s is the order of $\mathfrak{N}(G) \cap \mathfrak{N}(J)$. Then, for $i, j = 1$, (21) becomes

$$n_\lambda n_1^2 = gs\sum_\rho \chi_{\rho 1}^2\chi_{\rho\lambda}f_\rho^{-1}.$$

Since p and n_1 are relatively prime, there must exist a value of ρ such that

$$gn_\lambda^{-1}\chi_{\rho 1}^2\chi_{\rho\lambda}f_\rho^{-1} = \chi_{\rho 1}^2\omega_\rho(\mathrm{K}_\lambda)$$

is not divisible by a prime ideal divisor $\mathfrak{p}$ of p in the field of characters. Then

$$\chi_{\rho 1} \not\equiv 0 \pmod{p}, \qquad \omega_\rho(\mathrm{K}_\lambda) \not\equiv 0 \pmod{\mathfrak{p}}. \tag{24}$$

The second condition implies that χ_ρ belongs to a p-block of defect $d \leqq \nu$. Indeed, if χ_ρ belongs to a p-block B of defect d, there exists a character χ_μ in B with $g/(p^d f_\mu)$ prime to p. Now $\omega_\rho(\mathrm{K}_\lambda) \equiv \omega_\mu(\mathrm{K}_\lambda) \pmod{\mathfrak{p}}$. Since

$$\omega_\mu(\mathrm{K}_\lambda) = g\chi_{\mu\lambda}/(n_\lambda f_\mu)$$

is prime to $\mathfrak{p}$, n_λ must be divisible by p^d, that is, $\nu \geqq d$. The first condition (24) implies $\chi_{\rho 1} \neq 0$, and hence χ_ρ must be of positive defect for 2.

Corollary (5D). *If, in (5C), $n(G)$ is prime to p, then there exists a character of $\mathfrak{G}$ which is of defect 0 for p and of positive defect for 2.*

22. There is a second case in which we can prove that $\mathfrak{G}$ possesses characters of defect 0.

(5E). *Suppose that J is an involution and that for some odd prime p there exists a prime power group $\mathfrak{P}_0$ of order $p^c > 1$ such that no element of $\mathfrak{P}_0$ different from 1 is mapped on its inverse by any conjugate of J. If p divides $n(J)$ with the exact exponent ν, then there exists an irreducible character χ_ρ whose degree is divisible by $p^{c-\nu}$ and which is of positive defect for 2. In particular, if $\mathfrak{P}_0$ can be taken as a p-Sylow group of $\mathfrak{G}$ and if $\nu = 0$, then χ_ρ is of defect 0 for p.*

Proof. If J belongs to $\mathfrak{K}_1$, then the proof of (2A) shows that under our assumptions $a_{11\lambda} = 0$ for each class $\mathfrak{K}_\lambda$ with $\lambda > 0$ which contains elements of $\mathfrak{P}_0$. For $P \in \mathfrak{P}_0$, set

$$(25) \qquad \psi(P) = \sum_\rho \chi_\rho(J)^2 \chi_\rho(P) f_\rho^{-1}.$$

It follows from (21) that $\psi(P) = 0$ for every element $P \neq 1$ in $\mathfrak{P}_0$. For $P = 1$, we see from the orthogonality relations that

$$\psi(1) = \sum_\rho \chi_\rho(J)^2 = n(J).$$

If φ denotes the character of the regular representation of $\mathfrak{P}_0$, it follows that

$$(26) \qquad \psi = n(J)/p^c \varphi.$$

On the other hand, we can express the restriction of χ_ρ to $\mathfrak{P}_0$ in terms of the irreducible characters of $\mathfrak{P}_0$, in order to obtain an expression of $\psi(P)$ in terms of the irreducible characters of $\mathfrak{P}_0$. If the restriction of χ_ρ contains the 1-character of $\mathfrak{P}_0$ with the multiplicity b_ρ, then comparison of the multiplicity of this 1-character in (25) and (26) yields

$$\sum \chi_\rho(J)^2 b_\rho / f_\rho = n(J)/p^c.$$

Since the right hand side contains p with the exponent $c - \nu$ in the denominator, and since all $\chi_\rho(J)$ and b_ρ are rational integers, there must exist a value ρ such that $p^{c-\nu}$ divides f_ρ and $\chi_\rho(J) \neq 0$. This yields the statement.

An immediate consequence of (5E) is

(5F). *Let J be an involution and let p be an odd prime dividing g. If no element of order p is transformed into its inverse by J, and if p divides $n(J)$ with the exact exponent ν and g with the exact exponent a, then there exists an irreducible character whose degree is divisible by $p^{a-\nu}$ and which is of positive defect for 2.*

23. We conclude the paper with some remarks concerning the case that $\mathfrak{G}$ contains a subgroup $\mathfrak{H}$ which satisfies the assumptions of (4F). It is immaterial here whether the order g is even or odd, but we mention these remarks here since they can be applied in the case of (4G). If the notation is the same as in (4F), there exist t characters $\theta_1, \cdots, \theta_t$ of degree w of the group $\mathfrak{N}$. Then $\mathfrak{G}$ possesses t irreducible characters $\psi_1, \cdots, \psi_t$ all of the same degree z such that

$$\psi_j(H) = \gamma + \delta\theta_j(H)$$

for $H \in \mathfrak{H}$, $H \neq 1$. Here, γ and δ are independent of j and H, γ is a rational integer, and $\delta = \pm 1$. For elements G of $\mathfrak{G}$ which are not conjugate to such an element H, we have

$$\psi_1(G) = \psi_2(G) = \cdots = \psi_t(G).$$

If χ_j are the other characters, not of this "exceptional" kind, then $\chi_j(H)$ has a fixed rational integral value a_j for all $H \in \mathfrak{H}$, $H \neq 1$. The degree f_j of χ_j satisfies the congruence

$$f_j \equiv a_j \pmod{h},$$

while the degree z of the ψ_j satisfies

$$z \equiv \gamma + \delta w \pmod{h}.$$

These results are a special case of a more general result which was obtained originally as an application of a theorem on characters [1]. A direct simple proof using induced characters was given by M. Suzuki.

Using the orthogonality relations for group characters, one obtains the following additional relations:

$$(t-1)\gamma^2 + (\gamma - \delta)^2 + \sum a_j^2 = w + 1, \qquad t\gamma z + \sum f_j a_j = \delta z.$$

Harvard University
University of Arizona

References

[1] R. Brauer, *A characterization of the characters of groups of finite order*, Ann. of Math., 57 (1953), pp. 357–377.
[2] W. Burnside, Theory of Groups of Finite Order, Cambridge, 1897.
[3] H. Zassenhaus, Lehrbuch der Gruppentheorie, Leipzig, 1937.

Zur Darstellungstheorie der Gruppen endlicher Ordnung

Dem Andenken meines verehrten Lehrers ISSAI SCHUR gewidmet

Von

RICHARD BRAUER

§ 1. Einleitung

In seinen Vorlesungen über Gruppentheorie stellte I. Schur es gelegentlich als wünschenswert dar, die Darstellungen der Gruppen endlicher Ordnung von einem mehr zahlentheoretischen Standpunkt aus zu untersuchen. Die folgenden Ausführungen sind ein Versuch, ein solches Programm in Angriff zu nehmen.

Es sei $\mathfrak{G}$ eine Gruppe endlicher Ordnung g, und $\Gamma = \Gamma(\mathfrak{G}, \Omega)$ sei die zugehörige Gruppenalgebra über einem gegebenen Körper Ω. Das allgemeine Element von Γ hat dann die Form

$$\alpha = \sum_{G \in \mathfrak{G}} a_G G, \tag{1.1}$$

wo die Koeffizienten a_G Elemente aus Ω sind. Wählen wir Ω als algebraischen Zahlkörper endlichen Grades, und ist $\mathfrak{o}$ der Ring der ganzen algebraischen Zahlen in Ω, so bildet die Menge der Elemente (1.1) mit Koeffizienten a_G aus $\mathfrak{o}$ eine Ordnung J in Γ[1]). Man kann dann die Zahlentheorie in J untersuchen. Dabei ist zu beachten, daß J für $g > 1$ nicht Maximalordnung ist, so daß die Ergebnisse von Speiser, Artin und Hasse nicht ohne weiteres angewendet werden können. Zwar ist Γ halbeinfach, und daher kann man J in eine Maximalordnung einbetten, doch gehen dabei gerade die von unserem Standpunkt interessantesten arithmetischen Eigenschaften verloren.

Als erste Frage haben wir die Untersuchung der Primideale $\mathfrak{P}$ von J[2]). Der Durchschnitt von $\mathfrak{P}$ mit $\mathfrak{o}$ ist ein Primideal $\mathfrak{p}$ von $\mathfrak{o}$, und dann ist $J > \mathfrak{P} \geqq \mathfrak{p}J$. Die $\mathfrak{p}$ enthaltenden Ideale von J entsprechen eineindeutig den Idealen des Restklassenringes $J^* = J/\mathfrak{p}J$. Die Restklassenabbildung (mod $\mathfrak{p}J$) bildet den Teilring $\mathfrak{o}$ von J auf einen Körper Ω^* ab, der mit $\mathfrak{o}/\mathfrak{p}$ identifiziert werden darf. Man sieht dann, daß man J^* als die Gruppenalgebra $\Gamma(\mathfrak{G}, \Omega^*)$ von $\mathfrak{G}$ über Ω^* auffassen kann;

$$J^* = \Gamma(\mathfrak{G}, \Omega^*); \qquad \Omega^* = \mathfrak{o}/\mathfrak{p}. \tag{1.2}$$

[1]) Einige Begriffe und Resultate der Algebrentheorie, die im folgenden verwendet werden, sind im Anhang (§ 14) zusammengestellt.

[2]) Vgl. § 14, (a) und (c). Unter einem Ideal wird stets ein *zweiseitiges* Ideal verstanden. Ferner wird von einem Ideal einer Ordnung J vorausgesetzt, daß es von Null verschiedene Elemente von $\mathfrak{o}$ enthält.

Ein $\mathfrak{p}$ enthaltendes Ideal $\mathfrak{P}$ von J ist dann und nur dann Primideal von J, wenn sein Bild $\mathfrak{P}^*$ in J^* Primideal von J^* ist. Aus der Strukturtheorie der Algebren folgt, daß dies dann und nur dann der Fall ist, wenn $\mathfrak{P}^*$ maximal in J^* ist, d.h. wenn $J^*/\mathfrak{P}^*$ eine einfache Algebra ist.

Es seien $\mathfrak{F}_1, \mathfrak{F}_2, \ldots, \mathfrak{F}_l$ die wesentlich verschiedenen irreduziblen Darstellungen von $\mathfrak{G}$ durch lineare Transformationen mit Koeffizienten aus Ω^*. Jedes $\mathfrak{F}_i$ kann zu einer Darstellung von $\Gamma(\mathfrak{G}, \Omega^*)$ erweitert werden. Ist $\mathfrak{P}_i^*$ der Kern von $\mathfrak{F}_i$ in diesem Sinn, so sieht man leicht, daß $\mathfrak{P}_1^*, \mathfrak{P}_2^*, \ldots, \mathfrak{P}_l^*$ gerade die verschiedenen Primideale von $\Gamma(\mathfrak{G}, \Omega^*)$ sind. Das $\mathfrak{P}_i^*$ entsprechende Ideal $\mathfrak{P}_i$ von J besteht aus denjenigen Elementen α von J, für deren Restklasse α^* (mod $\mathfrak{p}J$) die Gleichung $\mathfrak{F}_i(\alpha^*) = 0$ gilt. Die so erhaltenen Ideale $\mathfrak{P}_1, \mathfrak{P}_2, \ldots, \mathfrak{P}_l$ sind dann die sämtlichen Primidealteiler von $\mathfrak{p}J$ in J. Es ist also die Untersuchung der zu einem festen Primideal $\mathfrak{p}$ von Ω gehörigen Primideale $\mathfrak{P}$ von J äquivalent mit der Untersuchung der irreduziblen Darstellungen von $\mathfrak{G}$ in Ω^*. Mit derartigen „modularen" Darstellungen von endlichen Gruppen haben sich C. J. Nesbitt und der Verfasser in einigen früheren Arbeiten befaßt[3]). In der vorliegenden Arbeit treten die modularen Gruppencharaktere nur als Hilfsmittel auf. Alles, was hier benötigt ist, ist mit kurzen Beweisskizzen in § 3 zusammengestellt.

Will man die Struktur des Ideals $\mathfrak{p}J$ von J weiter verfolgen, so liegt es nahe, die Restklassenalgebra $J^* = \Gamma(\mathfrak{G}, \Omega^*)$ als direkte Summe von direkt unzerlegbaren Idealen $\mathfrak{A}_i^*$ darzustellen;

$$\Gamma(\mathfrak{G}, \Omega^*) = \mathfrak{A}_1^* \oplus \mathfrak{A}_2^* \oplus \cdots \oplus \mathfrak{A}_t^*. \tag{1.3}$$

Setzt man dann

$$\mathfrak{B}_m^* = \sum_{\substack{i=1,2,\ldots,t\\ i\neq m}} \mathfrak{A}_i^*; \quad \mathfrak{A}_m^* = \bigcap_{\substack{i=1,2,\ldots,t\\ i\neq m}} \mathfrak{B}_i^*,$$

und ist $\mathfrak{B}_m$ das $\mathfrak{B}_m^*$ entsprechende Ideal von J, so ist

$$\mathfrak{p}J = \mathfrak{B}_1 \cap \mathfrak{B}_2 \cap \cdots \cap \mathfrak{B}_t. \tag{1.4}$$

Die Ideale $\mathfrak{B}_i$ und $\mathfrak{B}_j$ sind für $i \neq j$ teilerfremd; $(\mathfrak{B}_i, \mathfrak{B}_j) = J$; keins der Ideale $\mathfrak{B}_i$ läßt sich nicht-trivial als Durchschnitt teilerfremder Ideale darstellen. Die Darstellungen (1.3) und (1.4) sind eindeutig bestimmt. Wir nennen $\mathfrak{B}_1, \mathfrak{B}_2, \ldots, \mathfrak{B}_t$ die *Blockideale* von $\mathfrak{p}J$ in J.

Da $\mathfrak{B}_1\mathfrak{B}_2 \ldots \mathfrak{B}_t \subseteq \mathfrak{p}J$ ist, teilt jeder Primidealteiler $\mathfrak{P}$ von $\mathfrak{p}J$ genau eins der Blockideale $\mathfrak{B}_\tau$. Ist also B_τ die Menge der Primidealteiler von $\mathfrak{B}_\tau$, so sind die l Primideale $\mathfrak{P}_1, \mathfrak{P}_2, \ldots, \mathfrak{P}_l$ auf t *„Blöcke"* $B_1, B_2, \ldots, B_t$ verteilt. Nach dem oben Gesagten können wir dann auch von einer Verteilung der l modularen Darstellungen $\mathfrak{F}_1, \mathfrak{F}_2, \ldots, \mathfrak{F}_l$ auf die t Blöcke sprechen. Wie in § 4 genauer auseinandergesetzt werden wird, führt dies auch zu einer Verteilung der gewöhnlichen irreduziblen Darstellungen von $\mathfrak{G}$ auf die t Blöcke.

[3]) Man vgl. R. Brauer and C. Nesbitt, University of Toronto Studies, Math. Series No. 4, 1937; Annals of Math. (2) **42**, 556, Part I.

Diese Blöcke von Darstellungen sollen in der vorliegenden Arbeit untersucht werden[4]).

Kennt man die Charaktere von $\mathfrak{G}$, so kann man daraus weitgehende Information über die Blöcke und die zugehörigen Blockideale ablesen. Dies wird in § 4 und § 5 ausgeführt. Ist p die durch $\mathfrak{p}$ teilbare rationale Primzahl, und enthält g die Primzahl p genau in der Potenz p^a, so verstehen wir unter dem *Defekt* eines Blocks B die kleinste ganze Zahl d, für die p^{a-d} die Grade aller Darstellungen in B teilt (§ 6).

Das Hauptziel dieser Arbeit ist es, Zusammenhänge zwischen gruppentheoretischen Eigenschaften von $\mathfrak{G}$ und Eigenschaften der Blöcke aufzudecken. Daraus ergeben sich dann neue Eigenschaften der Charaktere von $\mathfrak{G}$. Diese Aufgabe wird in § 7 in Angriff genommen. Auf Grund einer einfachen gruppentheoretischen Bemerkung ergibt sich ein Zusammenhang zwischen den Blöcken gewisser Untergruppen von $\mathfrak{G}$ und den Blöcken von $\mathfrak{G}$. Wir können dann in § 8 jedem Block B von $\mathfrak{G}$ vom Defekt d eine Untergruppe $\mathfrak{D}$ von $\mathfrak{G}$ von der Ordnung p^d, seine *Defektgruppe*, zuordnen. Mit $\mathfrak{D}$ ist auch jede konjugierte Gruppe Defektgruppe von B; davon abgesehen ist $\mathfrak{D}$ eindeutig durch B bestimmt.

Ist $\mathfrak{D}$ eine beliebige Untergruppe von $\mathfrak{G}$, deren Ordnung eine Potenz p^d von p ist und ist $\mathfrak{N}$ der Normalisator von $\mathfrak{D}$ in $\mathfrak{G}$, so gibt es eine eineindeutige Zuordnung zwischen den Blöcken $\tilde{B}$ von $\mathfrak{N}$ vom Defekt d und den Blöcken B von $\mathfrak{G}$ mit der Defektgruppe $\mathfrak{D}$ (§ 10). Die Behandlung von $\mathfrak{N}$ läßt sich dann weiter auf Untersuchung von $\mathfrak{N}/\mathfrak{D}$ zurückführen. Für die Werte der Charaktere χ_i in B erhält man Kongruenzen nach Potenzen von $\mathfrak{p}$, vgl. (12A).

Interessiert man sich z.B. für die Bestimmung der Anzahl der Blöcke B von einem gegebenen Defekt d, so sieht man jetzt, daß man diese Frage im Fall eines positiven d vollständig auf die Untersuchung von Gruppen von kleinerer Ordnung als g zurückführen kann. Im Fall $d=0$ versagen diese Methoden, da hier $\mathfrak{N}=\mathfrak{G}$, $\mathfrak{D}=\{1\}$ ist. Als Ersatz leiten wir im § 13 ein andersartiges Resultat für die Anzahl der Blöcke vom Defekt 0 her.

Die Untersuchung soll in einer weiteren Mitteilung fortgesetzt werden, während Anwendungen an anderer Stelle behandelt werden sollen.

§ 2. Bezeichnungen

Im folgenden ist $\mathfrak{G}$ stets eine endliche Gruppe der Ordnung g. Ist $\mathfrak{A}$ eine beliebige Teilmenge von $\mathfrak{G}$, so bezeichnen wir mit $\mathfrak{N}(\mathfrak{A})$ den *Normalisator* von $\mathfrak{A}$, d.h. die Untergruppe von $\mathfrak{G}$, deren Elemente X die Gleichung $X\mathfrak{A}=\mathfrak{A}X$ erfüllen. Ferner sei $\mathfrak{C}(\mathfrak{A})$ der *Zentralisator* von $\mathfrak{A}$, das ist die Untergruppe von $\mathfrak{G}$, deren Elemente mit jedem Element von $\mathfrak{A}$ vertauschbar sind. Besteht $\mathfrak{A}$ aus einem einzigen Element A, so schreiben wir $\mathfrak{N}(A)$ $(=\mathfrak{C}(A))$ für $\mathfrak{N}(\mathfrak{A})$. Die Ordnung von $\mathfrak{N}(A)$ werde mit $n(A)$ bezeichnet.

---

[4]) Über einen Teil der Ergebnisse berichten ohne Beweis drei Noten, Proc. Nat. Acad. Sci. **30**, 109—114 (1944); **32**, 182—186, 215—219 (1946).

Es seien im folgenden $\mathfrak{K}_1, \mathfrak{K}_2, \ldots, \mathfrak{K}_k$ stets die Klassen konjugierter Elemente von $\mathfrak{G}$. Ist also $G_\alpha \in \mathfrak{K}_\alpha$, so besteht $\mathfrak{K}_\alpha$ aus $g/n(G_\alpha)$ Elementen. Die Anzahl k der Klassen $\mathfrak{K}_\alpha$ ist dann auch die Anzahl der irreduziblen Charaktere von $\mathfrak{G}$, die mit $\chi_1, \chi_2, \ldots, \chi_k$ bezeichnet werden sollen. Es sei x_i der Grad von χ_i; $x_i = \chi_i(1)$.

Ist Ξ ein beliebiger Körper, so kann man die Gruppenalgebra $\Gamma(\mathfrak{G}, \Xi)$ von $\mathfrak{G}$ über Ξ bilden. Dies ist eine assoziative Algebra der Dimension g über Ξ, deren g Basiselemente so gewählt werden können, daß sie bei Multiplikation eine zu $\mathfrak{G}$ isomorphe Gruppe bilden. Man kann dann die Basiselemente mit den Elementen von $\mathfrak{G}$ identifizieren, und dann ist das allgemeine Element von $\Gamma(\mathfrak{G}, \Xi)$ von der Form (1.1) mit Koeffizienten $a_G \in \Xi$. Offenbar ist das Einheitselement von $\mathfrak{G}$ auch Einheitselement von $\Gamma(\mathfrak{G}, \Xi)$. Mit K_α werde stets das Element von $\Gamma(\mathfrak{G}, \Xi)$ bezeichnet, das die Summe der Elemente der Klasse $\mathfrak{K}_\alpha$ ist,

$$K_\alpha = \sum_{G \in \mathfrak{K}_\alpha} G. \tag{2.1}$$

Bekanntlich bilden die k Elemente $K_1, K_2, \ldots, K_k$ eine Basis des Zentrums $Z(\mathfrak{G}, \Xi)$ der Gruppenalgebra. Wegen dieses Zusammenhangs nennen wir $Z(\mathfrak{G}, \Xi)$ die *Klassenalgebra* von $\mathfrak{G}$ über Ξ. Die Multiplikation der Basiselemente ist durch Formeln

$$K_\alpha K_\beta = \sum a_{\alpha\beta\gamma} K_\gamma \tag{2.2}$$

gegeben. Hier ist $a_{\alpha\beta\gamma}$ eine nicht-negative ganze rationale Zahl, die von der Wahl des Körpers unabhängig ist. Ist G_γ ein Element von $\mathfrak{K}_\gamma$, so ist nämlich $a_{\alpha\beta\gamma}$ die Anzahl der geordneten Paare (X, Y) von $\mathfrak{G}$, die die Bedingungen

$$XY = G_\gamma, \quad X \in \mathfrak{K}_\alpha, \quad Y \in \mathfrak{K}_\beta$$

erfüllen. Hat Ξ Primzahlcharakteristik p, so kann man natürlich $a_{\alpha\beta\gamma}$ (mod p) nehmen.

Ist Ξ der Körper der komplexen Zahlen, und ist $\mathfrak{X}_i$ die zum irreduziblen Charakter χ_i von $\mathfrak{G}$ gehörige Darstellung von $\mathfrak{G}$, so kann man $\mathfrak{X}_i$ als irreduzible Darstellung von $\Gamma(\mathfrak{G}, \Xi)$ auffassen. Ist ζ ein beliebiges Element von $Z(\mathfrak{G}, \Xi)$, so ist $\mathfrak{X}_i(\zeta)$ nach dem Schurschen Lemma ein skalares Vielfaches $\omega_i(\zeta)E$ der Identität E; $\omega_i(\zeta) \in \Xi$. Als Funktion von ζ stellt dann ω_i einen linearen Charakter der Klassenalgebra $Z(\mathfrak{G}, \Xi)$ dar. Natürlich ist ω_i vollständig durch die Werte $\omega_i(K_\alpha)$ für $\alpha = 1, 2, \ldots, k$ bestimmt. Diese kann man in bekannter Weise durch Berechnung der Spur von $\mathfrak{X}_i(K_\alpha)$ bestimmen. Ist G_α ein Element von $\mathfrak{K}_\alpha$, so wird es oft bequem sein, auch $\omega_i(G_\alpha)$ für $\omega_i(K_\alpha)$ zu schreiben; man hat dann

$$\omega_i(K_\alpha) = \omega_i(G_\alpha) = \frac{g}{n(G_\alpha)} \frac{\chi_i(G_\alpha)}{x_i} \qquad (x_i = \chi_i(1)). \tag{2.3}$$

Da $\chi_1, \chi_2, \ldots, \chi_k$ orthogonal und daher linear unabhängig sind, sind die k linearen Charaktere $\omega_1, \omega_2, \ldots, \omega_k$ von $Z(\mathfrak{G}, \Xi)$ voneinander verschieden. Da

$Z(\mathfrak{G}, \Xi)$ die Dimension k hat, sind dies sämtliche irreduziblen Darstellungen der Klassenalgebra. Man sieht auch jetzt noch leicht ein, daß $Z(\mathfrak{G}, \Xi)$ halbeinfach ist.

Jedes $\chi_i(G_\alpha)$ ist eine Summe von g-ten Einheitswurzeln. Ersetzt man also den Körper der komplexen Zahlen Ξ durch einen die g-ten Einheitswurzeln enthaltenden Teilkörper Ω, so zeigt (2.3), daß alle $\omega_i(K_\alpha)$ in Ω liegen. Daher besitzt auch $Z(\mathfrak{G}, \Omega)$ genau k verschiedene lineare Charaktere, die wir ebenfalls mit $\omega_1, \omega_2, \ldots, \omega_k$ bezeichnen.

Im folgenden ist Ω stets ein die g-ten Einheitswurzeln enthaltender algebraischer Zahlkörper endlichen Grades. Wie in § 1 sei $\mathfrak{o}$ der Ring der ganzen algebraischen Zahlen in Ω, und es sei $\mathfrak{p}$ ein festes Primideal von $\mathfrak{o}$ und p die durch $\mathfrak{p}$ teilbare rationale Primzahl. Die Restklasse eines Elements α von $\mathfrak{o}$ (mod $\mathfrak{p}$) wird stets mit α^* bezeichnet werden; es sei

$$\Omega^* = \mathfrak{o}/\mathfrak{p} \tag{2.4}$$

der Restklassenkörper. Ferner sei ν stets die zu $\mathfrak{p}$ gehörige exponentielle Bewertung von Ω, die wir so normieren, daß

$$\nu(p) = 1 \tag{2.5}$$

ist. Gelegentlich wird die Bezeichnung $\mathfrak{o}_\nu$ für den Ring der für $\mathfrak{p}$ ganzen Zahlen von Ω und $\mathfrak{p}_\nu$ für das Primideal von $\mathfrak{o}_\nu$ verwendet werden. Der Restklassenkörper $\mathfrak{o}_\nu/\mathfrak{p}_\nu$ darf mit Ω^* identifiziert werden.

Ist G ein Element von $\mathfrak{G}$, so schreiben wir zur Abkürzung $\nu(G)$ für $\nu(n(G))$ und nennen diese Zahl den *Defekt* von G. Die p-Sylowgruppen des Normalisators $\mathfrak{N}(G)$ haben dann also die Ordnung $p^{\nu(G)}$. Gehört G_α zur Klasse $\mathfrak{K}_\alpha$, so setzen wir noch

$$\nu(\mathfrak{K}_\alpha) = \nu(G_\alpha) = \nu(n(G_\alpha)), \qquad (G_\alpha \in \mathfrak{K}_\alpha); \tag{2.6}$$

wir nennen $\nu(\mathfrak{K}_\alpha)$ den *Defekt* der Klasse $\mathfrak{K}_\alpha$.

Ein Element G von $\mathfrak{G}$ heiße *p-regulär*, wenn die Ordnung von G zu p teilerfremd ist. Da das gleiche dann auch für alle konjugierten Elemente gilt, so können wir von *p-regulären Klassen* konjugierter Elemente sprechen.

§ 3. Eigenschaften von modularen Charakteren

Wir geben in diesem Abschnitt eine Schilderung der Theorie der modularen Gruppendarstellungen, soweit sie im folgenden benötigt wird.

(3 A) *Es sei Γ eine Algebra von endlicher Dimension über einem algebraisch abgeschlossenen Körper Ξ von Primzahlcharakteristik p. Es sei S der von den Elementen $\alpha\beta - \beta\alpha$ mit $\alpha, \beta \in \Gamma$ aufgespannte lineare Teilraum von Γ; es sei T die Menge der Elemente ϱ von Γ, für die es eine Potenz q von p gibt, derart daß ϱ^q zu S gehört. Dann ist T ein linearer Teilraum von Γ. Ferner ist die Anzahl der wesentlich verschiedenen irreduziblen Darstellungen von Γ gleich der Dimension des Restklassenraumes Γ/T.*

Beweis. Es seien ϱ und σ zwei Elemente von Γ; man bilde $(\varrho+\sigma)^p$ durch Ausmultiplizieren der Klammern. Zu jedem von ϱ^p und σ^p verschiedenen Glied gehören p Glieder, die aus ihm durch zyklische Permutation der Faktoren entstehen. Offenbar sind diese p Glieder untereinander kongruent (mod S). Daher ergibt sich

$$(\varrho+\sigma)^p \equiv \varrho^p + \sigma^p \pmod{S}. \tag{3.1}$$

Sind α, β Elemente von Γ und setzt man $\gamma=\beta(\alpha\beta)^{p-1}$, so folgt unter Verwendung von (3.1), daß

$$(\alpha\beta-\beta\alpha)^p \equiv (\alpha\beta)^p - (\beta\alpha)^p = \alpha\gamma - \gamma\alpha \equiv 0 \pmod{S}$$

ist. Jetzt sieht man, daß die p-te Potenz jedes Elementes von S wieder zu S gehört. Unter Verwendung von (3.1) folgt nun leicht, daß die in (3A) definierte Menge T wirklich ein linearer Teilraum von Γ ist, und ferner daß $T \geqq S$ ist.

Ist Γ einfach, also eine volle Matrizenalgebra über Ξ, so besteht S, wie man leicht sieht, gerade aus den Matrizen von der Spur Null. Ist η ein Idempotent mit von Null verschiedener Spur, so gehört η nicht zu S und auch nicht zu T. Da nun Γ/S die Dimension 1 hat und $\Gamma > T \geqq S$ ist, so folgt hier, daß Γ/T die Dimension 1 hat.

Im allgemeinen Fall folgt aus der Nilpotenz des Radikals N von Γ, daß N zu T gehört. Besitzt nun Γ genau l wesentlich verschiedene irreduzible Darstellungen, so ist Γ/N direkte Summe von l Idealen Γ_i, die einfache Algebren sind. Hat T_i die entsprechende Bedeutung für Γ_i wie T für Γ, so folgt jetzt leicht, daß die Dimension von Γ/T gleich der Summe der Dimensionen der Γ_i/T_i ist. Nach dem oben Gesagten ist aber jede der letzteren Dimensionen 1, und daher hat Γ/T die Dimension l wie behauptet war.

(3B) *Die Anzahl der wesentlich verschiedenen irreduziblen Darstellungen einer endlichen Gruppe $\mathfrak{G}$ in einem algebraisch abgeschlossenen Körper Ξ von Primzahlcharakteristik p ist gleich der Anzahl der p-regulären Klassen konjugierter Elemente in $\mathfrak{G}$.*

Beweis. Wir nehmen $\Gamma=\Gamma(\mathfrak{G},\Xi)$ und verwenden dieselben Bezeichnungen wie im Beweis von (3A). Ist G ein beliebiges Element von $\mathfrak{G}$, so können wir $G=RP$ setzen, wo R und P Potenzen von G sind und wo R p-regulär ist, während die Ordnung von P eine Potenz q von p ist. Aus (3.1) mit $\varrho=RP$, $\sigma=-R$ ergibt sich

$$(RP-R)^p \equiv R^p P^p - R^p \pmod{S}.$$

Diese Beziehung erhebe man wieder in die p-te Potenz und beachte, daß wie oben bemerkt, die p-te Potenz eines Elementes von S wieder zu S gehört. Fährt man so fort, so findet man schließlich

$$(RP-R)^q \equiv R^q P^q - R^q = R^q - R^q = 0 \pmod{S}.$$

Dies heißt aber, daß $RP-R \in T$ ist, also $G \equiv R \pmod{T}$.

28*

Jetzt folgt, daß (mod T) das allgemeine Element (1.1) von $\Gamma(\mathfrak{G}, \Xi)$ einem Element derselben Form kongruent ist, in dem nur die p-regulären Elemente von $\mathfrak{G}$ von Null verschiedene Koeffizienten haben. Sind ferner G und G_1 zwei konjugierte Elemente von $\mathfrak{G}$, so ist $G \equiv G_1$ (mod S) und also auch (mod T). Daher kann man setzen

$$\alpha \equiv \sum_R a_R R \qquad (\text{mod } T), \tag{3.2}$$

wo R ein Vertretersystem für die p-regulären Klassen durchläuft. Diese Elemente R bilden ein (mod T) linear unabhängiges System von Elementen von Γ. Ist nämlich α in (3.2) ein Element von T, so gibt es eine Potenz q von p, derart daß $\alpha^q \in S$ ist. Da die Elemente R p-regulär sind, kann man q so wählen, daß für jedes R die Gleichung $R^q = R$ gilt. Erhebt man nun α in (3.2) in die q-te Potenz, so ergibt sich mit Hilfe von wiederholter Anwendung von (3.1), daß

$$\sum_R a_R^q R \equiv 0 \qquad (\text{mod } S)$$

ist. Nun sieht man aber leicht, daß S aus den Elementen (1.1) besteht, für die (mod p) für jede Klasse $\mathfrak{K}_i$ die Summe der Koeffizienten der Elemente G in $\mathfrak{K}_i$ verschwindet. Da alle R in (3.2) zu verschiedenen Klassen gehören, sind (mod p) alle a_R^q und daher alle a_R in (3.2) Null, wie zu zeigen war.

Es bilden also die Elemente R in (3.2) eine Basis für Γ (mod T). Daher ist die Dimension l von Γ/T gleich der Anzahl der p-regulären Klassen $\mathfrak{K}_i$, und nun liefert (3 A) die Behauptung.

Wir beweisen weiter den für uns wichtigen Hilfssatz

(3 C) *Sind $\mathfrak{K}_1, \mathfrak{K}_2, \ldots, \mathfrak{K}_l$ die l p-regulären Klassen von $\mathfrak{G}$, und ist G_j ein Repräsentant für $\mathfrak{K}_j$, so kann man l (gewöhnliche) irreduzible Charaktere $\chi_1, \chi_2, \ldots, \chi_l$ von $\mathfrak{G}$ so auswählen, daß*

$$\text{Det}\,(\chi_i(G_j)) \not\equiv 0 \qquad (\text{mod } \mathfrak{p}) \tag{3.3}$$

ist, wobei in der Determinante der Zeilenindex i und der Spaltenindex j die Werte 1, 2, ..., l durchlaufen.

Beweis. Wäre (3 C) nicht richtig, so könnte man l ganze Zahlen $u_1, u_2, \ldots, u_l$ in Ω so wählen, daß für jeden irreduziblen Charakter χ_i die Kongruenz gilt

$$\sum_{j=1}^{l} u_j \chi_i(G_j) \equiv 0 \qquad (\text{mod } \mathfrak{p})$$

und etwa $u_m \not\equiv 0$ (mod $\mathfrak{p}$) ist. Ist Θ eine beliebige Linearverbindung der χ_i mit ganzen Koeffizienten in Ω, so ist dann

$$\sum_{j=1}^{l} u_j \Theta(G_j) \equiv 0 \qquad (\text{mod } \mathfrak{p}). \tag{3.4}$$

Es sei nun $\mathfrak{B} = \{G_m\}$ die von G_m erzeugte zyklische Gruppe, b sei die Ordnung und $\zeta_1, \zeta_2, \ldots, \zeta_b$ seien die b irreduziblen Charaktere von $\mathfrak{B}$. Dann wähle man eine p-Sylowgruppe $\mathfrak{Q}$ von $\mathfrak{N}(G_m)$ und setze $\mathfrak{E} = \{\mathfrak{B}, \mathfrak{Q}\} = \mathfrak{B} \times \mathfrak{Q}$. Ist BQ mit $B \in \mathfrak{B}$, $Q \in \mathfrak{Q}$ ein variables Element von $\mathfrak{E}$, so ist ein Charakter ϑ_i von $\mathfrak{E}$ durch

$$\vartheta_i(BQ) = \zeta_i(B)$$

definiert. Es sei Θ_i der von ϑ_i induzierte Charakter von $\mathfrak{G}$. Setzt man

$$\Theta = \sum_{i=1}^{b} \overline{\zeta_i(G_m)}\, \Theta_i,$$

so ist Θ eine Linearverbindung von Charakteren von $\mathfrak{G}$ mit ganzen Koeffizienten aus Ω. Aus der Definition der induzierten Charaktere ergibt sich

$$\Theta(G_j) = \frac{1}{(\mathfrak{E}:1)} \sum_{X \in \mathfrak{G}} \sum_{i=1}^{b} \overline{\zeta_i(G_m)}\, \vartheta_i(XG_jX^{-1}),$$

wo $\vartheta_i(XG_jX^{-1})$ gleich Null zu setzen ist, falls XG_jX^{-1} nicht zu $\mathfrak{E}$ gehört. Ist $XG_jX^{-1} \in \mathfrak{E}$ und $1 \leq j \leq l$, so ist XG_jX^{-1} p-regulär und also ein Element B von $\mathfrak{B}$. Dann ist $\vartheta_i(XG_jX^{-1}) = \zeta_i(B)$. Aus den Orthogonalitätsrelationen für die Charaktere von $\mathfrak{B}$ ergibt sich jetzt, daß der Wert der inneren Summe in der letzten Formel 0 oder b ist, je nachdem ob $B \neq G_m$ oder $B = G_m$ ist. Das letztere ist nur für $G_j = G_m$, d.h. für $j = m$ möglich. Jetzt folgt leicht, daß $\Theta(G_j) = 0$ für $j \neq m$ und $\Theta(G_m) = b(\mathfrak{N}(G_m):\mathfrak{E})$ ist. Wegen der Art, in der wir $\mathfrak{E}$ konstruierten, ist dieser Wert $\Theta(G_m)$ eine zu p teilerfremde ganze Zahl. Setzen wir dies in (3.4) ein, so erhalten wir einen Widerspruch, da $u_m \not\equiv 0 \pmod{\mathfrak{p}}$ war. Damit ist (3C) bewiesen.

Da der Körper Ω nach Voraussetzung die g-ten Einheitswurzeln enthält, kann man die irreduzible Darstellung $\mathfrak{X}_i$ von $\mathfrak{G}$ mit dem Charakter χ_i in Ω schreiben, $i = 1, 2, \ldots, k$[5]). Ebenso kann man die Koeffizienten der l absolut irreduziblen modularen Darstellungen $\mathfrak{F}_1, \mathfrak{F}_2, \ldots, \mathfrak{F}_l$ von $\mathfrak{G}$ im Restklassenkörper Ω^* wählen. Ist $\mathfrak{o}_\mathfrak{p}$ der Ring der für $\mathfrak{p}$ ganzen Zahlen in Ω, so darf man ferner annehmen, daß $\mathfrak{X}_i(G)$ für jedes $G \in \mathfrak{G}$ Koeffizienten in $\mathfrak{o}_\mathfrak{p}$ hat[6]). Die Restklassenabbildung führt dann $\mathfrak{X}_i$ in eine Darstellung $\mathfrak{X}_i^*$ mit Koeffizienten in $\mathfrak{o}_\mathfrak{p}/\mathfrak{p}_\mathfrak{p} = \Omega^*$ über. Tritt die irreduzible modulare Darstellung $\mathfrak{F}_j$ in $\mathfrak{X}_i$ mit der Vielfachheit d_{ij} auf, so nennen wir die nicht-negativen ganzen rationalen Zahlen d_{ij} die *Zerlegungszahlen* von $\mathfrak{G}$; $i = 1, 2, \ldots, k$; $j = 1, 2, \ldots, l$.

Wir definieren jetzt den Charakter φ einer modularen Darstellung $\mathfrak{F}$ mit Koeffizienten in Ω^*. Dabei beschränken wir uns auf p-reguläre Elemente G von $\mathfrak{G}$. Hat G die Ordnung h, so sind die charakteristischen Wurzeln $\varepsilon_1^*, \varepsilon_2^*, \ldots$ von $\mathfrak{F}(G)$ sämtlich h-te Einheitswurzeln. Da h zu p teilerfremd ist, geht

[5]) Vgl. etwa R. Brauer, Amer. J. Math. **69**, 709 (1947). Ersetzt man gegebenfalls Ω durch einen Erweiterungskörper endlichen Grades, so kann man die Verwendung dieser Tatsache vermeiden.

[6]) Vgl. § 14, (d) und (14B). Dort wird auch gezeigt, daß die im folgenden definierten Zahlen d_{ij} ungeändert bleiben, wenn man $\mathfrak{X}_i$ durch eine ähnliche Darstellung mit Koeffizienten in $\mathfrak{o}_\mathfrak{p}$ ersetzt.

jedes ε_i^* aus einer eindeutig bestimmten h-ten Einheitswurzel ε_i in Ω durch Restklassenabbildung hervor. Dann setzen wir $\varphi(G) = \sum \varepsilon_i$. Es ist also φ eine für p-reguläre Elemente $G \in \mathfrak{G}$ definierte Funktion mit Werten in Ω. Natürlich hängt $\varphi(G)$ nur von der Klasse konjugierter Elemente $\mathfrak{K}_\alpha$ ab, zu der G gehört.

(3D) *Haben die irreduziblen modularen Darstellungen* $\mathfrak{F}_1, \mathfrak{F}_2, \ldots, \mathfrak{F}_l$ *die Charaktere* $\varphi_1, \varphi_2, \ldots, \varphi_l$, *sind* d_{ij} *die Zerlegungszahlen, so ist für* p*-reguläre Elemente* G

$$\chi_i(G) = \sum_{j=1}^{l} d_{ij} \varphi_j(G). \tag{3.5}$$

In der Tat stimmt für p-reguläres G der Wert des Charakters von $\mathfrak{X}_i^*$ mit dem Wert $\chi_i(G)$ des Charakters χ_i von $\mathfrak{X}_i$ überein, und da $\mathfrak{F}_j$ genau d_{ij}-mal in $\mathfrak{X}_i^*$ auftritt, so erhält man sofort (3.5).

Setzt man die Werte (3.5) in (3.3) ein, so erhält man

(3E) *Die aus den Werten der* l *irreduziblen modularen Charaktere* φ_i *für die Repräsentanten* $G_1, G_2, \ldots, G_l$ *der* l p*-regulären Klassen von* $\mathfrak{G}$ *gebildete Determinante ist nicht durch* $\mathfrak{p}$ *teilbar;*

$$\operatorname{Det}\left(\varphi_i(G_j)\right) \not\equiv 0 \pmod{\mathfrak{p}}. \tag{3.6}$$

Sind ferner $\chi_1, \chi_2, \ldots, \chi_l$ *wie in* (3C) *gewählt, so ist*

$$\operatorname{Det}(d_{ij}) \not\equiv 0 \pmod{p}; \qquad (i, j = 1, 2, \ldots, l). \tag{3.7}$$

Wir setzen noch

$$c_{ij} = \sum_{\varrho=1}^{k} d_{\varrho i} d_{\varrho j} = c_{ji}, \qquad (i, j = 1, 2, \ldots, l); \tag{3.8}$$

$$\Phi_i = \sum_{j=1}^{k} d_{ji} \chi_j, \qquad (i = 1, 2, \ldots, l). \tag{3.9}$$

Die c_{ij} sind nicht-negative ganze rationale Zahlen, die sog. *Cartanschen Invarianten* von $\mathfrak{G}$ (oder eigentlich von $\Gamma(\mathfrak{G}, \Omega^*)$). Es erübrigt sich hier, ihre algebrentheoretische Bedeutung zu diskutieren. Die Φ_i sind gewöhnliche Charaktere von $\mathfrak{G}$, die in der modularen Theorie eine besondere Rolle spielen, worauf wir ebenfalls nicht näher einzugehen brauchen. Aus (3.9) und (3.5) ergibt sich unter Verwendung von (3.8) die Beziehung

$$\Phi_i(G) = \sum_{j=1}^{l} c_{ij} \varphi_j(G) \qquad (G\ p\text{-regulär}). \tag{3.10}$$

Die Orthogonalitätsbeziehungen für die Charaktere von $\mathfrak{G}$ besagen, daß

$$\sum_{i=1}^{k} \overline{\chi_i(G)}\, \chi_i(H) = n(G) \quad \text{oder} \quad = 0,$$

je nachdem ob die Elemente G und H von $\mathfrak{G}$ konjugiert sind oder nicht. Wählt man für H eins der p-regulären Elemente $G_1, G_2, \ldots, G_l$, sagen wir

$H = G_\alpha$, und drückt man $\chi_i(G_\alpha)$ nach (3.5) durch $\varphi_j(G_\alpha)$ aus, so folgt mit Hilfe von (3.9), daß

$$\sum_{j=1}^{l} \overline{\Phi_j(G)}\, \varphi_j(G_\alpha) = n(G) \quad \text{oder} \quad = 0 \tag{3.11}$$

ist, je nachdem ob G und G_α konjugiert sind oder nicht. Für p-singuläres G tritt der zweite Fall ein. Aus dem Nichtverschwinden der Determinante (3.6) ergibt sich dann, daß $\Phi_j(G) = 0$ sein muß. Für p-reguläres G folgt ebenso aus (3.6), daß $\overline{\Phi_j(G)}$ durch die höchste Potenz von $\mathfrak{p}$ teilbar ist, die in $n(G)$ aufgeht. Wegen $\overline{\Phi_j(G)} = \Phi_j(G^{-1})$, $n(G) = n(G^{-1})$ haben wir

(3F) *Die in* (3.9) *definierten Charaktere* Φ_j *verschwinden für alle p-singulären Elemente. Für p-reguläres G ist* $\nu(\Phi_j(G)) \geq \nu(G)$.

Ist G p-regulär, so kann man $\Phi_j(G)$ in (3.11) nach (3.10) durch $\varphi_i(G)$ ausdrücken. Schreibt man für G_α wieder H, so erhält man

(3G) *Für p-reguläre Elemente G und H von $\mathfrak{G}$ gilt*

$$\sum_{j=1}^{l} \overline{\Phi_j(G)}\, \varphi_j(H) = \sum_{i,j=1}^{l} c_{ij}\, \overline{\varphi_i(G)}\, \varphi_j(H) = n(G) \quad \text{oder} \quad 0, \tag{3.12}$$

je nachdem ob G und H konjugiert sind oder nicht.

§ 4. Blöcke von Charakteren

Wir schicken unserer Untersuchung einige Bemerkungen über die Klassenalgebra $Z = Z(\mathfrak{G}, \Omega)$ voraus. Wie bereits in § 2 bemerkt wurde, besitzt Z die k verschiedenen linearen Charaktere $\omega_1, \omega_2, \ldots, \omega_k$ und da Z die Dimension k hat, sind dies die sämtlichen irreduziblen Darstellungen von Z.

Offenbar ist der Durchschnitt $J \cap Z = J_0$ eine Ordnung von Z, die die $\mathfrak{o}$-Basis $K_1, K_2, \ldots, K_k$ besitzt; der Restklassenring $J_0/\mathfrak{p}J_0$ darf mit $Z^* = Z(\mathfrak{G}, \Omega^*)$ identifiziert werden. Anwendung der Restklassenabbildung auf $\omega_1, \omega_2, \ldots, \omega_k$ liefert dann k lineare Charaktere $\omega_1^*, \omega_2^*, \ldots, \omega_k^*$ von Z^*, die aber im allgemeinen nicht alle verschieden sind. Jeder Homomorphismus ψ der Algebra Z^* über Ω^* auf einen Erweiterungskörper von Ω^* ist ein linearer Charakter von Z^* und stimmt mit einem der ω_i^* überein[7]).

Wie in (1.3) betrachten wir die Zerlegung der modularen Gruppenalgebra $\Gamma(\mathfrak{G}, \Omega^*)$ in direkt unzerlegbare Ideale

$$\Gamma(\mathfrak{G}, \Omega^*) = \mathfrak{A}_1^* \oplus \mathfrak{A}_2^* \oplus \cdots \oplus \mathfrak{A}_t^*. \tag{4.1}$$

Setzt man dementsprechend

$$1 = \eta_1^* + \eta_2^* + \cdots + \eta_t^*, \qquad (\eta_\tau^* \in \mathfrak{A}_\tau^*), \tag{4.2}$$

so sind die η_τ^* orthogonale Idempotente des Zentrums $Z^* = Z(\mathfrak{G}, \Omega^*)$ von $\Gamma(\mathfrak{G}, \Omega^*)$. Das Ideal $\eta_\tau^* Z^*$ von Z^* ist direkt unzerlegbar und man hat

$$Z^* = \eta_1^* Z^* \oplus \eta_2^* Z^* \oplus \cdots \oplus \eta_t^* Z^*.$$

[7]) Vgl. § 14; (14A) und (14C).

Als kommutative direkt unzerlegbare Algebra ist $\eta_\tau^* Z^*$ primär. Ist also $\mathfrak{B}_\tau^*$ das Radikal von $\eta_\tau^* Z^*$, so ist $\eta_\tau^* Z^*/\mathfrak{B}_\tau$ ein Körper, der als Erweiterungskörper Ω_τ^* von Ω^* betrachtet werden kann. Man erweitere den natürlichen Homomorphismus von $\eta_\tau^* Z^*$ auf Ω_τ^* zu einem Homomorphismus ψ_τ von Z^* auf Ω_τ^*, indem man $\psi_\tau(\xi) = 0$ für $\xi \in \eta_\sigma^* Z^*$ mit $\sigma \neq \tau$ vorschreibt. Nach dem oben Gesagten erhält man so einen linearen Charakter ψ_τ von Z^*; jeder lineare Charakter von Z^* stimmt mit einem dieser ψ_τ überein. Man hat $\psi_\tau(\eta_\tau^*) = 1$.

Wie in § 1 sei $\mathfrak{B}_\tau^*$ die Summe der $\mathfrak{A}_\sigma^*$ mit $\sigma \neq \tau$. Da die Ideale $\mathfrak{B}_\tau^*$ von $\Gamma(\mathfrak{G}, \Omega^*)$ die Bilder der Ideale $\mathfrak{B}_\tau$ von $\Gamma(\mathfrak{G}, \Omega)$ in (1.4) sind, so haben wir das Resultat

(4A) *Hat das Ideal $\mathfrak{p} J$ in J die Blockideale $\mathfrak{B}_1, \mathfrak{B}_2, \ldots, \mathfrak{B}_t$ und bildet der natürliche Homomorphismus von J auf $J/\mathfrak{p} J = \Gamma(\mathfrak{G}, \Omega^*)$ das Ideal $\mathfrak{B}_\tau$ auf das Ideal $\mathfrak{B}_\tau^*$ von $\Gamma(\mathfrak{G}, \Omega^*)$ ab, so gibt es lineare Charaktere ψ_τ von $Z(\mathfrak{G}, \Omega^*)$, derart daß $\psi_\tau(\eta_\tau^*) = 1$ ist; $\tau = 1, 2, \ldots, t$. Die linearen Charaktere $\psi_1, \psi_2, \ldots, \psi_t$ sind die sämtlichen linearen Charaktere der Algebra $Z(\mathfrak{G}, \Omega^*)$.*

Ist wie immer $\mathfrak{p}$ ein festes Primideal von Ω, so ordnen wir jedem Blockideal $\mathfrak{B}_\tau$ von $\mathfrak{p} J$ einen „*Block*" B_τ zu. Darunter werden wir eine gewisse Menge von gewöhnlichen und modularen irreduziblen Darstellungen von $\mathfrak{G}$ verstehen. An Stelle von Darstellungen werden wir auch von den zu B_τ gehörigen gewöhnlichen und modularen Charakteren von $\mathfrak{G}$ sprechen. Den linearen Charakter ψ_τ von Z^* in (4A) nennen wir *den dem Block B_τ zugeordneten linearen Charakter der modularen Klassenalgebra $Z^* = Z(\mathfrak{G}, \Omega^*)$*.

Wir rechnen eine gewöhnliche irreduzible Darstellung $\mathfrak{X}_i$ von $\mathfrak{G}$ und ihren Charakter χ_i zum Block B_τ, wenn für den χ_i nach (2.3) entsprechenden linearen Charakter ω_i von Z die Gleichung gilt

(4.3) $$\omega_i^* = \psi_\tau.$$

Aus dem oben Gesagten ergibt sich

(4B) *Jeder irreduzible Charakter χ_i von $\mathfrak{G}$ gehört zu einem und nur einem Block B_τ und jeder Block enthält Charaktere χ_i. Zwei Charaktere χ_i und χ_j gehören dann und nur dann zum selben Block, wenn für alle $G \in \mathfrak{G}$ die Kongruenz gilt*

(4.4) $$\omega_i(G) \equiv \omega_j(G) \pmod{\mathfrak{p}}.$$

Wir zählen eine modulare irreduzible Darstellung $\mathfrak{F}_j$ und ihren Charakter φ_j zum Block B_τ, wenn der Kern $\mathfrak{P}_j^*$ von $\mathfrak{F}_j$ das Ideal $\mathfrak{B}_\tau^*$ enthält.

(4C) *Jede irreduzible modulare Darstellung $\mathfrak{F}_j$ gehört zu einem und nur einem Block B_τ und jeder Block B_τ enthält derartige $\mathfrak{F}_j$. Gehört die gewöhnliche irreduzible Darstellung $\mathfrak{X}_i$ zu B_τ, so gehören alle in $\mathfrak{X}_i^*$ als Bestandteil auftretenden $\mathfrak{F}_j$ zu B_τ.*

Beweis. Der erste Teil von (4C) ergibt sich unmittelbar aus dem in § 1 Gesagten. Gehört $\mathfrak{X}_i$ zu B_τ, so gilt (4.3). Wegen $\psi_\tau(\eta_\tau^*) = 1$ ist $\omega_i^*(\eta_\tau^*) = 1$. Als Element von Z^* hat η_τ^* die Form $\sum b_\lambda^* K_\lambda$ mit $b_\lambda^* \in \Omega^*$. Wählt man b_λ

in $\mathfrak{o}$ derart daß die Restklassenabbildung von $\mathfrak{o}$ auf Ω^* gerade b_λ in b_λ^* überführt und setzt man $\eta_\tau = \sum b_\lambda K_\lambda$, so ist η_τ ein Element von $J_0 = Z \cap J$, das durch die Restklassenabbildung von J gerade in η_τ^* übergeht. Dann ist $\mathfrak{X}_i(\eta_\tau) = \omega_i(\eta_\tau) E$, also $\mathfrak{X}_i^*(\eta_\tau^*) = \omega_i^*(\eta_\tau^*) E^* = E^*$. Tritt nun $\mathfrak{F}_j$ als Bestandteil in $\mathfrak{X}_i^*$ auf, so ist $\mathfrak{F}_j(\eta_\tau^*) = E^*$. Also ist $\eta_\tau^* \notin \mathfrak{P}_j^*$ und daher ist $\mathfrak{B}_\sigma$ für $\sigma \neq \tau$ nicht in $\mathfrak{P}_j$ enthalten. Dann muß $\mathfrak{B}_\tau \subseteqq \mathfrak{P}_j$ sein, und daher gehört $\mathfrak{F}_j$ zum Block B_τ, wie zu zeigen war.

Aus (4C) ersieht man, daß die Zerlegungszahl d_{ij} verschwinden muß, wenn $\mathfrak{X}_i$ und $\mathfrak{F}_j$ zu verschiedenen Blöcken gehören. Daraus folgt unter Verwendung von (3.8), daß $c_{ij} = 0$ ist, wenn $\mathfrak{F}_i$ und $\mathfrak{F}_j$ zu verschiedenen Blöcken gehören. Dies Resultat läßt sich folgendermaßen aussprechen

(4D) *Bildet man aus den Zerlegungszahlen d_{ij} und den Cartanschen Invarianten c_{ij} Matrizen D und C, so sind bei geeigneter Anordnung der Indizes diese Matrizen direkte Summen von den den einzelnen Blöcken B_τ entsprechenden Teilmatrizen D_τ bzw. C_τ.*

Besteht B_τ aus k_τ gewöhnlichen Darstellungen $\mathfrak{X}_i$ und aus l_τ modularen Darstellungen $\mathfrak{F}_j$, so hat D_τ gerade k_τ Zeilen und l_τ Spalten. Die Matrix C_τ ist quadratisch vom Grade l_τ. Aus (3.8) ergibt sich

$$C_\tau = D_\tau' D_\tau, \tag{4.5}$$

wo D_τ' die Transponierte von D_τ ist.

Beschränkt man sich ausschließlich auf die Betrachtung der gewöhnlichen Darstellungen, so hängt die Verteilung auf Blöcke allein von der rationalen Primzahl p und nicht von dem Primidealteiler $\mathfrak{p}$ von p ab. Beim Beweis dieser Behauptung darf man annehmen, daß Ω normal über dem Körper Q der rationalen Zahlen ist. Jeder Primidealteiler $\mathfrak{p}'$ von p in Ω geht aus dem festen Primideal $\mathfrak{p}$ durch Anwendung eines Elements T der Galoisschen Gruppe von Ω über Q hervor. Ist nun ε eine primitive g-te Einheitswurzel, so führt T dann ε in eine Potenz ε^s mit einem zu g teilerfremden Exponenten s über. Da $\chi_i(G)$ eine Summe von g-ten Einheitswurzeln ist, wird $\chi_i(G)$ in $\chi_i(G^s)$ transformiert, also $\omega_i(G)$ in $\omega_i(G^s)$. Wendet man T auf (4.4) an und ersetzt G^s durch G, so sieht man, daß (4.4) für $\mathfrak{p}'$ an Stelle von $\mathfrak{p}$ gilt. Daraus ergibt sich die Behauptung.

Kennt man die gewöhnlichen irreduziblen Charaktere χ_i von $\mathfrak{G}$, so kann man für jedes Primideal sofort die Anzahl t der Blöcke B_τ bestimmen, d.h. also feststellen, wieviel Blockideale $\mathfrak{B}_\tau$ in der Zerlegung (1.4) von $\mathfrak{p}J$ auftreten. Die Anzahl der in $\mathfrak{B}_\tau$ aufgehenden Primideale $\mathfrak{P}_j$ von J ist die Anzahl l_τ der modularen Darstellungen $\mathfrak{F}_j$ in B_τ. Sind die modularen Charaktere φ_j bekannt, so kann man leicht l_τ bestimmen. Die unten beim Beweis von (5A) verwendete Methode zeigt, daß man l_τ auch finden kann wenn man nur die χ_i kennt. Da unter unseren Voraussetzungen der Restklassenring $J/\mathfrak{P}_j$ ein voller Matrizenring über Ω^* ist, ist seine Struktur durch seine Gradzahl f_j bestimmt. Es ist aber f_j der Grad der zu $\mathfrak{P}_j$ gehörigen modularen Darstellung $\mathfrak{F}_j$, also $f_j = \varphi_j(1)$. Da sich die Cartanschen Invarianten c_{ij} aus den modularen Charakteren berechnen lassen[8]), so erhält man auch gewisse Information bezüglich des Restklassenrings $J/\mathfrak{B}_\tau$. Kennt man die modularen Charaktere φ_j, so kann man auch ein System von Kongruenzen für die Koeffizienten a_G des Elements α in (1.1) aufstellen, die die notwendigen und hinreichenden Bedingungen darstellen, daß α zu dem Primideal $\mathfrak{P}_j$ gehört. Es sei $\varphi_i^*(G)$ die Spur von $\mathfrak{F}_i(G)$ für $G \in \mathfrak{G}$. Setzt man wie gelegentlich in § 3 $G = RP$, wo R und P Potenzen von G sind, und wo R p-regulär ist, während die Ordnung von P eine Potenz von p ist, so sieht man leicht, daß $\varphi_i^*(G) = \varphi_i^*(R)$ ist, und daß $\varphi_i^*(R)$ aus dem Werte $\varphi_i(R)$ des Charakters φ_i von $\mathfrak{F}_i$ für das p-reguläre Element R durch Restklassenabbildung hervorgeht. Dann gilt

(4E) *Das zur irreduziblen modularen Darstellung $\mathfrak{F}_i$ von $\mathfrak{G}$ gehörige Primideal $\mathfrak{P}_i$ von J besteht aus denjenigen Elementen $\alpha = \sum a_G G$ von J, $a_G \in \mathfrak{o}$, für die die Gleichungen*

$$\sum_G a_G^* \varphi_i^*(GH) = 0 \tag{4.6}$$

für alle $H \in \mathfrak{G}$ gelten. Hier bezeichnet a_G^* das aus $a_G \in \mathfrak{o}$ durch Restklassenabbildung (mod $\mathfrak{p}$) *erhaltene Element von $\Omega^* = \mathfrak{o}/\mathfrak{p}$.*

Beweis. Man sieht leicht, daß die Elemente $\alpha \in J$, für die Gln. (4.6) gelten, ein Ideal $\mathfrak{M}$ von J bilden; $\mathfrak{P}_i \subseteqq \mathfrak{M} \subseteqq J$. Daher ist entweder $J = \mathfrak{M}$ oder $J = \mathfrak{P}_i$. Wäre $J = \mathfrak{M}$, so müßte die Spur von $\mathfrak{F}_i(\alpha^*)$ für alle $\alpha^* \in J^*$ verschwinden, was nicht der Fall ist. Daher ist $\mathfrak{M} = \mathfrak{P}_i$, wie behauptet war.

Der Beweis gilt auch dann noch, wenn Ω ein beliebiger algebraischer Zahlkörper ist, also nicht notwendig die g-ten Einheitswurzeln enthält. In der Tat ist dann Ω^* ein endlicher Körper, also vollkommen, und daher enthält J^* Elemente α^*, denen in der irreduziblen Darstellung $\mathfrak{F}_i$ von J^* lineare Transformationen $\mathfrak{F}_i(\alpha^*)$ mit von Null verschiedener Spur entsprechen.

§ 5. Zuordnung von p-regulären Klassen zu den Blöcken

Enthält der Block B_τ wie zuvor l_τ modulare Charaktere und ist l die Anzahl der p-regulären Klassen von $\mathfrak{G}$, so ist nach (3B)

$$l = l_1 + l_2 + \cdots + l_t. \tag{5.1}$$

Wir wollen jetzt die l p-regulären Klassen so den Blöcken B_τ zuordnen, daß dem Block B_τ gerade l_τ Klassen entsprechen, $\tau = 1, 2, \ldots, t$. Dabei wählen wir die Bezeichnungen so, daß dieselben Indizes j bei den $\varphi_j \in B_\tau$ und bei den B_τ zugeordneten Klassen auftreten. Genauer wollen wir zeigen

---

[8]) Sieht man die Zahlen $n(G)$ für p-reguläres G als bekannt an, so sieht man dies aus (3.12) und (3.6). Deutlicher noch ergibt sich diese Behauptung aus den Formeln $\gamma_{ij}\, g = \sum_R \overline{\varphi_i(R)}\, \varphi_j(R)$, wo R über alle p-regulären Elemente läuft und wo (γ_{ij}) die zu $C = (c_{ij})$ inverse Matrix C^{-1} bezeichnet. Für den Beweis der Formeln sei auf die in [3]) zitierten Arbeiten verwiesen.

(5 A) *Es sei S_τ die Menge der l_τ Indizes j, für die φ_j zu B_τ gehört, $\tau = 1, 2, \ldots, t$. Man kann die gewöhnlichen Charaktere χ_i von $\mathfrak{G}$ und die Klassen $\mathfrak{K}_i$ so numerieren, daß alle Klassen $\mathfrak{K}_i$ mit $i \in S_\tau$ p-regulär sind, daß die Charaktere χ_i mit $i \in S_\tau$ zu B_τ gehören und daß für jedes τ die folgenden Beziehungen gelten*

$$\left.\begin{array}{lll} (5.2) & \mathrm{Det}\,(\chi_i(G_j)) \not\equiv 0 & (\mathrm{mod}\,\mathfrak{p}), \\ (5.3) & \mathrm{Det}\,(\varphi_i(G_j)) \not\equiv 0 & (\mathrm{mod}\,\mathfrak{p}), \\ (5.4) & \mathrm{Det}\,(d_{ij}) \not\equiv 0 & (\mathrm{mod}\,p), \end{array}\right\} (i, j \in S_\tau),$$

wobei in allen drei Determinanten der Zeilenindex i und der Spaltenindex j über die l_τ Werte in S_τ laufen.

Beweis. Nach (3 C) kann man l Charaktere χ_i so auswählen, daß

$$\Delta = \mathrm{Det}\,(\chi_i(G_j)) \not\equiv 0 \quad (\mathrm{mod}\,\mathfrak{p}) \tag{5.5}$$

ist, wobei hier i und j die Werte $1, 2, \ldots, l$ durchlaufen und $G_1, G_2, \ldots, G_l$ Vertreter für die l p-regulären Klassen $\mathfrak{K}_1, \mathfrak{K}_2, \ldots, \mathfrak{K}_l$ sind. Entwickelt man Δ nach dem Laplaceschen Satz, so kann man Δ in der Form schreiben

$$\Delta = \sum_\varrho \pm \Delta_1^{(\varrho)} \Delta_2^{(\varrho)} \ldots \Delta_t^{(\varrho)},$$

wobei $\Delta_\tau^{(\varrho)}$ eine Unterdeterminante ist, in der die zu B_τ gehörigen unter den Charakteren χ_i auftreten. Es gebe etwa m_τ derartige Charaktere. Ist $m_\tau = 0$, so ist $\Delta_\tau^{(\varrho)} = 1$ zu setzen. Da in Δ gerade l Charaktere χ_i auftreten, hat man

$$l = m_1 + m_2 + \cdots + m_t. \tag{5.6}$$

Wegen (5.5) kann man ein Glied $\Delta_1^{(\varrho)} \Delta_2^{(\varrho)} \ldots \Delta_t^{(\varrho)}$ so auswählen, daß

$$\Delta_1^{(\varrho)} \Delta_2^{(\varrho)} \ldots \Delta_t^{(\varrho)} \not\equiv 0 \quad (\mathrm{mod}\,\mathfrak{p}) \tag{5.7}$$

ist. Drückt man die in $\Delta_\tau^{(\varrho)}$ auftretenden Charaktere nach (3.5) durch modulare Charaktere φ_j aus, so treten nur modulare Charaktere $\varphi_j \in B_\tau$ auf, vgl. (4 D). Wäre nun der Grad m_τ von $\Delta_\tau^{(\varrho)}$ größer als die Anzahl l_τ der modularen Charaktere in B_τ, so würde $\Delta_\tau^{(\varrho)}$ verschwinden, was im Widerspruch zu (5.7) steht. Daher ist $m_\tau \leq l_\tau$ für jedes τ. Vergleich von (5.1) und (5.6) zeigt jetzt, daß $m_\tau = l_\tau$ ist, $\tau = 1, 2, \ldots, t$.

Ändert man jetzt die Numerierung der Charaktere χ_i und der Klassen $\mathfrak{K}_j$ so ab, daß für jedes τ in der Determinante $\Delta_\tau^{(\varrho)}$ gerade die l_τ Charaktere χ_i mit $i \in S_\tau$ und die l_τ Klassen $\mathfrak{K}_j$ mit $j \in S_\tau$ auftreten, so folgt (5.2) aus (5.7). Drückt man wie oben die χ_i in dieser Determinante durch die φ_i aus, so sieht man, daß die Determinante in (5.2) das Produkt der Determinanten in (5.3) und (5.4) ist. Daher sind auch diese beiden Determinanten nicht durch $\mathfrak{p}$ teilbar. Da die d_{ij} ganze rationale Zahlen sind, kann man in (5.4) den Modul $\mathfrak{p}$ durch p ersetzen, und damit ist (5 A) bewiesen. Die Bedingungen von (5 A) können im allgemeinen auf mehrere Weisen erfüllt werden.

Nach (3.12) ist

$$\sum_{i,j} c_{\alpha\beta}\,\overline{\varphi_i(G_\alpha)}\,\varphi_j(G_\beta) = n(G_\alpha)\,\delta_{\alpha\beta},$$

wobei $\delta_{\alpha\beta}=0$ oder 1 zu setzen ist, je nachdem ob $\alpha \neq \beta$ oder $\alpha=\beta$ ist; $\alpha, \beta = 1, 2, \ldots, l$. Ist φ die Matrix $(\varphi_i(G_j))$, C die Matrix (c_{ij}) und N die Matrix $(n(G_i)\delta_{ij})$, $(i, j = 1, 2, \ldots, l)$, so kann man die Gleichung folgendermaßen in Matrizenform schreiben

$$\bar{\varphi}'\,C\,\varphi = N. \tag{5.8}$$

Arbeitet man im Ring $\mathfrak{o}_\mathfrak{p}$ der für $\mathfrak{p}$ ganzen Zahlen von Ω, so ist die Determinante von φ nach (3.6) eine Einheit. Daher zeigt (5.8), daß die Elementarteiler von C die in $n(G_1), n(G_2), \ldots, n(G_l)$ auftretenden Potenzen einer lokalen Primzahl von $\mathfrak{o}_\mathfrak{p}$ sind. Statt dessen kann man dann auch die Potenzen

$$p^{\nu(G_1)}, p^{\nu(G_2)}, \ldots, p^{\nu(G_l)} \tag{5.9}$$

nehmen. Da C ganze rationale Koeffizienten hat, sind dies auch die Elementarteiler von C im Ring I_p der für p ganzen, rationalen Zahlen. Tatsächlich ist die Determinante von C eine Potenz von p[9]), und daher sind die Zahlen (5.9) die Elementarteiler von C im Ring der ganzen rationalen Zahlen. Es wird aber im folgenden genügen, im Ring I_p zu operieren. Wir zeigen

(5B) *Die Elementarteiler der zum Block B_τ gehörigen Teilmatrix C_τ von C sind die Zahlen $p^{\nu(G_j)}$ mit $j \in S_\tau$.*

Beweis. Hat C_τ die Elementarteiler $e_1, e_2, \ldots, e_{l_\tau}$, so zeigen wir zunächst, daß bei geeigneter Anordnung der j in S_τ die Ungleichungen $e_j \geq p^{\nu(G_j)}$ gelten. Wir können die e_j als Potenzen von p wählen und setzen dementsprechend $e_j = p^{\varepsilon_j}$. Ist H_τ eine Diagonalmatrix, die in der Hauptdiagonale die Elementarteiler e_j enthält, so gibt es zwei in I_p unimodulare Matrizen $U=(u_{ij})$ und $V=(v_{ij})$, $(i, j \in S_\tau)$, vom Grade l_τ derart, daß

$$U\,C_\tau\,V^{-1} = H_\tau \tag{5.10}$$

ist.

Nach (3.10) und (4D) haben wir für p-reguläres G

$$\Phi_i(G) = \sum_{j\in S_\tau} c_{ij}\,\varphi_j(G); \qquad (i \in S_\tau). \tag{5.11}$$

Man setze

$$\tilde{\Phi}_i(G) = \sum_{j\in S_\tau} u_{ij}\,\Phi_j(G); \qquad \tilde{\varphi}_i(G) = \sum_{j\in S_\tau} v_{ij}\,\varphi_j(G); \qquad (i \in S_\tau). \tag{5.12}$$

Wegen (5.10) nimmt dann (5.11) die Form an

$$\tilde{\Phi}_j(G) = e_j\,\tilde{\varphi}_j(G) = p^{\varepsilon_j}\tilde{\varphi}_j(G), \qquad (j \in S_\tau).$$

---

[9]) Vgl. R. BRAUER, Ann. of Math. **42**, 53—61 (1941) oder Ann. of Math. **57**, 357—377 (1953), Satz 13.

Jetzt zeigt (3F), daß für alle p-regulären G und für alle $j \in S_\tau$ die Ungleichungen gelten

$$\nu(\tilde{\varphi}_j(G)) \geq \nu(G) - \varepsilon_j. \tag{5.13}$$

Aus (5.12) erhält man

$$\operatorname{Det}(\tilde{\varphi}_i(G_j)) = \operatorname{Det}(v_{ij}) \operatorname{Det}(\varphi_i(G_j)); \qquad (i, j \in S_\tau).$$

Da V unimodular ist, ist $\operatorname{Det}(v_{ij})$ eine Einheit von I_p. Aus (5.3) folgt daher, daß die Determinante auf der linken Seite eine Einheit von $\mathfrak{o}_\nu$ ist. Nimmt man also die l_τ Elemente G_j mit $j \in S_\tau$ in geeigneter Anordnung, so darf man annehmen, daß

$$\nu\Big(\prod_{j \in S_\tau} \tilde{\varphi}_j(G_j)\Big) = 0$$

ist. Dann ist $\nu(\tilde{\varphi}_j(G_j)) = 0$, und jetzt zeigt (5.13), daß $\varepsilon_j \geq \nu(G_j)$ ist, wie wir es zeigen wollten. Diese Ungleichungen gelten für jeden Block B_τ. Da C die direkte Summe der Matrizen C_τ ist, ist die Menge der Elementarteiler von C die Vereinigungsmenge der Mengen der Elementarteiler der C_τ. Würde auch nur in einem einzigen Fall in $\varepsilon_j \geq \nu(G_j)$ das Ungleichheitszeichen gelten, so könnten wir nicht in (5.9) die Elementarteiler von C haben. Also gilt stets $\varepsilon_j = \nu(G_j)$ und damit ist (5B) vollständig bewiesen.

Unsere Methode liefert auch noch den Hilfssatz

(5C) *Ist p^m der größte Elementarteiler von C_τ, so gelten für alle $\varphi_i \in B_\tau$, alle $\chi_i \in B_\tau$ und alle p-regulären Elemente G von $\mathfrak{G}$ die Beziehungen*

$$\nu(\varphi_i(G)) \geq \nu(G) - m; \qquad \nu(\chi_i(G)) \geq \nu(G) - m. \tag{5.14}$$

In der Tat folgt aus (5.13), daß $\nu(\tilde{\varphi}_j(G)) \geq \nu(G) - m$ ist, $j \in S_\tau$. Da V unimodular war, kann man nach den zweiten Gln. (5.12) φ_i als Linearverbindung der $\tilde{\varphi}_j$ mit Koeffizienten in $\mathfrak{o}_\nu$ darstellen. Daraus folgt die erste Behauptung (5.14). Die zweite Behauptung folgt aus der ersten unter Beachtung von (3.5) und (4D).

§ 6. Der Defekt eines Blockes

Es sei jetzt p^a die höchste in g aufgehende Potenz der Primzahl p, also in unseren Bezeichnungen $a = \nu(g)$. Wir definieren den Defekt d eines Blockes B_τ als die kleinste ganze rationale Zahl d, für die die Grade x_i aller Charaktere $\chi_i \in B_\tau$ durch p^{a-d} teilbar sind. Man kann dann also $x_i = p^{a-d} x_i'$ mit ganzem rationalem x_i' setzen; es gibt $\chi_i \in B_\tau$, für die x_i' zu p teilerfremd ist.

Wir haben bei der Definition des Defekts die gewöhnlichen Charaktere von B_τ verwendet. Wir hätten ebensogut die modularen Charaktere nehmen können. In der Tat gilt

(6A) *Hat B_τ den Defekt d, so sind die Grade f_i aller modularen irreduziblen Charaktere φ_i in B_τ sämtlich durch p^{a-d}, aber nicht alle durch p^{a-d+1} teilbar.*

Beweis. Man bestimme die ganze rationale Zahl δ so, daß die Grade f_i der $\varphi_i \in B_\tau$ alle durch $p^{a-\delta}$, aber nicht alle durch $p^{a-\delta+1}$ teilbar sind. Aus

(3.5) ergibt sich für $G=1$ unter Berücksichtigung von (4D), daß $p^{a-\delta}$ ein Teiler von p^{a-d} ist. Andererseits folgt aus (5.4) und (3.5), daß für p-reguläre Elemente die $\varphi_i \in B_\tau$ als Linearverbindungen von l_τ geeigneten Charakteren χ_i aus B_τ mit rationalen, für p ganzen, Koeffizienten darstellen kann. Für $G=1$ zeigt dies, daß p^{a-d} ein Teiler von $p^{a-\delta}$ ist. Daher muß $d=\delta$ sein, wie behauptet war.

Es seien jetzt Determinanten $\Delta_1^{(\varrho)}, \Delta_2^{(\varrho)}, \ldots, \Delta_t^{(\varrho)}$ wie in (5.7) ausgewählt. Verwenden wir dieselben Bezeichnungen wie in § 5, so haben wir also

$$\Delta_\tau^{(\varrho)} = \operatorname{Det}\left(\chi_i(G_j)\right) \not\equiv 0 \pmod{\mathfrak{p}}, \tag{6.1}$$

wo i und j die l_τ Werte in S_τ durchlaufen. Man wähle eine feste Spalte, etwa die r-te Spalte und multipliziere die Elemente dieser Spalte mit $g/n(G_r)$. Nach (2.3) ist $g\,\chi_i(G_r)/n(G_r) = x_i\,\omega_i(G_r)$. Da x_i durch p^{a-d} teilbar ist und $\omega_i(G_r)$ ganz algebraisch ist, so ist $g\,\Delta_\tau^{(\varrho)}/n(G_r)$ durch p^{a-d} teilbar. Aus (6.1) folgt dann

$$\nu(G_r) \leq d. \tag{6.2}$$

Da die χ_i mit $i \in S_\tau$ alle zum Block B_τ gehören, so sind nach (4.4) die l_τ Zahlen $\omega_i(G_r)$ für die l_τ Werte $i \in S_\tau$ alle untereinander $(\operatorname{mod} \mathfrak{p})$ kongruent. Daher kann das Gleichheitszeichen in (6.2) höchstens dann gelten, wenn $\omega_i(G_r) \not\equiv 0$ $(\operatorname{mod} \mathfrak{p})$ für $i \in S_\tau$ ist.

Nehmen wir nun an, daß für alle $r \in S_\tau$ das Ungleichheitszeichen in (6.2) gilt. Aus (5B) folgt dann, daß alle Elementarteiler von C kleiner als p^d sind. Dann ist $m \leq d-1$ in (5.14) und für $G=1$ ergibt sich, daß für die Grade aller $\chi_i \in B_\tau$ die Ungleichung $\nu(x_i) > a-d$ gilt. Dies widerspricht der Definition des Defekts d. Wir können also $r \in S_\tau$ so wählen, daß

$$\nu(G_r) = d, \qquad \omega_i(G_r) \not\equiv 0 \pmod{\mathfrak{p}} \tag{6.3}$$

ist.

Wählt man, was nach Definition von d möglich ist, den Charakter $\chi_i \in B_\tau$ zunächst so, daß $\nu(x_i) = a-d$ ist, so hat man

$$\nu\left(\omega_i(G)\right) = \nu\left(\frac{g\,\chi_i(G)}{n(G)\,x_i}\right) = a + \nu\left(\chi_i(G)\right) - \nu(G) - \nu(x_i) \geq d - \nu(G).$$

Insbesondere ist $\omega_i(G) \equiv 0 \pmod{\mathfrak{p}}$, falls $\nu(G) < d$ ist. Wegen (4.4) gilt dies für alle $i \in S_\tau$. Sagen wir wie in § 2, daß ein Element G den Defekt ν_0 hat, wenn $\nu(G) = \nu_0$ ist, so liefert (6.3) das folgende Resultat

(6B) *Ein irreduzibler Charakter χ_i von $\mathfrak{G}$ gehört dann und nur dann zu einem Block vom Defekt d, wenn für alle Elemente G von $\mathfrak{G}$ von einem Defekt kleiner als d die Kongruenz $\omega_i(G) \equiv 0 \pmod{\mathfrak{p}}$ gilt, während es p-reguläre Elemente G vom Defekt d gibt, für die $\omega_i(G) \not\equiv 0 \pmod{\mathfrak{p}}$ ist. (Wie immer ist ω_i der zu χ_i gehörige Charakter der Klassenalgebra $Z(\mathfrak{G}, \Omega)$.)*

Es sei jetzt B_τ ein Block, für den $l_\tau \geq 2$ ist und es seien r und s zwei verschiedene zu S_τ gehörige Indizes. Multipliziert man dann in (6.1) die r-te Spalte mit $g/n(G_r)$, die s-te Spalte mit $g/n(G_s)$ und drückt man in beiden

Spalten wieder die χ_i nach (2.3) durch die ω_i aus, so erhält man eine Formel

$$\frac{g}{n(G_r)}\,\frac{g}{n(G_s)}\,\Delta_\tau = |\ldots,\ x_i\,\omega_i(G_r),\ldots,\ x_i\,\omega_i(G_s),\ldots|,$$

wo auf der rechten Seite die i-te Zeile der Determinante angegeben ist, und wo in den nicht hingeschriebenen Spalten ganze algebraische Koeffizienten, nämlich Werte $\chi_i(G_j)$ stehen. Da alle $\chi_i \in B_\tau$ Grade x_i haben, die durch p^{a-d} teilbar sind, so kann man aus der r-ten und der s-ten Spalte je den Faktor p^{a-d} herausziehen. Ist dies geschehen, und wendet man auf die so entstehende Determinante die Restklassenabbildung (mod $\mathfrak{p}$) an, so werden nach (4.4) zwei Spalten proportional; man erhält also 0. Nach (6.1) hat man also $a - \nu(G_r) + a - \nu(G_s) > 2(a-d)$, d.h.

$$\nu(G_r) + \nu(G_s) < 2d. \tag{6.4}$$

Ist G_r wie in (6.3) gewählt, so hat man $\nu(G_s) < d$. Es ist also genau einer der Werte $\nu(G_j)$ mit $j \in S_\tau$ gleich d und alle anderen sind kleiner als d. Auf Grund von (5B) liefert dies

(6C) *Ist C_τ die zu einem Block vom Defekt d gehörige Teilmatrix der Matrix der Cartanschen Invarianten von $\mathfrak{G}$, so hat C_τ genau einen Elementarteiler p^d, während alle Elementarteiler kleiner als p^d sind.* anderen [R.B.]

Da nach (5.5) die Elementarteiler aller C_τ zusammen bekannt sind, liefert dies Ergebnis bereits gewisse Auskunft über die Anzahl der Blöcke von gegebenem Defekt[10]). Insbesondere kann eine Zahl p^a in (5.9) nur von einem Block vom Defekt a herrühren, und jeder solche Block liefert nur einmal p^a. Wir haben also

(6D) *Die Anzahl der Blöcke B vom maximalen Defekt $a = \nu(g)$ ist gleich der Anzahl der p-regulären Klassen $\mathfrak{K}_j$ vom Defekt a.*

Im Fall $d = 0$ kann C_τ überhaupt nur einen einzigen Elementarteiler haben, und dieser muß 1 sein. Man hat also $l_\tau = 1$, $C_\tau = (1)$. Nach (4.5) folgt dann, daß D_τ auch nur eine einzige Zeile haben kann, da alle d_{ij} ganze rationale Zahlen sind, und keine Zeile von D_τ ausschließlich aus Nullen besteht, wie man z.B. aus (3.5) für $G=1$ sieht. So erhalten wir das Resultat

(6E) *Jeder Block vom Defekt 0 besteht aus genau einem gewöhnlichen irreduziblen Charakter und einem irreduziblen modularen Charakter. Diese beiden Charaktere stimmen für p-reguläre Elemente überein, während der erstere für alle nicht p-regulären Elemente verschwindet.*

Der letzte Teil ergibt sich aus (3F) unter Verwendung von (3.9) und $D_\tau = (1)$.

Gehört χ_i zu einem Block B vom Defekt 0, so nennen wir χ_i einen *Charakter vom Defekt* 0. Nach Definition ist der Grad x_i durch die höchst mögliche

[10]) Die folgenden Ergebnisse (6D) und (6E) finden sich in der zweiten in [3]. erwähnten Arbeit. Der Satz (6C) wurde von C. Nesbitt und dem Verfasser kurz nach dem Entstehen der Arbeiten [3]. mit einem von dem hier gegebenen verschiedenen Beweis gefunden aber nicht veröffentlicht.

Potenz p^a von p teilbar. Man kann auch die Umkehrung zeigen, daß aus $p^a | x_i$ folgt, daß χ_i zu einem Block vom Defekt 0 gehört[11]). Für eine derartige Darstellung $\mathfrak{X}_i$ ist die entsprechende modulare Darstellung $\mathfrak{X}_i^*$ irreduzibel und nach den Formeln (3.9) gleichzeitig ein direkter Summand der modularen regulären Darstellung von $\mathfrak{G}$. Daraus folgt nach der allgemeinen Theorie der Darstellungen von Algebren auch noch, daß $\mathfrak{X}_i^*$ ein direkter Summand in allen modularen Darstellungen ist, in denen es als Bestandteil auftritt. Wir werden die letzteren Tatsachen im folgenden nicht verwenden. Man sieht, daß im Fall des Defekts 0 die modulare Darstellungstheorie mit der gewöhnlichen Darstellungstheorie zusammenfällt. Im allgemeinen sind die Abweichungen um so größer, je größer der Defekt ist. Es sei noch bemerkt, daß im Fall $g \not\equiv 0 \pmod{p}$ alle Blöcke natürlich den Defekt 0 haben müssen und daß daher überhaupt kein Unterschied zwischen der modularen und der gewöhnlichen Theorie besteht.

Von Wichtigkeit für uns ist das folgende Kriterium

(6F) *Es seien χ_α und χ_β zwei irreduzible Charaktere von $\mathfrak{G}$, die beide zu Blöcken vom Defekt d gehören. Die beiden Charaktere gehören dann und nur dann zu demselben Block, wenn für alle p-regulären Elemente G vom Defekt d die Kongruenz gilt*

$$\omega_\alpha(G) \equiv \omega_\beta(G) \pmod{\mathfrak{p}}.$$

Beweis. Die Notwendigkeit der Bedingung ist nach (4.4) klar. Um die Umkehrung zu zeigen, nehmen wir an, daß χ_α zum Block B_σ und χ_β zum Block $B_\tau \neq B_\sigma$ gehört. Wir verwenden jetzt eine Methode, die ganz analog der oben beim Beweis von (6C) verwendeten ist. Es sei $B_0 = B_\sigma \cup B_\tau$ die durch Zusammenwerfen der Blöcke B_σ und B_τ entstehende Menge von gewöhnlichen und modularen Charaktere. Geht man wieder von (5.5) aus und wendet man die Laplacesche Entwicklung an, wobei jetzt B_σ und B_τ durch die eine Menge B_0 ersetzt sind, so erhält man eine zu B_0 gehörige Unterdeterminante $\Delta_0 \not\equiv 0 \pmod{\mathfrak{p}}$. Durch weitere Entwicklung von Δ_0 sieht man, daß man annehmen darf, daß Δ_σ und Δ_σ in (5.7) komplementäre Unterdeterminanten von Δ_0 sind. Man darf also annehmen, daß die Indizes in Δ_0 gerade die Indizes in der Vereinigungsmenge $S_\sigma \cup S_\tau$ sind. Nach (6C) gibt es dann ein $r \in S_\sigma$ und ein $s \in S_\tau$ derart daß $\nu(G_r) = \nu(G_s) = d$ ist. Nach Voraussetzung ist nun $\omega_\alpha(G_r) \equiv \omega_\beta(G_r)$, $\omega_\alpha(G_s) \equiv \omega_\beta(G_s) \pmod{\mathfrak{p}}$. Daraus folgt nach (4.4), daß $\omega_i(G_r) \equiv \omega_j(G_r)$, $\omega_i(G_s) \equiv \omega_j(G_s) \pmod{\mathfrak{p}}$ für alle $\chi_i, \chi_j \in B_0 = B_\sigma \cup B_\tau$ ist. Drückt man jetzt in Δ_0 die Koeffizienten $\chi_i(G_r)$ und $\chi_i(G_s)$ in der r-ten und s-ten Spalte durch die entsprechenden ω_i aus, so erhält man genau wie oben die Ungleichung (6.4). Wegen $\nu(G_r) = \nu(G_s) = d$ ist dies ein Widerspruch, und damit ist (6F) bewiesen.

Schließlich zeigen wir noch:

(6G) *Ist eine Anzahl h von verschiedenen Charakteren χ_i vom Defekt 0 gegeben, so kann man h p-reguläre Elemente G_j vom Defekt 0 finden, derart daß*

$$\text{Det}\left(\chi_i(G_j)\right) \not\equiv 0 \pmod{\mathfrak{p}} \tag{6.5}$$

[11]) Vgl. § 13 der zweiten in [3]. genannten Arbeit.

ist, wo in der Determinante die Indizes i und j die h Werte durchlaufen, die zu den gegebenen Charakteren χ_i, bzw. zu den h gewählten Elementen G_j gehören.

Beweis. Nach (6E) ist jeder der m Charaktere χ_i der einzige gewöhnliche Charakter in seinem Block. Weiter folgt dann aus (5A), daß die h Charaktere χ_i in der Determinante (5.5) erscheinen. Durch Entwicklung nach den betreffenden Zeilen erhält man eine Unterdeterminante h-ten Grades, die die Bedingung (6.5) erfüllt. Wir haben zu zeigen, daß die hier auftretenden G sämtlich den Defekt 0 haben. Wäre eine der Zahlen $\nu(G_j)$ positiv, so folgte aus (2.3), daß $\chi_i(G_j) \equiv 0 \pmod{\mathfrak{p}}$ für alle unsere χ_i gilt, da $\omega_i(G_j)$ ganz algebraisch und g/x_i zu p teilerfremd ist. Dann wäre aber (6.5) unmöglich. Also haben alle G_j in (6.5) den Defekt 0.

§ 7. Zuordnung der Blöcke von $\mathfrak{G}$ zu Blöcken von Untergruppen

Es sei wie immer $\mathfrak{G}$ eine Gruppe endlicher Ordnung g, und p sei eine fest gewählte Primzahl. Wir betrachten Untergruppen $\mathfrak{T}$ und $\mathfrak{H}$ von $\mathfrak{G}$, die den folgenden Bedingungen genügen:

(*) Für jedes $T \in \mathfrak{T}$ gibt es eine p-Untergruppe $\mathfrak{P}_T$ von $\mathfrak{N}(T) \cap \mathfrak{N}(\mathfrak{T})$, derart daß $\mathfrak{C}(\mathfrak{P}_T) \subseteqq \mathfrak{T}$ gilt.

(**) $$\mathfrak{T} \subseteqq \mathfrak{H} \subseteqq \mathfrak{N}(\mathfrak{T}).$$

Die Existenz derartiger Untergruppen $\mathfrak{T}$ ergibt sich aus der Bemerkung

(7A) *Ist $\mathfrak{Q}$ eine p-Untergruppe von $\mathfrak{G}$, so genügt $\mathfrak{T} = \mathfrak{C}(\mathfrak{Q})$ der Bedingung* (*). *Ist $\mathfrak{H}$ eine Untergruppe von $\mathfrak{N}(\mathfrak{Q})$, die $\mathfrak{T}$ enthält, so ist* (**) *erfüllt.*

In der Tat, ist T ein Element von $\mathfrak{T} = \mathfrak{C}(\mathfrak{Q})$, so kann man $\mathfrak{P}_T = \mathfrak{Q}$ wählen. Da $\mathfrak{T}$ invariant in $\mathfrak{N}(\mathfrak{Q})$ ist, ist $\mathfrak{N}(\mathfrak{Q}) \subseteqq \mathfrak{N}(\mathfrak{T})$, also $\mathfrak{H} \subseteqq \mathfrak{N}(\mathfrak{T})$.

Die Gruppen $\mathfrak{T}$ und $\mathfrak{H}$ in (7A) sind die einzigen Gruppen, für die die folgenden Betrachtungen angewendet werden.

(7B) *Es sei Ξ ein Körper der Charakteristik p, und es seien $\mathfrak{T}$ und $\mathfrak{H}$ Untergruppen von $\mathfrak{G}$, die den Bedingungen* (*) *und* (**) *genügen. Für jede Klasse $\mathfrak{K}_\alpha$ konjugierter Elemente von $\mathfrak{G}$ sei wie früher K_α das Element der Gruppenalgebra $\Gamma(\mathfrak{G}, \Xi)$, das die Summe der Elemente in $\mathfrak{K}_\alpha$ ist; es sei $K_\alpha^{(1)}$ die Summe der Elemente in $\mathfrak{K}_\alpha \cap \mathfrak{T}$*[12]*. Dann gibt es einen eindeutig bestimmten Homomorphismus H der Klassenalgebra $Z(\mathfrak{G}, \Xi)$ von $\mathfrak{G}$ in die Klassenalgebra $Z(\mathfrak{H}, \Xi)$ von $\mathfrak{H}$, der K_α auf $K_\alpha^{(1)}$ abbildet.*

Beweis. Offenbar besteht $\mathfrak{K}_\alpha \cap \mathfrak{T}$ aus vollen Klassen von $\mathfrak{H}$, und daher kann $K_\alpha^{(1)}$ als Element von $Z(\mathfrak{H}, \Xi)$ betrachtet werden. Man setze $K_\alpha = K_\alpha^{(1)} + K_\alpha^{(2)}$, so daß also $K_\alpha^{(2)}$ die Summe der nicht in $\mathfrak{T}$ liegenden Elemente von $\mathfrak{K}_\alpha$ ist. Ist die Multiplikation der K_α durch die Formel (2.2) gegeben, so haben wir

$$(K_\alpha^{(1)} + K_\alpha^{(2)})\,(K_\beta^{(1)} + K_\beta^{(2)}) = \sum_\gamma a_{\alpha\beta\gamma}(K_\gamma^{(1)} + K_\gamma^{(2)}).$$

[12]) Gibt es kein derartiges Element, so ist $K_\alpha^{(1)} = 0$ zu setzen.

Man betrachte hier beide Seiten als Summen von Elementen von $\mathfrak{G}$ und vergleiche die Koeffizienten der in $\mathfrak{T}$ liegenden Elemente. Offenbar besteht $K_\alpha^{(1)}K_\beta^{(1)}$ ausschließlich aus solchen Elementen, während in $K_\alpha^{(1)}K_\beta^{(2)}$ und in $K_\alpha^{(2)}K_\beta^{(1)}$ nur außerhalb $\mathfrak{T}$ liegende Elemente auftreten. Können wir also zeigen, daß jedes in $K_\alpha^{(2)}K_\beta^{(2)}$ erscheinende Element von $\mathfrak{T}$ einen durch p teilbaren Koeffizienten hat, so ergibt der Koeffizientenvergleich, daß in $Z(\mathfrak{H}, \Xi)$ die Gleichung gilt

$$K_\alpha^{(1)}K_\beta^{(1)} = \sum_\gamma a_{\alpha\beta\gamma} K_\gamma^{(1)}. \tag{7.1}$$

Da die K_α eine Basis von $Z(\mathfrak{G}, \Xi)$ bilden, zeigt Vergleich mit (2.2), daß die durch $K_\alpha \to K_\alpha^{(1)}$ definierte lineare Abbildung H von $Z(\mathfrak{G}, \Xi)$ in $Z(\mathfrak{H}, \Xi)$ ein Homomorphismus ist.

Die Vielfachheit N, mit der ein Element T von $\mathfrak{T}$ in $K_\alpha^{(2)}K_\beta^{(2)}$ auftritt, ist offenbar die Anzahl der geordneten Paare (X, Y) von Elementen von $\mathfrak{G}$, die den Bedingungen genügen

$$XY = T; \quad X \in \mathfrak{K}_\alpha; \quad X \notin \mathfrak{T}; \quad Y \in \mathfrak{K}_\beta; \quad Y \notin \mathfrak{T}.$$

Ist nun P ein Element der Gruppe $\mathfrak{P}_T$ in $(*)$, so erfüllt mit (X, Y) auch das Paar $(P^{-1}XP,\ P^{-1}YP)$ die gleichen Bedingungen. Nennen wir also zwei Paare (X, Y) und (X', Y') äquivalent, wenn es ein P in $\mathfrak{P}_T$ mit $X' = P^{-1}XP$, $Y' = P^{-1}YP$ gibt, so sind unsere N Paare auf Äquivalenzklassen verteilt. Die (X, Y) enthaltende Äquivalenzklasse besteht offenbar aus $(\mathfrak{P}_T : \mathfrak{P}_T \cap \mathfrak{N}(X) \cap \mathfrak{N}(Y))$ verschiedenen Paaren. Dieser Gruppenindex ist eine Potenz von p. Wäre er 1, so wäre $\mathfrak{P}_T \subseteq \mathfrak{N}(X)$, also $X \in \mathfrak{C}(\mathfrak{P}_T)$, daher wegen $(*)$ $X \in \mathfrak{T}$. Da dies einen Widerspruch darstellt, ist der Gruppenindex und dann auch N durch p teilbar. Das ist aber, was wir zu zeigen hatten, und damit ist (7B) bewiesen.

Als Zusatz haben wir

(7C) *Ist $\mathfrak{U}$ der Kern des Homomorphismus H in* (7B), *so besteht eine Basis von $\mathfrak{U}$ aus denjenigen K_α, für die $\mathfrak{K}_\alpha$ keine Elemente von $\mathfrak{T}$ enthält. Eine Basis des Bildes $\mathfrak{B}$ von $Z(\mathfrak{G}, \Xi)$ in $Z(\mathfrak{H}, \Xi)$ besteht aus den von* 0 *verschiedenen $K_\alpha^{(1)}$.*

Dabei hat man nur zu beachten, daß die verschiedenen $K_\alpha^{(1)} \neq 0$ sicher linear unabhängig sind, da sich zwei verschiedene von ihnen aus gänzlich verschiedenen Elementen zusammensetzen.

In der eben eingeführten Bezeichnung haben wir

$$Z(\mathfrak{G}, \Xi)/\mathfrak{U} \cong \mathfrak{B} \subseteq Z(\mathfrak{H}, \Xi). \tag{7.2}$$

Wir wählen jetzt für Ξ den Körper Ω^* aus § 2. Ist ψ ein linearer Charakter von $Z(\mathfrak{G}, \Omega^*)$, dessen Kern das Ideal $\mathfrak{U}$ enthält, so kann man ψ als linearen Charakter von $Z(\mathfrak{G}, \Omega^*)/\mathfrak{U}$ betrachten. Durch Anwendung des Isomorphismus in (7.2) erhält man daraus einen linearen Charakter der Teilalgebra $\mathfrak{B}$ von $Z(\mathfrak{H}, \Omega^*)$, und dieser kann dann zu einem linearen Charakter $\tilde{\psi}$ von $Z(\mathfrak{H}, \Omega^*)$

erweitert werden[13]). Geht man umgekehrt von einem linearen Charakter $\tilde{\psi}$ von $Z(\mathfrak{H}, \Omega^*)$ aus, so erhält man durch Beschränkung einen linearen Charakter von $\mathfrak{B}$. Mit Hilfe von (7.2) ergibt sich daraus ein linearer Charakter von $Z(\mathfrak{G}, \Omega^*)/\mathfrak{U}$, und dieser kann als linearer Charakter von $Z(\mathfrak{G}, \Omega^*)$ gedeutet werden, dessen Kern $\mathfrak{U}$ enthält. In dieser Weise erkennen wir:

(7D) *Die linearen Charaktere $\tilde{\psi}$ von $Z(\mathfrak{H}, \Omega^*)$ entsprechen* ~~*eineindeutig*~~ *den linearen Charakteren ψ von $Z(\mathfrak{G}, \Omega^*)$, deren Kern das Ideal $\mathfrak{U}$ in* (7C) *(für $\Xi = \Omega^*$) enthält. Für zugeordnete Charaktere $\tilde{\psi}$ und ψ gilt* [R.B.]

$$\psi(K_\alpha) = \tilde{\psi}(K_\alpha^{(1)}) = \sum_\varrho \tilde{\psi}(L_\varrho), \tag{7.3}$$

wo auf der rechten Seite $\mathfrak{L}_\varrho$ über diejenigen Klassen von $\mathfrak{H}$ läuft, die in $\mathfrak{K}_\alpha \cap \mathfrak{T}$ liegen, und wo L_ϱ die Summe der Elemente in $\mathfrak{L}_\varrho$ ist.

Wie wir in § 4 gesehen haben, entsprechen die linearen Charaktere der Klassenalgebra $Z(\mathfrak{G}, \Omega^*)$ den Blöcken von $\mathfrak{G}$. Ebenso entsprechen die linearen Charaktere der Klassenalgebra $Z(\mathfrak{H}, \Omega^*)$ den Blöcken von $\mathfrak{H}$. Daher folgt:

(7E) *Auf Grund der Zuordnung $\tilde{\psi} \to \psi$ in* (7D) *entspricht jedem Block $\tilde{B}$ von $\mathfrak{H}$ ein Block B von $\mathfrak{G}$. Dabei ist $\tilde{\psi}$ der dem Block $\tilde{B}$ zugeordnete lineare Charakter der Klassenalgebra $Z(\mathfrak{H}, \Omega^*)$ und ψ der dem Block B zugeordnete lineare Charakter von $Z(\mathfrak{G}, \Omega^*)$.*

Ist χ_i ein zum Block B gehöriger Charakter von $\mathfrak{G}$, und ist ω_i wie immer durch (2.3) definiert, so ist $\psi = \omega_i^*$. Ähnlich kann man $\tilde{\psi}$ durch einen zu $\tilde{B}$ gehörigen Charakter $\tilde{\chi}_i$ von $\mathfrak{H}$ ausdrücken. Setzt man dies in (7.3) ein, so erhält man nach einfacher Umformung das Ergebnis:

(7F) *Ist χ_i vom Grad x_i ein irreduzibler Charakter des Blocks B von $\mathfrak{G}$ und $\tilde{\chi}_j$ vom Grad $\tilde{x}_j$ ein irreduzibler Charakter des Blocks $\tilde{B}$ von $\mathfrak{H}$ in* (7E), *so gilt für jedes Element G von $\mathfrak{G}$ die Kongruenz*[14])

$$\chi_i(G) \equiv \frac{n(G)}{g} \frac{x_i}{\tilde{x}_j} \sum \tilde{\chi}_j(H) \pmod{\mathfrak{p}, p^{\nu(G) - \nu(g/x_i)}}, \tag{7.4}$$

wo H über alle Elemente von $\mathfrak{T}$ läuft, die in $\mathfrak{G}$ zu G konjugiert sind. Gibt es keine derartigen Elemente, so ist die Summe gleich 0 zu setzen.

§ 8. Die Defektgruppe eines Blockes B

Es sei B ein Block von $\mathfrak{G}$ vom Defekt d. Es sei ψ der dem Block B zugeordnete lineare Charakter der Klassenalgebra $Z(\mathfrak{G}, \Omega^*)$. Aus (6B) folgt, daß es Klassen $\mathfrak{K}_\alpha$ vom Defekt d gibt, für die $\psi(K_\alpha) \neq 0$ ist. Ist G ein Element einer derartigen Klasse $\mathfrak{K}_\alpha$ und ist $\mathfrak{D}$ eine p-Sylowgruppe des Normalisators $\mathfrak{N}(G)$, so nennen wir $\mathfrak{D}$ eine *Defektgruppe* des Blocks B. Es ist dann klar,

---

13) Vgl. § 14, (14A).

14) Die Bedeutung dieser Kongruenz ist natürlich folgendermaßen gemeint: Ist δ die Differenz der linken und rechten Seite von (7.4), so ist

$$\nu(\delta) > \nu(G) - \nu(g/x_i).$$

daß jede zu $\mathfrak{D}$ konjugierte Gruppe ebenfalls eine Defektgruppe von B ist. Da $\nu(G)=d$ ist, hat jede Defektgruppe die Ordnung p^d.

(8A) *Ist* ψ *der dem Block* B *zugeordnete lineare Charakter von* $Z(\mathfrak{G}, \Omega^*)$, *und ist* $\psi(K_\beta) \neq 0$ *für eine Klasse* $\mathfrak{K}_\beta$, *so enthält der Normalisator* $\mathfrak{N}(G_\beta)$ *eines jeden Elementes* G_β *von* $\mathfrak{K}_\beta$ *eine Defektgruppe von* B.

Beweis. Es sei wie oben die Klasse $\mathfrak{K}_\alpha$ derart gewählt, daß $\psi(K_\alpha) \neq 0$ ist, und daß $\mathfrak{K}_\alpha$ den Defekt d hat. Ist die Multiplikation der K_i wie immer durch (2.2) definiert, so hat man

$$\psi(K_\alpha)\,\psi(K_\beta) = \sum_\gamma a_{\alpha\beta\gamma}\,\psi(K_\gamma).$$

Ist also $\psi(K_\beta) \neq 0$, so muß es einen Index γ geben, für den

$$a_{\alpha\beta\gamma} \not\equiv 0 \pmod{p}; \qquad \psi(K_\gamma) \neq 0 \tag{8.1}$$

ist. Es sei G_γ ein Element von $\mathfrak{K}_\gamma$ und $\mathfrak{Q}$ von der Ordnung p^r eine p-Sylowgruppe von $\mathfrak{N}(G_\gamma)$. Dann muß also $\nu(G_\gamma)=r$ sein. Aus der zweiten Bedingung (8.1) folgt nach (6B), daß $d \leq r$ ist.

Jetzt nehme man $\mathfrak{T} = \mathfrak{C}(\mathfrak{Q})$, $\mathfrak{H} = \mathfrak{N}(\mathfrak{Q})$. Wie wir in (7A) gesehen haben, sind dann die Bedingungen (*) und (**) erfüllt, und daher gelten die Resultate von § 7. Nach (7.1) ist also

$$K_\alpha^{(1)} K_\beta^{(1)} = \sum_\gamma a_{\alpha\beta\gamma} K_\gamma^{(1)}.$$

Nun ist $G_\gamma \in \mathfrak{T} \cap \mathfrak{K}_\gamma$ also $K_\gamma^{(1)} \neq 0$. Wegen der linearen Unabhängigkeit der von 0 verschiedenen $K_i^{(1)}$ müssen $K_\alpha^{(1)}$ und $K_\beta^{(1)}$ beide von 0 verschieden sein. Daher gibt es ein Element G_α in $\mathfrak{K}_\alpha$ und ein Element G_β in $\mathfrak{K}_\beta$, die zu $\mathfrak{T}$ gehören. Dann ist $\mathfrak{Q} \subseteqq \mathfrak{N}(G_\alpha)$ und $\mathfrak{Q} \subseteqq \mathfrak{N}(G_\beta)$. Da $\nu(G_\alpha)=d$ ist und $\mathfrak{Q}$ die Ordnung p^r hat, muß $d \geq r$ und also $d=r$ sein. Folglich ist $\mathfrak{Q}$ eine Sylowgruppe von $\mathfrak{N}(G_\alpha)$ und, da $\psi(K_\alpha) \neq 0$ war, ist $\mathfrak{Q}$ eine Defektgruppe von B. Damit haben wir gezeigt, daß $\mathfrak{N}(G_\beta)$ eine Defektgruppe von B enthält. Das gleiche gilt dann für den Normalisator jedes zu G_β konjugierten Elements, wie zu beweisen war.

Ist $\mathfrak{D}$ eine beliebige Defektgruppe von B, so gibt es nach Definition eine Klasse $\mathfrak{K}_\beta$ mit $\psi(K_\beta) \neq 0$, derart daß $\mathfrak{D}$ Sylowgruppe des Normalisators $\mathfrak{N}(G)$ eines Elements G von $\mathfrak{K}_\beta$ ist. Dann muß also $\nu(G)=d$ sein. Wählt man eine feste Klasse $\mathfrak{K}_\alpha$ vom Defekt d mit $\psi(K_\alpha) \neq 0$ und wendet die Methode im Beweis von (8A) an, so sieht man, daß es eine Sylowgruppe $\mathfrak{Q}$ eines Normalisators $\mathfrak{N}(G_\alpha)$ mit $G_\alpha \in \mathfrak{K}_\alpha$ gibt, derart daß $\mathfrak{Q}$ zum Normalisator von Elementen G_β von $\mathfrak{K}_\beta$ gehört. Dann muß $\mathfrak{Q}$ die Ordnung p^d haben und also Sylowgruppe von $\mathfrak{N}(G_\beta)$ sein. Da G_β und G als Elemente der Klasse $\mathfrak{K}_\beta$ konjugiert sind, sind die Sylowgruppen $\mathfrak{Q}$ und $\mathfrak{D}$ ihrer Normalisatoren konjugiert. Damit haben wir gezeigt, daß eine beliebige Defektgruppe $\mathfrak{D}$ zu den Sylowgruppen der Normalisatoren der Elemente einer festen Klasse $\mathfrak{K}_\alpha$ konjugiert sind. Wir haben also bewiesen:

(8B) *Je zwei Defektgruppen eines Blocks* B *sind konjugiert.*

Wir zeigen weiter:

(8C) *Ist B ein Block von $\mathfrak{G}$ mit der Defektgruppe $\mathfrak{D}$, und setzt man $\mathfrak{T} = \mathfrak{C}(\mathfrak{D})$, $\mathfrak{H} = \mathfrak{N}(\mathfrak{D})$, so gibt es einen Block $\tilde{B}$ von $\mathfrak{H}$, dem B im Sinn von (7E) zugeordnet ist.*

Beweis. Es sei ψ der zu B gehörige lineare Charakter der Klassenalgebra $Z(\mathfrak{G}, \Omega^*)$. Ist $\psi(K_\beta) \neq 0$ für eine Klasse $\mathfrak{K}_\beta$, so folgt aus (8A), daß $\mathfrak{K}_\beta$ Elemente G enthält, für die $\mathfrak{D} \leqq \mathfrak{N}(G)$ ist. Dann ist $G \in \mathfrak{C}(\mathfrak{D}) = \mathfrak{T}$, also $\mathfrak{K}_\beta \cap \mathfrak{T}$ nicht leer und $K_\beta^{(1)} \neq 0$. Ist daher $K_\beta^{(1)} = 0$, so muß $\psi(K_\beta)$ verschwinden. In der Bezeichnung von (7C) enthält also der Kern von ψ das Ideal $\mathfrak{U}$. Jetzt zeigt (7D), daß es einen linearen Charakter $\tilde{\psi}$ von $Z(\mathfrak{H}, \Omega^*)$ gibt, dem ψ zugeordnet ist, und daraus ergibt sich die Behauptung.

(8D) *Sind $\mathfrak{T}$ und $\mathfrak{H}$ zwei beliebige Untergruppen von $\mathfrak{G}$, die die Bedingungen (*) und (**) von § 7 erfüllen, und ist $\tilde{B}$ ein Block von $\mathfrak{H}$, dem der Block B von $\mathfrak{G}$ im Sinn von (7E) entspricht, so ist die Defektgruppe $\tilde{\mathfrak{D}}$ von $\tilde{B}$ in einer Defektgruppe $\mathfrak{D}$ von B enthalten. Insbesondere ist der Defekt $\tilde{d}$ von $\tilde{B}$ höchstens gleich dem Defekt d von B.*

Beweis. Nach Definition der Defektgruppe gibt es eine Klasse $\mathfrak{K}_\alpha$ vom Defekt d, für die $\psi(K_\alpha) \neq 0$ für den zu B gehörigen linearen Charakter ψ von $Z(\mathfrak{G}, \Omega^*)$ ist. Der zu $\tilde{B}$ gehörige lineare Charakter $\tilde{\psi}$ von $Z(\mathfrak{H}, \Omega^*)$ ist mit ψ durch die Formel (7.3) verbunden. Daraus folgt, daß es eine in $\mathfrak{K}_\alpha \cap \mathfrak{T}$ enthaltene Klasse $\mathfrak{L}_\varrho$ von $\mathfrak{H}$ gibt, für die $\tilde{\psi}(L_\varrho) \neq 0$ ist. Ist G ein Element von $\mathfrak{L}_\varrho$, so zeigt Anwendung von (8A) auf die Gruppe $\mathfrak{H}$, daß der Normalisator $\mathfrak{N}(G) \cap \mathfrak{H}$ von G in $\mathfrak{H}$ eine Defektgruppe $\tilde{\mathfrak{D}}$ von $\tilde{B}$ enthält. Dann gehört $\tilde{\mathfrak{D}}$ zu einer Sylowgruppe $\mathfrak{D}$ von $\mathfrak{N}(G)$, und da $G \in \mathfrak{K}_\alpha$ ist, ist $\mathfrak{D}$ eine Defektgruppe von B. Der letzte Teil der Behauptung folgt aus $(\mathfrak{D}:1) = p^d$, $(\tilde{\mathfrak{D}}:1) = p^{\tilde{d}}$.

§ 9. Gruppen mit invarianten p-Untergruppen

Wir unterbrechen jetzt unsere allgemeinen Betrachtungen und behandeln den speziellen Fall einer Gruppe $\mathfrak{G}$, die eine invariante Untergruppe $\mathfrak{M}$ enthält. Speziell sind wir an dem Fall interessiert, daß $\mathfrak{M}$ eine p-Gruppe ist, da die Ergebnisse später auf die Gruppe $\mathfrak{H} = \mathfrak{N}(\mathfrak{D})$ in (8C) an Stelle von $\mathfrak{G}$ angewendet werden sollen. Die Untersuchung ist dabei etwas weitergeführt, als es für unsere Zwecke unbedingt nötig ist.

Wir bezeichnen die Faktorgruppe von $\mathfrak{G}$ nach der invarianten Untergruppe $\mathfrak{M}$ mit $\overline{\mathfrak{G}}$;

$$\overline{\mathfrak{G}} = \mathfrak{G}/\mathfrak{M}. \tag{9.1}$$

Ist G ein Element von $\mathfrak{G}$, so sei stets $\overline{G} = G\mathfrak{M}$ das entsprechende Element von $\overline{\mathfrak{G}}$.

(9A) *Für $G \in \mathfrak{G}$ ist $\mathfrak{N}(G)\mathfrak{M}/\mathfrak{M}$ eine Untergruppe des Normalisators $\overline{\mathfrak{N}}(\overline{G})$ von $\overline{G}$ in $\overline{\mathfrak{G}}$. Die Ordnung $\bar{n}(\overline{G})$ von $\overline{\mathfrak{N}}(\overline{\mathfrak{G}})$ ist teilbar durch $n(G)/z$, wo z die Ordnung von $\mathfrak{N}(G) \cap \mathfrak{M} = \mathfrak{Z}$ ist.*

Der erste Teil der Behauptung ist klar. Der zweite Teil folgt daraus, daß $\mathfrak{N}(G)\mathfrak{M}/\mathfrak{M} \simeq \mathfrak{N}(G)/\mathfrak{Z}$ die Ordnung $n(G)/z$ hat.

Jeder Charakter von $\overline{\mathfrak{G}}$ kann als Charakter von $\mathfrak{G}$ aufgefaßt werden, dessen Kern die Gruppe $\mathfrak{M}$ enthält. In diesem Sinn behaupten wir:

(9B) *Jeder Block $\overline{B}$ von $\overline{\mathfrak{G}}$ ist in einem Block B von $\mathfrak{G}$ enthalten.*

Beweis. Wir setzen $m = (\mathfrak{M}:1)$. Ist χ_i ein gewöhnlicher irreduzibler Charakter von $\overline{\mathfrak{G}}$, so definieren wir $\overline{\omega}_i$ in Analogie mit (2.3) durch

$$\overline{\omega}_i(\overline{G}) = \frac{g}{m\,\overline{n}(\overline{G})}\,\frac{\chi_i(\overline{G})}{x_i}, \qquad (x_i = \chi_i(1)).$$

Betrachten wir χ_i jetzt als irreduziblen Charakter von $\mathfrak{G}$ und hat ω_i die alte Bedeutung (2.3), so ist also

$$\omega_i(G) = \frac{m\,\overline{n}(\overline{G})}{n(G)}\,\overline{\omega}_i(\overline{G}), \tag{9.2}$$

da wir $\chi_i(G) = \chi_i(\overline{G})$ zu setzen haben. Aus (9A) folgt, daß $\overline{n}(\overline{G})\,m/n(G)$ ganz ist, da ja z ein Teiler von m ist.

Gehören nun zwei irreduzible Charaktere χ_i und χ_j von $\overline{\mathfrak{G}}$ zum selben Block $\overline{B}$, so ist nach (4.4) $\overline{\omega}_i(\overline{G}) \equiv \overline{\omega}_j(\overline{G}) \pmod{\mathfrak{p}}$ für jedes G in $\mathfrak{G}$. Dann zeigt (9.2), daß $\omega_i(G) \equiv \omega_j(G) \pmod{\mathfrak{p}}$ für jedes G in $\mathfrak{G}$ ist. Daraus folgt, daß χ_i und χ_j als Charaktere von $\mathfrak{G}$ zum selben Block B gehören. Ist $\mathfrak{X}_i$ die Darstellung von $\overline{\mathfrak{G}}$ mit dem Charakter χ_i, so kann man die irreduziblen modularen Darstellungen von $\overline{B}$ als die in den Darstellungen $\mathfrak{X}_i^*$ für $\mathfrak{X}_i \in \overline{B}$ wirklich auftretenden irreduziblen Bestandteile $\mathfrak{F}_j$ definieren; vgl. (4D). Faßt man jetzt alle diese Darstellungen als Darstellungen von $\mathfrak{G}$ auf, so sieht man, daß alle $\mathfrak{F}_j \in \overline{B}$ zu B gehören, und damit ist (9B) bewiesen.

(9C) *Hat B in* (9B) *die Defektgruppe $\mathfrak{D}$, so enthält $\mathfrak{D}$ eine p-Sylowgruppe von $\mathfrak{M}$. Die Untergruppe $\mathfrak{M}\mathfrak{D}/\mathfrak{M}$ von $\overline{\mathfrak{G}}$ enthält eine Defektgruppe $\overline{\mathfrak{D}}$ von $\overline{B}$, und daher ist der Defekt $\overline{d}$ von $\overline{B}$ höchstens gleich $d - \nu(m)$.*

Beweis. Ist χ_i ein zu $\overline{B}$ und also zu B gehöriger Charakter, und ist ψ der B entsprechende lineare Charakter der Klassenalgebra $Z(\mathfrak{G}, \Omega^*)$, so ist $\omega_i^* = \psi$, § 4. Nach Definition der Defektgruppe $\mathfrak{D}$ gibt es Elemente $G \in \mathfrak{G}$ vom Defekt d, derart daß $\psi(K_\alpha) \neq 0$ ist, wobei $\mathfrak{K}_\alpha$ die G enthaltende Klasse ist. Dann ist $\omega_i(G) \not\equiv 0 \pmod{\mathfrak{p}}$. Aus (9.2) folgt dann

$$\nu(m) + \nu(\overline{n}(\overline{G})) = \nu(n(G)) = d; \qquad \overline{\omega}_i(\overline{G}) \not\equiv 0 \pmod{\mathfrak{p}}. \tag{9.3}$$

Da nach (9A) $\nu(\overline{n}(\overline{G})) \geq \nu(n(G)) - \nu(z) \geq \nu(n(G)) - \nu(m)$ ist, muß $\nu(z) = \nu(m)$ sein. Jede Sylowgruppe $\mathfrak{P}_0$ von $\mathfrak{Z}$ ist daher auch Sylowgruppe von $\mathfrak{M}$. Wegen $\mathfrak{Z} \subseteq \mathfrak{N}(G)$ kann man annehmen, daß $\mathfrak{P}_0$ in einer Sylowgruppe von $\mathfrak{N}(G)$, d.h. in einer Defektgruppe $\mathfrak{D}$ von B enthalten ist. Offenbar enthält dann auch jede zu $\mathfrak{D}$ konjugierte Gruppe eine Sylowgruppe von $\mathfrak{M}$. Aus dem zweiten Teil von (9.3) ergibt sich ferner nach (8A), daß jede Sylowgruppe von $\overline{\mathfrak{N}}(\overline{G})$ eine Defektgruppe $\overline{\mathfrak{D}}$ von $\overline{B}$ enthält. Nach dem ersten Teil von (9.3) hat eine solche Sylowgruppe die Ordnung $p^{d-\nu(m)}$. Offenbar hat die Untergruppe

$\mathfrak{M}\mathfrak{D}/\mathfrak{M}$ von $\mathfrak{N}(\overline{G})$ diese Ordnung, ist also Sylowgruppe, und daher kann man $\overline{\mathfrak{D}} \leqq \mathfrak{M}\mathfrak{D}/\mathfrak{M}$ annehmen, wie behauptet war.

Von jetzt an setzen wir voraus, daß die invariante Untergruppe von $\mathfrak{G}$ eine p-Gruppe ist.

(9D) *Enthält $\mathfrak{G}$ eine invariante Untergruppe $\mathfrak{M}$ der Ordnung p^μ, so ist $\mathfrak{M}$ in dem Kern jeder irreduziblen modularen Darstellung $\mathfrak{F}$ von $\mathfrak{G}$ enthalten.*

Beweis. Beschränkt man in der Darstellung $\mathfrak{F}$ das Argument auf $\mathfrak{M}$, so erhält man eine Darstellung von $\mathfrak{M}$, die nach einem Satz von A. H. Clifford[15]) vollständig reduzibel ist. Nun ist die einzige irreduzible modulare Darstellung einer p-Gruppe die Einsdarstellung [z.B. nach (3B)]. Daher ist $\mathfrak{F}(M)$ die Identität für jedes $M \in \mathfrak{M}$, und daraus ergibt sich die Behauptung.

(9E) *Enthält die Gruppe $\mathfrak{G}$ eine invariante Untergruppe $\mathfrak{M}$ der Ordnung p^μ, so enthält jeder Block B von $\mathfrak{G}$ vom Defekt d mindestens einen Block $\overline{B}$ von $\overline{\mathfrak{G}}$ vom Defekt $\bar{d} = d - \mu$. Alle in B enthaltenen Blöcke von $\overline{\mathfrak{G}}$ haben höchstens den Defekt $d - \mu$.*

Beweis. Nach (6A) kann man in B eine irreduzible modulare Darstellung $\mathfrak{F}$ finden, deren Grad f die Primzahl p genau mit dem Exponenten $\nu(g) - d$ enthält. Nach (9D) kann man $\mathfrak{F}$ als Darstellung von $\overline{\mathfrak{G}} = \mathfrak{G}/\mathfrak{M}$ betrachten. Gehört dann $\mathfrak{F}$ zu dem Block $\overline{B}$ vom Defekt $\bar{d}$, so muß f durch $p^{\nu(g)-\mu-\bar{d}}$ teilbar sein, da die Ordnung von $\overline{\mathfrak{G}}$ die Primzahl p mit dem Exponenten $\nu(g) - \mu$ enthält. Dies liefert $d \leqq \bar{d} + \mu$.

Andererseits ist nach (9C) der Defekt eines beliebigen in B enthaltenen Blocks von $\overline{\mathfrak{G}}$ höchstens gleich $d - \mu$. Dies liefert die letzte Behauptung in (9E). Für $\bar{d}$ ergibt sich jetzt, daß $\bar{d} = d - \mu$ ist, und damit ist auch der erste Teil von (9E) bewiesen.

Wir heben noch den folgenden speziellen Fall unserer Ergebnisse hervor:

(9F) *Enthält $\mathfrak{G}$ eine invariante Untergruppe $\mathfrak{M}$ der Ordnung p^μ, so ist $\mathfrak{M}$ in der Defektgruppe $\mathfrak{D}$ jedes Blocks B von $\mathfrak{G}$ enthalten, und daher ist der Defekt d von B mindestens gleich μ.*

Da nämlich der Block B von $\mathfrak{G}$ nach (9E) einen Block $\overline{B}$ von $\overline{\mathfrak{G}}$ enthält, kann man (9C) anwenden. Da $\mathfrak{M}$ seine eigene Sylowgruppe ist, ergibt sich $\mathfrak{M} \leqq \mathfrak{D}$ und daraus $\mu \leq d$.

§ 10. Der Hauptsatz

Wir brauchen den folgenden Hilfssatz:

(10A) *Es sei $\mathfrak{D}$ eine p-Untergruppe von $\mathfrak{G}$, und es werde $\mathfrak{T} = \mathfrak{C}(\mathfrak{D})$, $\mathfrak{H} = \mathfrak{N}(\mathfrak{D})$ gesetzt. Gibt es in einer Klasse $\mathfrak{K}_i$ von $\mathfrak{G}$ Elemente G, für die $\mathfrak{D}$ Sylowgruppe des Normalisators $\mathfrak{N}(G)$ ist, so enthält $\mathfrak{K}_i$ genau eine aus Elementen von $\mathfrak{T}$ bestehende Klasse $\mathfrak{L}_j$ von $\mathfrak{H}$. Ist H ein Element von $\mathfrak{H}$ und ist $\mathfrak{D}$ Sylowgruppe des Normalisators $\mathfrak{N}(H) \cap \mathfrak{H}$ von H in $\mathfrak{H}$, so ist $\mathfrak{D}$ Sylowgruppe von $\mathfrak{N}(H)$, und daher kann die erste Behauptung auf die Klasse $\mathfrak{K}_i$ von $\mathfrak{G}$ angewendet werden, in der H liegt.*

---

[15]) Clifford, A. H.: Ann. of Math. 38, 533—550 (1937). Vgl. auch H. Weyl, The classical groups. Princeton 1939.

Beweis. Ist $\mathfrak{D} \subseteqq \mathfrak{N}(G)$, so ist $G \in \mathfrak{T} = \mathfrak{C}(\mathfrak{D})$, und daher enthält die Klasse $\mathfrak{K}_i$ von G Elemente von $\mathfrak{T}$. Ist G' ein zweites Element von $\mathfrak{K}_i \cap \mathfrak{T}$, so kann man $G' = X^{-1}GX$ mit $X \in \mathfrak{G}$ setzen. Wegen $G' \in \mathfrak{T}$, ist $\mathfrak{D} \subseteqq \mathfrak{N}(G')$, und da $\mathfrak{D}$ Sylowgruppe von $\mathfrak{N}(G)$ ist, so ist $\mathfrak{D}$ ebenfalls Sylowgruppe von $\mathfrak{N}(G')$. Da $X^{-1}\mathfrak{D}X$ zu $\mathfrak{N}(X^{-1}GX) = \mathfrak{N}(G')$ gehört, ist $X^{-1}\mathfrak{D}X$ auch Sylowgruppe von $\mathfrak{N}(G')$ und daher zu $\mathfrak{D}$ in $\mathfrak{N}(G')$ konjugiert. Man kann also ein Element Y in $\mathfrak{N}(G')$ finden, so daß $X^{-1}\mathfrak{D}X = Y^{-1}\mathfrak{D}Y$ ist. Dann gehört XY^{-1} zu $\mathfrak{H} = \mathfrak{N}(\mathfrak{D})$, und da XY^{-1} das Element G in G' transformiert, so gehören G und G' zu derselben Klasse von $\mathfrak{H}$. Damit ist der erste Teil von (10A) bewiesen.

Es sei H ein Element von $\mathfrak{H}$, für das $\mathfrak{D}$ Sylowgruppe von $\mathfrak{N}(H) \cap \mathfrak{H}$ ist, und es sei $\mathfrak{D}_1$ eine $\mathfrak{D}$ enthaltende Sylowgruppe von $\mathfrak{N}(H)$. Wäre $\mathfrak{D}_1 > \mathfrak{D}$, so gäbe es nach bekannten Sätzen über p-Gruppen Elemente M in $\mathfrak{D}_1$, die zu $\mathfrak{N}(\mathfrak{D}) = \mathfrak{H}$, aber nicht zu $\mathfrak{D}$ gehören. Dann wäre $\{M, \mathfrak{D}\}$ eine von $\mathfrak{D}$ verschiedene p-Gruppe, die in $\mathfrak{N}(H) \cap \mathfrak{H}$ enthalten ist. Dies ist ein Widerspruch, da dann $\mathfrak{D}$ nicht Sylowgruppe von $\mathfrak{N}(H) \cap \mathfrak{H}$ wäre. Also ist $\mathfrak{D}$ Sylowgruppe von $\mathfrak{N}(H)$, wie behauptet war.

Jetzt sind wir in der Lage, den folgenden Satz zu beweisen, der das Hauptergebnis dieser Arbeit darstellt:

(10B) *Es sei $\mathfrak{D}$ eine Untergruppe von $\mathfrak{G}$, deren Ordnung eine Potenz p^d von p ist. Man setze $\mathfrak{T} = \mathfrak{C}(\mathfrak{D})$, $\mathfrak{H} = \mathfrak{N}(\mathfrak{D})$. Dann besteht eine eineindeutige Zuordnung zwischen den Blöcken B von $\mathfrak{G}$ mit der Defektgruppe $\mathfrak{D}$ und den Blöcken $\tilde{B}$ von $\mathfrak{H}$ vom Defekt d. Dabei ist B dem Block $\tilde{B}$ im Sinn von § 7 zugeordnet, so daß die Formeln (7.3) und (7.4) gelten. Die Blöcke $\tilde{B}$ von $\mathfrak{H}$ vom Defekt d besitzen $\mathfrak{D}$ als Defektgruppe.*

Beweis. Ist B ein Block von $\mathfrak{G}$ mit der Defektgruppe $\mathfrak{D}$, so gibt es nach (8C) einen Block $\tilde{B}$ von $\mathfrak{H}$, dem B im Sinn von § 7 zugeordnet ist. Aus (8D) ergibt sich, daß der Defekt $\tilde{d}$ von $\tilde{B}$ höchstens gleich dem Defekt von B ist. Da aber B eine Defektgruppe $\mathfrak{D}$ von der Ordnung p^d hat, hat B den Defekt d; es ist also $\tilde{d} \leq d$. Die Gruppe $\mathfrak{H} = \mathfrak{N}(\mathfrak{D})$ hat eine invariante Untergruppe $\mathfrak{D}$ der Ordnung p^d. Wendet man also (9F) auf die Gruppe $\mathfrak{H}$ an Stelle von $\mathfrak{G}$ an, so sieht man, daß der Defekt $\tilde{d}$ von $\tilde{B}$ mindestens gleich d ist. Damit ist gezeigt, daß $\tilde{d} = d$ ist. Der eben verwendete Hilfssatz (9F) zeigt auch noch, daß jeder Block von $\mathfrak{H}$ vom Defekt d die Defektgruppe $\mathfrak{D}$ besitzt.

Es sei $\tilde{B}$ ein beliebiger Block von $\mathfrak{H}$ mit der Defektgruppe $\mathfrak{D}$. Ist $\tilde{\psi}$ der zu $\tilde{B}$ gehörige lineare Charakter der Klassenalgebra $Z(\mathfrak{H}, \Omega^*)$, so kann man also eine Klasse $\mathfrak{L}_j$ von $\mathfrak{H}$ und ein Element H in $\mathfrak{L}_j$ so wählen, daß $\tilde{\psi}(L_j) \neq 0$ ist, und daß $\mathfrak{D}$ eine Sylowgruppe des Normalisators $\mathfrak{N}(H) \cap \mathfrak{H}$ von H in $\mathfrak{H}$ ist. Aus (10A) folgt dann, daß $\mathfrak{D}$ dann auch Sylowgruppe von $\mathfrak{N}(H)$ selbst ist, und weiter, daß wenn H in $\mathfrak{G}$ zur Klasse $\mathfrak{K}_i$ gehört, man $\mathfrak{L}_j = \mathfrak{K}_i \cap \mathfrak{T}$ hat.

Für gegebenes $\tilde{\psi}$ definieren die Formeln (7.3) einen linearen Charakter ψ der Klassenalgebra $Z(\mathfrak{G}, \Omega^*)$. Dieser gehört zu dem Block B von $\mathfrak{G}$, der dem Block $\tilde{B}$ von $\mathfrak{H}$ nach § 7 zugeordnet ist. Wendet man die Formeln (7.3) auf die Klasse $\mathfrak{K}_i$ an, so hat man nach dem eben Bewiesenen

$$\psi(K_i) = \tilde{\psi}(L_j). \tag{10.1}$$

Da $\tilde{\psi}(L_j) \neq 0$ war, ist $\psi(K_i) \neq 0$. Dann folgt aber aus (8A), daß $\mathfrak{N}(H)$ eine Defektgruppe von B enthält. Nach (8D) ist der Defekt von B mindestens gleich dem Defekt d von $\tilde{B}$. Da aber $\mathfrak{D}$ von der Ordnung p^d eine Sylowgruppe von $\mathfrak{N}(H)$ ist, so muß B den Defekt d haben und $\mathfrak{D}$ Defektgruppe von B sein.

Nehmen wir jetzt an, daß es noch einen zweiten Block $\tilde{B}_1$ von $\mathfrak{H}$ vom Defekt d gäbe, dem derselbe Block B von $\mathfrak{G}$ zugeordnet ist; es sei $\tilde{\psi}_1$ der zu $\tilde{B}_1$ gehörige lineare Charakter von $Z(\mathfrak{H}, \Omega^*)$. Dann hat auch $\tilde{B}_1$ die Defektgruppe $\mathfrak{D}$. Es sei $\mathfrak{L}_j$ eine Klasse von $\mathfrak{H}$ vom Defekt d in $\mathfrak{H}$. Gibt es in $\mathfrak{L}_j$ kein Element H, für das $\mathfrak{D}$ Sylowgruppe des Normalisators $\mathfrak{N}(H) \cap \mathfrak{H}$ von H in $\mathfrak{H}$ ist, so folgt aus (8A) daß $\tilde{\psi}(L_j) = 0$ und ebenso, daß $\tilde{\psi}_1(L_j) = 0$ ist. Gibt es in $\mathfrak{L}_j$ Elemente H, für die $\mathfrak{D}$ Sylowgruppe von $\mathfrak{N}(H) \cap \mathfrak{H}$ ist, so gilt (10.1) und ebenso $\psi(K_i) = \tilde{\psi}_1(L_j)$, da ja B ebenfalls dem Block $\tilde{B}_1$ von $\mathfrak{H}$ zugeordnet ist; beim Beweis von (10.1) wurde die Bedingung $\tilde{\psi}(L_j) \neq 0$ nicht verwendet. Damit haben wir gezeigt, daß für alle Klassen $\mathfrak{L}_j$ von $\mathfrak{H}$ vom Defekt d die Gleichung

$$\tilde{\psi}(L_j) = \tilde{\psi}_1(L_j)$$

gilt. Dann folgt aber durch Anwendung von (6F) auf $\mathfrak{H}$, daß $\tilde{\psi} = \tilde{\psi}_1$, also $\tilde{B} = \tilde{B}_1$ ist, und damit ist (10B) in allen Teilen bewiesen.

Im Fall $d = 0$ ist unser Resultat (10B) inhaltslos, da hier $\mathfrak{D} = \{1\}$, $\mathfrak{T} = \mathfrak{H} = \mathfrak{G}$ ist. Es sei $d > 0$. Will man alle Blöcke vom Defekt d finden, so hat man zunächst ein volles System von nicht-konjugierten Untergruppen $\mathfrak{D}$ von $\mathfrak{G}$ von der Ordnung p^d aufzustellen. Selbstverständlich kann man alle $\mathfrak{D}$ in einer festen Sylowgruppe von $\mathfrak{G}$ wählen. Für jedes $\mathfrak{D}$ muß man dann $\mathfrak{H} = \mathfrak{N}(\mathfrak{D})$ bilden und alle Blöcke $\tilde{B}$ von $\mathfrak{H}$ vom Defekt d finden. Ist $\mathfrak{D}$ nicht invariant in $\mathfrak{G}$, so handelt es sich bei $\mathfrak{H}$ um eine Gruppe von kleinerer Ordnung als g. Wir werden aber in § 12 sehen, daß man die Behandlung von $\mathfrak{H}$ stets auf die von $\mathfrak{H}/\mathfrak{D}$ zurückführen kann, das jedenfalls kleinere Ordnung als $\mathfrak{G}$ hat. Damit wird dann gezeigt sein, daß die Aufstellung aller Blöcke mit gegebener Defektgruppe $\mathfrak{D} \neq \{1\}$ auf eine Untersuchung von Gruppen kleinerer Ordnung zurückgeführt werden kann.

Nicht jede Untergruppe der Ordnung p^d kann als Defektgruppe auftreten. Wir zeigen:

(10C) *Ist die Untergruppe $\mathfrak{D}$ von der Ordnung p^d Defektgruppe, so gibt es p-reguläre Elemente G von $\mathfrak{G}$, für die $\mathfrak{D}$ Sylowgruppe von $\mathfrak{N}(G)$ ist. Ferner kann $\mathfrak{D}$ nicht in einer invarianten p-Untergruppe $\mathfrak{D}_1 \neq \mathfrak{D}$ von $\mathfrak{N}(\mathfrak{D})$ enthalten sein.*

Ist $\mathfrak{D}$ Defektgruppe des Blocks B, und verwenden wir dieselben Bezeichnungen wie früher, so zeigt (6B) daß es p-reguläre Klassen $\mathfrak{K}_\alpha$ vom Defekt d gibt, für die $\psi(K_\alpha) \neq 0$ ist. Dann ist $\mathfrak{D}$ Sylowgruppe von $\mathfrak{N}(G)$ für geeignete

$G \in \mathfrak{K}_\alpha$. Gibt es eine invariante p-Untergruppe $\mathfrak{D}_1 > \mathfrak{D}$ von $\mathfrak{H}$, so enthält jede Defektgruppe eines Blocks $\tilde{B}$ von $\mathfrak{H}$ nach (9F) die Gruppe $\mathfrak{D}_1$, und daher besitzt $\mathfrak{H}$ keine Blöcke vom Defekt d.

§ 11. Fortsetzung von § 9

Nehmen wir an, daß die Gruppe $\mathfrak{G}$ eine invariante Untergruppe $\mathfrak{W}$ enthält. Ist ϑ ein Charakter von $\mathfrak{W}$ und G ein festes Element von $\mathfrak{G}$, so stellt für variables W in $\mathfrak{W}$ die Funktion $\vartheta_1(W) = \vartheta(GWG^{-1})$ wieder einen Charakter ϑ_1 von $\mathfrak{W}$ dar. Mit ϑ ist auch ϑ_1 irreduzibel. Wir sagen, daß die Charaktere ϑ und ϑ_1 von $\mathfrak{W}$ in $\mathfrak{G}$ *assoziiert* sind. Offenbar ist dies eine Äquivalenzrelation. Die Elemente G von $\mathfrak{G}$, für die $\vartheta(W) = \vartheta(GWG^{-1})$ für alle W in $\mathfrak{W}$ gilt, bilden eine Untergruppe $\mathfrak{S} \supseteq \mathfrak{W}$ von $\mathfrak{G}$, die *Trägheitsgruppe* von ϑ in $\mathfrak{G}$. Die Anzahl r der verschiedenen zu ϑ in $\mathfrak{G}$ assoziierten Charaktere ist gleich dem Index $(\mathfrak{G}:\mathfrak{S})$.

Ist χ_i ein irreduzibler Charakter von $\mathfrak{G}$, so hat die Beschränkung $\chi_i^{(0)}$ von χ_i auf $\mathfrak{W}$ die Form

$$\chi_i^{(0)} = q \sum_{\rho=1}^{r} \vartheta_\rho, \tag{11.1}$$

wo $\vartheta_1, \vartheta_2, \ldots, \vartheta_r$ eine Klasse von in $\mathfrak{G}$ assoziierten Charakteren von $\mathfrak{W}$ bilden, und wo q eine natürliche Zahl ist. Hat ϑ den Grad t, so ergibt sich durch Gradvergleich die Gleichung

$$x_i = q r t. \tag{11.2}$$

Wir nehmen jetzt an, daß $\mathfrak{G}$ eine invariante Untergruppe $\mathfrak{D}$ der Ordnung p^d besitzt. Wir wollen die Blöcke B von $\mathfrak{G}$ untersuchen, die die Defektgruppe $\mathfrak{D}$ haben. Setzt man

$$\mathfrak{T} = \mathfrak{C}(\mathfrak{D}), \qquad \mathfrak{W} = \mathfrak{T}\mathfrak{D}, \tag{11.3}$$

so sind auch $\mathfrak{T}$ und $\mathfrak{W}$ invariante Untergruppen von $\mathfrak{G}$.

Wie aus (9E) folgt, ist in B ein Block $\bar{B}$ von $\bar{\mathfrak{G}} = \mathfrak{G}/\mathfrak{D}$ vom Defekt $d - d = 0$ enthalten. Es sei χ_i der gewöhnliche irreduzible Charakter von $\bar{\mathfrak{G}}$ in $\bar{B}$; wir betrachten wieder χ_i auch als Charakter von $\mathfrak{G}$. Ist $\nu(g) = a$, so teilt p die Ordnung von $\bar{\mathfrak{G}}$ mit dem Exponenten $a - d$, und daher hat man

$$\nu(x_i) = a - d. \tag{11.4}$$

Wir verwenden dieselben Bezeichnungen wie früher. Nach Definition der Defektgruppe gibt es Elemente G, für die $\mathfrak{D}$ Sylowgruppe von $\mathfrak{N}(G)$ ist, derart daß $\psi(K_\alpha) \not\equiv 0$ für die G enthaltende Klasse $\mathfrak{K}_\alpha$ von $\mathfrak{G}$ ist. Dann ist $\nu(G) = d$, und wir haben $G \in \mathfrak{C}(\mathfrak{D}) = \mathfrak{T}$. Da $\psi = \omega_i^*$ ist, so folgt daß $\omega_i(G) \not\equiv 0 \pmod{\mathfrak{p}}$ ist. Aus (2.3) folgt weiter wegen (11.4), daß $\chi_i(G) \not\equiv 0 \pmod{\mathfrak{p}}$ ist. Wendet man jetzt (11.1) an, so erhält man zunächst

$$q \not\equiv 0 \pmod{p}. \tag{11.5}$$

Da $\mathfrak{D}$ im Kern von χ_i enthalten ist, so ist $\mathfrak{D}$ auch im Kern von jedem ϑ_ϱ enthalten, und deswegen kann ϑ_ϱ als irreduzibler ~~Darstellung~~ von $\mathfrak{W}/\mathfrak{D}$ betrachtet werden. Daher teilt der Grad t von ϑ_ϱ die Ordnung von $\mathfrak{W}/\mathfrak{D}$. Hat $\mathfrak{W}$ die Ordnung w, so liefert dies $\nu(t) \leq \nu(w) - d$. Andererseits folgt aus (11.2) wegen (11.4) und (11.5), daß

⌊ Charakter [R.B.]

$$a - d = \nu(x_i) = \nu(r) + \nu(t)$$

ist. Also muß $a \leq \nu(r) + \nu(w)$ sein. Da $r = (\mathfrak{G}:\mathfrak{S})$ war und $\mathfrak{G} \geqq \mathfrak{S} \geqq \mathfrak{W}$ ist, ist dies nur möglich, wenn wir

$$(11.6) \qquad \nu((\mathfrak{S}:\mathfrak{W})) = 0; \quad \nu(t) = \nu(w) - d$$

haben. Aus der zweiten Gl. (11.6) folgt, daß jedes ϑ_ϱ als Charakter von $\mathfrak{W}/\mathfrak{D}$ zu einem Block vom Defekt 0 gehört.

Umgekehrt sei jetzt ϑ ein Charakter von $\mathfrak{W}/\mathfrak{D}$, der zu einem Block von $\mathfrak{W}/\mathfrak{D}$ vom Defekt 0 gehört. Es sei $\mathfrak{S}$ die Trägheitsgruppe von ϑ in $\mathfrak{G}$, und es werde vorausgesetzt, daß $(\mathfrak{S}:\mathfrak{W})$ zu p teilerfremd ist. Dann gilt also (11.6). Ist $X_1, X_2, \ldots, X_r$ ein rechtsseitiges Restsystem von $\mathfrak{G}$ nach $\mathfrak{S}$, und setzt man

$$(11.7) \qquad \vartheta_i(W) = \vartheta(X_i W X_i^{-1}) \qquad (\text{für } W \in \mathfrak{W}),$$

so sind $\vartheta_1, \vartheta_2, \ldots, \vartheta_r$ die verschiedenen, zu ϑ in $\mathfrak{G}$ assoziierten Charaktere. Bildet man den von ϑ induzierten Charakter ζ von $\mathfrak{G}$, so ist

$$\zeta(W) = \frac{1}{w} \sum_{X \in \mathfrak{G}} \vartheta(X W X^{-1}) = (\mathfrak{S}:\mathfrak{W}) \sum_{\varrho=1}^{r} \vartheta_\varrho(W) \qquad (\text{für } W \in \mathfrak{W}).$$

Ist χ_i ein irreduzibler, in ζ auftretender Charakter von $\mathfrak{G}$, so müssen also in der Formel (11.1) für χ_i gerade dieselben Charaktere $\vartheta_1, \vartheta_2, \ldots, \vartheta_r$ wie hier erscheinen. Da $\mathfrak{D}$ im Kern von ϑ und daher im Kern von allen ϑ_ϱ enthalten ist, gehört es auch zum Kern von ζ und dann auch von χ_i. Also kann χ_i auch als Charakter von $\mathfrak{G}/\mathfrak{D}$ betrachtet werden. Daher ist der Grad x_i von χ_i ein Teiler von g/p^d, also $\nu(x_i) \leq a - d$. Aus (11.1) folgt (11.2), und daraus erhalten wir unter Verwendung von (11.6) die Beziehung

$$\begin{aligned} \nu(x_i) &= \nu(q) + \nu(r) + \nu(t) = \nu(q) + \nu((\mathfrak{G}:\mathfrak{S})) + \nu(t) \\ &= \nu(q) + \nu((\mathfrak{G}:\mathfrak{W})) + \nu(w) - d = \nu(q) + a - d. \end{aligned}$$

Daher muß $\nu(x_i) = a - d$, $\nu(q) = 0$ sein. Es gelten also wieder (11.4) und (11.5).

Es sei B der Block von $\mathfrak{G}$, zu dem χ_i gehört. Wir wollen zeigen, daß B den Defekt d hat. Dazu brauchen wir den Hilfssatz:

(11A) *Es sei $\mathfrak{D}$ eine invariante Untergruppe der Ordnung p^d von $\mathfrak{G}$ und es sei $\mathfrak{T} = \mathfrak{C}(\mathfrak{D})$, $\mathfrak{W} = \mathfrak{T}\mathfrak{D}$. Ist $\overline{T} = T\mathfrak{D}$ ein p-reguläres Element von $\mathfrak{W}/\mathfrak{D}$, so kann man den Repräsentanten T in $T\mathfrak{D}$ als p-reguläres Element ⌊ wählen. Ist dies geschehen, so ist der Normalisator $\mathfrak{N}(\overline{T})$ von $\overline{T}$ in $\overline{\mathfrak{G}}$ gerade $\mathfrak{N}(T)/\mathfrak{D}$.*

Beweis. Ist $\overline{T}$ p-regulär, so gibt es eine zu p teilerfremde, natürliche Zahl c, für die $\overline{T}^c = 1$ ist. Ist T ein Element der Restklasse $\overline{T}$, so ist $T^c \in \mathfrak{D}$, und da c zur Ordnung von $\mathfrak{D}$ teilerfremd ist, kann man eine Potenz

⌊ l.-5: $T \in \mathfrak{T}$ [R.B.]

T_0 von T^e finden, für die $T_0^e = T^e$ ist. Da T und T_0 vertauschbar sind, ist $T_1 = TT_0^{-1}$ p-regulär. Offenbar ist $T_1\mathfrak{D} = T\mathfrak{D}$ und damit ist der erste Teil von (11 A) bewiesen.

Ist T schon selbst p-regulär, und gehört $\overline{V} = V\mathfrak{D}$ zum Normalisator von $\overline{T}$ in $\mathfrak{G}$, so ist $V^{-1}TV \in T\mathfrak{D}$; man setze $V^{-1}TV = TD$ mit $D \in \mathfrak{D}$. Hier ist das Element auf der linken Seite p-regulär. Auf der rechten Seite ist T p-regulär, während die Ordnung von D eine Potenz von p ist. Da D und T vertauschbar sind, kann TD nur dann p-regulär sein, wenn $D = 1$, d.h. wenn $V \in \mathfrak{N}(T)$ ist. Also ist $\overline{\mathfrak{N}}(\overline{T}) \subseteq \mathfrak{N}(T)/\mathfrak{D}$. Da $\mathfrak{N}(T)/\mathfrak{D}$ sicher zu $\overline{\mathfrak{N}}(\overline{T})$ gehört, ist auch der zweite Teil von (11 A) bewiesen.

Nach Voraussetzung ist ϑ ein Charakter vom Defekt 0 von $\mathfrak{W}/\mathfrak{D}$. Wendet man den inneren Automorphismus $G \to XGX^{-1}$ von $\mathfrak{G}$ an, der $\mathfrak{W}$ in sich und ϑ in ϑ_i überführt, so sieht man, daß das gleiche für die r Charaktere ϑ_i gilt. Wie in (6 G) gezeigt wurde, gibt es dann r p-reguläre Elemente $\overline{T}_1, \overline{T}_2, \ldots, \overline{T}_r$ von $\mathfrak{W}/\mathfrak{D}$, derart daß

(11.8) $$\operatorname{Det}\left(\vartheta_i(\overline{T}_j)\right) \not\equiv 0 \pmod{\mathfrak{p}}, \qquad (i, j = 1, 2, \ldots, r)$$

[R.B.] ist. ~~und daß die Ordnung von $\overline{\mathfrak{N}}(\overline{T}_j)$ zu p teilerfremd ist.~~ Dann zeigt (11 A), daß wir T_j in der Restklasse $\overline{T}_j$ so wählen können, daß

[R.B.] (11.9) $$\nu(T_j) \geq d$$

ist.

Nach (2.3) finden wir dann unter Verwendung von (11.1) und (11.2), daß

(11.10) $$\omega_i(W) = \frac{g}{n(W)\,t\,r} \sum_{\varrho=1}^{r} \vartheta_\varrho(W)$$

für $W \in \mathfrak{W}$ ist. Für $W = T_j$ haben wir $\nu\left(g/(n(W)\,t\,r)\right) = \nu(g/x_i) - \nu(T_j) = 0$. Aus (11.8) folgt, daß wir nicht $\omega_i(T_j) \equiv 0 \pmod{\mathfrak{p}}$ für alle T_j haben können. Auf Grund von (6 B) ergibt sich also aus (11.9), daß der Block B von $\mathfrak{G}$ höchstens den Defekt d haben kann. Nach (9 F) ist der Defekt aber nicht kleiner als d, und daher muß er genau gleich d sein, wie behauptet war. Die Defektgruppe von B ist folglich $\mathfrak{D}$.

Wie (6 F) zeigt, ist der Block eindeutig durch die Werte $\omega_i^*(G)$ für p-reguläre Elemente G vom Defekt d bestimmt. Dabei darf man sich auf Elemente G beschränken, die zu Klassen $\mathfrak{K}_\alpha$ gehören, die Elemente T von $\mathfrak{T}$ enthalten. Für die anderen Klassen ergibt sich nämlich durch Anwendung von (10 B) und (7.3) mit $\mathfrak{H} = \mathfrak{G}$, daß ω_i^* verschwindet. Jetzt zeigt (11.10), daß B nur von $\vartheta_1, \vartheta_2, \ldots, \vartheta_r$ abhängt.

Schließlich sei ϑ' ein Charakter von $\mathfrak{W}/\mathfrak{D}$, der zu einem Block vom Defekt 0 gehört, aber von $\vartheta_1, \vartheta_2, \ldots, \vartheta_r$ verschieden ist. Hat dann $\chi_{i'}$ die entsprechende Bedeutung für ϑ' wie χ_i für ϑ, und gehört $\chi_{i'}$ zum Block B' von $\mathfrak{G}$, so behaupten wir, daß B von B' verschieden ist. Um dies zu zeigen, hat man (6 G) zur Bildung einer Determinante zu verwenden, in der die verschiedenen zu ϑ und ϑ' assoziierten Charaktere von $\mathfrak{W}$ auftreten. Vergleicht man dann (11.10) mit der entsprechenden Formel für $\omega_{i'}$, so sieht man, daß $\omega_{i'}^*$ für geeignete Elemente T von ω_i^* verschieden ist, und dann muß $B \neq B'$ sein.

Hat der Charakter ϑ von $\mathfrak{W}/\mathfrak{D}$ als Charakter von $\mathfrak{W}$ die Trägheitsgruppe $\mathfrak{S}$ in $\mathfrak{G}$, so hat er als Charakter von $\mathfrak{W}/\mathfrak{D}$ die Trägheitsgruppe $\mathfrak{S}/\mathfrak{D}$ in $\mathfrak{G}/\mathfrak{D}$. Damit erhalten wir den Satz:

(11B) *Es sei $\mathfrak{G}$ eine Gruppe, die eine invariante Untergruppe $\mathfrak{D}$ der Ordnung p^d besitzt. Es sei $\mathfrak{T} = \mathfrak{C}(\mathfrak{D})$, $\mathfrak{W} = \mathfrak{T}\mathfrak{D}$. Um alle Blöcke von $\mathfrak{G}$ vom Defekt d zu erhalten, hat man alle Charaktere ϑ vom Defekt 0 von $\mathfrak{W}/\mathfrak{D}$ zu finden, für die $\mathfrak{W}/\mathfrak{D}$ in der Trägheitsgruppe $\mathfrak{S}/\mathfrak{D}$ von ϑ einen zu p teilerfremden Index hat. Verteilen sich diese Charaktere auf u Familien von in $\mathfrak{G}/\mathfrak{D}$ assoziierten Charakteren, so gibt es u Blöcke B von $\mathfrak{G}$ vom Defekt d. Ist B der durch die Familie $\vartheta_1, \vartheta_2, \ldots, \vartheta_r$ von assoziierten Charakteren vom Grad t von $\mathfrak{W}/\mathfrak{D}$ bestimmte Block, und ist χ_i ein Charakter von* [handwritten correction: B] *so ist für Elemente $W \in \mathfrak{W}$:*

$$\chi_i(W) = \frac{x_i}{g\,t} \sum_{X \in \mathfrak{G}} \vartheta(XWX^{-1}). \tag{11.11}$$

Dabei erhält man (11.11) aus (11.1), indem man X über alle g Elemente von $\mathfrak{G}$ an Stelle über das Restsystem X_i laufen läßt und dementsprechend, durch g/r dividiert; $qr/g = x_i/(gt)$.

§ 12. Kongruenzen für die Werte der Charaktere

Es sei wieder $\mathfrak{G}$ eine beliebige endliche Gruppe, und es sei B ein Block von $\mathfrak{G}$ vom Defekt d mit der Defektgruppe $\mathfrak{D}$. Setzt man $\mathfrak{H} = \mathfrak{N}(\mathfrak{D})$, $\mathfrak{T} = \mathfrak{C}(\mathfrak{D})$, $\mathfrak{W} = \mathfrak{T}\mathfrak{D}$, so gibt es nach (10B) einen Block $\tilde{B}$ von $\mathfrak{H}$ vom Defekt d, dem B zugeordnet ist. Ist χ_i ein Charakter von B und $\tilde{\chi}_i$ ein Charakter von $\tilde{B}$, so ist nach (7.4)

$$\chi_i(G) \equiv \frac{n(G)}{g} \frac{x_i}{\tilde{x}_j} \sum_T \tilde{\chi}_j(T) \qquad (\mathrm{mod}\ \mathfrak{p}_\nu, p^{\nu(G) - \nu(g/x_i)}), \tag{12.1}$$

wo T über alle Elemente von $\mathfrak{T}$ läuft, die in $\mathfrak{G}$ zu G konjugiert sind. Da $\mathfrak{H} = \mathfrak{N}(\mathfrak{D})$ die invariante Untergruppe $\mathfrak{D}$ hat, kann man auf $\mathfrak{H}$ das Resultat (11B) anwenden. Ist also der Block $\tilde{B}$ von $\mathfrak{H}$ durch den Charakter ϑ von $\mathfrak{W}/\mathfrak{D}$ vom Defekt 0 bestimmt, so hat man für [handwritten insertion: χ] $\tilde{\chi}_j$ an Stelle von χ_i eine Formel (11.11):

$$\tilde{\chi}_j(T) = \frac{\tilde{x}_j}{h\,t} \sum_{X \in \mathfrak{H}} \vartheta(XTX^{-1}),$$

wo h die Ordnung von $\mathfrak{H}$ ist. Setzt man dies in (12.1) ein, so hat man

$$\chi_i(G) \equiv \frac{n(G)\,x_i}{g\,h\,t} \sum_T \sum_{X \in \mathfrak{H}} \vartheta(XTX^{-1}) \qquad (\mathrm{mod}\ \mathfrak{p}_\nu, p^{\nu(G) - \nu(g/x_i)}).$$

Jedes Glied in der Summe auf der rechten Seite hat die Form $\vartheta(T')$ mit $T' = XTX^{-1}$; $T \in \mathfrak{T}$ zu G in $\mathfrak{G}$ konjugiert; $X \in \mathfrak{H}$. Daher ist auch T' zu G

↓ l. 14: B, dessen Kern $\mathfrak{D}$ enthält, [R.B.]
↓ l. −7: geeignetes [R.B.]

konjugiert und $T' \in \mathfrak{T}$. Ein festes T' wird genau h-mal in der Form XTX^{-1} erhalten. Schreibt man also für T' wieder T, so haben wir

$$\chi_i(G) \equiv \frac{n(G)}{g}\,\frac{x_i}{t}\sum_T \vartheta(T) \qquad (\mathrm{mod}\ \mathfrak{p}, p^{\nu(G)-\nu(g/x_i)}), \tag{12.2}$$

wo T dieselben Werte wie in (12.1) durchläuft.

Man kann dies Resultat noch in etwas anderer Form schreiben. Der von ϑ (als Charakter von $\mathfrak{W}$) in $\mathfrak{G}$ induzierte Charakter Θ ist durch die Formel

$$w\,\Theta(G) = \sum_{X \in \mathfrak{G}} \vartheta(XGX^{-1})$$

definiert, wo X über die Elemente von $\mathfrak{G}$ läuft, für die $XGX^{-1} \in \mathfrak{W}$ ist, und wo w die Ordnung von $\mathfrak{W}$ ist. Man betrachte hier zunächst die Summe Σ_1 der Glieder, für die XGX^{-1} ein Element von $\mathfrak{T}$ ist. Da jedes feste T $n(G)$-mal auftritt, ist $(1/n(G))\,\Sigma_1$ gerade die in (12.2) auftretende Summe.

Es sei weiter Σ_2 die Summe der übrigbleibenden Glieder,

$$\Sigma_2 = \Sigma\vartheta(XGX^{-1}); \qquad XGX^{-1} \in \mathfrak{W}, \quad XGX^{-1} \notin \mathfrak{T}.$$

Hier setze man $XGX^{-1} = W = TD_0$ mit $T \in \mathfrak{T}$, $D_0 \in \mathfrak{D}$ und schreibe $T = AB$, wo A und B Potenzen von T sind, derart daß die Ordnung von A zu p teilerfremd ist und die von B eine Potenz von p. Dann liegen A und B in $\mathfrak{T}$, sind also mit D_0 vertauschbar. Deutet weiterhin ein Querstrich Bildung der Restklasse nach $\mathfrak{D}$ an, so haben wir

$$W = ABD_0, \quad \overline{W} = \overline{AB}.$$

Ist nun $B \notin \mathfrak{D}$, also $\overline{B} \neq \overline{1}$, so ist $\overline{W}$ nicht p-regulär. Da ϑ ein Charakter von $\mathfrak{W}/\mathfrak{D}$ vom Defekt 0 war, ist $\vartheta(\overline{W}) = \vartheta(W) = 0$, vgl. (6E). Daher darf man sich in Σ_2 auf diejenigen Glieder beschränken, für die $XGX^{-1} = W$ die Form AD mit $D \in \mathfrak{D}$ und einem p-regulären $A \in \mathfrak{T}$ hat. Wegen $W \notin \mathfrak{T}$ muß $D \neq 1$ sein; wir haben $\vartheta(W) = \vartheta(A)$. Jedes derartige W tritt $n(G)$-mal auf. Mit jedem W fasse man die Glieder W' zusammen, die in $\mathfrak{W}$ zu W konjugiert sind. Diese erfüllen dieselben Bedingungen wie W; $W' = A'D'$ mit $D' \in \mathfrak{D}$ und p-regulären $A' \in \mathfrak{T}$, $D' \neq 1$. Ferner ist $\vartheta(W) = \vartheta(W')$. Zu jedem W gibt es $w/n_0(W)$ verschiedene W', wo $n_0(W)$ die Ordnung des Normalisators $\mathfrak{N}_0(W) = \mathfrak{N}(W) \cap \mathfrak{W}$ von W in $\mathfrak{W}$ ist. Da A und D vertauschbar sind und teilerfremde Ordnungen haben, ist A eine Potenz von W und also $\mathfrak{N}_0(W) \subseteqq \mathfrak{N}_0(A) = \mathfrak{N}(A) \cap \mathfrak{W}$. Wir behaupten, daß $\nu(n_0(W)) < \nu(n_0(A))$ ist, wo $n_0(A)$ die Ordnung von $\mathfrak{N}_0(A)$ bezeichnet.

In der Tat, würde Gleichheit bestehen, so würde eine Sylowgruppe $\mathfrak{P}_0$ von $\mathfrak{N}_0(W)$ gleichzeitig eine Sylowgruppe von $\mathfrak{N}_0(A)$ sein. Wegen $A \in \mathfrak{T} = \mathfrak{C}(\mathfrak{D})$ gibt es Sylowgruppen von $\mathfrak{N}_0(A)$, die $\mathfrak{D}$ enthalten. Da $\mathfrak{D}$ invariant in $\mathfrak{W}$ ist, enthalten also alle Sylowgruppen von $\mathfrak{N}_0(A)$ die Gruppe $\mathfrak{D}$. Man hätte also $\mathfrak{D} \subseteqq \mathfrak{P}_0 \subseteqq \mathfrak{N}_0(W)$, was $W \in \mathfrak{C}(\mathfrak{D}) = \mathfrak{T}$ zur Folge hätte; wir hatten aber $W \notin \mathfrak{T}$ in Σ_2. Also ist $\nu(n_0(W)) < \nu(n_0(A))$, wie behauptet war.

Schließlich ist $w\vartheta(A)/(n_0(A)t)$ ganz algebraisch, also

$$\nu(\vartheta(A)) + \nu(w) \geq \nu(n_0(A)) + \nu(t) > \nu(n_0(W)) + \nu(t).$$

Faßt man all dies zusammen, so ergibt sich, daß für die Summe Σ_2' der zu einem gegebenen W gehörigen Glieder $\vartheta(W')$ die Beziehung gilt

$$\nu(\Sigma_2') \geq \nu(G) + \nu(w) - \nu(n_0(W)) + \nu(\vartheta(A)) > \nu(G) + \nu(t).$$

Da Σ_2 eine Summe von Teilsummen Σ_2' ist, ist

$$\nu(\Sigma_2) > \nu(G) + \nu(t).$$

Wir können nun (12.2) in der Form schreiben

$$\chi_i(G) \equiv \frac{1}{g}\frac{x_i}{t}(w\Theta(G) - \Sigma_2) \quad (\mathrm{mod}\, \mathfrak{p}, p^{\nu(G)-\nu(g/x_i)}).$$

Jetzt sieht man, daß man hier das Glied Σ_2 weglassen darf. Wir haben damit das folgende Resultat bewiesen

(12A) *Es sei* $\mathfrak{D}$ *eine Untergruppe von* $\mathfrak{G}$, *deren Ordnung eine Potenz* p^d *der Primzahl* p *ist; es sei* $\mathfrak{T} = \mathfrak{C}(\mathfrak{D})$, $\mathfrak{H} = \mathfrak{N}(\mathfrak{D})$. *Man bestimme ein volles System von Charakteren* $\vartheta_1, \vartheta_2, \ldots$ *von* $\mathfrak{T}\mathfrak{D}/\mathfrak{D}$ *vom Defekt* 0, *derartig daß* ϑ_i *in* $\mathfrak{H}/\mathfrak{D}$ *eine Trägheitsgruppe* $\mathfrak{S}_i/\mathfrak{D}$ *hat, für die der Index* $(\mathfrak{S}_i/\mathfrak{T}\mathfrak{D})$ *zu* p *teilerfremd ist. In diesem System finde man ein vollständiges Teilsystem* $\vartheta^{(1)}, \vartheta^{(2)}, \ldots, \vartheta^{(u)}$ *von Charakteren* ϑ_j *von denen keine zwei in* $\mathfrak{H}/\mathfrak{D}$ *assoziiert sind. Jedem* $\vartheta^{(\varrho)}$ *kann man dann einen Block* $B^{(\varrho)}$ *von* $\mathfrak{G}$ *mit der Defektgruppe* $\mathfrak{D}$ *zuordnen, und alle Blöcke* B *von* $\mathfrak{G}$ *mit der Defektgruppe* $\mathfrak{D}$ *werden genau einmal in der Form* $B^{(\varrho)}$ *erhalten. Ist* χ_i *ein gewöhnlicher irreduzibler Charakter in* $B^{(\varrho)}$, *so ist*[16])

$$\chi_i(G) \equiv \frac{1}{(\mathfrak{G}:\mathfrak{T}\mathfrak{D})}\frac{x_i}{t^{(\varrho)}}\Theta^{(\varrho)}(G) \quad (\mathrm{mod}\, \mathfrak{p}, p^{\nu(G)-\nu(g/x_i)}). \tag{12.3}$$

Hier bezeichnet $\Theta^{(\varrho)}$ *den von* $\vartheta^{(\varrho)}$ *induzierten Charakter von* $\mathfrak{G}$, *wobei* $\vartheta^{(\varrho)}$ *wieder als Charakter von* $\mathfrak{T}\mathfrak{D}$ *aufzufassen ist;* x_i *ist der Grad von* χ_i *und* $t^{(\varrho)}$ *der von* $\vartheta^{(\varrho)}$.

§ 13. Über die Anzahl der Blöcke vom Defekt 0

Wie bereits bemerkt, versagen unsere Methoden im Fall der Blöcke vom Defekt 0. Daher entsteht die Aufgabe, die Bestimmung der Anzahl der Blöcke vom Defekt 0 zu diskutieren. Sind die Grade x_i der irreduziblen Charaktere χ_i bekannt, so kann man unmittelbar ablesen, welche von ihnen durch die volle Potenz p^a von p in g teilbar sind. Die Anzahl dieser x_i ist dann die gesuchte Anzahl M. Natürlich ist M vollständig durch die Klassenalgebra $Z(\mathfrak{G}, \Omega^*)$ bestimmt. Das folgende Resultat (13A) drückt dies in einer

[16]) Die Bedeutung der Kongruenz (12.3) ist wieder die folgende: Ist δ die Differenz der linken und rechten Seite von (12.3), so ist

$$\nu(\delta) > \nu(G) - \nu(g/x_i).$$

konkreteren Form aus. Wir sind aber nicht in der Lage, die Bestimmung von M ähnlich wie im Fall eines positiven Defekts durchzuführen.

(13 A) *Die Elemente K_α der Klassenalgebra $Z(\mathfrak{G}, \Omega^*)$, die zu Klassen $\mathfrak{K}'_\alpha$ vom Defekt* 0 *gehören, bilden die Basis einer Teilalgebra U. Die Anzahl der Blöcke von $\mathfrak{G}$ vom Defekt* 0 *ist gleich der Dimension von U^n für genügend großes n.*

Beweis. Für jede p-Untergruppe $\mathfrak{D}$ von $\mathfrak{G}$ liefert Anwendung von (7C) mit $\mathfrak{T}=\mathfrak{C}(\mathfrak{D})$, $\mathfrak{H}=\mathfrak{N}(\mathfrak{D})$ ein Ideal $\mathfrak{U}(\mathfrak{D})$ von $Z(\mathfrak{G}, \Omega^*)$, dessen Basis aus denjenigen K_α besteht, für die $\mathfrak{K}_\alpha \cap \mathfrak{T}$ leer ist. Offenbar ist dann U der Durchschnitt aller dieser $\mathfrak{U}(\mathfrak{D})$ mit $\mathfrak{D} \neq (1)$. Dies zeigt, daß U ein Ideal von $Z(\mathfrak{G}, \Omega^*)$ und daher jedenfalls eine Teilalgebra ist.

Jeder lineare Charakter ψ_0 von U kann zu einem linearen Charakter ψ von $Z(\mathfrak{G}, \Omega^*)$ erweitert werden. Gehörte hier ψ zu einem Block B von positivem Defekt, so würde nach (6B) ψ für alle K_α verschwinden, für die $\mathfrak{K}_\alpha$ Defekt 0 hat. Dann würde ψ für alle Elemente von U verschwinden, was unmöglich ist. Daher muß ψ zu einem Block vom Defekt 0 gehören.

Geht man umgekehrt von einem Block B vom Defekt 0 aus, so zeigt (6G), daß das zugehörige ψ nicht in U identisch verschwindet und ferner, daß zwei derartige ψ nicht dieselbe Beschränkung auf U haben können. Dies zeigt, daß die Anzahl M der Blöcke von $\mathfrak{G}$ vom Defekt 0 gleich der Anzahl der verschiedenen linearen Charaktere von U ist.

Es sei weiter ψ_0 ein linearer Charakter von U und ψ seine Erweiterung zu einem linearen Charakter von $Z(\mathfrak{G}, \Omega^*)$. Da ψ zu einem Block B von $\mathfrak{G}$ vom Defekt 0 gehört, so folgt aus (6E) und dem in § 4 Gesagten, daß ψ nur ein einziges Mal als irreduzibler Bestandteil in der regulären Darstellung $\mathfrak{R}$ von $Z(\mathfrak{G}, \Omega^*)$ auftritt. Ist $\mathfrak{R}_0$ die Beschränkung von $\mathfrak{R}$ auf U, so tritt dann ψ_0 nur ein einziges Mal in $\mathfrak{R}_0$ auf, da ψ eindeutig durch ψ_0 bestimmt war. Nun ist die reguläre Darstellung $\mathfrak{R}_U$ von U ein Bestandteil von $\mathfrak{R}_0$. Es folgt also, daß die reguläre Darstellung $\mathfrak{R}_U$ der kommutativen Algebra U jeden ihrer M linearen Charaktere ψ_0 mit der Vielfachheit 1 enthält. Dann zeigt aber die Strukturtheorie der Algebren, daß U direkte Summe des Radikals von U und von M Idealen ist, die als Algebren zu Ω^* isomorph sind. Daraus ergibt sich, daß U^n für genügend großes n in der Tat die Dimension M hat.

§ 14. Anhang: Bemerkungen über Algebren

Um die Arbeit möglichst unabhängig zu machen, geben wir hier einige Bemerkungen über Algebren, die wir aber nur so weit verfolgen, wie es in unserem Zusammenhang nötig ist. In § 14 bezeichnet Γ eine Algebra endlichen Ranges über einem Grundkörper Ω.

(a) Unter einem Ideal $\mathfrak{M}$ von Γ wird stets ein zweiseitiges Ideal verstanden. Ein Ideal $\mathfrak{M} \neq \Gamma$ ist ein *Primideal*, wenn ein Produkt $\mathfrak{U}\mathfrak{V}$ zweier Ideale $\mathfrak{U}$, $\mathfrak{V}$ von Γ nur dann in $\mathfrak{M}$ enthalten ist, falls mindestens einer der Faktoren in $\mathfrak{M}$ enthalten ist. Ist Γ halbeinfach und also direkte Summe $\Gamma = B_1 \oplus \cdots \oplus B_s$ von Idealen, die einfache Ringe sind, so sieht man sofort, daß man genau s Primideale hat. Dies sind die Ideale $\mathfrak{P}_j = \sum_{\lambda \neq j} B_\lambda$. Ist Γ nicht halbeinfach und ist N das Radikal, so folgt aus der Nilpotenz des Radikals leicht, daß N in jedem Primideal von Γ enthalten ist. Ist $\Gamma \neq N$, hat also z.B. Γ ein Einheitselement, so sind die Primideale von Γ gerade diejenigen Ideale, die bei dem natürlichen Homomorphismus von Γ auf Γ/N den Primidealen von Γ/N entsprechen. Daraus ergibt sich, daß jedes Primideal $\mathfrak{P}$ von Γ maximal in Γ ist; die Umkehrung ist trivial. Ist $\mathfrak{P}$ ein Primideal von Γ, so ist $\Gamma/\mathfrak{P}$ eine einfache Algebra und hat daher genau eine irreduzible Darstellung $\mathfrak{F}$ (wobei ähnliche Darstellungen als nicht verschieden betrachtet sind). Dann kann $\mathfrak{F}$ als irreduzible Darstellung von Γ mit dem Kern $\mathfrak{P}$ betrachtet werden. Umgekehrt ist der Kern einer irreduziblen Darstellung $\mathfrak{F}$ von Γ notwendig ein maximales Ideal $\mathfrak{P}$ in Γ. Es besteht also eine eineindeutige Zuordnung zwischen den irreduziblen Darstellungen $\mathfrak{F}$ von Γ und den Primidealen $\mathfrak{P}$ von Γ. Dabei ist jeder irreduziblen Darstellung $\mathfrak{F}$ der Kern $\mathfrak{P}$ von $\mathfrak{F}$ zugeordnet.

(b) Hat eine von der Nulldarstellung verschiedene Darstellung φ von Γ den Grad 1, so sprechen wir auch von einem *linearen Charakter* von Γ. Diese linearen Charaktere können dann also auch als die Homomorphismen der Algebra Γ auf Ω definiert werden.

Aus dem Schurschen Lemma ergibt sich leicht, daß jede irreduzible Darstellung $\mathfrak{F}$ von Γ als Bestandteil in der regulären Darstellung $\mathfrak{R}$ von Γ auftritt. Ist Γ kommutativ, so folgt aus dem Schurschen Lemma weiter, daß nach einer geeigneten Grundkörpererweiterung alle irreduziblen Darstellungen von Γ den Grad 1 haben. Gilt dies bereits für den Grundkörper Ω, so sagen wir, daß Ω *Zerfällungskörper* von Γ ist. Dann zerfällt also die reguläre Darstellung $\mathfrak{R}$ von Γ in Bestandteile ersten Grades (nicht notwendig vollständig).

(14A) *Ist Γ eine kommutative Algebra, für die der Grundkörper Zerfällungskörper ist, so bildet jeder von Null verschiedene Homomorphismus ψ von Γ in einen Erweiterungskörper von Ω notwendig Γ auf Ω ab; es ist also ψ ein linearer Charakter von Γ. Ist B eine Teilalgebra von Γ, so kann jeder lineare Charakter von B zu einem linearen Charakter von Γ erweitert werden.*

Beweis. Bildet ψ die Algebra Γ auf den Körper $\Omega_1 \geqq \Omega$ ab, so hat Ω_1 notwendig endlichen Grad m über Ω. Da Ω_1 als Algebra über Ω eine irreduzible Darstellung m-ten Grades hat, so kann man ψ selbst als irreduzible Darstellung m-ten Grades von Γ deuten. Nach Voraussetzung ist Ω Zerfällungskörper von Γ und also ist $m = 1$, d.h. $\Omega_1 = \Omega$.

Es sei $\mathfrak{R}^0$ die durch Beschränkung der regulären Darstellung $\mathfrak{R}$ von Γ auf die Teilalgebra B entstehende Darstellung von B. Es zerfällt also $\mathfrak{R}^0$ in Bestandteile ersten Grades. Diese Bestandteile sind die durch Beschränkung der Darstellungen ersten Grades von Γ entstehenden Darstellungen von B.

Ist $\tilde{\mathfrak{R}}$ die reguläre Darstellung von B, so sieht man sofort, daß $\tilde{\mathfrak{R}}$ als Bestandteil von $\mathfrak{R}^0$ aufgefaßt werden kann. Jeder lineare Charakter von B

ist ein Bestandteil von $\tilde{\mathfrak{K}}$, also von $\mathfrak{R}^0$, und ist daher die Beschränkung eines linearen Charakters φ von Γ auf B.

Damit ist (14A) bewiesen. **Natürlich kann man auch einen direkten Beweis geben.**

(c) Es sei Γ eine Algebra mit Einheitselement; wir fassen Ω als Teilkörper des Zentrums von Γ auf. In Ω sei ein fester Teilring $\mathfrak{i}$ gewählt, der das Einheitselement 1 enthält und Ω als Quotientenkörper besitzt. Unter einer *Ordnung* von Γ verstehen wir eine Teilmenge J, die die folgenden Bedingungen erfüllt:

(I) J ist ein Teilring von Γ; $1 \in J$.

(II) J ist ein $\mathfrak{i}$-Modul, der endlich erzeugt werden kann.

(III) J ist nicht in einer eigentlichen Teilalgebra von Γ enthalten.

Sprechen wir von einem Ideal von J, so meinen wir stillschweigend ein zweiseitiges Ideal, das von Null verschiedene Elemente von $\mathfrak{i}$ enthält.

Es sei $\mathfrak{T}$ eine Darstellung der Algebra Γ durch lineare Transformationen eines Vektorraumes V über Ω. Wählt man eine Basis $e_1, e_2, \ldots, e_n$ von V, so wird jede lineare Transformation $\mathfrak{T}(\alpha)$, $(\alpha \in \Gamma)$, durch eine Matrix $[\mathfrak{T}(\alpha)]$ mit Koeffizienten in Ω gegeben. Man erhält so eine Darstellung $[\mathfrak{T}]$ von Γ durch Matrizen. Man sagt, daß $[\mathfrak{T}]$ ganzzahlig für J ist, wenn jedes $[\mathfrak{T}(\alpha)]$ mit $\alpha \in J$ Koeffizienten in $\mathfrak{i}$ hat. Offenbar ist dies dann und nur dann der Fall, wenn der von $e_1, e_2, \ldots, e_n$ erzeugte $\mathfrak{i}$-Modul U durch alle linearen Transformationen $\mathfrak{T}(\alpha)$ mit $\alpha \in J$ in sich abgebildet wird. Schreibt man $v\alpha$ für das Bild eines Vektors $v \in V$ für die lineare Transformation $\mathfrak{T}(\alpha)$, so ist U ein J-Modul.

(d) Wir setzen weiter voraus, daß $\mathfrak{i}$ ein Hauptidealring ist, was für unsere Zwecke ausreicht. Dann besitzt jede Ordnung J eine $\mathfrak{i}$-Basis. Ist $\mathfrak{T}$ eine Darstellung von Γ und ist $e_1^{(0)}, \ldots, e_n^{(0)}$ eine Ω-Basis des zugrunde liegenden Vektorraumes V, so sei U der von den $e_i^{(0)}$ erzeugte J-Modul. Offenbar kann dann U auch als $\mathfrak{i}$-Modul endlich erzeugt werden und besitzt daher auch eine $\mathfrak{i}$-Basis $e_1, e_2, \ldots, e_n$. Diese bildet dann auch eine Ω-Basis von V. Legt man diese Basis bei der Bildung von $[\mathfrak{T}]$ zugrunde, so wird also $[\mathfrak{T}]$ eine ganzzahlige Darstellung von J. In jeder Klasse ähnlicher Darstellungen von Γ gibt es also ganzzahlige Darstellungen von J.

Es sei $\mathfrak{p}$ ein Primideal von $\mathfrak{i}$; der Restklassenkörper $\mathfrak{i}/\mathfrak{p}$ werde mit Ω^* bezeichnet. Man sieht leicht, daß $\mathfrak{i} \cap \mathfrak{p}J = \mathfrak{p}$ ist und daß $J^* = J/\mathfrak{p}J$ eine Algebra über Ω^* ist.

(e) Ist $[\mathfrak{T}]$ eine ganzzahlige Darstellung von J, so führt die Restklassenabbildung von $\mathfrak{i}\ (\mathrm{mod}\ \mathfrak{p})$ diese in eine Darstellung $\mathfrak{T}^*$ von J^* in Ω^* über. Wir zeigen

(14B) *Es sei $\mathfrak{i}$ der Ring $\mathfrak{o}_v$ der für die diskrete Bewertung v des Körpers Ω ganzen Zahlen, und es sei $\mathfrak{p}$ das zugehörige Primideal. Sind $[\mathfrak{T}_1]$ und $[\mathfrak{T}_2]$ zwei ähnliche ganzzahlige Darstellungen der Ordnung J, so haben die zugehörigen*

„modularen" Darstellungen $\mathfrak{T}_1^*$ und $\mathfrak{T}_2^*$ dieselben irreduziblen Bestandteile (natürlich als Darstellungen von J^* im Körper Ω^*).

Beweis. Es gehöre etwa $[\mathfrak{T}_1]$ zu dem J-Modul U_1 und $[\mathfrak{T}_2]$ zu dem J-Modul U_2, die beide demselben Vektorraum V angehören, und die beide den Ω-Modul V aufspannen. Es sei π eine zu $\mathfrak{p}$ gehörige Primzahl. Da man U_2 durch cU_2 mit c in Ω, $c \neq 0$ ersetzen kann, so dürfen wir annehmen, daß $U_2 \leqq U_1$ ist. Ferner gibt es eine natürliche Zahl r, derart daß $\pi^r U_1 \leqq U_2$ ist;

$$\pi^r U_1 \leqq U_2 \leqq U_1.$$

Der Darstellungsraum für $\mathfrak{T}_1^*$ kann als $U_1/\pi U_1$ und der für $\mathfrak{T}_2^*$ als $U_2/\pi U_2$ gewählt werden. Man setze

$$X_m = (U_2 \cap \pi^m U_1) + \pi U_2; \qquad Y_m = (U_1 \cap \pi^{1-m} U_2) + \pi U_1.$$

Dann ist

$$U_2 = X_0 \geqq X_1 \geqq \cdots \geqq X_{r+1} = \pi U_2,$$
$$\pi U_1 = Y_0 \leqq Y_1 \leqq \cdots \leqq Y_{r+1} = U_1.$$

Die Faktorgruppen X_m/X_{m+1} und Y_{m+1}/Y_m können als J^*-Moduln betrachtet werden. In diesem Sinne zerfällt die Darstellung $\mathfrak{T}_1^*$ in $U_1/\pi U_1$ in die Darstellungen von J^*, die zu den von (0) verschiedenen Moduln Y_{m+1}/Y_m gehören. Ebenso zerfällt $\mathfrak{T}_2^*$ in die Darstellungen, die zu den von (0) verschiedenen Moduln X_m/X_{m+1} gehören. Können wir also zeigen, daß die J-Moduln Y_{m+1}/Y_m und X_m/X_{m+1} isomorph sind,

(14.1) $$X_m/X_{m+1} \simeq Y_{m+1}/Y_m,$$

so sind wir fertig.

Wendet man nun das Zassenhaussche Lemma[17]) auf die J-Gruppen $\mathfrak{u} = \pi U_2$, $\mathfrak{U} = U_2$, $\mathfrak{v} = \pi^{m+1} U_1$, $\mathfrak{V} = \pi^m U_1$, an, so ergibt sich

$$\frac{X_m}{X_{m+1}} = \frac{(U_2 \cap \pi^m U_1) + \pi U_2}{(U_2 \cap \pi^{m+1} U_1) + \pi U_2} \simeq \frac{(\pi^m U_1 \cap U_2) + \pi^{m+1} U_1}{(\pi^m U_1 \cap \pi U_2) + \pi^{m+1} U_1}.$$

Bildet man weiter $(\pi^m U_1 \cap U_2) + \pi^{m+1} U_1$ durch den J-Isomorphismus $\xi \to \pi^{-m}\xi$ ab, so erhält man als Bild $(U_1 \cap \pi^{-m} U_2) + \pi U_1 = Y_{m+1}$, während der Teilmodul $(\pi^m U_1 + \pi U_2) + \pi^{m+1} U_1$ in $(U_1 \cap \pi^{1-m} U_2) + \pi U_1 = Y_m$ übergeht. Daher ist der Quotient auf der rechten Seite in unserer Beziehung zu Y_{m+1}/Y_m isomorph. Jetzt sieht man, daß (14.1) zunächst als J-Isomorphismus und dann auch als J^*-Isomorphismus gilt, und damit ist (14B) bewiesen.

Man kann den Beweis von (14B) auch leicht führen, indem man entweder zusammengehörige Basen der J-Moduln U_1 und U_2 verwendet oder auch, indem man die Elementarteilertheorie auf eine $[\mathfrak{T}_1]$ in $[\mathfrak{T}_2]$ transformierende Matrix anwendet.

(f) Schließlich zeigen wir

(14C) *Es sei Γ eine kommutative halbeinfache Algebra über dem Körper Ω, der Zerfällungskörper von Γ sei. Der Ring $\mathfrak{i}$ sei der Ring der für die diskrete*

[17]) ZASSENHAUS, H.: Lehrbuch der Gruppentheorie, S. 51. Leipzig u. Berlin 1937.

30*

Bewertung v *von* Ω *ganzen Zahlen, und es sei* $\mathfrak{p}$ *das zugehörige Primideal von* $\mathfrak{i}$. *Ist* J *eine Ordnung von* Γ, *so ist* $\Omega^* = \mathfrak{i}/\mathfrak{p}$ *Zerfällungskörper für die* Ω^**-Algebra* $J^* = J/\mathfrak{p}J$ *und jeder lineare Charakter von* J^* *geht aus einem linearen Charakter von* Γ *durch Restklassenbildung hervor.*

Beweis. Wählt man eine $\mathfrak{i}$-Basis von J zur Bildung der regulären Darstellung $\mathfrak{R}$ von Γ, so ist $[\mathfrak{R}]$ eine ganzzahlige Darstellung von J. Sind $\omega_1, \omega_2, \ldots, \omega_n$ die verschiedenen linearen Charaktere von Γ, so ist $[\mathfrak{R}]$ ähnlich zu der Darstellung $[\mathfrak{R}_1]$, die vollständig in $\omega_1, \omega_2, \ldots, \omega_n$ zerfällt. Da jede Darstellung $[\omega_i]$ ganzzahlig ist [z. B. nach dem in (d) Gezeigten], so gilt das gleiche für $[\mathfrak{R}_1]$. Jetzt zeigt (14B), daß $\mathfrak{R}^*$ die irreduziblen Bestandteile $\omega_1^*, \ldots, \omega_n^*$ besitzt, die alle Grad 1 haben. Offenbar ist $\mathfrak{R}^*$ die reguläre Darstellung von J^*. Da jeder lineare Charakter ψ von J^* in ihr als Bestandteil auftritt, stimmt ψ mit einem der ω_i^* überein. Damit ist (14C) vollständig gezeigt.

Dept. of Mathematics, Harvard University, Cambridge 38, Mass. (U.S.A.)

(Eingegangen am 24. August 1955)

Druck der Universitätsdruckerei H. Stürtz AG., Würzburg

Number Theoretical Investigations on Groups of Finite Order

Richard BRAUER

Consider a group G of finite order n. If K is a given field, we can form the group algebra Γ of G with regard to K. This is an associative algebra such that there exists a basis whose elements form a group G_1 isomorphic with G under multiplication. If we identify G_1 with G, then Γ consists of the elements

$$(1) \qquad \gamma=\sum_{g\in G} a_g g, \qquad a_g \in K.$$

If K is an algebraic number field, the elements γ with integral a_g form an 'order' J in Γ. Here, orders are defined by the following properties. (I) J is a subring of Γ, $1 \in J$. (II) If $\mathfrak{o}$ is the ring of integers in K, J is an $\mathfrak{o}$-module which can be generated finitely. (III) The ring J does not belong to any proper subalgebra Γ. We may then study the number theory in J, and in particular its relationship with group theoretical properties of G. If $n>1$, J is not a maximal order in Γ.

The main purpose of this talk is a discussion of a number of conjectures and open problems. I know the great interest Japanese mathematicians have taken as well in the theory of arithmetics of algebras as in group theory, and perhaps I should say that I have come here to ask for help, since I would like very much to know the answers to my questions.

In order to formulate the conjectures, I have to give some definitions and simple remarks. When I speak of an ideal of J, I shall always mean a two-sided ideal which contains elements different from 0 of K. If $\mathfrak{P}$ is a prime ideal of J, then $\mathfrak{P}\frown K=\mathfrak{p}$ is a prime ideal of $\mathfrak{o}$ and $J\supset\mathfrak{P}\supseteq\mathfrak{p}J$. The prime ideals $\mathfrak{P}$ dividing a fixed $\mathfrak{p}$ are in one-to-one correspondence with the prime ideals $\mathfrak{P}^*$ of $J/\mathfrak{p}J$. Now, $J/\mathfrak{p}J$ may be identified with the group algebra Γ^* of G formed over the residue class field $K^*=\mathfrak{o}/\mathfrak{p}$. Again, the prime ideals $\mathfrak{P}^*$ of Γ^* are in one-to-one correspondence with the irreducible representations F of Γ^* in the field K^* such that $\mathfrak{P}^*$ is the kernel of F. Each representation F is obtained from a representation F_0 of G in K^* by linear extension. If $\phi(g)$ is the trace of $F_0(g)$, $g \in G$, and if F_0 is absolutely irreducible, it is seen easily that the prime ideal $\mathfrak{P}$ of J corresponding to $\mathfrak{P}^*$ consists of the elements (1) such that

$$\sum a_g^* \phi(gh) = 0^* \qquad (\text{for all } h \in G).$$

Here, a_g^* is the residue class of $a_g \in \mathfrak{o}$ (mod $\mathfrak{p}$).

We next write $\mathfrak{p}J$ as intersection of ideals $\mathfrak{B}_\tau \neq J$,

$$\mathfrak{p}J = \mathfrak{B}_1 \frown \mathfrak{B}_2 \frown \cdots \frown \mathfrak{B}_t \tag{2}$$

such that $\mathfrak{B}_\sigma$ and $\mathfrak{B}_\tau$ are relatively prime for $\sigma \neq \tau$, i.e., $(\mathfrak{B}_\sigma, \mathfrak{B}_\tau) = J$. If none of the $\mathfrak{B}_\tau$ can be written as the intersection of two relatively prime ideals, we call $\mathfrak{B}_1, \mathfrak{B}_2, \cdots, \mathfrak{B}_t$ the block ideals of $\mathfrak{p}J$. They are uniquely determined. Actually, $\mathfrak{p}J$ is the product of the $\mathfrak{B}_\tau$, and any two of the $\mathfrak{B}_\tau$ commute. Each prime ideal $\mathfrak{P}$ dividing $\mathfrak{p}J$ divides $\mathfrak{B}_1\mathfrak{B}_2\cdots\mathfrak{B}_t$ and hence $\mathfrak{P}$ divides exactly one of the $\mathfrak{B}_\tau$. Let B_τ be the set of the prime ideals $\mathfrak{P}$ dividing $\mathfrak{B}_\tau$. Replace now each prime ideal $\mathfrak{P}$ by the corresponding modular representation F. Thus, the irreducible modular representations F of G are distributed into disjoint sets $B_1, B_2, \cdots, B_t$ which are called the 'blocks' of representations.[1]

We shall assume that the algebraic number field K is a splitting field of the semisimple algebra Γ, i.e., that the irreducible representations $X_1, X_2, \cdots, X_k$ of G in K are absolutely irreducible. This assumption is certainly satisfied, if K contains the n-th roots of unity.

If X_i is an ordinary irreducible representation of G in K, we can always find coordinate systems such that the coefficients of each $X_i(g)$, $g \in G$, are local integers for $\mathfrak{p}$. Then the residue class map carries X_i into a modular representation X_i^* of G in K^*. Here, X_i^* is not uniquely determined by X_i. However, if $F_1, F_2, \cdots, F_l$ are the different modular irreducible representations of G in K^*, then the multiplicity d_{ij} of F_j as irreducible constituent of X_i^* is uniquely determined. These d_{ij} are the decomposition numbers. For each X_i all F_j actually appearing in X_i^* belong to the same block B. We shall count X_i as a member of the block B in this case. Thus, the ordinary irreducible representations $X_1, X_2, \cdots, X_k$ of G in K are also distributed into our blocks $B_1, B_2, \cdots, B_t$.

Set

$$c_{ij} = \sum_{\rho=1}^{k} d_{\rho i} d_{\rho j}. \tag{3}$$

1) For a discussion of blocks, I may refer to the book by G. Azumaya and T. Nakayama, "Daisûgaku II" (Algebra II), Iwanami-Shoten, Tokyo (1954).

If F_i and F_j belong to different blocks, then $c_{ij}=0$. If i and j range over the indices for which F_i and F_j belong to a fixed block B_τ, the numbers c_{ij} represent the Cartan invariants of the algebra $J/\mathfrak{B}_\tau$.

Finally, for each block B we define the 'defect' δ of the block. If p is the rational prime divisible by $\mathfrak{p}$, δ is the largest integer such that p^δ divides $n/Dg\, X_i$ for some $X_i \in B$. Equivalently, we may say that δ is the largest integer such that p^δ divides some n/DgF_j with $F_j \in B$.

I now come to the discussion of the conjectures mentioned above.

(I). Given p and δ, consider all blocks B of defect δ for all finite groups G. The conjecture is that the decomposition numbers d_{ij} corresponding to the block lie below a bound depending only on p and δ. Because of (3), this is equivalent with the corresponding statement concerning the Cartan invariants c_{ij}. Our conjecture means that for given p and given defect δ, representations X_i^* cannot split into too many irreducible modular constituents.

Let me discuss some known results in support of this conjecture. If $\delta=0$, the block consists of one X_i and one F_j and $d_{ij}=1$, $c_{jj}=1$. The conjecture is still true for $\delta=1$. A number of years ago, I studied groups whose order is divisible only by the first power of a prime p.[2] The results obtained for blocks of defect 1 in this case hold in general for blocks of defect 1. We have here $d_{ij}\leqq 1$, $c_{ij}\leqq p$. We may make our conjecture more precise by asking: Is it true that

$$c_{ij}\leqq p^\delta$$

for the Cartan invariants of a block B of defect δ?

In the case $\delta=1$, we can describe exactly which systems (d_{ij}) can occur and can give relations between the types which occur and group theoretical properties of G. It would be highly desirable to have analogous results for higher defects. For $\delta=2$, it is no longer true that $d_{ij}\leqq 1$.

There is a result weaker than the conjecture above which can be proved for arbitrary defect. The Cartan invariants c_{ij} belonging to a block B of defect δ may be taken as the coefficients of a quadratic form H which is positive definite. Then it can be shown that, for given p and δ and all finite groups G, the form H belongs to a finite number of classes. This is proved by showing that if B consists of

2) R. Brauer, Investigations on group characters, Ann. of Math. **42** (1941), 936–958.

α ordinary representations X_i and β modular representations F_j, we have[3)]

(a)	$\det(c_{ij}) = p^{\mu}$	with	$\mu \leqq \delta + (\beta-1)(\delta-1)$.
(b)	$\beta < \alpha$	for	$\delta > 0$.
(c)	$\alpha \leqq p^{2\delta-2}$	for	$\delta > 1$.

For our present purposes, an inequality weaker than (c) would suffice which I have given a number of years ago. In the form stated here, it was obtained by W. Feit and myself improving the inequality $\alpha \leqq \frac{1}{4} p^{2\delta}$ which we had given recently.[4)]

For $\delta \leqq 2$, we have $\alpha \leqq p^{\delta}$. Probably, this is true for all δ. There is one way in which this inequality might be proved. The quadratic form belonging to the matrix $p^{\delta}(c_{ij})^{-1}$ has integral coefficients. If it could be shown that β is the smallest integer $\neq 0$ represented by this form, this would yield the inequality $\alpha \leqq p^{\delta}$. This method works for $\delta = 2$ and in some special cases; however I feel somewhat doubtful about the general case.

(II). We have the problem of determining the number of blocks of a fixed defect δ for a given group G. Let me mention first that this can be done at once when the characters of G are known. However, this is not what we want, since our aim is to obtain new properties of the group characters.

The situation is here opposite to that in (I), where small defects could be handled while the case of large defects seems difficult. For $\delta > 0$, we have a satisfactory treatment of (II) in form of a reduction to groups of smaller order.[5)] However, the problem remains to characterize the number of blocks of defect 0 of a group G by group theoretical properties. If a complete answer to the problem of the exact number of representations in a block could be given, this would answer our question. This process is feasible in the case $g \equiv 0 \pmod p$, $g \not\equiv 0$

3) The fact that $\det(c_{ij})$ is a power of p and the inequality in (b) are well known facts in the theory of modular representations of finite groups. The inequality in (a) is a consequence of the fact that δ is the largest elementary divisor of (c_{ij}) and that it appears with multiplicity 1.

Cf. R. Brauer, On the arithmetic in a group ring, Proc. Nat. Acad. Sci. **30** (1944), 109–114; M. Osima, Notes on blocks of group characters, Math. J. of Okayama Univ. **4** (1955), 175–188; R. Brauer, Zur Darstellungstheorie der Gruppen endlicher Ordnung, (to appear in Math. Zeitschr.).

4) Cf. R. Brauer, On the structure of groups of finite order, to appear in the Proc. Internat. Congress of Math. Amsterdam 1954.

5) Cf. the papers quoted in 3).

(mod p^2). Let P be a Sylow-subgroup of G of order p and let N be its normalizer. Then the number of blocks of defect 0 of G is $k-k'$ where k is the class number of G and k' that of N. In particular, $k \geqq k'$. I don't know a direct proof for this inequality which does not involve representation theory. Actually, it would be of interest to know the necessary and sufficient conditions for k and k' to be equal.

(III). Suppose that p^a is the exact power of p in g. By definition of the defect δ of a block B, the degrees of the ordinary representations in B are all divisible by $p^{a-\delta}$ while some of these degrees are not divisible by $p^{a-\delta+1}$. The problem arises to determine when degrees divisible by $p^{a-\delta+1}$ occur in a block B. Let us call this the case of a raised degree.

In the work mentioned in (II), a subgroup D of G of order p^δ is associated with a block B of defect δ. This is the defect group of the block. Each conjugate group may be taken; apart from this, the defect group D is uniquely determined. The simplest conjecture is that raised degrees occur, if and only if D is abelian. If this conjecture is true, it would furnish a generalization of the known fact that no raised degrees are possible, if $g \not\equiv 0 \pmod{p^2}$. The conjecture is trivially true in the case of p-groups. In order to test the conjecture further, I have studied the case $p=2$, $g \equiv 0 \pmod 8$, $g \not\equiv 0 \pmod{16}$ in which special methods are available. While the answer is not complete, I like to state the results obtained, since they are helpful in studying special groups. As above, let P denote the Sylow subgroup of G so that here P has order 8. Let B be a block of defect 3; $D=P$. We then have the following cases.

non [R.B.]

(A) *P the quaternion group of order* 8.

CASE 1. The block consists of four characters of odd degrees z_1, z_2, z_3, z_4 and three characters of even degress. We have an equation

$$\varepsilon_1 z_1 + \varepsilon_2 z_2 + \varepsilon_3 z_3 + \varepsilon_4 z_4 = 0$$

where each ε_i is ± 1 and where

$$\varepsilon_1 z_1 \equiv \varepsilon_2 z_2 \equiv \varepsilon_3 z_3 \equiv \varepsilon_4 z_4 + 4 \pmod 8.$$

The three even degrees are given by

$$z_5 = \pm(\varepsilon_1 z_1 + \varepsilon_2 z_2), \qquad z_6 = \pm(\varepsilon_1 z_1 + \varepsilon_3 z_3), \qquad z_7 = \pm(\varepsilon_1 z_1 + \varepsilon_4 z_4).$$

Actually, these are special cases of relations available for the values of the characters.

As an example, I mention the extended icosahedral group of order 120 which has a block B with the degrees

$$z_1=3,\ z_2=3,\ z_3=5,\ z_4=1,\ z_5=6,\ z_6=2,\ z_7=2.$$

CASE 2. The block consists of four equal odd degrees z and one even degree $2z$.

For instance, this case arises for $G=P$, $z=1$.

(B) *P the dihedral group of order* 8.

The block consists of four characters of odd degrees z_1, z_2, z_3, z_4 and one of an even degree z_5 and we have equations

$$\varepsilon_1 z_1+\varepsilon_2 z_2=\varepsilon_3 z_3+\varepsilon_4 z_4=z_5$$

with $\varepsilon_i=\pm 1$. For instance, for the alternating group A_6, we have $z_1=1$, $z_2=9$, $z_3=5$, $z_4=5$, $z_5=10$; for A_7: $z_1=1$, $z_2=15$, $z_3=21$, $z_4=35$, $z_5=14$; for the simple group of order 168: $z_1=1$, $z_2=7$, $z_3=3$, $z_4=3$, $z_5=6$.

(C) *P abelian of type* (2, 2, 2).

CASE 1. B consists of eight odd degrees z_i, $1\leqq i\leqq 8$ which satisfy a relation $\sum \varepsilon_i z_i=0$ with $\varepsilon_i=\pm 1$.

Example. The simple group of order 504 where we have the degrees 1, 7, 7, 7, 7, 9, 9, 9 in a block.

CASE 2. B consists of four characters of odd degrees z_1, z_2, z_3, z_4 and one even degree z_5 with an equation

$$2z_5=\varepsilon_1 z_1+\varepsilon_2 z_2+\varepsilon_3 z_3+\varepsilon_4 z_4, \qquad (\varepsilon_i=\pm 1).$$

(D) *and* (E) *P abelian of type* (4, 2) *or of type* (8).

Here, B consists of eight equal odd degrees.

I do not know whether the case 2 in (C) can occur. Apart from this, our conjecture is true for the groups considered here.

In principle, one can obtain a finite number of similar types for each given defect group by combining the known relations for characters. However, in general, the number of cases to be distinguished will be large and it will not be clear which cases can actually occur. Still it might be useful to work out further cases.

Our special results lead to the following question: If the characters of a finite group G are known, how much information about the structure of the Sylow groups P can be obtained from this knowledge of the characters? Even if the conjecture (III) is not true in general, it seems that results in this direction could be obtained.

(IV). In the case of defect $\delta=1$, the structure of the residue class ring $J/\mathfrak{B}$ of J modulo a block ideal $\mathfrak{B}$ can be described rather completely. The question arises whether analogous results exist for higher defects. Still more generally, it may be possible to investigate the rings $J/\mathfrak{B}^r$, $r \geqq 1$. This would lead to a deeper study of the number theory of J. In particular, from our point of view we would be interested in connections with group theorectical properties.

There are some known facts concerning the structure of the indecomposable components of the modular regular representation which might be understood more clearly by such an investigation. Also there would be hope that this work would lead to results in connection with the questions asked above.

(V). I mention some questions concerning modular representations which have remained unanswered. The degrees of the ordinary irreducible representations X_i divide the order n of the group. The corresponding fact is not true for the degrees f_j of the modular irreducible representations F_j. However, it seems possible that at least the power of the fixed prime p dividing f_j always divides n.

If Γ^* now denotes the group algebra of G with respect to an arbitrary field K^* of prime characteristic p, Γ^* is semi-simple, if and only if p does not divide n. If $p \mid n$, the problem arises to determine the radical N of Γ^*. It is not very difficult to give necessary and sufficient conditions that an element

$$\gamma = \sum_{g \in G} c_g g, \qquad c_g \in K^*$$

of Γ^* belongs to N. Let p^r denote the highest power of p for which there exist elements of order p^r in G. If h is an arbitrary element and t a p-regular element of G, let $Z(h, t)$ denote the set of all elements g of G for which the p^r-th powers of hg and of t are conjugate in G. Then necessary and sufficient conditions are given by the equations in K^*

$$\sum_{g \in Z(h,t)} c_g = 0$$

for all $h, t \in G$ and t p-regular.

However, in this form the equations are too complicated to be of much value. For instance, except in the simplest cases, they cannot be used to find the rank of the radical.[6] The question is whether it is possible to obtain simpler conditions which really enable us to study the radical. There would be some interest in having formulas for the ranks of the powers of N and for the exponent of N.

(VI). It must be emphasized that it is much easier to ask questions than to answer them. In this connection, I mention the

6) Of course, if the irreducible representations of G in K^* have the degrees $f_1, f_2, \cdots$, this rank is given by $n - f_1^2 - f_2^2 - \cdots$.

two well known problems: (1) What are the necessary and sufficient conditions that a square array A of complex numbers represents the table of characters of a finite group G? (2) How much additional information is required in order that G be uniquely determined? The answer to these questions seems to be very difficult. Perhaps, it may be worthwhile to study these questions for special types of groups.

(VII). There are various applications of the results obtained to group theoretical questions. However, these applications are still rather scattered and for this reason, I shall not attempt to give a summary. In many cases, we can get information about the characters of groups G of which only little is known. For instance, if the order n is given and if G is assumed to be simple, it is often possible to show that the known relations lead to contradictions and hence that no simple group of that order n can exist. For other values of n, the results enable us to find the characters and, on the basis of this, determine G. I think that it is possible to determine all simple groups up to some rather high order n in this manner, possibly with the exception of a few values.

In concluding, I would like to mention one particular type of problem. Given a group H of finite order and a set S of elements $h \neq 1$ of H. What are the groups G which contain H as a subgroup such that the centralizers of the elements $h \in S$ in G lie in H? For instance, we may take for S the elements $\neq 1$ of one or several Sylow groups of H and then the theory of blocks gives us at least some results concerning G. There are many ways in which our question can be modified.

Perhaps, I can describe the situation best by taking up a simple example. Let H be the cyclic group of order 3 and let S be the set consisting of a generating element of H. Here, the order n of G must have the form $n=3f(f+1)m^2$ where f and m are natural integers, $f \equiv 1 \pmod 3$, $m \not\equiv 0 \pmod 3$. There exist infinitely many groups G which satisfy the conditions. Each such G has a unique maximal normal subgroup $G_0 \neq G$ and the group G/G_0 is simple and satisfies the same conditions. Two simple groups G of this type are known, the simple groups of order 60 (with $f=4$ and $m=1$) and of order 168 (with $f=7$ and $m=1$). I don't know whether any further simple groups exist. In general, one can ask whether conditions of this type can ever be satisfied by infinitely many simple groups.

HARVARD UNIVERSITY

ON THE STRUCTURE OF GROUPS OF FINITE ORDER

RICHARD BRAUER

The theory of groups of finite order has been rather in a state of stagnation in recent years. This has certainly not been due to a lack of unsolved problems. As in the theory of numbers, it is easier to ask questions in the theory of groups than to answer them. If I present here some investigations on groups of finite order, it is with the hope of raising new interest in this field. The results which I am going to give will be incomplete and new unanswered questions will be added to the old ones. There is, however, hope that further progress will be possible along the lines suggested.

Let me start with the following question. If $\mathfrak{G}$ is a group of order g which is not cyclic of prime order, what can be said about the existence of proper subgroups $\mathfrak{H}$ of relatively large order h? It seems probable that the following conjecture is true. *There always exist proper subgroups $\mathfrak{H}$ of an order $h > \sqrt[3]{g}$.* For instance, if $\mathfrak{G}$ is soluble, it is easy to see that there exist proper subgroups $\mathfrak{H}$ even with $h \geqq \sqrt{g}$. There is a famous conjecture to the effect that all groups of odd order are soluble. Thus, if we accept this old conjecture, we have to consider only groups of even order. It is somewhat surprising that for groups of even order our conjecture can be proved. In other words, we have the following theorem

Theorem. If $\mathfrak{G}$ is a group of even order $g > 2$, there exists a proper subgroup $\mathfrak{H}$ of an order $h > \sqrt[3]{g}$.

The proof is quite elementary and I will sketch it, since it forms a preparation for some of the later developments. It will be convenient to work with the group algebra Γ of $\mathfrak{G}$ over the field of rational numbers. This is an associative algebra consisting of all formal expressions

$$\gamma = \sum_{\sigma} a_{\sigma}\sigma$$

where σ ranges over the elements of $\mathfrak{G}$ and where the a_{σ} are rational numbers. The operations in Γ are defined in the obvious manner.

By an *involution* of $\mathfrak{G}$, we shall mean an element J of $\mathfrak{G}$ of order 2. Let $\mathfrak{M}$ denote the set of all involutions in $\mathfrak{G}$. Since g is even, $\mathfrak{M}$ is not empty. Set

$$M = \sum_{J \in M} J$$

taken as an element of Γ. Then $M^2 \in \Gamma$ and we have a formula

$$M^2 = \sum_{\sigma} a_{\sigma}\sigma. \tag{1}$$

It is clear that a_σ here is the number of ordered pairs (X, Y) such that

(2) $$XY = \sigma, \qquad (X, Y \in \mathfrak{M}).$$

It is also clear that a_σ will depend only on the class of conjugate elements of $\mathfrak{G}$ to which σ belongs. If then the classes of conjugate elements of $\mathfrak{G}$ are denoted by $\mathfrak{K}_0, \mathfrak{K}_1, \ldots, \mathfrak{K}_{k-1}$ and if $K_i \in \Gamma$ is the sum of the elements of $\mathfrak{K}_i$, the formula (1) can be rewritten in the form

(1*) $$M^2 = \sum_{i=0}^{k-1} a_i K_i$$

where we set $a_i = a_\sigma$ for $\sigma \in \mathfrak{K}_i$.

If (2) holds, it follows at once that $X^{-1}\sigma X = \sigma^{-1}$. We shall say that an element σ is *real* in $\mathfrak{G}$, if σ and σ^{-1} are conjugate in $\mathfrak{G}$. Thus, if σ is non-real in $\mathfrak{G}$, (2) is impossible and $a_i = 0$ for non-real classes $\mathfrak{K}_i$. Moreover, if $\mathfrak{N}(\sigma)$ is the normalizer of σ in $\mathfrak{G}$, (i.e. the subgroup of $\mathfrak{G}$ consisting of the elements commuting with σ), and if $n(\sigma)$ is the order of $\mathfrak{N}(\sigma)$, there exist at most $n(\sigma)$ elements X with $X^{-1}\sigma X = \sigma^{-1}$. Thus, if σ_i is an element of $\mathfrak{K}_i$ we have $a_i \leqq n(\sigma_i)$. If σ_i is an involution, one sees easily that this inequality can be replaced by $a_i \leqq n(\sigma_i) - 2$. Actually, it will be clear that much more accurate statements concerning the values of a_i can be given, but this is not important for the moment.

Let us now compare the number of group elements appearing on both sides of (1*). If m denotes the total number of involutions in $\mathfrak{G}$, the left side of (1*) is the sum of m^2 elements of $\mathfrak{G}$. On the right, we have a sum of $\sum a_i g/n(\sigma_i)$ elements of $\mathfrak{G}$, since $\mathfrak{K}_i$ consists of $g/n(\sigma_i)$ elements. Hence

$$m^2 = \sum_{i=0}^{k-1} a_i g/n(\sigma_i).$$

Let $\mathfrak{K}_0$ denote the class containing the identity 1. Clearly, $a_0 = m$. Let $\mathfrak{K}_1, \ldots, \mathfrak{K}_r$ designate the classes consisting of involutions and let $\mathfrak{K}_{r+1}, \ldots, \mathfrak{K}_l$ designate the other classes consisting of real elements of $\mathfrak{G}$. If the estimates for the a_i are used, we find

$$m^2 \leqq m + \sum_{i=1}^{r} (n(\sigma_i) - 2)g/n(\sigma_i) + \sum_{j=r+1}^{l} n(\sigma_i)g/n(\sigma_i).$$

Each of the m involutions appears in one of the classes $\mathfrak{K}_1, \mathfrak{K}_2, \ldots, \mathfrak{K}_r$, and hence

$$m = \sum_{i=1}^{r} g/n(\sigma_i).$$

Thus,

(3) $$m^2 < \sum_{i=1}^{l} n(\sigma_i)g/n(\sigma_i) = lg.$$

On the other hand, the total number of elements in $\mathfrak{K}_1, \mathfrak{K}_2, \ldots, \mathfrak{K}_l$ is less than g, that is,

(4) $$\sum_{i=1}^{l} g/n(\sigma_i) < g.$$

Let h be the maximal value of $n(\sigma_i)$ appearing in (4). Then $g/h \leqq g/n(\sigma_i)$ and (4) yields

$$lg/h < g.$$

If this is used in (3), we have $m^2 < gh$. On the other hand, $m \geqq g/n(\sigma_1) \geqq g/h$. On combining these inequalities, we find $g < h^3$. If $h = n(\sigma_j) \neq g$, then $\mathfrak{N}(\sigma_j)$ is a proper subgroup $\mathfrak{H}$ of an order $h > \sqrt[3]{g}$. It remains to discuss the case that the maximal $n(\sigma_j)$ in (4) is equal to g. In this case, $\mathfrak{K}_j$ consists of one element only. For $r + 1 \leqq i \leqq l$ the class $\mathfrak{K}_i$ contains the two distinct elements σ_i and σ_i^{-1} Hence we must have $1 \leqq j \leqq r$, and then σ_j is an invariant involution. Now, induction can be used. If $G/\{\sigma_j\}$ has even order, the theorem holds for this quotient group and we see at once that it then holds for $\mathfrak{G}$. If $\mathfrak{G}/\{\sigma_j\}$ has odd order g', then the order g of $\mathfrak{G}$ is twice an odd number g'. It is well known [1]) that then $\mathfrak{G}$ has a subgroup or order $g/2$. Since $g/2 > \sqrt[3]{g}$ for $g \geqq 4$, the theorem holds again and this finishes the proof. Incidentally, it is possible to improve the theorem somewhat, but it does not seem to be worthwhile to go into this here. Instead I rather mention a related theorem which is proved by a similar method.

Theorem. Let m denote the number of involutions of the group $\mathfrak{G}$ of even order g and set $n = g/m$. There exists a subgroup $\mathfrak{H} \neq \mathfrak{G}$ of $\mathfrak{G}$ such that the index $t = (\mathfrak{G} : \mathfrak{H})$ is either 2·or t satisfies the inequality $t \leqq n(n+2)/2$.

This result has some interesting consequences. If the group $\mathfrak{G}$ has a subgroup $\mathfrak{H} \neq \mathfrak{G}$ of index t, then $\mathfrak{G}$ possesses a transitive permutation representation on t symbols. The kernel $\mathfrak{L}$ of this permutation representation is a normal subgroup of $\mathfrak{G}$ such that $\mathfrak{G}/\mathfrak{L}$ is isomorphic with a transitive subgroup of the symmetric group $\mathfrak{S}_t$ on t letters. This implies that

$$1 < (\mathfrak{G} : \mathfrak{L}) \leqq t! \leqq \text{Max } (2, [n(n+2)/2]!).$$

In particular, if $\mathfrak{G}$ is simple, we must have $\mathfrak{L} = \{1\}$ and hence

$$g \leqq [n(n+2)/2]!$$

If J is any involution in $\mathfrak{G}$, then $g/n(J) \leqq m = g/n$ and hence $n \leqq n(J)$, $g \leqq (n(J)(n(J)+2)/2)!$ This yields the result [2])

Theorem. There exist only a finite number of simple groups $\mathfrak{G}$ which contain an involution J such that the normalizer $\mathfrak{N}(J)$ of J is isomorphic to a given group.

It is not known whether a similar result holds, if, instead of an involution J, we take an element of a given odd prime order p.

The last theorem suggests the following problem: Given a group $\mathfrak{N}$ containing an involution J in its center. What are the groups $\mathfrak{G}$ containing $\mathfrak{N}$ as a subgroup such that $\mathfrak{N}$ is the normalizer of J in $\mathfrak{G}$? More generally, the case can

---

[1]) See, for instance, W. Burnside, The Theory of Groups of Finite Order, Cambridge 1911, p. 327, Theorem II.

[2]) A result of this kind with a different bound was first obtained by K. A. Fowler and the author by a somewhat different method.

be considered that a number of groups $\mathfrak{N}_1^*, \mathfrak{N}_2^*, \ldots, \mathfrak{N}_r^*$ are given such that each $\mathfrak{N}_i$ contains an invariant involution J_i. We then can study the groups $\mathfrak{G}$ containing subgroups $\mathfrak{N}_1^*, \mathfrak{N}_2^*, \ldots, \mathfrak{N}_r^*$ such that each $\mathfrak{N}_i^*$ is isomorphic to $\mathfrak{N}_i$ and that $\mathfrak{N}_i^*$ is the normalizer of the element J_i^* corresponding to J_i. We may further require that $J_1^*, J_2^* \ldots, J_r^*$ represent all the distinct classes of $\mathfrak{G}$ which contain involutions. Usually, it will be convenient to add further assumptions in order to characterize groups uniquely.

In this generality, the problem is far too difficult to handle. There are, however, some general methods which give at least some information and which we shall discuss next.

The first of these is closely related with what we have done above. If we use the same notation, the elements $K_1, K_2, \ldots, K_k$ form a basis of the center Λ of the group algebra Γ. We therefore have formulas

$$K_\alpha K_\beta = \Sigma\, c_{\alpha\beta\gamma} K_\gamma$$

with rational coefficients $c_{\alpha\beta\gamma}$. Actually, it is easy to see that the $c_{\alpha\beta\gamma}$ are non-negative integers. The $c_{\alpha\beta\gamma}$ can be expressed in terms of the irreducible characters of $\mathfrak{G}$ and, as a matter of fact, it is well known that a knowledge of the $c_{\alpha\beta\gamma}$ is equivalent with a knowledge of the characters.

Suppose now that g is even. The element M in (1*) is the sum of $K_1, K_2, \ldots K_r$. Hence the a_i in (1*) are sums of certain of the $c_{\alpha\beta\gamma}$. Since we can study the a_i directly and since we can express the $c_{\alpha\beta\gamma}$ in terms of the characters, we obtain new relations for the characters.

The second idea is of a more fundamental nature. It applies to groups of even or odd order and it can be used for other questions too. Let $\chi_1, \chi_2, \ldots, \chi_k$ designate the irreducible characters of $\mathfrak{G}$. Let p be a fixed prime number. In our application, we take $p = 2$, but this is not important now. The characters χ_i are distributed into disjoint sets, called the *p-blocks* of $\mathfrak{G}$. Here, χ_α and χ_β belong to the same block, if

$$\frac{g}{n(\sigma)} \frac{\chi_\alpha(\sigma)}{Dg\chi_\alpha} \equiv \frac{g}{n(\sigma)} \frac{\chi_\beta(\sigma)}{Dg\chi_\beta} \pmod{\mathfrak{p}}$$

for all $\sigma \in \mathfrak{G}$ where $\mathfrak{p}$ is a fixed prime ideal divisor of p in the field of the g-th roots of unity. This is a technical definition of the blocks. There is a better one which belongs to the theory of arithmetic of algebras, but it would take too long to discuss this here [3]).

[3]) Some of the results on blocks mentioned here have been stated without proof in the following papers: R. Brauer, On the arithmetic in a group ring, Proc. Nat. Acad. Sci. U.S.A. 30, 109—114 (1944) and On blocks of characters of groups of finite order I, II, Proc. Nat. Acad. Sci. U.S.A. 32, pp. 182—186, 215—219 (1946).

Suppose then we have a block B of characters, say, consisting of the characters $\chi_1, \chi_2, \ldots, \chi_s$. We form the highest power p^d which divides one of the numbers $g/Dg\chi_i$, $1 \leqq i \leqq s$. Then d is called the *defect* of B. The first basic result is, that for given p and d and all groups of finite order, there exist only a finite number of "types" of p-blocks of defect d. For each type, the number s of characters of the block is determined. Further, for each type, one can give explicitly all relations

$$\sum c_i \chi_i(\sigma) = 0$$

which hold simultaneously for all elements σ of $\mathfrak{G}$ of order prime to p. If p^a is the highest power dividing g, one may assume that, for each type, the values of the degrees $Dg\chi_i$ (mod p^{a-d+1}) are known.

In particular, for given p and d, there must exist an upper bound for the number s of characters in the block. It is probable that $s \leqq p^d$. However, this can only be proved for $d \leqq 2$. The best known result, proved recently by W. Feit and myself, is that $s < \frac{1}{4}p^{2d}$(for $d > 1$). Actually, for $d \leqq 1$, all possible types can be enumerated. For larger d, the number of types may be uncomfortably large. It would be very important for our work if the theory of blocks could be refined.

There is another result which I have to mention. If all subgroups $\mathfrak{D}$ of order p^d of a p-Sylow group $\mathfrak{P}$ of $\mathfrak{G}$ and their normalizers $\mathfrak{N}(\mathfrak{D})$, in $\mathfrak{G}$ are known, the number of p-blocks of defect d can be determined and information about the possible types obtained. For $d = 0$, we have $\mathfrak{G} = \mathfrak{N}(\mathfrak{D})$ and nothing can be learned about the number of p-blocks of defect $d = 0$. For $d > 0$, a real reduction can be achieved.

These results can be applied, if we have information about $\mathfrak{G}$ as outlined in the program above. The prime p is to be taken as 2. The method will give us a wealth of information about the characters of $\mathfrak{G}$ and this again can be applied to study the group $\mathfrak{G}$ further.

Let me now discuss some special cases. I first consider a relatively simple case. Consider a group $\mathfrak{G}$ of an order $g = 4g'$ where g' is odd. In order to have more definite results, we shall assume that no element outside a Sylow subgroup $\mathfrak{S}$ of order 4 commutes with all the elements of $\mathfrak{S}$. Further, we shall assume that $\mathfrak{G}$ does not have a normal subgroup of index 2. The normalizer $\mathfrak{N}(J)$ of an involution J has an order $n(J)$ of the form $n(J) = 4v$ with an odd v. If our methods are applied here, the result is obtained that the order g can be expressed by two integral rational parameters f_1, f_2 and a sign $\delta = \pm 1$. We have

$$g = 32v^3 \frac{f_1 f_2 f_3}{(f_1 + 1)(f_2 + \delta)(f_3 - 1)}$$

where we set $f_3 = 1 + f_1 + \delta f_2$. We must also have $f_1 \equiv 1 \pmod 4$, $f_2 \equiv \delta \pmod 4$. Actually, 1, f_1, d_2, f_3 are degrees of irreducible representations and all other such degrees are divisible by 4. As a strange consequence of our result we see that g will lie very close to $32v^3$. In particular, if $\mathfrak{G}$ is to be simple, then f_1, f_2, f_3 will increase with g and g is asymptotically equal to $32v^3$.

There do exist infinitely many simple groups of the type in question. They are groups $PSL(2, q)$ [4]) i.e. groups of unimodular projectivities of the projective line of a finite geometry belonging to a field with q elements. The prime power $q > 3$ has to satisfy the conditions $q \equiv 3$ or $5 \pmod 8$ in order that $\mathfrak{G}$ meets the requirements. Also, q may be taken equal to 4. In the latter case, the icosahedral group is obtained which is the only group possible for $v = 1$.

Probably, the groups mentioned here are the only simple groups of the type in question and, possibly, the only simple groups of an order $g = 4g'$ with an odd g'. However, this question has to be left open.

In any case, we come close here to a characterization of some of the simple groups $PSL(2, q)$. Actually, characterizations of all these groups can be given within the framework of our program. For instance, we have the following theorem first conjectured by M. Suzuki

Theorem. Suppose that $\mathfrak{G}$ is a group of even order g such that if two cyclic subgroups $\mathfrak{A}$ and $\mathfrak{B}$ of even order have an intersection different from $\{1\}$, then $\mathfrak{A}$ and $\mathfrak{B}$ both are subgroups of a cyclic subgroup of $\mathfrak{G}$. If $\mathfrak{G}$ does not have a normal subgroup of index 2, then $\mathfrak{G}$ is a group $PSL(2, q)$ (with $q > 2$) [5]).

Actually, under the assumptions of this theorem the characters of $\mathfrak{G}$ can be determined completely. In particular, this yields the order g of $\mathfrak{G}$ and now a characterization of the groups in question given by H. Zassenhaus [6]) can be used. There are various extensions and generalizations.

There is some hope that, for the other known simple groups, there exist similar characterizations. However, the continuation of this work becomes more and more difficult and, for this reason, I have been able to complete the work only in the case of the groups $PSL(3, q)$, the group of unimodular collineations of a projective plane under special assumptions for q. The result here is as follows

Theorem. Suppose that $\mathfrak{G}$ is a group of finite order which satisfies the following conditions

---

[4]) The notation here is that used in B. L. van der Waerden, Gruppen von linearen Transformationen, Ergebnisse der Math., vol. 4 (1935).

[5]) This theorem was proved more or less independently by M. Suzuki and the author and by G. E. Wall.

[6]) H. Zassenhaus, Abh. Math. Sem. Univ. Hamburg, 11, 17—40 (1936).

(I) $\mathfrak{G}$ *contains an involution* J *whose normalizer* $\mathfrak{N}(J)$ *is isomorphic to* $GL(2, q)$, *(the full linear group of degree* 2 *over a Galois field with* q *elements).*

(II) *If* C *is an element* $\neq 1$ *of the center* $\mathfrak{C}$ *of* $\mathfrak{N}(J)$, *the normalizer* $\mathfrak{N}(C)$ *of* C *in* $\mathfrak{G}$ *is* $\mathfrak{N}(J)$ *(and not larger than* $\mathfrak{N}(J)$*).*

(III) $\mathfrak{G}$ *is its own commutator subgroup.*

If $q \equiv -1$ *(mod* 4*),* $q \not\equiv 1$ (mod 3), *and* $q \neq 3$ *then* $\mathfrak{G}$ *is isomorphic to* $PSL(3, q)$. *If* $q = 3$, *we haven the additional case that* $\mathfrak{G}$ *can be the simple Mathieu group of order* 7912.

The latter group appears here as a kind of distorted version of the group $PSL(3, 3)$ of order 5616.

Since the group $GL(2, q)$ can be characterized by means of the preceding work, this theorem gives a characterization of the group $PSL(3, q)$ for the values of q in question. There seems to be little doubt that similar characterizations exist for the other values of q too. Some preliminary work indicates that the hyperorthogonal groups in three dimensions can be treated in a similar fashion. As was already stated, there is some hope that similar characterizations exist for the other simple groups though it is not clear whether these results are accessible to our methods.

I should like now to discuss briefly the ideas of the proof of the last theorem. If $\mathfrak{G}$ is actually equal to the group of collineations of a projective plane π, every involution J of $\mathfrak{G}$ leaves fixed a point P of π and all points of a line l not through P. Thus, to every J there corresponds a pair (P, l) of a point and a line. Instead of building up geometry using points as the fundamental elements, we can use such pairs (P, l) as the fundamental concept. The various incidence relations can then be described in terms of the involutions J corresponding to the pair (P, l). The axioms take the form of group theoretical statements. The projective plane is thus replaced by the set $\mathfrak{M}$ of all involutions $J \in \mathfrak{G}$. The projectivities are given by the transformations

$$T_\sigma : J \to \sigma^{-1} J \sigma \qquad (J \in \mathfrak{M})$$

of $\mathfrak{M}$ onto itself where σ is a fixed element of $\mathfrak{G}$. Of course, these T_σ form the full projective group $PGL(3, q)$ which for $q \not\equiv 1$ (mod 3) coincides with the special projective group.

Suppose now that $\mathfrak{G}$ is an arbitrary group which satisfies the conditions of the theorem and consider the set $\mathfrak{M}$ of involutions. If we can show that the axioms of projective geometry hold (in terms of involutions), then $\mathfrak{M}$ becomes a projective plane. If we can show further that the projectivities are given by the transformation T_σ, it follows that $\mathfrak{G}$ is isomorphic to $PSL(3, q)$ and we are finished.

The real difficulty then is to prove certain properties of the set $\mathfrak{M}$ of involutions which are in no way evident from the given assumptions. As a matter of fact, we have to use a long detour in order to obtain them. The methods described above suffice in our present case again to determine the characters of $\mathfrak{G}$ and, in particular, the order g. Now the knowledge of the characters provides us with the necessary information about the involutions and the proof can be finished.

The problem treated here of characterizing special simple groups is of interest in connection with the important unsolved problem of determining all simple groups of finite order. In order to be able to recognize the known simple groups, we need workable characterizations of these groups. Though we are certainly very far from a solution of the general problem of the simple groups of finite order, at least some kind of plan seems to evolve according to which the problem might be attacked. One hope would be that the only non-cyclic simple groups are the alternating groups, the finite analogues of the simple Lie groups and some distorted versions of a few of the latter groups. It would be necessary to see where these groups fit into our program and to show that no other cases can arise. It is not possible to say whether this plan will be workable, since some of the most important links are missing.

As a matter of fact, it is necessary to mention in this connection again the old conjecture that all groups of odd order are soluble. Of course, this would now obtain an added significance. No progress has been made on this question since it was first formulated.

Under these circumstances, it is perhaps of interest to state an equivalent conjecture. This is based on the following remark which can be proved rather easily. If $\mathfrak{G}$ is a group, if $\mathfrak{K}_0, \mathfrak{K}_1, \ldots, \mathfrak{K}_{k-1}$ are all the classes and if we use the same notation as above, it can be shown that $\mathfrak{G}$ is equal to its commutator group $\mathfrak{G}'$, if and only if the formula holds

$$g \,.\, K_0 K_1 \ldots K_{k-1} = \prod_{i=0}^{k-1} g_i \,.\, (K_0 + K_1 + \ldots + K_{k-1}) \tag{5}$$

where g_i is the number of elements in the class K_i. Thus the conjecture on groups of odd order is equivalent with the fact that the equation (5) can never hold for a group of odd order [7]).

Let me finish this talk with a number of remarks which are of some interest in connection with our problems. Let $\mathfrak{G}_1$ denote the set $\mathfrak{G} - \{1\}$ obtained from $\mathfrak{G}$ by removing the identity 1. Then the distance $d(\sigma, \tau)$ of two elements of $\mathfrak{G}_1$ can be defined as follows.For $\sigma = \tau$, we set $d(\sigma, \tau) = 0$. If σ and τ commute, but if $\sigma \neq \tau$, we set $d(\sigma, \tau) = 1$. If there exists a chain of elements σ_i of $\mathfrak{G}_1$,

$$\sigma_0 = \sigma, \sigma_1, \sigma_2, \ldots, \sigma_r = \tau,$$

---

[7]) I was informed by H. Wielandt that he also found this result.

of length r such that any two consecutive elements σ_{i-1} and σ_i commute, let $d(\sigma, \tau)$ be the length of the shortest such chain. Finally, if no such chain exists, set $d(\sigma, \tau) = \infty$. It is clear that $\mathfrak{G}_1$ then becomes a metric space (in which infinite distances are permitted).

Let $\mathfrak{G}$ now be an arbitrary group of even order g. The geometric part of the proof of the characterization of $PSL(3, q)$ shows that it will be of importance in the general case to consider the set $\mathfrak{M}$ of involutions (or the set $\mathfrak{G}_1$) as a kind of geometry in which the fundamental group is given by the transformations $T_\sigma : J \to \sigma^{-1}J\sigma$ where σ is a fixed element of $\mathfrak{G}$ while J ranges over $\mathfrak{M}$ (or over $\mathfrak{G}_1$). A very first step in this direction is given by the consideration of the distances. The following results can be proved without much difficulty.

Theorem. If $\sigma \neq 1$ is a real element of a group $\mathfrak{G}$ of even order and if the distance $d(\sigma, \mathfrak{M})$ of σ from the set of involutions is larger than 3, then the distance $d(\sigma, \mathfrak{M})$ is infinite. In this case, the normalizer $\mathfrak{N}(\sigma)$ of σ is an abelian group $\mathfrak{H}$ whose order is relatively prime to the index $(\mathfrak{G} : \mathfrak{H})$. The group $\mathfrak{H}$ consists only of real elements. It is the normalizer of each of its elements different from 1.

Theorem. If the group $\mathfrak{G}$ of even order contains more than one class of involutions, then any two involutions of $\mathfrak{G}$ have at most distance 3.

By combining these two statements, it is seen that if a group of even order contains more than one class of involutions, then any two real elements σ, $\tau \neq 1$ of $\mathfrak{G}$ either have infinite distance or distance $d(\sigma, \tau) \leqq 9$.

HARVARD UNIVERSITY.

Reprinted from ILLINOIS JOURNAL OF MATHEMATICS
Vol. 2, No. 4B, Miller Memorial Issue, 1958
Printed in U.S.A.

A CHARACTERIZATION OF THE ONE-DIMENSIONAL UNIMODULAR PROJECTIVE GROUPS OVER FINITE FIELDS[1]

In commemoration of G. A. Miller

BY
R. BRAUER, M. SUZUKI, AND G. E. WALL[2]

Let q be a power of a prime number, and denote by $LF(2, q)$ the group of all one-dimensional unimodular projectivities over the field Γ_q with q elements, i.e., of all linear fractional transformations

$$z' = \frac{az + b}{cz + d}$$

of determinant 1 with coefficients a, b, c, d in Γ_q. As is well known, $LF(2, q)$ is simple for $q \geqq 4$. The order of $LF(2, q)$ is $q(q + 1)(q - 1)/2$ for odd q and $q(q + 1)(q - 1)$ for even q.

It is our aim to give a group-theoretical characterization of these groups $LF(2, q)$. We shall prove

THEOREM. *Let $\mathfrak{G}$ be a group of finite order g which satisfies the following conditions:*

(I) *g is even;*

(II) *if $\mathfrak{A}$ and $\mathfrak{B}$ are two cyclic subgroups of $\mathfrak{G}$ of even orders, and if $\mathfrak{A} \cap \mathfrak{B} \neq \{1\}$, then there exists a cyclic subgroup $\mathfrak{Z}$ of $\mathfrak{G}$ which includes both $\mathfrak{A}$ and $\mathfrak{B}$;*

(III) *$\mathfrak{G}$ coincides with its commutator subgroup.*

Then $\mathfrak{G} \cong LF(2, q)$ where $q \geqq 4$ is a prime power.

We begin in I with an elementary discussion of groups which satisfy these conditions. Two cases, A and B, have to be considered. In Case A, the 2-Sylow group $\mathfrak{T}$ of $\mathfrak{G}$ is dihedral, and in Case B, $\mathfrak{T}$ is abelian of type $(2, 2, \cdots, 2)$. These two cases are treated in II and III respectively. The final result is obtained by applying a theorem of Zassenhaus [2] concerning doubly transitive permutation groups. We shall give some extensions of our theorem in a subsequent paper.

Received January 12, 1959.

[1] The results of this paper were obtained more or less independently by the three authors. Rather than publish three different papers, we preferred to combine our investigations.

The result for Case B has also been obtained by K. A. Fowler in his thesis, University of Michigan, 1952.

[2] Part of the work of the first two authors was done under an NSF contract.

I. Groups of type (S)

We shall say that a finite group $\mathfrak{G}$ is of type (S), if the following two conditions are satisfied:

(I) The group $\mathfrak{G}$ is of even order g.

(II) If $\mathfrak{A}$ and $\mathfrak{B}$ are two maximal cyclic subgroups of even orders, then either $\mathfrak{A} = \mathfrak{B}$ or $\mathfrak{A} \cap \mathfrak{B} = \{1\}$.

It follows from (II) that if two cyclic subgroups of even order of $\mathfrak{G}$ have an intersection different from $\{1\}$, then both are included in the same maximal cyclic subgroup of $\mathfrak{G}$. In particular, the elements of the two subgroups commute with each other. Moreover, if the two subgroups have the same order, they must be equal.

The following remarks are fairly obvious.

(I.A) *If $\mathfrak{G}$ has type (S), every subgroup $\mathfrak{H}$ of even order has type (S).*

Indeed, if $\mathfrak{A}_0$ and $\mathfrak{B}_0$ are two maximal subgroups of even order of $\mathfrak{H}$, and if $\mathfrak{A}_0 \cap \mathfrak{B}_0 \neq \{1\}$, then $\mathfrak{A}_0$ and $\mathfrak{B}_0$ are both subgroups of a cyclic subgroup $\mathfrak{Z}$ of $\mathfrak{G}$, and, because of the maximality of $\mathfrak{A}_0$ and $\mathfrak{B}_0$ in $\mathfrak{H}$, we have $\mathfrak{A}_0 = \mathfrak{H} \cap \mathfrak{Z}$, $\mathfrak{B}_0 = \mathfrak{H} \cap \mathfrak{Z}$.

(I.B) *Let $\mathfrak{G}$ be a group of type (S), and let G be an element of even order. If $G^r \neq 1$ for some exponent r, then the cyclic groups $\{G\}$ and $\{G^r\}$ have the same normalizer,*

$$\mathfrak{N}(\{G\}) = \mathfrak{N}(\{G^r\}). \tag{I1}$$

Proof. It is clear that $\mathfrak{N}(\{G\}) \subseteqq \mathfrak{N}(\{G^r\})$. Conversely, if $X \in \mathfrak{N}(\{G^r\})$, then $G^r \in \{G\} \cap X^{-1}\{G\}X$. Since the cyclic groups $\{G\}$ and $X^{-1}\{G\}X$ have a nontrivial intersection, and since they have the same order, they are equal. This means that $X \in \mathfrak{N}(\{G\})$, and hence we have (I1).

As a corollary, we have

(I.C) *Let $\mathfrak{G}$ be a group of type (S). Let G be an element of even order, and let I be the element of order 2 which is a power of G. Then*

$$\mathfrak{N}(\{G^r\}) = \mathfrak{C}(I) \tag{I2}$$

for all r for which $G^r \neq 1$.

(I.D) *Let $\mathfrak{G}$ be of type (S). Let I be an involution (i.e., an element of order 2) of $\mathfrak{G}$. If p is an odd prime dividing the order $c(I)$ of the centralizer $\mathfrak{C}(I)$ of I, then $\mathfrak{C}(I)$ includes a p-Sylow group $\mathfrak{P}$ of $\mathfrak{G}$.*

Proof. Let P be an element of order p of $\mathfrak{C}(I)$. If $X \in \mathfrak{C}(P)$, then $X \in \mathfrak{N}(\{IP\})$ and, by (I2), $X \in \mathfrak{C}(I)$. In particular, if P belongs to the p-Sylow group $\mathfrak{P}$ of $\mathfrak{G}$, the center of $\mathfrak{P}$ is included in $\mathfrak{C}(I)$. If we now replace P by an element of order p of the center of $\mathfrak{P}$, our argument shows that $\mathfrak{P} \subseteqq \mathfrak{C}(I)$.

(I.E) *Let $\mathfrak{G}$ be of type (S). Let I be an involution, and let p be an odd prime dividing $c(I)$. Then the p-Sylow groups of $\mathfrak{G}$ are cyclic.*

Indeed, by (I.D), a p-Sylow group $\mathfrak{P}$ of $\mathfrak{G}$ lies in $\mathfrak{C}(I)$. If $\mathfrak{P}$ were not cyclic, we could find two distinct subgroups $\{P_1\}$ and $\{P_2\}$ of $\mathfrak{P}$ of order p, and then

$$I \in \{IP_1\} \cap \{IP_2\}.$$

Since $\{IP_1\}$ and $\{IP_2\}$ both have the same order $2p$, this implies $\{IP_1\} = \{IP_2\}$, and hence $\{P_1\} = \{P_2\}$. This is a contradiction.

We next study the 2-Sylow groups of $\mathfrak{G}$.

(I.F) *If $\mathfrak{G}$ is of type (S), the 2-Sylow group $\mathfrak{T}$ of $\mathfrak{G}$ is of one of the following types:*

(a) *$\mathfrak{T}$ is cyclic.*

(b) *$\mathfrak{T}$ is abelian of type $(2, 2, \cdots, 2)$.*

(c) *$\mathfrak{T}$ is dihedral: $\mathfrak{T} = \{A, B\}$ with $A^{2^m} = 1$, $B^2 = 1$, $B^{-1}AB = A^{-1}$.*

Proof. By (I.A), $\mathfrak{T}$ itself is of type (S). If all elements $T \neq 1$ of $\mathfrak{T}$ have order 2, we have case (b). Suppose that $\mathfrak{T}$ contains elements R of order $r \geqq 4$. If I is an involution in the center of $\mathfrak{T}$, then

$$R^2 \in \{R\} \cap \{IR\},$$

and since R and IR have the same order r, and since $R^2 \neq 1$, it follows that $\{R\} = \{IR\}$, whence $I \in \{R\}$. Thus, $I = R^{r/2}$.

Let A be an element of maximal order r of $\mathfrak{T}$; $r \geqq 4$. If B is an element of order $\geqq 4$ of $\mathfrak{T}$, then as just shown, $I \in \{A\}$ and $I \in \{B\}$. Since $\{A\}$ is a maximal cyclic subgroup of $\mathfrak{T}$, it follows that $B \in \{A\}$. In particular, $\{A\}$ is the only maximal cyclic subgroup of $\mathfrak{T}$ of order $\geqq 4$. It follows that $\{A\}$ is normal in $\mathfrak{T}$. If $\mathfrak{T} = \{A\}$, we have case (a).

Suppose then that $\{A\} \neq \mathfrak{T}$. If $B \notin \{A\}$, B has order 2, and, as $AB \notin \{A\}$, AB also has order 2. Thus, $B^2 = 1$, $ABAB = 1$, whence $B^{-1}AB = A^{-1}$. Hence, if A has order 2^m, we have

$$A^{2^m} = 1, \qquad B^2 = 1, \qquad B^{-1}AB = A^{-1}.$$

If B_1 is any element of $\mathfrak{T}$ with $B_1 \notin \{A\}$, then we also have $B_1^{-1}AB_1 = A^{-1}$. Then $B_1B^{-1} \in \mathfrak{C}(A)$. If we had $B_1B^{-1} \notin \{A\}$, we would have $(B_1B^{-1})^{-1}A(B_1B^{-1}) = A^{-1} \neq A$; $B_1B^{-1} \notin \mathfrak{C}(A)$. Thus, $B_1B^{-1} \in \{A\}$, $B_1 \in \{A\}B$. It follows that $\mathfrak{T} = \{A, B\}$. Hence $\mathfrak{T}$ is dihedral.

We now study the centralizers in $\mathfrak{G}$ of the involutions in the center of a 2-Sylow group $\mathfrak{T}$ of $\mathfrak{G}$. It will not be necessary to consider the case that $\mathfrak{T}$ is cyclic.

(I.G) *Let $\mathfrak{G}$ be of type (S). Let $\mathfrak{T}$ be a 2-Sylow group of $\mathfrak{G}$, and let T be an involution in the center of $\mathfrak{T}$.*

(1) *If $\mathfrak{T}$ is abelian of type $(2, 2, \cdots, 2)$ and of order $2^r > 4$, then $\mathfrak{C}(T) = \mathfrak{T}$.*

(2) *If $\mathfrak{T}$ is abelian of type* (2, 2), *then* $\mathfrak{C}(T) = \mathfrak{T}\mathfrak{B}$ *where* $\mathfrak{B}$ *is cyclic of odd order* v. *Moreover*, $B^{-1}VB = V^{-1}$ *for* $B \in \mathfrak{T}$, $B \neq 1, T$; $V \in \mathfrak{B}$.

(3) *If* $\mathfrak{T} = \{A, B\}$ *is dihedral of order* $2^{m+1} \geqq 8$, $A^{2^m} = 1$, $B^2 = 1$, $B^{-1}AB = A^{-1}$, *then* $\mathfrak{C}(T) = \{A, B, \mathfrak{B}\}$ *where* $\mathfrak{B}$ *is cyclic of odd order* v, $\mathfrak{B} \subseteqq \mathfrak{C}(A)$, *and* $B^{-1}VB = V^{-1}$ *for* $V \in \mathfrak{B}$.

Proof. (α) Suppose that V is an element of odd order of $\mathfrak{C}(T)$; $V \neq 1$. By (I.B), $\mathfrak{N}(\{V\}) = \mathfrak{N}(\{TV\}) = \mathfrak{C}(T)$. If J is an involution of $\mathfrak{C}(T)$, we must have $J^{-1}VJ \in \{V\}$, say $J^{-1}VJ = V^j$. Since J has order 2, then $V^{j^2} = V$. Assume that the order of V is a prime power p^s. Since $j^2 \equiv 1 \pmod{p^s}$ and p is odd, $j \equiv \pm 1 \pmod{p^s}$. If $j \equiv 1 \pmod{p^s}$, $J^{-1}VJ = V$, and then $V \in \{JV\} \cap \{TV\}$, which implies $J = T$. Thus, for $J \neq T$, we have $J^{-1}VJ = V^{-1}$. Since this holds for all elements V of odd prime power of $\mathfrak{C}(T)$, it holds for all elements V of $\mathfrak{C}(T)$ of odd order.

$$\text{(I3)} \qquad J^{-1}VJ = V^{-1} \quad \text{if} \begin{cases} V \in \mathfrak{C}(T), & V \text{ of odd order,} \\ J \in \mathfrak{C}(T), & J \text{ an involution, } J \neq T. \end{cases}$$

(β) If $\mathfrak{T}$ is abelian of type $(2, 2, \cdots, 2)$ and order $2^r > 4$, then (I3) would hold for all $2^r - 2$ elements $J \neq 1, T$ of $\mathfrak{T}$. If we choose two such elements J and J_1 such that $J_1 \neq J, JT$, we have a contradiction.

(γ) Suppose that $\mathfrak{T}$ is abelian of type (2, 2). Then $\mathfrak{T}$ is a 2-Sylow subgroup of $\mathfrak{C}(T)$. If $J \in \mathfrak{T}$, $J \neq 1, T$, then J and JT are not conjugate in $\mathfrak{C}(T)$. Indeed, if we had $X^{-1}JX = JT$ with $X \in \mathfrak{C}(T)$, then $X^{-1}JTX = J$. Thus, $X^2 \in \mathfrak{C}(J)$, and then $X^2 \in \mathfrak{C}(\{T, J\}) = \mathfrak{C}(\mathfrak{T})$. Since the order of $\{X, \mathfrak{C}(\mathfrak{T})\}$ cannot be divisible by 8, and since $X \in \mathfrak{N}(\mathfrak{T})$, we find $X \in \mathfrak{C}(\mathfrak{T})$, a contradiction. Thus, no two distinct elements of $\mathfrak{T}$ are conjugate in $\mathfrak{C}(T)$. It follows from Burnside's Theorem that $\mathfrak{C}(T)$ has a normal subgroup $\mathfrak{B}$ of index 4. Then $\mathfrak{C}(T) = \mathfrak{T}\mathfrak{B}$. Since J maps every element V of $\mathfrak{B}$ on its inverse, $\mathfrak{B}$ is abelian. But the Sylow groups of $\mathfrak{B}$ are cyclic by (I.E), and $\mathfrak{B}$ itself is cyclic.

(δ) Suppose now that $\mathfrak{T}$ is dihedral and of order $\geqq 8$. Clearly, $\{A\}$ is a 2-Sylow subgroup of $\mathfrak{C}(A)$. Since no two distinct elements of $\{A\}$ are conjugate in $\mathfrak{C}(A)$, Burnside's Theorem shows that we may set $\mathfrak{C}(A) = \{A\}\mathfrak{B}$ where $\mathfrak{B}$ is a normal subgroup of $\mathfrak{C}(A)$ of odd order v. Then

$$\mathfrak{C}(A) = \{A\} \times \mathfrak{B}.$$

If W is an element of odd order of $\mathfrak{C}(T)$, then $T \in \{A\} \cap \{TW\}$. It follows that A commutes with TW and hence with W. Thus, $W \in \mathfrak{C}(A)$. Thus, every p-Sylow group of $\mathfrak{C}(T)$ lies in $\mathfrak{C}(A)$ for odd p. Since $\mathfrak{C}(A) \subseteqq \mathfrak{C}(T)$, it follows that $(\mathfrak{C}(T) : \mathfrak{C}(A))$ is a power of 2. But $\mathfrak{T} = \{A, B\} \in \mathfrak{C}(T)$, $A \in \mathfrak{C}(A)$, $B \notin \mathfrak{C}(A)$, and we see that

$$\mathfrak{C}(T) = \{\mathfrak{T}, \mathfrak{B}\} = \{B, \mathfrak{C}(A)\}.$$

Again, (I3) implies that $\mathfrak{B}$ is abelian, and then (I.E) shows that $\mathfrak{B}$ is cyclic. This completes the proof of (I.G).

We shall consider the case (2), $r = 2$ as a special case of (3) taking $m = 1$, $A = T$. In this sense, the dihedral group of order 4 is the abelian group of type (2, 2).

If we assume that $\mathfrak{G}$ is a group of type (S) such that $\mathfrak{G}$ does not have a normal subgroup of index 2, the case (a) of (I.F) is excluded (by Burnside's Theorem). We then have to consider the following cases:

Case A. $\mathfrak{T}$ is dihedral of order 2^{m+1}, $m \geqq 1$. We set $\mathfrak{V} = \{V\}$ and $H = AV$. Then $\mathfrak{C}(T) = \{H, B\}$ with

$$H^h = 1, \qquad B^2 = 1, \qquad B^{-1}HB = H^{-1},$$

where $h = 2^m v$.

Case B. $\mathfrak{T}$ is abelian of type $(2, 2, \cdots, 2)$, $(\mathfrak{T}:1) \geqq 4$. Here $\mathfrak{C}(T) = \mathfrak{T}$ for $T \in \mathfrak{T}$, $T \neq 1$.

If $m = 1$, $v = 1$ in Case A, we may consider this as Case B with $(\mathfrak{T}:1) = 4$. Consequently, in dealing with Case A, we may assume that $h \geqq 4$.

II. The Case A

1. Assumptions

We assume that $\mathfrak{G}$ is a group of type (S) which does not have a normal subgroup of index 2 and that we have Case A in the notation of I. Fully stated our assumptions are

(I) $\mathfrak{G}$ is a finite group of even order g.

(II) If $\mathfrak{A}$ and $\mathfrak{B}$ are two maximal cyclic subgroups of $\mathfrak{G}$ of even order, then either $\mathfrak{A} = \mathfrak{B}$ or $\mathfrak{A} \cap \mathfrak{B} = \{1\}$.

(III) There does not exist a normal subgroup of index 2 in $\mathfrak{G}$.

(IV) The 2-Sylow subgroups of $\mathfrak{G}$ are dihedral.

Let $\mathfrak{T}$ be a 2-Sylow subgroup of $\mathfrak{G}$. If its order is 2^{m+1} with $m \geqq 1$, we can set $\mathfrak{T} = \{A, B\}$ where

$$A^{2^m} = 1, \qquad B^2 = 1, \qquad B^{-1}AB = A^{-1}. \tag{1}$$

We set

$$T = A^{2^{m-1}} \tag{2}$$

Then T has order 2.

2. The centralizer $\mathfrak{C}(T)$ of T

As shown by (I.G), the centralizer $\mathfrak{C}(T)$ is dihedral too. We shall denote its order by $2h$. Then

$$(\mathfrak{C}(T):\{1\}) = 2h = 2^{m+1}v; \qquad (v, 2) = 1. \tag{3}$$

If V is an element of order v in $\mathfrak{C}(T)$, and if we set

$$H = AV, \tag{4}$$

then $AV = VA$, and $\mathfrak{C}(T) = \{H, B\}$;

$$(5) \qquad H^h = 1, \qquad B^{-1}HB = H^{-1}.$$

Since $\mathfrak{T}$ is a 2-Sylow subgroup of $\mathfrak{G}$, 2^{m+1} is the exact power of 2 in g. By (I.D), we can set

$$(6) \qquad g = 2^{m+1}vg_0, \qquad (g_0, 2^{m+1}v) = 1.$$

The number h is even, and we shall set

$$(7) \qquad s = h/2 - 1.$$

The case $h = 2$ may be considered as part of Case B so that we may assume

$$(8) \qquad s \geqq 1.$$

(II.A) *In $\mathfrak{G}$, every element of $\mathfrak{C}(T)$ is conjugate to a power H^α of H. Two powers H^α and H^β are conjugate in $\mathfrak{G}$ if and only if $H^\beta = H^{\pm\alpha}$. Thus, the elements*

$$(9) \qquad 1, \quad H^{s+1} = T; \qquad H, \quad H^2, \quad \cdots, \quad H^s$$

form a full system of representatives for those classes of conjugate elements of $\mathfrak{G}$ in which the order is not relatively prime to $2h = 2^{m+1}v$. We have

$$(10) \qquad c(1) = g, \qquad c(T) = 2h; \qquad c(H^\alpha) = h \quad \text{for} \quad H^\alpha \neq 1, T.$$

Proof. In the dihedral group $\mathfrak{C}(T) = \{H, B\}$, the elements (9) together with the elements B and BH form a full system of representatives for the classes of conjugate elements. The elements 1 and T form classes by themselves, while the class of H^α for $H^\alpha \neq 1, T$ consists of H^α and $H^{-\alpha}$. The class of B consists of all elements BH^ρ with even ρ and the class of BH of all elements BH^σ with odd σ.

Since $T = H^{h/2} = H^{s+1}$ is a power of H, it follows from (I.C) that $\mathfrak{N}(\{H^r\}) = \mathfrak{N}(\{T\}) = \mathfrak{C}(T)$ for $H^r \neq 1$. Consequently, two powers H^α and H^β can be conjugate in $\mathfrak{G}$ only if they are conjugate in $\mathfrak{C}(T)$, and this is the case if and only if $H^\beta = H^{\pm\alpha}$. Moreover, for $H^r \neq 1$, we have $\mathfrak{C}(H^r) \subseteqq \mathfrak{N}(\{H^r\}) \subseteqq \mathfrak{C}(T)$. This implies that the order of the centralizer of $H^r \neq 1$ in $\mathfrak{G}$ is the order of the centralizer of H^r in $\mathfrak{C}(T)$. This order is $2h$ for $H^r = T$ and h for $H^r \neq 1, T$. Since $c(1) = g$, the equations (10) are true.

We show next that the elements T, B, $BH^v = BA^v$ are conjugate in $\mathfrak{G}$. Let $\mathfrak{T}^*$ denote the subgroup of $\mathfrak{T}$ generated by all quotients XY^{-1} of elements X, Y of $\mathfrak{T}$ which are conjugate in $\mathfrak{G}$. If X is a power $H^r \neq 1$, T, then $Y = H^{\pm r}$, since the order of H^r is different from 2 and H^r is not conjugate to the involutions B, BH^v. Thus, $XY^{-1} = 1$, or $XY^{-1} = H^{2r}$. For $X = 1$, we have $Y = 1$, $XY^{-1} = 1$. If no two different ones of the elements T, B, BH^v are conjugate in $\mathfrak{G}$, then $\mathfrak{T}^* = \{A^2\}$. If two, but not all three, elements T, B, BH^v are conjugate in $\mathfrak{G}$, then $\mathfrak{T}^*$ is one of the groups $\{A^2, BT^{-1}\}$, $\{A^2, BH^vT^{-1}\}$, $\{A^2, BH^vB^{-1}\}$. Since $\mathfrak{T}/\{A^2\}$ is abelian of type (2, 2), the group $\mathfrak{T}$ cannot be generated by A^2 and one further element. Hence $\mathfrak{T}^* \neq \mathfrak{T}$. Now, the generalization of Burnside's Theorem shows that the 2-Sylow group of $\mathfrak{G}/\mathfrak{G}'$ is isomorphic with $\mathfrak{T}/\mathfrak{T}^*$. Hence $\mathfrak{G}/\mathfrak{G}'$ has an order divisible by 2. Then $\mathfrak{G}$ has a normal subgroup of index 2. But this is a contradiction to condition (III). Hence T, B, and BH^v are conjugate in $\mathfrak{G}$. Since v is odd, then T, B, BH are conjugate in $\mathfrak{G}$.

If X has an order divisible by a prime factor p of $2^{m+1}v$, and if P is a power of X whose order is p, then $X \in \mathfrak{C}(P)$. If $p = 2$, the preceding argument shows that P is conjugate to T in $\mathfrak{G}$. If $p \neq 2$, it follows from (I.D) and (I.E) that P is conjugate to an element of the form H^r. Then X is conjugate to an element of $\mathfrak{C}(T)$ or even of $\mathfrak{C}(H^r) \subset \mathfrak{C}(T)$. This concludes the proof of (II.A).

3. The restrictions of the irreducible characters of $\mathfrak{G}$ to $\mathfrak{H} = \{H\}$

Set $\mathfrak{H} = \{H\}$. If ε is a fixed primitive h^{th} root of unity, let ε_j denote the irreducible character of $\mathfrak{H}$ defined by

$$\varepsilon_j(H^r) = \varepsilon^{jr} \qquad (r = 0, 1, \cdots, h-1). \tag{11}$$

Then $\varepsilon_i = \varepsilon_j$ if and only if $i \equiv j \pmod h$. Clearly,

$$\varepsilon_{-s}, \varepsilon_{-s+1}, \cdots, \varepsilon_{s-1}, \varepsilon_s, \varepsilon_{s+1} \tag{12}$$

are all the irreducible characters of $\mathfrak{H}$.

If $\chi^{(1)}, \chi^{(2)}, \cdots, \chi^{(k)}$ are the irreducible characters of $\mathfrak{G}$, we can set

$$\chi^{(\mu)} \mid \mathfrak{H} = \textstyle\sum_{j=-s}^{s+1} a_{\mu j}\, \varepsilon_j \qquad (\mu = 1, 2, \cdots, k) \tag{13}$$

with nonnegative rational integers $a_{\mu j}$. Since $\chi^{(\mu)}(H^r) = \chi^{(\mu)}(H^{-r})$, we see that

$$a_{\mu j} = a_{\mu,-j}\,. \tag{14}$$

It follows from (13) that

$$a_{\mu j} = (1/h)\textstyle\sum_{X\in\mathfrak{H}} \chi^{(\mu)}(X)\bar{\varepsilon}_j(X). \tag{15}$$

(II.B) *If* $u = (g - 2h)/h^2$, *then, for* $0 \leqq i, j \leqq s + 1$, *we have*

$$\sum_{\mu=1}^{k} a_{\mu i}\, a_{\mu j} = \begin{cases} u & \text{for } i \neq j, \\ u + 1 & \text{for } i = j;\quad i \neq 0, s+1, \\ u + 2 & \text{for } i = j;\quad i = 0 \text{ or } i = s+1. \end{cases} \tag{16}$$

Proof. Since $a_{\mu j}$ is rational, we have, by (15)

$$\textstyle\sum_{\mu=1}^{k} a_{\mu i}\, a_{\mu j} = \sum_{\mu=1}^{k} a_{\mu i}\, \bar{a}_{\mu j} = (1/h^2)\sum_{X,Y} \bar{\varepsilon}_i(X)\varepsilon_j(Y)\sum_{\mu=1}^{k} \chi^{(\mu)}(X)\bar{\chi}^{(\mu)}(Y).$$

Here, the inner sum on the right is 0, if X and Y are not conjugate in $\mathfrak{G}$, while in the other case, the value is $c(X)$. On account of (II.A), we find

$$\sum_{\mu=1}^{k} a_\mu a_{\mu j} = \frac{1}{h^2} g + \frac{1}{h^2} 2h\bar{\varepsilon}_i(T)\varepsilon_j(T) + \frac{h}{h^2} \sum_{X \in \mathfrak{H}, X \neq 1, T} \bar{\varepsilon}_i(X)(\varepsilon_j(X) + \varepsilon_j(X^{-1})).$$

By the orthogonality relation for the characters of $\mathfrak{H}$, we have

$$\frac{1}{h} \sum_{X \in \mathfrak{H}} \bar{\varepsilon}_i(X)(\varepsilon_j(X) + \bar{\varepsilon}_j(X)) = \begin{cases} 0 & \text{for} \quad i \neq j, \\ 1 & \text{for} \quad i = j; \quad i \neq 0, s+1, \\ 2 & \text{for} \quad i = j; \quad i = 0 \text{ or } i = s+1, \end{cases}$$

where we assumed $0 \leqq i, j \leqq s + 1$. If this is subtracted from the preceding equation, and if $2\bar{\varepsilon}_i(T)\varepsilon_j(T) = \bar{\varepsilon}_i(T)(\varepsilon_j(T) + \bar{\varepsilon}_j(T))$ is taken into account, we obtain

$$\sum_{\mu=1}^{k} a_{\mu i} a_{\mu j} - \begin{cases} 0 \\ 1 \\ 2 \end{cases} = \frac{1}{h^2} g - \frac{2}{h}.$$

This yields (16).

If with each of the characters $\chi^{(\mu)}$, there is associated a complex number z_μ, we arrange these k numbers z_μ in the form of a column. We define the inner product of two such columns in the usual manner. In particular, let $\mathfrak{a}_i$ denote the column consisting of the numbers $a_{\mu i}$ with fixed i. Because of (14), it will be sufficient to consider the columns

$$\mathfrak{a}_0, \quad \mathfrak{a}_{s+1}; \qquad \mathfrak{a}_1, \quad \mathfrak{a}_2, \quad \cdots, \quad \mathfrak{a}_s.$$

Now, (16) can be written in the form

$$\begin{aligned} &\mathfrak{a}_i\, \mathfrak{a}_j = u && \text{for} \quad i \neq j; \quad 0 \leqq i, j \leqq s+1, \\ (17) \qquad &\mathfrak{a}_i^2 = u + 1 && \text{for} \quad i = 1, 2, \cdots, s, \\ &\mathfrak{a}_i^2 = u + 2 && \text{for} \quad i = 0, \quad \text{and for} \quad i = s+1. \end{aligned}$$

It follows from (15) that

$$a_{\mu j} - a_{\mu 1} = \frac{1}{h} \sum_{X \in \mathfrak{H}, X \neq 1} \chi^{(\mu)}(X)(\bar{\varepsilon}_j(X) - \bar{\varepsilon}_1(X)).$$

If Y is either 1 or an element of $\mathfrak{G}$ which is not conjugate to a power H^r of H, and if we multiply here by $\bar{\chi}^{(\mu)}(Y)$ and add over $\mu = 1, 2, \cdots, k$, the orthogonality relations for the $\chi^{(\mu)}$ show that we obtain 0. Replacing Y by Y^{-1}, we have

(II.C) *Let Y be an element of $\mathfrak{G}$ which is not conjugate in $\mathfrak{G}$ to a power $H^r \neq 1$. If $\chi(Y)$ denotes the column consisting of the numbers $\chi^{(\mu)}(Y)$, we have*

$$(18) \qquad (\mathfrak{a}_j - \mathfrak{a}_1)\cdot\chi(Y) = 0 \qquad (j = 0, 1, \cdots, s+1).$$

In particular,

$$(19) \qquad (\mathfrak{a}_j - \mathfrak{a}_1)\cdot\chi(1) = 0 \qquad (j = 0, 1, \cdots, s+1).$$

4. The exceptional characters of $\mathfrak{G}$

Assume first that $s > 1$. We shall say that an irreducible character $\chi^{(\mu)}$ of $\mathfrak{G}$ is *exceptional*, if not all the coefficients in $\mathfrak{a}_2 - \mathfrak{a}_1, \mathfrak{a}_3 - \mathfrak{a}_1, \cdots, \mathfrak{a}_s - \mathfrak{a}_1$ belonging to $\chi^{(\mu)}$ are 0.

(II.D) *Suppose that $s \geqq 2$. There exist exactly s exceptional characters of $\mathfrak{G}$. These can be taken for $\chi^{(1)}, \chi^{(2)}, \cdots, \chi^{(s)}$ in such an order that*

$$a_{ij} = a + \tau\delta_{ij} \qquad (\text{for } i, j = 1, 2, \cdots, s) \tag{20}$$

where a is a fixed rational integer and where $\tau = \pm 1$.

Proof. Let j range over $2, 3, \cdots, s$. It follows from (17) that

$$(\mathfrak{a}_j - \mathfrak{a}_1)^2 = 2.$$

Since the coefficients of $\mathfrak{a}_j - \mathfrak{a}_1$ are rational integers, only two of these coefficients have values different from 0, and these values are ± 1. Now (19) shows that one of the values is $+1$ and the other is -1, since the coefficients of $\chi(1)$ are the degrees of the $\chi^{(\mu)}$ and hence positive.

Assume that $s \geqq 3$. If j' also is one of the values $2, 3, \cdots, s$, then (17) shows that

$$(\mathfrak{a}_j - \mathfrak{a}_1)(\mathfrak{a}_{j'} - \mathfrak{a}_1) = 1 \qquad \text{for } j \neq j'.$$

Hence $\mathfrak{a}_j$ and $\mathfrak{a}_{j'}$ must have an equal coefficient ± 1 in one and the same row, while in all other rows at least one of them has the coefficient 0.

Choose $\chi^{(1)}$ such that $\mathfrak{a}_2 - \mathfrak{a}_1$ and $\mathfrak{a}_3 - \mathfrak{a}_1$ have the same coefficient ± 1 in the first row, and denote the value of this coefficient by $-\tau$.

We claim that we can arrange the characters $\chi^{(1)}, \chi^{(2)}, \quad, \chi^{(k)}$ in such an order that the matrix M of the columns $\mathfrak{a}_2 - \mathfrak{a}_1, \mathfrak{a}_3 - \mathfrak{a}_1, \cdots, \mathfrak{a}_s - \mathfrak{a}_1$ has the form

$$M = \begin{pmatrix} -\tau & -\tau & -\tau & \cdots & -\tau \\ \tau & 0 & 0 & \cdots & 0 \\ 0 & \tau & 0 & \cdots & 0 \\ \cdots & \cdots & \cdots & \cdots & \cdots \\ 0 & 0 & 0 & \cdots & \tau \\ 0 & 0 & 0 & \cdots & 0 \\ 0 & 0 & 0 & \cdots & 0 \\ \cdots & \cdots & \cdots & \cdots & \cdots \\ 0 & 0 & 0 & \cdots & 0 \end{pmatrix}. \tag{21}$$

Since the coefficient τ must appear in $\mathfrak{a}_2 - \mathfrak{a}_1$ and in $\mathfrak{a}_3 - \mathfrak{a}_1$ in different rows, we can choose $\chi^{(2)}, \chi^{(3)}$ such that the first two columns $\mathfrak{a}_2 - \mathfrak{a}_1$ and $\mathfrak{a}_3 - \mathfrak{a}_1$ have the form given in (21). If $s = 3$, we are finished. If $s \geqq 4$, then $\mathfrak{a}_4 - \mathfrak{a}_1$ must have either the coefficient $-\tau$ in the first row or the coefficient τ both in the third and fourth row. The latter case is impossible, since the two nonvanishing coefficients of $\mathfrak{a}_4 - \mathfrak{a}_1$ have different values. Thus, we have $-\tau$ in the first row of $\mathfrak{a}_4 - \mathfrak{a}_1$, and if $\chi^{(4)}$ is chosen suitably,

the coefficient τ appears in the fourth row. Continuing in this manner we see that the $s - 1$ columns $\mathfrak{a}_j - \mathfrak{a}_1$ can be assumed to have the form given in (21).

We see at once that this result is true for $s = 2$ too. Here, we may take $\tau = 1$, choosing $\chi^{(1)}, \chi^{(2)}$ such that $\mathfrak{a}_2 - \mathfrak{a}_1$ has coefficient 1 in the first row and -1 in the second row.

It is evident from (21) that we have exactly s exceptional characters, and, in our notation, they are the characters $\chi^{(1)}, \chi^{(2)}, \cdots, \chi^{(s)}$. Moreover, by (17), $\mathfrak{a}_1(\mathfrak{a}_i - \mathfrak{a}_1) = -1$ for $i = 2, 3, \cdots, s$. Using the form (21) of $\mathfrak{a}_j - \mathfrak{a}_1$, we obtain $a_{11}(-\tau) + a_{j1}\tau = -1$, whence $a_{i1} = a_{11} - \tau$ for $i = 2, 3, \cdots, s$. Adding $\mathfrak{a}_1$ and $\mathfrak{a}_j - \mathfrak{a}_1$ (given in (21)), $j = 2, 3, \cdots, s$, we see that $a_{ii} = a_{11}, a_{ij} = a_{11} - \tau$ for $i \neq j$. This completes the proof of (II.D); $a = a_{11} - \tau$.

(II.E) *The s exceptional characters $\chi^{(1)}, \chi^{(2)}, \cdots, \chi^{(s)}$ have the same degree f. Moreover, we have*

$$\chi^{(1)}(Y) = \chi^{(2)}(Y) = \cdots = \chi^{(s)}(Y)$$

for every element Y of $\mathfrak{G}$ which is not conjugate to a power of H.

This is an immediate consequence of (18) and (19) combined with (21).

5. The values of the characters of $\mathfrak{G}$ for the elements $H^r \neq 1$

We now determine the columns $\mathfrak{a}_0 - \mathfrak{a}_1$ and $\mathfrak{a}_{s+1} - \mathfrak{a}_1$. It follows from (17) that we have

$$(22)\quad (\mathfrak{a}_0 - \mathfrak{a}_1)^2 = 3, \qquad (\mathfrak{a}_{s+1} - \mathfrak{a}_1)^2 = 3, \qquad (\mathfrak{a}_0 - \mathfrak{a}_1)(\mathfrak{a}_{s+1} - \mathfrak{a}_1) = 1,$$

$$(23)\quad \mathfrak{a}_0(\mathfrak{a}_j - \mathfrak{a}_1) = 0, \qquad \mathfrak{a}_{s+1}(\mathfrak{a}_j - \mathfrak{a}_1) = 0 \qquad \text{for } j = 2, 3, \cdots, s.$$

Then (22) shows that $\mathfrak{a}_0 - \mathfrak{a}_1$ and $\mathfrak{a}_{s+1} - \mathfrak{a}_1$ both have three nonvanishing coefficients, and these have the values ± 1. At least in one row, both columns have an equal coefficient ± 1. If there were another row in which both had nonvanishing coefficients, then the three nonvanishing coefficients in both columns would appear in the same three rows, and then $(\mathfrak{a}_0 - \mathfrak{a}_1) - (\mathfrak{a}_{s+1} - \mathfrak{a}_1)$ would have only one nonvanishing coefficient ± 2. Then again (19) leads to a contradiction, since all $\chi^{(\mu)}(1)$ are positive.

Thus, we have exactly one row in which $\mathfrak{a}_0 - \mathfrak{a}_1$ and $\mathfrak{a}_{s+1} - \mathfrak{a}_0$ have a nonvanishing coefficient, and these two coefficients are equal and ± 1. There are two other rows in which $\mathfrak{a}_1 - \mathfrak{a}_0$ has a nonvanishing coefficient and two further rows in which $\mathfrak{a}_{s+1} - \mathfrak{a}_0$ has a nonvanishing coefficient. It follows from (19) that the three nonvanishing coefficients of $\mathfrak{a}_1 - \mathfrak{a}_0$ cannot all have the same sign. The same is true for $\mathfrak{a}_{s+1} - \mathfrak{a}_0$.

The equations (23) combined with (21) show that the first s coefficients of $\mathfrak{a}_0$ all have the same value a_{10}. By (20), the first s coefficients of $\mathfrak{a}_0 - \mathfrak{a}_1$ then are

$$(24) \qquad a_{10} - a - \tau, \quad a_{10} - a, \quad a_{10} - a, \quad \cdots, \quad a_{10} - a.$$

Since these coefficients are all equal to 0, +1, or −1, we have either $a_{10} = a$ or $a_{10} = a + \tau$. A similar argument applies to $\mathfrak{a}_{s+1} - \mathfrak{a}_1$. The first s coefficients here are

$$(25) \quad a_{1,s+1} - a - \tau, \quad a_{1,s+1} - a, \quad \cdots, \quad a_{1,s+1} - a, \text{ and } a_{1,s+1} = a \text{ or } a + \tau.$$

We wish to show that $a_{10} = a_{1,s+1} = a$. This is clear for $s \geqq 4$ since otherwise in (24) or in (25), there appear three or more coefficients $\tau = \pm 1$, which has been seen to be impossible.

Before dealing with the cases $s = 3$ and $s = 2$, let us first observe that if $\chi^{(j)}$ is the unit character, then (13) shows that $a_{j0} = 1$, $a_{jv} = 0$ for $v \neq 0$. Hence $\chi^{(j)}$ is not exceptional. We may then choose our notation such that $j = s + 1$, that is, such that $\chi^{(s+1)}$ is the unit character.

Assume now that $s = 3$. If we had $a_{10} = a_{1,s+1} = a + \tau$, then (24) and (25) show that both columns $\mathfrak{a}_0 - \mathfrak{a}_1$ and $\mathfrak{a}_{s+1} - \mathfrak{a}_0$ would have nonzero coefficients in rows 2 and 3, which is impossible. If one of a_{10}, $a_{s+1,0}$ is a and the other is $a + \tau$, the first three coefficients of $\mathfrak{a}_0 - \mathfrak{a}_{s+1}$ are equal, and their value is $\tau = \pm 1$. Moreover, the coefficient in the fourth row is ± 1, and as $(\mathfrak{a}_0 - \mathfrak{a}_{s+1})^2 = 4$ by (17), all other coefficients vanish. By (II.E), $\chi^{(1)}, \chi^{(2)}, \chi^{(3)}$ have the same degree f, while the unit character $\chi^{(4)}$ has degree 1. By (19), $(\mathfrak{a}_0 - \mathfrak{a}_{s+1})\chi(1) = (\mathfrak{a}_0 - \mathfrak{a}_1)\chi(1) - (\mathfrak{a}_{s+1} - \mathfrak{a}_1)\chi(1) = 0$, and hence $\pm 3f \pm 1 = 0$. This is impossible. Again, we must have $a_{10} = a_{1,s+1} = a$ for $s = 3$.

Take $s = 2$. If $a_{10} = a_{1,s+1} = a + \tau$, we simply replace $a + \tau$ by a, τ by $-\tau$, and we interchange $\chi^{(1)}$ and $\chi^{(2)}$. It is seen easily that (20) remains valid and that we have the desired case $a_{10} - a_{s+1,0} = a$. If $a_{10} = a + \tau$, $a_{s+1,0} = a$, it follows from (13) and (14) that

$$\chi^{(1)} \mid \mathfrak{H} = \tau(\varepsilon_0 + \varepsilon_1 + \varepsilon_1^{-1}) + a \textstyle\sum_{i=-s}^{s+1} \varepsilon_i .$$

Since here $h = 6$, we see that

$$\chi^{(1)}(H) = 2\tau, \qquad \chi^{(1)}(H^2) = 0.$$

Now, H^2 is an element of order 3. Since the degree f of $\chi^{(1)}$ is congruent to $\chi^{(1)}(H^2)$ modulo a prime ideal divisor of 3, we have $f \equiv 0 \pmod 3$. Then $g\chi^{(1)}(H)/c(H)f = 2\tau g/3h = \tau g/9$, and since we have an algebraic integer, we find $g \equiv 0 \pmod 9$. This is inconsistent with (6).

Similarly, if $a_{10} = a$, $a_{1,s+1} = a + \tau$, a contradiction is obtained by considering $\chi^{(2)}$. We find here

$$\chi^{(2)} \mid \mathfrak{H} = \tau(\varepsilon_3 + \varepsilon_2 + \varepsilon_2^{-1}) + a \textstyle\sum_{i=-2}^{3} \varepsilon_i ,$$

$$\chi^{(2)}(H) = -2\tau, \qquad \chi^{(2)}(H^2) = 0,$$

which again leads to $g \equiv 0 \pmod 9$, in contradiction to (6). Thus we may again assume that we have $a_{10} = a_{1,s+1} = a$. This is then true for $s \geqq 2$.

It now follows from (24), (25) that the first coefficient in $\mathfrak{a}_0 - \mathfrak{a}_1$ and $\mathfrak{a}_{s+1} - \mathfrak{a}_1$ is $-\tau$ and the next $s - 1$ coefficients are 0. In the row $s + 1$ belonging to the unit characters, we have the coefficients 1 and 0 respectively. Our previous results show that we may choose $\chi^{(s+2)}$, $\chi^{(s+3)}$, $\chi^{(s+4)}$ such that $\mathfrak{a}_0 - \mathfrak{a}_1$ has a coefficient ± 1 in the row $s + 2$ and $\mathfrak{a}_{s+1} - \mathfrak{a}_1$ has coefficients ± 1 in the rows $s + 3$, $s + 4$, while all other coefficients in the two columns vanish. Collecting these results (cf. (13), (14), (20)) we find

$$
\begin{aligned}
&\chi^{(i)} \mid \mathfrak{H} = \tau(\varepsilon_i + \varepsilon_{-i}) + a\textstyle\sum_{j=-s}^{s+1} \varepsilon_j, \qquad i = 1, 2, \cdots, s,\\
&\chi^{(s+1)} \mid \mathfrak{H} = \varepsilon_0,\\
&\chi^{(s+2)} \mid \mathfrak{H} = \delta_2\, \varepsilon_0 + a_{s+2,1} \textstyle\sum_{j=-s}^{s+1} \varepsilon_j,\\
(26)\qquad &\chi^{(s+3)} \mid \mathfrak{H} = \delta_3\, \varepsilon_{s+1} + a_{s+3,1} \textstyle\sum_{j=-s}^{s+1} \varepsilon_j,\\
&\chi^{(s+4)} \mid \mathfrak{H} = \delta_4\, \varepsilon_{s+1} + a_{s+4,1} \textstyle\sum_{j=-s}^{s+1} \varepsilon_j,\\
&\chi^{(s+\mu)} \mid \mathfrak{H} = a_{s+\mu,1} \textstyle\sum_{j=-s}^{s+1} \varepsilon_j,
\end{aligned}
$$

where δ_2, δ_3, $\delta_4 = \pm 1$. We set $\delta_1 = 1$.

We note that this result remains valid for $s = 1$. Indeed, we have here only the columns $\mathfrak{a}_0$, $\mathfrak{a}_2$, $\mathfrak{a}_1 = \mathfrak{a}_{-1}$. As before, there will be exactly one row in which $\mathfrak{a}_0 - \mathfrak{a}_1$ and $\mathfrak{a}_2 - \mathfrak{a}_1$ have a nonvanishing coefficient. The coefficient in this row will be the same in both columns and will be denoted by τ; the corresponding character is taken as $\chi^{(1)}$. The characters $\chi^{(2)}$, $\chi^{(3)}$ can be chosen such that $\mathfrak{a}_0 - \mathfrak{a}_1$ has coefficients ± 1 in the corresponding rows, and the characters $\chi^{(4)}$, $\chi^{(5)}$ such that $\mathfrak{a}_2 - \mathfrak{a}_1$ has coefficients ± 1 in the corresponding rows. Then (26) holds again.

We write χ_μ for $\chi^{(s+\mu)}$, $\mu > 0$. We then have the result

(II.F) *The irreducible characters of* $\mathfrak{G}$ *can be denoted by* $\chi^{(1)}, \chi^{(2)}, \cdots, \chi^{(s)}$, $\chi_1, \chi_2, \cdots, \chi_m$ *with* $m \geqq 4$ *in such a manner that we have for* $H^r \neq 1$

$$
\begin{aligned}
&\chi^{(i)}(H^r) = \tau(\varepsilon^{ir} + \varepsilon^{-ir}), && i = 1, 2, \cdots, s,\\
&\chi_j(H^r) = \delta_j, && j = 1, 2,\\
(27)\qquad &\chi_j(H^r) = \delta_j(-1)^r, && j = 3, 4,\\
&\chi_j(H^r) = 0, && j \geqq 5.
\end{aligned}
$$

Indeed, this follows from (26), if we note that $\sum_{i=-s}^{s+1} \varepsilon_i$ vanishes for all $H^r \neq 1$.

6. Congruences for the degrees of the characters of $\mathfrak{G}$

If ψ is any character of $\mathfrak{C}(T)$, it follows from the orthogonality relations that

$$\psi(1) + (h/2)\psi(B) + (h/2)\psi(BH) + \textstyle\sum_{H^r \neq 1} \psi(H^r) \equiv 0 \pmod{2h}.$$

If ψ is the restriction of a character of $\mathfrak{G}$, then, by (II.A), $\psi(B) = \psi(BH) = \psi(T)$. Combining this with (27), we find

(II.G) *If f is the degree of the exceptional characters $\chi^{(i)}$, and if f_j is the degree of χ_j, $j = 1, 2, \cdots, m$, then*

$$(28) \qquad \begin{aligned} f &\equiv 2\tau && \pmod{2h} \\ f_j &\equiv \delta_j && \pmod{2h} && (j = 1, 2) \\ f_j &\equiv h + \delta_j && \pmod{2h} && (j = 3, 4) \\ f_j &\equiv 0 && \pmod{2h} && (j \geqq 5). \end{aligned}$$

Proof. For $\psi = \chi^{(i)} \mid \mathfrak{C}(T)$, our congruence reads

$$f + h\tau((-1)^{h/2} + (-1)^{h/2}) + \tau \textstyle\sum_{H^r \neq 1} (\varepsilon_i(H^r) + \bar{\varepsilon}_i(H^r)) \equiv 0 \pmod{2h}.$$

The orthogonality relations for the characters of $\mathfrak{H}$ show that

$$2 + \textstyle\sum_{H^r \neq 1} (\varepsilon_i(H^r) + \bar{\varepsilon}_i(H^r)) = 0.$$

This yields $f \equiv 2\tau \pmod{2h}$. Similarly, for $j = 1, 2$

$$f_j + \delta_j h + \delta_j \textstyle\sum_{H^r \neq 1} 1 \equiv 0 \pmod{2h}.$$

Here, $\sum_{H^r \neq 1} 1 = h - 1$, and hence $f_j \equiv \delta_j \pmod{2h}$. The proof of the last 2 congruences (28) is analogous.

(II.H) *We have*

$$1 + \delta_2 \chi_2(X) = \tau\chi^{(j)}(X), \qquad \delta_3 \chi_3(X) + \delta_4 \chi_4(X) = \tau\chi^{(j)}(X)$$

for elements X of $\mathfrak{G}$ which are not conjugate to element $H^r \neq 1$. In particular,

$$(29) \qquad 1 + \delta_2 f_2 = \tau f, \qquad \delta_3 f_3 + \delta_4 f_4 = \tau f.$$

This follows from $(\mathfrak{a}_0 - \mathfrak{a}_1)\chi(X) = 0$, $(\mathfrak{a}_{s+1} - \mathfrak{a}_1)\chi(X) = 0$ in conjunction with the values of the coefficients of the columns $\mathfrak{a}_0 - \mathfrak{a}_1$ and $\mathfrak{a}_{s+1} - \mathfrak{a}_1$ obtained above (χ_1 was the unit character, $\delta_1 = 1$).

7. The class relation

Let $\mathfrak{K}_1, \mathfrak{K}_2, \cdots, \mathfrak{K}_k$ denote the classes of conjugate elements where we choose that notation such that $1 \in \mathfrak{K}_1$, $T \in \mathfrak{K}_2$, $H^j \in \mathfrak{K}_{2+j}$ for $j = 1, 2, \cdots, s$ (cf. (II.A)), and where $\mathfrak{K}_1, \mathfrak{K}_2, \cdots, \mathfrak{K}_t$ $(t \geqq 2 + s)$ are the classes consisting of the "real" elements G of $\mathfrak{G}$, i.e., the elements G which are conjugate to their reciprocals G^{-1}.

We work in the group algebra Γ of $\mathfrak{G}$ over the field of rational numbers or rather in the center Z of Γ. If K_j denotes the sum of the elements in $\mathfrak{K}_j$, then $K_1, K_2, \cdots, K_k$ form a basis of Z. In particular, we have an equation

$$(30) \qquad K_2^2 = \textstyle\sum_{j=1}^{k} c_j K_j.$$

Here, c_j denotes the number of ordered pairs (X, Y) of elements of $\mathfrak{K}_2$ such that XY is equal to a fixed element $G_j \in \mathfrak{K}_j$. Since X, Y have order 2, the equation $XY = G_j$ implies $G_j^{-1} = YX = Y^{-1}G_jX$. Conversely, if X has order 2, and if $XG_jX^{-1} = G_j^{-1}$, then for $Y = XG_j$, we have $XY = G_j$, and $Y^2 = XG_jXG_j = G_j^{-1}G_j = 1$. Thus Y has order 2, except when $Y = 1$, $G_j = X$. This latter case arises only if G_j has order 2, i.e., if $j = 2$. Thus

(II.I) *If G_j is a representative of $\mathfrak{K}_j$, then for $f \neq 2$ the number c_j in* (30) *denotes the number of elements X of order 2 which satisfy the equation*

$$X^{-1}G_jX = G_j^{-1}.$$

For $j = 2$, c_j is one less than the number of X of order 2 which satisfy the corresponding equation.

If $j = 1$, c_1 is simply the number of elements of order 2, i.e., the number of elements of $\mathfrak{K}_2$. By (II.A), $c_1 = g/2h$. For $j = 2$, we may take $G_j = T$, and $c_2 + 1$ is the number of elements of order 2 which commute with T. It follows from (II.A) that $c_2 + 1 = h + 1$, $c_2 = h$. For $3 \leqq j \leqq 2 + s$, we may choose $G_j = H^{j-2}$. Since

$$X^{-1}G_jX = G_j^{-1} \quad \text{implies} \quad X \in \mathfrak{N}(\{H^{j-2}\}) = \mathfrak{C}(T),$$

(cf. (I.C)), the number c_j denotes the number of elements of order 2 of the dihedral group $\mathfrak{C}(T)$ which transform H^{j-2} into its reciprocal $(H^{j-2})^{-1}$. Hence $c_j = h$ for $j = 3, \cdots, 2 + s$. If $2 + s < j \leqq t$, the elements G_j and G_j^{-1} are conjugate in $\mathfrak{G}$. Hence there exist exactly $c(G_j)$ elements X in $\mathfrak{G}$ such that $X^{-1}G_jX = G_j^{-1}$. Then X^2 commutes with G_j. If the order of X^2 contained a prime factor of $2^{m+1}v = 2h$, by (II.A), X^2 would belong to one of the classes $\mathfrak{K}_1, \mathfrak{K}_2, \cdots, \mathfrak{K}_{2+s}$. Moreover, for $X^2 \neq 1$, $\mathfrak{C}(X^2)$ would consist only of elements which lie in the same $s + 2$ classes. This is impossible for $j > s + 2$ since $G_j \in \mathfrak{C}(X^2)$. Hence $X^2 = 1$. Thus we have exactly $c(G_j)$ elements X of order 2 for which $X^{-1}G_jX = G_j^{-1}$ and $c_j = c(G_j)$ for $2 + s < j \leqq t$. Finally, for $j > t$, the elements G_j and G_j^{-1} are not conjugate, and $c_j = 0$.

If we count the number of elements of $\mathfrak{G}$ appearing as summands on both sides of (30), we have

$$(g/c(T))^2 = \textstyle\sum_{j=1}^{k} c_j(g/c(G_j)).$$

Substituting the values of c_j just found, this yields

$$\frac{g^2}{4h^2} = \frac{g}{2h} + \frac{g}{2h}h + \sum_{j=3}^{s+2}\frac{g}{h}h + \sum_{j=s+3}^{t}\frac{g}{c(G_j)}c(G_j).$$

This yields

$$g^2/4h^2 = g/2h + g/2 + g(t - 2),$$

whence

$$t - 2 = (g - 2h - 2h^2)/4h^2. \tag{31}$$

8. The degrees of the irreducible characters of $\mathfrak{G}$

It will be necessary to separate the cases $\tau = 1$ and $\tau = -1$.

The case $\tau = 1$. It follows from (29) that $\delta_2 = 1$ since $\tau f = f > 0$. As shown by (29), the character χ_2 must be real, since $\bar{\chi}_2$ cannot be equal to any of the characters except χ_2. If $f_2 = 1$, then $\mathfrak{G}$ would have a linear character $\chi_2 \neq \chi_1$, $\chi_2^2 = \chi_1$. Since the multiplicative group of linear characters is isomorphic with $\mathfrak{G}/\mathfrak{G}'$, it would follow that $\mathfrak{G}/\mathfrak{G}'$ has even order. This is impossible, if $\mathfrak{G}$ does not have a normal subgroup of index 2. Hence $f_2 \neq 1$, and (28) shows that $f_2 \geqq 2h + 1$. Now (29) shows that $f \geqq 2h + 2$. Interchanging χ_3 and χ_4 if necessary, we see from (29) that we may assume $\delta_3 = 1$. By (28), $f_3 \geqq h + 1$, $f_4 \geqq h + \delta_4$, $f_j \geqq 2h$ for $j \geqq 5$. Thus

$$(32)\qquad \begin{aligned} &f \geqq 2h + 2,\quad f_1 = 1,\quad f_2 \geqq 2h + 1,\quad f_3 \geqq h + 1,\\ &\qquad\qquad\qquad f_4 \geqq h + \delta_4,\quad f_j \geqq 2h \quad \text{for} \quad j \geqq 5. \end{aligned}$$

Since we have $s = h/2 - 1$ characters $\chi^{(i)}$, we shall have $k - s - 4 = k - h/2 - 3$ characters χ_j with $j \geqq 5$.

Now, the order g is the sum of the squares of the degrees of the irreducible characters. This yields

$$(33)\qquad \begin{aligned} g \geqq (h/2 - 1)(2h + 2)^2 + 1 + (2h + 1)^2 + (h + 1)^2\\ + (h + \delta_4)^2 + (k - h/2 - 3)4h^2. \end{aligned}$$

On account of (31), g can be written in the form

$$g = 4h^2(t - 2) + 2h + 2h^2.$$

Substituting this in (33), we have, after simplification,

$$(34)\qquad 4h^2t \geqq 4kh^2 + 2\delta_4 h - 2h.$$

If $\delta_4 = 1$, this yields $t \geqq k$. Since $k \geqq t$, we have $k = t$, and we must have the equality sign everywhere in (32). If $\delta_4 = -1$, we still can conclude $k = t$, and we see that the left side in (33) exceeds the right side by $4h$. Thus, we must have an inequality in (32) for some j. If we write the inequality in (32) in the form $f_j > f_j^*$, then (28) shows that

$$f_j \geqq f_j^* + 2h.$$

Since $f_j^2 \geqq f_j^{*2} + 4hf_j^* + 4h^2$, we can add $4hf_j^* + 4h^2$ on the right-hand side of (33) and (34). This leads to a contradiction. Thus,

(II.J) *If* $\tau = 1$, *we have* $g = 4h^2k - 6h^2 + 2h$, $\delta_2 = \delta_3 = \delta_4 = 1$,

$$f = 2h + 2,\qquad f_1 = 1,\qquad f_2 = 2h + 1,$$

$$f_3 = f_4 = h + 1,\qquad f_j = 2h \quad \text{for} \quad j \geqq 5.$$

The Case $\tau = -1$. Here $\delta_2 = -1$ by (29) and, after interchanging χ_3 and χ_4 if necessary, we may assume that $\delta_3 = -1$. Then (29) reads

$$(29^*) \qquad f_2 - 1 = f, \qquad f_3 - \delta_4 f_4 = f.$$

We separate the cases $\delta_4 = 1$ and $\delta_4 = -1$.

Subcase $\delta_4 = +1$. It follows from (28) that $f \geqq 2h - 2$, and hence $f_2 \geqq 2h - 1$. Moreover $f_4 \geqq h + 1$, and (29*) shows that $f_3 \geqq 3h - 1$. Also $f_j \geqq 2h$ for $j \geqq 5$.

Instead of (33), we find here

$$(33^*) \qquad g \geqq (h/2 - 1)(2h - 2)^2 + 1 + (2h - 1)^2 + (3h - 1)^2 + (h + 1)^2 + (k - h/2 - 3)4h^2.$$

Then (34) can be replaced by

$$(34^*) \qquad 4h^2 t \geqq 4h^2 k.$$

It follows that we must have $k = t$, and that we must have equalities in all estimates

$$f = 2h - 2, \quad f_1 = 1, \quad f_2 = 2h - 1, \quad f_3 = 3h - 1, \quad f_4 = h + 1, \quad f_j = 2h \quad \text{for} \quad j \geqq 5.$$

Subcase $\delta_4 = -1$. Here,

$$f \geqq 2h - 2, \quad f_2 \geqq 2h - 1, \quad f_3 \geqq h - 1, \quad f_4 \geqq h - 1, \quad f_j \geqq 2h \text{ for } j \geqq 5.$$

Then

$$g \geqq (h/2 - 1)(2h - 2)^2 + 1 + (2h - 1)^2 + (h - 1)^2 + (h - 1)^2 + (k - h/2 - 3)4h^2.$$

This leads to

$$(34^{**}) \qquad 4th^2 \geqq 4kh^2 - 8h^2.$$

If one of the degrees f_j has a value larger than the estimate f_j^* used here, by (28), $f_j \geqq f_j^* + 2h$, and we can add a term $4hf_j^* + 4h^2$ on the right-hand side of (34**). If $j \geqq 5$, this is an additional term $12h^2$, which is impossible as $t \leqq k$. If $1 \leqq j \leqq 4$, it follows from (29) that we must have inequality for two values of j, and again, this gives a contradiction. Finally, if $f > f^*$, we have an additional term $(h/2 - 1)(4h(2h - 2) + 4h^2)$. Again, we have a contradiction. It follows that the degrees have the values used in the estimates and that the equality sign holds in (34**), that is, that $t = k - 2$. Thus

(II.J*) *If* $\tau = -1$, *then we can assume* $\delta_2 = \delta_3 = -1$,

$$f = 2h - 2, \qquad f_1 = 1, \qquad f_2 = 2h - 1, \qquad f_j = 2h \quad \textit{for} \quad j \geqq 5.$$

Case (a) *If* $\delta_4 = 1$, *then*

$$f_3 = 3h - 1, \quad f_4 = h + 1, \quad k = t, \quad \text{and} \quad g = 4kh^2 - 6h^2 + 2h.$$

Case (b) *If* $\delta_4 = -1$, *then*

$$f_3 = f_4 = h - 1, \quad t = k - 2, \quad g = 4kh^2 \quad 14h^2 + 2h.$$

9. The order g

It follows easily from the basic properties of the group characters that the coefficients c_j in (30) are given by the formula

$$(35) \qquad c_j = \frac{g}{c(T)^2}\sum_{\mu=1}^{k}\frac{\chi^{(\mu)}(T)^2\chi^{(\mu)}(G_j)}{\chi^{(\mu)}(1)},$$

where $\chi^{(\mu)}$ ranges over all irreducible characters of $\mathfrak{G}$, and where $G_j \in \mathfrak{K}_j$. For $G_i = H$, we had $c_i = h$, $c(T) = 2h$. Now (27) yields $\chi^{(j)}(T) = \pm 2$, and we find from (27), (II.J), (II.J*), if either $\tau = 1$, or $\tau = -1$, $\delta_4 = -1$, that

$$4h^3 = g\left(\frac{4}{2h + 2\tau}\left(\tau\sum_{j=1}^{s}(\varepsilon^j + \varepsilon^{-j})\right) + \frac{1}{1} + \frac{\tau}{2h + \tau} + 2\frac{-\tau}{h + \tau}\right).$$

Now, $\sum_{j=1}^{s}(\varepsilon^j + \varepsilon^{-j}) + 1 + (-1) = 0$, whence $\sum_{j=1}^{s}(\varepsilon^j + \varepsilon^{-j}) = 0$, and

$$4h^3 = g\,\frac{2h^2 + 3h\tau + 1 + h\tau + 1 - 4h\tau - 2}{(2h + \tau)(h + \tau)} = \frac{2h^2}{(2h + \tau)(h + \tau)}\,g,$$

whence

$$(36) \qquad g = 2h(2h + \tau)(h + \tau).$$

On the other hand, if $\tau = -1$, $\delta_4 = 1$, the same method yields

$$4h^3 = g\left(1 - \frac{1}{2h - 1} + \frac{1}{3h - 1} - \frac{1}{h + 1}\right),$$

whence we find mod $h - 1$ that

$$4 \equiv g(1 - 1 + \tfrac{1}{2} - \tfrac{1}{2}) \equiv 0.$$

(Note that $h - 1$ is relatively prime to $2h - 1$, $3h - 1$, and $h + 1$ since h is even.) But then $h - 1$ divides 4, which is impossible for $h \neq 2$, and $h = 2$ was excluded. Thus, this case is impossible; in (II.J*), the subcase $\delta_4 = 1$ is excluded, and we have

$$(\text{II.J**}) \qquad \delta_2 = \tau, \quad \delta_3 = \tau, \quad \delta_4 = \tau.$$

10. The elements R and the elements S

Let R denote any element whose order contains a prime factor p' of $h + \tau$, and let S denote any element whose order contains a prime factor p of $2h + \tau$. Since any two of $2h$, $2h + \tau$, $h + \tau$ are relatively prime, it follows from (36) that f_2 is divisible by the full power of p dividing g; cf. (II.J), (II.J*). Hence

$$\chi_2(S) = 0. \tag{37}$$

Similarly, f_3, f_4 are divisible by the full power of p' dividing g and we have

$$\chi_3(R) = \chi_4(R) = 0. \tag{38}$$

It follows from (II.H) that $\chi^{(j)}(S) = \tau$, $\chi^{(j)}(R) = 0$, $\chi_2(R) = -\tau$. Clearly, $R \neq S$. Hence no element can have an order divisible by primes p and p'. In conjunction with (II.A), this yields

(II.K) *For every element R, $c(R)$ divides $h + \tau$, and for every element S, $c(S)$ divides $2h + \tau$.*

Apply now (35) taking $G_j = R$. We find (cf. (36))

$$c_j = \frac{g}{4h^2}\left(1 - \frac{\tau}{2h + \tau}\right) = \frac{h + \tau}{2h} 2h = h + \tau.$$

Since $c_j \neq 0$, the class $\mathfrak{R}_j$ is real, and we have

$$c(R) = h + \tau \tag{39}$$

for every R.

Similarly, taking $G_j = S$ in (34), we find

$$c_j = \frac{g}{4h^2}\left(\frac{4s}{2h + 2\tau}\chi^{(1)}(S) + \frac{1}{h + \tau}(\chi_3(S) + \chi_4(S)) + 1\right);$$

cf. (II.E), (II.J), (II.J*), (II.J**). But $\chi_3(S) + \chi_4(S) = \chi^{(1)}(S)$ by (II.H), and we have

$$c_j = \frac{g}{4h^2}\left(\frac{(h - 2)\tau + \tau}{h + \tau} + 1\right) = \frac{2h + \tau}{2h} h(1 + \tau).$$

Hence, if $\tau = 1$, the class $\mathfrak{R}_j$ is real, and

$$c(S) = 2h + 1 \qquad \text{for} \quad \tau = 1.$$

If $\tau = -1$, the class $\mathfrak{R}_j$ is not real. Since there exist only two nonreal classes (as $k = t + 2$), we have exactly two classes containing elements S. This implies that $2h - 1$ is a prime power in this case.[3]

Actually this result holds for $\tau = 1$ too. Indeed, by (II.J) and (36), $2hk - 3h + 1 = 2h^2 + 3h + 1$ for $\tau = 1$, whence $k = h + 3$. Since we have $s + 2 = h/2 + 1$ classes containing elements 1, T, and H^j, we have $h/2 + 2$ classes containing elements R and S. The orthogonality relation

[3] If $2h - 1$ were divisible by two distinct primes p_1 and p_2, we would have elements S of orders p_1 and of orders p_2. For $\tau = 1$, S and S^{-1} are not conjugate. We would have at least four classes of elements S.

for χ_2, χ_0 yields

$$(2h + 1) + g/2h + s(g/h) - \textstyle\sum_R 1 = 0.$$

This shows that the number of elements R is equal to

$$2h + 1 + (2h + 1)(h + 1)(1 + 2s)$$
$$= (2h + 1)(1 + (h + 1)(h - 1)) = (2h + 1)h^2.$$

By (39) each class $\mathfrak{R}_j$ consisting of elements R contains

$$g/(h + 1) = (2h + 1)(2h)$$

such elements, and hence there are $h/2$ classes consisting of elements R. Thus there are exactly two classes consisting of elements S. As already remarked,[4] this is only possible if $2h + \tau$ is a power p^n of a prime,

$$2h + \tau = p^n. \tag{40}$$

We had shown above that $c(S) = 2h + \tau$ for $\tau = 1$. We can now prove that this holds for $\tau = -1$. Indeed, since $k = t + 2$, we have two nonreal classes in this case, and these must be the classes containing the elements S. If S is chosen such that S occurs in the center of the p-Sylow group $\mathfrak{P}$, then $c(S) = p^n = 2h + \tau = 2h - 1$. Since the other class is represented by S^{-1}, we have $c(S^{-1}) = 2h - 1$ for the elements of this class. Hence

$$c(S) = 2h + \tau. \tag{41}$$

11. The groups $\mathfrak{N}$ and $\mathfrak{S}$ and their normalizers

(II.L) *The group $\mathfrak{G}$ has an abelian subgroup $\mathfrak{N}$ of order $h + \tau$. The centralizer $\mathfrak{C}(R)$ of each element R can be taken for $\mathfrak{N}$.*

Proof. Our previous results show that for each element R, we have

$$c(R) = h + \tau,$$

that the elements of $\mathfrak{C}(R)$ different from 1 are of the type R again, and that R is real. Now, Theorem (4D) of [1] shows that $\mathfrak{N} = \mathfrak{C}(R)$ is abelian.

It follows from Theorem (4F) of [1] that if the order of the normalizer $\mathfrak{N}(\mathfrak{N})$ is $w(h + \tau)$, then w divides $h + \tau - 1$, and we can set

$$g = w(h + \tau)(1 + N(h + \tau))$$

with integral $N \geqq 0$. Hence $w(1 + N(h + \tau)) = 2h(2h + \tau)$, whence $w \equiv 2h(2h + \tau) \pmod{h + \tau}$. Hence $w \equiv -2\tau(-\tau) \equiv 2 \pmod{h + \tau}$. Since $w \leqq h + \tau - 1$, we must have $w = 2$. Thus

[4] Since the case $\tau = -1$ has been settled above, we may assume $\tau = 1$. Here, $c(S) = 2h + 1$. If $2h + 1$ were divisible by two distinct primes p_1 and p_2, we would have classes of elements S of each of the orders p_1, p_2, and $p_1 p_2$.

(II.M) *The normalizer of the subgroup* $\mathfrak{R}$ *in* (II.L) *has the order* $2(h + \tau)$.

As a consequence, we have

(II.N) *If an element* R *is conjugate to a power* R^μ *in* $\mathfrak{G}$, *then* $R^\mu = R^{\pm 1}$.

Indeed, if $G^{-1}RG = R^\mu$, and if we take $\mathfrak{R} = \mathfrak{C}(R)$, then $G^{-1}\mathfrak{R}G = \mathfrak{R}$. Hence $G \in \mathfrak{N}(\mathfrak{R})$, and by (II.M) $G^2 \in \mathfrak{R}$, $R^{\mu^2} = R$, $\mu^2 \equiv 1 \pmod{h + \tau}$. Since $h + \tau$ is odd, then $\mu \equiv \pm 1 \pmod{h + \tau}$.

If $\tau = 1$, the element S belongs to a real class, and the same argument as used in (II.L) shows that $\mathfrak{C}(S) = \mathfrak{S}$ is abelian. In order to have the same result for $\tau = -1$, we have to use a more complicated procedure.

(II.O) Lemma. *Let* $\mathfrak{G}$ *be any finite group. Let* p *be a prime dividing the order* g *of* $\mathfrak{G}$, *and make the following assumptions*:

(a) *If* Q *is an element of prime power order* $q^\alpha > 1$, $p \neq q$, *the order of* $\mathfrak{N}(\{Q\})$ *is not divisible by* p.

(b) *A generalized quaternion group does not appear as a subgroup of* $\mathfrak{G}$.

Then either the p-*Sylow subgroup of* $\mathfrak{G}$ *is normal in* $\mathfrak{G}$, *or any two distinct* p-*Sylow subgroups have intersection* $\{1\}$.

Proof. Suppose that there exist two distinct p-Sylow subgroups $\mathfrak{P}$ and $\mathfrak{P}_1$ with $\mathfrak{P} \cap \mathfrak{P}_1 = \mathfrak{D} \neq \{1\}$. Choose $\mathfrak{P}$ and $\mathfrak{P}_1$ so that $\mathfrak{D}$ has maximal order. Then $\mathfrak{P} \cap \mathfrak{N}(\mathfrak{D}) \supset \mathfrak{D}$, $\mathfrak{P}_1 \cap \mathfrak{N}(\mathfrak{D}) \supset \mathfrak{D}$, and

$$(\mathfrak{P} \cap \mathfrak{N}(\mathfrak{D})) \cap (\mathfrak{P}_1 \cap \mathfrak{N}(\mathfrak{D})) = \mathfrak{D}.$$

Hence $\mathfrak{N}(\mathfrak{D})$ has two distinct p-Sylow subgroups whose intersection is not $\{1\}$.

Choose a principal series of $\mathfrak{N}(\mathfrak{D})$ through $\mathfrak{D}$, and let $\mathfrak{B}$ be the last group different from $\{1\}$ in this series. Since $\mathfrak{B} \subseteqq \mathfrak{D}$, $\mathfrak{B}$ is a p-group and hence abelian of type $(p, p, \cdots, p)$. Choose a normal subgroup $\mathfrak{A}$ of $\mathfrak{N}(\mathfrak{D})$ such that (1) $\mathfrak{A} \supseteqq \mathfrak{B}$, (2) $\mathfrak{A}$ has at least two distinct p-Sylow subgroups, and (3) $\mathfrak{A}$ has minimal order, subject to the previous conditions.

Since $\mathfrak{B}$ is normal in $\mathfrak{A}$, the transformation of $\mathfrak{B}$ by an element $A \in \mathfrak{A}$ can be described by a matrix $[A]$ with coefficients in the Galois field with p elements, and the mapping $A \to [A]$ is a homomorphism. Let q denote the smallest prime factor $\neq p$ of $(\mathfrak{A}:1)$. Since $\mathfrak{A}$ cannot be a p-group, such a prime q exists. Let $\mathfrak{Q}$ be a q-Sylow group of $\mathfrak{A}$, and let $Q \in \mathfrak{Q}$, $Q \neq 1$. The assumption (a) shows that Q cannot commute with an element $B \neq 1$ of $\mathfrak{B}$. Hence the linear transformation belonging to $[Q]$ does not leave a vector $\neq 0$ fixed. It follows that $\mathfrak{Q}$ cannot have a subgroup of type (q, q). It follows that $\mathfrak{Q}$ is cyclic, $\mathfrak{Q} = \{Q_0\}$ say. Suppose that two distinct powers Q and Q^μ are conjugate in $\mathfrak{A}$. We may then find an element A of prime power order p_0^γ such that $A^{-1}QA = Q^\nu \neq Q$, $A \in \mathfrak{A}$. Then p_0 must be different from q. Since p_0 must divide $q - 1$, it follows from our choice of q that p_0 can only be p itself. But this is excluded by the assumption (a).

Hence no two distinct elements of $\mathfrak{Q}$ are conjugate in $\mathfrak{A}$. Now, Burnside's Theorem shows that $\mathfrak{A}$ has a normal subgroup $\mathfrak{A}_0$ consisting of the elements of $\mathfrak{A}$ of orders prime to q. Clearly, $\mathfrak{A}_0$ is even normal in $\mathfrak{N}(\mathfrak{D})$. Moreover, all p-Sylow subgroups of $\mathfrak{A}$ appear in $\mathfrak{A}_0$ and $\mathfrak{B} \subseteqq \mathfrak{A}_0$. This shows that $\mathfrak{A}_0$ satisfies the conditions (1), (2) imposed on $\mathfrak{A}$, and as $\mathfrak{A}_0 \subset \mathfrak{A}$, we have a contradiction. The lemma (II.O) has been proved.

We use (II.O) to show that for any element S, the group $\mathfrak{C}(S)$ is abelian. If p is as in (40), and if $\mathfrak{P}$ is the p-Sylow subgroup of $\mathfrak{G}$, we have to show that $\mathfrak{P}$ is abelian, since we know that $c(S)$ divides $2h + \tau$; cf. (II.K). As already remarked, we may assume that $\tau = -1$. If we choose $S_0 \neq 1$ in the center of $\mathfrak{P}$, we certainly have $c(S_0) = p^n$. Since every element S is conjugate to S_0 or S_0^{-1} in the case $\tau = -1$, $c(S) = p^n$. Hence every element S appears in the center of some p-Sylow subgroup. If $\mathfrak{P}$ is not abelian, there must exist two distinct p-Sylow subgroups whose intersection is different from $\{1\}$. Since the condition (b) of (II.O) is satisfied for our $\mathfrak{G}$, it remains to check condition (a) of (II.O). Since $q \neq p$, an element Q of order $q^\alpha > 1$ is either conjugate to an element of H or to an element R. If we had $P^{-1}QP = Q^\nu$, $Q^\nu = Q^{\pm 1}$; cf. (II.A), (II.N). But since

$$c(P) = 2h + \tau,$$

we cannot have $P^{-1}QP = Q$. If $P^{-1}QP = Q^{-1}$, P would have even order, which is equally impossible. Hence we have a contradiction.

Thus,

(II.P) *The p-Sylow group $\mathfrak{P}$ of order $2h + \tau = p^n$ of $\mathfrak{G}$ is abelian.*

Two elements of $\mathfrak{P}$ are conjugate in $\mathfrak{G}$ if and only if they are conjugate in $\mathfrak{N}(\mathfrak{P})$. Since $c(P) = 2h + \tau$ for every $P \neq 1$ in $\mathfrak{P}$ (P being of type (S)), we see that the number of conjugates of P belonging to $\mathfrak{P}$ is equal to $(\mathfrak{N}(\mathfrak{P}):\mathfrak{P})$. But since we have two classes of elements S, it follows that

$$2(\mathfrak{N}(\mathfrak{P}):\mathfrak{P}) = p^n - 1.$$

Hence $\mathfrak{N}(\mathfrak{P})$ has the order

$$\tfrac{1}{2}p^n(p^n - 1) = \begin{cases} (2h + 1)h & \text{for} \quad \tau = 1, \\ (2h - 1)(h - 1) & \text{for} \quad \tau = -1. \end{cases}$$

12. Proof of the main theorem

Since $\mathfrak{G}$ has a subgroup $\mathfrak{N}(\mathfrak{P})$ of order $(2h + 1)h$ for $\tau = 1$ and of order $(2h - 1)(h - 1)$ for $\tau = -1$, it follows that $\mathfrak{G}$ has a representation $\mathfrak{Z}$ as a transitive group of permutations in $2(h + 1)$ or $2h$ letters respectively.

The case $\tau = 1$. After removing the unit character from the character of $\mathfrak{Z}$, we have a character of degree $2h + 1$ which no longer contains the unit character. It follows from (II.J) that this character must be irreducible and equal to χ_2. Hence $\mathfrak{Z}$ has the character $\chi_1 + \chi_2$. Since two distinct irreducible constituents appear, $\mathfrak{Z}$ is doubly transitive. The number

$$\chi_1(G) + \chi_2(G)$$

gives the number of symbols left fixed by $\mathfrak{Z}(G)$. It follows from (27) that this is 2 for $G = H^r \neq 1$. For $G = S$, it is 1 by (37), and since

$$\chi_2(R) = -\tau$$

as remarked in connection with (38), it is 0 for R. Hence no $\mathfrak{Z}(G)$ with $G \neq 1$ leaves three letters fixed. Because of the double transitivity, the subgroup leaving two letters fixed has order $g/(2h + 2)(2h + 1) = h$, and the subgroup leaving one letter fixed has order $h(2h + 1)$. The elements of order 2 leave two letters fixed.

The case $\tau = -1$. Here, the character of $\mathfrak{Z}$ has the form $\chi_1 + \chi$ where χ is a character of degree $2h - 1$. It follows from (II.J*) that $\chi = \chi_2$.

Again, $\mathfrak{Z}$ is doubly transitive. It follows here from (27) that the elements $\mathfrak{Z}(H^r)$ for $H^r \neq 1$ do not leave any letter fixed. The elements $\mathfrak{Z}(R)$ leave two letters fixed, and the elements $\mathfrak{Z}(S)$ leave one letter fixed. No element $\mathfrak{Z}(G)$, $G \neq 1$, then leaves three letters fixed.

The subgroup for which $\mathfrak{Z}(G)$ leaves two letters fixed has order

$$(2h - 1)(h - 1),$$

and the subgroup for which $\mathfrak{Z}(G)$ leaves two letters fixed has order $h - 1$.

In both cases, $\mathfrak{Z}$ is faithful. Indeed, the degree is at least 3 and only $\mathfrak{Z}(1)$ leaves three letters fixed.

We now apply Zassenhaus' method; cf. [2]. We have a group of permutations of $N + 1$ letters, doubly transitive, such that only the identity leaves three letters fixed. The order of the group is

$$\tfrac{1}{2}(N + 1)N(N - 1), \qquad N = 2h + \tau.$$

In the case $\tau = -1$, Zassenhaus' assumptions are not quite satisfied, since the subgroup leaving two letters fixed does not contain elements of order 2. However, the method still works. This yields the result:

$$\mathfrak{G} \cong LF(2, 2h + \tau) \qquad (\tau = \pm 1).$$

III. The Case B

1. Assumptions

We assume here

(I) $\mathfrak{G}$ is a finite group of type (S).

(II) The 2-Sylow subgroup $\mathfrak{T}$ of $\mathfrak{G}$ is abelian of type $(2, 2, \cdots, 2)$, of order $2^a > 2$.[5]

(III) $\mathfrak{G}$ does not have a proper normal subgroup which includes $\mathfrak{T}$, and $\mathfrak{G} \neq \mathfrak{T}$.

[5] For $a = 2$ assume also that $\mathfrak{C}(T) = \mathfrak{T}$ for $T \in \mathfrak{T}$, $T \neq 1$; cf. (1).

As shown in I, it follows from (I) and (II) that for $a \geqq 3$, we have

$$\mathfrak{C}(T) = \mathfrak{T}$$

for $T \in \mathfrak{T}$, $T \neq 1$. The case $a = 2$, $\mathfrak{C}(T) \neq \mathfrak{T}$ for some $T \in \mathfrak{T}$, $T \neq 1$ has been treated in II. Hence we assume that if $a = 2$, we still have

$$\mathfrak{C}(T) = \mathfrak{T} \qquad \text{for} \quad T \in \mathfrak{T}, \quad T \neq 1. \tag{1}$$

2. The classes of involutions

(III.A) *All elements of order* 2 *of* $\mathfrak{G}$ *belong to the same class of conjugate elements.*

Proof. Suppose that X and Y are two involutions which belong to different classes. Then by Lemma (3A) of [1], there exists an involution Z such that $Z \in \mathfrak{C}(X)$, $Z \in \mathfrak{C}(Y)$. If $X \in \mathfrak{T}$, it follows from (1) that

$$Z \in \mathfrak{C}(X) = \mathfrak{T}, \qquad Y \in \mathfrak{C}(Z) = \mathfrak{T}.$$

Hence all the elements of the class of Y belong to $\mathfrak{T}$. By reasons of symmetry, the same is true for the class of X. It follows that $\mathfrak{T}$ consists of full classes of conjugate elements. Hence $\mathfrak{T}$ is normal in $\mathfrak{G}$, and this has been excluded.

3. The normalizer of $\mathfrak{T}$

Let $\mathfrak{N} = \mathfrak{N}(\mathfrak{T})$. It is clear that two elements of $\mathfrak{T}$ are conjugate in $\mathfrak{G}$ if and only if they are conjugate in $\mathfrak{N}$. Hence the class of T ($T \in \mathfrak{T}$, $T \neq 1$) in $\mathfrak{N}$ consists of all elements $\neq 1$ of $\mathfrak{T}$. Thus,

$$2^a - 1 = (\mathfrak{N}:(\mathfrak{N} \cap \mathfrak{C}(T)) = (\mathfrak{N}:\mathfrak{T}).$$

It follows that

$$(\mathfrak{N}:1) = (2^a - 1)2^a.$$

Any two different 2-Sylow groups $\mathfrak{T}$ and $\mathfrak{T}_1$ have intersection $\{1\}$. Indeed, if $T_0 \in \mathfrak{T} \cap \mathfrak{T}_1$, and if we had $T_0 \neq 1$, we would find

$$\mathfrak{C}(T_0) = \mathfrak{T} \quad \text{and} \quad \mathfrak{C}(T_0) = \mathfrak{T}_1.$$

It is now clear that the number of 2-Sylow groups of $\mathfrak{G}$ is congruent to 1 (mod 2^a). If we denote this number by $1 + 2^aN$, we have $N \geqq 1$, since $\mathfrak{T}$ is not normal in $\mathfrak{G}$. Hence $(\mathfrak{G}:\mathfrak{N}) = 1 + 2^aN \geqq 1 + 2^a$, and we have

$$g = 2^a(2^a - 1)(1 + 2^aN) \geqq (2^a + 1)2^a(2^a - 1). \tag{2}$$

4. The class relation

Let $\mathfrak{K}_1, \mathfrak{K}_2, \cdots, \mathfrak{K}_k$ denote the classes of conjugate elements of $\mathfrak{G}$ where we choose the notation such that $1 \in \mathfrak{K}_1$, $T \in \mathfrak{K}_2$ for T of order 2, and such that $\mathfrak{K}_1, \mathfrak{K}_2, \cdots, \mathfrak{K}_t$ are real. Let K_j denote the sum of the elements of $\mathfrak{K}_j$ taken in the group algebra of $\mathfrak{G}$ over the field of rational numbers. It follows from [1], (2A), (4B), that

(3) $$K_2^2 = (g/2^a)K_1 + (2^a - 2)K_2 + \sum_{j=3}^{t} c(G_j)K_j,$$

where G_j denotes a representative of $\mathfrak{K}_j$. Comparing the number of elements of $\mathfrak{G}$ appearing on both sides of (3), we find

$$g^2/2^{2a} = g/2^a + 2^a(g/2^a) - 2(g/2^a) + (t-2)g,$$

whence

(4) $$g = (t-1)2^{2a} - 2^a.$$

5. The degrees of the irreducible characters of $\mathfrak{G}$

Let $\chi_1, \chi_2, \cdots, \chi_k$ denote the irreducible characters of $\mathfrak{G}$; let

$$f_j = \chi_j(1)$$

be the degree of χ_j. Take χ_1 as the unit character.

If we set $\chi_j(T) = z_j$, then z_j is a rational integer. The orthogonality relations for $\chi_j \mid \mathfrak{T}$ yield

(5) $$f_j + (2^a - 1)z_j = b_j 2^a,$$

where b_j is a nonnegative rational integer, the multiplicity of the unit character in $\chi_j \mid \mathfrak{T}$. We choose our notation so that $z_j > 0$ for $j = 1, 2, \cdots, r$; $z_j < 0$ for $j = r+1, r+2, \cdots, r+s$; $z_j = 0$ for $j = r+s+1, \cdots, k$. Except for $j = 1$, we do not have $z_j = f_j$, since otherwise $\mathfrak{T}$ would belong to the kernel of χ_j, and this is excluded by the assumption (III). It follows from (5) that we can set

(6) $$f_j = z_j + 2^a c_j$$

with rational integers c_j. Since $z_j < f_j$ for $j \neq 1$, we have here

(6a) $$c_j \geqq 1 \qquad \text{for} \quad j = 2, 3, \cdots, r.$$

For $j = r+1, \cdots, r+s$, we have $c_j = b_j - z_j \geqq -z_j$, that is,

(6b) $$c_j \geqq |z_j| \qquad \text{for} \quad j = r+1, \cdots, r+s.$$

Finally

(6c) $$c_j \geqq 1 \qquad \text{for} \quad j = r+s+1, \cdots, k.$$

Now $g = \sum_{j=1}^{k} f_j^2$. Since $k \geqq t$, it follows from (4) that some of the f_j with $j \geqq 2$ must be smaller than 2^a. It follows from (6) that this can only be so in the case of (6b). Since $\sum_{j=1}^{k} |\chi_j(T)|^2 = c(T) = 2^a$, we have

(7) $$\sum_{j=1}^{r+s} z_j^2 = 2^a.$$

Thus, $|z_j| < 2^a$, and we must have $c_j = 1$, $z_j = -1$.

Suppose that we have b values of j for which $z_j = -1$, $c_j = 1$, that is,

$f_j = 2^a - 1$. Since $z_1 = 1$, it follows from (7) that

$$b \leqq 2^a - 1. \tag{8}$$

The term $f_1 = 1$ and the b terms $2^a - 1$ contribute

$$1 + b(2^a - 1)^2 = b2^{2a} - 2^{a+1}b + b + 1$$

to $\sum_{j=1}^{k} f_j^2 = g$. Hence, for the remaining $k - b - 1$ terms, we have, by (4),

$$\begin{aligned} \textstyle\sum' f_j^2 &= g - b2^{2a} + 2^{a+1}b - b - 1 \\ &= (t - b - 1)2^{2a} - 2^a + 2^{a+1}b - b - 1. \end{aligned} \tag{9}$$

Since k is the number of all classes of conjugate elements of $\mathfrak{G}$, and t the number of "real" classes, we have $k \geqq t$. If $k > t$, then $k \geqq t + 2$, as the non-real classes appear in pairs. Then, there appear

$$k - b - 1 \geqq (t - b - 1) + 2$$

terms $f_j^2 \geqq 2^{2a}$ on the left-hand side of (9), and this side is at least

$$(t - b - 1)2^{2a} + 2 \cdot 2^{2a}.$$

By (8), $2^{a+1}b \leqq 2 \cdot 2^{2a}$, and (9) leads to a contradiction. Thus, $t = k$ and we have

(III.B) *All classes of* $\mathfrak{G}$ *are real.*

Suppose that some of the f_j in (9) were at least $2^{a+1} - 2$. Then

$$f_j^2 = 2^{2a+2} - 2^{a+3} + 4 = 3 \cdot 2^{2a} + 2^{2a} - 8 \cdot 2^a + 4,$$

and we see that the left side of (9) would be at least

$$(k - b - 1)2^{2a} + 3 \cdot 2^{2a} - 8 \cdot 2^a + 4.$$

Because $k = t$, (9) yields

$$3 \cdot 2^{2a} - 8 \cdot 2^a + 4 \leqq -2^a + 2^{a+1}b - b - 1.$$

Using (8), we obtain $2^{a+1}b \leqq 2 \cdot 2^{2a} - 2 \cdot 2^a$, and hence

$$2^{2a} + 6 \leqq 5 \cdot 2^a.$$

This is certainly false for $a \geqq 3$, since $2^{2a} \geqq 8 \cdot 2^a$. It is also false for $a = 2$. Hence all f_j satisfy $f_j < 2 \cdot 2^a - 2$. Now (6) shows that we must have $c_j = 1$ in the case of (6a) and (6c). If $|z_j| \geqq 3$ in the case of (6b), then $c_j \geqq 3$ and $f_j = z_j + 3 \cdot 2^a \geqq -2^a + 3 \cdot 2^a$, since $|z_j| < 2^a$ by (7). This is impossible. If $z_j = -2$, then $c_j \geqq 2$ and $f_j \geqq 2 \cdot 2^a - 2$, which was also excluded. If $z_j = -1$ and $c_j \geqq 2$, we have likewise a contradiction. Hence we must have

$z_j = -1$, $c_j = 1$. This yields the result:

(III.C) *The degrees of the irreducible representations of* $\mathfrak{G}$ *have the following values:*

$$f_1 = 1, \qquad f_j = z_j + 2^a, \qquad j = 2, 3, \cdots, r,$$

with $z_j > 0$,

$$f_j = 2^a - 1, \qquad j = r+1, \cdots, r+s,$$

and $s = b$,

$$f_j = 2^a, \qquad \text{for } j > r+s+1.$$

Moreover, (*cf.* (7)),

$$\sum_{j=1}^{r} z_j^2 + s = 2^a. \tag{10}$$

By the orthogonality relations, we also have

$$0 = \sum_j f_j\, \chi_j(T) = \sum_j f_j\, z_j = 1 + \sum_{j=2}^{r}(2^a + z_j)z_j - s(2^a - 1).$$

This yields $2^a(\sum_{j=2}^{r} z_j - s) + 2^a = 0$, and hence

$$\sum_{j=2}^{r} z_j = s - 1. \tag{11}$$

The coefficient $2^a - 2$ of K_2 in (3) can be expressed by the characters in the form

$$2^a - 2 = \frac{g}{2^{2a}} \sum_j \frac{\chi_j(T)^3}{f_j}. \tag{12}$$

Because of the values obtained for the f_j and $z_j = \chi_j(T)$, the sum here is

$$1 + \sum_{j=2}^{r} \frac{z_j^3}{2^a + z_j} - s\frac{1}{2^a - 1}.$$

Now,

$$\frac{z_j^3}{2^a + z_j} = \frac{z_j^2}{(2^a/z_j) + 1} \geqq \frac{z_j^2}{2^a + 1},$$

and the sum is at least equal to

$$1 + \frac{1}{2^a + 1}\sum_{j=2}^{r} z_j^2 - s\frac{1}{2^a - 1} = 1 + \frac{2^a - s - 1}{2^a + 1} - s\frac{1}{2^a - 1};$$

cf. (10). Thus, (12) yields

$$(2^a - 2)2^{2a} \geqq g\,\frac{2^{2a} - 1 + 2^{2a} - s2^a - 2^a - 2^a + s + 1 - s2^a - s}{(2^a + 1)(2^a - 1)},$$

$$(2^a - 2)2^{2a}(2^a + 1)(2^a - 1) \geqq g(2^{2a+1} - 2^{a+1}s - 2^{a+1}). \tag{13}$$

Combining this with (2), we find

$$2^a - 2 \geqq 2^{a+1} - 2s - 2, \qquad 2s \geqq 2^a.$$

On the other hand, by (11) and (10)

$$s = \sum_{j=1}^{r} z_j \leqq \sum_{j=1}^{r} z_j^2 = 2^a - s, \tag{14}$$

whence $2s \leqq 2^a$. It follows that $2s = 2^a$; moreover, in (13) and (14) the equality sign must hold. This implies that $g = (2^a + 1)2^a(2^a - 1)$, that $z_1 = z_2 = \cdots = z_r = 1$, and, finally, that $r = s$. This yields the results

(III.D) *$\mathfrak{G}$ has the order $g = (2^a + 1)2^a(2^a - 1)$.*

(III.E) *$\mathfrak{G}$ has exactly $2^{a-1} - 1$ degrees $2^a + 1$, and 2^{a-1} degrees $2^a - 1$, and one degree 1. All other degrees are 2^a.*

It remains to find the number of degrees 2^a. Combining (4) with the value of g, and the equation $t = k$, we have $(k - 1)2^a - 1 = (2^a - 1)(2^a + 1)$, whence $k - 1 = 2^a$. Since we have 2^a degrees $1, 2^a + 1, 2^a - 1$, we have exactly one degree 2^a.

(III.E*) *There is exactly one degree 2^a; $k = 2^a + 1$.*

6. The main result

Since $\mathfrak{G}$ has a subgroup $\mathfrak{N}$ of order $(2^a - 1)2^a$, that is, of index $2^a + 1$, it follows that $\mathfrak{G}$ has a transitive representation $\mathfrak{Z}$ by permutations of $2^a + 1$ objects. If the character of $\mathfrak{Z}$ is $\chi_1 + \chi$, then χ is a character of $\mathfrak{G}$ of degree 2^a which no longer contains χ_1. Comparison with (III.E), (III.E*) shows that $\chi = \chi_k$. Since χ_k is irreducible, $\mathfrak{Z}$ is doubly transitive.

If R is an element of $\mathfrak{G}$ whose order is divisible by a prime factor p of $2^a + 1$, then all characters χ_j of degree $2^a + 1$ vanish for R. Likewise, if S is an element of $\mathfrak{G}$ whose order is divisible by a prime factor p' of $2^a - 1$, then $\chi_l(S) = 0$ for all χ_l of degree $2^a - 1$. Thus $\chi_j(R)\chi_j(S) = 0$ for $1 < j < k$. Now the orthogonality relations for group characters yield $\chi_k(R)\chi_k(S) + 1 = 0$. Since χ_k is the only irreducible character of its degree, its values are rational integers. It follows that $\chi_k(R) = \pm 1$, $\chi_k(S) = \mp 1$. Thus $\chi_k(X)$ for $X \neq 1$ is $0, +1$, or -1, and the character of $\mathfrak{Z}$ for $X \neq 1$ has only the values 1, 2, 0. In particular, the representation $\mathfrak{Z}$ is faithful. Moreover, no $\mathfrak{Z}(X)$ with $X \neq 1$ leaves three objects fixed. It follows that $\mathfrak{Z}$ is triply transitive: The subgroup leaving one letter fixed has order $2^a(2^a - 1)$; the subgroup leaving two letters fixed has order $2^a - 1$; the subgroup leaving three letters fixed has order 1.

Now, Zassenhaus' results apply. It follows that $\mathfrak{G} = LF(2, 2^a)$.

7. Groups $\mathfrak{G}$ which satisfy the assumptions (I), (II), but not the assumption (III)

If $\mathfrak{G}$ satisfies the assumptions (I) and (II), but not the assumption (III), let $\mathfrak{G}_0$ be a seminormal subgroup of $\mathfrak{G}$ of minimal order which includes the 2-Sylow subgroup $\mathfrak{T}$. Then $\mathfrak{G}_0 \neq \mathfrak{G}$. Again $\mathfrak{G}_0$ satisfies the assumptions (I) and (II). If $\mathfrak{G}_0 \neq \mathfrak{T}$, then $\mathfrak{G}_0$ will satisfy the assumptions (I), (II), (III). Hence $\mathfrak{G}_0 = LF(2, 2^a)$.

Let $\mathfrak{G}_1$ be a group which precedes $\mathfrak{G}_0$ in a composition series from $\mathfrak{G}$ to $\mathfrak{G}_0$. Then $\mathfrak{G}_0$ is normal in $\mathfrak{G}_1$. If $T \in \mathfrak{T}$, $T \neq 1$, and if $X \in \mathfrak{G}_1$, then $X^{-1}TX$ is an involution of $\mathfrak{G}_0$ and hence conjugate to T in $\mathfrak{G}_0$. Thus $X^{-1}TX = Y^{-1}TY$ with $Y \in \mathfrak{G}_0$. It follows that $XY^{-1} \in \mathfrak{C}(T)$. Since $c(T) = 2^a$, $\mathfrak{C}(T) = \mathfrak{T}$, and we find $X \in \mathfrak{T}Y \subseteqq \mathfrak{G}_0$. Hence $\mathfrak{G}_1 = \mathfrak{G}_0$, a contradiction.

Thus, $\mathfrak{G}_0 = \mathfrak{T}$. Suppose

$$\mathfrak{G} \supset \mathfrak{H}_1 \supset \cdots \supset \mathfrak{H}_r = \mathfrak{T}$$

is a composition series from $\mathfrak{G}$ to $\mathfrak{T}$. Suppose we know already that $\mathfrak{T}$ is normal in $\mathfrak{H}_l$ for some l. Then $\mathfrak{T}$ is characteristic in $\mathfrak{H}_l$ and hence normal in $\mathfrak{H}_{l-1}$. This shows that $\mathfrak{T}$ is normal in $\mathfrak{G}$,

$$\mathfrak{G} = \mathfrak{N}(\mathfrak{T}).$$

Since $\mathfrak{N}(\mathfrak{T})/\mathfrak{C}(\mathfrak{T})$ is isomorphic with a subgroup $\mathfrak{M}$ of $LH(a, 2)$ in the usual manner, we have here $\mathfrak{G}/\mathfrak{T} \cong \mathfrak{M}$. In our case, no element $M \neq 1$ of $\mathfrak{M}$ has a fixed point. Also, $\mathfrak{M}$ has odd order. It follows that all Sylow subgroups of $\mathfrak{M}$ are cyclic, and this implies that $\mathfrak{M}$ is soluble. Hence $\mathfrak{G}$ is soluble too. Thus, we have

(III.F) *Let $\mathfrak{G}$ satisfy the assumptions*: (I) $\mathfrak{G}$ *is of type* (S). (II) *The* 2-*Sylow subgroup* $\mathfrak{T}$ *of* $\mathfrak{G}$ *is abelian of type* $(2, 2, \cdots, 2)$, *order* $2^a \geqq 4$. *For* $a = 2$ *assume also that* $\mathfrak{C}(T) = \mathfrak{T}$ *for* $T \in \mathfrak{T}$, $T \neq 1$.

If $\mathfrak{G}$ *does not satisfy the assumption* (III), *then* $\mathfrak{T}$ *is normal in* $\mathfrak{G}$, *and* $\mathfrak{G}$ *is soluble.*

References

1. R. Brauer and K. A. Fowler, *On groups of even order*, Ann. of Math. (2), vol. 62 (1955), pp. 565–583.
2. H. Zassenhaus, *Kennzeichnung endlicher linearer Gruppen als Permutationsgruppen*, Abh. Math. Sem. Univ. Hamburg, vol. 11 (1936), pp. 17–40.

Harvard University
Cambridge, Massachusetts
University of Illinois
Urbana, Illinois
University of Sydney
Sydney, Australia

Annals of Mathematics
Vol. 68, No. 3, November, 1958
Printed in Japan

ON A PROBLEM OF E. ARTIN

By Richard Brauer and W. F. Reynolds *

(Received March 14, 1958)

To E. Artin on His Sixtieth Birthday

1. Introduction

E. Artin, in two important papers [1], has studied the orders of the known types of simple finite groups. His results indicate a new approach to the investigation of the simple finite groups G. In this connection, it is of interest to study the largest prime power which divides the order g of such a group. The simplest question of this type is the following one which Artin asked us in a letter: what are the simple finite groups G of an order g which is divisible by a prime $p > g^{1/3}$? He conjectured that all such groups G belong to the known types. We shall show that this is actually true. Following Artin, we shall denote by $L_2(m)$ the group of unimodular linear collineations of a Desarguean projective line which belongs to a finite field with m elements. We will prove

THEOREM 1. *Let G be a finite simple non-cyclic group whose order g is divisible by a prime $p > g^{1/3}$. Then G is isomorphic either to $L_2(p)$ where $p > 3$ is a prime or to $L_2(p-1)$ where $p>3$ is a Fermat prime, $p=2^n+1$, with integral n.*

It is seen easily that the assumptions of Theorem 1 imply that g is divisible by p with the exact exponent 1. We shall investigate groups G of an order of this type. Sylow's theorem asserts that G has subgroups P of order p and that the number of such subgroups is of the form $1+rp$ where $r \geqq 0$ is an integer. It seems of interest to study this number r further and to obtain refinements of Sylow's theorem for our class of groups. We prove the following result from which Theorem 1 can be obtained as a corollary.

THEOREM 2. *Let G be a finite group whose order contains a prime number p with the exact exponent 1. Assume that G is its own commutator subgroup, $G = G'$. If the number of subgroups of G of order p is $1 + rp$ then r is of the form*

$$r = (hup + u^2 + u + h)/(u + 1) \tag{1}$$

with positive integers u and h, except when G has a homomorphic image

* This research was supported by the National Science Foundation under Grant NSF-G-1123.

G^* *of one of the following types*:

(I) $G^* \cong L_2(p)$, $p > 3$ *a prime.*

(II) $G^* \cong L_2(p-1)$, $p > 3$ *a Fermat prime.*

In the case (I), *we have* $r = 1$ *and in the case* (II), $r = (p-3)/2$.

Our proof of Theorem 2 is based on methods which were developed in two previous papers [2], [3] in which similar results were obtained for a more restricted class of groups. In a later paper, we will improve our results further.

It is perhaps of interest to mention that Theorem 1 is of importance if one wants to determine all simple groups whose order g lies below a given numerical bound, since we have here a bound for the prime factors p of g for which $g \not\equiv 0 \pmod{p^2}$. If $p^2 \mid g$, elementary methods show that $p^4 < g$ and if $p^a \mid g$ with $a \geqq 3$, then $p^{a+3} < g$. This imposes strong conditions on the orders g.

2. Group-theoretical preliminaries

Let G be a group of finite order g and assume that there exists a prime number p such that $g \equiv 0 \pmod p$, $g \not\equiv 0 \pmod{p^2}$. Then a p-Sylow subgroup P of G has order p. It follows easily (for instance from Burnside's theorem [4, p. 133]) that the centralizer $C(P)$ of P is a direct product $C(P) = P \times W$ of P with a group W of an order w prime to p. If π is a generating element of P and if σ is an element of the normalizer $N(P)$ of P, we have an equation $\sigma^{-1}\pi\sigma = \pi^j$ where the integer $j = j(\sigma)$ depends on σ. It is seen at once that the mapping $\sigma \rightarrow j(\sigma)$ is a homomorphism of $N(P)$ into the multiplicative group of residue classes of integers prime to $p \pmod p$. Since the kernel is $C(P)$, it follows that $N(P)/C(P)$ is cyclic of an order q which divides $p-1$. Then there exist exactly q distinct powers of π which are conjugate to π in G.

By Sylow's theorem, the number of distinct subgroups of order p of G has the form $1 + rp$ with integral $r \geqq 0$ and $1 + rp = (G : N(P))$. Combining these facts, we have

$$g = pqw(1 + rp) . \tag{2}$$

We call (q, w, r) the *system of p-invariants* of G.

If $r = 0$ and $w = 1$, then G has order pq and G can be defined by the relations $\pi^p = 1$, $\kappa^q = 1$, $\kappa^{-1}\pi\kappa = \pi^j$ where j is an integer which belongs to the exponent $q \pmod p$. If j is changed, we obtain an isomorphic group. We call this group G the *p-metacylic group* of order pq.

PROPOSITION 1. *If G is a group whose order g contains the prime number p with the exact exponent* 1 *and if G has the p-invariants* (q, w, r) *with*

$r=0$, then there exists a normal subgroup H of G such that G/H is isomorphic with the p-metacyclic group of order pq. In particular, $G \neq G'$ in this case.

Indeed, if $r = 0$, then P is normal in G, that is, $G = N(P)$. Since W is a characteristic subgroup of $C(P)$, W is normal in G. It is seen easily that G/W is isomorphic with the p-metacyclic group of order pq. Since G/W is soluble, we have $G \neq G'$.

PROPOSITION 2. *Let G be a group whose order g contains the prime number p with the exact exponent* 1 *and let (q, w, r) be the system of p-invariants of G. If S is a normal subgroup of G of an order $s \not\equiv 0$ (mod p) and if $G_1 = G/S$ has the p-invariants (q_1, w_1, r_1), then $q = q_1$. Moreover, there exists an integer $z \geqq 0$ such that $r = z + r_1 + zr_1p$.*

PROOF. Let θ be the natural homomorphism of G onto $G_1 = G/S$. The image $P_1 = \theta(P)$ of P is a p-Sylow subgroup of G_1. It is clear that the image $\theta(N(P))$ of $N(P)$ is included in the normalizer $N_1(P_1)$ of P_1 in G_1. On the other hand, if $\theta(\tau) \in N_1(P_1)$ for some $\tau \in G$, then $\tau^{-1}P\tau \subseteqq PS$. It follows from Sylow's theorem applied to PS, that there exists an element $\rho \in PS$ with $\tau^{-1}P\tau = \rho^{-1}P\rho$. Then $\tau\rho^{-1} \in N(P)$, $\tau \in N(P)PS = N(P)S$. It is now clear that $\theta(N(P)) = N_1(P_1)$.

Let π again denote a generator of P. Then $\pi_1 = \theta(\pi)$ is a generator of P_1. If an element τ_1 of G_1 transforms π_1 into a power π_1^j, then $\tau_1 \in N_1(P_1)$ and, as was just shown, we can set $\tau_1 = \theta(\tau)$ with $\tau \in N(P)$. It follows that we have an equation $\tau^{-1}\pi\tau = \pi^i$ with integral i. Application of θ shows that τ_1 transforms π_1 into π_1^i. Hence $\pi_1^i = \pi_1^j$. Since S has an order $s \not\equiv 0$ (mod p), two powers of π_1 are equal only if the corresponding powers of π are equal. It is now clear that we may take $i = j$ and that two powers of π_1 are conjugate in G_1 if and only if the corresponding powers of π are conjugate in G. This implies $q = q_1$. Moreover, if τ_1 belongs to the centralizer $C_1(P_1)$ of P_1 in G_1, we may take $j = 1$ and our argument shows that we can set $\tau_1 = \theta(\tau)$ with $\tau \in C(P)$. It follows that $\theta(C(P)) = C_1(P_1)$. If d is the order of $C(P) \cap S$, then since $C(P)$ has order pw, the order pw_1 of $C_1(P_1)$ is equal to pw/d. This yields $w = dw_1$.

As in (2), we have $g/s = pq_1w_1(1+r_1p)$. Substituting $q=q_1$ and $w=dw_1$ and using (2), we find $d(1 + rp) = s(1 + r_1p)$. It follows that $d \equiv s$ (mod p). Since d divides s, we may set $s = d(1+zp)$ with integral $z \geqq 0$ and then our equation yields $r = z + r_1 + zr_1p$. This completes the proof.

PROPOSITION 3. *Let G be a group whose order g contains the prime number p with the exact exponent* 1 *and let (q, w, r) be the system of p-invari-*

ants of G. Assume that no quotient group of G is isomorphic with the p-metacyclic group of order pq. Let G_0 be the unique minimal normal subgroup of G for which G/G_0 is soluble. Then the order g_0 of G_0 is divisible by p and if G_0 has the p-invariants (q_0, w_0, r_0), we have $r = r_0$.

PROOF. Suppose that p does not divide g_0. Using a principal series from G to G_0, we can find normal subgroups S and T of G such that $G_0 \subseteqq S \subset T \subseteqq G$ and that $(T : S) = p$. Then $G_1 = G/S$ has a normal subgroup T/S of order p. Proposition 2 shows that the first p-invariant of G_1 is q and it now follows from Proposition 1 that G_1 has a quotient group isomorphic with the p-metacyclic group of order pq. However, this is impossible since G did not have such a quotient group. Consequently, p divides g_0. Since every p-Sylow subgroup of G is included in G_0, we have $r = r_0$.

It is easy to show that in the case of Proposition 3, we have $q_0 \mid q$, $w_0 \mid w$.

3. Proof of Theorem 2

We now prove Theorem 2 in a slightly refined form. As a convenient abbreviation, we introduce the rational function

$$F(p, u, h) = (hup + u^2 + u + h)/(u + 1)\,. \tag{3}$$

THEOREM 2*. *Let G be a group whose order g contains the prime p with the exact exponent* 1. *Suppose that G has the p-invariants (q, w, r). Let G_0 be the minimal normal subgroup of G for which G/G_0 is soluble and let S be the maximal normal subgroup of G_0 of an order prime to p. Then either r can be represented in the form $r = F(p, u, h)$ with positive integers u and h or we have one of the cases:*

(I) *$G_0/S \cong L_2(p)$ with $p > 3$ a prime; $r = 1$.*

(II) *$G_0/S \cong L_2(p-1)$ with $p > 3$ a Fermat prime; $r = (p-3)/2$.*

(III) *G can be mapped homomorphically on the p-metacyclic group of order pq.*

PROOF. Suppose that we do not have case (III). It is clear that S is characteristic in G_0 and hence normal in G. Proposition 3 shows that p divides the order g_0 of G_0 and, if G_0 has the p-invariants (q_0, w_0, r_0), we have $r = r_0$. Obviously, G_0 is its own commutator-subgroup and, if we set $G^* = G_0/S$, the analogous fact holds for G^*. Moreover, G^* does not have a non-trivial normal subgroup of an order prime to p. If G^* has the p-invariants (q^*, w^*, r^*), it follows from Proposition 2 that there exists a non-negative integer z such that

$$r = r_0 = z + r^* + zr^*p. \tag{4}$$

Since G^* is its own commutator group, we have $r^*>0$, cf. Proposition 1. Set now $h=(z+1)r^*$, $u=z$. Then, by (3) and (4), $F(p,u,h)=r$. If $z>0$, r can be represented in the form $r=F(p,u,h)$ with positive integers u, h.

Assume that $z=0$. It follows here that $r=r^*$. If we write G for G^*, it is clear that it will suffice to prove Theorem 2* for groups G which are equal to their commutator subgroups and which do not possess non-trivial normal subgroups of order prime to p. Under these additional assumptions, the theorem states that if r cannot be represented in the form $F(p,u,h)$ with positive integers u, h, then either $G\cong L_2(p)$ with $p>3$ or $G\cong L_2(p-1)$ with $p>3$ a Fermat prime. If $w=1$, this follows at once from Theorem 10 of [3]. Hence we may assume that $w>1$.

Consider the p-block B of irreducible characters of G to which the unit character χ_1 of G belongs. As shown in [2], B consists of q non-exceptional characters $\chi_1, \chi_2, \cdots, \chi_q$ and $t=(p-1)/q$ exceptional characters $\chi_0^{(\lambda)}$, $\lambda=1,2,\cdots,t$. For each $\chi_j\in B$, we have a sign $\delta_j=\pm 1$ such that

$$\chi_j(\sigma)=\delta_j \qquad (\text{for } \sigma\in P\times W, \sigma\notin W). \tag{5}$$

Similarly, there is a sign $\delta_0=\pm 1$ such that

$$\textstyle\sum_{\lambda=1}^{t}\chi_0^{(\lambda)}(\sigma)=\delta_0 \qquad (\text{for } \sigma\in P\times W, \sigma\notin W). \tag{6}$$

Moreover, all $\chi_0^{(\lambda)}$ have the same degree f_0 and if χ_j has degree f_j, we have

$$\textstyle\sum_{i=0}^{q}\delta_i f_i=0\,. \tag{7}$$

Take σ as a generating element π of P. Then $\chi_j(\sigma)$ is a sum of f_j p^{th} roots of unity. If $\mathfrak{p}$ is a fixed prime ideal divisor of p in the field of the pth roots of unity over the rationals, every p^{th} root of unity is congruent to 1 (mod $\mathfrak{p}$). Hence (5) implies $f_j\equiv\delta_j\ (\text{mod } \mathfrak{p})$ whence

$$f_j\equiv\delta_j \qquad (\text{mod } p) \tag{8}$$

for $j=1,2,\cdots,q$. Similarly, (6) yields

$$tf_0\equiv\delta_0 \qquad (\text{mod } p). \tag{9}$$

Let $c(\sigma)$ denote the order of the centralizer $C(\sigma)$ of an element σ of G. As is well known, the number $g\chi_j(\sigma)/(c(\sigma)f_j)$ is an algebraic integer. For $\sigma=\pi$, we have $c(\sigma)=pw$ and it follows from (5) and (2) that $f_j\,|\,q(1+rp)$ whence

$$tf_j\,|\,(p-1)(1+rp) \qquad (j=1,2,\cdots,q). \tag{10}$$

Similarly, $g\chi_0^{(\lambda)}(\sigma)/c(\sigma)f_0$ is an algebraic integer for $\lambda=1,2,\cdots,t$. Adding over λ and using (6), we obtain the analogous relation

(11) $$tf_0 \mid (p-1)(1+rp)\ .$$

Assume that r cannot be written in the form $F(p, u, h)$ with positive integers u and h. An elementary divisibility consideration [3, Lemmas 1 and 2] applied to (8) and (10) shows that we must have one of the cases

(12) $$f_j = 1\ , \qquad f_j = 1 + rp \qquad (\text{for } \delta_j = 1)\ ;$$

(13) $$f_j = p - 1\ , \quad f_j = p(pr - r + 1) - 1 \quad (\text{for } \delta_j = -1).$$

This applies for $j = 1, 2, \cdots, q$.

Similarly, (9) and (11) imply that tf_0 also can have only one of the values in (12), (13),

(14) $$tf_0 = 1\ , \qquad tf_0 = 1 + rp \qquad (\text{for } \delta_0 = 1)\ ;$$

(15) $$tf_0 = p - 1\ , \quad tf_0 = p(pr - r + 1) - 1 \quad (\text{for } \delta_0 = -1).$$

In (7), we have $q + 1$ terms $\delta_i f_i$ with $i = 0, 1 \cdots, q$. For at most q of them, δ_i can have the value $+1$. For each of these terms, $f_i \leqq 1 + rp$ by (12) and (14). For $i = 1$, χ_1 is the unit character of degree 1 and here $\delta_1 = 1$, as shown by (5). Since we could assume $G = G'$ and hence $r > 0$ (cf. Proposition 1), $\delta_1 f_1 = 1 < 1 + rp$. Thus, the sum of the terms of (7) with $\delta_i = 1$ is less than $q(1+rp) = \{(p-1)(1+rp)\}/t = \{p(pr-r+1)-1\}/t$. It now follows from the equation (7) that the cases $f_j = p(pr-r+1)-1$ in (13) and $tf_0 = p(pr - r + 1) - 1$ in (15) are impossible. Since terms with $\delta_i = -1$ must appear in (7), one of the degrees $p - 1$ in (13) or $(p - 1)/t$ in (15) occurs. If we have $f_j = p - 1$ for some j with $1 \leqq j \leqq q$, let Z be the representation of G which belongs to the character χ_j. If $f_j \neq p - 1$ for $j = 1, 2, \cdots, q$, take Z as the (reducible) representation of G whose character is the sum of the t exceptional characters $\chi_0^{(\lambda)}$ of B. In either case, we have a representation Z of degree $p - 1$ of G. If ζ is the character of Z, the formulas (5) and (6) show that

(16) $$\zeta(\sigma) = -1$$

for $\sigma \in P \times W, \sigma \notin W$.

Of course, the restriction of ζ to $C(P) = P \times W$ is a character of $P \times W$. Hence we can find distinct irreducible characters $\psi_1, \psi_2, \cdots, \psi_n$ of W and characters $\varphi_1, \varphi_2, \cdots, \varphi_n$ of P such that

(17) $$\zeta(\tau\rho) = \sum_{i=1}^{n} \psi_i(\tau)\varphi_i(\rho)$$

for $\tau \in W, \rho \in P$. Take ρ as a generating element π of P. By (16), $\zeta(\tau\pi) = -1$. Because of the linear independence of the irreducible characters of W, it follows from (17) that the unit character of W must appear among $\psi_1, \psi_2, \cdots, \psi_n$. If, say, ψ_1 is the unit character of W, our argument shows that $\varphi_1(\pi) = -1$, $\varphi_2(\pi) = 0, \cdots, \varphi_n(\pi) = 0$. If we had $n > 1$, again $\varphi_2(1) \equiv \varphi_2(\pi) \pmod{\mathfrak{p}}$. Since $\varphi_2(\pi)=0$, the degree of φ_2 would be divisible by p. This is impossible, because ζ had only the degree $p-1$. Thus, $n = 1$ and (17) reads $\zeta(\tau\rho) = \psi_1(\tau)\varphi_1(\rho) = \varphi_1(\rho)$. In particular, $\zeta(\tau)$ has the value $\varphi_1(1)$ for all $\tau \in W$. Taking $\tau = 1$, we see that $\zeta(\tau)$ is equal to the degree $p - 1$ for all these τ. Then all characteristic roots of the linear transformation $Z(\tau)$ are 1. Since τ is an element of a finite group, this implies that Z represents τ by the identity.

Let S denote the kernel of Z. Then $W \subseteq S$. If the normal subgroup S of G had an order divisible by p, we would have $P \subseteq S$ and hence $\pi \in S$, $\zeta(\pi) = p - 1$. However, (16) for $\sigma = \pi$ shows that $\zeta(\pi) = -1$. Thus, S has an order prime to p. Since we had $w>1$, S is a non-trivial normal subgroup of G of an order prime to p. As shown above, we could assume that no such subgroup exists. We have a contradiction and Theorem 2* is proved.

Theorem 2 is an immediate consequence of Theorem 2* since the assumption $G = G'$ implies that $G_0 = G$ and that case (III) of Theorem 2* is excluded.

4. Proof of Theorem 1

Assume now that G is a finite group and suppose that there exists a prime factor p of the order g of G such that

$$g < p^3. \tag{18}$$

If $g \equiv 0 \pmod{p^2}$ and if $1 + rp$ is the number of p-Sylow groups of G, then $g \geqq p^2(1 + rp)$ and (18) implies that $r = 0$. Hence a p-Sylow group P is normal in G.

Suppose then that $g \not\equiv 0 \pmod{p^2}$. Let (q, w, r) denote the p-invariants of G. If we assume that $G = G'$, Theorem 2 can be applied. The minimum of $F(p, u, h)$ in (3) for positive integers u and h is $(p + 3)/2$. Burnside's theorem shows that $q \geqq 2$. Hence if r can be represented in the form $F(p, u, h)$, (2) yields

$$g = pqw(1 + rp) \geqq 2p(1 + p(p + 3)/2) = p(p + 1)(p + 2)\,, \tag{19}$$

contrary to (18).

If r does not have the form $F(p, u, h)$ with integral $u > 0$, $h>0$, G has a quotient group $G/H = G^*$ of one of the forms

(I) $G^* \cong L_2(p)$ with $p > 3$;

(II) $G^* \cong L_2(p - 1)$ with $p > 3$ a Fermat prime.

It follows from (18) that in the case (I), H can have at most order 2. The same is true in the case (II) when $p = 5$. If $p > 5$, we must have

$G = G^*$ in the case (II).

This yields the result:

THEOREM 1*. *Let G be a finite group of order g and assume that g has a prime factor $p > g^{1/3}$. Assume further*

(α) *that $G = G'$,*

(β) *that the p-Sylow group P of G is not normal in G,*

(γ) *that G does not have a normal subgroup of order 2.*

Then $G \cong L_2(p)$ where $p > 3$ is a prime or $G \cong L_2(p-1)$ where $p > 3$ is a Fermat prime.

Theorem 1 is a special case of Theorem 1*.

It is of interest to replace (18) by weaker inequalities. As suggested by (19), the next class of simple groups encountered are the groups $L_2(p+1)$ where p is a Mersenne prime.

HARVARD UNIVERSITY
TUFTS UNIVERSITY

REFERENCES

1. E. ARTIN, *The orders of the linear groups*, Comm. Pure Appl. Math., 8 (1955), 355–365.

———, *The orders of the classical simple groups*, Comm. Pure Appl. Math., 8 (1955), 455–472.

2. R. BRAUER, *On groups whose order contains a prime number to the first power* I, Amer. J. of Math., 64 (1942), 401–420.

3. ———, *On permutation groups of prime degree and related classes of groups*, Ann. of Math., 44 (1943), 57–79.

4. H. ZASSENHAUS, Lehrbuch der Gruppentheorie, 1937.

Reprinted from the Proceedings of the NATIONAL ACADEMY OF SCIENCES
Vol. 45, No. 3, pp. 361–365. March, 1959.

ON THE NUMBER OF IRREDUCIBLE CHARACTERS OF FINITE GROUPS IN A GIVEN BLOCK

BY RICHARD BRAUER AND WALTER FEIT

HARVARD UNIVERSITY AND CORNELL UNIVERSITY

Communicated January 16, 1959

Let $\mathfrak{G}$ be a group of finite order g and let p be a fixed prime number. The blocks of irreducible characters of $\mathfrak{G}$ with regard to p and their significance for the arithmetic in the group ring of $\mathfrak{G}$ have been discussed in several previous publications.[1] In particular, it has been stated that the number m of ordinary irreducible characters in a p-block of defect d is at most equal to $p^{d(d+1)/2}$. In the present note, this result will be improved. We shall show:

THEOREM 1. *Let B be a* p*-block of defect* d *of a finite group* $\mathfrak{G}$. *The number* m *of ordinary irreducible characters* $\chi_1, \chi_2, \ldots \chi_m$ *in B satisfies the inequality*

$$m \leq {}^1/_4 p^{2d} + 1. \tag{1}$$

Our new proof which we shall now sketch is far more elementary than the previous proof of the weaker result. Let B consist of the modular irreducible characters $\varphi_1, \varphi_2, \ldots \varphi_n$. For p-regular elements R of $\mathfrak{G}$, i.e., for elements R of $\mathfrak{G}$ whose order is prime to p, we have formulas

$$\chi_i(\mathrm{R}) = \sum_{j=1}^{n} d_{ij}\varphi_j(R), \qquad (1 \leq i \leq m) \tag{2}$$

where the d_{ij} are nonnegative rational integers, the decomposition numbers of $\mathfrak{G}$ belonging to B. Set

$$a_{ij} = (p^d/g) \sum_{R} \chi_i(R)\overline{\chi_j(R)}, \qquad (1 \leq i, j \leq m) \tag{3}$$

where R ranges over all p-regular elements of $\mathfrak{G}$. Let A denote the $m \times m$ matrix (a_{ij}) and let D denote the $m \times n$ matrix (d_{ij}). If D' denotes the transpose of D, then $C = D'D$ is the $n \times n$ matrix of Cartan invariants of $\mathfrak{G}$ belonging to B. It follows from (2), (3), and the orthogonality relations for the modular group characters that

$$A = p^d DC^{-1}D'. \tag{4}$$

It is known that C has integral coefficients and that its largest elementary divisor is p^d. Now (4) shows that A has integral coefficients. Clearly, A is a symmetric matrix.

The nth determinantal divisor of the $m \times n$ matrix D is 1.[2] Using the normal form of D, we deduce from (4) that there exists an $m \times m$ matrix U with integral coefficients and determinant 1 such that

$$UAU' = p^d \begin{pmatrix} C^{-1} & 0 \\ 0 & 0 \end{pmatrix}.$$

It follows that not all coefficients of UAU' are divisible by p and hence not all coefficients of A are divisible by p.

Let $\{R_1, R_2, \ldots, R_h\}$ be a system of representatives for the classes of p-regular conjugate elements of $\mathfrak{G}$. If the class of R_α consists of g_α elements, and if $x_i = \chi_i(1)$ is the degree of χ_i, then it is well known that the number

$$\omega_i(R_\alpha) = g_\alpha X_i(R_\alpha)/x_i$$

is an algebraic integer. Introducing these quantities in (3), we obtain

$$a_{ij} = (p^d/g)x_i \sum_{\alpha=1}^{h} \omega_i(R_\alpha)\bar{\chi}_j(R_\alpha). \tag{5}$$

If χ_r is another irreducible character in B and if ω_r has the analogous significance as ω_i, then modulo a suitable prime ideal divisor $\mathfrak{p}$ of p, we have

$$\omega_i(R_\alpha) \equiv \omega_r(R_\alpha) \qquad (\text{mod } \mathfrak{p}).$$

Now, (5) yields easily the congruence

$$(g/p^d)\ (a_{ij}/x_i) \equiv (g/p^d)\ (a_{rj}/x_r) \qquad (\text{mod } \mathfrak{p}).$$

Let ν denote the p-adic exponential valuation, $\nu(p) = 1$. By the definition of the defect d of a block B, we can set

$$\nu(x_i) = \nu(g) - d + \lambda_i \tag{6}$$

with $\lambda_i \geq 0$ for $1 \leq i \leq m$. We shall term the integer λ_i the height of the character χ_i in B. There exist characters of height 0 in B. We choose χ_r as such a character, $\lambda_r = 0$.

If congruences are taken as congruences in the ring of local integers for p, our result becomes

$$a_{ij} \equiv (x_i/x_r)a_{rj} \qquad (\text{mod } p^{1+\lambda_i}).$$

Using the symmetry of a_{ij}, we find

$$a_{ij} \equiv (x_i x_j/x_r^2)a_{rr} \qquad (\text{mod } p^{1+\lambda_i}). \tag{7}$$

It follows that a_{rr} is not divisible by p, since otherwise all a_{ij} would be divisible by p, and we have already seen that this is not true. Taking $j = r$ in (7), we obtain

$$\nu(a_{ir}) = \lambda_i. \tag{8}$$

In particular, $a_{ir} \neq 0$ for $1 \leq i \leq m$.

It follows from (4) together with $D'D = C$ that

$$A^2 = p^d A. \tag{9}$$

Hence

$$\sum_{i=1}^{m} a_{ir}^2 = p^d a_{rr}. \tag{10}$$

Since $a_{ir} \neq 0$, this yields

$$(m - 1) + a_{rr}^2 \leq p^d a_{rr}.$$

The maximum of $p^d x - x^2$ is obtained for $x = p^d/2$ and hence $m - 1 \leq p^{2d}/4$. This completes the proof of Theorem 1.

The formulas developed here can be used to obtain some further results. It follows from (3) in conjunction with the orthogonality relations for group characters that $0 < a_{ii} \leq p^d$. By (9),

$$\sum_{j=1}^{m} a_{ji}^2 = p^d a_{ii} \tag{11}$$

for $i = 1, 2, \ldots, m$. If $a_{ii} = p^d$, all a_{ji} with $j \neq i$ vanish. Then (8) implies that $i = r$. Since $\nu(a_{rr}) = 0$, it follows that $d = 0$. Thus $a_{ii} < p^d$ for $d > 0$. This can also be seen easily directly.

We defined the height λ_i of a character χ_i of B by the equation (6). If $\lambda_i > 0$, then (7) shows that a_{ii} is divisible by $p^{\lambda_i + 1}$. Hence $\lambda_i + 1 < d$. We thus have a new proof of the following result.[3]

THEOREM 2. *If B is a block of defect $d \geq 2$, the height λ_i of a character χ_i in B is at most $d - 2$. For $d = 0, 1$, and 2, we have $\lambda_i = 0$ for all χ_i in B.*

There exist examples of blocks of arbitrary defect $d \geq 2$ which contain characters of height $d - 2$.

Let m_λ denote the number of characters χ_i of B of height λ;

$$m = \sum_{\lambda} m_\lambda. \tag{12}$$

Then as shown by (8), m_λ terms a_{ir} in (10) contain p with the exact exponent p^λ. Hence the method used above yields

$$(m_0 - 1) + m_1 p^2 + m_2 p^4 + \ldots \leq {}^1/_4\, p^{2d}. \tag{13}$$

In particular

$$m_\lambda < {}^1/_4 p^{2d-2\lambda} \tag{14}$$

for $\lambda > 0$, as it is easy to see that $m_0 > 1$.

On the other hand, if B contains characters χ_i of positive height $\lambda_i = \lambda > 0$, then m_0 terms a_{ji} in (11) contain p with the exact exponent λ. Hence

$$m_0 \leq {}^1/_4 p^{2d-2\lambda}. \tag{15}$$

On combining (12), (13), and (15) we obtain

THEOREM 3. *If the p-block B of defect d contains irreducible characters χ_i of positive height, the number m of irreducible characters in B is less than ${}^1/_2 p^{2d-2}$.*

So far we have only used p-regular elements R of $\mathfrak{G}$. If P now is an element of $\mathfrak{G}$ whose order is a power $p^\alpha \geq 1$ of p, and if R is a p-regular element of the centralizer $\mathfrak{C}(P)$ of P, the formula (2) can be generalized to

$$\chi_i(PR) = \sum_j d_{ij}^P \varphi_j^P(R),$$

where the φ_j^P are the modular irreducible characters of $\mathfrak{C}(P)$ and where the d_{ij}^P are algebraic integers of the field Ω_α of the p^αth roots of unity. If we consider only χ_i belonging to a fixed p-block B of $\mathfrak{G}$, only characters φ_j^P have to be taken which belong to a well-determined set of p-blocks $\tilde{B}$ of $\mathfrak{C}(P)$. We set

$$a_{ij}^P = (p^d/n(P)) \textstyle\sum_R \chi_i(PR)\bar{\chi}_j(PR),$$

where $n(P)$ is the order of $\mathfrak{C}(P)$ and where R ranges over the p-regular elements of $\mathfrak{C}(P)$. For $P = 1$, we have $a_{ij}^P = a_{ij}$. Let A^P be the matrix of the a_{ij}^P, D^P the matrix of the d_{ij}^P, and $\mathfrak{C}^P$ the direct sum of the matrices $\tilde{C}$ of the Cartan invariants of the blocks $\tilde{B}$ of $\mathfrak{C}(P)$ associated with B. Then (4) can be generalized to

$$A^P = p^d D^P (C^P)^{-1} (\bar{D}^P)'.$$

It follows that the a_{ij}^P are algebraic integers of the field Ω_α. Similarly to (9), we have $(A^P)^2 = p^d A^P$. There is also a form in which (7) can be generalized. If P ranges over a system of representatives for the classes of conjugate elements whose order is a power of p, we have

$$\sum_P a_{ij}^P = p^d \delta_{ij}.$$

Finally, there exists a simple connection between the p-blocks of $\mathfrak{C}(P)$ and those of $\mathfrak{C}(P)/\{P\}$.

We now state a number of results without proof which can be shown by means of these formulas. If we use known results[4] on blocks of defect 0 and 1 and Theorem 3, we obtain the following improvement of Theorem 1.

THEOREM 1*. *If B, p, d, m have the same significance as in Theorem 1, then $m \leq p^d$ for $d = 0, 1, 2$ and $m < p^{2d-2}$ for $d > 2$.*

Probably, the inequality $m \leq p^d$ holds for all d, but we have not been able to prove this. However, for certain types of defect groups $\mathfrak{D}$ of the block B, our results can be improved. For instance, we can prove:

THEOREM 4. *If the defect group $\mathfrak{D}$ of the block B in Theorem 1 is cyclic, then $m \leq p^d$.*

Actually, there is a connection between the structure of the defect group $\mathfrak{D}$ of the block B and the maximal height λ of characters in B.

THEOREM 5. *Let $\mathfrak{D}$ of order p^d be the defect group of the p-block B of $\mathfrak{G}$ and assume that there exist u elements in the center of $\mathfrak{D}$ such that no two of them are conjugate in $\mathfrak{G}$. If B contains characters χ of height $\lambda > 0$, then*

$$u < p^{d-\lambda}. \tag{16}$$

If $p = 2$, then (16) can be replaced by

$$u \leq 2^{d-\lambda-1}. \tag{16*}$$

COROLLARY. *If B is a block of full defect $d = \nu(g)$, i.e., if the defect group $\mathfrak{D}$ is a* *p-Sylow group of $\mathfrak{G}$ and if the center of $\mathfrak{D}$ contains an element P of order p^α, then B can contain characters of height λ only if*

$$\lambda \leq d - \alpha.$$

For $p = 2$ and $\lambda > 0$, we even have

$$\lambda \leq d - \alpha - 1.$$

[1] Brauer, R., these PROCEEDINGS, **30**, 109–114 (1944); **32**, 182–186 (1944); and **32**, 215–219 (1946). Cf. also Brauer, R. *Math. Zeitschr.*, **63**, 406–444 (1956).

[2] See, for instance, Brauer, R., *Ann. Math.*, **57**, 357–377 (1953).

[3] Brauer, R., *Ann. Math.*, **42**, 936–958 (1941).

[4] Cf. the paper quoted in reference 3.

BRAUER, R.
Math. Zeitschr. 72, 25–46 (1959)

Zur Darstellungstheorie der Gruppen endlicher Ordnung. II

LEON LICHTENSTEIN zum Gedächtnis

Von
RICHARD BRAUER

§ 1. Einleitung[1]

In Fortführung der Untersuchungen des ersten Teils[2]) betrachten wir die Blöcke von Charakteren einer endlichen Gruppe $\mathfrak{G}$ für eine fest gewählte Primzahl p. In § 2 werden in Erweiterung der Ergebnisse von I Beziehungen zwischen Blöcken von $\mathfrak{G}$ und Blöcken geeigneter Untergruppen von $\mathfrak{G}$ hergeleitet.

Sind $\chi_1, \chi_2, \ldots, \chi_k$ die gewöhnlichen irreduziblen Charaktere von $\mathfrak{G}$ und sind $\varphi_1, \varphi_2, \ldots, \varphi_l$ die irreduziblen modularen Charaktere von $\mathfrak{G}$, so gelten nach I (3.5) für p-reguläre Elemente $G \in \mathfrak{G}$ die Gleichungen

$$\chi_i(G) = \sum_j d_{ij} \varphi_j(G), \tag{1.1}$$

wo die d_{ij} die Zerlegungszahlen von $\mathfrak{G}$ (für p) sind. Man erhält ohne Schwierigkeit ähnliche Formeln für p-singuläre Elemente von $\mathfrak{G}$, wobei aber rechts anstelle der φ_j die irreduziblen modularen Charaktere gewisser Untergruppen $\mathfrak{N}_1, \mathfrak{N}_2, \ldots, \mathfrak{N}_r$ von $\mathfrak{G}$ auftreten (§ 3). Das Hauptergebnis der vorliegenden Arbeit (§ 6) besagt, daß, wenn man in den verallgemeinerten Formeln (1.1) den Charakter χ_i in einem festen Block B wählt, rechts nur die modularen Charaktere $\varphi_j^{(\varrho)}$ gewisser B zugeordneter Blöcke der Gruppen $\mathfrak{N}_\varrho$ auftreten können. Die Gesamtzahl dieser $\varphi_j^{(\varrho)}$ ist gleich der Zahl $\varkappa$ der Charaktere χ_i in B, und die in den Formeln auftretende Matrix der Koeffizienten ist quadratisch vom Grade $\varkappa$, nicht-singulär, und die Determinante ist im wesentlichen eine Potenz von p. Für die genaue Formulierung muß auf § 6 und § 7 verwiesen werden. In § 4 werden einige Hilfsbetrachtungen über Klassenalgebren gegeben, § 5 ist dem Beweis einiger Hilfssätze gewidmet. Schließlich werden in § 7 eine Reihe von Folgerungen gegeben.

Die Bezeichnungen sind wie in I gewählt. Hier sei nur erwähnt, daß Ω ein die g-ten Einheitswurzeln enthaltender algebraischer Zahlkörper endlichen Graues ist, wo $g = (\mathfrak{G}:1)$ die Ordnung von $\mathfrak{G}$ bezeichnet. Es sei ν eine Erweiterung der p-adischen exponentiellen Bewertung der rationalen Zahlen auf Ω,

[1]) This research was supported in part by the United States Air Force under Contract No. AF 49 (638)-287 monitored by the AF Office of Scientific Research of the Air Research and Development Command.

[2]) Die erste Mitteilung ist in der Math. Z. 63, 406–444 (1956) erschienen; sie wird im folgenden mit I zitiert.

$\nu(p)=1$; es sei $\mathfrak{o}_\nu$ der Ring der für ν ganzen Zahlen von Ω, $\mathfrak{p}_\nu$ das zugehörige Primideal und schließlich

$$\Omega^* = \mathfrak{o}_\nu/\mathfrak{p}_\nu$$

der Restklassenkörper.

Schreiben wir eine Matrix $A=(a_{ij})$, so ist *stets* der Zeilenindex mit i, der Spaltenindex mit j bezeichnet. Der Wertebereich der Indizes i und j ist in der Regel aus dem Zusammenhang zu entnehmen. Ist z.B. $R_1, R_2, \ldots, R_l$ ein Vertretersystem für die p-regulären Klassen von $\mathfrak{G}$, so ist $\boldsymbol{\varphi}=(\varphi_j(R_j))$ die Matrix l-ten Grades der irreduziblen modularen Charaktere von $\mathfrak{G}$.

Wir geben hier noch eine Ergänzung zu der in I § 3 entwickelten Theorie der modularen Charaktere. Es sei $(n(R_i)\delta_{ij})=N$ die aus den Ordnungen $n(R_i)$ der Normalisatoren $\mathfrak{N}(R_i)$ der Elemente R_i gebildete Diagonalmatrix. Es sei $C=(c_{ij})$ die Matrix der Cartanschen Invarianten von $\mathfrak{G}$, vgl. I (3.8). Aus I (3 G) folgt dann

$$\bar{\boldsymbol{\varphi}}' C \boldsymbol{\varphi} = N, \tag{1.2}$$

also

$$C \boldsymbol{\varphi} N^{-1} \bar{\boldsymbol{\varphi}}' = E = (\delta_{ij}). \tag{1.3}$$

Ich benütze die Gelegenheit, einige Ungenauigkeiten in I zu berichtigen. Auf S. 416, Zeile 21 ersetze man Z durch Z^*. Auf S. 423, Zeile 17 ist vor „Elementarteiler" das Wort „anderen" hinzuzufügen. Herr H. WIELANDT macht mich darauf aufmerksam, daß beim Beweis von (6E) die Elementarteiler in bezug auf den Ring der ganzen rationalen Zahlen zu nehmen sind (vgl. dazu Anmerkung [9]) von I). Auf S. 427, Zeile 6 muß das Wort „eindeutig" gestrichen werden. In (11.1) ist die Summe durch $\sum_{\varrho=1}^{r} \vartheta_\varrho$ zu ersetzen. Auf S. 434, Zeile 27 muß es $\bar{B}$ anstatt B heißen. Auf S. 435, Zeile 2 werde das Wort „Darstellung" durch „Charakter" ersetzt. Auf S. 436, Zeile 17 streiche man die Worte „und daß die Ordnung von $\overline{\mathfrak{N}}(\overline{T}_j)$ zu p teilerfremd ist". Die Beziehung (11.9) muß $\nu(T_j) \geqq d$ lauten, in (11.10) werde die Summe durch $\sum_{\varrho=1}^{r} \vartheta_\varrho(W)$ ersetzt und in der folgenden Zeile muß es $\nu(g/x_i) - \nu(T_j) \leqq 0$ (anstelle von Gleichheit) heißen. Auf S. 437 muß die Zeile 14 ersetzt werden durch „Block, und ist χ_i ein Charakter von B, dessen Kern $\mathfrak{D}$ enthält, so ist für Elemente $W:\mathfrak{W}$:". Schließlich füge man auf S. 437, Zeile 29 vor $\tilde{\chi}_j$ das Wort „geeignetes" hinzu.

§ 2. Beziehungen zwischen Blöcken von $\mathfrak{G}$ und Blöcken von Untergruppen

Wie in I § 2 sei $Z(\mathfrak{G}, \Xi)$ die Klassenalgebra der Gruppe $\mathfrak{G}$ über dem beliebigen Körper Ξ. Sind $\mathfrak{K}_1, \mathfrak{K}_2, \ldots, \mathfrak{K}_k$ die Klassen konjugierter Elemente von $\mathfrak{G}$ und setzt man

$$K_\alpha = \sum_{\sigma \in \mathfrak{K}_\alpha} \sigma \qquad (\text{für } \alpha = 1, 2, \ldots, k),$$

so ist $Z(G, \Xi)$ ein Vektorraum mit der Basis $K_1, K_2, \ldots, K_k$. Es ist klar, was man unter einer *linearen Funktion* f in $Z(\mathfrak{G}, \Xi)$ zu verstehen hat. Dabei ist f eindeutig durch die Werte $f(K_1), f(K_2), \ldots, f(K_k)$ bestimmt, und man kann diese k Werte als beliebige k Elemente von Ξ vorschreiben.

Es sei $\tilde{\mathfrak{G}}$ eine Untergruppe von $\mathfrak{G}$. Wir verwenden für $\tilde{\mathfrak{G}}$ durchgehend dieselben Bezeichnungen wie für $\mathfrak{G}$ unter Hinzufügung eines $\tilde{}$-Zeichens. So werden z.B. die Klassen von $\tilde{\mathfrak{G}}$ mit $\tilde{\mathfrak{K}}_1, \tilde{\mathfrak{K}}_2, \ldots$ bezeichnet. Ist h eine lineare Funktion in $Z(\tilde{\mathfrak{G}}, \Xi)$, so definieren wir eine mit $h^{\mathfrak{G}}$ bezeichnete lineare Funktion in $Z(\mathfrak{G}, \Xi)$ durch die Festsetzung

$$h^{\mathfrak{G}}(K_\alpha) = \sum_{\tilde{\mathfrak{K}}_\beta \leq \mathfrak{K}_\alpha} h(\tilde{K}_\beta), \tag{2.1}$$

wo also $\tilde{\mathfrak{K}}_\beta$ die in $\mathfrak{K}_\alpha$ enthaltenen Klassen von $\tilde{\mathfrak{G}}$ durchläuft. Ist $\mathfrak{A}$ eine zwischen $\mathfrak{G}$ und $\tilde{\mathfrak{G}}$ gelegene Gruppe, so ist offenbar

$$(h^{\mathfrak{A}})^{\mathfrak{G}} = h^{\mathfrak{G}}. \tag{2.2}$$

Wir wählen jetzt $\Xi = \Omega^*$. Dann hat man eine eineindeutige Zuordnung zwischen den linearen Charakteren $\psi_1, \psi_2, \ldots, \psi_t$ von $Z(\mathfrak{G}, \Omega^*)$ (d.h. zwischen den Algebra-Homomorphismen ψ_τ von $Z(\mathfrak{G}, \Omega^*)$ auf Ω^*) und den p-Blöcken $B_1, B_2, \ldots, B_t$ von $\mathfrak{G}$.

Es sei $\tilde{\psi}$ ein linearer Charakter von $Z(\tilde{\mathfrak{G}}, \Omega^*)$, der zu dem p-Block $\tilde{B}$ von $\tilde{\mathfrak{G}}$ gehöre. Ist dann $\psi = \tilde{\psi}^{\mathfrak{G}}$ ein linearer Charakter von $Z(\mathfrak{G}, \Omega^*)$, so gehört ψ zu einem p-Block B von $\mathfrak{G}$. Dann wollen wir

$$B = \tilde{B}^{\mathfrak{G}}$$

schreiben. Es ist aber zu beachten, daß ψ kein linearer Charakter von $\mathfrak{G}$ zu sein braucht, so daß $\tilde{B}^{\mathfrak{G}}$ nicht immer definiert ist.

(2A) *Es sei $\mathfrak{Q}$ eine p-Untergruppe von $\mathfrak{G}$, es sei $\mathfrak{T} = \mathfrak{C}(\mathfrak{Q})$ der Zentralisator von $\mathfrak{Q}$ und es sei $\mathfrak{H}$ eine Untergruppe des Normalisators $\mathfrak{N}(\mathfrak{Q})$ von $\mathfrak{Q}$, derart daß $\mathfrak{H} \geqq \mathfrak{Q}\mathfrak{T}$ ist. Ist $\tilde{B}$ ein Block von $\mathfrak{H}$, so ist der p-Block $B = \tilde{B}^{\mathfrak{G}}$ stets definiert (und es ist B gerade der $\tilde{B}$ nach* I (7E) *zugeordnete Block von* $\mathfrak{G}$).

Beweis. Man wähle $\tilde{\mathfrak{G}} = \mathfrak{H}$. Ist $\tilde{\psi}$ der zu $\tilde{B}$ gehörige lineare Charakter von $Z(\mathfrak{G}, \Omega^*)$, so zeigt Vergleich von (2.1) für $h = \tilde{\psi}$ mit I (7.3), daß es genügt $\tilde{\psi}(\tilde{K}_\beta) = 0$ für diejenigen Klassen $\tilde{\mathfrak{K}}_\beta$ von $\tilde{\mathfrak{G}}$ zu zeigen, die nicht in $\mathfrak{T}$ liegen. Da $\mathfrak{T} \lhd \tilde{\mathfrak{G}}$ ist (d.h. $\mathfrak{T}$ eine invariante Untergruppe von $\mathfrak{G}$), ist dann $\tilde{\mathfrak{K}}_\beta \cap \mathfrak{T} = \emptyset$. Wäre nun $\tilde{\psi}(\tilde{K}_\beta) \neq 0$, so würde aus I (8A) folgen, daß der Zentralisator $\tilde{\mathfrak{N}}(\tilde{G})$ eines Elementes $\tilde{G} \in \tilde{\mathfrak{K}}_\beta$ in $\tilde{\mathfrak{G}}$ eine Defektgruppe $\tilde{\mathfrak{D}}$ von $\tilde{B}$ enthält. Wegen I (9F) gilt $\mathfrak{Q} \leqq \tilde{\mathfrak{D}}$. Also folgt $\tilde{G} \in \mathfrak{C}(\tilde{\mathfrak{D}}) \leqq \mathfrak{C}(\mathfrak{Q}) = \mathfrak{T}$, was einen Widerspruch ergibt.

Aus (2A) ergibt sich, daß die Zuordnung $\tilde{B} \to B$ in I (7E) nicht von $\mathfrak{Q}$ abhängt, falls $\mathfrak{H}$ die Voraussetzungen für verschiedene Wahlen von $\mathfrak{Q}$ erfüllt.

(2B) ***Ist $\tilde{B}$ ein Block einer beliebigen Untergruppe $\tilde{\mathfrak{G}}$ von $\mathfrak{G}$ und ist $\tilde{B}^{\mathfrak{G}} = B$ definiert, so kann die Defektgruppe $\mathfrak{D}$ von B so gewählt werden, daß sie eine gegebene Defektgruppe $\tilde{\mathfrak{D}}$ von $\tilde{B}$ enthält.***

Dies kann in derselben Weise wie I (8D) bewiesen werden.

Aus (2.2) ergibt sich:

(2C) *Es seien $\mathfrak{A}$ und $\tilde{\mathfrak{G}}$ zwei Untergruppen von $\mathfrak{G}$ mit $\mathfrak{A} \geqq \tilde{\mathfrak{G}}$ und es sei $\tilde{B}$ ein Block von $\tilde{\mathfrak{G}}$. Sind $\tilde{B}^{\mathfrak{A}}$ und $\tilde{B}^{\mathfrak{G}}$ definiert, so ist $(\tilde{B}^{\mathfrak{A}})^{\mathfrak{G}}$ definiert und $(\tilde{B}^{\mathfrak{A}})^{\mathfrak{G}} = \tilde{B}^{\mathfrak{G}}$. Sind $\tilde{B}^{\mathfrak{A}}$ und $(\tilde{B}^{\mathfrak{A}})^{\mathfrak{G}}$ definiert, so ist $\tilde{B}^{\mathfrak{G}}$ definiert.*

Weiterhin zeigen wir:

(2D) *Es sei B ein Block von $\mathfrak{G}$ mit der Defektgruppe $\mathfrak{D}$. Im Sinne von* I (12A) *werde B die Klasse $\{\vartheta\}$ von in $\mathfrak{N}(\mathfrak{D})$ assoziierten Charakteren von $\mathfrak{D}\mathfrak{C}(\mathfrak{D})$ zugeordnet. Liegt ϑ im Block $\tilde{B}$ von $\mathfrak{D}\mathfrak{C}(\mathfrak{D})$, so ist $B = \tilde{B}^{\mathfrak{G}}$. Auch $\tilde{B}$ hat die Defektgruppe $\mathfrak{D}$.*

Beweis. Es sei $\chi_i \in B$. Aus I (12.2) ergibt sich, daß der zu χ_i gehörige lineare Charakter ω_i von $Z(\mathfrak{G}, \Omega)$ die Kongruenz

$$\omega_i(G) \equiv \frac{1}{\vartheta(1)} \sum_T \vartheta(T) \pmod{\mathfrak{p}_\nu}$$

erfüllt, wo T die Elemente der Klasse $\mathfrak{K}_\alpha$ von G durchläuft, die in $\mathfrak{C}(\mathfrak{D})$ liegen. Ist also $\tilde{\omega}$ der zu ϑ gehörige lineare Charakter von $Z(\mathfrak{D}\mathfrak{C}(\mathfrak{D}), \Omega)$, so ist

$$\omega_i(K_\alpha) \equiv \sum_{\tilde{\mathfrak{K}}_\beta \subseteq \mathfrak{K}_\alpha} \tilde{\omega}(\tilde{K}_\beta) \pmod{\mathfrak{p}_\nu}, \tag{2.3}$$

wo hier die Bezeichnung $\tilde{\mathfrak{K}}_\beta$ für die Klassen von $\tilde{\mathfrak{G}} = \mathfrak{D}\mathfrak{C}(\mathfrak{D})$ verwendet ist. Entspricht B dem linearen Charakter ψ von $Z(\mathfrak{G}, \Omega^*)$ und $\tilde{B}$ dem linearen Charakter $\tilde{\psi}$ von $Z(\tilde{\mathfrak{G}}, \Omega^*)$, so führt die Restklassenabbildung ω_i in ψ und $\tilde{\omega}$ in $\tilde{\psi}$ über. Jetzt zeigt (2.3), daß $\psi = \tilde{\psi}^{\mathfrak{G}}$, also $B = \tilde{B}^{\mathfrak{G}}$ ist. Da $\mathfrak{D} \triangleleft \mathfrak{D}\mathfrak{C}(\mathfrak{D})$ ist, zeigt I (9F), daß $\mathfrak{D}$ in der Defektgruppe $\tilde{\mathfrak{D}}$ von $\tilde{B}$ enthalten ist. Dann folgt aus (2B), daß $\mathfrak{D} = \tilde{\mathfrak{D}}$ ist.

Jetzt können wir (2A) folgendermaßen verallgemeinern:

(2E) *Es sei $\tilde{\mathfrak{G}}$ eine Untergruppe von $\mathfrak{G}$ und $\tilde{B}$ ein Block von $\tilde{\mathfrak{G}}$ mit der Defektgruppe $\tilde{\mathfrak{D}}$. Enthält $\tilde{\mathfrak{G}}$ den Zentralisator $\mathfrak{C}(\tilde{\mathfrak{D}})$ von $\tilde{\mathfrak{D}}$ in $\mathfrak{G}$, so ist $\tilde{B}^{\mathfrak{G}}$ definiert.*

Beweis. Aus der Voraussetzung folgt, daß $\mathfrak{C}(\tilde{\mathfrak{D}})$ auch der Zentralisator von $\tilde{\mathfrak{D}}$ in $\tilde{\mathfrak{G}}$ ist. Wendet man (2D) auf $\tilde{\mathfrak{G}}$ und $\tilde{B}$ an, so folgt, daß es einen Block $B^{\#}$ von $\tilde{\mathfrak{D}}\mathfrak{C}(\tilde{\mathfrak{D}})$ gibt, für den $\tilde{B} = (B^{\#})^{\tilde{\mathfrak{G}}}$ gilt. Andererseits ist $(B^{\#})^{\mathfrak{G}} = B$ nach (2A) definiert. Aus (2C) folgt, daß $\tilde{B}^{\mathfrak{G}}$ definiert und gleich B ist.

Sind $\tilde{B}$ und B wie in (2B) gewählt, so ergibt sich für die Defekte $\tilde{d}$ und d die Ungleichung

$$\tilde{d} \leqq d. \tag{2.4}$$

Wir behandeln einen Fall, in dem (2.4) verschärft werden kann:

(2F) *Es sei* $\mathfrak{Q}$ *eine* p*-Untergruppe von* $\mathfrak{G}$, *und es sei* $\tilde{\mathfrak{G}} = \mathfrak{Q}\mathfrak{C}(\mathfrak{Q})$. *Es sei* $\tilde{\chi}$ *ein irreduzibler Charakter von* $\tilde{\mathfrak{G}}$ *und* $\mathfrak{S}$ *seine Trägheitsgruppe in* $\mathfrak{N}(\mathfrak{Q})$, *und es sei* $\varrho = \nu(\mathfrak{S}:\tilde{\mathfrak{G}})$. *Gehört* $\tilde{\chi}$ *zu dem Block* $\tilde{B}$ *von* $\tilde{\mathfrak{G}}$ *mit der Defektgruppe* $\tilde{\mathfrak{D}}$, *so ist* $B = \tilde{B}^{\mathfrak{G}}$ *definiert, und seine Defektgruppe* $\mathfrak{D}$ *kann so gewählt werden, daß* $\mathfrak{D} \geqq \tilde{\mathfrak{D}}$ *und* $(\mathfrak{D} \cap \mathfrak{S} : \mathfrak{D} \cap \tilde{\mathfrak{G}}) = p^{\varrho}$ *ist. Hat* B *den Defekt* d *und* $\tilde{B}$ *den Defekt* $\tilde{d}$, *so ist*

$$d \geqq \tilde{d} + \varrho. \tag{2.5}$$

Beweis. Da $\mathfrak{Q} \lhd \tilde{\mathfrak{G}}$ ist, ist $\mathfrak{Q} \subseteqq \tilde{\mathfrak{D}}$, also $\mathfrak{C}(\tilde{\mathfrak{D}}) \subseteqq \mathfrak{C}(\mathfrak{Q}) \subseteqq \tilde{\mathfrak{G}}$. Daher ist $\tilde{B}^{\mathfrak{G}} = B$ nach (2E) definiert.

Nach dem Sylowschen Satz gibt es eine Untergruppe $\mathfrak{A} \geqq \tilde{\mathfrak{G}}$ von $\mathfrak{S}$ mit $(\mathfrak{A}:\tilde{\mathfrak{G}}) = p^{\varrho}$. Wir nehmen zunächst an, daß $\mathfrak{G} = \mathfrak{A}$ ist. Ist χ ein irreduzibler Charakter in B, ist ω der zu χ gehörige lineare Charakter von $Z(\mathfrak{G}, \Omega)$ und $\tilde{\omega}$ der zu $\tilde{\chi}$ gehörige lineare Charakter von $Z(\tilde{\mathfrak{G}}, \Omega)$, so gilt für alle α

$$\omega(K_\alpha) \equiv \sum_{\tilde{\mathfrak{K}}_\beta \subseteqq \mathfrak{K}_\alpha} \tilde{\omega}(\tilde{K}_\beta) \pmod{\mathfrak{p}_\nu}. \tag{2.6}$$

Ist $X \in \mathfrak{G}$ und ist $\tilde{\mathfrak{K}}_\beta \subseteqq \mathfrak{K}_\alpha$ eine Klasse von $\tilde{\mathfrak{G}} \lhd \mathfrak{A} = \mathfrak{G}$, so ist $X^{-1}\tilde{\mathfrak{K}}_\beta X$ wieder eine Klasse $\tilde{\mathfrak{K}}_\beta^X \subseteqq \mathfrak{K}_\alpha$ von $\tilde{\mathfrak{G}}$. Da jetzt $\mathfrak{G}$ die Trägheitsgruppe von $\tilde{\chi}$ ist, gilt $\tilde{\omega}(\tilde{K}_\beta) = \tilde{\omega}(\tilde{K}_\beta^X)$. Die Elemente X mit $\tilde{\mathfrak{K}}_\beta^X = \tilde{\mathfrak{K}}_\beta$ bilden eine Untergruppe $\mathfrak{N}_\beta \geqq \tilde{\mathfrak{G}}$ von $\mathfrak{G}$. Dann ist die Anzahl der verschiedenen $\tilde{\mathfrak{K}}_\beta^X$ gleich $(\mathfrak{G}:\mathfrak{N}_\beta)$, also ein Teiler von $(\mathfrak{G}:\tilde{\mathfrak{G}}) = p^{\varrho}$ und daher eine Potenz von p. Es genügt also, $\tilde{\mathfrak{K}}_\beta$ in (2.6) über diejenigen Klassen laufen zu lassen, für die $\mathfrak{N}_\beta = \mathfrak{G}$ ist. Entweder gibt es kein derartiges $\tilde{\mathfrak{K}}_\beta$ und dann ist $\omega(K_\alpha) \equiv 0 \pmod{\mathfrak{p}_\nu}$, oder $\mathfrak{K}_\alpha$ stimmt mit einem derartigen $\tilde{\mathfrak{K}}_\beta$ überein und wir haben

$$\omega(K_\alpha) \equiv \tilde{\omega}(\tilde{K}_\beta) \pmod{\mathfrak{p}_\nu}. \tag{2.7}$$

Aus der Definition der Defektgruppe $\mathfrak{D}$ von B ergibt sich, daß es Klassen $\mathfrak{K}_\alpha$ gibt, für die $\omega(K_\alpha) \not\equiv 0 \pmod{p_\nu}$ ist, und für die $\mathfrak{D}$ p-Sylowuntergruppe des Normalisators $\mathfrak{N}(Y)$ von Elementen Y von $\mathfrak{K}_\alpha$ ist. Dann muß also $\mathfrak{K}_\alpha$ eine Klasse $\tilde{\mathfrak{K}}_\beta$ von $\tilde{\mathfrak{G}}$ sein. Insbesondere ist $Y \in \tilde{\mathfrak{G}}$. Sind $n(Y)$ und $\tilde{n}(Y)$ die Ordnungen der Normalisatoren $\mathfrak{N}(Y)$ und $\tilde{\mathfrak{N}}(Y) = \mathfrak{N}(Y) \cap \tilde{\mathfrak{G}}$ von Y in $\mathfrak{G}$ und in $\tilde{\mathfrak{G}}$, so folgt aus $\mathfrak{K}_\alpha = \tilde{\mathfrak{K}}_\beta$, daß $g/n(Y) = \tilde{g}/\tilde{n}(Y)$ ist. Wegen $(\mathfrak{G}:\tilde{\mathfrak{G}}) = (\mathfrak{A}:\tilde{\mathfrak{G}}) = p^{\varrho}$ ist also

$$n(Y) = p^{\varrho}\tilde{n}(Y). \tag{2.8}$$

Nach (2.7) ist $\tilde{\omega}(\tilde{K}_\beta) \not\equiv 0 \pmod{\mathfrak{p}_\nu}$. Daher enthält $\tilde{\mathfrak{N}}(Y)$ nach I (8A) eine Defektgruppe $\tilde{\mathfrak{D}}$ von $\tilde{B}$. Ersetzt man nötigenfalls $\mathfrak{D}$ durch eine Konjugierte, so kann man annehmen, daß $\tilde{\mathfrak{D}}$ zu der Sylowgruppe $\mathfrak{D}$ von $\mathfrak{N}(Y)$ gehört.

Dann zeigt (2.8), daß $(\mathfrak{D}:\tilde{\mathfrak{D}}) \geqq p^{\varrho}$ ist, daß also für die Defekte d und $\tilde{d}$ die Ungleichung $d \geqq \tilde{d} + \varrho$ gilt. Aus (2.8) folgt leicht, daß $(\mathfrak{D}:\mathfrak{D} \frown \tilde{\mathfrak{G}}) = p^{\varrho}$ ist. Da $\mathfrak{D}\tilde{\mathfrak{G}}/\tilde{\mathfrak{G}} \cong \mathfrak{D}/\mathfrak{D} \frown \tilde{\mathfrak{G}}$ ist, haben wir $\mathfrak{G} = \mathfrak{D}\tilde{\mathfrak{G}}$.

Kehren wir zu dem Fall eines beliebigen $\mathfrak{G} \geqq \mathfrak{A}$ zurück! Da $\tilde{\mathfrak{G}} \geqq \mathfrak{C}(\tilde{\mathfrak{D}})$ war, enthält $\tilde{\mathfrak{G}}$ den Zentralisator $\mathfrak{C}(\tilde{\mathfrak{D}}) \frown \mathfrak{A}$ von $\tilde{\mathfrak{D}}$ in $\mathfrak{A}$ und nach (2E) ist der Block $\hat{B} = \tilde{B}^{\mathfrak{A}}$ von $\mathfrak{A}$ definiert. Wendet man die vorangehenden Betrachtungen auf $\mathfrak{A}$ anstelle von $\mathfrak{G}$ an, so zeigt sich, daß die Defektgruppe $\hat{\mathfrak{D}}$ von $\hat{B}$ so gewählt werden kann, daß sie $\tilde{\mathfrak{D}}$ enthält und daß $(\hat{\mathfrak{D}}:\tilde{\mathfrak{D}}) \geqq p^{\varrho}$ und $\mathfrak{A} = \hat{\mathfrak{D}}\tilde{\mathfrak{G}}$ ist.

Nach (2E) und (2C) ist $\hat{B}^{\mathfrak{G}}$ definiert und $\tilde{B}^{\mathfrak{G}} = \hat{B}^{\mathfrak{G}}$. Wegen (2B) kann die Defektgruppe $\mathfrak{D}$ von B so gewählt werden, daß $\mathfrak{D} \geqq \hat{\mathfrak{D}}$ ist. Daher ist $(\mathfrak{D}:\tilde{\mathfrak{D}}) \geqq p^{\varrho}$. Ferner haben wir $\hat{\mathfrak{D}} \leqq \mathfrak{D} \frown \mathfrak{S}$, also $((\mathfrak{D} \frown \mathfrak{S})\tilde{\mathfrak{G}}:\tilde{\mathfrak{G}}) \geqq (\hat{\mathfrak{D}}\tilde{\mathfrak{G}}:\tilde{\mathfrak{G}}) = (\mathfrak{A}:\tilde{\mathfrak{G}}) = p^{\varrho}$. Andererseits ist

$$((\mathfrak{D} \frown \mathfrak{S})\tilde{\mathfrak{G}} : \tilde{\mathfrak{G}}) = (\mathfrak{D} \frown \mathfrak{S} : \mathfrak{D} \frown \mathfrak{S} \frown \tilde{\mathfrak{G}}) = (\mathfrak{D} \frown \mathfrak{S} : \mathfrak{D} \frown \tilde{\mathfrak{G}}).$$

Dieser Index ist eine Potenz von p, die $(\mathfrak{S}:\tilde{\mathfrak{G}})$ teilt und daher höchstens gleich p^{ϱ} ist. Also ist $(\mathfrak{D} \frown \mathfrak{S}:\mathfrak{D} \frown \tilde{\mathfrak{G}}) = p^{\varrho}$. Damit ist (2F) bewiesen.

Wir behandeln noch den Zusammenhang zwischen den Blöcken von $\mathfrak{Q}\mathfrak{C}(\mathfrak{Q})$ und denen von $\mathfrak{Q}\mathfrak{C}(\mathfrak{Q})/\mathfrak{Q}$.

(2G) *Ist $\mathfrak{Q}$ eine p-Untergruppe von $\mathfrak{G}$, ist $\tilde{\mathfrak{G}} = \mathfrak{Q}\mathfrak{C}(\mathfrak{Q})$ und $\bar{\mathfrak{G}} = \tilde{\mathfrak{G}}/\mathfrak{Q}$, so hat man eine eineindeutige Zuordnung zwischen den Blöcken $\tilde{B}$ von $\tilde{\mathfrak{G}}$ und den Blöcken $\bar{B}$ von $\bar{\mathfrak{G}}$. Ist dabei $\bar{\chi}$ ein irreduzibler Charakter von $\bar{\mathfrak{G}}$ in $\bar{B}$, so gehört $\bar{\chi}$ als Charakter von $\tilde{\mathfrak{G}}$ betrachtet zu dem zugeordneten Block $\tilde{B}$. Ist $\tilde{\mathfrak{D}}$ die Defektgruppe von $\tilde{B}$, so kann die Defektgruppe von $\bar{B}$ als $\bar{\mathfrak{D}} = \tilde{\mathfrak{D}}/\mathfrak{Q}$ gewählt werden.*

Beweis. Aus I § 9 ergibt sich, daß jedem Block $\bar{B}$ von $\bar{\mathfrak{G}}$ ein Block $\tilde{B}$ von $\tilde{\mathfrak{G}}$ entspricht, derart daß wenn $\bar{\chi}$ ein irreduzibler Charakter von $\bar{\mathfrak{G}}$ in $\bar{B}$ ist, $\bar{\chi}$ als Charakter von $\tilde{\mathfrak{G}}$ in $\tilde{B}$ liegt. Jeder Block $\tilde{B}$ ist dabei mindestens einem $\bar{B}$ zugeordnet. Ist $(\mathfrak{Q}:1) = p^{\mu}$ und hat $\tilde{B}$ den Defekt $\tilde{d}$, so kann man $\bar{B}$ als Block vom Defekt $\tilde{d} - \mu$ wählen.

Es seien $\tilde{\chi}$ und $\bar{\chi}$ zwei Charaktere von $\tilde{\mathfrak{G}}$ und $\bar{\mathfrak{G}}$, die in der genannten Weise zusammenhängen; es seien $\tilde{\omega}$ und $\bar{\omega}$ die zugehörigen linearen Charaktere von $Z(\tilde{\mathfrak{G}}, \Omega)$ bzw. $Z(\bar{\mathfrak{G}}, \Omega)$. Durchläuft $\tilde{R}$ die p-regulären Elemente von $\tilde{\mathfrak{G}}$ und bezeichnet $\bar{R} = \tilde{R}\mathfrak{Q}$ das entsprechende Element von $\bar{\mathfrak{G}} = \tilde{\mathfrak{G}}/\mathfrak{Q}$, so folgt aus I (11A), daß $\bar{R}$ die p-regulären Elemente von $\bar{\mathfrak{G}}$ durchläuft, und daß der Normalisator von $\bar{R}$ in $\bar{G}$ gerade $\tilde{\mathfrak{N}}(\tilde{R})/\mathfrak{Q}$ ist (wo $\tilde{\mathfrak{N}}(\tilde{R})$ der Normalisator von $\tilde{R}$ in $\tilde{\mathfrak{G}}$ ist). Daraus ergibt sich

$$\tilde{\omega}(\tilde{R}) = \bar{\omega}(\bar{R}).$$

Jetzt zeigt I (6F), daß ein gegebener Block $\tilde{B}$ von $\tilde{G}$ nur als Bild eines einzigen Blocks $\bar{B}$ von $\bar{\mathfrak{G}}$ auftritt. Wählt man, was möglich ist, $\tilde{R}$ so, daß $\tilde{\omega}(\tilde{R}) \not\equiv 0$ (mod $\mathfrak{p}_{\nu}$) und daß die p-Sylowgruppe $\tilde{\mathfrak{D}}$ von $\tilde{\mathfrak{N}}(\tilde{R})$ die Ordnung $p^{\tilde{d}}$ hat, so ist

$\tilde{\mathfrak{D}}$ eine Defektgruppe von $\tilde{B}$. Da dann $\tilde{\mathfrak{D}}/\mathfrak{Q} = \mathfrak{D}$ eine p-Sylowgruppe von $\tilde{\mathfrak{N}}(\tilde{R})/\mathfrak{Q}$ ist, ist $\mathfrak{D}$ von der Ordnung $p^{d-\mu}$ eine Defektgruppe von $\overline{B}$.

Bemerkung. Soll in (2F) der Fall $\mathfrak{D} = \mathfrak{Q}$ eintreten, d.h. $d = \mu$ sein, so ergibt sich aus (2.5) und (2G), daß $\nu(\mathfrak{S}:\tilde{\mathfrak{G}}) = 0$ sein muß, und daß das zu $\tilde{B}$ gehörige $\overline{B}$ den Defekt 0 haben muß. Dies zeigt erneut die Notwendigkeit der in I (12A) auftretenden Bedingungen.

§ 3. Verallgemeinerte Zerlegungszahlen[3])

Unter einem *p-Element* P von $\mathfrak{G}$ wollen wir ein Element verstehen, dessen Ordnung eine Potenz von p ist. Ein beliebiges Element G von $\mathfrak{G}$ läßt sich dann eindeutig als Produkt

$$G = PR \tag{3.1}$$

eines p-Elements P mit einem p-regulären Element R des Normalisators $\mathfrak{N}(P)$ von P schreiben. Dabei sind P und R geeignete Potenzen von G. Wir nennen P den *p-Faktor* von G und R den *p-regulären Faktor* von G.

Es seien $\varphi_1^P, \varphi_2^P, \ldots, \varphi_{l(P)}^P$ die irreduziblen modularen Charaktere von $\mathfrak{N}(P)$, also $l(P)$ die Anzahl der p-regulären Klassen von $\mathfrak{N}(P)$. Ist $\tilde{\chi}_\lambda$ ein irreduzibler Charakter von $\mathfrak{N}(P)$, so ist $\tilde{\omega}_\lambda(P) = \varepsilon_\lambda$ eine p^α-te Einheitswurzel, wo p^α die Ordnung von P bezeichnet. Weiter ist $\tilde{\chi}_\lambda(PR) = \tilde{\omega}_\lambda(P)\,\tilde{\chi}_\lambda(R) = \varepsilon_\lambda \tilde{\chi}_\lambda(R)$, und hier läßt sich $\tilde{\chi}_\lambda(R)$ nach (1.1) durch die $\varphi_j^P(R)$ ausdrücken.

Da die Beschränkung $\chi_i | \mathfrak{N}(P)$ sich aus den irreduziblen Charakteren $\tilde{\chi}_\lambda$ von $\mathfrak{N}(P)$ zusammensetzt, so hat man für p-reguläres $R \in \mathfrak{N}(P)$ Formeln

$$\chi_i(PR) = \sum_j d_{ij}^P \varphi_j^P(R), \tag{3.2}$$

wo die d_{ij}^P ganz algebraisch im Körper der p^α-ten Einheitswurzeln sind und nicht von R abhängen. Wegen der linearen Unabhängigkeit von $\varphi_1^P, \varphi_2^P, \ldots$ sind die d_{ij}^P eindeutig bestimmt; wir nennen sie die zu P gehörigen *verallgemeinerten Zerlegungszahlen* von $\mathfrak{G}$. Für $P = 1$ kommt man auf die Formeln (1.1), also auf die gewöhnlichen Zerlegungszahlen von $\mathfrak{G}$ zurück.

(3A) *Ist c_{ij}^P die zu φ_i^P, φ_j^P gehörige Cartansche Invariante von $\mathfrak{N}(P)$, so genügen die verallgemeinerten Zerlegungszahlen den Orthogonalitätsrelationen*

$$\sum_{\mu=1}^{k} \bar{d}_{\mu i}^P d_{\mu j}^P = c_{ij}^P. \tag{3.3}$$

Ist ferner Q ein nicht zu P konjugiertes p-Element von $\mathfrak{G}$, so ist

$$\sum_{\mu=1}^{k} \bar{d}_{\mu i}^P d_{\mu j}^Q = 0. \tag{3.4}$$

Beweis. Es sei $R_1^P, R_2^P, \ldots, R_{l(P)}^P$ ein Vertretersystem für die p-regulären Klassen von $\mathfrak{N}(P)$. Setzt man

$$\boldsymbol{\chi}^P = \left(\chi_i(PR_j^P)\right), \quad D^P = (d_{ij}^P), \quad \boldsymbol{\varphi}^P = \left(\varphi_i^P(R_j^P)\right), \quad C^P = (c_{ij}^P),$$

[3]) Zu den Ergebnissen von § 3 vgl. R. Brauer, On the connection between the ordinary and the modular characters of groups of finite order. Ann. of Math. **42**, 926–935 (1941).

so zeigt (3.2), daß $\boldsymbol{\chi}^P = D^P \boldsymbol{\varphi}^P$ ist, also

$$(3.5)\qquad (\bar{\boldsymbol{\chi}}^P)' \boldsymbol{\chi}^Q = (\bar{\boldsymbol{\varphi}}^P)'(\bar{D}^P)' D^Q \boldsymbol{\varphi}^Q.$$

Für $P = Q$ ist nach den Orthogonalitätsrelationen für Gruppencharaktere die linke Seite die Diagonalmatrix $N^P = \big(n(PR_i^P)\delta_{ij}\big)$, also wegen (1.2) (für $\mathfrak{N}(P)$ anstelle von $\mathfrak{G}$) gleich $(\bar{\boldsymbol{\varphi}}^P)' C^P \boldsymbol{\varphi}^P$. Da $\boldsymbol{\varphi}^P$ nicht singulär ist, folgt $C^P = (\bar{D}^P)' D^P$ was mit (3.3) gleichwertig ist. Sind P und Q nicht in $\mathfrak{G}$ konjugiert, so sind PR_α^P und QR_β^Q nicht in $\mathfrak{G}$ konjugiert. Dann verschwindet die linke Seite von (3.5), was $(\bar{D}^P)' D^Q = 0$ und damit (3.4) ergibt.

Aus $(\bar{\boldsymbol{\varphi}}^P)' C^P \boldsymbol{\varphi}^P = N^P$ folgt auch noch

$$\boldsymbol{\varphi}^P (N^P)^{-1} (\bar{\boldsymbol{\varphi}}^P)' C^P = E,$$

also wegen $\boldsymbol{\chi}^P = D^P \boldsymbol{\varphi}^P$ die Gleichung

$$\boldsymbol{\chi}^P (N^P)^{-1} (\bar{\boldsymbol{\varphi}}^P)' C^P = D^P.$$

Explizit geschrieben gibt dies

$$(3.6)\qquad d_{ij}^P = \sum_{\alpha,\beta=1}^{l(P)} \chi_i(PR_\alpha^P)\, n(PR_\alpha^P)^{-1} \bar{\varphi}_\beta^P(R_\alpha^P)\, c_{\beta j}^P.$$

Es sei $g = p^a g_0$ mit ganzem, zu p teilerfremdem g_0, $a = \nu(g)$. Es bezeichne Ω_0 den Körper der g_0-ten und Ω_1 den Körper der p^a-ten Einheitswurzeln. Ist σ ein Element der Galoisschen Gruppe von Ω über Ω_0 und ist χ_i ein irreduzibler Charakter von $\mathfrak{G}$, so ist auch die algebraisch Konjugierte χ_i^σ ein irreduzibler Charakter χ_{i^*} von $\mathfrak{G}$. Da die modularen Charaktere von $\mathfrak{N}(P)$ Werte in Ω_0 haben, bleiben sie bei der Anwendung von σ fest. Daher folgt aus (3.2), daß

$$\chi_{i^*}(PR) = \big(\chi_i(PR)\big)^\sigma = \sum_j (d_{ij}^P)^\sigma \varphi_j^P(R)$$

ist; man hat also

$$(3.7)\qquad d_{i^*j}^P = (d_{ij}^P)^\sigma \quad \text{für} \quad \chi_{i^*} = \chi_i^\sigma.$$

Wir sagen, daß χ_i und $(\chi_i)^\sigma$ *p-konjugierte Charaktere* sind. Für $P = 1$ sieht man, daß sie dieselben modularen Bestandteile haben. Also gehören p-konjugierte Charaktere stets zum selben Block (z.B. nach I(4D)).

Unter der *Sektion* $\mathfrak{S}(P)$ eines p-Elements P wollen wir die Gesamtheit aller $G \in \mathfrak{G}$ verstehen, deren p-Faktor zu P in $\mathfrak{G}$ konjugiert ist. Offenbar ist $\mathfrak{S}(P)$ die Vereinigungsmenge der Klassen $\mathfrak{K}_\alpha$, die von den Elementen PR_1^P, $PR_2^P, \ldots, PR_{l(P)}^P$ repräsentiert werden; das sind genau $l(P)$ verschiedene Klassen. Ist $P_1, P_2, \ldots, P_r$ ein Vertretersystem für die p-Klassen von $\mathfrak{G}$, so gehört jedes $G \in \mathfrak{G}$ zu genau einer der Sektionen $\mathfrak{S}(P_1), \mathfrak{S}(P_2), \ldots, \mathfrak{S}(P_r)$. Daher ergibt sich für Klassenzahl von $\mathfrak{G}$ die Formel

$$(3.8)\qquad k = \sum_{j=1}^{r} l(P_j).$$

Da Charaktere χ_i für konjugierte Elemente denselben Wert haben, genügt es, in (3.2) P das System $P_1, P_2, \ldots, P_r$ und bei festem P das Element R das System $R_1^P, R_2^P, \ldots, R_{l(P)}^P$ durchlaufen zu lassen.

Wir wollen noch sehen, wie sich d_{ij}^{P} ändert, wenn man P durch ein konjugiertes p-Element $Q=G^{-1}PG$ (mit $G\in\mathfrak{G}$) ersetzt. Dann ist $\mathfrak{N}(Q)=G^{-1}\mathfrak{N}(P)G$. Durchläuft R die p-regulären Elemente von $\mathfrak{N}(P)$, so durchläuft $R^G=G^{-1}RG$ die p-regulären Elemente von $\mathfrak{N}(Q)$. Setzt man

$$\varphi_i^Q(R^G)=\varphi_i^P(R), \tag{3.9}$$

so sind offenbar $\varphi_{1'}^Q, \varphi_{2'}^Q, \ldots, \varphi_{l(P)'}^Q$ die irreduziblen modularen Charaktere von $\mathfrak{N}(Q)$. Aus

$$\chi_i(QR^G)=\chi_i(G^{-1}PRG)=\chi_i(PR)$$

ergibt sich mit Hilfe von (3.2) und (3.9), daß

$$\chi_i(QR^G)=\sum_j d_{ij}^P\varphi_j^Q(R^G)$$

ist, daß wir also

$$d_{ij'}^Q=d_{ij}^P \tag{3.10}$$

haben.

§ 4. Ganzwertige lineare Funktionen in $Z(\mathfrak{G},\Omega)$

Wir wollen ein Element $\Sigma a_i K_i$ von $Z=Z(\mathfrak{G},\Omega)$ *ganz* nennen, wenn die Koeffizienten a_i zu $\mathfrak{o}_\nu$ gehören. Diese ganzen Elemente bilden einen Teilring $J=J(\mathfrak{G})$ von Z. Eine lineare Funktion f in $Z(\mathfrak{G},\Omega)$ (vgl. § 2) soll *ganzwertig* genannt werden, wenn $f(\alpha)\in\mathfrak{o}_\nu$ für alle $\alpha\in J$ gilt. Diese ganzwertigen linearen Funktionen bilden einen $\mathfrak{o}_\nu$-Modul $\mathfrak{M}$ vom Rang k.

Wir können $Z(\mathfrak{G},\Omega^*)$ mit $J/\mathfrak{p}_\nu J$ identifizieren. Ist $\zeta\in J$, so wird das durch Restklassenbildung erhaltene Element von $Z(\mathfrak{G},\Omega^*)$ durchgehend mit ζ^* bezeichnet werden.

Es seien $B_1, B_2, \ldots, B_t$ die p-Blöcke von $\mathfrak{G}$. Ist $G_i\in\mathfrak{K}_i$ und setzt man nach Osima[4])

$$\eta_\tau=\sum_{j=1}^{k} b_j^{(\tau)}K_j \tag{4.1}$$

mit

$$b_j^{(\tau)}=\frac{1}{g}\sum_{\chi_\mu\in B_\tau}\chi_\mu(1)\bar{\chi}_\mu(G_j), \tag{4.2}$$

so sind $\eta_1, \eta_2, \ldots, \eta_t$ ein System orthogonaler Idempotente von $Z(\mathfrak{G},\Omega)$ mit der Summe 1. Ist ω_i der zu χ_i gehörige lineare Charakter von $Z(\mathfrak{G},\Omega)$, so folgt nämlich aus (4.1) und (4.2) und den Orthogonalitätsrelationen für Gruppencharaktere, daß

$$\omega_i(\eta_\tau)=\begin{cases}1 & \text{für } \chi_i\in B_\tau\\ 0 & \text{für } \chi_i\notin B_\tau\end{cases} \tag{4.3}$$

ist, woraus sich leicht die Behauptungen ergeben.

[4]) Osima, Masaru: Notes on blocks of group characters. Math. J. Okayama Univ. 4, 175—188 (1955), p. 178.

Ferner gilt nach I (3.5) für die durch I (3.9) erklärten Charaktere Φ_i die Gleichung

$$b_j^{(\tau)} = \frac{1}{g} \sum_{\varphi_i \in B_\tau} \varphi_i(1)\,\bar{\Phi}_i(G_j).$$

Aus I (3 F) folgt, daß $b_j^{(\tau)} = 0$ für p-singuläre G_j ist. Für p-reguläre G_j hat man nach (4.2), I (3.5) und I (3.9)

$$b_j^{(\tau)} = \frac{1}{g} \sum_{\varphi_i \in B_\tau} \Phi_i(1)\,\bar{\varphi}_i(G_j). \tag{4.4}$$

Aus I (3 F) folgt, daß $b_j^{(\tau)} \in \mathfrak{o}_\nu$, daß also $\eta_\tau \in J$ ist. Aus (4.3) folgt auch noch, daß η_τ^* das in I § 4 auftretende Idempotent von $Z(\mathfrak{G}, \Omega^*)$ ist.

Wir wollen sagen, daß eine ganzwertige lineare Funktion f mit dem Block B_τ *assoziiert* ist, falls

$$f(\zeta \eta_\sigma) \equiv 0 \pmod{\mathfrak{p}_\nu} \tag{4.5}$$

für alle $\zeta \in J$ und alle $\sigma \neq \tau$ gilt. Da ω_i ein linearer Charakter von Z ist und $\omega_i(\zeta) \in \mathfrak{o}_\nu$ ist, folgt aus (4.3), daß für $\chi_i \in B_\tau$ die ganzwertige lineare Funktion ω_i mit B_τ assoziiert ist.

Man setze

$$\mathfrak{M}_\tau = \mathfrak{M} \frown \sum_{\chi_i \in B_\tau} \Omega\,\omega_i. \tag{4.6}$$

Offenbar ist $\mathfrak{M}_\tau$ ein $\mathfrak{o}_\nu$-Modul, dessen Rang gleich der Anzahl k_τ der $\chi_i \in B_\tau$ ist.

(4 A) *Sind $f_1^{(\tau)}, f_2^{(\tau)}, \ldots, f_{k_\tau}^{(\tau)}$ eine $\mathfrak{o}_\nu$-Basis von $\mathfrak{M}_\tau$ $(1 \leqq \tau \leqq t)$, so bilden die Funktionen $f_j^{(\tau)}$ für $\tau = 1, 2, \ldots, t$; $j = 1, 2, \ldots, k_\tau$ eine $\mathfrak{o}_\nu$-Basis von $\mathfrak{M}$. Dann und nur dann ist die lineare Funktion*

$$f = \sum_\tau \sum_j a_j^{(\tau)} f_j^{(\tau)} \qquad (a_j^{(\tau)} \in \mathfrak{o}_\nu) \tag{4.7}$$

mit B_σ assoziiert, wenn $a_j^{(\tau)} \equiv 0 \pmod{\mathfrak{p}_\nu}$ für alle $\tau \neq \sigma$ und $j = 1, 2, \ldots, k_\tau$ ist.

Beweis. Da wir im ganzen k Charaktere χ_i haben, ist

$$k = \sum_{\tau=1}^{t} k_\tau. \tag{4.8}$$

Also besteht das System $f_j^{(\tau)}$ $(1 \leqq \tau \leqq t,\ 1 \leqq j \leqq k_\tau)$ aus k Funktionen. Um zu zeigen, daß es eine $\mathfrak{o}_\nu$-Basis von $\mathfrak{M}$ bildet, genügt es nachzuweisen, daß wenn ein Ausdruck (4.7) zu $\mathfrak{p}_\nu \mathfrak{M}$ gehört, alle Koeffizienten $a_j^{(\tau)}$ in $\mathfrak{p}_\nu$ liegen. Man setze

$$f_\tau = \sum_j a_j^{(\tau)} f_j^{(\tau)}, \tag{4.9}$$

also

$$f = \sum_\tau f_\tau. \tag{4.10}$$

Aus (4.6) und (4.3) folgt, daß dann

$$f_\varrho(\eta_\lambda \zeta) = 0 \quad \text{für} \quad \varrho \neq \lambda \text{ und } \zeta \in J \tag{4.11}$$

gilt. Also ist

$$f(\eta_\tau \zeta) = f_\tau(\eta_\tau \zeta). \tag{4.12}$$

Gehört nun f zu $\mathfrak{p}_\nu \mathfrak{M}$ und ist π ein Primelement von $\mathfrak{p}_\nu$, so ist $f(\eta_\tau \zeta)/\pi \in \mathfrak{o}_\nu$. Jetzt zeigen (4.11) und (4.12), daß $f_\tau/\pi \in \mathfrak{M}$ ist, nach (4.6) ist also $f_\tau/\pi \in \mathfrak{M}_\tau$. Da $f_1^{(\tau)}, f_2^{(\tau)}, \ldots$ eine $\mathfrak{o}_\nu$-Basis von $\mathfrak{M}_\tau$ bilden, ist nach (4.9) $a_j^{(\tau)} \in \mathfrak{p}_\nu$ für alle j. Da dies für alle τ gilt, bilden die k Funktionen $f_j^{(\tau)}$ (für alle τ und j) in der Tat eine $\mathfrak{o}_\nu$-Basis von $\mathfrak{M}$.

Nimmt man an, daß f in (4.7) mit B_σ assoziiert ist, so gilt $f(\eta_\tau \zeta)/\pi \in \mathfrak{o}_\nu$ für $\tau \neq \sigma$, und die vorangehende Betrachtung liefert $f_\tau/\pi \in \mathfrak{M}_\tau$ für $\tau \neq \sigma$, also $a_j^{(\tau)} \in \mathfrak{p}_\nu$ für $\tau \neq \sigma$. Sind umgekehrt diese Bedingungen erfüllt, so folgt aus (4.11), daß f mit B_σ assoziiert ist.

(4B) *Ist τ fest und durchläuft i die k_τ Werte, für die $\chi_i \in B_\tau$ ist, so hat man Formeln*

$$\omega_i = \sum_j q_{ij} f_j^{(\tau)} \tag{4.13}$$

mit $q_{ij} \in \mathfrak{o}_\nu$. Die zugehörige Determinante verschwindet nicht.

Beweis. Die Existenz von Formeln (4.13) mit $q_{ij} \in \mathfrak{o}_\nu$ folgt aus $\omega_i \in \mathfrak{M}_\tau$. Die Determinante kann nicht verschwinden, da die k_τ Funktionen ω_i mit $\chi_i \in B_\tau$ linear unabhängig sind.

Wir benötigen später noch das folgende Ergebnis.

(4C) *Es sei $\mathfrak{Q}$ eine p-Untergruppe von $\mathfrak{G}$, es sei $\mathfrak{T} = \mathfrak{C}(\mathfrak{Q})$ und $\tilde{\mathfrak{G}}$ eine die Bedingungen $\mathfrak{T} \leqq \tilde{\mathfrak{G}} \leqq \mathfrak{N}(\mathfrak{Q})$ erfüllende Untergruppe. Es sei $\tilde{B}$ ein Block von $\tilde{\mathfrak{G}}$ und $\tilde{f}$ eine mit $\tilde{B}$ assoziierte ganzwertige lineare Funktion in $Z(\tilde{\mathfrak{G}}, \Omega)$. Definiert man eine ganzwertige lineare Funktion f in $Z(\mathfrak{G}, \Omega)$ durch die Bedingungen*

$$f(K_\alpha) = \sum_{\tilde{\mathfrak{K}}_\beta \subseteq \mathfrak{K}_\alpha \cap \mathfrak{T}} \tilde{f}(\tilde{K}_\beta), \tag{4.14}$$

so ist f mit dem Block $B = \tilde{B}^{\mathfrak{G}}$ von $\mathfrak{G}$ assoziiert[5]).

Beweis. Man definiere eine lineare Abbildung L von $Z(\mathfrak{G}, \Omega) \to Z(\tilde{\mathfrak{G}}, \Omega)$ derart, daß

$$L(K_\alpha) = \sum_{\tilde{\mathfrak{K}}_\beta \subseteq \mathfrak{K}_\alpha \cap \mathfrak{T}} \tilde{K}_\beta \tag{4.15}$$

ist. Ist $\tilde{J}$ die Menge der ganzen Elemente von $Z(\tilde{\mathfrak{G}}, \Omega)$, so gilt $J \to \tilde{J}$. Aus (4.14) folgt, daß f die aus $\tilde{f}$ und L zusammengesetzte Funktion ist:

$$f = \tilde{f} \circ L.$$

Durch Restklassenabbildung entsteht aus L eine lineare Abbildung H von $Z(\mathfrak{G}, \Omega^*) \to Z(\tilde{\mathfrak{G}}, \Omega^*)$. Dabei ist H nach I(7B) ein Algebra-Homomorphismus. Insbesondere ist also $H(\eta_\tau^*)$ entweder ein Idempotent von $Z(\tilde{\mathfrak{G}}, \Omega^*)$ oder 0.

[5]) Ist $\tilde{\mathfrak{G}} = \mathfrak{C}(\mathfrak{Q})$ gewählt, so kann man f in der Form $f = \tilde{f}^{\mathfrak{G}}$ schreiben.

3*

Haben $\tilde{\eta}_1, \tilde{\eta}_2, \ldots$ die analoge Bedeutung für $Z(\tilde{\mathfrak{G}}, \Omega)$ wie oben $\eta_1, \eta_2, \ldots$ für $Z(\mathfrak{G}, \Omega)$, so folgt, daß $H(\eta_\tau^*)$ eine Summe $\sum \tilde{\eta}_\lambda^*$ ist, wobei λ über eine (möglicherweise leere) Menge $M(\tau)$ von Indizes läuft. Daher hat man

$$(4.16) \qquad L(\eta_\tau) \equiv \sum_{\lambda \in M(\tau)} \tilde{\eta}_\lambda \pmod{\mathfrak{p}_\nu \tilde{J}}.$$

Ist nun $\tilde{\chi}$ ein irreduzibler Charakter des Blockes $\tilde{B}_\sigma$ von $\tilde{\mathfrak{G}}$, ist $\tilde{\omega}$ der zugehörige lineare Charakter von $Z(\tilde{G}, \Omega)$ und ist $\omega = \tilde{\omega} \circ L$, so ergibt Vergleich von (4.15) mit I (7D), daß ω mod $\mathfrak{p}_\nu \mathfrak{M}$ den zu $\tilde{B}_\sigma^{\mathfrak{G}}$ gehörigen linearen Charakteren von $Z(\mathfrak{G}, \Omega)$ kongruent ist. Ist $\tilde{B}_\sigma^{\mathfrak{G}} = B_\tau$, so ist nach (4.3) $\omega(\eta_\tau) \equiv 1 \pmod{\mathfrak{p}_\nu}$, also nach (4.16)

$$\sum_{\lambda \in M(\tau)} \tilde{\omega}(\tilde{\eta}_\lambda) \equiv 1 \pmod{\mathfrak{p}_\nu}.$$

Jetzt folgt aus (4.3) (auf $\tilde{\mathfrak{G}}$ angewendet), daß σ dann zu $M(\tau)$ gehören muß. Ist umgekehrt $\sigma \in M(\tau)$, so ist $\omega(\eta_\tau) \equiv 1 \pmod{\mathfrak{p}_\nu}$ und daher $\tilde{B}_\sigma^{\mathfrak{G}} = B_\tau$. Also besteht $M(\tau)$ aus den Indizes σ, für die $\tilde{B}_\sigma^{\mathfrak{G}} = B_\tau$ ist.

Ist nun $\tilde{f}$ eine mit $\tilde{B} = \tilde{B}_\sigma$ assoziierte ganzwertige lineare Funktion in $Z(\tilde{\mathfrak{G}}, \Omega)$, so folgt für $f = \tilde{f} \circ L$ und $\zeta \in J$, daß

$$\begin{aligned} f(\zeta\eta_\varrho) &\equiv \tilde{f}(L(\zeta\eta_\varrho)) \equiv \tilde{f}(L(\zeta)\,L(\eta_\varrho)) \\ &\equiv \sum_{\lambda \in M(\varrho)} \tilde{f}(L(\zeta)\tilde{\eta}_\lambda) \end{aligned} \qquad \pmod{\mathfrak{p}_\nu}$$

ist. Ist $B_\varrho \neq \tilde{B}_\sigma^{\mathfrak{G}}$, also $\sigma \notin M(\varrho)$, so liegt der letzte Ausdruck in $\mathfrak{p}_\nu$ und also ist f mit $\tilde{B}_\sigma^{\mathfrak{G}}$ assoziiert, wie behauptet war.

§ 5. Hilfssätze

In § 5 wird angenommen, daß P ein p-Element im Zentrum $\mathfrak{C}(\mathfrak{G})$ von $\mathfrak{G}$ ist. Wir werden später bei beliebigem $\mathfrak{G}$ die Ergebnisse auf die Untergruppe $\mathfrak{N}(P)$ von $\mathfrak{G}$ anwenden. Aus der Voraussetzung folgt, daß für die irreduzible Darstellung $\mathfrak{X}_i$ von $\mathfrak{G}$ mit dem Charakter χ_i die Gleichung $\mathfrak{X}_i(PG) = \varepsilon_i \mathfrak{X}(G)$ für $G \in \mathfrak{G}$ gilt, wo ε_i eine Einheitswurzel ist. Daher ist

$$(5.1) \qquad \chi_i(PG) = \varepsilon_i \chi_i(G).$$

Da jetzt $\mathfrak{N}(P) = \mathfrak{G}$ ist, kann man in § 3 $\varphi_j^P = \varphi_j$ wählen. Vergleich von (5.1) und (1.1) mit (3.2) für p-reguläres G liefert dann

$$(5.2) \qquad d_{ij}^P = \varepsilon_i d_{ij}.$$

(5A) *Es sei P ein p-Element von $\mathfrak{C}(\mathfrak{G})$, und es sei Q ein von P verschiedenes p-Element von $\mathfrak{G}$. Ist U ein Element der Sektion $\mathfrak{S}(P)$ von P und V ein Element der Sektion $\mathfrak{S}(Q)$, so gilt für jeden Block B von $\mathfrak{G}$ die Gleichung*

$$\sum_{\chi_i \in B} \chi_i(U)\bar{\chi}_i(V) = 0.$$

Beweis. Wir können $U = PR$, $V = QT$ setzen, wo R und T p-regulär sind und $T \in \mathfrak{N}(Q)$ ist. Aus (5.1) folgt

$$\chi_i(U)\bar{\chi}_i(V) = \chi_i(PR)\chi_i(T^{-1}Q^{-1}) = \varepsilon_i\chi_i(R)\chi_i(Q^{-1}T^{-1})$$
$$= \chi_i(R)\chi_i(PQ^{-1}T^{-1}).$$

Wegen $P \neq Q$ ist das Element $S = PQ^{-1}T^{-1}$ p-singulär. Summiert man über alle $\chi_i \in B$, so erhält man nach I(3.5), I(4D) und I(3.9) die Summe

$$\sum_i \chi_i(R)\chi_i(S) = \sum_i \sum_{\varphi_j \in B} d_{ij}\varphi_j(R)\chi_i(S) = \sum_{\varphi_j \in B} \varphi_j(R)\Phi_j(S),$$

die wegen I(3F) in der Tat verschwindet.

(5B) *Es sei P ein p-Element von $\mathfrak{C}(\mathfrak{G})$, es sei B_τ ein Block von $\mathfrak{G}$ und man nehme an, daß in den Bezeichnungen von* (4A) *für alle i die Gleichung*

$$\sum_{j=1}^{k} c_j f_i^{(\tau)}(K_j) = 0 \tag{5.3}$$

mit von i unabhängigen Koeffizienten $c_j \in \mathfrak{o}_\nu$ gilt. Dann ist

$$\sum_\alpha{}' c_\alpha f_i^{(\tau)}(K_\alpha) = 0, \tag{5.4}$$

wo in $\sum'$ nur über die Klassen $\mathfrak{K}_\alpha$ in der Sektion $\mathfrak{S}(P)$ zu summieren ist.

Beweis. Ist $\chi_i \in B$ und ist ω_i der zugehörige lineare Charakter von $Z(\mathfrak{G}, \Omega)$, so folgt aus (4B), daß (5.3) richtig bleibt, wenn man $f_i^{(\tau)}$ durch ω_i ersetzt. Drückt man nach I(2.3) ω_i durch χ_i aus, so erhält man

$$\sum_{j=1}^{k} c_j \frac{1}{n(G_j)} \chi_i(G_j) = 0,$$

wo G_j ein Vertreter der Klasse $\mathfrak{K}_j$ ist. Es sei $\mathfrak{K}_\beta$ eine zur Sektion $\mathfrak{S}(P)$ gehörige Klasse; man multipliziere mit $\bar{\chi}_i(G_\beta)$ und addiere über alle $\chi_i \in B$. Auf Grund von (5A) findet man

$$\sum_\alpha{}' c_\alpha n(G_\alpha)^{-1} \sum_{\chi_i \in B} \chi_i(G_\alpha)\bar{\chi}_i(G_\beta) = 0.$$

Multipliziert man hier mit $\bar{c}_\beta n(G_\beta)^{-1}$ und addiert man über alle $\mathfrak{K}_\beta \in \mathfrak{S}(P)$, so hat man

$$\sum_{\chi_i \in B} \left| \sum_\alpha{}' c_\alpha n(G_\alpha)^{-1} \chi_i(G_\alpha) \right|^2 = 0.$$

Also gilt für jedes $\chi_i \in B$ die Gleichung

$$\sum_\alpha{}' c_\alpha n(G_\alpha)^{-1} \chi_i(G_\alpha) = 0.$$

Dies liefert für dieselben i die Beziehung

$$\sum_\alpha{}' c_\alpha \omega_i(G_\alpha) = 0.$$

Erneute Anwendung von (4B) liefert die Behauptung (5.4).

(5C) *Die Behauptung von* (5B) *bleibt richtig, wenn man in* (5.3) *und* (5.4) *Gleichheit durch Kongruenz* (mod $\mathfrak{p}_\nu$) *ersetzt.*

Beweis. Bildet man aus den Werten der k Funktionen $f_i^{(\sigma)}$ ($\sigma = 1, 2, \ldots, t$; $i = 1, 2, \ldots, k_\sigma$) für die k Elemente K_j eine Determinante Δ, so zeigen wir zunächst

$$(5.5) \qquad \Delta \in \mathfrak{o}_\nu, \quad \Delta \not\equiv 0 \pmod{\mathfrak{p}_\nu}.$$

Die erste Behauptung folgt aus $f_i^{(\sigma)}(K_j) \in \mathfrak{o}_\nu$. Wäre $\Delta \equiv 0 \pmod{\mathfrak{p}_\nu}$, so könnte man eine Linearverbindung f der $f_i^{(\sigma)}$ mit Koeffizienten in Ω bilden, derart, daß nicht alle Koeffizienten zu $\mathfrak{o}_\nu$ gehören, daß aber alle $f(K_j) \in \mathfrak{o}_\nu$ sind. Dies widerspricht (4A), da dann $f \in \mathfrak{M}$ wäre.

Aus (5.5) folgt für festes τ, daß man k_τ Klassen $\mathfrak{K}_j$ so auswählen kann, daß die entsprechende Unterdeterminante von Δ vom Grad k_τ,

$$(5.6) \qquad \operatorname{Det}\left(f_i^{(\tau)}(K_j)\right) \not\equiv 0 \pmod{\mathfrak{p}_\nu}$$

ist. Werden die so ausgewählten k_τ Klassen als die Klassen $\mathfrak{K}_\gamma$ bezeichnet, wobei also γ geeignete Indizes durchläuft, so hat man für $j = 1, 2, \ldots, k$ und für $i = 1, 2, \ldots, k_\tau$ Formeln

$$(5.7) \qquad f_i^{(\tau)}(K_j) = \sum_\gamma b_{j\gamma} f_i^{(\tau)}(K_\gamma),$$

wo die $b_{j\gamma}$ in $\mathfrak{o}_\nu$ liegen und nicht von i abhängen. Ist hier $\mathfrak{K}_j$ eine Klasse der Sektion $\mathfrak{S}(P)$, so folgt aus (5B), daß man annehmen kann, daß in (5.7) rechts nur Klassen $\mathfrak{K}_\gamma \in \mathfrak{S}(P)$ auftreten. Ebenso folgt für $\mathfrak{K}_j \notin \mathfrak{S}(P)$, daß man annehmen kann, daß in (5.7) rechts nur Klassen $\mathfrak{K}_\gamma \notin \mathfrak{S}(P)$ auftreten.

Gilt nun für $i = 1, 2, \ldots, k_\tau$ die Kongruenz

$$\sum_j a_j f_i^{(\tau)}(K_j) \equiv 0 \pmod{\mathfrak{p}_\nu}$$

mit $a_j \in \mathfrak{o}_\nu$, so trenne man zunächst die Glieder für die $\mathfrak{K}_j \in \mathfrak{S}(P)$ und die Glieder mit $\mathfrak{K}_j \notin \mathfrak{S}(P)$. Drückt man in jedem Fall $f_i^{(\tau)}(K_j)$ nach (5.7) durch die $f_i^{(\tau)}(K_\gamma)$ aus, so erhält man eine Kongruenz

$$\sum_\gamma c_\gamma f_i^{(\tau)}(K_\gamma) \equiv 0 \pmod{\mathfrak{p}_\nu}$$

mit von i unabhängigen $c_\gamma \in \mathfrak{o}_\nu$. **Wegen (5.6) müssen dann aber alle c_γ in $\mathfrak{p}_\nu$** liegen. Insbesondere ist also

$$(5.8) \qquad \sum_\gamma{}' c_\gamma f_i^{(\tau)}(K_\gamma) \equiv 0 \pmod{\mathfrak{p}_\nu},$$

wo in $\sum'$ der Index nur die Werte von γ durchläuft, für die $\mathfrak{K}_\gamma \in \mathfrak{S}(P)$ ist. Aus der im Anschluß an (5.7) gemachten Bemerkung folgt aber, daß (5.8) mit

$$\sum_\alpha{}' a_\alpha f_i^{(\tau)}(K_\alpha) \equiv 0 \pmod{\mathfrak{p}_\nu}$$

gleichwertig ist, wo $\mathfrak{K}_\alpha$ alle Klassen aus $\mathfrak{S}(P)$ durchläuft. Dies liefert (5C).

Auf Grund von (4A) kann man (5C) auch in der folgenden Weise **formulieren:**

(5C*) ***Es sei P ein p-Element von $\mathfrak{C}(\mathfrak{G})$. Gibt es k Koeffizienten $a_1, a_2, \ldots, a_k$ in $\mathfrak{o}_\nu$, derart daß für alle ganzwertigen linearen, mit B_τ assoziierten Funktionen f die Kongruenz***

$$\sum_j a_j f(K_j) \equiv 0 \pmod{\mathfrak{p}_\nu} \tag{5.9}$$

gilt, so bleibt (5.9) *richtig, wenn man K_j nur die Klassen von $\mathfrak{S}(P)$ durchlaufen läßt.*

In § 6 wird von den bisherigen Ergebnissen nur (5C*) verwendet werden. Wir brauchen noch einen weiteren speziellen Hilfssatz:

(5D) *Es sei P ein p-Element von $\mathfrak{C}(\mathfrak{G})$, es sei B_τ ein Block von $\mathfrak{G}$. Es sei $R_1, R_2, \ldots, R_l$ ein Vertretersystem für die p-regulären Klassen von $\mathfrak{G}$, und die Klasse von PR_λ sei mit $\mathfrak{K}_\lambda$ bezeichnet. Ist jedem irreduziblen modularen Charakter $\varphi_j \in B_\tau$ ein Element $y_j \in \mathfrak{o}_\nu$ zugeordnet, derart daß für alle ganzwertigen linearen mit B_τ assoziierten Funktionen f die Kongruenz gilt*

$$\sum_{\alpha=1}^{l} \sum_j f(K_\alpha)\, \overline{\varphi}_j(R_\alpha)\, y_j \equiv 0 \pmod{\mathfrak{p}_\nu}, \tag{5.10}$$

so gehören alle y_j zu $\mathfrak{p}_\nu$.

Beweis. Offenbar sind $\mathfrak{K}_1, \mathfrak{K}_2, \ldots, \mathfrak{K}_l$ gerade die Klassen von $\mathfrak{S}(P)$.

Es sei $\chi_i \in B_\sigma \neq B_\tau$. Nach (5.2) und I(4D) gilt dann $d_{ij}^P = 0$ für alle $\varphi_j \in B_\tau$. Wegen (3.6) ergibt sich

$$\sum_{\alpha=1}^{l} \sum_{\beta=1}^{l} \chi_i(PR_\alpha)\, n(PR_\alpha)^{-1}\, \overline{\varphi}_\beta(R_\alpha)\, c_{\beta j}^P = 0$$

für alle diese j. Wegen $\mathfrak{N}(P) = \mathfrak{G}$ ist $c_{\beta j}^P = c_{\beta j}$. Nach I(4D) genügt es, β über die Werte laufen zu lassen, für die $\varphi_\beta \in B_\tau$ ist. Da

$$\mathrm{Det}(c_{ij}) \neq 0 \qquad (\varphi_i, \varphi_j \in B_\tau)$$

ist, folgt

$$\sum_{\alpha=1}^{l} \chi_i(PR_\alpha)\, n(PR_\alpha)^{-1} \overline{\varphi}_j(R_\alpha) = 0$$

für $\varphi_j \in B_\tau$, $\chi_i \in B_\sigma \neq B_\tau$. Also ist

$$\sum_{\alpha=1}^{l} \omega_i(K_\alpha)\, \overline{\varphi}_j(R_\alpha) = 0$$

und nach (4B)

$$\sum_{\alpha=1}^{l} f_i^{(\sigma)}(K_\alpha)\, \overline{\varphi}_j(R_\alpha) = 0$$

für $\sigma \neq \tau$. Daher gilt (5.10) nicht nur für $f = f_i^{(\tau)}$, sondern auch für $f = f_i^{(\sigma)}$ mit $\sigma \neq \tau$. Wegen (5.5) hat man dann

$$\sum_j \overline{\varphi}_j(R_\alpha)\, y_j \equiv 0 \pmod{\mathfrak{p}_\nu}$$

für alle α. Da die φ_j linear unabhängig mod $\mathfrak{p}_\nu$ sind, folgt, daß alle $y_j \in \mathfrak{p}_\nu$ sind.

§ 6. Der Hauptsatz

Wir beweisen jetzt den folgenden Satz:

(6A) *Es sei P ein p-Element einer endlichen Gruppe $\mathfrak{G}$. Ist φ_j^P ein irreduzibler modularer Charakter der Gruppe $\mathfrak{N}(P)$, der zum Block $\tilde{B}$ von $\mathfrak{N}(P)$ gehört, und ist χ_i ein irreduzibler Charakter von $\mathfrak{G}$, so kann die Zerlegungszahl d_{ij}^P nur dann von 0 verschieden sein, wenn χ_i zum Block $B = \tilde{B}^{\mathfrak{G}}$ von $\mathfrak{G}$ gehört.*

Der Beweis wird in einer Reihe von Schritten geführt werden.

1. Ist $\tilde{B}$ ein Block von $\tilde{\mathfrak{G}} = \mathfrak{N}(P)$, so ist $B = \tilde{B}^{\mathfrak{G}}$ nach (2A) mit $\mathfrak{Q} = \{P\}$ definiert. Wir definieren eine Matrix D_0 durch

$$D_0 = (d_{ij}^P), \tag{6.1}$$

wo i die Werte durchläuft, für die $\chi_i \in B$ ist, und j die Werte, für die $\varphi_j^P \in \tilde{B}$. Enthält also $\tilde{B}$ etwa h irreduzible modulare Charaktere, so hat D_0 gerade h Spalten. Wir wollen zunächst zeigen, daß D_0 den Rang h hat. Wäre das nicht richtig, so gäbe es h nicht sämtlich verschwindende Elemente $z_j \in \Omega$, derart daß für alle $\chi_i \in B$

$$\sum_j d_{ij}^P z_j - 0 \tag{6.2}$$

ist. Setzt man den Wert von d_{ij}^P aus (3.6) ein, drückt χ_i durch ω_i aus und verwendet (4B), so sieht man, daß für $i = 1, 2, \ldots, k_\tau$ die Gleichung

$$\sum_\alpha \sum_\beta \sum_j f_i^{(\tau)}(K_\alpha)\,\bar{\varphi}_\beta^P(R_\alpha^P)\, c_{\beta j}^P z_j = 0$$

gilt, wo $\mathfrak{R}_\alpha$ die Klasse von PR_α^P ist. Da $\varphi_j^P \in \tilde{B}$ ist, genügt es nach I (4D), β die h Werte durchlaufen zu lassen, für die $\varphi_\beta^P \in \tilde{B}$ ist. Da die aus den h^2 Werten c_{ij}^P mit φ_i^P, $\varphi_j^P \in \tilde{B}$ gebildete Determinante nicht verschwindet, so sind nicht alle h Zahlen

$$y_\beta = \sum_j c_{\beta j}^P z_j$$

Null. Dann ist

$$\sum_\alpha \sum_\beta f_i^{(\tau)}(K_\alpha)\,\bar{\varphi}_\beta^P(R_\alpha^P)\, y_\beta = 0 \qquad (i = 1, 2, \ldots, k_\tau).$$

Multipliziert man alle y_β mit einem geeigneten festen Element aus $\mathfrak{o}_\nu$, so darf man annehmen, daß alle y_β in $\mathfrak{o}_\nu$, aber nicht alle in $\mathfrak{p}_\nu$ liegen. Nach (4A) gilt dann für eine beliebige ganzwertige, mit B_τ assoziierte lineare Funktion f in $Z(\mathfrak{G}, \Omega)$ die Kongruenz

$$\sum_\alpha \sum_\beta f(K_\alpha)\,\bar{\varphi}_\beta^P(R_\alpha^P)\, y_\beta \equiv 0 \pmod{\mathfrak{p}_\nu}. \tag{6.3}$$

Ist $\tilde{f}$ eine ganzwertige Funktion in $Z(\tilde{\mathfrak{G}}, \Omega)$, die mit $\tilde{B}$ assoziiert ist, und setzt man $f = \tilde{f}^{\mathfrak{G}}$, so folgt aus (4C) mit $\mathfrak{Q} = \{P\}$, $\mathfrak{T} = \tilde{\mathfrak{G}} = \mathfrak{N}(P)$, daß (6.3) für dieses f gilt. Hier ist

$$f(K_\alpha) = \sum_{\tilde{\mathfrak{R}}_\varrho \subseteq \mathfrak{R}} \tilde{f}(\tilde{K}_\varrho). \tag{6.4}$$

Ist etwa $\breve{\mathfrak{K}}_\alpha$ die Klasse von PR_χ^P in $\widetilde{\mathfrak{G}} = \mathfrak{N}(P)$, so tritt $\widetilde{\mathfrak{K}}_\alpha$ unter den $\widetilde{\mathfrak{K}}_\varrho$ auf. Offenbar ist $\widetilde{\mathfrak{K}}_\alpha$ die einzige unter den Klassen $\widetilde{\mathfrak{K}}_\varrho \subseteq \mathfrak{K}_\alpha$, die zur Sektion $\widetilde{\mathfrak{S}}(P)$ von P in $\widetilde{\mathfrak{G}}$ gehört. Setzt man daher (6.4) in (6.3) ein und wendet man (5 C*) (für $\widetilde{\mathfrak{G}} = \mathfrak{N}(P)$ anstelle von $\mathfrak{G}$) an, so findet man

$$\sum_{\alpha,\beta} \tilde{f}(\widetilde{K}_\alpha)\, \bar{\varphi}_\beta^P(R_\alpha^P)\, y_\beta \equiv 0 \pmod{\mathfrak{p}_\nu}. \tag{6.5}$$

Dies gilt für alle ganzwertigen linearen, mit $\widetilde{B}$ assoziierten Funktionen $\tilde{f}$ in $Z(\widetilde{\mathfrak{G}}, \Omega)$. Jetzt folgt aus (5 D), daß alle $y_\beta \in \mathfrak{p}_\nu$ sind; dies ist ein Widerspruch.

Also hat D_0 in der Tat den Rang h.

2. Man setze

$$\bar{D}_0' D_0 = U_0 = (u_{ij}). \tag{6.6}$$

Offenbar ist U_0 eine Matrix h-ten Grades, die zu einer nicht negativen Hermiteschen Form H_0 gehört. Verschwindet H_0 für eine Spalte $\mathfrak{z}$, die aus h komplexen Zahlen besteht, und ist $\mathfrak{x}$ die Spalte $\mathfrak{x} = D_0\mathfrak{z}$, so folgt aus $\bar{\mathfrak{z}}' U_0 \mathfrak{z} = 0$, daß $\bar{\mathfrak{x}}'\mathfrak{x} = 0$. Also ist $\mathfrak{x}$ die Nullspalte 0. Da $D_0\mathfrak{z} = 0$ ist und D_0 den Rang h hat, ist $\mathfrak{z} = 0$. Daher ist H_0 positiv definit und folglich Det $U_0 > 0$.

Man setze $D_1 = (d_{ij}^P)$, wo hier i im Gegensatz zu (6.1) die Werte durchläuft, für die $\chi_i \notin B$ ist, und j dieselben Werte wie in (6.1). Dann ist

$$U_1 = \bar{D}_1' D_1 \tag{6.7}$$

ebenfalls die Matrix einer nicht negativen Hermiteschen Form H_1. Ferner ergibt sich aus (6.6), (6.7) und (3 A) für den Koeffizienten in der i-ten Zeile, j-ten Spalte von $U_0 + U_1$ der Wert

$$\sum_{\chi_\mu \in B} \bar{d}_{\mu i}^P d_{\mu j}^P + \sum_{\chi_\mu \notin B} \bar{d}_{\mu i}^P d_{\mu j}^P = c_{ij}^P. \tag{6.8}$$

Ordnet man die Charaktere $\varphi_1^P, \varphi_2^P, \ldots$ so, daß die ersten h zu $\widetilde{B}$ gehören, so ist nach I(4D) für $\widetilde{\mathfrak{G}}$ die Matrix C^P direkte Summe

$$C^P = \begin{pmatrix} C^P(\widetilde{B}) & 0 \\ 0 & * \end{pmatrix} \tag{6.9}$$

der zu $\widetilde{B}$ gehörigen Teilmatrix $C^P(B)$ und einer Matrix vom Grad $l(P) - h$. Jetzt zeigt (6.8), daß

$$U_0 + U_1 = C^P(\widetilde{B}) \tag{6.10}$$

ist. Die Behauptung (6A) wird bewiesen sein, wenn wir $U_1 = 0$ zeigen können, da dies $D_1 = 0$ zur Folge hat. Wäre $U_1 \neq 0$, so würde aus (6.10) nach bekannten Sätzen über Hermitesche Formen folgen, daß

$$\text{Det } U_0 < \text{Det } C^P(\widetilde{B}) \tag{6.11}$$

ist. Nun werden wir in **3** zeigen, daß U_0 eine Matrix mit ganzen rationalen Koeffizienten ist, die links im Bereich der ganzen rationalen Zahlen durch

$C^P(\tilde{B})$ **teilbar ist. Da Det $U_0 > 0$ war, ist dann Det $U_0 \geqq$ Det $C^P(\tilde{B})$.** Dies ist **ein Widerspruch zu (6.11), und (6A) wird bewiesen sein.**

3. Wir setzen jetzt

$$(6.12) \qquad \Big(\sum_{\chi_\mu \in B} \bar{d}^P_{\mu i} d^P_{\mu j}\Big) = (u_{ij}) = U,$$

wo hier i und j alle Werte $1, 2, \ldots, l(P)$ durchlaufen. Dann ist U_0 die Teilmatrix in den ersten h Zeilen und Spalten von U. Können wir zeigen, daß U ganze rationale Koeffizienten hat und links durch C^P teilbar ist, so ergeben sich daraus wegen (6.9) die am Ende von **2.** bezüglich U_0 ausgesprochenen Behauptungen, und (6A) wird bewiesen sein.

Es mögen Ω_0 und Ω_1 dieselbe Bedeutung wie in § 3 haben. Wie dort gezeigt wurde, ist dann jedes $d^P_{\mu i}$ ganz algebraisch in Ω_1, und daher gilt dasselbe für alle u_{ij}. Wendet man ferner ein Element σ der Galoisschen Gruppe von Ω über Ω_0 an, so werden die Charaktere $\chi_\mu \in B$ permutiert, und nach (3.7) bleibt jedes u_{ij} fest. Also gilt $u_{ij} \in \Omega_0$, d.h. $u_{ij} \in \Omega_0 \cap \Omega_1$, und daher sind in der Tat alle u_{ij} ganz rational.

Man setze

$$(6.13) \qquad (C^P)^{-1} U = W = (w_{ij}).$$

Da Det C^P eine Potenz p^ϱ von p ist[6]), sind alle w_{ij} rationale Zahlen mit dem Nenner p^ϱ. Können wir zeigen, daß $w_{ij} \in \mathfrak{o}_\nu$ ist, so sind alle w_{ij} ganz rational, und wir sind fertig.

Es mögen $\boldsymbol{\varphi}^P$ und N^P dieselbe Bedeutung wie in § 3 haben. Ferner sei jetzt $D = (d^P_{ij})$ mit $\chi_i \in B$, $j = 1, 2, \ldots, l(P)$. Aus (6.12) und (6.13) folgt

$$(\boldsymbol{\varphi}^P)^{-1} W \boldsymbol{\varphi}^P = (C^P \boldsymbol{\varphi}^P)^{-1} \bar{D}' D \boldsymbol{\varphi}^P.$$

Nun ist $(\bar{\boldsymbol{\varphi}}^P)' C^P \boldsymbol{\varphi}^P = N^P$. Also haben wir

$$(\boldsymbol{\varphi}^P)^{-1} W \boldsymbol{\varphi}^P = (N^P)^{-1} (\bar{\boldsymbol{\varphi}}^P)' \bar{D}' D \boldsymbol{\varphi}^P.$$

Da die Koeffizienten von $\boldsymbol{\varphi}^P$ in $\mathfrak{o}_\nu$ liegen und Det $\boldsymbol{\varphi}^P$ nach I(3E) eine Einheit von $\mathfrak{o}_\nu$ ist, genügt es zu zeigen, daß die Koeffizienten der rechten Seite zu $\mathfrak{o}_\nu$ gehören. Nach (3.2) ist der Koeffizient in der i-ten Zeile, j-ten Spalte rechts gleich

$$n(PR_i^P)^{-1} \sum_{\chi_\mu \in B} \bar{\chi}_\mu(PR_i^P)\, \chi_\mu(PR_j^P).$$

Damit ist der ganze Beweis auf den Beweis des folgenden Hilfssatzes zurückgeführt:

HILFSSATZ. *Es sei B ein Block von $\mathfrak{G}$. Sind $L, M \in \mathfrak{G}$, so ist*

$$\sum_{\chi_i \in B} \bar{\chi}_i(L)\, \chi_i(M) \in n(L)\, \mathfrak{o}_\nu.$$

[6]) Vgl. etwa R. BRAUER, Ann. of Math. **57**, 357—377 (1953), Theorem 13. Für das Theorem 1 dieser Arbeit, aus dem der fragliche Satz hergeleitet wird, findet sich ein einfacherer Beweis bei R. BRAUER und J. TATE, Ann. of Math. **63**, 1—7 (1955).

Beweis des Hilfssatzes. Wie in I(2.2) setze man

$$K_\alpha K_\beta = \sum_\gamma a_{\alpha\beta\gamma} K_\gamma. \tag{6.14}$$

Für jedes j sei G_j ein Element aus $\mathfrak{K}_j$. Dann ist $a_{\alpha\beta\gamma}$ die Anzahl der geordneten Paare (ϱ, σ) von Elementen von $\mathfrak{G}$, die die Bedingungen

$$\varrho\sigma = G_\gamma, \quad \varrho \in \mathfrak{K}_\alpha, \quad \sigma \in \mathfrak{K}_\beta$$

erfüllen. Da $\mathfrak{K}_\gamma$ aus $g/n(G_\gamma)$ Elementen besteht, gibt es $g a_{\alpha\beta\gamma}/n(G_\gamma)$ geordnete Paare (ϱ, σ) von Elementen, für die

$$\varrho\sigma \in \mathfrak{K}_\gamma, \quad \varrho \in \mathfrak{K}_\alpha, \quad \sigma \in \mathfrak{K}_\beta$$

ist. Für jedes j sei $\mathfrak{K}_{j'}$ die Klasse, die aus den Inversen der Elemente von $\mathfrak{K}_j$ besteht. Da $\varrho\sigma = \tau$ mit $\sigma\tau^{-1} = \varrho^{-1}$ äquivalent ist, erhält man $g a_{\alpha\beta\gamma}/n(G_\gamma) = g a_{\beta\gamma'\alpha'}/n(G_\alpha)$, woraus

$$a_{\alpha\beta\gamma} \in \frac{n(G_\gamma)}{n(G_\alpha)}\, \mathfrak{o}, \tag{6.15}$$

folgt.

Da ω_i ein linearer Charakter von $Z(\mathfrak{G}, \Omega)$ ist, ergibt sich aus (6.14)

$$\omega_i(K_\alpha)\, \omega_i(K_\beta) = \sum_\gamma a_{\alpha\beta\gamma}\, \omega_i(K_\gamma).$$

Drückt man ω_i durch χ_i aus, so hat man nach einfacher Umformung

$$\chi_i(G_\alpha)\, \chi_i(G_\beta) = \frac{n(G_\alpha)\, n(G_\beta)}{g} \sum_\gamma \frac{a_{\alpha\beta\gamma}}{n(G_\gamma)}\, \chi_i(G_\gamma)\, \chi_i(1).$$

Summiert man jetzt über alle $\chi_i \in B$, so ergibt (4.2) für $B = B_\tau$ die Gleichung

$$\sum_{\chi_i \in B} \chi_i(G_\alpha)\, \chi_i(G_\beta) = \sum_\gamma \frac{n(G_\alpha)\, n(G_\beta)}{n(G_\gamma)}\, a_{\alpha\beta\gamma}\, b_{\gamma'}^{(\tau)}.$$

Wegen $b_{\gamma'}^{(\tau)} \in \mathfrak{o}$, und (6.15) liegt die Summe in $n(G_\beta)\,\mathfrak{o}$. Wählt man $\mathfrak{K}_\alpha$ und $\mathfrak{K}_\beta$ so, daß $L^{-1} \in \mathfrak{K}_\beta$, $M \in \mathfrak{K}_\alpha$, so erhält man wegen $n(L) = n(G_\beta^{-1}) = n(G_\beta)$ den Hilfssatz.

Damit ist (6A) vollständig bewiesen.

§ 7. Folgerungen

Es sei P ein p-Element von $\mathfrak{G}$ und $R \in \mathfrak{N}(P)$ p-regulär. Gehört χ_i zum Block B_τ, so zeigt (6A), daß es genügt, in den Formeln (3.2):

$$\chi_i(PR) = \sum_j d_{ij}^P \varphi_j^P(R) \tag{7.1}$$

φ_j^P über diejenigen irreduziblen modularen Charaktere von $\mathfrak{N}(P)$ laufen zu lassen, die zu Blöcken $\tilde{B}$ von $\mathfrak{N}(P)$ mit $\tilde{B}^{\mathfrak{G}} = B$ gehören. Gibt es kein derartiges $\tilde{B}$, so ist $\chi_i(PR) = 0$. Ist $\tilde{B}^{\mathfrak{G}} = B$, so gehört die Defektgruppe $\tilde{\mathfrak{D}}$ von $\tilde{B}$ nach (2B) und I(8B) zu einer Konjugierten der Defektgruppe $\mathfrak{D}$ von B.

Andererseits ist nach I (9F) $P \in \tilde{\mathfrak{D}}$, da P im Zentrum von $\mathfrak{N}(P)$ liegt. Also gehört dann P zu einer Konjugierten von $\mathfrak{D}$. Dies liefert das folgende Ergebnis:

(7A) *Es sei B ein Block von $\mathfrak{G}$ mit der Defektgruppe $\mathfrak{D}$. Ist P ein p-Element von $\mathfrak{G}$, dessen Klasse kein Element von $\mathfrak{D}$ enthält, so verschwindet jedes $\chi_i \in B$ für alle Elemente der Sektion $\mathfrak{S}(P)$.*

Als Spezialfall ist darin das frühere Ergebnis enthalten, daß ein Charakter vom Defekt 0 für alle p-singulären Elemente verschwindet.

(7B) *Sind P und Q zwei nicht konjugierte p-Elemente von $\mathfrak{G}$, so gilt für jeden Block B die Gleichung*

$$\sum_{\chi_\mu \in B} \bar{d}^P_{\mu i} d^Q_{\mu j} = 0. \tag{7.2}$$

Beweis. Wegen (3.4) genügt es, (7.2) für alle Blöcke B von $\mathfrak{G}$ mit Ausnahme eines festen Blockes B_0 zu beweisen. Gehört φ^P_i zum Block $\tilde{B}$ von $\mathfrak{N}(P)$, so wähle man dabei $B_0 = \tilde{B}^{\mathfrak{G}}$. Für $B \neq B_0$ sind dann alle in (7.2) auftretenden $d^P_{\mu i}$ gleich Null, und (7.2) ist trivial.

Sind R und S p-reguläre Elemente von $\mathfrak{N}(P)$ bzw. $\mathfrak{N}(Q)$, multipliziert man (7.2) mit $\bar{\varphi}^P_i(R)\, \varphi^Q_j(S)$ und addiert über alle i und j, so findet man die folgende Verschärfung der Orthogonalitätsrelationen für Gruppencharaktere:

(7C) *Es sei B ein Block von $\mathfrak{G}$, es seien L und M zwei Elemente von $\mathfrak{G}$, die zu verschiedenen Sektionen gehören. Dann ist*

$$\sum_{\chi \in B} \bar{\chi}(L)\, \chi(M) = 0.$$

(7D) *Es seien $B_1, B_2, \ldots, B_t$ die Blöcke von $\mathfrak{G}$, und es sei $P_1, P_2, \ldots, P_r$ ein Vertretersystem für die p-Klassen von $\mathfrak{G}$. Für jedes p-Element P bezeichne $l_\tau(P)$ die Anzahl der irreduziblen Charaktere φ^P_ϱ von $\mathfrak{N}(P)$, die zu Blöcken $\tilde{B}$ von $\mathfrak{N}(P)$ mit $\tilde{B}^{\mathfrak{G}} = B_\tau$ gehören. Dann ist die Anzahl k_τ der Charaktere $\chi_i \in B_\tau$ gegeben durch*

$$k_\tau = \sum_{i=1}^{r} l_\tau(P_i). \tag{7.3}$$

Beweis. Die Summe auf der rechten Seite von (7.3) sei mit k'_τ bezeichnet. Für jedes φ^P_j definiere man eine Klassenfunktion ψ^P_j in $\mathfrak{G}$ durch die Festsetzung

$$\begin{aligned} \psi^P_j(PR) &= \varphi^P_j(R) \qquad \text{für } p\text{-reguläres } R \in \mathfrak{N}(P), \\ \psi^P_j(G) &= 0 \qquad \text{für alle } G \notin \mathfrak{S}(P). \end{aligned}$$

Dann besagen die Formeln (7.1), daß die k_τ Charaktere $\chi_i \in B_\tau$ Linearkombinationen von k'_τ geeigneten Funktionen ψ^P_j sind. Da die χ_i linear unabhängig sind, ist also

$$k_\tau \leq k'_\tau. \tag{7.4}$$

Andererseits ist $\sum_\tau l_\tau(P)$ die Anzahl aller irreduziblen modularen Charaktere

φ_j^P von P, also gleich der Anzahl $l(P)$ der p-regulären Klassen von $\mathfrak{N}(P)$. Unter Verwendung von (3.8) und (4.8) ergibt sich

$$\sum_\tau k'_\tau = \sum_\tau \sum_i l_\tau(P_i) = \sum_i l(P_i) = k = \sum_\tau k_\tau .$$

Also muß in (7.4) für alle τ Gleichheit gelten. Dies liefert (7.3).

Man betrachte die Zahlen d_{ij}^P mit festem P und j als eine Spalte $\mathfrak{d}_j^P$ der Länge k; $i = 1, 2, \ldots, k$. Dann besagen die Formeln (3.2), daß bei geeigneter Anordnung der Klassen die Matrix $(\chi_i(G_j)) = \boldsymbol{\chi}$ der Gruppencharaktere $(i, j = 1, 2, \ldots, k)$ sich als Produkt

$$\boldsymbol{\chi} = \boldsymbol{D}\,\boldsymbol{\varphi}$$

schreiben läßt, wo $\boldsymbol{D}$ die aus den k Spalten $\mathfrak{d}_j^P$ mit $P = P_1, P_2, \ldots, P_r$ gebildete Matrix ist, während $\boldsymbol{\varphi}$ die direkte Summe der Matrizen $\boldsymbol{\varphi}^P$ für $P = P_1, P_2, \ldots, P_r$ ist. Hierbei ist $\boldsymbol{\varphi}^P$ wie in § 3 definiert. Übrigens folgt aus (3.3) und (3.4) ohne Schwierigkeiten, daß $\pm(\mathrm{Det}\,\boldsymbol{D})^2$ eine Potenz von p ist. Andererseits ist leicht zu sehen, daß $(\mathrm{Det}\,\boldsymbol{\varphi})^2$ eine zu p teilerfremde ganze Zahl ist.

Bildet man aus den Zahlen d_{ij}^P mit festem P und j und $\chi_i \in B_\tau$ eine Spalte $\mathfrak{d}_{\tau j}^P$ der Länge k_τ, so hat man nach (7D) k_τ Spalten $\mathfrak{d}_{\tau j}^P$, für die P eins der Elemente $P_1, P_2, \ldots, P_r$ ist und φ_j^P zu einem Block $\tilde{B}$ von $\mathfrak{N}(P)$ mit $\tilde{B}^{\mathfrak{G}} = B$ gehört. Es sei $\boldsymbol{D}_\tau$ die aus diesen k_τ Spalten gebildete Matrix vom Grade k_τ. Dann besagt (6A), daß man die Spalten von $\boldsymbol{D}$ so permutieren kann, daß $\boldsymbol{D}$ in die direkte Summe von $\boldsymbol{D}_1, \boldsymbol{D}_2, \ldots, \boldsymbol{D}_t$ übergeht. Insbesondere ist $\pm(\mathrm{Det}\,\boldsymbol{D}_\tau)^2$ eine Potenz von p.

Wir nennen $\boldsymbol{D}_\tau$ *die zum Block* B_τ *gehörige Matrix von verallgemeinerten Zerlegungszahlen.*

(7E) *Ist* $\boldsymbol{D}_\tau$ *die zum Block* B_τ *gehörige Matrix von verallgemeinerten Zerlegungszahlen, so ist* $\boldsymbol{D}_\tau$ *eine quadratische Matrix und* $\pm(\mathrm{Det}\,\boldsymbol{D}_\tau)^2$ *eine Potenz von* p.

Gelegentlich ist es zweckmäßig, zum Block B_τ alle irreduziblen modularen Charaktere von $\mathfrak{N}(P_1), \mathfrak{N}(P_2), \ldots, \mathfrak{N}(P_r)$ zu rechnen, die den Spalten von $\boldsymbol{D}_\tau$ entsprechen. Dann besteht also B_τ aus k_τ gewöhnlichen irreduziblen Charakteren χ_i und k_τ irreduziblen modularen Charakteren φ_j^P. Dabei kann aber derselbe Charakter φ mehrfach, nämlich für mehrere verschiedene P auftreten. Für $P = 1$ erhält man die im ursprünglichen Sinn zu B_τ gerechneten modularen Charaktere.

Haben Ω_0 und Ω_1 dieselbe Bedeutung wie in § 3, und ist σ ein Element der Galoisschen Gruppe $\mathfrak{g}$ von Ω über Ω_0, so folgt aus (3.7), daß bei Anwendung von σ auf $\boldsymbol{D}_\tau$ die Zeilen der Matrix permutiert werden. Ist ε eine primitive p^a-te Einheitswurzel, so hat man weiterhin

$$\varepsilon^\sigma = \varepsilon^s$$

mit ganzem rationalem, zu p teilerfremdem s. Daraus folgt, daß für p-Elemente P und p-reguläres $R \in \mathfrak{N}(P)$ die Gleichung gilt

$$\chi_i(PR)^\sigma = \chi_i(P^s R). \tag{7.5}$$

Für jedes $P = P_1, P_2, \ldots, P_r$ ist natürlich P^s zu einem Element desselben Vertretersystems konjugiert. Jetzt zeigt (6A) zusammen mit (3.2) und (3.10), daß man die Anwendung von σ auf $\boldsymbol{D}_\tau$ auch als eine Permutation der Spalten ansehen kann. Wegen Det $\boldsymbol{D}_\tau \neq 0$ kann die Methode aus § 6 und § 7 der unter [3]) zitierten Arbeit angewendet werden. Dies liefert die folgenden Ergebnisse:

(7F) *Es sei Λ ein Teilkörper von Ω; es sei ε_0 eine primitive g_0-te Einheitswurzel (für $g = p^a g_0$ mit $(g_0, p) = 1$). Besteht der Block $B = B_\tau$ aus m Familien von Charakteren χ_i, die bezüglich $\Lambda(\varepsilon_0)$ algebraisch konjugiert sind, so verteilen sich die Spalten der zugehörigen Matrix $\boldsymbol{D}_\tau$ auf m Familien, die bezüglich Λ algebraisch konjugiert sind.*

(7G) *Es sei $p \neq 2$. Bestehen die m Familien von Charakteren χ_i in (7F) aus $q_1, q_2, \ldots,$ bzw. q_m Mitgliedern, so bestehen die m Familien von Spalten von $\boldsymbol{D}_\tau$ ebenfalls gerade aus $q_1, q_2, \ldots,$ bzw. q_m Mitgliedern. Dies gilt auch für $p = 2$, $a \leqq 2$. Ist $p = 2$, $a \geqq 3$, so gilt es, falls $\sqrt{-1} \in \Lambda$ ist.*

Die $l_\tau = l_\tau(1)$ zur Sektion $\mathfrak{S}(1)$ gehörigen Spalten von $\boldsymbol{D}_\tau$ haben rationale Koeffizienten, da die d_{ij} in I § 3 rational sind. Es sei $p \neq 2$, Λ der Körper der rationalen Zahlen. Dann müssen mindestens l_τ Charaktere $\chi_i \in B_\tau$ in Ω_0 liegen. Das liefert das Ergebnis:

(7H) *Ist $p \neq 2$ und besteht der Block B_τ aus l_τ modularen Charakteren φ (im ursprünglichen Sinn), so liegen mindestens l_τ Charaktere χ_i von B_τ im Körper der g_0-ten Einheitswurzeln.*

Ähnliche Aussagen kann man bei beliebiger Wahl von Λ machen. Enthält Λ die p^e-ten Einheitswurzeln, so liegen jedenfalls alle Spalten $\mathfrak{d}_j^P$ in Λ, für die P höchstens die Ordnung p^e hat.

Dept. of Mathematics, Harvard University, Cambridge 38, Mass. (U.S.A.)
und
Tata Institute for Fundamental Research, Bombay

(Eingegangen am 7. März 1959)

Druck der Universitätsdruckerei H. Stürtz AG., Würzburg

Reprinted from the Proceedings of the NATIONAL ACADEMY OF SCIENCES
Vol. 45, No. 12, pp. 1757–1759. December, 1959.

ON FINITE GROUPS OF EVEN ORDER WHOSE 2-SYLOW GROUP IS A QUATERNION GROUP

BY RICHARD BRAUER* AND MICHIO SUZUKI†

HARVARD UNIVERSITY AND UNIVERSITY OF ILLINOIS

Communicated October 28, 1959

We shall sketch a proof of the following theorem:[1]

THEOREM 1. *Let G be a group of finite even order. If the 2-Sylow group P of G is a quaternion group (ordinary or generalized), then G is not simple.*

Proof: If P has order 2^n, it can be generated by two elements α and β with

$$\alpha^{2^{n-1}} = 1, \qquad \beta^2 = \alpha^{2^{n-2}}, \qquad \beta^{-1}\alpha\beta = \alpha^{-1}.$$

The element $\mu = \alpha^{2^{n-2}}$ is the only element of order 2 in P and hence all elements of order 2 in G are conjugate to μ in G. Let $\chi_1 = 1, \chi_2, \ldots, \chi_k$ denote the irreducible characters of G and set $x_i = \chi_i(1)$. If σ is an element of G of even order, it can be shown without difficulty[2] that

$$\sum_{i=1}^{k} \chi_i(\mu)^2\chi_i(\sigma)/x_i = 0. \qquad (1)$$

Let $\pi \neq 1$ be a fixed element of P and let ρ be an element of odd order of the centralizer $C(\pi)$ of π. Applying the theory of modular character for the prime $p = 2$, we have

$$\chi_i(\pi\rho) = \sum_j d_{ij}^\pi \varphi_j^\pi(\rho) \qquad (2)$$

where the d_{ij}^π are the decomposition numbers of G for $p = 2$ and where $\varphi_1^\pi = 1$, φ_2^π, φ_3^π, ... are the irreducible modular characters of $C(\pi)$ for $p = 2$. On substituting (2) in (1) for $\sigma = \pi\rho$ and using the linear independence of the φ_j^π, we obtain

$$\sum_i \chi_i(\mu)^2 d_{ij}^\pi / x_i = 0, \; (j = 1,2,\ldots). \qquad (3)$$

Assume first that P is a generalized quaternion group, $n > 3$. We let π range over the elements of $\{\alpha\}$ of orders 2^{n-1} and 2^{n-2}. It can be seen that there exists a linear combination of the corresponding equations (3) with $j = 1$ of the form

$$\sum_i \chi_i(\mu)^2 t_i / x_i = 0 \qquad (4)$$

such that the t_i are rational integers, $t_i = 1$, and

$$\sum_i t_i^2 = 3. \qquad (5)$$

It follows from the properties of the decomposition[3] numbers that

$$(a) \;\; \sum_i t_i x_i = 0; \quad (b) \;\; \sum_i t_i \chi_i(\mu) = 0. \qquad (6)$$

Because of (5), only three of the t_i are different from 0. On combining (4) and (6), we can deduce $\chi_i(\mu) = \chi_i(1)$ for these three values of i. Hence μ belongs to the kernels of the corresponding representations and G cannot be simple.

If P is the ordinary quaternion group, the argument is more complicated. We may assume that G does not have a normal subgroup of index 2. By the *principal block* of characters of a group, we mean the block which contains the principal character $\chi_1 = 1$ (for the given prime p, here $p = 2$). We take $\pi = \alpha$ and $\pi = \mu$ choosing for φ_j^π the irreducible modular characters of the principal block of $C(\pi)$. For $\pi = \alpha$, we have only $\varphi_1^\mu = 1$ while for $\pi = \mu$ we have three characters φ_j^μ. This gives us four equations (3). In addition, we use (2) and the properties of the decomposition numbers, in particular, the orthogonality relations.[3] If the characters χ_i are taken in a suitable order, it can be shown that there exist signs $\delta_1 = 1$, $\delta_2 = \pm 1$, $\delta_3 = \pm 1$, ... such that the following relations hold:

(*a*) The equations (4) and (6) with $t_i = \delta_i$ for $1 \leqq i \leqq 4$ and $t_i = 0$ for $i > 4$.

(*b*) The equations (4) and (6a) in the following three cases:

1) $t_1 = \delta_1$, $t_4 = -\delta_4$, $t_6 = \delta_6$, $t_7 = \delta_7$, all other $t_i = 0$;
2) $t_2 = \delta_2$, $t_4 = -\delta_4$, $t_5 = \delta_5$, $t_7 = \delta_7$, all other $t_i = 0$;
3) $t_3 = \delta_3$, $t_4 = -\delta_4$, $t_5 = \delta_5$, $t_6 = \delta_6$, all other $t_i = 0$;

(*c*) $\delta_5\chi_5(\mu) = \delta_2\chi_2(\mu) + \delta_3\chi_3(\mu)$, $\delta_6\chi_6(\mu) = \delta_1\chi_1(\mu) + \delta_3\chi_3(\mu)$, $\delta_7\chi_7(\mu) = \delta_1\chi_1(\mu) + \delta_2\chi_2(\mu)$.

An elementary, but somewhat messy, computation allows to deduce $\chi_2(\mu) = x_2$. Again, G cannot be simple.

A more detailed discussion of the decomposition numbers leads to the following refinement of Theorem 1:

THEOREM 2. *Let G be a finite group whose 2-Sylow group is a quaternion group (ordinary or generalized). If H is the unique maximal normal subgroup of odd order, then G/H has a center of order 2.*

An example is given by the extended icosahedral group of order 120.

* The research of the first author was supported by the United States Air Force under Contract No. AF 49(638)-287 monitored by the Air Force Office of Scientific Research of the Air Research and Development Command.

† The work of the second author was supported by a Faculty Summer Fellowship from the University of Illinois.

[1] Theorem 1 in the case that P is the ordinary quaternion group was obtained by R. Brauer and communicated to M. Suzuki who then also gave a proof. Both authors independently obtained the extension to the case that P is a generalized quaternion group.

[2] Cf. Brauer, R., and K. A. Fowler, *Annals of Mathematics*, **62**, 565–583 (1955), in particular, Theorem (2A) and equation (23).

[3] For the results used here, cf. Brauer, R., *Annals of Mathematics*, **42**, 926–935 (1941); these PROCEEDINGS, **32**, 215–219 (1946).

Reprinted from the Proceedings of the NATIONAL ACADEMY OF SCIENCES
Vol. 47, No. 12, pp. 1888–1890. December, 1961.

ON BLOCKS OF REPRESENTATIONS OF FINITE GROUPS

BY RICHARD BRAUER*

HARVARD UNIVERSITY

Communicated October 20, 1961

The blocks of representations of a finite group G have been studied in several previous papers.[1] Here, a number of further results are given which are needed for application of this theory to an investigation of the structure of groups of even order.

1. Let G be a group of finite order g and let p be a fixed prime number. The irreducible characters $\chi_1, \chi_2, \ldots, \chi_k$ are distributed into disjoint sets, the p-blocks of G. We shall denote by G° the set of p-regular elements of G and by $\chi_\mu{}^\circ$ the restriction of χ_μ to G°. If B is a fixed block, the functions $\chi_\mu{}^\circ$ with $\chi_\mu \in B$ generate a module M_B with regard to the ring $\mathbf{Z}$ of integers. By a *basic set* φ_B for B, we mean any basis $\{\varphi_\rho\}$ of M_B. Thus,

$$\chi_\mu{}^\circ = \sum_\rho d_{\mu\rho}\, \varphi_\rho \quad \text{for} \quad \chi_\mu \in B \tag{1}$$

with $d_{\mu\rho} \in \mathbf{Z}$. For $\varphi_\rho, \varphi_\sigma \in \varphi_B$, set

$$c_{\rho\sigma} = \sum_\mu d_{\mu\rho}\, d_{\mu\sigma}, \tag{2}$$

the sum extending over all μ with $\chi_\mu \in B$. Hence $(c_{\rho\sigma})$ is the matrix of a quadratic form Q. If φ_B is replaced by another basic set, Q is replaced by an equivalent quadratic form.

It is well known that the irreducible modular characters in B form a basic set. In this case, the $d_{\mu\rho}$ are the *decomposition numbers* and the $c_{\rho\sigma}$ are the *Cartan invariants* of B. We shall use the same terms in the case of an arbitrary basic set.

Let d be the defect of B. We consider p^d as fixed and shall say that a quantity is bounded if it is bounded for all p-blocks of defect d of all finite groups with bounds depending only on p^d. In this sense, the dimension r of M_B and $\det(c_{ij})$ are bounded.

Hence, Q belongs to one of a finite number of classes of positive definite quadratic forms. As a consequence, we have

THEOREM 1. *There exist bounds $\gamma = \gamma(p^d)$ depending only on p^d such that for each p-block B of defect d of a finite group, a basic set can be chosen such that the Cartan invariants are at most equal to γ.*

We remark that the union of basic sets φ_B for all p-blocks B af G is still linearly independent.

2. Let P be a p-element of G, i.e., an element whose order p^α is a power of p. By the *section* S_P of P, we mean the set of elements of G conjugate to an element PR, where R is a p-regular element of the centralizer $C(P)$ of P in G; $R \in C(P)^\circ$. Thus, if P ranges over a set of representatives for the conjugate classes of p-elements in G, each element of G belongs to exactly one S_P. If $R \in C(P)^\circ$, we can set

$$\chi_\mu (PR) = \sum_b \sum_\rho d_{\mu\rho}^P \varphi_\rho^P (R). \tag{3}$$

Here, b ranges over the p-blocks of $C(P)$ and, for each b, ρ ranges over the indices of the elements φ_ρ^P of a basic set φ_b. Moreover, the $d_{\mu\rho}^P$ are algebraic integers of the field of p^αth roots of unity which do not depend on R. We call the $d_{\mu\rho}^P$ *the generalized decomposition numbers.*

Each p-block b of $C(P)$ determines a p-block $B = b^G$, cf. the papers quoted in reference 1. If we consider only $\chi_\mu \in B$ in (3), it suffices to let b range over those blocks of $C(P)$ for which $b^G = B$. For φ_ρ^P, $\varphi_\sigma^P \in \varphi_b$,

$$\sum_\mu \bar{d}_{\mu\rho}^P d_{\mu\sigma}^P = c_{\rho\sigma}^P, \tag{4}$$

where μ ranges over the indices for which $\chi_\mu \in B$ and where $c_{\rho\sigma}^P$ is the Cartan invariant of b. On the other hand, for the same range of μ,

$$\sum_\mu \bar{d}_{\mu\rho}^P d_{\mu\sigma}^{P'} = 0 \tag{5}$$

when P and P' are two nonconjugate p-elements or when $P = P'$ but φ_ρ^P, φ_σ^P belong to basic sets of two distinct p-blocks of $C(P)$.

If $b^G = B$, the defect d_0 of b is at most equal to the defect d of B. If the basic set φ_b is chosen in accordance with Theorem 1, $|c_{\rho\sigma}^P| \leqq \gamma(p^d)$. It follows from (4) that we have only finitely many possibilities for the matrix $(d_{\mu\rho}^P)$. This leads to

THEOREM 2. *For a given p^d, there exist a finite number of possible types of p-blocks B of defect d. For each type, the set of generalized decomposition numbers $\{d_{\mu\rho}^P\}$ is completely determined assuming that suitable basic sets are used. If the group $C(P)$ is given, the values of the $\chi_\mu \in B$ for elements of the section of P are determined.*

We may assume that, for a fixed type, the defect group D of the block is given as an abstract p-group and that it is also given which conjugate classes of D are "fused" in G, i.e., are included in the same conjugate class of G. If P is not conjugate to an element of D, then each $\chi_\mu \in B$ vanishes on S_P. In the last part of Theorem 2, it is assumed that we know to which element of D (if any) P is conjugate.

3. Estimates for the number of types for given p^d obtainable by the previous method would be extremely large. There are other methods available which also give Theorems 1 and 2 and which yield better results. These are based on the follow-

ing remarks: 1. The discussion of the Cartan invariants of $C(P)$ can be reduced to the same discussion for $C(P)/\{P\}$. If $P \neq 1$, the defect is reduced. We can therefore use an inductive procedure to obtain the possibilities for the $d^P_{\mu\rho}$ with a fixed $P \neq 1$. 2. Consider the $d^P_{\mu\rho}$ with fixed ρ and P as the coefficients of a column $\mathfrak{d}^P_\rho$ with μ as row index. The columns with coefficients in **Z** which are orthogonal to all columns $\mathfrak{d}^P_\rho$ with $P \neq 1$ form a **Z**-module X. By (5), $\mathfrak{d}_\rho^1 \in X$. Any **Z**-basis of X can be used for the set of columns $\mathfrak{d}_\rho^1$, assuming a suitable choice of the basic set. 3. In addition to (4) and (5), there are a number of other results which facilitate the discussion. In particular, congruences for the columns $\mathfrak{d}^P_\rho$ can be established. Also, Theorems 4 and 5 of the third paper quoted in reference 1 can be used; these results can be refined further.

4. One of the p-blocks of G must contain the principal character $\chi_0 = 1$. We term this block the *principal block* B_0 of G and state a number of results for B_0.

THEOREM 3. *Let b be a block of a subgroup H of G, let T be its defect group in H and assume that the centralizer of T in G is included in H so that b^G is defined as a block of G. Then b^G is the principal block B_0 of G if and only if b is the principal block b_0 of H.*

It follows from this that for $B = B_0$, the sum $\sum_b$ in (3) consists only of the term $b = b_0$. This simplification is of importance for the applications.

By the *p-regular core* $K_p(G)$ of a group G, we mean the unique maximal normal subgroup of G of an order prime to p.

THEOREM 4. *The intersection of the kernels of the irreducible representations in the principal p-block B_0 of G is the p-regular core $K_p(G)$ of G.*

This means that the principal p-block of G can be identified with that of $G/K_p(G)$.

COROLLARY. If *n is the sum of the degrees of the irreducible characters $\chi_r \in B_0$,* then,

$$n \leqq (G:K_p(G)) \leqq [2n]!. \tag{6}$$

Indeed, the algebraic conjugates of $\chi_\mu \in B_0$ lie again in B_0 and hence $\sum \chi_\mu \in B_0$ is a rational character belonging to a faithful representation of $G/K_p(G)$. Now the right-hand part of (6) is obtained from a theorem of I. Schur[3] (even in a somewhat sharper form). The other part is trivial.

THEOREM 5. *If B is the principal block, the basic set φ_B can be chosen such that the constant 1 appears in it and that all Cartan invariants $c_{\rho\sigma}$ belonging to it lie below a bound $\gamma_0(p^n)$ where p^n is the highest power of p dividing g.*

* This research was supported by the United States Air Force under Contract No. AF 49 (638)-287 monitored by the Air Force Office of Scientific Research of the Air Research and Development Command.

[1] Brauer, R., these PROCEEDINGS, **30**, 109–114 (1944), **32**, 182–186 (1946), **32**, 215–219 (1946); *Mathematische Zeitschrift*, **63**, 406–414 (1956), **72**, 25–46(1959). Brauer, R. ,and W. Feit, these PROCEEDINGS, **45**, 361–365 (1959).

[2] Using results from the theory of quadratic forms, explicit estimates for $\gamma(p^d)$ can be given, but they are probably much too large.

[3] Schur, I., *Sitzungsberichte der Preussischen Akademie Berlin, Mathematisch-Naturwissenschaftliche Klasse*, 77–91 (1905).

Reprinted from the Proceedings of the NATIONAL ACADEMY OF SCIENCES
Vol. 47, No. 12, pp. 1891–1893. December, 1961.

INVESTIGATION ON GROUPS OF EVEN ORDER, I*

BY RICHARD BRAUER

HARVARD UNIVERSITY

Communicated October 23, 1961

1. Let G be a group of finite order g and let $\tilde{G}$ be a subgroup of order $\tilde{g}$. If f is a complex-valued class function defined on G, i.e. a function whose value remains constant on each of the conjugate classes $K_1, K_2, \ldots K_k$ of G, then f can be expressed by the values of the irreducible characters $\chi_1, \chi_2, \ldots, \chi_k$ of G. If we can find a class function f on G, a class function $\tilde{f}$ on $\tilde{G}$, and an element σ of $\tilde{G}$ such that $f(\sigma) = \tilde{f}(\sigma)$, this yields a relation between the values of $\chi_1, \chi_2, \ldots$, and the values of the irreducible characters $\tilde{\chi}_1, \tilde{\chi}_2, \ldots$ of $\tilde{G}$ for σ.

2. We shall assume that the order g is even. Then G contains involutions, i.e. elements of order 2. Let K_α and K_β be two fixed conjugate classes of involutions. For any $\sigma \in G$, let $f_{\alpha\beta}(\sigma)$ denote the number of ordered pairs $\langle \xi, \eta \rangle$ of elements of G such that

$$\xi\eta = \sigma, \qquad \xi \in K_\alpha, \quad \eta \in K_\beta. \tag{1}$$

Then $f_{\alpha\beta}$ is a class function. Let J_λ be a representative for the class K_λ. If $c(\tau)$ denotes the order of the centralizer $C(\tau)$ of an element $\tau \in G$, then,

$$c(J_\alpha)c(J_\beta)f_{\alpha\beta}(\sigma) = g \sum_{i=1}^{k} \chi_i(J_\alpha)\chi_i(J_\beta)\chi_i(\sigma)/x_i, \tag{2}$$

where $x_i = \chi_i(1)$ is the degree of χ_i.

The elements ζ of G for which $\zeta^{-1}\sigma\zeta$ is either σ or σ^{-1} form a subgroup $C^*(\sigma)$, say, of order $c^*(\sigma)$. Thus, $c^*(\sigma)$ is $c(\sigma)$ or $2c(\sigma)$. Since ξ and η in (1) are involutions, it follows at once that $\xi^{-1}\sigma\xi = \sigma^{-1}$ and $\eta^{-1}\sigma\eta = \sigma^{-1}$, which implies $\xi, \eta \in C^*(\sigma)$. Moreover, we have

LEMMA 1. *If no element of K_α transforms σ into σ^{-1}, then,*

$$f_{\alpha\beta}(\sigma) = 0. \tag{3}$$

3. We shall use the same notation in $\tilde{G}$ as introduced in G with tilde signs added. Thus, $\tilde{K}_1, \tilde{K}_2, \ldots$ will denote the conjugate classes of $\tilde{G}$, etc. The remarks above yield at once the result:

LEMMA 2. *Suppose that $\tilde{G}$ satisfies the condition*

$$\tilde{G} \supseteq C^*(\sigma). \tag{4}$$

Let K_α, K_β be classes of involutions of G. We can write

$$K_\alpha \cap \tilde{G} = \bigcup_\lambda \tilde{K}_\lambda, \; K_\beta \cap G = \bigcup_\mu \tilde{K}_\mu. \tag{5}$$

If $\tilde{f}_{\lambda\mu}$ has the same significance for $\tilde{G}$, $\tilde{K}_\lambda$, $\tilde{K}_\mu$ as $f_{\alpha\beta}$ has for G, K_α, K_β, then,

$$f_{\alpha\beta}(\sigma) = \sum_\lambda \sum_\mu \tilde{f}_{\lambda\mu}(\sigma), \tag{6}$$

where the range of λ and μ is the same in (5) and (6).

4. It follows from §1 that (3) represents an equation between $\chi_1(\sigma), \chi_2(\sigma), \ldots$, while (6) represents an equation between $\chi_1(\sigma), \chi_2(\sigma), \ldots$ and $\tilde{\chi}_1(\sigma), \tilde{\chi}_2(\sigma), \ldots$

We can use the theory of blocks[1] to derive relations which can be handled more easily.

Let p be a fixed prime and let P be a p-element of G. We shall take $\sigma = PR$, where R is an arbitrary p-regular element of $C(P)$. If no element of K_α transforms P into P^{-1}, no element of K_α transforms σ into σ^{-1} and (3) applies. In the case of Lemma 2, we assume $\tilde{G} \supseteq C^*(P)$. This implies (4) for $\sigma = PR$ and (6) holds. We can now use the formulas (3) of *Bl* to express $\chi_\mu(PR)$ by means of the generalized decomposition numbers $d^P_{\mu\rho}$ and the values $\varphi_\rho^P(R)$ of the functions of a basic set φ_b for the p-blocks b of $C(P)$. In the case of (6), the corresponding formulas can be applied in $\tilde{G}$; the same basic set φ_b can be used. Because of the linear independence of the union of the sets φ_b for all p-blocks b of $C(P)$, the coefficients of each φ_ρ^P in the resulting equations must be equal. This yields the following propositions:

PROPOSITION 1. *Let G be a finite group of even order. Let p be a prime, let P be a p-element of G, and let φ_ρ^P be a member of a basic set φ_b for a p-block b of $C(P)$. If J_α and J_β are two involutions of G and if no conjugate of J_α in G transforms P into P^{-1}, then for all ρ,*

$$\sum_\mu \chi_\mu(J_\alpha)\,\chi_\mu(J_\beta) d^P_{\mu\rho}/x_\mu = 0. \tag{7}$$

Here, χ_μ ranges over the irreducible characters of the block $B = b^G$, the $d^P_{\mu\rho}$ are the corresponding generalized decomposition numbers, and $x_\mu = \chi_\mu(1)$.

PROPOSITION 2. *Let G be a group of even order g. Let p be a prime and let P be a p-element of G. Suppose that $\tilde{G}$ is a subgroup of order $\tilde{g}$ of G such that $\tilde{G} \supseteq C^*(P)$. If J_α and J_β are two involutions of G and if φ_ρ^P is a member of a basic set φ_b for a p-block b of $C(P)$, then,*

$$g \sum_\mu \chi_\mu(J_\alpha)\,\chi_\mu(J_\beta)\, d^P_{\mu\rho}/x_\mu = \tilde{g} c(J_\alpha)\, c(J_\beta) \sum_\mu h_{\mu\alpha} h_{\mu\beta}\, \tilde{d}^P_{\mu\rho}/\tilde{x}_\mu. \tag{8}$$

Here, the notation on the left is the same as in (7) while on the right, μ ranges over the indices for which the irreducible character $\tilde{\chi}_\mu$ of $\tilde{G}$ belongs to $\tilde{B} = b^{\tilde{G}}$, $\tilde{x}_\mu = \tilde{\chi}_\mu(1)$, and $\tilde{d}^P_{\mu\rho}$ is the corresponding generalized decomposition number. Finally,

$$h_{\mu\alpha} = \sum_\lambda \tilde{\chi}_\mu(\tilde{J}_\lambda)/\tilde{c}(\tilde{J}_\lambda), \tag{9}$$

with $\{\tilde{J}_\lambda\}$ ranging over a system of representatives for the $\tilde{G}$-classes included in the G-class of J_α. The definition of $h_{\mu\beta}$ is analogous.

The advantage of (7) and (8) compared to (3) and (6) is that if B is a p-block of defect d, then μ here ranges over at most p^{2d} values. Moreover, if p^d is given, only a finite number of possibilities for the $d^P_{\mu\rho}$, $\tilde{d}^P_{\mu\rho}$ occur. Of course, the sum on the right depends only on $\tilde{G}$.

5. If $b = b_0$ is the principal block of $C(P)$, then $B = b_0{}^G$ is the principal block of G. Its defect is the highest exponent n for which p^n divides g. Then we may assume that $1 \in \varphi_b$. It follows from *Bl*, Theorem 5, that we may assume that all $|d^P_{\mu\rho}|$ lie below $\gamma_0(p^n)^{1/2}$. If $\varphi_\rho^P = 1$, then for the principal character $\chi_0 \in B_0$, $d^P_{\mu\rho} = 1$. If the sum in (7) for $\alpha = \beta$ is denoted by u, either u cannot differ too much from 1, or there must exist a $\chi_\mu \neq \chi_0$ in B such that $\chi_\mu(J_\alpha)^2/x_\mu$ lies above a positive bound depending only on p^n.

The following theorems can be obtained by this method.

THEOREM 1. *Let G be a group of even order g and let J be an involution of G. Let p be a prime,* $g = p^n g_0$ *with* $g_0 \in \mathbf{Z}$, $(g_0, p) = 1$ *and let P be a p-element of G. Finally, let* $A > 1$ *be a given real number. Then, we have at least one of the following two cases:*

Case I. The value of g is determined approximately by $C^*(P)$ *and* $c(J)$;

$$\frac{A}{A+1}\, c(J)^2 c^*(P) v \leqq g \leqq \frac{A}{A-1}\, c(J)^2 c^*(P) v, \tag{10}$$

where v is a positive real number depending only on $C^*(P)$ *and the set of elements of* $C^*(P)$ *conjugate to J in G.*

Case II. There exists a character $\chi_\mu \neq \chi_0$ *in the principal p-block* B_0 *such that*

$$x_\mu < A p^{2n} \gamma_0(p^n)^{1/2} \chi_\mu(J)^2, \tag{11}$$

where $\gamma_0(p^n)$ *is the bound in Bl, Theorem 5.*

Remarks: 1. For $0 < A$, either the left inequality in (11) holds or the statement of Case II is true.

2. In Case I, we have $|v| \leqq \quad \gamma_0(p^n)^{1/2} l$. where l is the number of classes of $C(P^*)$ included in the G-class of J. For $p = 2$, we have $|v| \leqq \gamma_0(p^n)^{1/2}$.

THEOREM 2. *If P and* P^{-1} *are not conjugate in G, then* $v = 0$ *and we must have Case II.*

In Case II, the kernel of the irreducible representation belonging to χ_μ is a normal subgroup $N \neq G$ of G. This leads to

THEOREM 3. *In case II of Theorem 1, there exists a normal subgroup* $N \neq G$ *of G which includes the p-regular core* $K_p(G)$ *of G and is such that*

$(G:N) \leqq [2Ap^{2n}\gamma_0(p^n)^{1/2}c(J)]!$;

$(G:N) \leqq [2Ap^{2n}\gamma_0(p^n)^{1/2}(C(J):\ K_2C(J))]!$ for $p = 2$.

5. We shall say that a group G is *corefree with regard to p*, if $\{1\}$ is the only normal subgroup of order prime to p, i.e. if $K_p(G) = \{1\}$.

Let H be a given p-group. The question arises whether or not there exist infinitely many corefree groups G with H as their p-Sylow group. We can answer this question when $p = 2$ and H is abelian.

THEOREM 4. *Let H be an abelian 2-group. There exist infinitely many groups G, corefree with regard to 2 and with H as their 2-Sylow group, if and only if H has a direct factor which is elementary abelian of type (2,2).*

THEOREM 5. *Let H be an abelian 2-group of order* $2^n \geqq 8$ *which has no direct factor which is elementary abelian of type (2,2,2). There exist only finitely many simple groups G with H as their 2-Sylow subgroup.*

* This research was supported by the United States Air Force under Contract No. 49 (638)-287 monitored by the Air Force Office of Scientific Research of the Air Research and Development Command.

[1] Cf. the preceding paper, "On blocks of representations of finite groups," these PROCEEDINGS, 47, 1888 (1961). This paper will be referred to as *Bl.* We shall use the same notation here as in *Bl.*

ON FINITE GROUPS WITH AN ABELIAN SYLOW GROUP

RICHARD BRAUER AND HENRY S. LEONARD, JR.

1. Introduction. We shall consider finite groups $\mathfrak{G}$ of order of g which satisfy the following condition:

(*) *There exists a prime p dividing g such that if $P \neq 1$ is an element of a p-Sylow group $\mathfrak{P}$ of $\mathfrak{G}$ then the centralizer $\mathfrak{C}(P)$ of P in $\mathfrak{G}$ coincides with the centralizer $\mathfrak{C}(\mathfrak{P})$ of $\mathfrak{P}$ in $\mathfrak{G}$.*

This assumption is satisfied for a number of important classes of groups. It also plays a role in discussing finite collineation groups in a given number of dimensions.

Of course (*) implies that $\mathfrak{P}$ is abelian. It is possible to obtain rather detailed information about the irreducible characters of groups $\mathfrak{G}$ in this class (§ 4). The method used here is that of considering the exceptional characters of $\mathfrak{G}$ with regard to the normalizer $\mathfrak{N}(\mathfrak{P})$ of $\mathfrak{P}$ (§ 3) and comparing the results with those obtained by studying the p-blocks of characters of $\mathfrak{G}$. It may be mentioned that the same method can be applied under wider assumptions than (*), but the results are less definite.

If g is divisible by p but not by p^2, it is clear that (*) will always be satisfied. The corresponding class of groups has been studied by one of us previously **(3; 4)**. Our results contain many but not all of the previous results. On the other hand, our new method is more elementary.

As a consequence of our results we show in § 5 that if a group $\mathfrak{G}$ satisfying condition (*) has a faithful representation of degree less than $(p^n - 1)^{\frac{1}{2}}$, where p^n is the order of the p-Sylow group $\mathfrak{P}$, then $\mathfrak{P}$ is normal in $\mathfrak{G}$. This is a generalization of a theorem of Blichfeldt **(1)**.

In order to make the paper more self-contained, we start with a specially elementary treatment of the theory of exceptional characters in the form needed here.† The method used can be applied under wider assumptions.

2. Exceptional characters. Let $\mathfrak{G}$ be a group of finite order g and let $\mathfrak{H}$ be a subgroup of order h. We assume that there exists a non-empty set $\mathscr{L}$

Received August 12, 1961. The second author's research was supported in part by National Science Foundation grants NSF-G2268 and NSF-G9656 and by the Office of Scientific Research of the United States Air Force. An earlier version of part of this paper appeared in the second author's doctoral dissertation at Harvard University in 1958 and was presented to the American Mathematical Society April 25, 1959, under the title *On the order of primitive linear groups.*

†Exceptional characters were first used by the first author in his original version of the work published in the joint paper **(9)**. The proof was based on **(5,** Theorem **1)**. M. Suzuki **(10)** then gave a much more elementary proof. For refinements cf. **(11)**.

of conjugate classes $\mathfrak{L}$ of $\mathfrak{H}$ such that if $L \in \mathfrak{L}$, $G \in \mathfrak{G}$, and $G^{-1}LG \in \mathfrak{H}$, then $G \in \mathfrak{H}$, that is, that only the elements of $\mathfrak{H}$ transform an element of $\mathfrak{L}$ into an element of $\mathfrak{H}$. Clearly this condition on $\mathfrak{L}$ is equivalent to the following two conditions: (I) If $L \in \mathfrak{L}$ then the centralizer $\mathfrak{C}(L)$ of L in $\mathfrak{G}$ lies in $\mathfrak{H}$. (II) If $\mathfrak{L}$ is included in the conjugate class $\mathfrak{K}$ of $\mathfrak{G}$ then $\mathfrak{K} \cap \mathfrak{H} = \mathfrak{L}$. It is not required that $\mathscr{L}$ consist of all classes of $\mathfrak{H}$ with these properties.

Let $\psi_1, \psi_2, \ldots, \psi_l$ denote the irreducible characters of $\mathfrak{H}$. We arrange them in "families" $F_1, F_2, \ldots$, such that ψ_i and ψ_j belong to the same family if and only if $\psi_i(H) = \psi_j(H)$ for all $H \in \mathfrak{H}$ which do not belong to a class $\mathfrak{L} \in \mathscr{L}$. We shall say that a family F is *proper* if it has more than one member.

In the following, $\mathfrak{G}$, $\mathfrak{H}$, and $\mathscr{L}$ will be fixed. The notion of exceptional characters of $\mathfrak{G}$ will depend on $\mathfrak{H}$ and $\mathscr{L}$. For convenience we shall call the classes $\mathfrak{L} \in \mathscr{L}$ *special classes* and their elements *special elements.*

(2A) *Let $F = \{\psi_1, \psi_2, \ldots, \psi_m\}$ be a proper family. There exist m irreducible characters $\chi_1, \chi_2, \ldots, \chi_m$ of $\mathfrak{G}$ such that*

$$\chi_i|\mathfrak{H} = \epsilon\psi_i + S, \qquad (i = 1, 2, \ldots, m), \tag{2.1}$$

where ϵ is a sign ± 1 independent of i, and S is a character of $\mathfrak{H}$ independent of i in which $\psi_1, \psi_2, \ldots, \psi_m$ appear with the same multiplicity. If χ is an irreducible character of $\mathfrak{G}$ other than $\chi_1, \chi_2, \ldots, \chi_m$ then $\psi_1, \psi_2, \ldots, \psi_m$ appear with the same multiplicity in $\chi|\mathfrak{H}$.

Proof. If $\chi_1, \chi_2, \ldots, \chi_k$ are the irreducible characters of $\mathfrak{G}$, we can set

$$\chi_i|\mathfrak{H} = \sum_{j=1}^{l} a_{ij}\psi_j, \qquad (i = 1, 2, \ldots, k), \tag{2.2}$$

where the a_{ij} are non-negative integers. Then

$$ha_{ij} = \sum_{H \in \mathfrak{H}} \chi_i(H)\bar{\psi}_j(H). \tag{2.3}$$

We shall view the k numbers a_{ij} with fixed j as the coefficients of a column A_j with i serving as row index. Likewise, if $G \in \mathfrak{G}$, we arrange the k numbers $\chi_i(G)$ in a column $X(G)$. Then (2.3) becomes

$$hA_j = \sum_{H \in \mathfrak{H}} X(H)\bar{\psi}_j(H), \qquad (j = 1, 2, \ldots, l). \tag{2.3*}$$

The inner product of two columns is defined in the usual manner. If $G \in \mathfrak{G}$, we have

$$h(X(G), A_j - A_1) = \sum_{H \in \mathfrak{H}} (X(G), X(H))(\psi_j(H) - \psi_1(H)).$$

Assume here that $1 \leqslant j \leqslant m$. It will suffice to let H range over the special elements of $\mathfrak{H}$ since otherwise $\psi_j(H) = \psi_1(H)$. By the orthogonality relations $(X(G), X(H)) = 0$ if G and H are not conjugate in $\mathfrak{G}$. If G and H are conjugate, $(X(G), X(H))$ is the order $c(H)$ of the centralizer $\mathfrak{C}(H)$. For a special element H, $\mathfrak{C}(H)$ is also the centralizer of H in $\mathfrak{H}$ and the class $\mathfrak{L}$ of H in $\mathfrak{H}$ consists of $h/c(H)$ elements. Thus, for $j = 1, 2, \ldots, m$,

$$(2.4) \qquad (X(G), A_j - A_1) = \begin{cases} \psi_j(H) - \psi_1(H) \\ 0, \end{cases}$$

where the first case applies when G is conjugate in $\mathfrak{G}$ to special elements $H \in \mathfrak{H}$ and the second case when G is not conjugate to such elements.

Combining (2.3*) and (2.4), we have

$$h(A_r, A_j - A_1) = \sum_{H \in \mathfrak{H}} \bar{\psi}_r(H)(\psi_j(H) - \psi_1(H)), \qquad (1 \leqslant r \leqslant l; 1 \leqslant j \leqslant m).$$

Hence, if δ_{ij} has the usual significance,

$$(2.5) \qquad (A_r, A_j - A_1) = \delta_{jr} - \delta_{1r}, \qquad (1 \leqslant r \leqslant l; 1 \leqslant j \leqslant m).$$

This implies that $(A_j - A_1, A_j - A_1) = 2$ for $2 \leqslant j \leqslant m$. Hence $A_j - A_1$ contains two coefficients ± 1 while all other coefficients vanish. If $\mathfrak{H} \neq \mathfrak{G}$, the unit element 1 is not special and (2.4) shows that $(X(1), A_j - A_1) = 0$. Since $X(1)$ has positive coefficients, the two non-vanishing coefficients in $A_j - A_1$ have opposite signs. If $\mathfrak{H} = \mathfrak{G}$, this is still true, since here the matrix $[a_{ij}]$ in (2.2) is the unit matrix. If $1 < i < j \leqslant m$ then $(A_i - A_1, A_j - A_1) = 1$ by (2.5). It now follows easily that the characters $\chi_1, \chi_2, \ldots, \chi_k$ can be taken in such an order that the first m rows in the $m - 1$ columns $A_2 - A_1, A_3 - A_1, \ldots, A_m - A_1$ are given by

$$\begin{matrix} -\epsilon & -\epsilon & \ldots & -\epsilon \\ \epsilon & 0 & \ldots & 0 \\ 0 & \epsilon & \ldots & 0 \\ \cdot & \cdot & \ldots & \cdot \\ 0 & 0 & \ldots & \epsilon \end{matrix} \qquad \text{with } \epsilon = \pm 1$$

while the remaining rows vanish. In other words

$$(2.6) \qquad a_{ij} - a_{i1} = \begin{cases} -\epsilon \text{ for } i = 1, 2 \leqslant j \leqslant m \\ \epsilon \text{ for } i = j, 2 \leqslant j \leqslant m \\ 0 \text{ otherwise.} \end{cases}$$

(For $j = 1$, the left side of (2.6) vanishes trivially.)

It now follows from (2.5) that

$$(2.7) \qquad -a_{1r}\epsilon + a_{jr}\epsilon = \delta_{jr} - \delta_{1r}, \qquad (1 \leqslant r \leqslant l, 1 \leqslant j \leqslant m).$$

In particular

$$(2.7^*) \qquad a_{jr} = a_{1r}, \qquad (1 \leqslant j \leqslant m, r > m).$$

We find from (2.7) and (2.6) that

$$(2.8) \qquad a_{jr} - \epsilon\delta_{jr} = a_{1r} - \epsilon\delta_{1r} = a_{11} - \epsilon, \quad (1 \leqslant j \leqslant m, 1 \leqslant r \leqslant m).$$

Here $a_{11} - \epsilon = a_{12} \geqslant 0$. On combining (2.8) and (2.7*) with (2.2) we see that

$$\chi_j|\mathfrak{H} - \epsilon\psi_j = (a_{11} - \epsilon) \sum_{r=1}^{m} \psi_r + \sum_{r>m} a_{1r}\psi_r, \qquad (1 \leqslant j \leqslant m).$$

Thus this expression is a character S of $\mathfrak{H}$ which is independent of j and which contains $\psi_1, \psi_2, \ldots, \psi_m$ with the same multiplicity. This yields (2.1). The last statement of (2A) is immediate from (2.2) and (2.6) with $i > m$.

We call χ_j the *exceptional character* of $\mathfrak{G}$ corresponding to ψ_j, $(1 \leqslant j \leqslant m)$. It is clear from (2A) that χ_j is uniquely determined by ψ_j.

On combining (2.4) and (2.6) we have

(2B) *If G is an element of $\mathfrak{G}$ which is not conjugate to a special element then the exceptional characters χ_j of $\mathfrak{G}$ corresponding to the characters of $\mathfrak{H}$ in a proper family F all take the same value for G. In particular, if $\mathfrak{G} \neq \mathfrak{H}$ then these characters χ_j all have the same degree.*

The last statement is obtained by taking $G = 1$ which is permitted for $\mathfrak{G} \neq \mathfrak{H}$.

Let F^* be a proper family of characters of $\mathfrak{H}$ different from the family F in (2A). Then $\psi_r \in F^*$ appears with the same multiplicity in $\chi_1|\mathfrak{H}, \chi_2|\mathfrak{H}, \ldots, \chi_m|\mathfrak{H}$. This shows that $\chi_1, \chi_2, \ldots, \chi_m$ are different from the exceptional characters of $\mathfrak{G}$ associated with the characters in F^*. We shall say that an irreducible character χ of $\mathfrak{G}$ is *exceptional*, if χ is the exceptional character of $\mathfrak{G}$ corresponding to a character of H in some *proper* family. By definition, exceptional characters of $\mathfrak{G}$ are irreducible. Combining the preceding remark with (2A), we have

(2C) *If χ is an exceptional character of $\mathfrak{G}$ and ψ the irreducible character of $\mathfrak{H}$ with which it is associated, then the correspondence $\chi \to \psi$ is a $(1-1)$ correspondence between the exceptional characters of $\mathfrak{G}$ and the irreducible characters of $\mathfrak{H}$ belonging to proper families. If ψ belongs to the proper family F_0 then we have formulae*

$$\chi|\mathfrak{H} = \epsilon(F_0)\psi + \sum_{F} a(F_0, F) \sum_{\psi \in F} \psi \tag{2.9}$$

where F ranges over all families of characters of $\mathfrak{H}$, proper or improper. Here $\epsilon(F_0)$ is a sign depending only on F_0, and $a(F_0, F)$ is a non-negative rational integer depending only on F_0 and F.

Likewise, if χ_μ is an irreducible character of $\mathfrak{G}$ which is non-exceptional then we have a formula

$$\chi_\mu|\mathfrak{H} = \sum_{F} a(\mu, F) \sum_{\psi \in F} \psi \tag{2.10}$$

where the $a(\mu, F)$ are non-negative rational integers.

3. Application of the exceptional characters. Let p be a prime number. We consider groups $\mathfrak{G}$ of finite order g divisible by p which satisfy the following assumption:

(*) *If* $\mathfrak{P}$ *is a p-Sylow subgroup of* $\mathfrak{G}$ *and* $P \neq 1$ *is an element of* $\mathfrak{P}$, *then the centralizer* $\mathfrak{C}(P)$ *of* P *in* $\mathfrak{G}$ *coincides with the centralizer* $\mathfrak{C}(\mathfrak{P})$ *of* $\mathfrak{P}$ *in* $\mathfrak{G}$.

It follows from (*) that $\mathfrak{P}$ is abelian. The order of $\mathfrak{P}$ will be denoted by p^n. Set $\mathfrak{C} = \mathfrak{C}(\mathfrak{P})$. As is well known,

$$\mathfrak{C} = \mathfrak{P} \times \mathfrak{V} \tag{3.1}$$

where $\mathfrak{V}$ is a group of order v prime to p. We choose $\mathfrak{H}$ as the normalizer $\mathfrak{N}(\mathfrak{P})$ of $\mathfrak{P}$ in $\mathfrak{G}$. Both $\mathfrak{P}$ and $\mathfrak{V}$ are normal in $\mathfrak{H}$. Since $\mathfrak{P}$ is the only p-Sylow subgroup of $\mathfrak{H}$, every p-singular class $\mathfrak{K}$ of $\mathfrak{H}$ consists of elements of the form PV with $P \in \mathfrak{P}, P \neq 1, V \in \mathfrak{V}$. It follows from (*) that $\mathfrak{P} \subseteq \mathfrak{C}(PV) \subseteq \mathfrak{C} \subseteq \mathfrak{H}$. Now Sylow's theorem shows that if $G^{-1}PVG \in \mathfrak{H}$ for some $G \in \mathfrak{G}$, then $G \in \mathfrak{H}$. Hence we may choose for $\mathscr{L}$ in § 2 the set of all p-singular classes $\mathfrak{K}$ of $\mathfrak{H}$.

We next discuss the irreducible characters of $\mathfrak{H}$. If $\vartheta_1, \vartheta_2, \ldots, \vartheta_a$ are the irreducible characters of $\mathfrak{V}$ and $\lambda_1, \lambda_2, \ldots, \lambda_b$ those of $\mathfrak{P}$, and if we define $\lambda_\beta\vartheta_\alpha$ by

$$(\lambda_\beta\vartheta_\alpha)(PV) = \lambda_\beta(P)\vartheta_\alpha(V), \qquad (P \in \mathfrak{P},\ V \in \mathfrak{V}),$$

then the ab products $\lambda_\beta\vartheta_\alpha$ are the ab distinct irreducible characters of the direct product $\mathfrak{C}$ in (3.1). Every irreducible character ψ of $\mathfrak{H}$ appears as a constituent in a character $(\lambda_\beta\vartheta_\alpha)^*$ of $\mathfrak{H}$ induced by an irreducible character $\lambda_\beta\vartheta_\alpha$ of $\mathfrak{C}$. Here

$$(\lambda_\beta\vartheta_\alpha)^*(H) = \begin{cases} 0, & (H \notin \mathfrak{C}), \\ \sum_N (\lambda_\beta^N\vartheta_\alpha^N)(H), & (H \in \mathfrak{C}), \end{cases} \tag{3.2}$$

where N ranges over a complete residue system R of $\mathfrak{H}$ (mod $\mathfrak{C}$). It follows that we have $(\lambda_\beta\vartheta_\alpha)^* = (\lambda_{\beta'}\vartheta_{\alpha'})^*$ if and only if there exists an $N \in \mathfrak{H}$ such that $\lambda_{\beta'} = \lambda_\beta^N$ and $\vartheta_{\alpha'} = \vartheta_\alpha^N$. In particular, to obtain all characters $(\lambda_\beta\vartheta_\alpha)^*$ it will suffice to let ϑ_α range over a set $\mathfrak{S}$ of irreducible characters of $\mathfrak{V}$ such that each irreducible character of $\mathfrak{V}$ is an associate in $\mathfrak{H}$ of exactly one element of $\mathfrak{S}$.

If $N \in \mathfrak{H} - \mathfrak{C}$ then N transforms the abelian group $\mathfrak{P}$ into itself leaving only the unit element fixed. Hence every $P \neq 1$ in $\mathfrak{P}$ has $(\mathfrak{H} : \mathfrak{C})$ conjugates in $\mathfrak{H}$. Thus if w conjugate classes of $\mathfrak{H}$ contain such elements P and if $(\mathfrak{P} : 1) = p^n$, then

$$p^n - 1 = w(\mathfrak{H} : \mathfrak{C}). \tag{3.3}$$

Moreover for $N \in \mathfrak{H} - \mathfrak{C}$, the mapping $\lambda \to \lambda^N$ permutes the irreducible characters λ of $\mathfrak{P}$ leaving only the principal character 1 fixed.† Hence if $\lambda_\beta \neq 1$ in (3.2) then the characters λ_β^N appearing in (3.2) are distinct. If ψ is an irreducible constituent of $(\lambda_\beta\vartheta_\alpha)^*$, then $\psi|\mathfrak{C}$ is a sum of terms $\lambda_\beta^N\vartheta_\alpha^N$ such that with each term its distinct associates appear. Since we have just seen that all $(\mathfrak{H} : \mathfrak{C})$ associates are distinct, we have $(\lambda_\beta\vartheta_\alpha)^* = \psi$; that is,

†Cf., e.g., the "permutation lemma" in **(2)**.

$(\lambda_\beta\vartheta_\alpha)^*$ is irreducible if $\lambda_\beta \neq 1$. We shall have $(\lambda_\beta\vartheta_\alpha)^* = (\lambda_{\beta'}\vartheta_\alpha)^*$ if and only if there exists an $N \in \mathfrak{H}$ such that $\lambda_{\beta'} = \lambda_\beta^N$ and $\vartheta_\alpha = \vartheta_\alpha^N$. The latter condition means that N must belong to the inertial group $\mathfrak{J}(\vartheta_\alpha) \supseteq \mathfrak{C}$ of ϑ_α in $\mathfrak{H}$.

If we set

$$(3.4) \qquad (\mathfrak{H} : \mathfrak{J}(\vartheta_\alpha)) = r_\alpha, \qquad (\mathfrak{J}(\vartheta_\alpha) : \mathfrak{C}) = s_\alpha,$$

it follows from our remarks that we have $(p^n - 1)/s_\alpha$ irreducible characters ψ of $\mathfrak{H}$ of the form $(\lambda_\beta\vartheta_\alpha)^*$ with fixed α and $\lambda_\beta \neq 1$. We shall denote this set of characters by $F(\vartheta_\alpha)$. Then, by (3.3) and (3.4), $F(\vartheta_\alpha)$ consists of wr_α distinct irreducible characters ψ of $\mathfrak{H}$. As shown by (3.2), each $\psi \in F(\vartheta_\alpha)$ vanishes for the elements $H \in \mathfrak{H} - \mathfrak{C}$. If V is a p-regular element of $\mathfrak{C}$ then $V \in \mathfrak{B}$ and we have

$$(3.5) \qquad \psi(V) = \sum_{N \in R} \vartheta_\alpha^N(V), \qquad (V \in \mathfrak{B}).$$

Hence all characters in $F(\vartheta_\alpha)$ have the same value for elements of $\mathfrak{H}$ which are not special in the sense of § 2. Thus the family $F(\vartheta_\alpha)$ is included in one of the families F in § 2. We note that (3.5) can be written

$$(3.5^*) \qquad \psi(V) = s_\alpha\zeta_\alpha(V) \text{ with } \zeta_\alpha(V) = \sum_M \vartheta_\alpha^M(V) \qquad (V \in \mathfrak{B}),$$

where M ranges over a complete residue system of $\mathfrak{H}(\text{mod } \mathfrak{J}(\vartheta_\alpha))$.

It follows from (3.2) that the kernel of a character $\psi \in F(\vartheta)$ cannot include $\mathfrak{P}$. If we let ϑ range over the set $\mathfrak{S}$, the families $F(\vartheta)$ obtained will be disjoint. The remarks above show that all irreducible characters ψ of $\mathfrak{H}$ will belong to such a family $F(\vartheta)$ except those which appear as a constituent of some $(1\vartheta_\rho)^*$ with $\vartheta_\rho \in \mathfrak{S}$. Let us denote the set of distinct irreducible constituents ψ of $(1\vartheta_\rho)^*$ by $U(\vartheta_\rho)$. It follows from (3.2) with $\lambda_\beta = 1$ that all these ψ have kernels which include $\mathfrak{P}$. Consequently no $\psi \in U(\vartheta_\rho)$ can belong to an $F(\vartheta)$.

We summarize some of the results in the following proposition.

(3A) *Let G be a group satisfying condition* (*), *$\mathfrak{P}$ a p-Sylow group, $\mathfrak{C} = \mathfrak{C}(\mathfrak{P}) = \mathfrak{P} \times \mathfrak{B}$, $\mathfrak{N}(\mathfrak{P}) = \mathfrak{H}$. Let $\mathfrak{S}$ be a system of irreducible characters ϑ_α of $\mathfrak{B}$ such that every irreducible character of $\mathfrak{B}$ is an associate in $\mathfrak{H}$ of exactly one $\vartheta_\alpha \in \mathfrak{S}$. The irreducible characters ψ of $\mathfrak{H}$ whose kernel does not include $\mathfrak{P}$ are distributed in the disjoint families $F(\vartheta_\alpha)$, $\vartheta_\alpha \in \mathfrak{S}$. Here $F(\vartheta_\alpha)$ consists of the characters of $\mathfrak{H}$ induced by characters $\lambda\vartheta_\alpha$ of $\mathfrak{C}$ where λ is a non-principal irreducible character of $\mathfrak{P}$. The number of members of $F(\vartheta_\alpha)$ is $(p^n - 1)/s_\alpha = wr_\alpha$ where $s_\alpha = (\mathfrak{J}(\vartheta_\alpha) : \mathfrak{C})$, $r_\alpha = (\mathfrak{H} : \mathfrak{J}(\vartheta_\alpha))$, $p^n - 1 = w(\mathfrak{H} : \mathfrak{C})$, and where $\mathfrak{J}(\vartheta_\alpha)$ is the inertial group of ϑ_α in $\mathfrak{H}$.*

We now study the characters in the set $U(\vartheta_\rho)$. It follows from (3.2) that for $\psi \in U(\vartheta_\rho)$, $P \in \mathfrak{P}$, $V \in \mathfrak{B}$,

$$(3.6) \qquad \psi(PV) = e(\psi) \sum_M \vartheta_\rho^M(V) = e(\psi)\zeta_\rho(V),$$

where M ranges over a residue system of $\mathfrak{H}$ (mod $\mathfrak{F}(\vartheta_\rho)$) and where $e(\psi)$ is a natural integer depending only on ψ. In particular this shows that, for two distinct characters ϑ_ρ and $\vartheta_{\rho'}$ in $\mathfrak{S}$, the sets $U(\vartheta_\rho)$ and $U(\vartheta_{\rho'})$ are disjoint.

(3B) *Let $U(\vartheta_\rho)$ denote the set of irreducible constituents of the character of $\mathfrak{H}$ induced by the character $1\vartheta_\rho$ of $\mathfrak{C}$. Each irreducible character of $\mathfrak{H}$ whose kernel includes $\mathfrak{P}$ belongs to exactly one set $U(\vartheta_\rho)$ with $\vartheta_\rho \in \mathfrak{S}$.*

In applying the lemma of § 2 each ψ in a set $U(\vartheta_\rho)$ will be considered as forming a family of one element; the remaining families will be the sets $F(\vartheta_\alpha)$. We claim that if we sum over all ψ in $F(\vartheta_\alpha)$, we have

$$\sum_\psi \psi(H) = \begin{cases} 0, & (H \notin \mathfrak{C}), \\ -\zeta_\alpha(V), & (H \in \mathfrak{C} - \mathfrak{V}), \\ (p^n - 1)\zeta_\alpha(V), & (H = V \in \mathfrak{V}), \end{cases} \tag{3.7}$$

where $H = PV$ with $P \in \mathfrak{P}$, $V \in \mathfrak{V}$, for $H \in \mathfrak{C}$. This is immediate from (3.2) for $H \notin \mathfrak{C}$ and from (3.5*) for $H \in \mathfrak{V}$. Assume that $H \in \mathfrak{C} - \mathfrak{V}$. If we let λ range over the $p^n - 1$ non-principal irreducible characters of $\mathfrak{P}$, it follows from (3.4) and the remarks preceding it that each of the characters ψ in $F(\vartheta_\alpha)$ is obtained s_α times in the form $(\lambda\vartheta_\alpha)^*$, and we have

$$s_\alpha \sum_\psi \psi|\mathfrak{C} = \sum_{\lambda\neq 1} \sum_{N\in R} \lambda^N \vartheta_\alpha^N .$$

For $H = PV$ with $P \neq 1$, we have $\sum_{\lambda\neq 1}\lambda^N(P) = -1$. Thus

$$\sum_\psi \psi(H) = -(1/s_\alpha) \sum_{N\in R} \vartheta_\alpha^N(V) = -\zeta_\alpha(V)$$

by (3.5*). This completes the proof of (3.7).

The family $F(\vartheta_\alpha)$ is proper except when $w = 1$ and $r_\alpha = 1$, that is, when $(\mathfrak{H} : \mathfrak{C}) = p^n - 1$ and ϑ_α is associated only with itself. If $F(\vartheta_\alpha)$ is a proper family, there exist exactly wr_α exceptional characters χ corresponding to the characters $\psi \in F(\vartheta_\alpha)$. If $\chi_\beta^{(\alpha)}$ is the exceptional character of $\mathfrak{G}$ corresponding to $(\lambda_\beta\vartheta_\alpha)^* \in F(\vartheta_\alpha)$, then by (2.9)

$$\chi_\beta^{(\alpha)}|\mathfrak{H} = \epsilon(F(\vartheta_\alpha))(\lambda_\beta\vartheta_\alpha)^* + \sum_F a(F(\vartheta_\alpha),F) \sum_{\psi\in F} \psi. \tag{3.8}$$

For $P \in \mathfrak{P}$, $P \neq 1$, $V \in \mathfrak{V}$, using (3.2), (3.6), (3.7), we have

$$\chi_\beta^{(\alpha)}(PV) = \epsilon(F(\vartheta_\alpha)) \sum_{N\in R} \lambda_\beta^N(P)\vartheta_\alpha^N(V) + \sum_\rho b_\rho^{(\alpha)}\zeta_\rho(V), \tag{3.9}$$

where ρ ranges over the values for which $\vartheta_\rho \in \mathfrak{S}$, and where the $b_\rho^{(\alpha)}$ are rational integers independent of P. Applying the same method for the element $V \in \mathfrak{V}$, we see that

$$\chi_\beta^{(\alpha)}(V) - \epsilon(F(\vartheta_\alpha))(\lambda_\beta\vartheta_\alpha)^*(V) \equiv \chi_\beta^{(\alpha)}(PV) - \epsilon(F(\vartheta_\alpha))(\lambda_\beta\vartheta_\alpha)^*(PV) \tag{3.10}$$

mod p^n in the ring of all algebraic integers.

If χ_μ is a non-exceptional character of $\mathfrak{G}$, we apply (2.10). It follows that we have formulae

$$\chi_\mu(PV) = \sum_\rho b_{\mu\rho}\zeta_\rho(V) \tag{3.11}$$

for $P \neq 1$ in $\mathfrak{P}$ and V in $\mathfrak{V}$, where ρ has the same range as in (3.9) and where the $b_{\mu\rho}$ are rational integers. Also

$$\chi_\mu(PV) \equiv \chi_\mu(V) \qquad (\text{mod } p^n). \tag{3.12}$$

4. The decomposition numbers. We now study the p-blocks B of groups $\mathfrak{G}$ which satisfy the assumption (*). If $\mathfrak{D}$ is the defect group of B, we may assume $\mathfrak{D} \subseteq \mathfrak{P}$. There exist p-regular elements $G \in \mathfrak{G}$ such that $\mathfrak{D}$ is a p-Sylow subgroup of $\mathfrak{C}(G)$ **(6, § 8)**. It follows from (*) that either $\mathfrak{D} = \mathfrak{P}$ or $\mathfrak{D} = \{1\}$. Hence

(4A) *If $\mathfrak{G}$ satisfies assumption* (*) *for a prime p then every p-block B of $\mathfrak{G}$ has either full defect n or defect* 0.

We investigate the blocks B of full defect further. Here $\mathfrak{D} = \mathfrak{P}$, $\mathfrak{D}\mathfrak{C}(\mathfrak{D}) = \mathfrak{C} = \mathfrak{P} \times \mathfrak{V}$ by (3.1), and $\mathfrak{D}\mathfrak{C}(\mathfrak{D})/\mathfrak{D} \cong \mathfrak{V}$ is of order prime to p. The number s of blocks B of full defect is equal to the number of classes $\{\vartheta\}$ of irreducible characters of $\mathfrak{V}$ associated in $\mathfrak{N}(\mathfrak{D})/\mathfrak{D} = \mathfrak{H}/\mathfrak{P}$ **(6, § 12)**, that is, classes $\{\vartheta\}$ of irreducible characters associated in $\mathfrak{H}$. Thus s is the number of $\vartheta_\alpha \in \mathfrak{S}$.

On the other hand we use the main result of **(7)**. As is easily seen, each block $\mathfrak{b}$ of $\mathfrak{C}$ consists of the p^n irreducible characters $\lambda_\beta\vartheta_j$ for some fixed j. We shall denote this block by $\mathfrak{b}(\vartheta_j)$. There is only one modular irreducible character in $\mathfrak{b}(\vartheta_j)$, and it may be identified with ϑ_j. If χ_i is an irreducible character of $\mathfrak{G}$ belonging to a block B, then

$$\chi_i(PV) = \sum_j d_{ij}^P\vartheta_j(V), \qquad (P \in \mathfrak{P}, P \neq 1, V \in \mathfrak{V}), \tag{4.1}$$

where the $d_{ij}{}^P$ are the generalized decomposition numbers and where $d_{ij}{}^P \neq 0$ only for values of j for which

$$\mathfrak{b}(\vartheta_j)^{\mathfrak{G}} = B. \tag{4.2}$$

For each block B of defect n, there exist ϑ_j for which (4.2) holds. Indeed, if this were not so, then by (4.1) $\chi_i(P) = 0$ for all $P \neq 1$. If χ_i is non-exceptional then by (3.12) $\chi_i(1) \equiv 0$ (mod p^n), that is, χ_i would be of defect 0, a contradiction. If χ_i is exceptional then taking $V = 1$ and summing in (3.10) over the classes relative to $\mathfrak{H}$ of elements $P \neq 1$ in $\mathfrak{P}$, again we would have $\chi_i(1) \equiv 0$ (mod p^n).

Let $F(\vartheta_\alpha)$ be a proper family, take χ_i as an exceptional character $\chi_\beta{}^{(\alpha)}$ and let B be the block to which $\chi_\beta{}^{(\alpha)}$ belongs. Comparison of (4.1) with (3.9) yields

$$\epsilon(F(\vartheta_\alpha)) \sum_{N \in \bar{\mathfrak{N}}} \lambda_\beta^N(P)\vartheta_\alpha^N(V) + \sum_\rho b_\rho^{(\alpha)}\zeta_\rho(V) = \sum_j d_{ij}^P\vartheta_j(V), \tag{4.3}$$

where ζ_ρ is given by (3.5*). Since this holds for all $V \in \mathfrak{B}$, each irreducible character ϑ of $\mathfrak{B}$ appears with the same coefficient on both sides. The coefficient of $\vartheta_\alpha{}^N$ is

$$\epsilon(F(\vartheta_\alpha)) \sum_T \lambda_\beta^{TN}(P) + b_\alpha^{(\alpha)},$$

where T ranges over a residue system of $\mathfrak{F}(\vartheta_\alpha)$ (mod $\mathfrak{C}$). This expression would vanish for all $P \neq 1$ in $\mathfrak{P}$ only if all irreducible characters $\lambda \neq 1$ appeared among the $\lambda_\beta{}^{TN}$. This would mean that $s_\alpha = p^n - 1$. But then (3.3) and (3.4) would imply that $wr_\alpha = 1$, a contradiction since $F(\vartheta_\alpha)$ consists of wr_α elements and would not be a proper family. Hence each $\vartheta_\alpha{}^N \in \{\vartheta_\alpha\}$ must appear among the ϑ_j on the right in (4.3), that is, (4.2) holds for all $\vartheta_j \in \{\vartheta_\alpha\}$.

If $F(\vartheta_\alpha)$ is not a proper family, we have $r_\alpha = 1$ (in addition to $w = 1$). Then the class $\{\vartheta_\alpha\}$ consists only of ϑ_α. We may therefore say for all classes $\{\vartheta_\alpha\}$ that if ϑ_j in (4.2) ranges over $\{\vartheta_\alpha\}$, the block B on the right is the same for all j. Thus (4.2) establishes a mapping of the set of classes $\{\vartheta_\alpha\}$ into the set of blocks B of full defect. We have already noted that the mapping is onto and that the number of classes $\{\vartheta_\alpha\}$ and of blocks B of full defect are equal. Hence we have a $(1 - 1)$ correspondence. We shall now denote the block B corresponding to the class $\{\vartheta_\alpha\}$ by $B(\vartheta_\alpha)$. Then we have shown:

(4B) *Every block B of $\mathfrak{G}$ of full defect is obtained exactly once in the form $B(\vartheta_\alpha)$ when ϑ_α ranges over $\mathfrak{S}$. If $wr_\alpha > 1$, the wr_α exceptional characters $\chi_\beta{}^{(\alpha)}$ belong to $B(\vartheta_\alpha)$.*

(4C) *For $\chi_i \in B(\vartheta_\alpha)$, we have formulae*

$$\chi_i(PV) = \sum_j d_{ij}^P \vartheta_j(V), \qquad (P \in \mathfrak{P}, P \neq 1, V \in \mathfrak{B}),$$

where ϑ_j ranges over the class $\{\vartheta_\alpha\}$.

It follows from (4C) that in (3.9) and (4.3) $b_\rho{}^{(\alpha)} = 0$ for $\rho \neq \alpha$. Thus

(4D) *If χ_i is an exceptional character $\chi_\beta{}^{(\alpha)}$ then for $\vartheta_j = \vartheta_\alpha{}^M$ we have*

$$d_{ij}^P = \epsilon_\alpha \sum_T \lambda_\beta^{TM}(P) + b_\alpha^{(\alpha)}, \qquad (P \in \mathfrak{P}, P \neq 1). \tag{4.4}$$

Here T ranges over a residue system of $\mathfrak{F}(\vartheta_\alpha)$ (mod $\mathfrak{C}$), $b_\alpha{}^{(\alpha)}$ is a rational integer which does not depend on P, and $\epsilon_\alpha = \epsilon(F(\vartheta_\alpha))$ is a sign ± 1 depending only on α. Hence

$$\chi_\beta^{(\alpha)}(PV) = \sum_M \Big(\epsilon_\alpha \sum_T \lambda_\beta^{TM}(P) + b_\alpha^{(\alpha)}\Big) \vartheta_\alpha^M(V)$$

for $P \in \mathfrak{P}$, $P \neq 1$, $V \in \mathfrak{B}$, with M ranging over a residue system of $\mathfrak{H}$ (mod $\mathfrak{F}(\vartheta_\alpha)$).

We may also take for χ_i a non-exceptional irreducible character of $\mathfrak{G}$ of positive defect and apply (4C) to (3.11). It follows that if $\chi_i \in B(\vartheta_\alpha)$ then $b_{i\rho} = 0$ for $\rho \neq \alpha$. Hence

(4E) *If χ_i is a non-exceptional irreducible character of $\mathfrak{G}$ belonging to $B(\vartheta_\alpha)$ and $\vartheta_j = \vartheta_\alpha^M$, then d_{ij}^P is a rational integer $b_i = b_{i\alpha}$ which depends neither on P for $P \neq 1$ nor on M. Hence*

$$\chi_i(PV) = b_i \sum_M \vartheta_\alpha^M(V), \qquad (P \in \mathfrak{P}, P \neq 1, V \in \mathfrak{V}),$$

where M ranges over a residue system of $\mathfrak{H}$ (mod $\mathfrak{F}(\vartheta_\alpha)$, *that is, ϑ_α^M ranges over the distinct associates of ϑ_α in $\mathfrak{H}$.*

If $w = 1$, that is, if $(\mathfrak{N}(\mathfrak{P}) : \mathfrak{C}(\mathfrak{P}))$ has the maximal value $p^n - 1$ possible when assumption (*) is satisfied, we may have blocks $B(\vartheta_\alpha)$ which do not contain any exceptional character of $\mathfrak{G}$. Indeed, this will be so if $r_\alpha = 1$ since, by (4B), the exceptional characters belong to the blocks $B(\vartheta_\alpha)$ with $wr_\alpha > 1$. If $w = 1$ and $r_\alpha = 1$ we shall now pick an irreducible character arbitrarily from $B(\vartheta_\alpha)$, denote it by $\chi^{(\alpha)}$, and from now on count it as the exceptional character of $B(\vartheta_\alpha)$. We show that (4D) remains valid. Since $r_\alpha = 1$, we have $\mathfrak{F}(\vartheta_\alpha) = \mathfrak{H}$. Then T in (4.4) ranges over a complete residue system of $\mathfrak{H}$ (mod $\mathfrak{C}$), λ_β^{TM} ranges over all irreducible characters $\neq 1$ of $\mathfrak{P}$, and $\sum_T \lambda_\beta^{TM}(P) = -1$ for every $P \neq 1$ in $\mathfrak{P}$. By (4E) applied to $\chi_i = \chi^{(\alpha)}$, d_{ij}^P is a rational integer which for $\vartheta_j = \vartheta_\alpha^M$ is independent of P ($P \neq 1$) and M. Hence (4.4) is true, if we take $\epsilon_\alpha = 1$ and

$$b_\alpha^{(\alpha)} = d_{ij}^P + 1.$$

(4F) *If $B(\vartheta_\alpha)$ is a block of $\mathfrak{G}$ of full defect and if $B(\vartheta_\alpha)$ contains l_α modular irreducible characters of $\mathfrak{G}$, then $B(\vartheta_\alpha)$ contains $wr_\alpha + l_\alpha$ ordinary irreducible characters: wr_α exceptional ones and l_α non-exceptional ones.*

Proof. As shown in (7, (7D)), the total number of irreducible characters in $B(\vartheta_\alpha)$ is obtained by letting P range over a system of representatives of the conjugate classes of order a power of p, determining for each the number $l_\alpha(P)$ of modular irreducible characters of $\mathfrak{C}(P)$ corresponding to $B(\vartheta_\alpha)$ and taking $\sum_P l_\alpha(P)$. We know that for $P \neq 1$ the modular irreducible characters of $\mathfrak{C}(P)$ corresponding to $B(\vartheta_\alpha)$ are the r_α characters $\vartheta_\alpha^N \in \{\vartheta_\alpha\}$. By (3.3) there are w such representatives P. For $P = 1$ we have $l_\alpha = l_\alpha(1)$ by definition. Thus $\sum_P l_\alpha(P) = wr_\alpha + l_\alpha$. This yields the first statement. According to (4B), wr_α of these characters are exceptional; in the case $wr_\alpha = 1$ this was a matter of definition. The remaining l_α characters then must be non-exceptional.

In particular, (4F) shows that every block $B(\vartheta_\alpha)$ contains non-exceptional irreducible characters.

As a corollary of the results of §§ 2 and 3, we have

(4G) *All exceptional characters in a block $B(\vartheta_\alpha)$ take the same value for p-regular elements $G \in \mathfrak{G}$. In particular, they all have the same degree.*

For the degrees of the characters, we have congruences mod p^n:

(4H) *If χ_i is a non-exceptional character in $B(\vartheta_\alpha)$ and b_i has the same significance as in* (4E), *then* $\mathrm{Dg}\,\chi_i \equiv b_i r_\alpha \,\mathrm{Dg}\,\vartheta_\alpha \pmod{p^n}$. *In particular* $b_i \neq 0$. *If the notation is as in* (4D), *then*

$$\mathrm{Dg}\,\chi_\beta^{(\alpha)} \equiv \epsilon_\alpha r_\alpha s_\alpha \,\mathrm{Dg}\,\vartheta_\alpha + b_\alpha^{(\alpha)} r_\alpha \,\mathrm{Dg}\,\vartheta_\alpha \qquad \pmod{p^n}.$$

Proof. For the non-exceptional character, χ_i, it follows from (4E) for $V = 1$, $P \in \mathfrak{P}$, $P \neq 1$ that $\chi_i(P) = b_i r_\alpha \,\mathrm{Dg}\,\vartheta_\alpha$. Now (3.12) can be applied. Since $\mathrm{Dg}\,\chi_i$ cannot be divisible by p^n, we must have $b_i \neq 0$.

In the case of $\chi_\beta^{(\alpha)}$ we apply (3.10) for $V = 1$. Since $\mathrm{Dg}\,(\lambda_\beta \vartheta_\alpha)^* = (\mathfrak{H} : \mathfrak{C})\,\mathrm{Dg}\,\vartheta_\alpha$, the left side becomes $\mathrm{Dg}\,\chi_\beta^{(\alpha)} - \epsilon_\alpha r_\alpha s_\alpha \,\mathrm{Dg}\,\vartheta_\alpha$. Using (3.9) and the fact that $b_\rho^{(\alpha)} = 0$ for $\rho \neq \alpha$, we can write the right side in the form $b_\alpha^{(\alpha)} \zeta_\alpha(1) = b_\alpha^{(\alpha)} r_\alpha \,\mathrm{Dg}\,\vartheta_\alpha$, and the last part of (4H) is obtained.

We may also use (7, (3A)). Since for $P \in \mathfrak{P}$, $P \neq 1$, the Cartan matrix of each block of $\mathfrak{C}(P) = \mathfrak{P} \times \mathfrak{B}$ is (p^n), we find for $\phi_j{}^P = \vartheta_\alpha{}^M$,

$$\sum_i |d_{ij}^P|^2 = p^n,$$

and because of the orthogonality of these d-columns, we find

$$\sum_i \Big| \sum_P \sum_j d_{ij}^P \Big|^2 = w r_\alpha p^n, \tag{4.5}$$

where P ranges over a system of representatives of the p-classes $\neq 1$ of $\mathfrak{G}$ and for each P, j ranges over the r_α values belonging to the r_α characters $\vartheta_\alpha{}^M \in \{\vartheta_\alpha\}$. If χ_i is non-exceptional then by (4E)

$$\sum_P \sum_j d_{ij}^P = w r_\alpha b_i. \tag{4.6}$$

For $\chi_i = \chi_\beta^{(\alpha)}$, by (4.4)

$$\sum_P \sum_j d_{ij}^P = \epsilon_\alpha \sum_P \sum_{N \in R} \lambda_\beta^N(P) + w r_\alpha b_\alpha^{(\alpha)} \tag{4.7}$$

since T in (4.4) ranges over a residue system of $\mathfrak{F}(\vartheta_\alpha) \pmod{\mathfrak{C}}$ while M has to range over a residue system of $\mathfrak{H} \pmod{\mathfrak{F}(\vartheta_\alpha)}$. As N ranges over a residue system R of $\mathfrak{H} \pmod{\mathfrak{C}}$, NPN^{-1} ranges over the set of elements of $\mathfrak{P}$ conjugate to P in $\mathfrak{G}$, and the first term on the right in (4.7) can be written as $\epsilon_\alpha \sum_Q \lambda_\beta(Q)$, where Q ranges over $\mathfrak{P} - \{1\}$. Thus (4.7) becomes

$$\sum_P \sum_j d_{ij}^P = -\epsilon_\alpha + w r_\alpha b_\alpha^{(\alpha)}. \tag{4.8}$$

Since we have $w r_\alpha$ characters $\chi_\beta^{(\alpha)}$ in $B(\vartheta_\alpha)$, (4.5) becomes

$$w^2 r_\alpha^2 \sum_i{}' b_i^2 + w r_\alpha (w r_\alpha b_\alpha^{(\alpha)} - \epsilon_\alpha)^2 = w r_\alpha p^n$$

where in $\sum'$ the index i ranges over the l_α values belonging to the non-exceptional characters in $B(\vartheta_\alpha)$. Using (3.3) and (3.4) we can write this in the form

$$\sum_i{}' b_i^2 + w r_\alpha b_\alpha^{(\alpha)2} - 2\epsilon_\alpha b_\alpha^{(\alpha)} = s_\alpha$$

or

$$\sum_i{}' b_i^2 + (wr_\alpha - 1)b_\alpha^{(\alpha)2} + (b_\alpha^{(\alpha)} - \epsilon_\alpha)^2 = s_\alpha + 1. \tag{4.9}$$

At least one of the numbers $b_\alpha^{(\alpha)}$, $b_\alpha^{(\alpha)} - \epsilon_\alpha$ is not zero. Since by (4H) $b_i \neq 0$, it follows for $wr_\alpha \neq 1$ that $\sum_i' b_i^2$ has at most s_α terms. If $wr_\alpha = 1$ then $s_\alpha = p^n - 1 \equiv -1 \pmod{p^n}$, and since $\mathrm{Dg}\,\chi^{(\alpha)} \not\equiv 0 \pmod{p^n}$, (4H) shows that $b_\alpha^{(\alpha)} - \epsilon_\alpha \not\equiv 0 \pmod{p^n}$. Hence the first sum again contains at most s_α terms. Thus

(4I) *The number l_α of non-exceptional characters χ_i in $B(\vartheta_\alpha)$ is at most s_α.*

It follows from (4.9) that if $b_\alpha^{(\alpha)} \neq 0$ then

$$wr_\alpha - 1 \leqslant 1 - l_\alpha + (p^n - 1)/(wr_\alpha).$$

Finally, it follows from the orthogonality relations for decomposition numbers (7, (3A)) and (4.6), (4.8), and (4G) that we have

(4J) *If G is a p-regular element of $\mathfrak{G}$, then for every β,*

$$\sum{}' b_i\chi_i(G) + (wr_\alpha b_\alpha^{(\alpha)} - \epsilon_\alpha)\chi_\beta^{(\alpha)}(G) = 0$$

where χ_i ranges over the non-exceptional characters of $B(\vartheta_\alpha)$.

5. Estimates for the degrees of the representations of $\mathfrak{G}$. We use our results to give lower estimates for the degrees of the irreducible representations of a group $\mathfrak{G}$ which satisfies condition (*). The following statement is obvious from (4A).

(5A) *If the irreducible character χ of $\mathfrak{G}$ does not belong to a block of defect n, the degree $\mathrm{Dg}\,\chi$ is divisible by p^n.*

We next consider the irreducible characters χ_i in the block $B(\vartheta_\alpha)$ of full defect.

(5B) *Let χ_i be a non-exceptional character belonging to $B(\vartheta_\alpha)$. Then either $\mathrm{Dg}\,\chi_i \geqslant p^n - 1$ or $\mathfrak{P}$ is included in the kernel $\mathfrak{K}(\chi_i)$ of χ_i (that is, the kernel of the representation of $\mathfrak{G}$ with the character χ_i).*

Proof. Let b_i have the same significance as in (4E). By (4.9), $|b_i| < p^n$. It follows from (4E) that

$$\chi_i(P) = b_i r_\alpha \mathrm{Dg}\,\vartheta_\alpha, \qquad (P \in \mathfrak{P}, P \neq 1).$$

Now the orthogonality relations for group characters show that

$$\chi_i(1) + (p^n - 1)b_i r_\alpha \mathrm{Dg}\,\vartheta_\alpha = ap^n \tag{5.1}$$

where a is the multiplicity of the principal character in $\chi_i|\mathfrak{P}$; $a \geqslant 0$. If $b_i < 0$, it follows from (5.1) that

$$\chi_i(1) \geqslant (p^n - 1)|b_i| r_\alpha \mathrm{Dg}\,\vartheta_\alpha \geqslant p^n - 1.$$

If $b_i > 0$, we have $\chi_i(1) \geqslant \chi_i(P) > 0$ for all $P \in \mathfrak{P}$. According to (4H), $\chi_i(1) \equiv \chi_i(P) \pmod{p^n}$. If $\chi_i(1) > \chi_i(P)$ for some P, we have $\mathrm{Dg}\,\chi_i > p^n$. On the other hand, if $\chi_i(P) = \chi_i(1)$ for all $P \in \mathfrak{P}$, then, as is well known, $\mathfrak{P}$ is included in the kernel $\mathfrak{K}(\chi_i)$.

We now turn to a discussion of the exceptional characters $\chi_\beta{}^{(\alpha)}$. If $wr_\alpha = 1$, then $\chi_\beta{}^{(\alpha)}$ is not truly exceptional and (5B) still applies.

(5C) *Let* $\chi = \chi_\beta{}^{(\alpha)}$ *be an exceptional character of* $B(\vartheta_\alpha)$, *with* $wr_\alpha > 1$. *If* $\epsilon_\alpha = -1$, *then* $\mathrm{Dg}\,\chi \geqslant \frac{1}{2}(p^n - 1)$. *If* $\epsilon_\alpha = 1$ *and* $\mathrm{Dg}\,\chi < p^n$ *then* $b_\alpha{}^{(\alpha)} \geqslant 0$ *and*

$$\chi \,|\, \mathfrak{C} = \sum_{N \in R} \lambda_\beta^N \vartheta_\alpha^N + b_\alpha^{(\alpha)} \sum_M \vartheta_\alpha^M \tag{5.2}$$

where M *ranges over a residue system of* $\mathfrak{H}$ (mod $\mathfrak{F}(\vartheta_\alpha)$). *In particular*

$$\mathrm{Dg}\,\chi = (s_\alpha + b_\alpha^{(\alpha)}) r_\alpha \,\mathrm{Dg}\,\vartheta_\alpha. \tag{5.3}$$

Proof. If $\epsilon_\alpha = -1$ then $a(F(\vartheta_\alpha), F(\vartheta_\alpha)) \geqslant 1$ in (3.8). Indeed, in (2.1) S must contain ψ_i as a constituent when $\epsilon = -1$. Since $(\lambda_\beta \vartheta_\alpha)^*$ has degree $r_\alpha s_\alpha \,\mathrm{Dg}\,\vartheta_\alpha$ and since $F(\vartheta_\alpha)$ consists of wr_α characters of this degree,

$$\mathrm{Dg}\,\chi \geqslant (wr_\alpha - 1) r_\alpha s_\alpha \,\mathrm{Dg}\,\vartheta_\alpha \geqslant \tfrac{1}{2} wr_\alpha r_\alpha s_\alpha \,\mathrm{Dg}\,\vartheta_\alpha = \tfrac{1}{2}(p^n - 1) r_\alpha \,\mathrm{Dg}\,\vartheta_\alpha$$

whence the first statement.

Assume now $\epsilon_\alpha = 1$. If in (3.8) $a(F(\vartheta_\alpha), F(\vartheta_{\alpha'})) \neq 0$ for some α', an argument similar to the one used for $\epsilon_\alpha = -1$ shows that $\mathrm{Dg}\,\chi \geqslant 1 + (p^n - 1) = p^n$. If this is not so, then in (3.8) $a(F(\vartheta_\alpha), F)$ vanishes except for families F consisting of one irreducible character belonging to a set $U(\vartheta_\rho)$. If $b_\rho{}^{(\alpha)}$ has the same significance as in (3.9), we have here

$$\chi \,|\, \mathfrak{C} = \sum_{N \in R} \lambda_\beta^N \vartheta_\alpha^N + \sum_\rho b_\rho^{(\alpha)} \zeta_\rho$$

and all $b_\rho{}^{(\alpha)}$ are non-negative. Since $b_\rho{}^{(\alpha)} = 0$ for $\rho \neq \alpha$, this yields (5.2) and (5.3).

(5D) *Let* $\mathfrak{G}$ *be a group which satisfies assumption* (*) *for a prime* p *and let* p^n *be the order of the* p*-Sylow group. If* $\mathfrak{G}$ *has an irreducible character* $\chi \neq 1$ *of degree* $x < (p^n - 1)^{\frac{1}{2}}$, *then either* $\mathfrak{G}$ *has a proper normal subgroup* $\mathfrak{K} \supseteq \mathfrak{P}$, *or* $\mathfrak{G}$ *has a normal subgroup* $\mathfrak{M}$ *of index* p^n *and* $x = 1$.

Proof. It follows from (5A), (5B), and (5C), that we may assume χ is of the form $\chi = \chi_\beta{}^{(\alpha)}$ and $wr_\alpha > 1$. Then by (4G) $\chi_1{}^{(\alpha)} - \chi_2{}^{(\alpha)}$ vanishes for all p-regular elements. Let ϑ_1 denote the principal character of $\mathfrak{V}$. It follows from (4C) that the principal character 1 of $\mathfrak{G}$ belongs to $B(\vartheta_1)$. We distinguish two cases:

Case I. $B(\vartheta_1)$ contains an irreducible non-exceptional character $\chi_i \neq 1$. If b_i has the same significance as in (4E), then $\chi_i - b_i$ vanishes for all p-singular elements of $\mathfrak{G}$. Hence

$$(\chi_1^{(\alpha)} - \chi_2^{(\alpha)})(\chi_t - b_t) = 0$$

identically; that is,

$$\chi_1^{(\alpha)}\chi_t + b_t\chi_2^{(\alpha)} = \chi_2^{(\alpha)}\chi_t + b_t\chi_1^{(\alpha)}. \tag{5.4}$$

If $b_t > 0$ then $\chi_1^{(\alpha)}$ is a constituent of $\chi_1^{(\alpha)}\chi_t$. Using the orthogonality relations, we see that χ_t is a constituent of $\chi_1^{(\alpha)}\bar{\chi}_1^{(\alpha)}$. Hence

$$\mathrm{Dg}\,\chi_t \leqslant (\mathrm{Dg}\,\chi)^2 = x^2 < p^n - 1. \tag{5.5}$$

Hence by (5B) the kernel $\mathfrak{K}(\chi_t)$ includes $\mathfrak{P}$, and, as $\chi_t \neq 1$, $\mathfrak{K}(\chi_t) \neq \mathfrak{G}$, we have the first alternative in (5D).

If $b_t < 0$, it follows from (5.4) that $\chi_1^{(\alpha)}$ is a constituent of $\chi_2^{(\alpha)}\chi_t$. Then χ_t is a constituent of $\chi_1^{(\alpha)}\bar{\chi}_2^{(\alpha)}$. This implies (5.5) and again (5D) holds. (Actually, the proof of (5B) shows that this case is impossible under our assumptions.)

Case II. The only irreducible non-exceptional character in $B(\vartheta_1)$ is the principal character of $\mathfrak{G}$. Then (4F) shows that $B(\vartheta_1)$ contains only one irreducible modular character. This implies that $\mathfrak{G}$ has a normal subgroup $\mathfrak{M}$ of index p^n (**8**, p. 587). It follows at once that $\mathfrak{N}(\mathfrak{P}) = \mathfrak{C}(\mathfrak{P})$, that is, that $\mathfrak{H} = \mathfrak{C}$. In particular,

$$r_\rho = s_\rho = 1 \text{ for all } \rho, \qquad w = p^n - 1.$$

Since $s_\alpha = 1$, (4.9) shows that for $p^n > 3$, we must have $b_\alpha^{(\alpha)} = 0$. Moreover $B(\vartheta_\alpha)$ then contains only one non-exceptional character χ_t, and for this χ_t we have $b_t = \pm 1$. As shown by (4J) then $\mathrm{Dg}\,\chi_t = \mathrm{Dg}\,\chi_\beta^{(\alpha)} = x$. If $\vartheta_\alpha \neq \vartheta_1$, χ_t is not the principal character. Since $\mathrm{Dg}\,\chi_t < (p^n - 1)^{\frac{1}{2}}$, (5B) shows that $\mathfrak{G}$ has a proper normal subgroup $\mathfrak{K}(\chi_t) \supseteq \mathfrak{P}$. If $\vartheta_\alpha = \vartheta_1$ then $\chi_t = 1$ and we have $x = 1$.

It remains to discuss the cases $p^n \leqslant 3$. If $p^n = 2$ then $wr_\alpha = 1$, a contradiction. If $p^n = 3$ and $b_\alpha^{(\alpha)} = 0$, the argument above applies. If $b_\alpha^{(\alpha)} \neq 0$ then (4.9) shows that $b_\alpha^{(\alpha)} = \epsilon_\alpha$ and that $B(\vartheta_\alpha)$ still contains only one non-exceptional χ_t and that $b_t = \pm 1$. Then (4J) shows that $\mathrm{Dg}\,\chi_t = x$ and we can conclude the proof as in the case $p^n > 3$.

(5E) *Let $\mathfrak{G}$ be a finite group which satisfies the assumption (*) for some prime p. If the p-Sylow group $\mathfrak{P}$ has order p^n and if $\mathfrak{G}$ has a faithful representation $\mathfrak{X}$ of degree $x < (p^n - 1)^{\frac{1}{2}}$, then $\mathfrak{P}$ is normal in $\mathfrak{G}$.*

Proof. We may assume that (5E) has been proved for groups of smaller order than $\mathfrak{G}$. Since $\mathfrak{X}$ must have an irreducible constituent which is not the principal representation, (5D) applies. If $\mathfrak{G}$ has a proper normal subgroup $\mathfrak{K} \supseteq \mathfrak{P}$, application of (5E) to $\mathfrak{K}$ shows that $\mathfrak{P}$ is normal in $\mathfrak{K}$. Since $\mathfrak{P}$ then is characteristic in $\mathfrak{K}$, it is normal in $\mathfrak{G}$.

It remains to deal with the case that $\mathfrak{G}$ has a normal subgroup $\mathfrak{M}$ of index p^n and that all irreducible constituents of $\mathfrak{X}$ have degree 1. Since $\mathfrak{X}$ is a faithful representation, $\mathfrak{G}$ then is abelian, and, of course, $\mathfrak{P}$ is normal in $\mathfrak{G}$.

The case $n = 1$ has been studied elsewhere **(4)**. If $n \geqslant 2$, it follows from (5E) that if the order g of a finite group $\mathfrak{G}$ is divisible by the square p^2 of a prime p and if $\mathfrak{G}$ has a faithful representation $\mathfrak{X}$ of degree $x < p$, then either the p-Sylow group $\mathfrak{P}$ is normal in $\mathfrak{G}$, or there exist elements $P \neq 1$ in $\mathfrak{P}$ such that $\mathfrak{C}(P) \neq \mathfrak{C}(\mathfrak{P})$.

References

1. H. F. Blichfeldt, *On the order of linear homogeneous groups (supplement)*, Trans. Amer. Math. Soc., *7* (1906), 523–529.

2. R. Brauer, *On the connection between the ordinary and the modular characters of groups of finite order*, Ann. Math., *42* (1941), 926–935.

3. ——— *On groups whose order contains a prime number to the first power*, I, Amer. J. Math., *64* (1942), 401–420.

4. ——— *On groups whose order contains a prime number to the first power*, II, Amer. J. Math., *64* (1942), 421–440.

5. ——— *A characterization of the characters of groups of finite order*, Ann. Math., *57* (1953), 357–377.

6. ——— *Zur Darstellungstheorie der Gruppen endlicher Ordnung*, I, Math. Z., *63* (1956), 406–444.

7. ——— *Zur Darstellungstheorie der Gruppen endlicher Ordnung*, II, Math. Z., *72* (1959), 25–46.

8. R. Brauer and C. Nesbitt, *On the modular characters of groups*, Ann. Math., *42* (1941), 556–590.

9. R. Brauer, M. Suzuki, and G. E. Wall, *A characterization of the one-dimensional unimodular projective groups over finite fields*, Illinois J. Math., *2* (1958), 718–745.

10. M. Suzuki, *On finite groups with cyclic Sylow subgroups for all odd primes*, Amer. J. Math., *77* (1955), 657–691.

11. ——— *Applications of group characters*, Proceedings of Symposia in Pure Mathematics, Amer. Math. Soc., *1* (1959), 88–99.

Harvard University
and
Carnegie Institute of Technology

Sonderabdruck aus
ARCHIV DER MATHEMATIK
Vol. XIII, 1962 BIRKHÄUSER VERLAG, BASEL UND STUTTGART Fasc. 1—3

On Groups of Even Order with an Abelian 2-Sylow Subgroup

To Reinhold Baer on his sixtieth birthday

By

Richard Brauer

1. Introduction. Let $\mathfrak{P}$ be a given 2-group of order 2^n. It is rather surprising that it is possible to obtain information about arbitrary finite groups $\mathfrak{G}$ which contain $\mathfrak{P}$ as their 2-Sylow subgroup though at present these results cannot be given in an explicit form.

We consider here the special case that $\mathfrak{P}$ is abelian. By the *2-regular core* $\mathfrak{K}(\mathfrak{G})$ of a finite group $\mathfrak{G}$, we mean the unique maximal normal subgroup of $\mathfrak{G}$ of odd order. Of course, $\mathfrak{K}(\mathfrak{G})$ is characteristic in $\mathfrak{G}$. We shall then prove:

Theorem 1. *Let $\mathfrak{P}$ be an abelian group of order 2^n which does not possess a direct factor elementary abelian of order 4. Then there exists a bound b depending only on n such that for every finite group $\mathfrak{G}$ with $\mathfrak{P}$ as its 2-Sylow subgroup, we have*

$$(\mathfrak{G} : \mathfrak{K}(\mathfrak{G})) \leqq b. \tag{1}$$

This is of course equivalent with saying that there exist only finitely many $\mathfrak{G}$ with $\mathfrak{P}$ as their 2-Sylow subgroup and $\mathfrak{K}(\mathfrak{G}) = \{1\}$. On the other hand, if $\mathfrak{P} = \mathfrak{P}_1 \times \mathfrak{P}_2$ is an abelian 2-group and if $\mathfrak{P}_1$ is elementary abelian of order 4, there exist infinitely many finite groups $\mathfrak{G}$ whose 2-Sylow subgroup is $\mathfrak{P}$ and which have core $\mathfrak{K}(\mathfrak{G}) = \{1\}$. We only have to take the groups $\mathfrak{G} = L_2(q) \times \mathfrak{P}_2$ where q is a prime power with $q \equiv \pm 3 \bmod 8$. This shows that the assumption in Theorem 1 is essential.

It follows from Theorem 1 that if $\mathfrak{P}$ satisfies the assumptions of the Theorem, there can exist at most finitely many simple groups $\mathfrak{G}$ with $\mathfrak{P}$ as their 2-Sylow subgroup. Our method also yields:

Theorem 2. *Let $\mathfrak{P}$ be a given abelian group of order $2^n \geqq 8$ which does not possess a direct factor elementary abelian of order 8. There exist at most finitely many simple groups $\mathfrak{G}$ with $\mathfrak{P}$ as their 2-Sylow subgroup.*

2. Proof of Theorem 1. We use induction on n. For $n = 0$, $\mathfrak{K}(\mathfrak{G}) = \mathfrak{G}$ and we can take $b = 1$ in (1). Assume then that $n \geqq 1$ and that we have a bound b_1 such that

$$(\mathfrak{G}_1 : \mathfrak{K}(\mathfrak{G}_1)) \leqq b_1 \tag{2}$$

for all groups $\mathfrak{G}_1$ with an abelian 2-Sylow group $\mathfrak{P}_1$ of an order less than 2^n for which $\mathfrak{P}_1$ does not have an elementary abelian direct factor of order 4. Let $\mathfrak{P}$ have order 2^n and satisfy the assumptions of Theorem 1 and let $\mathfrak{G}$ be a finite group with $\mathfrak{P}$ as its 2-Sylow group.

Let $\mathfrak{P}^*$ denote the subgroups of $\mathfrak{P}$ generated by all quotients $P_1^{-1}P_2$ of elements P_1, P_2 of $\mathfrak{P}$ which are conjugate in $\mathfrak{G}$. This *"focal"* group $\mathfrak{P}^*$ of $\mathfrak{P}$ in $\mathfrak{G}$ is the image of $\mathfrak{G}$ under the transfer map into $\mathfrak{P}$. As $\mathfrak{P}$ is abelian, by Burnside's remark, P_1 and P_2 are conjugate in the normalizer $\mathfrak{N}(\mathfrak{P})$ of $\mathfrak{P}$ in $\mathfrak{G}$. Every element of $\mathfrak{N}(\mathfrak{P})$ defines an automorphism of $\mathfrak{P}$. If $\mathfrak{C}(\mathfrak{P})$ is the centralizer of $\mathfrak{P}$ in $\mathfrak{G}$, then $\mathfrak{N}(\mathfrak{P})/\mathfrak{C}(\mathfrak{P})$ is isomorphic with a group $\mathfrak{A}$ of automorphisms of $\mathfrak{P}$. Since $\mathfrak{P} \subseteqq \mathfrak{C}(\mathfrak{P})$, $\mathfrak{A}$ has odd order. It can be seen easily that $\mathfrak{P}^*$ has the following two properties[1]):

(a) $\mathfrak{P}^*$ is a direct factor of $\mathfrak{P}$.

(b) There does not exist an element different from 1 in $\mathfrak{P}^*$ which is invariant under all elements of $\mathfrak{A}$.

We shall distinguish several cases in our proof of Theorem 1.

Case I. The group $\mathfrak{P}^*$ is different from $\mathfrak{P}$.

By the basic properties of the focal group $\mathfrak{P}^*$, it follows that there exists a normal subgroup $\mathfrak{G}_1$ of $\mathfrak{G}$ with the 2-Sylow group $\mathfrak{P}^*$ such that $\mathfrak{G}/\mathfrak{G}_1 \simeq \mathfrak{P}/\mathfrak{P}^*$ is a 2-group. Since $\mathfrak{K}(\mathfrak{G}_1)$ is characteristic in $\mathfrak{G}_1$, it is normal in $\mathfrak{G}$ and hence $\mathfrak{K}(\mathfrak{G}_1) = \mathfrak{K}(\mathfrak{G})$. It now follows from (2) that

$$(\mathfrak{G} : \mathfrak{K}(\mathfrak{G})) \leqq 2^n b_1. \tag{3}$$

Case II. $\mathfrak{P}^* = \mathfrak{P}$.

Let J be an element of order 2 in $\mathfrak{P}$. Let $\mathfrak{G}_0 = \mathfrak{C}(J)$ be the centralizer of J in $\mathfrak{G}$. Then $\mathfrak{G}_0$ has the 2-Sylow subgroup $\mathfrak{P}$. If $\mathfrak{P}_0^*$ is the focal group of $\mathfrak{P}$ in $\mathfrak{G}_0$, the property (b) of the focal group stated above shows that $J \notin \mathfrak{P}_0^*$. Indeed, J is invariant under the action of all elements of $\mathfrak{G}_0$. Since then $\mathfrak{P}_0^* \neq \mathfrak{P}$, (3) applies for $\mathfrak{G}_0$ and we have

$$(\mathfrak{G}_0 : \mathfrak{K}(\mathfrak{G}_0)) \leqq 2^n b_1.$$

If $\mathfrak{P}$ does not contain elements of order 4, then $\mathfrak{P}$ is elementary abelian. By the assumption of Theorem 1, $\mathfrak{P}$ has to be of order 2. In this case, $\mathfrak{P}^* = \{1\}$ and we have Case I. Hence $\mathfrak{P}$ must contain elements P of order 4. If P was conjugate to P^{-1} in $\mathfrak{G}$, say $G^{-1}PG = P^{-1}$, replacing G be a suitable power with odd exponent, we may assume that G is a 2-element. Then $\{G, P\}$ would be a non-abelian 2-subgroup of $\mathfrak{G}$ which is impossible, if the 2-Sylow group $\mathfrak{P}$ is abelian.

---

[1]) A short direct proof of (a) and (b) has been given by J. E. McLaughlin and H. N. Ward, [4] p. 12. In fact, there the following proposition is proved in a few lines. Let $\mathfrak{A}$ be a group of finite order a and let $\mathfrak{P}$ be an $\mathfrak{A}$-module such that the map $P \to aP$ of $\mathfrak{P}$ is an automorphism of $\mathfrak{P}$. Let $\mathfrak{P}^*$ be the submodule of $\mathfrak{P}$ generated by the elements $\sigma P - P$, $\sigma \in \mathfrak{A}$, $P \in \mathfrak{P}$. Then $\mathfrak{P}$ is the direct sum of $\mathfrak{P}^*$ and the module consisting of the elements of $\mathfrak{P}$ left fixed by all $\sigma \in \mathfrak{A}$.

One can also use a recent theorem of Grün [3], which states that if $\mathfrak{P}$ is an abelian p-group (for any prime p) and $\mathfrak{A}$ a group of automorphisms of $\mathfrak{P}$ of an order prime to p, then $\mathfrak{P}$ can be written as a direct product $\mathfrak{P} = \mathfrak{P}_1 \times \mathfrak{P}_2 \times \cdots \times \mathfrak{P}_r$ such that each $\mathfrak{P}_\varrho$ is left fixed by $\mathfrak{A}$ and is a direct product of cyclic groups of the same order p^m. If $\overline{\mathfrak{P}}_\varrho$ is the quotient group of $\mathfrak{P}_\varrho$ modulo the group of elements of $p^{m_\varrho - 1}$-st powers of elements of $\mathfrak{P}_\varrho$, then $\overline{\mathfrak{P}}_\varrho$ is a vector space over a field with p elements and the action of $\mathfrak{A}$ on $\overline{\mathfrak{P}}_\varrho$ defines a modular representation T_ϱ of $\mathfrak{A}$. Now, Grün's theorem can be refined as follows: The $\mathfrak{P}_\varrho$ can be chosen such that all T_ϱ are irreducible in the field with p elements. (Several of the P_ϱ may have the same exponent p^{m_ϱ}.) This also yields (a) and (b).

Hence, P and P^{-1} are not conjugate in $\mathfrak{G}$. Now a recent result[2]) of the author shows that $\mathfrak{G}$ possesses a subgroup $\mathfrak{M}$ such that

(4) $$\mathfrak{M} \lhd \mathfrak{G}, \quad \mathfrak{M} \neq \mathfrak{G},$$

(5) $$\mathfrak{M} \supseteqq \mathfrak{K}(\mathfrak{G}),$$

(6) $$(\mathfrak{G} : \mathfrak{M}) \leqq b_2,$$

where b_2 is a bound depending only on n.

Let $\mathfrak{P}_1$ be a 2-Sylow subgroup of $\mathfrak{M}$. We may assume that $\mathfrak{P}_1 \subseteqq \mathfrak{P}$. We separate two subcases:

Case IIA. $\mathfrak{P}_1 \neq \mathfrak{P}$.

Set $\mathfrak{G}_1 = \mathfrak{P}\mathfrak{M}$. Then $\mathfrak{G}_1 \neq \mathfrak{M}$ and since $\mathfrak{G}_1/\mathfrak{M} \cong \mathfrak{P}/(\mathfrak{P} \cap \mathfrak{M})$ is a 2-group of order at least 2, the focal group $\mathfrak{P}_1^*$ of the 2-Sylow group $\mathfrak{P}$ of $\mathfrak{G}_1$ in $\mathfrak{G}_1$ is not $\mathfrak{P}$. Hence we have Case I for $\mathfrak{G}_1$. By (3)

(3*) $$(\mathfrak{G}_1 : \mathfrak{K}(\mathfrak{G}_1)) \leqq 2^n b_1.$$

Moreover, $\mathfrak{K}(\mathfrak{G}_1) = \mathfrak{K}(\mathfrak{M})$. On the other hand, by (5), $\mathfrak{K}(\mathfrak{G}) \subseteqq \mathfrak{K}(\mathfrak{M})$. Since $\mathfrak{M}$ is normal in $\mathfrak{G}$, $\mathfrak{K}(\mathfrak{M})$ is normal in $\mathfrak{G}$ and hence $\mathfrak{K}(\mathfrak{M}) \subseteqq \mathfrak{K}(\mathfrak{G})$. Thus, $\mathfrak{K}(\mathfrak{G}) = \mathfrak{K}(\mathfrak{M}) = \mathfrak{K}(\mathfrak{G}_1)$. It now follows from (6) and (3) that if we set $b_3 = 2^n b_1 b_2$,

(7) $$(\mathfrak{G} : \mathfrak{K}(\mathfrak{G})) \leqq b_3.$$

Case II B. $\mathfrak{P}_1 = \mathfrak{P}$, i.e. $\mathfrak{P} \subseteqq \mathfrak{M}$.

If we replace $\mathfrak{G}$ by $\mathfrak{G}/\mathfrak{K}(\mathfrak{G})$, we may assume that $\mathfrak{K}(\mathfrak{G}) = \{1\}$. We have to find an upper bound for the order g of $\mathfrak{G}$.

Let $\mathfrak{P}^{(1)}, \mathfrak{P}^{(2)}, \ldots, \mathfrak{P}^{(r)}$ denote the distinct 2-Sylow groups of $\mathfrak{G}$. One of them, viz. $\mathfrak{P} = \mathfrak{P}_1$ lies in $\mathfrak{M} \lhd \mathfrak{G}$. Hence all $\mathfrak{P}^{(\varrho)}$ lie in $\mathfrak{M}$. Let $\mathfrak{G}_2$ be the group generated by $\mathfrak{P}^{(1)}, \mathfrak{P}^{(2)}, \ldots, \mathfrak{P}^{(r)}$,

(8) $$\mathfrak{G}_2 = \{\mathfrak{P}^{(1)}, \mathfrak{P}^{(2)}, \ldots, \mathfrak{P}^{(r)}\}$$

so that $\mathfrak{G}_2 \subseteqq \mathfrak{M}$ and $\mathfrak{G}_2 \lhd \mathfrak{G}$. Since $\mathfrak{K}(\mathfrak{G}) = \{1\}$, we have $\mathfrak{K}(\mathfrak{G}_2) = \{1\}$.

The subgroup $\mathfrak{G}_2$ of $\mathfrak{G}$ has the Sylow group $\mathfrak{P}$. We wish to show that its order g_2 lies below the bound b_3. This will follow from (3) and (7) and $b_1 \leqq b_3$, if we show that if $\mathfrak{G}_2$ is considered instead of $\mathfrak{G}$, we have Case I or Case IIA. Suppose that we have Case IIB. Let $\mathfrak{M}_2$ have the same significance for $\mathfrak{G}_2$ as $\mathfrak{M}$ had for $\mathfrak{G}$. But then in the Case IIB, $\mathfrak{P} \subseteqq \mathfrak{M}_2$. Thus the 2-Sylow groups $\mathfrak{P}^{(\varrho)}$ of $\mathfrak{G}_2$ all lie in $\mathfrak{M}_2$, $\varrho = 1, 2, \ldots, r$, and (8) implies $\mathfrak{G}_2 = \mathfrak{M}_2$ which contradicts (4) for $\mathfrak{G}_2$ instead of $\mathfrak{G}$. Hence, as stated

(9) $$g_2 \leqq b_3.$$

Since r is the number of distinct 2-Sylow groups of both $\mathfrak{G}$ and $\mathfrak{G}_2$, we have

(10) $$(\mathfrak{G} : \mathfrak{N}(\mathfrak{P})) = r = (\mathfrak{G}_2 : (\mathfrak{G}_2 \cap \mathfrak{N}(\mathfrak{P}))) \leqq g_2 \leqq b_3.$$

If b_4 is the odd factor of the order of the group of automorphisms of $\mathfrak{P}$, the group $\mathfrak{A}$

2) See [1]. Since P and P^{-1} are not conjugate in $\mathfrak{G}$, theorem 2 of the paper shows that Case II of Theorem 1 applies and then Theorem 3 yields the desired result.

above has at most order b_4, i.e.

(11) $$(\mathfrak{N}(\mathfrak{P}):\mathfrak{C}(\mathfrak{P})) \leqq b_4.$$

Finally, by Burnside's theorem applied to $\mathfrak{C}(\mathfrak{P})$, we have

(12) $$\mathfrak{C}(\mathfrak{P}) = \mathfrak{P} \times \mathfrak{V}$$

where $\mathfrak{V}$ is a group of odd order v,

(13) $$(\mathfrak{C}(\mathfrak{P}):1) = 2^n v.$$

It follows from (10), (11) and (13) that

(14) $$g/v \leqq b_3 b_4 2^n.$$

Replacing $\mathfrak{P}$ by its conjugate $\mathfrak{P}^{(\varrho)}$ in $\mathfrak{G}$, we can set

$$\mathfrak{C}(\mathfrak{P}^{(\varrho)}) = \mathfrak{P}^{(\varrho)} \times \mathfrak{V}^{(\varrho)}$$

where $\mathfrak{V}^{(1)}, \mathfrak{V}^{(2)}, \dots, \mathfrak{V}^{(r)}$ are conjugates of $\mathfrak{V}$ in $\mathfrak{G}$. In particular,

(15) $$\mathfrak{V}^{(1)} \cap \mathfrak{V}^{(2)} \cap \dots \cap \mathfrak{V}^{(r)} = \{1\}$$

since the left side is a normal subgroup of $\mathfrak{G}$ of odd order and we assumed $\mathfrak{K}(\mathfrak{G}) = \{1\}$.
For $\varrho = 1, 2, \dots, r$, set

$$\mathfrak{D}_\varrho = \mathfrak{V}^{(1)} \cap \mathfrak{V}^{(2)} \dots \cap \mathfrak{V}^{(\varrho)}$$

and let d_ϱ be the order of $\mathfrak{D}_\varrho$. If $\varrho < r$, $\mathfrak{D}_\varrho \mathfrak{V}^{(\varrho+1)}$ consists of $d_\varrho v/d_{\varrho+1}$ elements since $\mathfrak{D}_\varrho \cap \mathfrak{V}^{(\varrho+1)} = \mathfrak{D}_{\varrho+1}$. Hence $d_\varrho v/d_{\varrho+1} \leqq g$ and by (14),

(16) $$d_\varrho/d_{\varrho+1} \leqq b_3 b_4 2^n$$

for $\varrho = 1, 2, \dots, r-1$. Multiply (16) for $\varrho = 1, 2, \dots, r-1$. As $d_1 = v$ and, by (15), $d_r = 1$, we find

$$v = \prod_{\varrho=1}^{r-1} (d_\varrho/d_{\varrho+1}) \leqq (b_3 b_4 2^n)^{r-1}.$$

Since $r \leqq b_3$, v and then g lie below a bound b_5 depending only on n and this finishes the proof of Theorem 1.

3. Proof of Theorem 2. Assume now that $\mathfrak{P}$ is an abelian 2-group of order $2^n \geqq 8$ which does not have a direct factor elementary abelian of order 8. If there should exist infinitely many simple groups $\mathfrak{G}$ with $\mathfrak{P}$ as their 2-Sylow group, by Theorem 1, $\mathfrak{P}$ has to have a direct factor $\mathfrak{P}_1$ which is elementary abelian of order 4, say

$$\mathfrak{P} = \mathfrak{P}_1 \times \mathfrak{P}_2.$$

Here $\mathfrak{P}_2$ cannot have a direct factor of order 2. For each $\mathfrak{G}$, we must have $\mathfrak{P}^* = \mathfrak{P}$. Let J be an element of order 2 in $\mathfrak{P}_1$. Take $\mathfrak{H} = \mathfrak{C}(J)/\{J\}$. Since the 2-Sylow group $\mathfrak{Q}$ of $\mathfrak{H}$ is isomorphic with $(\mathfrak{P}_1/\{J\}) \times \mathfrak{P}_2$, $\mathfrak{Q}$ does not have a direct factor which is elementary abelian of order 4. Hence Theorem 1 shows that $(\mathfrak{H} : \mathfrak{K}(\mathfrak{H}))$ lies below a bound b^* depending only on n. Then $(\mathfrak{C}(J) : \mathfrak{K}(\mathfrak{C}(J))) \leqq 2b^*$. As in section 2, $\mathfrak{P}$ contains elements P of order 4 and P and P^{-1} are not conjugate in $\mathfrak{G}$. This implies again the existence of a subgroup $\mathfrak{M}$ of $\mathfrak{G}$ satisfying (4), (5), (6) with some constant b_2 depending on n. Since $\mathfrak{G}$ is simple, $\mathfrak{M} = \{1\}$ and hence $g \leqq b_2$. But this means that we can have only finitely many simple $\mathfrak{G}$ and Theorem 2 is proved.

4. Concluding Remarks. Our proof of Theorem 1 depends on the use of characters, since the results of [1] depend on them. However, it is quite possible that direct proofs of Theorem 1 can be given. It is also not unlikely that our Theorem is only a first form of a far more explicit and precise statement.

Actually, by a study of the characters of $\mathfrak{G}$ for a fixed group $\mathfrak{P}$, much more detailed information concerning $\mathfrak{G}$ can be obtained. For instance, if $\mathfrak{P}$ is the direct product of two cyclic groups of order 4, then $\mathfrak{G}/\mathfrak{K}(\mathfrak{G})$ is either $\mathfrak{P}$ or a definite group of order 48.

No simple groups seem to be known whose 2-Sylow group $\mathfrak{P}$ is abelian and is not elementary abelian. If $\mathfrak{P}$ is elementary abelian of order $2^n \geqq 4$, we always have one simple group $\mathfrak{G}$, the group $L_2(2^n)$. In addition, for $n = 2$, we have the groups $L_2(q)$ where q is a prime power with $q \equiv \pm 3 \bmod 8$, $q \neq 3$. For $n = 3$, we have the infinite class of simple groups discovered recently by Ree [5]. It seems to be an open question whether there exist other simple groups with an elementary abelian 2-Sylow group $\mathfrak{P}$.

Our results suggest that an investigation of the following questions may be worthwhile. Let $\mathfrak{P}$ be a given 2-group, abelian or non-abelian.

(I) For which $\mathfrak{P}$ is it true that any finite group $\mathfrak{G}$ with $\mathfrak{P}$ as its 2-Sylow subgroup has a normal subgroup of index 2 ? In other words: For which $\mathfrak{P}$ is the focal group $\mathfrak{P}^*$ of $\mathfrak{P}$ in $\mathfrak{G}$ always different from $\mathfrak{P}$? As is well known, the cyclic groups $\mathfrak{P}$ have this property. There are infinitely many other such $\mathfrak{P}$.

(II) For which $\mathfrak{P}$ is it true that every $\mathfrak{G}$ with $\mathfrak{P}$ as its 2-Sylow subgroup has a normal subgroup of odd prime number index ? As was mentioned above, the direct product $\mathfrak{P}$ of two cyclic groups of order 4 has this property.

(III) For which $\mathfrak{P}$ is it true that for every $\mathfrak{G}$ with $\mathfrak{P}$ as its 2-Sylow subgroup, the group $\mathfrak{G}/\mathfrak{K}(\mathfrak{G})$ has a non-trivial center ? The quaternion groups (ordinary and generalized) are of this nature [2].

(IV) For which $\mathfrak{P}$ do there exist only finitely many groups $\mathfrak{G}$ with $\mathfrak{P}$ as their 2-Sylow group and with $\mathfrak{K}(\mathfrak{G}) = \{1\}$? This is the question which was answered for abelian $\mathfrak{P}$ by Theorem 1.

(V) For which $\mathfrak{P}$ do there exist simple $\mathfrak{G}$, but only finitely many such $\mathfrak{G}$? For instance, the 2-Sylow group $\mathfrak{P}$ of order 64 of Suzuki's simple group of order 29120 can be shown to have this property [5].

(VI) For which $\mathfrak{P}$ do there exist infinitely many simple groups with $\mathfrak{P}$ as their 2-Sylow subgroup ?

These questions can still be modified in various ways.

References

[1] R. BRAUER, Investigation on groups of even order, I. Proc. Nat. Acad. Sci., USA **47**, 1891 to 1893 (1961).

[2] R. BRAUER and M. SUZUKI, On finite groups of even order whose 2-Sylow group is a quaternion group. Proc. Nat. Acad. Sci., USA **45**, 1757–1759 (1959).

[3] O. GRÜN, Einige Sätze über Automorphismen abelscher p-Gruppen. Abh. Math. Sem. Univ. Hamburg **24**, 54–58 (1960).

[4] J. E. MCLAUGHLIN and H. N. WARD, Mimeographed Notes of a Seminar on Groups. Harvard University 1960/61.

[5] R. REE, A family of simple groups associated with the simple Lie algebra of type (G_2). Amer. J. Math. **83**, 432–462 (1961).

[6] M. SUZUKI, A new type of simple groups of finite order. Proc. Nat. Acad. Sci., USA **46**, 868 to 870 (1960).

Eingegangen am 23. 2. 1962

Anschrift des Autors:
Richard Brauer
Department of Mathematics
Harvard University
Cambridge 38 (Mass.), USA

8

On Some Conjectures Concerning Finite Simple Groups

Richard Brauer

One of the most interesting open problems in the theory of finite groups is that of constructing all finite simple groups. At present, we have no clue to the number of groups missing in the list of known simple groups.* As a matter of fact, hardly any properties of finite simple groups are known that are helpful in attacking the problem.

It is our aim to state a number of conjectures concerning finite simple groups. Although these are perhaps not of a very sweeping nature, there is some reason to believe that an answer to these questions may lie within reach. At least some special cases of these conjectures can be proved, and our discussion of the conjectures will enable us to present these results in a more unified form.

1. Our starting point is the celebrated theorem of Burnside which states that the order of a non-cyclic simple group is divisible by at least three distinct primes. Our first conjecture asserts:

(A) *There exist only finitely many simple groups G whose order g is divisible by exactly three distinct primes.*

Eight simple groups of this type are known. Their orders are

(1) 60, 168, 360, 2448, 5616, 6048, 504, 25920.

As a result in the direction of this conjecture, we have

THEOREM 1. *Let G be a simple group of an order $g = 2^n p^a q^b$, where p and q are two distinct primes. If the 2-Sylow group S of G of order 2^n contains elements of order 2^{n-1}, then G is one of the groups $L_2(5)$, $L_2(7)$, $L_2(9)$, $L_2(17)$, $L_3(3)$.*

The proof is based on a method developed in a recent paper [5]. There

This research was supported in part by the United States Air Force Office of Scientific Research of the Air Research and Development Command under Contract AF 49 (638)-287.

* For a list of the known simple finite groups, see Artin [1]. To this should be added the groups discovered more recently by Suzuki [14] and Ree [11, 12].

Reprinted from *Studies in Mathematical Analysis and Related Topics,* edited by Gabor Szegö and others, with the permission of the publishers, Stanford University Press.

are other classes of 2-Sylow groups S for which a similar theorem can be obtained. For instance, if $m \geqq 2$, we may consider the groups $S = \{\sigma_1, \sigma_2, \gamma\}$ defined by

$$\sigma_1^{2m} = \sigma_2^{2m} = 1,\ \gamma^2 = 1;\ \sigma_1\sigma_2 = \sigma_2\sigma_1,\ \gamma^{-1}\sigma_1\gamma = \sigma_2 .$$

The only simple group G of an order $2^n p^a q^b$ with such a 2-Sylow group S is the group $U_3(3)$ of order 6048.

The groups in Theorem 1 correspond to the first five orders.

Another result in the direction of the conjecture is:

THEOREM 2. *If G is a simple group of an order $g = p^a q^b r$, where p, q, and r are distinct primes and $a = 1$ or 2, then G is one of the groups*

$$L_2(5),\ L_2(7),\ L_2(9),\ L_2(17),\ \text{or}\ L_2(8) .$$

The case $a = 1$ is treated in a paper by H. F. Tuan and the author [8]. Somewhat similar but more complicated methods yield the result for $a = 2$.

2. The case $a = 1$ of Theorem 2 is a consequence of the following theorem:

THEOREM 3 (R. Brauer and H. F. Tuan). *Let G be a simple group of order $g = pq^b g_0$, where p and q are primes and b and g_0 are positive integers such that $g_0 < p - 1$. Then either $G \approx L_2(p)$ with $p = 2^m \pm 1 > 3$ or $G \approx L_2(2^m)$ with $p = 2^m + 1 > 3$.*

Incidentally, the question whether there exist only finitely many simple groups G satisfying the conditions of the theorem is equivalent to the question whether there exist only finitely many Fermat and Mersenne primes.

Theorem 3 suggests the following conjectures:

(B) *Let G be a simple group of order $g = p^a q^b g_0$, where p and q are distinct primes ($a > 0$, $b > 0$) and where g_0 is an integer prime to p.* (i) *For fixed g_0 but arbitrary p^a and q^b, the group G is one of a finite set of groups.* (ii) *We have $g_0 \geqq (p - 1)/2$.*

The following special cases can be handled:

THEOREM 4. *Let G be a simple group of the form mentioned in conjecture* (B). *If it is assumed either that $a \leqq 2$ or that the p-Sylow group P is cyclic, then* (B) *holds true.*

PROOF. Let $\chi_1 = 1, \chi_2, \cdots, \chi_m$ denote the irreducible characters of G that belong to the principal p-block. If $\pi \neq 1$ is a p-element of G,

$$\sum_{j=1}^{m} \chi_j(1)\chi_j(\pi) = 0 . \tag{2}$$

It follows that there must exist $\chi_j \neq 1$ whose degree $x_j = \chi_j(1)$ is not divisible by q. Under the assumptions of Theorem 4, x_j is not divisible by p [6]. Hence $x_j \mid g_0$. It now follows from the Theorem of Camille Jordan that g lies below a bound depending only on g_0. Moreover, a recent theorem of W. Feit and J. G. Thompson [10] shows that $x_j \geqq (p - 1)/2$ and hence $g_0 \geqq (p - 1)/2$. Q.E.D. Actually the same proof yields $g_0 \geqq (q - 1)/2$.

Remark. There is a conjecture to the effect that if the p-Sylow group P of G is Abelian, the degrees of the irreducible characters in the principal p-block of G are prime to p [3]. If this conjecture is true, then (B) is true if the p-Sylow group or the q-Sylow group of G is Abelian.

There are some results in the direction of conjecture (B ii) for groups G with certain types of non-Abelian p-Sylow groups P:

THEOREM 5. *Let G be a simple group of the form mentioned in conjecture* (B). *Assume that $a = 3$ and that the p-Sylow group P of G is non-Abelian and does not contain elements of order p^2. Then $g_0 > \sqrt{p/2}$.*

PROOF. As in the proof of Theorem 4, there exist irreducible characters $\chi_\mu \neq 1$ in the principal p-block whose degree x_μ is not divisible by q. If $(x_\mu, p) = 1$, the proof of Theorem 4 applies. If $p \mid x_\mu$, we have $x_\mu \not\equiv 0 \pmod{p^2}$ and hence $x_\mu/p \leqq g_0$. Let X_μ denote the irreducible representation of G with the character χ_μ. If π is an element of the center $Z(P)$ of P, it is seen without difficulty that $X_\mu(\pi)$ has at most g_0 distinct characteristic roots. On the other hand, if $\pi \in P$, $\pi \notin Z(P)$, all pth roots of unity appear as characteristic roots of $X_\mu(\pi)$. Now the methods developed by H. F. Blichfeldt [2] and A. Speiser [13] apply and yield the statement.

The other type of non-Abelian groups of order p^3 is covered by the following Theorem:

THEOREM 6. *Let G be a group of order $g = p^a q^b g_0$, where p and q are distinct primes $[a > 0,\ b > 0$, and $(g_0, pq) = 1]$ such that G does not possess a normal subgroup different from G of an order divisible by p. Assume that the p-Sylow group P of G is not cyclic, and suppose that there exists an order p^α such that there exist elements π of order p^α in P and such that for all such π the centralizer $C_P(\pi)$ of π in P is $\{\pi\}$. Then*

$$g_0 > p^{(3\alpha/2)-a+1} \text{ for } \alpha \text{ even}; \quad g_0 > p^{[(3\alpha+1)/2]-a} \text{ for } \alpha \text{ odd}.$$

PROOF. It follows from the assumption that the centralizer $C_G(\pi)$ of π in G has a normal p-complement. If the notation is as in the proof of Theorem 4, again (2) holds. In addition, we have*

$$\sum_{\mu=1}^{m} |\chi_\mu(\pi)|^2 = p^\alpha . \tag{3}$$

Let ν denote an exponential valuation of the field Ω of algebraic numbers with $\nu(p) = 1$. If we set $x_\mu = \chi_\mu(1)$ and $\nu(x_\mu) = h_\mu$, we find

$$\nu(\chi_\mu(\pi)) > h_\mu + \alpha - a \qquad (\mu = 1, 2, \cdots, m), \tag{4}$$

since χ_μ belongs to the principal p-block. It can be deduced without difficulty from (2), (3), and (4) that there exists a μ $(1 < \mu \leqq m)$ such that

$$x_\mu \not\equiv 0 \pmod{q}, \qquad h_\mu < a - \frac{1}{2}\alpha - 1 .$$

* This follows from the results in [4].

Set

$$\rho = a - \frac{\alpha}{2} - 2 + \log g_0/\log p \quad \text{for } \alpha \text{ even},$$

$$\rho = a - \frac{\alpha}{2} - \frac{3}{2} + \log g_0/\log p \quad \text{for } \alpha \text{ odd}.$$

Since x_μ/p^{h_μ} divides g_0, it follows that $x_\mu \leqq p^\rho$.

Let X_μ again denote the irreducible representation belonging to the character χ_μ. Write X_μ so that its coefficients lie in Ω and are local integers with regard to ν. Then a modular representation X_μ^* is obtained from X_μ by the residue class map. If N is the kernel of X_μ and N^* that of X_μ^*, then $N \subseteq N^*$ and it is seen without difficulty that N^*/N is a p-group. It follows from the assumptions of Theorem 6 that $N^* = N$ and that the order of N is prime to p. Hence $X_\mu^*(\pi)$ has order p^α and this implies $p^{\alpha-1} < x_\mu$. Hence $\alpha - 1 < \rho$, and this yields the statement.

If P is a non-Abelian group of order p^3 that possesses elements of order p^2, then the assumptions of the theorem are satisfied with $a = 3$ and $\alpha = 2$, and we obtain $g_0 > p^{1/2}$.

3. Our next conjecture states:

(C) *If the order g of a non-cyclic simple finite group G is divisible by a prime power p^a, then $g > p^{2a}$.*

Obviously, Burnside's Theorem (Sec. 1) would follow from (C). On the other hand, (C) follows from the slightly stronger conjecture:

(C*) *If the p-Sylow groups of a simple (non-cyclic) group G have order p^a, the number of these p-Sylow groups exceeds p^a.*

This conjecture (C*) is true, and hence (C) is true, if the p-Sylow groups of G are Abelian. This follows from the following Theorem:

THEOREM 7. *Let G be a finite group of order g whose p-Sylow groups are Abelian for some prime p. Then the intersection of all p-Sylow groups of G can be represented as the intersection of two p-Sylow groups.*

A proof of Theorem 7 can be found in a paper of J. M. Brodkey [9]. I am indebted to Garrett Birkhoff for the remark that Theorem 7 fails to hold if the p-Sylow groups of G are non-Abelian.

In the case of non-Abelian p-Sylow groups, we can prove:

THEOREM 8. *Assume that the p-Sylow groups of the finite group G are non-Abelian. If G has no normal subgroup different from G of an order divisible by p, then $g > p^{a+3}$.*

In particular, (C) holds for simple groups in the case $a = 3$. The proof of Theorem 8 is elementary but complicated, and we shall not give it here.

We shall deal with another special case:

THEOREM 9. *Assume that G is a group of finite even order g. Suppose that there is a prime p such that* (i) *the p-Sylow group P of G is non-Abelian of order p^a;* (ii) *There are no elements of order $2p$ in G. Then $g > p^{2a}$.*

PROOF. We must have $p > 2$. Let $\pi \neq 1$ be an element of P and suppose that there exists an involution $\eta \in G$ such that $\eta^{-1}\pi\eta = \pi^{-1}$. Let $\gamma \neq 1$ be an element of the center $Z(P)$ of P, and let P_0 be a p-Sylow group of the centralizer $C_G(\pi)$ of π in G with $P_0 \supseteq \{\pi, \gamma\}$. Then there exists an element $\sigma \in C_G(\pi)$ such that $\eta\sigma$ tranforms $\pi \to \pi^{-1}$, $P_0 \to P_0$. Replacing $\eta\sigma$ by a suitable power, we obtain a 2-element τ in the normalizer $N_G(P_0)$ of P_0 that carries $\pi \to \pi^{-1}$. Since τ^2 then commutes with π, it follows from the assumptions that $\tau^2 = 1$. If we map $\xi \to \tau^{-1}\xi\tau$ for $\xi \in P_0$, P_0 is mapped onto P_0 and only the unit element of P_0 remains fixed. Since P_0 has odd order, it follows that P_0 is Abelian and that $\tau^{-1}\xi\tau = \xi^{-1}$ for all $\xi \in P_0$. In particular, $\tau^{-1}\gamma\tau = \gamma^{-1}$.

We now apply the same argument, replacing π and η by γ and τ. We see that a p-Sylow group P_1 of $C_G(\gamma)$ is Abelian. Since $\gamma \in Z(P)$, it follows that P_1 is a p-Sylow subgroup of G. This contradicts assumption (i).

Hence no element $\pi \neq 1$ of P can be transformed into its inverse by an involution. Now, a theorem of K. A. Fowler and the author [7] shows that there exist irreducible characters χ_μ of G of p-defect 0. But this implies $g > \chi_\mu(1)^2 \geqq p^{2a}$. Q.E.D.

Remark (1). In a recent letter, J. A. Green communicated to me the theorem that the defect group of a p-block can always be represented as intersection of two p-Sylow subgroups. This shows that actually not only (C) but also (C*) holds under the assumptions of Theorem 9.

Remark (2). J. G. Thompson raised the question whether there exist groups G that satisfy the assumptions made in conjectures (B i) and (B ii).

4. In concluding, we remark that the problem concerning simple groups whose order g is divisible by exactly four primes leads to difficult number-theoretical questions. For instance, if p were to be a prime of the form $p = 3 \cdot 2^n - 1$ with $n \geqq 2$ for which $q = 3 \cdot 2^{n-1} - 1 = (p-1)/2$ is also a prime, then $G = L_2(p)$ would satisfy our requirements. It seems to be unknown whether there exist infinitely many such primes p.

The first case one may consider is that of an order $g = p^a p_1 p_2 p_3$, where p, p_1, p_2, and p_3 are primes ($p_1 > p_2 > p_3$). It seems advantageous to distinguish three cases:

(I) There exists an irreducible representation of degree p_2.

(II) There does not exist an irreducible representation of degree p_2, but there exists an irreducible representation of degree p_1 in the principal p_2-block or in the principal p_3-block.

(III) We have neither Case I nor Case II.

In each of these cases, results can be obtained. For instance, in Case (III) the primes p_1, p_2, and p_3 must satisfy the equation $u_1 p_1 + u_2 p_2 + u_3 p_3 = 1$, where u_j is the unique integer such that

$$u_j p_1 p_2 p_3 / p_j \equiv 1 \pmod{p_j}\,; \quad |u_j| < p_j/2; \qquad (j = 1, 2, 3)\,.$$

In addition, u_j has to divide $p_j - 1$. Suzuki's simple group of order 29120 forms an example of this type of groups.

I don't know the complete answer to our question.

Of course, the orders discussed here are of a special type. However, the order g of $L_2(m)$ for any odd prime power m can be written in the form $g = 2^a p_1 p_2 p_3$ so that, although p_1, p_2, and p_3 are not necessarily primes, the group behaves in many ways as if they were primes. In the case in which m is a power of 2, we even have the simpler case $g = 2^a p_1 p_2$ of Theorem 2.

Coming now to the final conjecture (D), I have to be rather vague. There is some evidence that there is a situation similar to that mentioned for $L_2(m)$ for each class of classical groups of a fixed degree n over a finite field. There should be finitely many types, depending on the number m of elements of the finite field, and for each type, we should have a factorization

$$g = p_1^{a_1} p_2^{a_2} \cdots p_r^{a_r} \tag{5}$$

of the group order into powers of integers $p_1, \cdots, p_r$ so that the characters of the group behave as if these factors $p_1, p_2, \cdots, p_r$ were primes. If this question can be settled, we may think of extending conjecture (A) as follows:

(A*) *There exist only finitely many simple groups G whose order g is divisible by exactly r distinct primes and which do not belong to a class of classical groups for which the number r in* (5) *does not exceed n.*

Harvard University

References

[1] Artin, E., The Orders of the Classical Simple Groups, *Comm. Pure Appl. Math.*, **8** (1955), 455-72.

[2] Blichfeldt, H. F., *Finite Collineation Groups.* Chicago: Univ. of Chicago Press, 1917, chap. IV.

[3] Brauer, R., Number Theoretical Investigations on Groups of Finite Order, *Proc. Internat. Symp. on Algebr. Number Theory, 1955.* Tokyo and Nikko: The Science Council of Japan, 1956, pp. 55-64. (On p. 59, line 18, the word "abelian" should be replaced by the word "non-abelian.")

[4] Brauer, R., Zur Darstellungstheorie der Gruppen endlicher Ordnung, II, *Math. Z.*, **72** (1959), 25-46.

[5] Brauer, R., Investigation on Groups of Even Order, I, *Proc. Nat. Acad. Sci., U.S.A.*, **47** (1961), 1891-93.

[6] Brauer, R., and W. Feit, On the Number of Irreducible Characters of Finite Groups in a Given Block, *Proc. Nat. Acad. Sci., U.S.A.*, **45** (1959), 361-65.

[7] Brauer., R., and K. A. Fowler, On Groups of Even Order, *Ann. Math.*, **62** (1955), 565-83, Th. (5E).

[8] Brauer, R., and H. F. Tuan, On Simple Groups of Finite Order, I, *Bull. Amer. Math. Soc.*, **51** (1945), 756-66.

[9] Brodkey, J. M. (to be published in *Proc. Amer. Math. Soc.*).

[10] Feit, W., and J. G. Thompson (to be published in *Pacific J. Math.*).

[11] Ree, R., A Family of Simple Groups Associated with the Simple Lie Algebra of Type (F_4), *Amer. J. Math.*, **83** (1961), 401-20.

[12] Ree, R., A Family of Simple Groups Associated with the Simple Lie Algebra of Type (G_2), *Amer. J. Math.*, **83** (1961), 432-62.

[13] Speiser, A., *Theorie der Gruppen von endlicher Ordnung*, 3d ed. Berlin: Springer, 1937, Sec. 70.

[14] Suzuki, M., A New Type of Simple Groups of Finite Order, *Proc. Nat. Acad. Sci., U.S.A.*, **46** (1960), 868-70.

ON FINITE GROUPS AND THEIR CHARACTERS

RICHARD BRAUER

The idea of a presidential address seems to require a lecture delivered in the most refined and dignified scientific atmosphere yet understandable to the layman, a lecture which treats a difficult field of mathematics in such a complete manner that the audience has the excitement, the aesthetic enjoyment of seeing a mystery resolved, perhaps only with the slightly bitter feeling, of asking afterwards: Why did I not think of that myself?

Well, I don't know how my predecessors did it, but I know that I can't do it. Since the founding fathers of the Society have placed the presidential address at the time in the life of the president when he disappears into anonymity among the ranks of the Society, I shall not even try it. The choice of the field about which I am going to speak was a natural one for me, not only because of my own work in the theory of groups of finite order, but because of the new life which has appeared in this field in recent years. However, in spite of all our efforts, we know very little about finite groups. The mystery has not been resolved, we cannot even say for sure whether order or chaos reigns. If any excitement can be derived from what I have to say, it should come from the feeling of being at a frontier across which we can see many landmarks, but which as a whole is unexplored, of planning ways to find out about the unknown, even if the pieces we can put together are few and far apart. My hope then is that some of you may go out with the idea: "Now let me think of something better myself."

Let me first mention one difficulty of the theory. We have not learned yet how to describe properties of groups very well; we lack an appropriate language. One of the things we can do is to speak about the characters of a group G. I cannot define characters here. Let me only mention that we have a partitioning of the group G into disjoint sets $K_1, K_2, \cdots, K_k$, the classes of conjugate elements. The characters then are k complex-valued functions $\chi_1, \cdots, \chi_k$, each constant on each class K_i. They have a number of properties which connect them with properties of the group. These characters can be used to prove general theorems on groups, but we seem to have little control about what can be done and what not. You will see this more clearly later when I discuss specific results. I can give two reasons

Presidential address delivered before the Annual Meeting of the Society, January 28, 1960; received by the editors November 19, 1962.

for this particular behavior of the characters. The first is that I believe that we don't know all about characters that we should know. My reason for this statement is the following: If you take groups of a special type, groups whose order g contains some prime number p only to the first power

$$g = pg_0, \qquad (p, g_0) = 1,$$

some powerful results can be proved. The groups G have a subgroup P of order p. If N is the normalizer of P, i.e. the subgroup of all σ in G with $\sigma P = P\sigma$, and if the characters of N are known, the values of the characters of G for the so-called p-singular classes of G can be given (apart from certain $\pm$ signs). It is this type of connection between groups and subgroups we are looking for. Where our assumptions are satisfied, this result can be used very well to explore groups. Now it seems natural to assume that this result must be a special case of some general result where

$$g = p^a g_0 \quad \text{with} \quad (p, g_0) = 1 \quad \text{and} \quad a \geqq 1.$$

Actually, this idea has motivated a good deal of my work for quite a number of years. While many things can be proved, the part which would be most desirable for applications is lacking; there are not even conjectures about what one should try to prove. I may perhaps mention in passing that the results in the special case $a = 1$ in some sense are not quite as special as it may seem. Conditions can be given for groups of an order $g = pg_0$, $(p, g_0) = 1$, with p no longer a prime number where many of the results for the special case still hold. It may not be unreasonable to conjecture that all or almost all simple groups are of this type; there are some heuristic reasons for such a statement.

There is a second reason which may explain why the characters may fall short of our expectations. This is that there is a notion, more general than that of a group, where we have characters. I may call these objects pseudo-groups just to have a name. Perhaps, I should explain briefly how a group G qualifies as a pseudo-group. Suppose you observe somebody who has a group and who picks elements σ, τ and forms the product $\sigma\tau$. You don't see actually what the elements are, but you observe to which classes K_α the elements σ, τ, $\sigma\tau$ belong. You may then be able to determine what the probability is that the product of an element of K_α with an element of K_β lies in a given class K_γ. Whenever you have a setup of this kind and some further conditions are satisfied, you have a pseudo-group. (This description is not quite honest, because the further conditions do not look very natural in this language.) Theorems which we may want to prove for groups may not hold for pseudo-groups, and in such cases, characters are not the appropriate tool.

We shall now turn to a discussion of groups themselves. Since we assume that the order is finite, an inductive approach seems indicated. Suppose we know certain subgroups H_i of a group G, what can we say about G? If we ask the question in this form, the answer is: Not very much. Obviously, a group H can be imbedded in many different groups G which have little in common. What we have to do is to assume that the given subgroups H_i play a particular rôle in G. For instance, in the theory of characters, the centralizers $C(\sigma)$ of elements σ are important. This is the set of all elements τ of G which commute with σ. To formulate a definite question, let me ask:

Given the centralizers

$$C(\sigma_1),\ C(\sigma_2),\ \cdots,\ C(\sigma_r)$$

of all elements of prime orders, and suppose we know the intersections of any two of them. If G has center 1, these are proper subgroups of G. If we could say what G is, we would have an inductive approach. Actually, we can say something about G. For instance, it is trivial to see that the order of G is determined. It is not so trivial to see that the characters of G are determined, and this shows that we can do even a bit in putting the pieces together. However, we are far from being able to prove an isomorphism theorem stating that our pieces determine G up to isomorphism.

I should mention that the question posed can be modified in many ways. The form given above is not the best possible one, but one which can be stated quickly.

Since we are not able to put the pieces together, one may ask: How much does information about a single $C(\sigma)$ help with a discussion of G? I will show later that in some concrete cases, we can say amazingly much about G.

I shall next discuss a general principle by which relations between groups G and subgroups H can be discussed. I mentioned earlier that many group theoretical properties or quantities can be expressed in terms of characters. Clearly, there are many such quantities which have the same value for G and for suitable subgroups. Each will then give a relation between the characters of G and those of suitable subgroups H.

For instance, let σ be an element of G, let n be an integer and denote by $A_G^{(n)}(\sigma)$ the number of solutions of $\xi^n=\sigma$ in G. Clearly each such ξ lies in $C(\sigma)$. So if $H\supseteq C(\sigma)$,

$$A_G^{(n)}(\sigma) = A_H^{(n)}(\sigma).$$

On the other hand, $A_G^{(n)}(\sigma)$ can be expressed in terms of the characters of G and we obtain relations between characters of G and H. Unfortunately, they are too complicated to be exploited except in special cases.

Another application of the same principle yields some of the results mentioned above, but an actual statement would require too many definitions and be too technical.

I come now to still another application of the same principle which leads to the program I mentioned in the beginning. This application works only for groups of even order. There is an old conjecture to the effect that all groups of odd order are solvable. The program would require a proof of this so that groups of odd order could be discarded. We are far from a proof, but recently, more progress has been made on this problem first by John Thompson and later by him, M. Hall and W. Feit so that it may not be daydreaming to say that some day this question will be cleared up.[1]

So then I will concentrate on groups of even order. The quantity which we use here is the following one. If σ is an element of G, let $B_G(\sigma)$ denote the number of elements ξ of order 2 which satisfy the equation

$$\xi^{-1}\sigma\xi = \sigma^{-1}.$$

Clearly, the elements y of G satisfying $y^{-1}\sigma y = \sigma^{\pm 1}$ form a group $C^*(\sigma)$ which is either $C(\sigma)$ or has twice the order. Then if H is any subgroup of G with $H \supseteq C^*(\sigma)$, we have

$$B_G(\sigma) = B_H(\sigma).$$

Again, $B_G(\sigma)$ can be expressed by the characters of G. The method then yields relations

$$g \sum_{\chi_\mu} \frac{h_\mu^2}{\chi_\mu^{(1)}} \chi_\mu(\sigma) =$$

where g is the order of G, the h_μ are certain rational integers formed by means of the characters and where on the right the same expressions for the group H appear. The sum extends over all characters of G. The method is to use combinations of these formulas for various elements σ in which only relatively few characters appear. This re-

[1] See the note added at the end of the paper.

quires some deep results on characters which I cannot describe in detail without becoming too technical. The outcome is that we obtain estimates for the order g by means of certain subgroups. As I mentioned, the general result requires concepts from the theory of group representations which are beyond the scope of this lecture. Rather than tiring you with them, I shall describe some particular cases in which the results can be brought into an explicit form. In principle, we have corresponding statements whenever we know the 2-Sylow group P of G. For the sake of simplicity, I assume that G does not have a normal subgroup H of half its order, since in the excluded case more direct methods for constructing G out of H are available.

Let first P be a dihedral group of order 2^n

$$P = \{\rho, \sigma\}, \qquad \rho^{2^{n-1}} = 1, \qquad \sigma^2 = 1, \qquad \sigma^{-1}\rho\sigma = \rho^{-1}.$$

Here, set $\tau = \rho^{2^{n-2}}$. This is an element of order 2. If $n \geqq 3$, g has the form

$$g = 2\alpha c(\tau)^3 \left(\frac{1}{c(\tau, \sigma)} + \frac{1}{c(\tau, \rho\sigma)} \right)^2$$

where $c(\xi)$, $c(\xi, \eta)$ denote the number of elements commuting with ξ or with ξ and η. Here, α is a rational number which satisfies the inequalities

$$\left(1 - \frac{1}{2^{n-1}}\right)\left(1 - \frac{1}{2^n}\right) \leqq \alpha \leqq \left(1 + \frac{1}{2^{n-1}}\right)\left(1 + \frac{1}{2^n}\right),$$

i.e. which is close to 1. In fact, α has the form $x(x+\delta)/(x-\delta)^2$ where $\delta = \pm 1$ and where x is the degree of an irreducible representation, $x \equiv \delta \pmod{2^n}$, $x > 1$. Of course, if $C(\tau)$ is known, $c(\tau, \sigma)$ and $c(\tau, \rho\sigma)$ are known. We say that we have a regular case, if $c(\tau, \sigma) = c(\tau, \rho\sigma)$. Here, g can be brought into the form

$$g = \beta \frac{x(x+1)(x-1)}{2}, \qquad \beta \text{ integral.}$$

The groups $L_2(x)$, x a prime power, form an example. It is not known whether there exist examples where x is not a prime power.

In the irregular case, $c(\tau, \sigma)/c(\tau, \rho\sigma)$ can be shown to lie close to 1. The only known example of an irregular group is the simple group of order 2520.

There are similar results, somewhat more complicated in the case $n = 2$.

As a second example, I mention the 2-Sylow group

$$P = \{\sigma, \tau, \rho\},$$
$$\rho^{-1}\sigma\rho = \tau, \qquad \rho^{-1}\tau\rho = \sigma, \qquad \tau\sigma = \sigma\tau,$$
$$\sigma^{2^m} = 1, \qquad \tau^{2^m} = 1, \qquad \rho^2 = 1, \qquad m \geqq 2.$$

Then P is somewhat more complicated, the discussion is far more difficult, but the results are far simpler. It turns out that g is completely determined, if we know $C(\rho)$, in fact when we know the orders of certain subgroups of $C(\rho)$. As an example, I mention the projective groups $LF(3, q)$ of a projective geometry over a finite field with $q \equiv 1 \pmod 4$. Actually the group can be characterized within this framework. The other cases of q have been treated before. There also is the unitary group $HO(3, q^2)$.

Returning to the general case, let me say what the general program of treating groups of even order is to which these methods lead. We consider a group H which contains an element σ of order 2 in its center and we consider groups $G \supset H$ for which $C(\sigma) = H$ and which have the same 2-Sylow group P as H. The construction of H can be reduced to that of $H/\{\sigma\}$, a group of smaller order. On the other hand as I have indicated, once H is known, properties of G can be obtained, in particular, the order g of G can be estimated. Of course, this is not enough, but my examples above indicate that it seems feasible to characterize the simple groups of finite order in this manner.

Added November, 1962. The hope which I expressed in my address has been fulfilled. Recently, W. Feit and John G. Thompson announced that they succeeded in proving the famous conjecture stating that groups of odd order are solvable (Proc. Nat. Acad. Sci. U.S.A. **48** (1962), 968–970). It is already evident that the deep methods developed by these authors will have far reaching consequences for other problems in group theory. For instance, D. Gorenstein and J. Walter have made progress with the question of groups with dihedral 2-Sylow groups discussed as an example in this lecture. There is every indication that this case will be cleared up soon.

Originally, I had not intended to submit the manuscript of my talk for publication. I have changed my mind since the hope for progress on the problem is now far greater.

HARVARD UNIVERSITY

6

Representations of Finite Groups

Richard Brauer

INTRODUCTION

It has been said by E. T. Bell that "wherever groups disclosed themselves or could be introduced, simplicity crystallized out of comparative chaos." This may often be true, but, strangely enough, it does not apply to group theory itself, not even when we restrict ourselves to groups of finite order. We are reminded of the educators who want to educate the world and cannot handle their own children. A tremendous effort has been made by mathematicians for more than a century to clear up the chaos in group theory. Still, we cannot answer some of the simplest questions.

This may sound as if I am critical of the theory of finite groups. On the contrary, this is the reason why I am fascinated by it. It is the unknown and the mysterious which attracts our attention. With this as my basic attitude, you will understand that I will talk less about what has been achieved, but rather discuss questions which cannot be answered today. Some of these problems look rather hopeless, and perhaps there is no clear answer to them. But it seems unfair to crow about the successes of a theory and to sweep all its failures under the rug.

Why are we interested in groups? It has been said that implicitly the idea of a group is as old as mathematical thinking. What is

Reprinted from *Lectures on Modern Mathematics, Volume I,* edited by Dr. Thomas Saaty. Published by John Wiley & Sons, Inc., 1963.

meant is that the question of symmetry was one of the oldest questions connected with mathematics to attract man's mind. This seems natural because man found symmetry in nature.*

Groups are the mathematical concept with which we describe symmetry. One of the outstanding achievements of Greek mathematics is the discovery of the five regular polyhedra, the Platonic solids. Each of them is closely associated with a finite group. While groups do not appear in Euclid's treatment of geometry, Poincaré could say: "Euclidean space is simply a group."

We have to confess that it took mathematicians more than 2000 years to achieve a mathematical formulation of the group concept. In its abstract form, it was given by Cayley, after permutation groups had been used by Lagrange, Cauchy, Abel, Galois, and others in their work on the solution of algebraic equations by radicals.

Once we have the concept of an abstract group, we have two ways to study it. We can either approach groups directly in their abstract form using combinatorial methods, or we can utilize concrete realizations of groups for the investigations. The latter is done in the theory of representations with which we are concerned here. This has the advantage that results from algebra and geometry can be applied. It must be expected that, if we are to achieve our final goals, both methods will have to work together.

I believe that the combinatorial approach to group theory will form the subject of some other lectures in this series. Even with the restriction to the representation theory of finite groups, it was not possible to include all pertinent topics. In particular, permutation representations, integral representations, and projective representations are not treated. Questions of homological algebra have not been considered. To some extent, the selection of the material was dictated by my own familiarity with the questions.

The original lectures are incorporated in the next section. The "Supplements," page 159, contain some further results which may serve as illustrations for the general discussion. In order to make the report self-contained, I have also included the basic definitions and theorems in an Appendix page, 170. The material in the Supplements is arranged in the order in which the various points come up in the next section. For further reference, see C. W. Curtis and I. Reiner [19] which has just appeared. Since a rather exhaustive

* A beautiful account of the role of symmetry has been given by H. Weyl [69].

bibliography is given there, I have dispensed with this here. I also mention a set of mimeographed notes by T. Tsuzuku in cooperation with N. Ito, A. Mizutani, and T. Nakayama, to be published soon by Nagoya University. They are based to some extent on lectures given by the author.

A reader who is interested in getting acquainted with the basic facts of the theory will find an introduction in most books on finite groups. I mention here only M. Hall, Jr. [34] Chapter 16. One of the original papers on the subject, by I. Schur [51], is still enjoyable reading. Finally, the following books may be mentioned: W. Burnside [17]: A. Speiser [57]; and B. L. van der Waerden [67].

SURVEY OF THE THEORY OF REPRESENTATIONS OF FINITE GROUPS

§1. We start by considering the group algebra Γ of a given finite group G. This Γ consists of all formal sums (see the Appendix, §1)

$$\gamma = \Sigma a_g g \tag{1}$$

with g ranging over G and the coefficients a_g taken from a given field Ω. If the operations are defined in a natural manner, Γ becomes an associative algebra. This concept of group algebra also goes back to Cayley; it is as old as that of an abstract group and it seems to be an equally worthy object of study.

What is the advantage? We have two operations, addition and multiplication, and this gives us more elbow room to work than we have in a group. More precisely, we can apply the results of the theory of algebras. If we have a subring R of Ω with a natural number theory, we can restrict the a_g in (1) to R and study "number theory" in the corresponding subset of Γ. There is a marked disadvantage: if we just look at Γ, we do not know which elements γ are the group elements g. Two questions come to our mind:

Problem 1. Which (finite-dimensional associative) algebras Γ *are group algebras?*

Problem 2. When do nonisomorphic groups have isomorphic group algebras?

I cannot answer either question. I am not even sure that they have good complete answers. Still, they will be important in giving some direction to our work.

§2. In order to have the simplest possible case, let us assume that Ω is the field $\mathbf{C}$ of complex numbers. There is a necessary condition for Problem 1: the algebra Γ must be semisimple. In our present case, this means that Γ is a direct sum of a certain number k of complete matrix algebras:

$$\Gamma = \Xi_1 \oplus \Xi_2 \oplus \cdots \oplus \Xi_k \tag{2}$$

Here, each Ξ_i consists of all matrices of a fixed degree x_i. If we only take the elements $g \in G \subseteq \Gamma$ and associate with each the corresponding matrix $X_i = X_i(g)$ in Ξ_i for fixed i, we have a matrix representation (Appendix, §2)

$$X_i\colon g \rightarrow X_i(g)$$

The matrix representation X_i is irreducible, and apart from equivalence, all irreducible representations in G are obtained exactly once for $i = 1, 2, \ldots, k$. We see that, as an algebra, Γ is completely determined by the condition of semisimplicity, the number k, and the k degrees $x_1, x_2, \ldots, x_k$.

The number k can be described easily. We form the center Z of Γ. The decomposition (2) carries over from Γ to Z and becomes

$$Z = \Omega_1 \oplus \Omega_2 \oplus \cdots \oplus \Omega_k \tag{3}$$

where each Ω_i here is simply an isomorphic copy of the field Ω. Hence k is the dimension of Z and this is simply the class number of G.

One might ask for an equally direct description of the k degrees $x_1, x_2, \ldots, x_k$ in terms of G. However, no such description is known. This is unfortunate because otherwise we might have a kind of answer to our Problem 2.

Representations are described in terms of their characters. In §4 of the Appendix, the basic properties of characters are listed. In particular, we have k irreducible characters $\chi_1, \chi_2, \ldots, \chi_k$. The existence of k such functions χ_i is a fundamental property of groups from which a considerable number of theorems on groups have been derived. Sometimes this has been considered as the use of an alien tool. I should perhaps then state exactly how much group theoretical information we need in order to be able to find the χ_i. This can be answered completely. The class algebra Z has scalars of multiplication $a_{\alpha\beta\gamma}$ with regard to its natural basis. Knowledge of the characters χ_i is equivalent with knowledge of the

$a_{\alpha\beta\gamma}$. Now, the $a_{\alpha\beta\gamma}$ have an immediate group theoretical interpretation (Appendix, §3). Thus, there is nothing alien about the characters. As a corollary of this, we see that for each given group G, we can actually compute the k degrees $x_1, x_2, \ldots, x_k$.

§3. Let us go back to Problem 1. The results on characters give the following information on the degrees x_i.

(a) $$x_1^2 + x_2^2 + \cdots + x_k^2 = n$$

where n is the order $|G|$ of G.

(b) $$x_i \text{ divides } n$$

The following fact is proved easily:

(c) The number of degrees $x_i = 1$ is a divisor of n. It is equal to the index of the commutator group G' of G.

If n is fairly small, the systems $\{x_1, x_2, \ldots, x_k\}$ satisfying conditions (a), (b), and (c) can be discussed without difficulty. A discussion of a numerical example in the supplements, §2, shows already that we shall have far more systems $\{x_1, \ldots, x_k\}$ than we have group algebras so that we are far from having a set of sufficient conditions for group algebras.

We add a remark which can be interpreted as a necessary condition for Problem 1. E. Landau [42] proved that the group order n lies below a bound depending on the class number k only. For instance, his method can be used to show that (for $k \geqq 3$),

$$n \leqq (2k)^{2^{k-3}}(k-1)^{2^{k-3}}(k-2)^{2^{k-4}}(k-3)^{2^{k-5}} \cdots 5^{2^3}4^{2^2}3^2 2 \; *$$

This implies that if n is very large, k cannot be small. Incidentally, this means that for growing n, a complete discussion of (a) will become increasingly messy.

Since very little group theory is used in Landau's method, there is every hope that much better inequalities can be obtained. I propose, therefore, as my next problem:

Problem 3. Give upper bounds for the order n *of a group with a given class number* k, *which lie substantially below the bounds obtainable by Landau's method.*

If we restrict ourselves to the group algebras of simple groups G, further conditions can be added. It follows from a theorem of

* For $k = 1$, of course, $n = 1$, and for $k = 2$, $n = 2$.

I. Schur [52] that if $r > 1$ is an integer and if exactly m degrees x_i are equal to r, then a prime number p divides $n = |G|$ at most with the exponent

$$e = \left[\frac{mr}{p-1}\right] + \left[\frac{mr}{p(p-1)}\right] + \left[\frac{mr}{p^2(p-1)}\right] + \cdots^*$$

In particular, $e \leqq mrp/(p-1)^2$.

Later, other conditions will be given which apply only for values n of a special kind, but which are far more suitable for an actual discussion. Their existence shows that certainly the last word has not been said about Problem 1.

§4. The simplest examples show that nonisomorphic groups may have isomorphic group algebras over **C**. For instance, this is always true for two nonisomorphic Abelian groups of the same order n, since for both all degrees x_i are 1. If we wish to obtain a criterion for the isomorphism of two groups, we will have to strengthen our assumptions. Since we know that the degrees x_i are determined by the scalars of multiplication $a_{\alpha\beta\gamma}$ of the class algebra Z, we may require that these numbers $a_{\alpha\beta\gamma}$ are the same for both groups, if the classes are indexed suitably. This is the same as to require that both groups have the same character table $[\chi]$ provided, of course, that characters and classes are indexed suitably. Still, this is not enough as the two non-Abelian groups of order p^3 for primes p have the same character table.

We strengthen our assumption further. If K is a conjugate class of G and m an integer, there is a conjugate class $K^{[m]}$ of G which consists of the mth powers of the elements of K. We now ask:

Problem 4. Let G *and* G* *be two finite groups and assume that there exists a one-to-one mapping* $K_j \to K_j^*$ *of the set of conjugate classes of* G *onto the set of conjugate classes of* G*, *and a one-to-one mapping* $\chi_i \to \chi_i^*$ *of the set of irreducible characters of* G *onto the set of irreducible characters of* G* *such that*

(a) $\chi_i(K_j) = \chi_i^*(K_j^*)$

(b) $(K_j^{[m]})^* = (K_j^*)^{[m]}$ for all integers m.

Are G *and* G* *isomorphic?*

No counterexamples seem to be known, but nobody seems to have

* Here, $[t]$ denotes the largest integer which does not exceed the real number t.

looked closely. Perhaps it would be worthwhile to do some checking; it may not even be difficult to find counterexamples. On the other hand, if the answer was affirmative, this would be wonderful.

My next problem is somewhat vague.

Problem 5. Investigate connections between the values of the characters and the mappings $\mathrm{K} \to \mathrm{K}^{[m]}$.

There are a number of obvious remarks one can make here (Supplements, §3). The question is of practical importance, if characters are used to study properties of particular groups.

We can now state all we can ever hope for in the theory of characters in form of a problem.

Problem 6. Give necessary and sufficient conditions for a square matrix (with operation on the columns, corresponding to the mapping $\mathrm{K} \to \mathrm{K}^{[m]}$ *thrown in) to be the character table of a finite group.*

There are good reasons to believe that important properties of characters have not yet been discovered. This means of course that any attempt to solve Problem 6 now would be futile. One may even doubt whether a satisfactory concrete solution can ever be given.

§5. I continue my discussion of group characters by stating a more recent result. A subgroup E of a group G will be called elementary, if it is a direct product of a cyclic group A and a p-group P (that is, a group of prime power order p^m). Necessary and sufficient conditions can be given for a function θ defined on G to be an irreducible character of G. They are as follows.

1. The function θ is constant on each class K_j.
2. For every elementary subgroup E of G, the restriction $\theta|E$ of θ to E is a character of E.*
3. We have $\sum_{g \in G} |\theta(g)|^2 = |G|$.

In order to apply this result to a construction of the irreducible characters of G, it will be sufficient to consider a set of elementary subgroups

$$E_1, E_2, \ldots, E_t \tag{4}$$

* It is sufficient to require here only that $\theta|E$ is a generalized character of E.

of G such that each elementary subgroup of G is conjugate to one of the groups (4). If we know the irreducible characters of the groups E_j and, if for each conjugate class of the E_j, it is known to which of the k conjugate classes $K_1, K_2, \ldots, K_k$ of G it belongs, then the irreducible characters χ_i of G can be constructed. For the first part, it would be highly desirable to have more information on the following problem:

Problem 7. Study the irreducible characters of p-*groups.*

We say that two conjugate classes of a subgroup E of G are fused in G, if both lie in the same conjugate class of G. For the second part in our construction of the χ_i we have to know what the fusions of E-classes in G are. We would like to know more about the ways in which the classes of a given group E can be fused in larger groups. For instance, if we are interested in all finite groups which contain a given p-group P as their p-Sylow group, we are led to the question:

Problem 8. Given a p-*group* P, *can we find constructively all possible ways in which the conjugate classes of* P *can be fused in finite groups* G *with* P *as their* p-*Sylow group?*

There are a number of ways in which the above procedure can be modified. In particular the groups (4) can be replaced by a set of subgroups $H_1, H_2, \ldots$ of G such that every elementary subgroup of G is conjugate in G to a subgroup of some H_i. The following special case seems to deserve attention. Pick a representative σ_i in each of the k conjugate classes K_i, $(i = 1, 2, \ldots, k)$ of G and let C_i be its centralizer in G; $C_i = C(\sigma_i)$. If, say, $\sigma_k = 1$, then $C_k = G$. It is obvious that the remaining groups

$$C_1, C_2, \ldots C_{k-1} \tag{5}$$

satisfy the requirements imposed above on $H_1, H_2, \ldots$ We now ask:

Problem 9. How much information about the group G *can be obtained if all or some of the groups* C_i *in (5) are given (possibly combined with information of the type mentioned above in connection with the* E_i)*?*

For some comments, see the Supplements, §4. We shall see later that in some special cases amazingly much can be said, if a single C_i is known.

§6. We shall conclude our survey of the classical theory of characters with a brief discussion of some of the questions about groups G which can be answered if the characters of G are known. First, we can find all normal subgroups N and determine the characters of G/N. In particular, we can recognize whether or not G is simple and whether or not G is soluble. Among the normal subgroups N of G, we can identify the members of the ascending and descending central series of G. Also, we can determine the Frattini subgroup of G.

There is more difficulty about getting information concerning the structure of normal subgroups N of G. For instance, I do not know the answer to the following question.

Problem 10. Given the character table of a group G *and the set of conjugate classes of* G *which make up a normal subgroup* N *of* G, *can it be decided whether or not* N *is Abelian?*

This would be of importance if we want to identify the members of the derived series of G.

If G possesses an arbitrary subgroup H of index r, then G possesses a transitive permutation representation of degree r, and, correspondingly, a reducible character of degree r. If we were able to recognize which characters of G belong to transitive permutation representations, we could determine the orders of all subgroups of G. However, no criterion for permutation characters is available, only a number of necessary conditions. For this reason, we are usually only able to show that G cannot have subgroups of certain orders. In a more refined form, we can use Frobenius' idea of induced characters (see Appendix, §6).

Suppose we have two groups G and H such that the order of H divides that of G and suppose we know the character tables $[\chi]$ of G and $[\psi]$ of H. If H is a subgroup of G, we have a mapping θ: $L_i \rightarrow K_j$ of the set of conjugate classes of H into the set of conjugate classes of G determined by $L_i \subseteq K_j$. Moreover, we have formulas

$$(6) \qquad \chi_\rho(K_j) = \sum_\sigma a_{\rho\sigma}\psi_\sigma(L_i) \qquad \text{for } K_j = \theta(L_i)$$

with non-negative integral coefficients $a_{\rho\sigma}$ which are independent of i.

If H is no longer assumed to be a subgroup of G, we may still have a mapping θ and formulas (6) such that all conditions known for the case of a subgroup H of G are satisfied. We should include here the conditions which are connected with the mappings $K \to K^{[m]}$. It would be a generalization of Problem 4 if we ask whether H then has to be isomorphic with a subgroup of G such that θ and the formulas (6) have the meaning given them in the case $H \subseteq G$.

In a less ambitious mood, we formulate here the questions:

Problem 11. Given the character table of a group G, *how much information about existence of subgroups can be obtained?*

Problem 12. Given the character table of a group G *and a prime* p *dividing* n = |G|, *how much information about the structure of the* p-*Sylow group* P *can be obtained? In particular, can it be decided whether or not* P *is Abelian?*

For a number of comments, see the Supplements, §6.

Every automorphism of a group G produces a permutation of the irreducible characters of G. Because of this, we may ask:

Problem 13. How much information about the automorphism group A(G) *of a group* G *can be obtained if the characters of* G *are known?*

Some remarks are given in the Supplements, §7.

§7. So far we have taken the underlying field Ω as the field **C** of complex numbers. It would not make any difference if instead we use an algebraically closed field Ω of characteristic 0. We now drop the assumption that Ω be algebraically closed, but, for the time being, we shall stick to fields of characteristic 0. Then the group algebra Γ of G with respect to Ω is still semisimple. As in (2) we can write Γ as a direct sum

$$\Gamma = \Lambda_1 \oplus \Lambda_2 \oplus \cdots \oplus \Lambda_r \tag{7}$$

of simple algebras. Each Λ_ρ corresponds to a class of equivalent representations of G, now with coefficients in Ω and irreducible in Ω. All irreducible representations L of G in Ω are obtained in this fashion. In order to investigate such representations L of G, we may see how they behave in the algebraic closure $\bar{\Omega}$ of Ω. Of course, in $\bar{\Omega}$ we have the situation studied above. This approach was first carried out by I. Schur who found that each Λ_ρ is in one-to-one correspondence with a family T_ρ of irreducible characters χ_j, taken in $\bar{\Omega}$ and algebraically conjugate with regard to Ω. If L_ρ is an irreducible representation of G in Ω, which is associated with the term Λ_ρ in (7), the character λ_ρ has the form

$$\lambda_\rho = m_\rho \sum_j \chi_j \tag{8}$$

where χ_j ranges over the characters in T_ρ and where the m_ρ are natural integers. This formula (8) corresponds to the breaking up of the representation L_ρ in the algebraic closure $\bar{\Omega}$ of Ω. At this stage, we know little about the integers m_ρ, except that they divide the degree x_j of $\chi_j \in T_\rho$. We call the m_ρ the Schur indices.

So far we have disregarded any connections with Wedderburn's theory of simple algebras. According to this theory, each Λ_ρ is isomorphic with a complete matrix algebra of a certain degree q_ρ with coefficients in a division algebra Δ_ρ over Ω. Let $\Omega(\chi_j)$ denote the field obtained from Ω by adjoining all values $\chi_j(g)$ with $g \in G$ of the character χ_j. It turns out that the center Z_ρ of Δ_ρ is isomorphic to $\Omega(\chi_j)$ for $\chi_j \in T_\rho$. This implies that the degree $z_\rho = [Z_\rho : \Omega]$ is equal to the number of characters χ_j in T_ρ. Moreover, the dimension of Δ_ρ over Z_ρ is exactly $m_\rho{}^2$. Finally, the Wedderburn degree q_ρ is equal to x_j/m_ρ.

If the irreducible characters $\chi_1, \chi_2, \ldots, \chi_k$ of G are known, we know of course the number r of terms in (7), the fields Z_ρ, and their degrees. It is more difficult to determine the Schur indices m_ρ. Still, a great deal of information is available. It is possible to reduce the same question to a treatment for the case of solvable groups. If the field Ω contains the nth roots of unity for $n = |G|$, then all m_ρ are equal to 1, and it follows that the irreducible representations of G in Ω remain irreducible in the algebraic closure $\bar{\Omega}$. They can then be used as our old $X_1, X_2, \ldots, X_k$. As has been shown by L. Solomon [55], all m_ρ are already 1, if Ω contains a primitive pth root of 1 for every odd prime divisor p of $n = |G|$ and, in addition, a primitive fourth root of 1, if n is even.

Since in the case of an algebraic number field Ω a complete characterization of all central division algebras by arithmetical invariants has been obtained by H. Hasse [37], the problem arises to develop methods by which the Hasse invariants of the Λ_ρ in (7) can be determined. Such methods have been given by E. Witt [71].

It is clear that the preceding results can be used to discuss special cases of Problems 1 and 2, cf. Supplements, §9.

I conclude §7 by formulating a rather old problem which is still open.

Problem 14. Characterize by means of group theoretical properties of G *the number of irreducible representations* X_i *of* G *in* **C** *which are equivalent to representations with coefficients in the field* **R** *of real numbers.*

§8. As our next generalization, we consider fields Ω of arbitrary characteristic. If we go back to Problem 2 for a moment, we may have a better chance of success if we admit fields of a different type. At least, no counterexample is known for the following version.

Problem 2. If two groups* G_1 *and* G_2 *have isomorphic group algebras over every ground field* Ω, *are* G_1 *and* G_2 *isomorphic?*

Little work seems to have been done on this question. Incidentally, the assumption is not weakened if only prime fields Ω are considered. If it is assumed that the group rings of G_1 and G_2 over the ring of integers are isomorphic, this implies isomorphism of the group algebras, and we should have a better chance of success if we try to prove isomorphism of G_1 and G_2 under this assumption. The theory developed for fields Ω of characteristic p enables us to deal at least with some special cases.

Let Ω now be a field of prime characteristic p. The interesting case is that p divides the order n of the group G. The first marked difference is that in this case the group algebra Γ is no longer semisimple, the radical N of Γ is not (0).

It seems natural to formulate the following problem.

Problem 15. Characterize by group theoretical properties of G *the dimensions of the radical* N *and of its powers* $N^2, N^3, \ldots$. *In particular, determine the smallest exponent* e *for which* $N^e = 0$.

Unfortunately, nothing is known about Problem 15 except in the case of a p-group where p is the characteristic of Ω. In this particular case, a satisfactory theory has been given by S. A. Jennings [41].

Returning for a moment to Problem 1, we observe that, in the case of fields of characteristic p, a new type of necessary condition for group algebras Γ can be formulated: Γ belongs to a special class of algebras; it is a symmetric algebra.

It would be interesting to find other necessary conditions for Problem 1 of a similar nature, that is, to find general classes of algebras which include the group algebras. A variation of this leads to the problem:

Problem 16. Obtain classes of groups G *by imposing group theoretical conditions which can also be characterized by algebra-theoretic conditions imposed on the group algebra* Γ *(Supplements, §10).*

§9. While it is possible to start an investigation of the irreducible representations of G in the field Ω of characteristic p directly, we prefer a different approach. If p is a given prime, we denote by $\nu_p(a)$ the exact exponent with which p divides the integer $a \neq 0$. If we set $\nu_p(b/a) = \nu_p(b) - \nu_p(a)$ for two integers $b, a \neq 0$ and set $\nu_p(0) = \infty$, we have the p-adic (exponential) valuation ν_p of the field Q of rational numbers. It follows from a basic theorem of algebra* that ν_p can be extended to a valuation of the field $\bar{Q} = A$ of all algebraic numbers. We shall denote a fixed such extension again by ν_p. The set of elements α of A, with $\nu_p(\alpha) \geqq 0$ forms a subring $\mathfrak{o}$ of A, the ring of local integers with regard to ν_p, and A is the quotient field of $\mathfrak{o}$. The elements α of p with $\nu_p(\alpha) > 0$ form a maximal ideal $\mathfrak{p}$ in $\mathfrak{o}$. The residue class ring $\mathfrak{o}/\mathfrak{p} = \Omega$ is a field algebraically closed of characteristic p. For each $\alpha \in \mathfrak{o}$, we have a corresponding element $\alpha^* \in \Omega$; the residue class map $\alpha \rightarrow \alpha^*$ will be denoted by θ.

We now start from a matrix representation X of G in A. Actually, since only finitely many coefficients appear in the $X(g)$, $g \in G$, the representation X will lie in a subfield A_0 of finite degree over Q. It can be seen easily that after replacing X by an equivalent representation we may assume that all coefficients of all $X(g)$ lie in $\mathfrak{o}$. Now the residue class map θ of $\mathfrak{o} \rightarrow \Omega$ changes X into a representation X^θ in Ω. It is quite possible that X is irreducible in A and that X^θ is reducible in Ω.

Let again $X_1, X_2, \ldots, X_k$ denote the irreducible representations of G in the algebraically closed field A of characteristic 0 and let $F_1, F_2, \ldots, F_l$ denote the different (that is, nonequivalent) irreducible representations of G in Ω. The number l of distinct representations F_i turns out to be the number of conjugate classes K_λ of G consisting of elements ι whose order is prime to p, or, as we shall say, consisting of p-regular elements. We can then speak of the multiplicity $d_{\mu j}$ with which F_j appears in the representation $X_\mu{}^\theta$ obtained from X_μ by means of the residue class map θ. Actually, θ may transform equivalent representations X_i and $X_i^{(1)}$ with coeffi-

* See, for instance, Zariski-Samuel [72, Chap. 6] or van der Waerden [67, Chap. 10].

cients in $\mathfrak{o}$ into inequivalent representations X_i^θ and $X_i^{(1)\theta}$ in Ω. However, it can be shown that in spite of this the decomposition numbers $d_{\mu j}$ are well defined.

Each F_j appears in some X_μ^θ. We shall refer to the representations of G in A as to the ordinary representations and to the representations of G in Ω as to the modular representations of G.

We now define the character ϕ of a representation Y of G in Ω. We restrict ourselves to p-regular elements g of G. If g has order m prime to p, the characteristic roots of the matrix $Y(g)$ are mth roots $\eta_1, \eta_2, \ldots$ of unity in Ω. For each such η_λ, there exists a unique mth root of unity ϵ_λ in $\mathfrak{o}$ mapped by θ on η_λ. We set $\phi(g) = \epsilon_1 + \epsilon_2 + \cdots$. Thus, ϕ is a complex-valued function defined on the set G^0 of p-regular elements in G. Two representations of G in Ω have the same character if and only if they have the same irreducible constituents.

Let ϕ_j denote the character of the irreducible representation F_j. If restriction of a function defined on G to the set G^0 is indicated by a superscript zero, the definition of the decomposition numbers d_{ij} yields immediately the formulas

$$\chi_\mu{}^0 = \sum_{j=1}^{l} d_{\mu j}\phi_j \qquad (\mu = 1, 2, \ldots, k) \tag{9}$$

In the further development of the theory, the integers c_{ij} defined by the formulas

$$c_{ij} = \sum_{\mu=1}^{k} d_{\mu i}\, d_{\mu j} \qquad (i, j = 1, 2, \ldots, l) \tag{10}$$

are of importance. They are the Cartan invariants which play a role in the investigation of the algebra Γ.

Now orthogonality relations for the modular characters ϕ_j can be given. In accordance with the general spirit of the lectures, I shall talk about two questions which I cannot answer:

Problem 17. If $f_i = \phi_i(1)$ *is the degree of the irreducible representation* F_i, *is it true that* $\nu_p(f_i) \leqq \nu_p(n)$, *that is, that* n/f_i *is a local integer? (In general,* f_i *does not divide the group order* n.)

For p-solvable groups, an affirmative answer to our question has been given by P. Fong [30].

Problem 18. For each modular irreducible character ϕ_i*, there exist ordinary generalized characters* ψ_i *of* G*, such that* $\psi_i{}^0 = \phi_i$*. If* ψ *is any ordinary generalized character of* G*, how can we decide whether* $\psi_i{}^0$ *is a modular character of* G?

Stated in a slightly different manner: we would like to have a characterization of the modular irreducible characters ϕ_i similar to that given above for the ordinary irreducible characters χ_i.

§10. The formulas (9) give the main reason for our interest in the modular characters ϕ_j: the ϕ_j are considered as the building stones out of which the ordinary characters χ_i are built. To be true, in (9) only the values of χ_i for p-regular elements appear. However, there exist similar formulas for the other elements in which instead of $\phi_1, \ldots, \phi_l$ the modular irreducible characters of certain subgroups of G and "generalized decomposition" numbers occur. We refer to R. Brauer [6, 9] for details.

We give some additional motivation for our interest in modular representations. All the known simple groups with the exception of the alternating groups and the five Mathieu groups can be defined in a natural way by means of linear groups with coefficients in a finite field. If we formulate the analogue of Problem 4 for modular characters, then, for the groups mentioned here, the problem appears no longer inaccessible. If we knew the decomposition numbers $d_{\mu j}$, the same would be true for Problem 4 in its original form for the groups in question. Perhaps we are not overoptimistic if we hope that the ideas may play an important part in investigating the problem of characterizing the known and unknown finite simple groups by group theoretical properties.

Finally, it may be mentioned that the problem of extension of elementary Abelian p-groups by means of G leads immediately to an investigation of modular representations of G.

Actually, we are here interested in the representations of G in a subfield of Ω, the prime field. The relation between the irreducible representations of G in Ω and in subfields of Ω can be treated in much the same way as was done earlier for ordinary representations. According to a celebrated theorem of Wedderburn, all finite division algebras are commutative. This implies that the Schur indices are 1; we have a simpler situation than in the former case.

In the other direction, nothing is really changed if Ω is replaced by an extension field W_0. Since every algebraically closed field W of characteristic p is isomorphic to such a field W_0, nothing is gained by studying the irreducible representations of G in W.

§11. In continuing our discussion of group algebras Γ, we now turn our attention to the center Z of Γ, the class algebra of G. Let us go back for a moment to the case of an algebraically closed field of characteristic 0, or in our present notation, let us replace Ω by the field A of all algebraic numbers. As explained in the Appendix, §5, we have k algebra homomorphism ω_i of Z onto the field A. If we consider only "integral" elements of Z, (that is, elements γ for which in (1) all a_g lie in $\mathfrak{o}$), the residue class map θ changes each ω_i into an algebra homomorphism $\omega_i{}^\theta$ of the class algebra of G over Ω onto Ω. Every such homomorphism is obtained, but in general $\omega_1{}^\theta, \omega_2{}^\theta, \ldots, \omega_k{}^\theta$ are not distinct. We distribute $\omega_1, \ldots, \omega_k$ into subsystems called blocks* where ω_i and ω_j belong to the same block, if and only if $\omega_i{}^\theta = \omega_j{}^\theta$. Because of the connection between the χ_i and the ω_i, we can then also speak of the irreducible characters χ_i or of the irreducible representations X_i of G in A belonging to a block. In particular (by the Appendix, §5), χ_i and χ_j belong to the same block, if and only if

$$\frac{n\chi_i(g)}{c(g)x_i} \equiv \frac{n\chi_j(g)}{c(g)x_j} \quad \bmod \mathfrak{p}$$

for all $g \in G$. Thus, we can find the distribution of the χ_i into blocks if the values of the characters are known.

If we write the algebra Γ as a direct sum

$$\Gamma = \Psi_1 \oplus \Psi_2 \oplus \cdots \oplus \Psi_s \tag{11}$$

of indecomposable algebras Ψ_i, we have a one-to-one correspondence between the set of Ψ_α and the set of algebra homomorphisms of the center Z of Γ onto Ω. It follows from this that each Ψ_α corresponds exactly to one block B_α and that the number of blocks is exactly the number s occurring in (11). Moreover, each irreducible modular representation F_j will represent all but one of the Ψ_ρ by the zero matrix. If the exceptional Ψ_ρ is Ψ_α, we say that the character ϕ_j of F_j belongs to the block B_α. Now, both the ordinary irreducible characters $\chi_1, \chi_2, \ldots, \chi_k$ of G and the modular irreducible

* If it is necessary to indicate the prime p we speak of p-blocks instead of blocks. It should be mentioned that the distribution of $\chi_1, \chi_2, \ldots, \chi_k$ into blocks depends only on p and not on the special choice of the valuation ν_p.

characters $\phi_1, \phi_2, \ldots, \phi_l$ are distributed into the blocks $B_1, B_2, \ldots, B_s$. It turns out that the decomposition number $d_{\mu j}$ in (9) can be different from 0 only if χ_μ and ϕ_j belong to the same block. Likewise, c_{ij} in (10) can be different from 0 only if ϕ_i and ϕ_j belong to the same block. The block containing the principal character $\chi_i = 1$ will be called the principal block.

§12. We define the defect $d(B)$ of a block B as the maximal value of the numbers $\nu_p(n/\chi_i(1))$ for $\chi_i \in B$. It can then be shown that to each B there corresponds a class of conjugate subgroups of G of order $p^{d(B)}$, the defect groups of B. If D is a p-subgroup of G, say of order p^d, we have a natural one-to-one correspondence between the blocks B of G with defect group D, and blocks b of the normalizer $N(D)$ of D with the defect group D. On the other hand, we can find the blocks b if we know the blocks of $N(D)/D$ of defect 0. Thus, the question of determining the number of blocks B of G of a given positive defect d can be answered if we know the blocks of defect 0 of groups $N(D)/D$ which have smaller order than g. In the case of defect 0, we are faced with the problem:

Problem 19. Characterize the number of blocks of defect 0 *of* G *by group theoretical properties of* G.

A number of results concerning blocks of defect 0 are given in the Supplements, §13. If the prime p does not divide the group order n, these results show that there is really no difference between the ordinary and the modular theory in this case. This had first been shown by A. Speiser. The interesting case is that in which p divides n.

We formulate a number of further questions on blocks which we cannot answer.

Problem 20. Is it true that a p*-block* B *of defect* d *consists of at most* p^d *ordinary characters?*

Problem 21. If $q > 1$ *is an arbitrary prime power, can functions* $f(q)$ *be given such that* $f(q) \to \infty$ *for* $q \to \infty$, *and that for all finite groups* G *the number of irreducible characters in a* p*-block of defect* $d > 0$ *is at least* $f(p^d)$*?*

Problem 22. If ϕ_i *and* ϕ_j *are two modular irreducible characters of a* p*-block* B *of defect* d, *is it true that the Cartan invariant* c_{ij} *is at most equal to* p^d*?*

If χ_i belongs to a p-block B of defect d, it follows from the definition of defect, that the integer $h_i = d - \nu_p(n/\chi_i(1))$ is non-negative. We call h_i the height of χ_i (for given p). Our next question then is:

Problem 23. Is it true that a block B *has Abelian defect group* D, *if and only if all* $\chi_i \in$ B *have height* 0*?*

For some comments, see §14. The results mentioned there indicate that rather complete results are available for blocks of defect 1.

§13. If the order n of G contains the prime p with the exact exponent 1,

$$n = pn_0 \qquad \text{with } (n_0, p) = 1 \tag{12}$$

then only blocks of defects 1 and 0 occur. We can obtain a great deal of information about the characters of G, see R. Brauer [5]. We state some of the results.

(a) Let P of order p be a p-Sylow subgroup of a group G which satisfies the condition. Let $C = C(P)$ be the centralizer of P in G and let $N = N(P)$ denote the normalizer of P in G. The index $(N: C)$ has the form $m = (p - 1)/t$ where the integer t divides $p - 1$. Then the principal p-block B_0 of G contains m irreducible characters $\chi_1 = 1, \chi_2, \ldots \chi_m$ such that each χ_μ takes a fixed value $\delta_\mu = \pm 1$ for all p-singular $g \in G$. The full number of irreducible characters χ_i in B_0 is $m + t$, and the t additional characters are all algebraically conjugate and hence have all the same degree x^*. The degrees $x_i = \chi_i(1)$ satisfy the relations

$$\sum_{\mu=1}^{m} \delta_\mu x_\mu \pm x^* = 0 \tag{13}$$

$$x_\mu \equiv \delta_\mu \qquad (\text{mod } p) \text{ for } 1 \leqq \mu \leqq m \tag{14}$$

(b) If the values of the irreducible characters of N for p-singular elements are known, the values of the irreducible characters of G for p-singular elements can be given, except that certain $\pm$ signs remain in doubt.

(c) If $k(G)$ is the class number of G and $k(N)$ that of N, then

$$k(G) \geqq k(N) \tag{15}$$

Since (15) is stated without reference to characters, we may ask:

Problem 24. Prove (15) directly. In particular, determine when the equality sign holds.

The reason for our interest in the second part lies in the fact that $k(G) - k(N)$ is the number of irreducible characters χ_i of defect 0.

Problem 25. Give an inequality for the class number of an arbitrary finite group G *which in the case (12) specializes to (15).*

In the case (12), the conditions (13), (14), and the analogous conditions for the other p-blocks represent very strange necessary conditions for our original Problem 1. They are much more suited for the actual discussion of the problem than anything mentioned before. For instance, R. G. Stanton has used the results on blocks and especially (13), (14) to investigate the simple groups of order 244,823,040 (see the Supplements, §6). He could show first that the degrees x_i of the irreducible characters are uniquely determined, then that the characters are uniquely determined, and finally that G is unique.

As an application of a more general nature, I mention that it can be shown for groups G satisfying the condition (12) and equal to their commutator group G' that the values possible for the number $(G: N)$ of p-Sylow groups are of a rather restricted nature; see R. Brauer and W. F. Reynolds [15]. This leads me to ask:

Problem 26. Let p *be a given prime. What can be said about the values* r *which can occur as the number of* p*-Sylow groups in finite groups* G, *in particular in groups* G *with* G = G'?

In the case (12), the principal block B_0 of G consists of $(p - 1)/t = m$ modular irreducible representations. It has been shown by H. F. Tuan [66] that these can all be written in the prime field of characteristic p. Again, it would be of interest to see how this can be generalized. For instance, we have the question:

Problem 27. Let G *be a group whose order* n *is divisible by the prime* p. *When does the principal block* B_0 *contain irreducible modular representations besides the principal representation which can be written in the prime field of characteristic* p?

Related considerations lead to the following remark. Let G be a group satisfying (12) and let t have the same significance as above. If A is the group of automorphisms of G, then A leaves at least

$m = (p - 1)/t$ irreducible characters of G (in fact $\chi_1, \chi_2, \ldots, \chi_m$) invariant.

Problem 28. Can a result be given concerning the irreducible characters of a finite group G *left invariant by the automorphism group* A *of* G *which in the case (12) yields the result stated?*

§14. In a rather weak sense, the results for defect 1 can be generalized to the case of an arbitrary defect d. Let p^d be a given power of p. Then for all finite groups G, only finitely many "types" are possible for p-blocks of defect d. For each type, the defect group D of B is determined as an abstract group, and the numbers of irreducible ordinary characters χ_i and modular characters ϕ_j in B are determined. Moreover, the values of the degrees x_i of $\chi_i \in B$ are determined mod p^d. Let π be an element of D and assume that the centralizer $C(\pi)$ of π in G is known. If ρ is a p-regular element of $C(\pi)$, for each fixed type, the values of $\chi_i \in B$ for $g = \pi\rho$ can be found. If π is a p-element of G, which is not conjugate to an element of D in G, and if ρ is a p-regular element of $C(\pi)$ then $\chi_i(\pi\rho) = 0$ for $\chi_i \in B$.

In principle, this gives us all we had in the case $d = 1$. There are cases with $d > 1$ where we have all we can wish for (see, for instance, the Supplements, §15). The trouble is that usually the number of possible types is so tremendously large that the actual discussion is out of question. In fact, we do not even know the p-groups of order p^d except when d is small. It is certain that most of the types which we cannot exclude on the basis of our present knowledge may never occur. Thus, the following rather vague problem is important.

Problem 29. To obtain information on p*-blocks of defect* d *for given* p^{d} *which restrict the possibilities for types.*

§15. The results outlined here must appear complicated, but this has to be expected. The characters of many classes of groups have been determined and few, if any, regularities are evident. It was my wish to show that in spite of this, there is a strong interrelation between properties of the representations and properties of the abstract group. This frequently makes it possible to obtain information about the characters, if we know rather little about the groups. Based on this, we can often study groups about which we have made only rather weak assumptions.

The more connections we have between properties of the representations and properties of the abstract group, the better we can expect this to work. Most of the questions I have asked above can be summarized essentially in one single problem: to find new connections.

§16. In the next sections, I shall leave the general theory of representations and discuss some topics of a more concrete nature where the results discussed so far can be applied.

Recently, W. Feit and J. G. Thompson [28] succeeded in proving the famous old conjecture which states that all finite groups of odd order n are solvable; the complete proof is to appear in the *Pacific Journal of Mathematics.* If we are interested in nonsolvable groups, we can now concentrate on groups of even order.

Assume that $n = |G|$ is even. Then the group G contains elements j of order 2 or, as we shall say, involutions j. If j belongs to the conjugate class K_α of G, the scalars of multiplication $a_{\alpha\alpha\gamma}$ of the class algebra Z of G have some special properties. Expressing the $a_{\alpha\alpha\gamma}$ by means of the characters and using the blocks to investigate the resulting relations, results on groups can be obtained which we shall now discuss.

It will be necessary to introduce some notation. Let p be a fixed prime dividing n and let π be an element of order $p^a > 1$ of G. The generalized centralizer $C^*(\pi)$ of π is the subgroup of G consisting of all $\sigma \in G$ for which either $\sigma^{-1}\pi\sigma = \pi$ or $\sigma^{-1}\pi\sigma = \pi^{-1}$.

A rather crude application of the method yields results of the following kind.

Choose an arbitrary real number $A > 1$. We then have one of two cases.

Case 1. There exists an approximate formula for n.

$$\frac{A}{A+1}\,\xi c(j)^2 \leqq n \leqq \frac{A}{A-1}\,\xi c(j)^2 \tag{16}$$

where ξ depends only on $C^*(\pi)$, and the manner in which the intersection $K_\alpha \cap C^*(\pi)$ of the conjugate class K_α of j with $C^*(\pi)$ breaks up into conjugate classes of $C^*(\pi)$. If k_0 such $C^*(\pi)$ classes appear in K_α, then

$$n \leqq c(\pi)c(j)^2\lambda k_0 \tag{17}$$

where the real number λ depends on A, p, and $\nu_p(c(j))$.

Case 2. There exists a normal subgroup N $\neq$ G *of* G *such that* (G: N) *lies below* $[\beta c(j)]!$ *with* β *depending on* A, *p*, *and* $\nu_p(n)$.*

If π and π^{-1} are not conjugate in G, only Case 2 has to be considered.

For $p = 2$, the result can be improved somewhat. In (17) the factor k_0 can be deleted. If H is any finite group, we define its 2-regular core $K_2(H)$ as the maximal normal subgroup of H of odd order. In the Case 2, $(G: N)$ lies below $[\beta(C(j): K_2C(j))]!$.

Let us restrict our attention to the case $p = 2$. We can obtain much more precise results, if we know the type to which the principal 2-block B_0 of G belongs. We shall then assume that we know the defect group of B_0, that is, the 2-Sylow subgroup P of G. This leads to the following problem.

Problem 30. Given a 2-group P, *investigate the finite groups* G *with* P *as their Sylow subgroup. In particular, investigate the groups* G *in which in addition the centralizer* C(j) *of an involution is given.*

If j belongs to the center of P, then P is determined by $C(j)$ and we have a special type of Problem 9 with only one C_i given. It may seem surprising that it is possible to say anything about G. Still, this is the case for a number of classes of 2-groups (Supplements, §16) and, in principle, the method can be used for any P. As a more precise question in the direction of Problem 30, we may ask:

Problem 31. For which 2-groups P *can there exist only finitely many (nonisomorphic) groups* G *with 2-Sylow group* P *and with 2-regular core* $K_2(G) = \{1\}$ *(that is, without normal subgroup of odd order larger than 1)?*

This can at least be answered for Abelian P; see R. Brauer [10]. In §4 of this paper a number of other possibilities are mentioned which can arise in connection with Problem 30.

The answer to the following question seems to be unknown.

Problem 32. If G *is a simple (noncyclic) group of order* n, *does there exist an involution* j *such that* $n \leqq c(j)^3$?

If G has at least two conjugate classes of involutions, this can be

* A similar result is proved by elementary methods in R. Brauer and K. A. Fowler [13].

proved by quite elementary methods even without the assumption of simplicity of G (see R. Brauer and K. A. Fowler [13]). The formula (17) with $\pi = j$ shows that the methods discussed here might be applicable. It would be useful if we had the answer to the following question.

Problem 33. For which 2-groups P *can it happen that there exist finite groups* G *with* P *as their 2-Sylow group and with only one conjugate class consisting of involutions?*

§17. One of the principal outstanding problems in the theory of finite groups is that of determining all simple finite groups. Families of simple groups and a few individual simple groups are known (Supplements, §17). However, we have no idea how many of the simple finite groups have not yet been discovered. While it is not unthinkable that some day a direct and uniform approach to the study of all finite simple groups might be found, we do not know when this day will come.

In the meantime, there are at least a few things we can try if we are much more modest. In some of the more approachable cases, one can find group theoretical characterizations of particular simple groups by using the idea of Problem 30 and imposing further conditions on $C(j)$. If we knew more about the 2-Sylow subgroups of the simple groups and more about Problem 30, we would have at least an opening. Of course, for the simple groups of Lie type, it is a different prime, the characteristic of the field, which plays a distinguished role. But at least in some ways, the prime $p = 2$ seems to be second best.

I am aware that I am going far out on a limb if I now make some other suggestions about things one might try to do. The answers may not be what I hope them to be, and the direction may not be the right one. But at least the first steps do not look hopeless, and, without trying, we do not know where they may lead.

As my starting point, I use a well-known theorem of Burnside: there do not exist any finite simple groups whose order n is divisible by exactly two distinct primes. I ask:

Problem 34. Is it true that there exist only finitely many simple groups whose order is divisible by exactly three distinct primes?

If we wish to go on to the case of four primes, we have to change the question somewhat.

Problem 35. Is it true that there exist only finitely many simple groups whose order n *is divisible by exactly four distinct primes and which are not of the type* $L_2(q)$?

Before turning to group orders with larger number of prime factors, we note that the orders of the simple groups in a given family $\mathfrak{F}$ (Supplements, §17) with fixed dimension can be written in the form

$$n(F) = q^a \cdot \prod_r f_r(q)^{a_r} \tag{18}$$

where $f_r(q)$ denotes the rth cyclotomic polynomial, where a and a_r are positive integral exponents, and where r ranges over a certain finite set of positive integers. Let $M(\mathfrak{F}) - 1$ denote the number of values r appearing here. We now can pose the question:

Problem 36. Given an integer $s \geqq 3$, *is the number of simple groups* G *finite whose order* n *is divisible by exactly* s *distinct primes and which do not belong to families* $\mathfrak{F}$ *with* $M(\mathfrak{F}) \leqq s$?

A number theoretical theorem of C. L. Siegel [54] shows that the known simple groups will not provide counterexamples. Also, for given s, there are only finitely many families with $m(\mathfrak{F}) \leqq s$.

It is very likely that before much progress is made on questions concerning simple groups, more attention will have to be paid to the known families of simple groups. We discuss some points which fit in with the general program discussed in this report. First of all, we would like to know the characters of the groups in the various families. By operating rather recklessly, it is often possible to make some good guesses. It seems not unreasonable to hope that the structure of the groups G in a family $\mathfrak{F}$ depends somewhat regularly on the parameter q. A first question in this direction is:

Problem 37. Consider the groups G *in a fixed family* $\mathfrak{F}$ *of simple groups. Do there exist polynomials*

$$x_1(q), x_2(q), \ldots, x_t(q),$$
$$n_1(q), n_2(q), \ldots, n_t(q)$$

with the following property: for any fixed value of q, *the group* G *possess exactly* $n_i(q)$ *irreducible characters* χ_i *of degree* $x_i(q)$, $(i = 1, 2, \ldots, t)$?

λ ℓ.7: const. [R.B.]

The primes p dividing the values r occurring in (18) play a somewhat exceptional role. If they are removed from $f_r(q)$ (for some fixed q, or, more generally, for some class of values of q determined by congruence conditions), the remaining part h_r of $f_r(q)$ often seems to behave as if it was a prime factor of n. This is specially useful, if the exponent a_r is 1, since we can then apply tentatively the strong results known for the case where the group order is divisible by a prime with the exact exponent 1. For all families $\mathfrak{F}$, such factors $f_r(q)$ with $a_r = 1$ occur.

If we have guessed the values of some irreducible character of G, it may be possible to prove that we actually have a character by showing that the necessary and sufficient conditions for characters are satisfied (see §5 in this section).

Of course, ideas of the same nature can be tried for other group theoretical properties of G.

It is not necessary for our discussion that the polynomial (18) actually belongs to a family of simple groups. We ask:

Problem 38. Consider a polynomial n(q) *and assume that for all* q *in a certain set of integers there exists a group* $\mathrm{G_q}$ *depending on* q *of order* n(q). *Give necessary and sufficient conditions of a group theoretical nature which guarantee that the heuristic principles described above are true statements.*

§18. As my last topic, I wish to discuss briefly the finite linear groups of a fixed degree m. By a linear group G of degree m over a field Ω, we mean a group G which has a faithful representation of degree m in Ω. Given Ω, the question then is: what are these linear groups of a given degree?

If $m = 3$ and if Ω is the field **R** of real numbers, this problem is essentially equivalent to a study of the Platonic solids. We only have to add some degenerate solids, the dihedra. The linear transformations mapping the solid onto itself form a finite group. Given a finite linear group G, we can find the vertices of an associated solid by looking for points for which the total number of distinct images under transformations of G is relatively small. Our general question about the linear groups is an extension of the old question of Greek mathematics.

Again, it is natural to work in the field $\Omega = \mathbf{C}$ of complex numbers. The basic result is a theorem of Camille Jordan (1878) which states that, for a given degree m, there exist bounds β_m

depending on m only such that a linear group G of degree m has a normal Abelian subgroup A for which the index $(G: A)$ lies below β_m. In particular, there are only finitely many simple linear groups of degree m. Explicit bounds for β_m have been given by Blichfeldt, Bieberbach, and Frobenius; see H. F. Blichfeldt [3] for references. The small dimensions $m = 2, 3, 4$ have been studied, and, with the newer results on characters, the case $m = 5$ and probably the next cases can be handled.

On the whole, we know very little about this subject. It would be of interest to obtain substantially better values for the bound β_m. Since for large m, more and more linear groups of degree m occur, it would be futile to try to list them all. Rather, we should concentrate on the interesting cases. Before this can be done, it would be necessary to decide which groups deserve special attention.

It may also be worthwhile to look more closely at the case where the dimension m is a prime power. It has been shown by R. Steinberg [59] that if G is a finite simple Lie group and if p^a is the highest power of the characteristic dividing the group order n, then G has an irreducible representation of degree p^a. These Steinberg linear groups seem to play an important role, and it would be of interest if they could be singled out from all the linear groups of prime power degree.

With the whole field wide open, there may not be too much point in placing emphasis on any particular question. I am mentioning the next problem because it plays a role in Blichfeldt's work and because recent progress has been made on it.

Problem 39. For which linear groups G *of degree* m *does it happen that there exist primes* $p > m + 1$ *dividing the group order* n *such that* G *does not have a normal subgroup of order* $p^a > 1$?

Recently, W. Feit and J. G. Thompson [26] proved that if the prime p divides the order n of the linear group G of degree m, and if the p-Sylow subgroup of G is not normal in G, then $p \leqq 2m + 1$. Their proof is based on a result of the author dealing with the case $n \not\equiv 0 \pmod{p^2}$ which depends on properties of group characters of the type discussed here.

I have been informed by W. Feit that he can now show that if the linear group G of degree m has order $n = n_1 n_2$, where all prime factors of n_1 are larger than $m + 1$ while the prime factors of n_2 are at most equal to $m + 1$, then G has a normal subgroup either of order n_1 or of order n_1/p with p a prime. This disposes of a question which I asked at the original lectures. For some other results, see the Supplements, §18.

There is no reason to exclude fields of characteristic p. The celebrated Theorem B of P. Hall and G. Higman [35] deals with a property of p-soluble linear groups over fields of characteristic p. This theorem, which has already proved its value in J. G. Thompson's work, opens a number of new questions.

If Ω is a finite field, say with q elements, we are dealing with the subgroups G of a particular finite group, the general linear group $GL(m, q)$. Still, we know very little about the linear groups of degree m. If the case $m = 4$ had been studied systematically, Suzuki's simple groups would have been discovered earlier. It may then not be superfluous to ask:

Problem 40. Determine all linear groups of small degrees m *over finite fields.*

No result in direction of Jordan's theorem can exist for algebraically closed fields of characteristic p instead of **C**. For some remarks, see the Supplements, §19.

§19. I could go on. There are many topics about which we would want to know more, for example, invariants of linear groups and linear groups with integral coefficients. But the questions formulated here will suffice to show that a great deal of work remains to be done in the theory of group representations, even if we restrict ourselves to finite groups. If more mathematicians become interested and help to answer the questions, I have accomplished what I set out to do.

SUPPLEMENTS

§1. *Introductory Remarks.* It is our aim to demonstrate that characters form a powerful tool for the study of finite groups. It is essential that it is often possible to say quite a bit about the characters, if only very scanty information about the group is available. On the other hand, a great deal of information about the group is hidden in the characters. Thus, the interrelations between properties of characters and of group theoretical properties is of great interest. I have already said that I believe that for further progress combinatorial group theoretical methods will have to be used together with methods from the theory of representations. At first glance, it may sometimes appear as if we try to let characters do all kinds of tricks just for the love of it. What we have in mind are the first steps of the program outlined here. There are indications that for the eventual success of the method, it will not only be the general principles which will matter, but also the ability to adapt ourselves to a particular situation.

§2. *Numerical Examples.* It may be instructive to mention an example in connection with Problem 1. I have chosen the case $n = 60$. The conditions (a), (b), (c) in Survey* §3 for the degrees $x_1, x_2, \ldots$ can be satisfied in 65 ways. Actually, there are only 9 group algebras as can be seen by the methods developed in the later sections of the part on Survey. On the other hand, the number of (nonisomorphic) groups of order 60 is 13. If we are interested only in groups G for which G is its own commutator group G', exactly one of the x_i is 1. With this condition added, there are 6 systems satisfying (a), (b), (c). Only one of these, the system $\{1, 3, 3, 4, 5\}$ corresponds to a group algebra, that of the icosahedral group.

In order to show the mysterious nature of the degrees $x_1, x_2, \ldots$, I give another example, that of the Mathieu group M_{24} of order 244,823,040. The characters of M_{24} were obtained by Frobenius in 1904. There are 26 irreducible characters. The degrees are 1, 23, 45, 45, 231, 231, 252, 253, 483, 770, 770, 990, 990, 1035, 1035, 1035, 1265, 1771, 2024, 2277, 3312, 3520, 5313, 5544, 5796, 10395. There are all kinds of hidden relations, for instance, of the type discussed in finite Groups, §13.

§3. *Relations between the Characters and the Mappings $K \to K^{[m]}$.* As a first remark, we observe that if we know the characters of G, we can determine the orders $c(\sigma)$ of the centralizers of the elements $\sigma \in K$; we also write $c(K)$ for $c(\sigma)$. It is obvious that $c(K)$ divides $c(K^{[m]})$ and this gives a necessary condition for the class $K^{[m]}$ for given K. Moreover, if m is relatively prime to $n = |G|$, the values $\chi_i(K^{[m]})$ are algebraically conjugate to $\chi_i(K)$ in a definite manner and then $K^{[m]}$ is uniquely determined in this case. More generally, we can use the fact that if σ has order r and if χ is a character of

* "Survey" refers to section headed SURVEY OF THE THEORY OF REPRESENTATIONS OF FINITE GROUPS.

degree x, then $\chi(K)$ is a sum of x rth roots of unity ϵ_i and $\chi(K^{[m]})$ is the sum of the ϵ_i^m. The number r itself is a divisor of $c(K)$.

A result of a related nature is a theorem of Blichfeldt which gives a relationship between the irrationalities appearing in the values of the characters and the orders of the elements of G. A very short proof is given in a paper of the author, forthcoming in the *Proceedings of the American Mathematical Society*. I mention a rather special question which I cannot answer.

Problem 41. Let p, q, r *be three distinct primes and suppose that* G *has an irreducible character* χ *for which there exist elements* α, β, γ *of* G *such that* $\chi(\alpha)$, $\chi(\beta)$, $\chi(\gamma)$ *are irrational and* $\chi(\alpha) \in Q(\sqrt{\pm qr})$, $\chi(\beta) \in Q(\sqrt{\pm pr})$, $\chi(\gamma) \in Q(\sqrt{\pm pq})$. *Does* G *contain elements of order* pqr?

If the answer should be affirmative, various generalizations could be tried.

If the characters of G are known and if p is a given prime, we may be interested in finding out which classes K contain p-elements (for instance, as a first step in attempting to determine the structure of a p-Sylow group of G). If $\chi_i(K)$ is rational, a necessary condition is given by $\chi_i(K) \equiv \chi_i(1) \pmod p$. If $\chi_i(K)$ is irrational, this has to be replaced by a congruence modulo a suitable prime ideal.

§4. *Remarks on Problems 3 and 4.* If the notation is as in Finite Groups §5, and if we write c_i for $c(\sigma_i)$, the k numbers c_i are connected by the equation

$$\frac{1}{c_1} + \frac{1}{c_2} + \cdots + \frac{1}{c_k} = 1 \tag{1}$$

For $\sigma_k = 1$, we have $c_k = n$. Landau's method mentioned in connection with Problem 3 is based on a simple discussion of this Diophantine equation (1). Of course, (1) represents a condition which the groups C_i in Survey, (5) have to satisfy.

If $n > 1$, none of the c_i can be 1. The case where some c_i is 2 can be discussed easily. Recently, W. Feit and J. G. Thompson [27] determined all G for which some c_i is 3. For the case that some c_i is 4, see M. Suzuki [60].

The answer to the following question seems to be unknown.

Problem 42. Let p *be a fixed prime. Can there exist infinitely many simple groups for which some* $C(\sigma)$ *is cyclic of order* p?

Groups G in which all groups C_i in Finite Groups (5) are nilpotent have been called *CN* groups. They have been studied rather exhaustively by M. Suzuki [63, 64]. An earlier paper of W. Feit, M. Hall, Jr., and J. G. Thompson [25] should be mentioned here. Though the results have now been superseded it has played an important role in recent developments. Suzuki in his work also determined all simple groups (of even order) for which the groups C_i in Survey (5) are either 2-groups or of odd order. It was in this connection that he discovered his new class of simple groups.

§5. *Remarks on Induced Characters.* In many applications of character theory to investigation of group theoretical properties, Frobenius' relations between characters of groups and subgroups (Appendix, §6) play a fundamental role. A generalization of Frobenius' reciprocity theorem has been given by G. W. Mackey [44].

In certain cases quite definite information on the characters of G can be obtained easily on the basis of Frobenius' idea. This led M. Suzuki and the author to a discussion of the so-called exceptional characters. For a very simple proof see R. Brauer and H. S. Leonard [14]. Refined versions appear in Suzuki's papers and in W. Feit [24].

A most ingenious use of ideas of this general nature is of importance in W. Feit's and J. G. Thompson's proof of the solvability of the groups of odd orders. Since the intermediate results are geared to the special situation encountered there, it is impossible to give a taste of them here.

§6. *Information about Subgroups Available If the Characters of* G *Are Known.* If H is a subgroup of G of index $(G: H) = r$ the principal representation $h \rightarrow 1$ of H induces a representation X of G of degree r. Actually, X is the permutation representation of G belonging to the subgroup H. The character χ is easily seen to have the following properties (see, for instance, D. E. Littlewood [43]), Chap. 9, §3.

(a) The degree r of χ divides $n = |G|$.
(b) The values of χ are non-negative rational integers.
(c) $\chi(K^{[m]}) \geqq \chi(K)$ for all conjugate classes of G and all exponents m.

(d) The irreducible character χ_i of G appears in χ at most with the multiplicity $x_i = \chi_i(1)$. In particular, the principal character of G appears exactly once in χ.
(e) For each class K of G, the number $n\chi(K)/(rc(K))$ is an integer.
(f) If $\sigma \in G$ has an order m which does not divide n/r, then $\chi(\sigma) = 0$.

If the irreducible characters $\chi_1, \chi_2, \ldots, \chi_k$ of G are known, we can form the reducible characters χ of G which satisfy these conditions. There can exist a subgroup H of G of index r only if r is the degree of such a character χ.

The permutation representation X belonging to a subgroup H is doubly transitive, if and only if $\chi - 1$ is irreducible.

It has already been mentioned that if the characters of G are known, we can find the orders $c(\sigma)$ of the centralizers of elements $\sigma \in G$. It is also not difficult to obtain the order of the normalizer of the cyclic group generated by σ.

In some special cases, methods are available which allow us to obtain new subgroups from given ones. For instance, it may happen that we know two subgroups A, B of G and $D = A \cap B$. If χ is an irreducible character of G, we can find the multiplicity α with which the principal character of A appears in $\chi|A$. Likewise, if β and δ have analogous significance for B and D respectively, these integers can be determined. If $\alpha + \beta > \delta$, then A and B generate a proper subgroup H of G. For instance, in his paper on simple groups of order 244, 823, 040, R. G. Stanton [58] succeeded in constructing the character tables of these groups. There exists exactly one irreducible character χ of degree 23. There are normalizers A and B of cyclic groups of orders 253 and 55 respectively, and we may assume that $D = A \cap B$ has order 11. Here, $\alpha = 1$, $\beta = \delta = 3$ and the method applies. It can now be seen that H has index 24 in G. Then G has a permutation representation of degree 24, and it is not very difficult to show that G must be isomorphic to the Mathieu group M_{24}. We have here a case where the character table of a group, actually a very small part of it, determines a group uniquely.

There exist other methods to construct subgroups by means of characters, but they all are of a rather special nature, and we shall not discuss them.

As another example of an application of characters in a numerical case, a paper of E. T. Parker [46] may be mentioned. It seems to have been conjectured that every (noncyclic) simple group can be represented as a multiply transitive permutation group. Parker showed that this is not true for the simple group of order 25920.

If the conjecture stated in Problem 23 is true, this would allow us to decide whether the p-Sylow group P of G is Abelian. If we know the p-classes K, we have a necessary condition for this question: $c(K)$ for these classes must be divisible by the full power of p dividing n. It does not seem to be known whether this condition is sufficient for P to be Abelian.

§7. *Remarks Concerning Automorphisms of Groups.* The automorphism τ of G which leave every irreducible character χ_i of G fixed form a normal subgroup $B(G)$ of the automorphism group $A(G)$ of G. Thus, $A(G)/B(G)$ appears as a permutation group on the set $\{\chi_1, \chi_2, \ldots, \chi_k\}$. If p is a fixed prime, the p-blocks of G are permuted among themselves. In many cases, this can be used to study $A(G)/B(G)$.

The elements of $B(G)$ can also be characterized as the automorphisms τ of G which map every conjugate class of G onto itself. Of course, all inner automorphisms of G belong to $B(G)$.

§8. *Some Comments on Survey,* §7. The number r of (nonequivalent) irreducible representations of the group G in the field Ω of characteristic 0 can also be described in more direct group theoretical terms. Let ϵ denote a primitive nth root of unity and let g denote the Galois group of $\Omega(\epsilon)$ over Ω. Every $\sigma \in g$ maps ϵ on a power $\epsilon^{s(\sigma)}$ where $s(\sigma)$ is a rational integer prime to n. For every $\sigma \in g$, we can form the permutation $K \to K^{[s(\sigma)]}$ of the set $\{K_1, \ldots, K_k\}$ and g is then represented as a permutation group Π on this set. The number r is equal to the number of transivity classes of Π.

This result probably was known to I. Schur. It is a corollary of a simple lemma of the author [6]. It was obtained independently by S. D. Berman [2].

There are extensions to the case of fields of characteristic p.

For references to work of P. Roquette, E. Witt, S. D. Berman, and the author on Schur indices and related questions, see the paper of L. Solomon [55] quoted in Finite Groups, §7. Solomon's method provides a good approach to much of this work.

Another result of interest was given by A. Speiser [56]; see also I. Schur [53]. Speiser proved (with $\Omega \subseteq \mathbf{C}$) that if the values of

χ_j are real, and if the degree x_j is odd, the corresponding Schur index is 1. For real χ_j of even degree x_j, m_ρ is 1 or 2, see R. Brauer [4] in conjunction with later results of H. Hasse [36] on the exponents of central simple algebras. There are connections with questions concerning invariants.

As to the behavior of the irreducible representation X_j of G in C in the field R, it was shown by G. Frobenius and I. Schur [31] that the number

$$\epsilon_j = \frac{1}{n} \sum_{g \in G} \chi_j(g^2)$$

can have only one the values 1, −1, 0. We have $\epsilon_j = 1$, if X_j can be written in R; we have $\epsilon_j = -1$, if this is not possible, but if χ_j is real. Finally, $\epsilon_j = 0$, if χ_j is nonreal. If the notation is as in Survey, §7, we have $\epsilon_j = 1$ if the division algebra Δ_ρ is R, and $\epsilon_j = -1$ if Δ_ρ is the quaternion algebra. Finally, $\epsilon_j = 0$, if $\Delta_\rho = \mathbf{C}$. For any $\sigma \in G$, the number

$$N(\sigma) = \sum_{j=1}^{k} \epsilon_j \chi_j(\sigma)$$

represents the number of solutions of the equation

$$\xi^2 = \sigma \qquad \text{with } \xi \in G$$

While it is easy to characterize the number of characters χ_j with $\epsilon_j = 1$ or $\epsilon_j = -1$, no direct characterization of the number of χ_j with $\epsilon_j = 1$ alone is known. This is the point of Problem 14.

§9. *Some Remarks on Problems 2 and 2**. It follows from the remark on the centers Z_ρ of the algebras Λ_ρ in Survey, §7(7) that nonisomorphic Abelian groups have nonisomorphic group algebras over the field Q of rational numbers. The case of Abelian groups has been studied further by S. Perlis and G. Walker [47] and W. E. Deskins [20].

The two (nonisomorphic) non-Abelian groups of order p^3 have isomorphic group algebras over every field Ω of characteristic 0 if p is odd. It follows from Jennings' results (Survey, §8) that for fields Ω of characteristic p, the algebras are not isomorphic. For $p = 2$, this is already true for the field $\Omega = \mathbf{Q}$.

Jennings' results allow us to define many numerical invariants of group algebras of p-groups over the prime fields of characteristic p by counting the number of elements of a power of the radical which satisfy given conditions. This suggests that it may be much easier to study Problem 2 for this particular case.

§10. *Groups with Cyclic p-Sylow Group.* An interesting theorem of D. G. Higman [38] gives a class of groups of the kind required in Problem 16. A finite group G has a cyclic p-Sylow group if and only if the degrees of all indecomposable representations of the group algebra Γ over a field Ω of characteristic p have degrees which lie below a fixed bound.

§11. *A Remark on the Dimension of the Radical of the Modular Group Algebra.* If we know the degrees f_j of the modular irreducible representations F_j in Finite Groups, §9, we can easily find the dimension h of the radical of Γ. Indeed, $h = n - \sum_j f_j^2$. If we could characterize h directly as required in Problem 15, this would give an interesting connection.

As an example, we mention the groups $G = L_2(p)$. Here, the f_j are known and we find $h = \frac{1}{6}(p + 1)(2p^2 - 5p)$ for $p > 2$.

§12. *Blocks.* For properties of blocks, see R. Brauer [7, 9]; a continuation is being prepared for publication. Modifications and simplifications are given in M. Osima [45], K. Iizuka [39], and A. Rosenberg [50].

In this connection, I wish to mention the interesting investigations of J. A. Green [32, 33].

§13. *Blocks of Defect 0.* Blocks of defect 0 have properties much simpler than those encountered in the general case. Each block B of defect 0 consists of one ordinary irreducible character χ_i and one modular irreducible character which happens to be the restriction χ_i^0 of χ_i to the set of p-regular elements of G. An ordinary irreducible character χ_i belongs to a block of defect 0, if and only if its degree x_i is divisible by the full power of p dividing n. An equivalent condition is that χ_i vanishes for all p-singular elements. If the representation X_i belonging to χ_i is a constituent of a reducible representation, it is a direct summand.

If the characteristic p of Ω does not divide the group order n, all p-blocks have defect 0. It follows from the results stated here that there is really no difference between the ordinary and the modular theory in this case. This has first been proved by A. Speiser.

As to Problem 19, the situation is of a rather mysterious nature. For instance, if G is the symmetric group $\mathfrak{S}_m$ of all permutations of m objects and if $p = 2$, then G has blocks of defect 0, if and only if $m = r(r + 1)/2$ with $r = 1, 2, 3 \ldots.$ In this case we have exactly one such block. We remark that for the symmetric groups the blocks are known for all primes p. The result in question has been conjectured by T. Nakayama and proved by G. de B. Robinson and the author [49].

In the case of a solvable group G, sufficient conditions for the existence of blocks of defect 0 have been given by N. Ito [40]. For a large class of group orders n including all odd n, these conditions are necessary.

For sufficient conditions of a different nature for arbitrary finite G of even order, see R. Brauer and K. A. Fowler [13].

§14. *Comments on Problems 20–23.* The answer to Problem 20 is affirmative for $d \leqq 2$. For $d \geqq 3$ the number of ordinary irreducible characters in a p-block of defect d is less than p^{2d-2} (R. Brauer and W. Feit [12]).

For blocks B of defect 1, the number of $\chi_i \in B$ is of the form $\tau + (p - 1)/\tau$ where τ is a divisor of $p - 1$. In particular, this number is at least $2\sqrt{p-1}$. If p but not p^2 divides n, the principal p-block (that is, the block B_0 which contains the principal character $\chi_0 = 1$ of G) has defect 1. It follows that if G has class number k, then $p \leqq 1 + (k^2/4)$. This is a partial result in direction of Problem 3 for primes p with $\nu_p(n) = 1$. An answer to Problem 21 might lead to an answer to Problem 3 in a similar fashion.

Problem 22 can also be answered in the affirmative for blocks of defect 1. In fact, much more precise results on the values of the decomposition numbers and the Cartan invariants are known in this case.

If the answer to Problem 23 is affirmative, this would yield an answer to the last part of Problem 12. In the case of p-solvable groups, it has been shown by P. Fong [30] that if the center of the defect group D of a block B has index p^c in D, then all $\chi_i \in B$ have height at most c. In particular, if D is Abelian, all heights are 0. Conversely, if the principal block of G contains only characters

χ_i of height 0, then the p-Sylow group of G is Abelian. Fong has also solved Problem 22 for p-solvable G. For arbitrary finite G, it can be shown that blocks with cyclic defect group contain only χ_i of height 0 and the answer to the question raised in Problem 20 is affirmative in this case. We ask:

Problem 43. Can a theory be developed for blocks with cyclic defect group D *which for* $|\mathrm{D}| = \mathrm{p}$ *yields the known results for defect 1?*

§15. *Remarks Concerning Types of Blocks.* There are cases where rather complete information is available about the possible types of blocks. The discussion is somewhat easier for $p = 2$, since additional methods are available in this case.

As an example, we mention the 2-groups P of order 2^{2m+1} with $m \geqq 2$ generated by three elements σ_1, σ_2, τ such that

$$\sigma_1^{2^m} = \sigma_2^{2^m} = \tau^2 = 1$$

$$\tau^{-1}\sigma_1\tau = \sigma_2, \ \sigma_1\sigma_2 = \sigma_2\sigma_1$$

Each such P occurs as 2-Sylow group of infinitely many simple groups.

Let G be a finite group with P as its 2-Sylow subgroup such that G does not have a normal subgroup of index 2. Then the principal block B_0 of G can be shown to consist of $\frac{1}{3}(2^{2m-1} + 3 \cdot 2^{m+1} + 1)$ ordinary characters χ_i. There exists an integer

$$f \equiv 1 + 2^m (\mathrm{mod}\ 2^{m+1})$$

but not necessarily positive such that for $\delta = \operatorname{sign} f$, the degrees of the $\chi_i \in B_0$ are as follows: 1, $f(f + 1)$, δf^3, 2^{m-1} equal degrees $\delta(f - 1)(f^2 + f + 1)$, $\frac{1}{3}(2^{2m-1} - 3 \cdot 2^{m-1} + 1)$ equal degrees $\delta(f + 1)(f^2 + f + 1)$, $2^m - 1$ equal degrees $f^2 + f + 1$, $2^m - 1$ equal degrees $\delta f(f^2 + f + 1)$. A great deal of further information is available concerning the values of these χ_i.

§16. *Remarks on Problem 30.* For results of this kind, see R. Brauer [8], R. Brauer and M. Suzuki [16], M. Suzuki [61], R. Brauer [10], and H. N. Ward [68]. I have been informed that D. Gorenstein and J. Walter succeeded in determining the structure of all finite groups with dihedral 2-Sylow subgroup using the deep methods of J. G. Thompson.

As another example, the groups P of order 2^{2m+1} in §15 of this section may be mentioned. If G has P as its 2-Sylow subgroup and if G does not have a normal subgroup of index 2, the order n of G has the form

$$n = \alpha^3\lambda f^3(f + 1)(f - 1)(f^2 + f + 1)$$

where f is the same number as in §15 of this section and where α and λ are integers. If j is an involution in G, then $c(j) = \alpha\lambda(f + 1)f(f + 1)$ and α, λ, f and hence n are completely determined by $C(j)$. The following groups fall under the groups discussed here: (1) The projective groups of a projective Desarguean plane over a field with q elements, $q \equiv 1 \bmod 4$. (2) The full unitary projective group of degree 3 over a ground field with q elements, $q \equiv -1 (\bmod 4)$. In the first case, $f = q$ and in the second $f = -q$.

§17. *The Known Simple Finite Groups.* The following simple finite groups are known. First, we have the alternating permutation groups A_m with $m \geqq 5$. Next, there are the classical groups, the linear, unitary, symplectic, and orthogonal groups distributed into families L_m, U_m, S_m, $O_m{}^+$, $O_m{}^-$ where the groups in each family depend on a parameter q which is a prime power. The index m is the dimension of the underlying vector space. Then there are the families $E^{(2)}$, $E^{(4)}$, $E^{(6)}$, $E^{(7)}$, $E^{(8)}$ which are analogues of the exceptional Lie groups. There are three more families related to exceptional Lie groups, the Suzuki groups and the two families of Ree groups. In the first two of these families the parameter q is a power of 2 with odd exponent larger than 1, and in the last family q is a power of 3 with odd exponent larger than 1. Finally, there are five individual groups, the Mathieu groups M_{11}, M_{12}, M_{22}, M_{23}, M_{24} which play a mysterious role.

The classical groups were discussed in L. E. Dickson [21]; and J. Dieudonné [22, 23].

The connection with Lie groups was studied by C. Chevalley [18] in a much more general context. Chevalley's paper contains a wealth of information concerning these groups. Chevalley also discovered the groups $E^{(4)}$, $E^{(6)}$, $E^{(7)}$, $E^{(8)}$. The family $E^{(2)}$ had already been found by Dickson.

For the Suzuki and Ree groups, see M. Suzuki [62] and R. Ree [48]. See also the papers of R. Steinberg and J. Tits quoted in Ree. For the Mathieu groups, see E. Witt [70].

The orders of the known simple groups were considered from a number theoretical point of view by E. Artin [1].

§18. *Remarks on Linear Groups.* Combining W. Feit's new theorem quoted in Survey, §18 with older results, we know now all cases in Problem 39 where $p = 2m + 1$ (R. Brauer) and where $2m + 1 > p > (3m - 1)/2$ (H. F. Taun [66]). Very recently, S. Hayden and the author have shown that the case $(3m - 1)/2 \geqq p > (6m - 1)/5$ is impossible.

The case $m = 2$ in Problem 40 has been treated by L. E. Dickson in his book on linear groups [21]. Using methods related to those discussed in Survey, §16, D. Bloom has recently handled the case $m = 3$. Nothing seems to be known for higher dimensions.

APPENDIX

§1. While a definition of the elements of the group algebra as formal sums $\Sigma a_g g$ as in Finite Groups (1) is hardly satisfactory from a logical point of view, the reader will have no difficulty in constructing an associative algebra Γ over the given field Ω with a basis whose elements form a group G_0 under multiplication isomorphic with the given group G. If we identify corresponding elements of the groups G_0 and G, the elements of Γ appear in the form Finite Groups (1).

§2. We speak of a matrix representation X of a group G in the field Ω, if with every $g \in G$, there is associated a nonsingular square matrix $X(g)$ of a fixed degree x such that the mapping $g \rightarrow X(g)$ is a homomorphism, that is, that

$$X(g_1)X(g_2) = X(g_1g_2)$$

for all $g_1, g_2 \in G$. We should keep in mind that square matrices really are the analytic expressions for linear transformations of a vector space V into itself, if a fixed coordinate system (basis) has been chosen. Two representations X and X^* of the same "degree" x are equivalent if there exists a fixed nonsingular matrix P of degree x such that $P^{-1}X(g)P = X^*(g)$ for all $g \in G$. This means that $X(g)$ and $X^*(g)$ describe the same linear transformation of V in two different coordinate systems. A representation X is irreducible if there does not exist a subspace W of V with $W \neq (0)$, $W \neq V$ such that W is mapped into itself by all $X(g)$ with $g \in G$.

By a matrix representation of an algebra Γ with unit element 1, we mean an algebra homomorphism of Γ onto a set of square matrices such that 1 corresponds to the unit matrix I. Every representation X of G can be extended to a representation of the group algebra Γ by linearity, and every representation of Γ is obtained in this fashion.

The radical N of an algebra Γ is the set of those elements $\gamma \in \Gamma$ represented by the zero matrix in every irreducible representation of Γ. Then N is a two-sided ideal of Γ. If it consists only of 0, we say that Γ is semisimple.

§3. The center Z of an algebra Γ consists of those elements ζ of Γ which commute with all $\gamma \in \Gamma$, that is, for which $\zeta\gamma = \gamma\zeta$ for all $\gamma \in \Gamma$. Of course, Z is a subalgebra of Γ. In the case of the group algebra Γ of G, we can easily construct a basis of Z. Let $K_1, K_2, \ldots, K_k$ denote the conjugate classes of G. For each K_i, let $\{K_i\}$ denote the sum of the elements of K_i taken of course as element of Γ. It is seen immediately that $\{K_1\}, \{K_2\}, \ldots, \{K_k\}$ form a basis of Z. We have formulas

$$(1) \qquad \{K_\alpha\} \cdot \{K_\beta\} = \sum_{\gamma=1}^{k} a_{\alpha\beta\gamma}\{K_\gamma\}, \ (\alpha, \beta = 1, 2, \ldots, k)$$

where the $a_{\alpha\beta\gamma}$ are non-negative integers. (It will be clear what is meant by this in the case that Ω has characteristic $p \neq 0$.) If g_γ is a fixed element of K_γ, then $a_{\alpha\beta\gamma}$ is simply the number of ordered pairs (ξ, η) of elements $\xi \in K_\alpha$, $\eta \in K_\beta$ such that $\xi\eta = g_\gamma$.

Because of the connection with the conjugate classes of G, the algebra Z is called the class algebra of G over Ω. Its dimension is the class number k of G. The basis $\{K_1\}, \{K_2\}, \ldots, \{K_k\}$ will be called the natural basis of the class algebra of G.

§4. Assume that the underlying field Ω has characteristic 0. If X is a representation of G in Ω and if $g \in G$, denote by $\chi(g)$ the trace of the matrix $X(g)$, that is, the sum of the coefficients in the main diagonal of $X(g)$. The resulting function χ defined on G with values in Ω is called the character of X. Since the trace of a matrix depends only on the linear transformation represented by the matrix, equivalent representations have the same character. Moreover, each character χ is a class function, that is, the value of χ is constant on each conjugate class K of G. Occasionally, this constant value will be denoted by $\chi(K)$.

The product of two characters is again a character. By a generalized character, we mean the difference of two characters. Then the generalized characters form a ring.

We speak of an irreducible character if the corresponding representation is irreducible. One of these irreducible characters is the constant 1. This particular character is called the principal character. The generalized characters are the linear combinations

$$\chi = a_1\chi_1 + a_2\chi_2 \cdots \tag{2}$$

of the irreducible characters $\chi_1, \chi_2, \ldots$ with integral coefficients a_i. We have a character χ in (2) if all a_i here are non-negative and not all are 0.

Assume now that Ω is the field $\mathbf{C}$ of complex numbers. We have exactly k irreducible characters $\chi_1, \chi_2, \ldots, \chi_k$ where k is the class number of G. If θ and η are two complex valued functions defined on the group G of finite order n, we define the inner product

$$(\theta, \eta) = \frac{1}{n}\sum_g \theta(g)\overline{\eta(g)}$$

Then, $\chi_1, \chi_2, \ldots, \chi_k$ form an orthonormal system:

$$(\chi_i, \chi_j) = \delta_{ij} = \begin{cases} 1 & i = j \\ 0 & i \neq j \end{cases} \tag{3}$$

It follows that the a_i in (2) are given by $a_i = (\chi, \chi_i)$. Moreover, $(\chi, \chi) = \Sigma a_i^2$.

There is an equivalent way in which we can write (3). If $\sigma \in G$, we denote its centralizer in G by $C(\sigma)$, that is, $C(\sigma)$ consists of all $\tau \in G$ with $\sigma\tau = \tau\sigma$. The order $|C(\sigma)|$ will be denoted by $c(\sigma)$. Then, for $\sigma, \sigma_1 \in G$,

$$\sum_{i=1}^{k} \chi_i(\sigma)\overline{\chi_i(\sigma_1)} = \begin{cases} c(\sigma) \\ 0 \end{cases} \tag{4}$$

where the upper case applies, if and only if σ and σ_1 are conjugate in G.

By the degree χ of a character, we mean the degree x of the matrices of the corresponding representation X. Then $x = \chi(1)$. The degrees $x_1, x_2, \ldots, x_k$ of the k irreducible characters all divide n.

The values of each character χ lie in the field A of all algebraic numbers, and nothing is changed if we replace the field $\mathbf{C}$ by A. On the other hand, if Ω is any algebraically closed field of characteristic 0, then A can be identified with a subfield of Ω. There is no change, if we replace A by Ω.

§5. Assume further that Ω is an algebraically closed field of characteristic 0. The class algebra Z of G is a direct sum

$$Z = \Omega_1 \oplus \cdots \oplus \Omega_k$$

(see Finite Groups, §3, (3)) where each Ω_i is a field isomorphic with Ω. Consider a fixed i, $1 \leqq i \leqq k$. Using a fixed isomorphism, identify Ω_i with Ω. Let ω_i denote the mapping which maps the arbitrary element $\zeta \in Z$ on the corresponding element of Ω_i. Then, ω_i is clearly an algebra homomorphism of Z onto Ω. Moreover, $\omega_1, \omega_2, \ldots, \omega_k$ are distinct, and every homomorphism of Z onto Ω is obtained in this manner.

Each ω_i corresponds to an irreducible character χ_i of G. Explicitly, this correspondence is given by

$$\omega_i(\{K_\lambda\}) = n\chi_i(\sigma_\lambda)/(\chi_i(1)c(\sigma_\lambda)) \tag{5}$$

where σ_λ is an element of K_λ.

§6. If H is a subgroup of G, it is obvious that the restriction of a character χ of G to H is a character of H. Thus, if $\chi_1, \chi_2, \ldots, \chi_k$ are the irreducible characters of G and $\psi_1, \psi_2, \ldots, \psi_m$ are the irreducible characters of H, we have formulas (as in Finite Groups §6 (6))

$$\chi_i|H = \sum_{j=1}^{m} a_{ij}\psi_j \tag{6}$$

with non-negative rational integers a_{ij}. If ψ is a character of H, Frobenius constructed a character ψ^* of G by means of ψ. This is the character ψ^* of G induced by ψ. It is given by

$$\psi^*(g) = \frac{1}{|H|}\sum{}' \psi(xgx^{-1})$$

with the sum extending over those $x \in G$ for which $xgx^{-1} \in H$. For $\psi = \psi_j$,

$$\psi_j^* = \sum_{i=1}^{k} \chi_i a_{ij} \tag{7}$$

where the a_{ij} are the same as in (6).

Actually, if Y is the representation of H with the character ψ,

Frobenius gave an explicit construction of the representation Y^* of G with the character ψ^*.

This method plays an important role in all applications of the theory.

REFERENCES

1. E. Artin, *Comm. Pure Appl. Math.* **8** (1955) 355–365; 455–472.
2. S. D. Berman, *Dokl. Akad. Nauk SSSR* (N. S.) **86** (1952) 885–888 University of Chicago Press.
3. H. F. Blichfeldt, *Finite collineation groups*, Chicago (1917).
4. R. Brauer, *Sitzber. Preuss. Akad. Wiss* (1926) 410–416.
5. R. Brauer, *Amer. J. Math.* **64** (1941) 401–420.
6. R. Brauer, *Ann. of Math.* **42** (1941) 926–935.
7. R. Brauer, *Math. Zeit.* **63** (1956) 406–444.
8. R. Brauer, Séminaire P. Dubreil, M. L. Dubreil et C. Pisot (1958–1959) 6.1–6.16.
9. R. Brauer, *Math. Zeit.* **72** (1959) 25–46.
10. R. Brauer, *Arch. Math.* **13** (1960) 55–60.
11. R. Brauer, *Proc. Amer. Math. Soc.*, in press.
12. R. Brauer and W. Feit, *Proc. Nat. Acad. Sci. U.S.A.* **45** (1959) 361–365.
13. R. Brauer and K. A. Fowler, *Ann. of Math.* **62** (1955) 565–583.
14. R. Brauer and H. S. Leonard, *Can. J. Math.* **14** (1962) 436–450.
15. R. Brauer and W. F. Reynolds, *Ann. of Math.* **68** (1958) 713–720.
16. R. Brauer and M. Suzuki, *Proc. Nat. Acad. Sci. U.S.A.* **45** (1959) 1757–1759.
17. W. Burnside, *The theory of groups of finite order*, Cambridge (1911) Cambridge University Press, 2nd edition.
18. C. Chevalley, *Tôhoku Math. J.* (2) **7** (1955) 14–66.
19. C. W. Curtis and I. Reiner, *Representation theory of finite groups and associative algebras*, New York (1962) Wiley.
20. W. E. Deskins, *Duke Math. J.* **23** (1956) 35–40.
21. L. E. Dickson, *Linear groups*, Leipzig (1901) Teubner.
22. J. Dieudonné, Hermann (1948) Paris.
23. J. Dieudonné, *Ergebnisse der Mathematik und ihrer Grenzgebiete* (N. F.), Heft 5. Springer-Verlag. Berlin-Göttingen-Heidelberg, 1955.
24. W. Feit, *Proc. Symp. in Pure Math.* **6** (1962) 69–70.
25. W. Feit, M. Hall, Jr., and J. G. Thompson, *Math. Zeit.* **74** (1960) 1–17.
26. W. Feit and J. G. Thompson, *Pacific J. Math.* **11** (1961) 1257–1262.
27. W. Feit and J. G. Thompson, *Nagoya Math. J.* **21** (1962) 185–197.
28. W. Feit and J. G. Thompson, *Proc. Nat. Acad. Sci. U.S.A.* **48** (1962) 968–970.
29. W. Feit and J. G. Thompson, *Pacific J. Math.*, in press.
30. P. Fong, *Trans. Amer. Math. Soc.* **98** (1961) 263–284.
31. G. Frobenius and I. Schur, *Sitzber. Preuss. Akad. Wiss* (1906) 186–208.
32. J. A. Green, *Math. Zeit.* **70** (1959) 430–445.
33. J. A. Green, *Math. Zeit.* **79** (1962) 100–115.

34. M. Hall, Jr., *The theory of groups*, New York (1959) MacMillan.
35. P. Hall and G. Higman, *Proc. London Math. Soc.* (3)**7** (1956) 1–42.
36. H. Hasse, *J. Reine Angew. Math.* **167** (1931) 399–404.
37. H. Hasse, *Math. Ann.* **107** (1933) 731–760.
38. D. G. Higman, *Duke Math. J.* **21** (1954) 377–381.
39. K. Iizuka, *Math. Zeit.* **75** (1960–1961) 299–304.
40. N. Ito, *Nagoya J. Math.* **3** (1951) 31–48.
41. S. A. Jennings, *Trans. Amer. Math. Soc.* **50** (1941) 175–185.
42. E. Landau, *Math. Ann.* **56** (1903) 671–676.
43. D. E. Littlewood, *The theory of group characters*, New York (1940) Oxford Univ. Press, Chap. 9, §3.
44. G. W. Mackey, *Amer. J. Math.* **73** (1951) 576–592.
45. M. Osima, *Math. J. Okayama Univ.* **4** (1955) 175–188.
46. E. T. Parker, *Proc. Amer. Math. Soc.* **5** (1954) 606–611.
47. S. Perlis and G. Walker, *Trans. Amer. Math. Soc.*, **68** (1950) 420–426.
48. R. Ree, *Amer. J. Math.* **83** (1961) 401–420; 432–462.
49. G. de B. Robinson and R. Brauer, *Trans. Roy. Soc. Can.* Sect. III **41** (1947) Springer, 3rd 709–716.
50. A. Rosenberg, *Math. Zeit.* **76** (1961) 209–216.
51. I. Schur, *Sitzber. Preuss. Akad. Wiss.* (1905) 406–432.
52. L. Schur, *Sitzber. Preuss. Akad. Wiss.* (1905) 77–91.
53. I. Schur, *Math. Zeit.* **5** (1919) 6–10.
54. C. L. Siegel, *Math. Zeit.* **10** (1921) 173–213, Theor. 7.
55. L. Solomon, *Jap. J. Math.* **13** (1961) 144–164.
56. A. Speiser, *Math. Zeit.* **5** (1919) 1–6.
57. A. Speiser, *Die Theorie der Gruppen endlicher Ordnung*, Berlin (1937) Springer, 3rd edition.
58. R. G. Stanton, *Can. J. Math.* **3** (1951) 164–174.
59. R. Steinberg, *Can. J. Math.* **8** (1956) 580–591; **9** (1957) 347–351.
60. M. Suzuki, *Ill. J. Math.* **3** (1959) 255–271.
61. M. Suzuki, *Proc. Symp. in Pure Math.* **1** (1959) 88–100.
62. M. Suzuki, *Proc. Nat. Acad. Sci. U.S.A.* **46** (1960) 868–870.
63. M. Suzuki, *Trans. Amer. Math. Soc.* **99** (1961) 425–470.
64. M. Suzuki, *Ann. of Math.* **75** (1962) 105–145.
65. T. Tsuzuku, N. Ito, A Mizutani, and T. Nakayama, mineographed notes to be published by Nagoya University.
66. H. F.Tuan, *Ann. of Math.* **45** (1944) 110–140.
67. B. L. van der Waerden, *Modern algebra*, New York (1949) Ungar, Vol. 2.
68. H. N. Ward, *Bull. Amer. Math. Soc.* **69** (1962) 113–114.
69. H. Weyl, *Symmetry*, Princeton, N.J. (1952) Princeton University Press.
70. E. Witt, *Abh. Math. Sem. Univ. Hamburg* **12** (1937–1938) 256–264.
71. E. Witt, *J. Reine Angew, Math.* **190** (1952) 231–245.
72. O. Zariski and P. Samuel, *Commutative algebra*, New York (1960) Van Nostrand, Vol. II.

On quotient groups of finite groups

By

RICHARD BRAUER

§ 1. Introduction

Let G be a group and H a subgroup. Every normal subgroup G_0 of G gives rise to a normal subgroup

$$H_0 = G_0 \cap H \tag{1}$$

of H. If, in addition,

$$G = G_0 H, \tag{2}$$

we have isomorphic quotient groups $G/G_0 \simeq H/H_0$.

Conversely, given a subgroup H of G and a normal subgroup H_0 of H, we can ask: Does there exist a normal subgroup G_0 of G such that (1) and (2) hold?

Theorems guaranteeing the existence of such normal subgroups G_0 under suitable assumptions can be used to obtain criteria for simplicity of groups.

The problem has been studied by a number of authors. I mention here only an interesting paper of WIELANDT [9].

It is my aim to give some new contributions to this question. As in the preceding work, the group G is always supposed to be finite. The main result (Theorem 1, § 3) is somewhat complicated. It is proved in § 5. A partial converse is given in § 7. In § 8, it is shown that Wielandt's theorems [9] can be deduced from Theorem 1. § 9 contains a set of necessary and sufficient conditions for the existence of a normal π-complement in a finite group. In § 10 a further corollary of Theorem 1 is obtained, and in § 11 a generalization of the Theorem of P. HALL and GRÜN, (cf. M. HALL [4], Theorem 14.4.6) is deduced from it.

In order to have a convenient terminology, we give a definition.

Definition. Let G be a group, H a subgroup, H_0 a normal subgroup of H, $H_0 \lhd H \subseteqq G$. A normal subgroup G_0 of G is a *normal complement of H over H_0*, if

$$G = G_0 H, \qquad H_0 = G_0 \cap H$$

and hence $G/G_0 \cong H/H_0$.

§ 2. Notation. Preliminaries

If G is a finite group and π a set of primes, we say that G is a *π-group*, if the order $|G|$ of G is divisible only by primes in π. We call G *π-regular*, if $|G|$ is not divisible by primes in π. Similarly we say that an element g

of G is a *π-element*, if its order is divisible only by primes in π and we speak of a *π-regular element* g when the order of g is not divisible by primes in π.

Every element g has a unique decomposition

$$g = g_\pi g_{\pi'} = g_{\pi'} g_\pi$$

into a commuting product of a π-element g_π and a π-regular element $g_{\pi'}$. Both g_π and $g_{\pi'}$ are powers of g. We then call g_π the *π-factor* of g and $g_{\pi'}$ the *π-regular factor* of g. If π consists only of one prime p, we write g_p for g_π and term g_p the *primary factor* of g belonging to the prime p. Each g is the commuting product of its primary factors for the different primes.

If M is a subset of G, the number of elements in M will be designated by $|M|$. The notation $\mathfrak{C}_G(M)$ and $\mathfrak{N}_G(M)$ is used for the centralizer and normalizer of M in G. We use $\mathfrak{C}_G(m)$ for the centralizer of an element m.

We shall consider the following situation: We have a finite group G, a subgroup H of G, a normal subgroup H_0 of H. In symbols,

$$H_0 \lhd H \leqq G.$$

We form the conjugate classes of H/H_0 and take their reciprocal images

$$H_0, H_1, \ldots, H_n \tag{3}$$

in H. Of course

$$H = \bigcup_{i=0}^{n} H_i \quad \text{(disjoint)} \tag{4}$$

and hence

$$|H| = \sum_{i=1}^{n} |H_i|. \tag{5}$$

The set π will be taken as the set of primes p which divide the index $(H:H_0)$. Then all π-regular elements of H lie in H_0. Each H_i contains π-elements, since $H_i H_0 = H_i$.

Corresponding to the sets (3), we define subsets

$$G_0, G_1, G_2, \ldots, G_n \tag{6}$$

of G. For $i \neq 0$, G_i is to consist of those $g \in G$ whose π-factor g_π is conjugate in G to an element of H_i. For G_0, we take the complement

$$G_0 = G - \bigcup_{i=1}^{n} G_i \tag{7}$$

of the union of the G_i with $i \neq 0$.

In general, the sets G_i need not to be disjoint, but in the cases considered in this paper, this will always be true.

§ 3. Statement of Theorem 1

Theorem 1. *Let G be a finite group, H a subgroup of G, H_0 a normal subgroup of H and let π be the set of primes dividing $(H:H_0)$. Assume the following four conditions:*

(AI) *The indices* $(G:H)$ *and* $(H:H_0)$ *are coprime.*

(AII) *If two elements* h_1, h_2 *of* H *are conjugate in* G, *then* h_1H_0 *and* h_2H_0 *are conjugate in* H/H_0.

(AIII) ***If** h **is a** π**-element of** $H-H_0$ **and if** p **is a prime in** π **which does not divide the order of** h, **then** p **does not divide the index***

$$(\mathfrak{C}_G(h):\mathfrak{C}_H(h)).$$

(AIV) *If* $G_0, G_1, \ldots, G_n$ *are the sets defined in* (6) *there do not exist indices* i, j *with* $0 \leqq i, j \leqq n$ *such that*

$$|G_i| > (G:H)\cdot|H_i|, \qquad |G_j| < (G:H)\cdot|H_j|.$$

Then H *has a normal complement in* G *over* H_0.

This will be proved in § 5.

§ 4. Consequences of the hypotheses (AI), (AII), (AIII)

Lemma 1. *If* $H_0 \triangleleft H \leqq G$, *if* π *is the set of primes dividing* $(H:H_0)$, *and if* (AI), (AII), (AIII) *hold, then each* G_i *is a union of conjugate classes of* G *and*

$$G = \bigcup_{i=0}^{n} G_i \quad (disjoint) \tag{8}$$

$$H_i = G_i \cap H \quad for \quad i = 0, 1, \ldots, n. \tag{9}$$

Proof. It is clear from the definition of the G_i that if $g \in G_i$, all the conjugates of g lie in G_i. As shown by (7), $G_0 \cap G_i = \emptyset$ for $i \neq 0$. If $g \in G_i$ with $i > 0$, the conjugate class K of g_π meets H_i. If also $g \in G_j$, $j > 0$, then K meets H_j and hence there exist $h_i \in H_i$ and $h_j \in H_j$ which are conjugate in G. Then (AII) implies that h_iH_0 and h_jH_0 are conjugate in H/H_0. Consequently, $i = j$. Hence the sets (6) are disjoint. Since, by (7), their union is G, we have (8).

It is clear from the definition of G_i that $H_i \subseteqq G_i \cap H$ for $i > 0$. If $h \in G_i \cap H$ and $i > 0$, then h_π is conjugate in G to some element of H_i and (AII) yields $h_\pi \in H_i$.

The π-regular element $h_{\pi'}$ of H belongs to H_0 and hence $h = h_\pi h_{\pi'} \in H_iH_0 = H_i$. Consequently, (9) holds for $i > 0$. Now (8) implies that

$$G_0 \cap H = H - \bigcup_{i=1}^{n} (G_i \cap H) = H - \bigcup_{i=1}^{n} H_i.$$

It follows from (4) that (9) holds for $i = 0$ too.

On account of (AI), H contains a p-Sylow group P of G for every $p \in \pi$. Then Sylow's theorem shows that every p-element x of G is conjugate to an element $y \in P \subseteqq H$. We say that x is of the *first kind*, if $y \in H - H_0$ and that x is of the *second kind*, if $y \in H_0$. It follows from (AII) that for given x, this does not depend on the particular choice of y in the intersection of H with the conjugate class of x.

For any $g \in G$, let $\alpha(g)$ be the product of all primary factors of g_π of the first kind, and let $\beta(g)$ be the product of all primary factors of the second kind. Then

$$g=\alpha(g)\,\beta(g)\,g_{\pi'}.$$

Lemma 2. *Assumptions as in Lemma 1. If $\alpha(g)\neq 1$ for an element g of G, then g_π is conjugate in G to an element of $H-H_0$.*

Proof. If g_π is a p-element for some prime $p\in\pi$, this follows from a remark above. We use induction on the number of primary factors different from 1 of g_π. If g_π is not a p-element, we may set $g_\pi=xg_p$ where g_p is a primary factor of g_π for some prime $p\in\pi$ and where x is the product of the remaining primary factors. Here, p may be chosen such that a primary factor of the first kind occurs in x. Then $\alpha(x)\neq 1$ and by induction x is conjugate to an element of $H-H_0$. Replacing g by a conjugate, we may assume that $x\in H-H_0$. Now (AIII) implies that $\mathfrak{C}_H(x)$ contains a p-Sylow subgroup P_0 of $\mathfrak{C}_G(x)$. Since g_p is a p-element of $\mathfrak{C}_G(x)$, it is conjugate in $\mathfrak{C}_G(x)$ to an element of $P_0\subseteq H$ and then $g_\pi=xg_p$ is conjugate in G to an element $h\in H$. If here $h\in H_0$, then each primary factor of g_π (being a power of g_π) would be conjugate to a power of h, that is, to an element of H_0. Then each primary factor of g_π would be of the second kind which would mean $\alpha(g)=1$, a contradiction. This proves Lemma 2.

Lemma 3. *The same assumptions.*

(*a*) *We have $g\in G_0$ if and only if $\alpha(g)=1$.*

(*b*) *For every $g\in G$, the elements g, g_π, and $\alpha(g)$ lie in the same G_i.*

(*c*) *If $g\in G_0$, all powers of g lie in G_0.*

(*d*) *All π-regular elements g lie in G_0.*

Proof. It follows from Lemma 2 that if $\alpha(g)\neq 1$, then g lies in some G_i with $i\neq 0$. Conversely, if $g\in G_i$ with $i\neq 0$, then g_π is conjugate to some $h\in H_i$. It follows easily that $\alpha(g)$ is conjugate to $\alpha(h)$ and $\beta(g)$ to $\beta(h)$. Since every primary factor y of $\beta(h)$ lies in H and is conjugate in G to an element of H_0, it follows from (AII) that $y\in H_0$ and then $\beta(h)\in H_0$. Thus, $h=\alpha(h)\,\beta(h)\in\alpha(h)\,H_0$ and $\alpha(h)\in H_iH_0=H_i$. In particular, $\alpha(h)\neq 1$ whence $\alpha(g)\neq 1$. This completes the proof of (*a*).

At the same time, we have shown that if $g\in G_i$ with $i\neq 0$, then $\alpha(g)$ is conjugate to an element $\alpha(h)\in H_i$ and as $\alpha(g)=(\alpha(g))_\pi$, we have $\alpha(g)\in G_i$. For $i=0$, this is trivial by (*a*). It is clear from the definition of G_i that g_π and g lie in the same G_i. Thus, g, g_π and $\alpha(g)$ lie in the same G_i.

The statements (*c*) and (*d*) are immediate consequences of (*a*).

Lemma 4. *The same assumptions. If E is a π-subgroup of the form*

$$E=U\times P,$$

where $U=\{u\}$ is a cyclic group and P a p-group for some prime $p\in\pi$, not dividing $|U|$, and if $\alpha(u)\neq 1$, then E is conjugate in G to a subgroup of H.

Proof. It follows from Lemma 2 that u is conjugate to an element of $H-H_0$. After replacing E by a conjugate, we may assume $u\in H-H_0$. Now (AIII) shows that $\mathfrak{C}_H(u)$ contains a p-Sylow group P^* of $\mathfrak{C}_G(u)$. By Sylow's theorem, P is conjugate in $\mathfrak{C}_G(u)$ to a subgroup of P^*, and hence E is conjugate to a subgroup of $U \times P^* \subseteqq H$. This proves the lemma.

Lemma 5. *The same assumptions. The hypothesis* (AIV) *is equivalent with the condition*

(AIV*) $$|G_i| = (G:H) \cdot |H_i| \quad \textit{for} \quad i = 0, 1, \ldots, n.$$

Proof. It follows from (8) and (4) that

$$|G| = \sum_{i=0}^{n} |G_i|, \qquad |H| = \sum_{i=0}^{n} |H_i|.$$

If, say, $|G_i| \geqq (G:H) \cdot |H_i|$ for all i, this shows that the equality sign must hold for all i. The same is true, if $|G_i| \leqq (G:H) \cdot |H_i|$ for all i.

§ 5. Proof of theorem 1

Assume now all four hypotheses of Theorem 1. Let ψ be a generalized character of H/H_0 and let ψ also denote the corresponding generalized character of H. Then ψ is constant on all sets H_i. We define a function ϑ_ψ on G by setting

(10) $$\vartheta_\psi | G_i = \psi | H_i.$$

Clearly, ϑ_ψ is a class function on G and

(11) $$\vartheta_\psi | H = \psi.$$

By an *elementary subgroup* E of G, we mean a subgroup which is a direct product $U \times V$ of a cyclic subgroup U and a p-subgroup V for some prime p. It does not mean a restriction to require that p does not divide $|U|$. If we can show that $\vartheta_\psi | E$ is a generalized character of E for all elementary subgroups E of G, it follows from a theorem of mine[1]) that ϑ_ψ is a generalized character of G. Clearly, it will suffice to consider only one E in each class of conjugate subgroups.

Let E be an elementary group. We can write it as a direct product

$$E = E_0 \times E_1,$$

where E_0 is a π-group and E_1 is π-regular. Let ϱ be the projection of E onto E_0. Then $x \in E$ has the π-factor $x_\pi = \varrho(x)$ and by Lemma 3 (b) and (10), $\vartheta_\psi(x) = \vartheta_\psi(\varrho(x))$, i.e. $\vartheta_\psi | E = \vartheta_\psi \circ \varrho$. It will therefore suffice to prove that $\vartheta_\psi | E_0$ is a generalized character of E_0.

The π-group E_0 is elementary again, say

$$E_0 = U \times P,$$

where $U = \{u\}$ is cyclic and P is a p-group for some prime $p \in \pi$ which does not divide $|U|$. If $\alpha(u) \neq 1$, then Lemma 4 shows that after replacing E_0 by a conjugate, we may assume $E_0 \subseteqq H$. Now (11) shows that $\vartheta_\psi | E_0$ is a

[1]) Since now simple proofs (e.g. in BRAUER and TATE [2]) are available, there is no longer any reason to avoid the use of the theorem. — Of course, elementary groups in the sense used here must not be confused with elementary abelian p-groups.

generalized character of E_0. Assume then that $\alpha(u)=1$. Then $\alpha(u^m)=1$ for all powers u^m of u. If σ is the projection of $E_0=U\times P$ onto P, we have $\alpha(x)=\alpha(\sigma(x))$ for $x\in E_0$. Now Lemma 3(b) and (10) show that

$$\vartheta_\psi | E_0 = \vartheta_\psi \circ \sigma.$$

On account of (AI) and Sylow's theorem, we may assume that $P\subseteqq H$ and (11) shows that ϑ_ψ is a generalized character of E_0.

This completes the proof that ϑ_ψ is a generalized character of G.

If ϑ and η are class functions defined on G, we define the inner product as usual by

$$(\vartheta, \eta)_G = \frac{1}{|G|} \sum_{x\in G} \vartheta(x)\, \overline{\eta(x)}.$$

Then by (10), if h_i is an element in H_i,

$$(\vartheta_\psi, \vartheta_\psi)_G = \frac{1}{|G|} \sum_{i=0}^{n} |G_i| \cdot |\psi(h_i)|^2.$$

By Lemma 5, (AIV*) holds and we find

$$(\vartheta_\psi, \vartheta_\psi)_G = \frac{1}{|H|} \sum_{i=0}^{n} |H_i| \cdot |\psi(h_i)|^2 = (\psi, \psi)_H.$$

In particular, if ψ is an irreducible character of H, we have $(\vartheta_\psi, \vartheta_\psi)_G = (\psi, \psi)_H = 1$ and since $\vartheta_\psi(1)=\psi(1)>0$, ϑ_ψ is an irreducible character of G. Now, (11) shows that every irreducible character ψ of H/H_0 can be extended to a character ϑ_ψ of G. The same then is true for every character ψ of H/H_0.

Choose ψ as the character of the regular representation of H/H_0.

$$\psi | H_0 = \psi(1), \qquad \psi | H_i = 0 \quad \text{for} \quad i>0;$$

By (10),

$$\vartheta_\psi | G_0 = \vartheta_\psi(1), \qquad \vartheta_\psi(G_i)=0 \quad \text{for} \quad i>0.$$

Hence G_0 is the kernel of the representation of G with the character ϑ_ψ. This proves that $G_0 \lhd G$. By Lemma 3(d), all π-regular elements of G lie in G_0 and hence G/G_0 is a π-group. Now (AI) implies $G=G_0H$ while (9) yields $H_0=G_0\cap H$. Hence G_0 is a normal complement of H in G over G_0, q.e.d.

§ 6. Remarks concerning the condition (AIV)

In applying Theorem 1, we may often have difficulties in checking the assumption (AIV). The following remarks may facilitate this in some cases.

1. If we can prove $|G_i| = (G:H)\cdot|H_i|$ for all but one i, say for all $i\neq 0$, (AIV) holds.

2. Suppose that for fixed i, the set G_i is represented as disjoint union $G_i = \bigcup_\lambda M_\lambda$ of subsets M_λ and that for each M_λ we can prove $|M_\lambda| \geqq (G:H)\cdot|M_\lambda \cap H|$. Then, by (9)

$$|G_i| \geqq (G:H)\cdot|G\cap H_i| = (G:H)\cdot|H_i|.$$

Likewise, if $|M_\lambda| \leqq (G:H)\cdot|M_\lambda\cap H|$ for all λ, we have $|G_i| \leqq (G:H)\cdot|H_i|$.

Since G_i is a union of conjugate classes of G, we can try to take for the M_λ the distinct conjugate classes contained in G_i.

3. If x is a π-element of G, the *π-section* $S(x)$ of x is defined as the set of all $g \in G$ such that g_π is conjugate to x in G. Every G_i is a disjoint union of π-sections and we can try to take these π-sections for the M_λ in 2.

We determine $|S(x)|$. For any finite group X, let $r_\pi(X)$ denote the number of π-regular elements in X. If g is to belong to $S(x)$, we have $(G:\mathfrak{C}_G(x))$ possibilities for g_π. If x' is one of them, we have $g_\pi = x'$, if and only if $g = x'y$ where y is a π-regular element of $\mathfrak{C}_G(x') \simeq \mathfrak{C}_G(x)$. Hence we have $r(\mathfrak{C}_G(x))$ possibilities for y. It follows that

$$|S(x)| = (G:\mathfrak{C}_G(x)) \cdot r_\pi(\mathfrak{C}_G(x)). \tag{12}$$

Of course, $|S(x) \cap H|$ is a disjoint union of π-sections of H.

§ 7. A partial converse of Theorem 1

Theorem 2. *Assume that $H_0 \lhd H \subseteqq G$ and that there exist normal complements G_0 of H in G over H_0. Then* (A II) *holds. If in addition* (A I) *and* (A III) *hold with π denoting the set of prime divisors of $(H:H_0)$, then* (A IV*) *and hence* (A IV) *hold. Moreover, G_0 is uniquely determined by G, H, H_0.*

Proof. If h_1 and h_2 are two elements of H and if there exists a $g \in G$ such that $g^{-1}h_1 g = h_2$, we can set $g = g_0 h$ with $g_0 \in G_0$, $h \in H$. Then $h^{-1}h_1 h \equiv h_2 \pmod{G_0}$. Since $h^{-1}h_1 h$ and h_2 lie in H, we have $h^{-1}h_1 h \equiv h_2 \bmod (G_0 \cap H)$, i.e. mod H_0, and this yields (A II).

Assume now that (A I) and (A III) hold too. We shall use the same notation as in § 2, but since it is not clear yet that the set G_0 defined in (7) coincides with the given normal subgroup G_0, we shall reserve the notation G_0 for the latter.

Since $G/G_0 \simeq H/H_0$ is a π-group, all π-regular elements of G lie in G_0. If x is a p-element of G for some $p \in \pi$, then by (A I) and Sylow's theorem, x is conjugate in G to an element $y \in H$. If $x \in G_0$, the normality of G_0 implies $y \in G_0 \cap H = H_0$ and x is a primary element of the second kind. It follows that if g is an element of G_0, then $\alpha(g) = 1$. Conversely, if $\alpha(g) = 1$ for some $g \in G$, every primary factor x of g_π is conjugate to an element of $H_0 \subseteqq G_0$. Hence $x \in G_0$ and then $g_\pi \in G_0$. It follows that $g = g_\pi g_{\pi'} \in G_0$. Since we now know that the elements g of G_0 are characterized by the condition $\alpha(g) = 1$, comparison with Lemma 3 shows that G_0 coincides with the set defined by (7). In particular, G_0 is uniquely determined by the conditions.

If $g \in G_i$ with $i > 0$, then g_π is conjugate in G to some element $h_i \in H_i$. There exist then elements $g_0 \in G_0$, $h \in H$, such that $g_\pi = g_0^{-1}h^{-1}h_i h g_0$. Since $h^{-1}h_i h \in H_i$, this implies $g_\pi \in H_i G_0$. As the π-regular element $g_{\pi'}$ lies in G_0, then $g = g_\pi g_{\pi'} \in H_i G_0$. Thus, $G_i \subseteqq H_i G_0$. This holds for $i = 0$, too. Moreover, if $h_i g_0 = h_j' g_0'$ with $h_i \in H_i$, $h_j' \in H_j$ and $g_0, g_0' \in G_0$, we have $h_i^{-1}h_j' \in G_0 \cap H = H_0$ and $i = j$. Set $h_i^{-1}h_i' = h \in H_0$. Then $h_i' = h_i h$, $g_0' = h^{-1}g_0$ and we see that $|H_i G_0| = |H_i| \cdot |G_0| / |H_0| = (G_0 : H_0) \cdot |H_i|$.

Our argument shows that the sets $H_i G_0$ with $i=0, 1, \ldots, n$ are disjoint. Since $G_i \subseteqq H_i G_0$, (8) implies that $G_i = H_i G_0$. Hence $|G_i| = (G:H) \cdot |H_i|$ and this is (AIV*).

§ 8. A Theorem of H. Wielandt

In [9], H. WIELANDT proved an interesting theorem which is itself a generalization of a classical theorem of Frobenius:

Theorem of Wielandt. *Let G be a finite group, H a subgroup of G and H_0 a normal subgroup of H. Assume that*

$$H \cap H^x \subseteqq H_0 \quad \textit{for all} \quad x \in G - H. \tag{13}$$

There exists exactly one normal complement G_0 of H in G over H_0 and

$$G_0 = G - \bigcup_{x \in G} (H - H_0)^x. \tag{14}$$

We wish to point out that this theorem can be deduced from Theorem 1. Indeed, if the assumptions of Wielandt's theorem are satisfied, a Sylow group argument shows that (AI) holds. As is seen easily, (13) implies the following stronger forms of (AII) and (AIII).

(AII*) If two elements h and h' of H are conjugate in G and if $h \notin H_0$, then h and h' are conjugate in H.

(AIII*) If $h \in H - H_0$, then $\mathfrak{C}_G(h) = \mathfrak{C}_H(h)$.

If a conjugate class K of G lies in G_i with $i > 0$, then it follows easily from (AIII*) that K meets $H - H_0$. By (AII*), $K \cap H$ is a conjugate class of H. If $h \in K \cap H$, (AIII*) shows that

$$|K| = (G:\mathfrak{C}_H(h)) = (G:H)(H:\mathfrak{C}_H(h)) = (G:H)\,|K \cap H|.$$

Now, § 6, Remarks 1 and 2 apply. This shows that (AI), (AII*), (AIII*) imply (AIV). The existence of G_0 follows from Theorem 1 and the uniqueness and (14) are obtained from Theorem 2 and (7) since here $\bigcup_{i \neq 0} G_i = \bigcup_{x \in G} (H - H_0)^x$.

We may note that, on account of (AIII*), it suffices to require (AII*) for π-elements h, h'. Indeed, if $h \in H - H_0$ and if $g^{-1} h g = h' \in H$ for some $g \in G$, then $g^{-1} h_\pi g = h'_\pi$. If there exists an $y \in H$ with $y^{-1} h_\pi y = h'_\pi$, then $g y^{-1} \in \mathfrak{C}_G(h_\pi)$ and by (AIII*), $g y^{-1} \in H$ and then $g \in H$.

Using this remark, we can also obtain Wielandt's second theorem ([9], Satz 2) directly.

§ 9. Necessary and sufficient conditions for the existence of normal π-complements

Let G be a finite group and let π be a set of primes. We say that a subgroup G_0 of G is a *normal π-complement* in G, if G_0 is a normal π-regular subgroup of G such that G/G_0 is a π-group. The question when such normal π-complements exist has been studied by a number of authors, e.g. BAER [1], SAH [7]; cf. the literature quoted in [7]. We prove

Theorem 3. *Let G be a finite group and let π be a set of primes. The following conditions are necessary and sufficient for the existence of normal π-complements in G:*

(B I) *There exist S_π-subgroups H of G, i.e. subgroups H whose order is exactly the product of the contributions of the primes $p \in \pi$ to $|G|$.*

(B II) *If two elements of H are conjugate in G, they are conjugate in H.*

(B III) *If E is an elementary π-subgroup of G then E is conjugate in G to a subgroup of H.*

Proof. (*a*) Assume that (B I), (B II), (B III) hold. Primes $p \in \pi$ which do not divide $|G|$ may be removed from π. We apply Theorem 1 with $H_0 = 1$. Since H is an S_π-subgroup of G, (A I) holds. In the case $H_0 = 1$, (A II) becomes identical with (B II).

Let $h \neq 1$ be an element of H. Let P be a p-Sylow group of $\mathfrak{C}_G(h)$ for some prime $p \in \pi$. Then $E = \{h, P\}$ is an elementary π-subgroup of G. By (B III), there exist $g \in G$ such that $E^g \subseteqq H$. In particular $h^g \in H$ and by (B II), there exist $y \in H$ such that $h^{gy} = h$. Then $E^{gy} \subseteqq H$ and hence $P^{gy} \subseteqq H$. Since P^{gy} lies in the centralizer of $h^{gy} = h$, we see that there exist p-Sylow subgroups P^{gy} of $\mathfrak{C}_G(h)$ which lie in H. Hence

$$((\mathfrak{C}_G(h) : \mathfrak{C}_H(h)), |H|) = 1. \tag{15}$$

This is trivially true for $h = 1$. It is now clear that (AIII) holds.

In order to prove (AIV), we use Remarks 2 and 3 of § 6. Since $H_0 = 1$, the sets H_i are simply the conjugate classes of H. Choose $h_i \in H_i$. Then h_i is a π-element of G_i and for $i \neq 0$, G_i is here the π-section $S(h_i)$. This is true for $i = 0$ too, since Lemma 3 shows that G_0 consists exactly of the π-regular elements of G; $G_0 = S(1)$.

It now follows from (12) that

$$|G_i| = (G : \mathfrak{C}_G(h_i)) \cdot r_\pi(\mathfrak{C}_G(h_i)).$$

Let a_i be the contribution of the primes in π to $|\mathfrak{C}_G(h_i)|$ and let b_i be the contribution of the other primes. The π-regular elements of $\mathfrak{C}_G(h_i)$ are the solutions of the equation $x^{b_i} = 1$ in $\mathfrak{C}_G(h_i)$ and a theorem of FROBENIUS (cf. [*4*], Theorem 9.1.2) shows that the number $r(\mathfrak{C}_G(h_i))$ of these solutions is at least b_i. Hence

$$|G_i| \geqq \frac{|G|}{a_i b_i} b_i = \frac{|G|}{a_i}.$$

On the other hand, $|H_i| = (H : \mathfrak{C}_H(h_i))$. As shown by (15), $|\mathfrak{C}_H(h_i)|$ is exactly the contribution a_i of the primes $p \in \pi$ to $|\mathfrak{C}_G(h_i)|$. Hence $|H_i| = |H|/a_i$ and we have proved $|G_i| \geqq (G:H)\,|H_i|$ for all i. Thus, (A IV) holds and Theorem 1 applies and yields the existence of a normal π-complement G_0 in G.

(*b*) Conversely, assume that normal π-complements G_0 in G exist. It follows from the theorem of ZASSENHAUS-SCHUR ([*10*], Theorem 25, p. 162) that there exist S_π-groups H of G. Clearly, $H \cap G_0 = 1$. By the first part of Theorem 2, (A II) holds and, for $H_0 = 1$, this is (B II).

If $g \in G$, set

$$g = w(g)\, h(g)$$

with $w(g) \in G_0$, $h(g) = H$. Then $w(g)$ and $h(g)$ are uniquely determined by g and the mapping $g \to h(g)$ is a homomorphism ϑ of G onto H with the kernel G_0.

If E is an elementary π-subgroup of G, then $\vartheta | E$ is an isomorphism of E onto $\vartheta(E) \subseteqq H$ and

$$G_0 E = G_0 \vartheta(E).$$

Since E is soluble, a theorem of ZASSENHAUS ([*10*], Theorem 27, p. 162), shows that E and $\vartheta(E)$ are conjugate in $G_0 E$ and hence in G. Thus (B III) holds and the proof is complete.

Remarks

1. The condition (B III) in Theorem 1 can be replaced by (A III). Indeed in part (*a*) of the proof, (B II) and (B III) were shown to imply (A III) and no other use was made of (B III) in part (*a*).

2. If π consists of a single prime p, then (BI) and (BIII) are consequences of Sylow's theorem. Hence (B II) alone represents a necessary and sufficient condition for the existence of a normal p-complement. This fact is well known[2]).

3. We know from FEIT-THOMPSON [*3*] that the last part of the proof of Theorem 4 works for arbitrary non-soluble π-subgroups E of G. Hence the word "elementary" can be removed in (BIII). In the terminology of P. HALL [*5*], (B I) and (B III) then can be stated in the form: (D_π) holds for G.

§ 10. Another sufficient condition for the existence of normal complements of H in G over H_0

Theorem 4. *Let G be a finite group, H a subgroup, H_0 a normal subgroup of H and let π be the set of primes dividing $(H:H_0)$. The following condition implies the existence of a normal complement to H in G over H_0:*

(C) *If U is an abelian π-subgroup of H, every conjugate class K of $\mathfrak{C}_G(U)$ consisting of π-elements meets H in a conjugate class of $\mathfrak{C}_H(U)$. (In particular, this is to hold for $U=1$, $\mathfrak{C}_G(U)=G$, $\mathfrak{C}_H(U)=H$.)*

Proof. (*a*) We first note that if G and $H \subseteqq G$ are groups which satisfy (C) for any set π of primes and if h is a π-element of H, then

$$G^* = \mathfrak{C}_G(h), \qquad H^* = \mathfrak{C}_H(h) = G^* \cap H$$

[2]) Actually, a much deeper theorem has been obtained by HUPPERT [*6*] in the case $\pi = (p)$, p odd. J. G. THOMPSON has shown that this theorem remains true for $p=2$. Recently, TATE [*8*] has given an elegant proof of the theorem by homological methods.

again satisfy (C) for the same π. Indeed, let U^* be an abelian π-subgroup of H^*. Then $U=\{h, U^*\}$ is an abelian π-subgroup of H and

$$\mathfrak{C}_G(U)=\mathfrak{C}_G(h)\cap\mathfrak{C}_G(U^*)=G^*\cap\mathfrak{C}_G(U^*)=\mathfrak{C}_{G^*}(U^*),$$
$$\mathfrak{C}_H(U)=\mathfrak{C}_G(U)\cap H=\mathfrak{C}_{G^*}(U^*)\cap H=\mathfrak{C}_{H^*}(U^*).$$

If K^* is a conjugate class of $\mathfrak{C}_{G^*}(U^*)=\mathfrak{C}_G(U)$ consisting of π-elements, then $K^*\subseteqq G^*$ and hence

$$K^*\cap H=K^*\cap G^*\cap H=K^*\cap H^*.$$

Now, (C) for G and H shows that $K^*\cap H^*$ is a conjugate class of $\mathfrak{C}_{H^*}(U^*)$.

(*b*) Assume further that G and H satisfy (C) for a set of primes π. We claim that

$$r_\pi(G)/|G|=r_\pi(H)/|H| \tag{16}$$

where r_π is defined in § 6.

In order to prove (16), let $h_0=1, h_1, \ldots, h_m$ be a set of representatives for the conjugate classes of H consisting of π-elements. It follows from (C) with $U=1$ that G is a disjoint union of the π-sections (§ 6) $S(h_i)$:

$$G=\bigcup_{j=0}^{m} S(h_i) \quad \text{(disjoint)}. \tag{17}$$

Likewise,

$$H=\bigcup_{j=0}^{m} S_H(h_i) \quad \text{(disjoint)}, \tag{18}$$

where $S_H(x)$ designates the π-section in H. On comparing the number of elements on both sides in (17) and (18) and using (12), we find

$$\sum_{j=0}^{m} r_\pi(\mathfrak{C}_G(h_i))/|\mathfrak{C}_G(h_i)|=\sum_{j=0}^{m} r_\pi(\mathfrak{C}_H(h_i))/|\mathfrak{C}_H(h_i)|. \tag{19}$$

Both sides in (19) are equal to 1.

Assume now that (16) is true if G and H are replaced by groups G^* and H^* which satisfy (C) and for which $|G^*|<|G|$. By part (*a*), we may take here $G^*=\mathfrak{C}_G(h_j)$, $H^*=\mathfrak{C}_H(h_j)$, provided that h_i does not lie in the center $\mathfrak{Z}(G)$ of G. Thus (16) yields

$$r_\pi(\mathfrak{C}_G(h_j))/|\mathfrak{C}_G(h_j)|=r_\pi(\mathfrak{C}_H(h_j))/|\mathfrak{C}_H(h_j)|. \tag{20}$$

Remove the corresponding terms from (19). If k of the h_j lie in $\mathfrak{Z}(G)$, we find

$$k r_\pi(G)/|G|=k r_\pi(H)/|H|.$$

Since $h_0=1\in\mathfrak{Z}(G)$, we have $k\neq 0$ and obtain (16).

(*c*) We need one more lemma.

Lemma 6. *If X is a finite group and p a prime, the number $r_p(X)$ of p-regular elements of X is not divisible by p.*

Proof. Let P be a p-Sylow group of X and distribute the set of p-regular elements of X into classes of elements which are conjugate with regard to P. The number of elements in such a class L is a power of p and we have $|L|=1$ if and only if L consists of a p-regular element of $\mathfrak{C}_X(P)$. Hence

$$r_p(X) \equiv r_p(\mathfrak{C}_X(P)) \pmod{p}.$$

Now, $\mathfrak{C}_X(P)$ has the p-Sylow group $\mathfrak{Z}(P)$. Since no two distinct elements of $\mathfrak{Z}(P)$ are conjugate in $\mathfrak{C}_X(P)$, Burnside's theorem ([*4*], Theorem 14.3.1), or if we prefer, Theorem 3, shows that $\mathfrak{C}_X(P)$ has a normal p-complement W. Then $r_p(\mathfrak{C}_X(P)) = |W|$. Since $|W|$ is not divisible by p, neither is $r_p(X)$.

(*d*) We can now prove Theorem 4 quickly. Let p be a fixed prime in π and let h be a π-element in H. Then (C) holds for $G^* = \mathfrak{C}_G(h)$, $H^* = \mathfrak{C}_H(h)$ and π. This remains true, if π is replaced by the subset $\pi_0 = \{p\}$, and (16) can be applied to this case. In conjunction with Lemma 6, this shows that $(\mathfrak{C}_G(h):\mathfrak{C}_H(h))$ cannot be divisible by any prime p in π. It follows that (AIII) holds. If we take $h=1$ (which was not excluded here), we obtain (AI). If we take $U=1$ in (C), we see that (AII) holds.

It remains to prove (AIV). We use here remarks 2 and 3 of § 6. Every π-section $S(x)$ in G coincides with one of the $S(h_j)$ in (17). On account of (C) with $U=1$, we have $S(h_j) \cap H = S_H(h_j)$ and (20) and (12) yield $|S(h_j)| = (G:H) \cdot |S_H(h_j)|$. Hence

$$|S(x)| = (G:H) \cdot |S(x) \cap H|$$

for every π-section $S(x)$ of G. As shown in § 6, this implies (AIV*) and hence Theorem 1 applies and yields Theorem 5.

§ 11. A generalization of the Theorem of Hall and Grün

As an application of Theorem 4, we prove

Theorem 5. *Let G be a finite group, $\pi \neq \emptyset$ a set of primes and assume*

(DI) *There exists an S_π-group R of G and a subgroup T of the center $\mathfrak{Z}(R)$ which is weakly closed in R with regard to G.*

(DII) *If $U \geqq 1$ is an abelian π-subgroup of $H = \mathfrak{N}_G(T)$, then (D_π) holds for $\mathfrak{C}_G(U)$.*

If H/H_0 is a maximal π-quotient group of H then H has a normal complement in G over H_0.

Proof. (*a*) Let U be an abelian π-subgroup of H and set

$$G^* = \mathfrak{C}_G(U), \qquad H^* = \mathfrak{C}_H(U) = G^* \cap H.$$

Then $\{U, T\}$ is a π-subgroup of G and since (D_π) holds for G [by (DII) with $U=1$], $\{U, T\}$ lies in a conjugate R^y of R, $y \in G$. It follows from the weak closure of T in R that $T = T^y \leqq \mathfrak{Z}(R^y)$. Now $U \leqq R^y$ implies

(21) $$T \leqq \mathfrak{C}_G(U) = G^*.$$

By (DII) there exists an S_π-group $R^* \geqq T$ of G^*. Again, R^* belongs to some R^w with $w \in G$ and we conclude $T \leqq \mathfrak{Z}(R^w)$ and then $T \leqq \mathfrak{Z}(R^*)$. It is

6*

seen easily that T is weakly closed in R^*. Hence (DI) holds again for G^*, R^*, T (instead of G, R, T).

(*b*) If U^* is an abelian π-subgroup of H^*, then $U_1=\{U, U^*\}$ is an abelian π-subgroup of H and

$$\mathfrak{C}_G(U_1)=\mathfrak{C}_G(U)\cap\mathfrak{C}_G(U^*)=G^*\cap\mathfrak{C}_G(U^*)=\mathfrak{C}_{G^*}(U^*).$$

By (DII), (D_π) holds for $\mathfrak{C}_G(U_1)=\mathfrak{C}_{G^*}(U^*)$. Since $\mathfrak{N}_{G^*}(T)=H\cap G^*=H^*$, we now see that (DII) holds for G^*, R^*, and T.

(*c*) We show that (DI) and (DII) imply that if K is a conjugate class of G consisting of π-elements, then $K\cap H$ is a conjugate class of H. First of all, it follows from (D_π) for G, that K meets R and hence H. Let h_1 and h_2 be two elements of $K\cap H$. Taking $U=\{h_i\}$ with $i=1$ or 2 in (*a*), we see from (21) that $T\leqq\mathfrak{C}_G(h_i)$. By (D_π) for $\mathfrak{C}_G(h_i)$, there exists an S_π group $Q_i\geqq\{T, h_i\}$ of $\mathfrak{C}_G(h_i)$. If $g^{-1}h_1g=h_2$ with $g\in G$, then $g^{-1}Q_1g$ and Q_2 are both S_π-groups of $\mathfrak{C}(h_2)$ and again by (D_π), we may assume that g is chosen such that $g^{-1}Q_1g=Q_2$. It follows that Q_2 includes both T and T^g and as Q_2 belongs to an S_π-group R^x of G, $x\in G$, the weak closure implies that $T=T^g$, that is, that $g\in H$. Hence h_1 and h_2 are conjugate in H.

We have now shown that $K\cap H$ is a conjugate class of H.

If U is an abelian π-subgroup of H, then by (*a*) and (*b*), the corresponding result holds for $G^*=\mathfrak{C}_G(U)$ and $H^*=\mathfrak{C}_H(U)$: If K is a conjugate class consisting of π-elements of G^*, then $K\cap H^*$ is a conjugate class of H^*.

This shows that G satisfies the hypothesis (C) of Theorem 4. This remains true if we replace π by the subset of primes which actually appear in $(H:H_0)$. Now, Theorem 4 can be applied and yields the desired result.

Remark. If π consists of only one prime, then by Sylow's theorem, (DII) is satisfied. Theorem 5 then yields the theorem of Hall-Grün, M. HALL [*4*], Theorem 14.4.6.

References

[*1*] BAER, R.: Die Existenz Hallscher Normalteiler. Arch. Math. **11**, 77–87 (1960).

[*2*] BRAUER, R., and J. TATE: On the characters of finite groups. Ann. Math. **62**, 1–7 (1955).

[*3*] FEIT, W., and J. G. THOMPSON: A solvability criterion for finite groups and some consequences. Proc. Nat. Acad. Sci. U.S.A. **48**, 968–970 (1962).

[*4*] HALL, M.: The Theory of Groups. New York: MacMillan 1959.

[*5*] HALL, P.: Theorems like Sylow's. Proc. London Math. Soc. (3) **6**, 286–304 (1956).

[*6*] HUPPERT, B.: Subnormale Untergruppen und p-Sylow-Gruppen. Acta Math. Szeged **22**, 46–61 (1961).

[*7*] SAH, C. H.: Normal complements in finite groups. Illinois J. Math. **6**, 282–291 (1962).

[*8*] TATE, J.: Nilpotent Quotient Groups. Forthcoming in Topology.

[*9*] WIELANDT, H.: Die Existenz von Normalteilern in endlichen Gruppen. Math. Nachr. **18**, 274–280 (1959).

[*10*] ZASSENHAUS, H.: The Theory of Groups, 2nd ed. New York: Chelsea 1958.

Dept. of Math., Harvard University, Cambridge 38, Mass. (U.S.A.)

(Received July 23, 1963)

A NOTE ON THEOREMS OF BURNSIDE AND BLICHFELDT

RICHARD BRAUER

1. The irreducible constituents of the tensor powers of a representation of a group. In his book [3], W. Burnside proved the following theorem (Theorem IV of Chapter XV):

THEOREM 1. *Let G be a finite group and let X be a faithful representation of G over the field* **C** *of complex numbers. Each irreducible representation X_λ of G appears as a constituent in some tensor power of X.*

Recently, R. Steinberg [5] has given a very simple proof of this theorem generalizing it at the same time. I shall give still another proof of Theorem 1. While this proof is less conceptual than Steinberg's proof, it is very short, and it refines the theorem in another direction.

THEOREM 1*. *Assume that the character χ of the representation X in Theorem 1 takes on a total of r distinct values $a_1, a_2, \cdots, a_r$ on G. Each irreducible character χ_λ of G appears as a constituent of one of the characters $\chi^0 = 1, \chi, \chi^2, \cdots, \chi^{r-1}$.*

PROOF. Let A_j be the set of elements $g \in G$ for which $\chi(g) = a_j$. Choose $g_j \in A_j$. If χ_λ is not contained in χ^i, then

$$|G|\,(\chi^i, \chi_\lambda) = \sum_j \chi^i(g_j) \sum_{g \in A_j} \bar{\chi}_\lambda(g) = 0.$$

If this holds for $i = 0, 1, \cdots, r-1$, it follows from the nonvanishing

Received by the editors December 1, 1962.

of the Vandermonde determinant that the inner sum vanishes for all j. Since X is faithful, one of the A_j consists only of the unit element, and we would have $D_g\, \chi_\lambda = \chi_\lambda(1) = 0$, which is absurd.

2. **Remarks.** 1. If $n = D_g\, \chi$, it follows from Theorem 1* that the sum of the degrees of the irreducible representations of G is at most equal to

$$s = 1 + n + n^2 + \cdots + n^{r-1}.$$

This implies that the order $|G|$ of G is at most equal to s^2.

2. If it is known that all values of χ lie in a fixed algebraic number field K of finite degree over the field Q of rational numbers, then since $|\chi(g)|$ and all its algebraic conjugates are at most equal to n for $g \in G$, it follows that r lies below a bound depending only on K and n. Hence $|G|$ also lies below bounds depending only on K and n. The existence of such bounds has first been observed by I. Schur [4]. We do not obtain here the sharp values given by Schur.

3. The proof of Theorem 1* shows that instead of $1, \chi, \cdots, \chi^{r-1}$, we may take any r consecutive powers of χ. If χ assumes the value 0, even $r-1$ consecutive powers of χ with positive exponents will do.

4. If C is replaced by an algebraically closed field K of prime characteristic p, the same result will hold for the irreducible representations X_λ of G in K. If we take for χ_λ the character of the indecomposable component of the regular representation, associated with X_λ, the same proof applies. Here, $a_1, a_2, \cdots, a_r$ are to be taken as the values assumed by χ for p-regular elements of G. (See [2] for the basic properties of "modular" characters.)

5. The proof of Theorem 1* remains the same if C is replaced by an algebraically closed field K of characteristic 0. It is immediate that here and in Remark 4 the condition that K be algebraically closed can be dropped.

3. **A result on the conjugates of a character.** We show now

THEOREM 2. *Let χ be an irreducible character of a finite group G. Let Ω be the field obtained by adjoining the values of χ to the field Q of rational numbers and let Γ be the Galois group of Ω over Q. If $\theta_1, \theta_2, \cdots, \theta_n$ is a system of n elements of Γ, we have one of the following two cases.*

Case A. *There exists an element $g \in G$ such that $\theta_i\chi(g) \neq \chi(g)$ for $i = 1, 2, \cdots, n$.*

Case B. *There exists a product $\theta_\alpha\theta_\beta \cdots \theta_\rho$ with $1 \leqq \alpha < \beta < \cdots < \rho \leqq n$ with an odd number of factors which leaves χ invariant.*

PROOF. If ξ is an element of the group ring of Γ over the ring of rational integers, we define $\xi\cdot\chi$ in the natural manner by linearity. Note that $\theta_i\theta_j=\theta_j\theta_i$. Form

$$P = \prod_i (1-\theta_i)\cdot\chi.$$

If there exists an element $g\in G$ with $P(g)\neq 0$, we must have $\theta_i\chi(g)\neq\chi(g)$ for all i and this leads to Case A. If P vanishes identically, this yields

$$(1)\qquad \chi+\sum_{i<j}\theta_i\theta_j\chi+\cdots=\sum_i\theta_i\chi+\sum_{i<j<k}\theta_i\theta_j\theta_k\chi+\cdots.$$

It follows that the irreducible character χ appears on the right and this means that we have the result of Case B.

REMARK. It is trivial that Theorem 2 remains valid, if Ω is replaced by a larger field normal over Q.

4. Existence of elements of certain orders in G. It will be convenient to say that a number ω "requires" the mth roots of unity, if ω lies in the field of the mth roots of unity over Q, and if m is the least positive integer for which this is true. If m here is even, it will be divisible by 4. We shall also say that a character χ *"requires"* the mth roots of unity, if all values of χ lie in the field of the mth roots of unity, and if again m cannot be replaced by a smaller integer.

We now give some corollaries of Theorem 2. The first one is due to Blichfeldt [1] and Burnside [3, Theorem X of Chapter XVI].

COROLLARY 1. *Let χ be an irreducible character of a finite group. Let $p_1, p_2, \cdots, p_n$ be distinct primes, and assume that there exist elements $g_1, g_2, \cdots, g_n$ of G such that $\chi(g_i)$ requires the $p_i^{a_i}$th roots of unity for some $a_i>0$, $i=1, 2, \cdots, n$. Then G contains elements of order $p_1^{a_1}p_2^{a_2}\cdots p_n^{a_n}$.*

PROOF. Let Ω be the field of the $|G|$th roots of unity over Q (see the Remark in 3). If p_i divides $|G|$ with the exact exponent r_i, $(i=1, 2, \cdots, n)$, then $a_i\leqq r_i$. Let Ω_i be the field of the $(|G|/p_i^{r_i-a_i+1})$st roots of unity over Q. Since $\chi(g_i)\notin\Omega_i$, we can find an element θ_i of the Galois group of Ω over Ω_i such that $\theta_i\chi(g_i)\neq\chi(g_i)$. On the other hand, by construction $\theta_i\chi(g_j)=\chi(g_j)$ for $j\neq i$, $1\leqq j\leqq n$. If we now apply Theorem 2, we cannot have Case B since a product $\theta_\alpha\theta_\beta\cdots$ will not leave $\chi(g_\alpha)$ fixed. Hence there exist $g\in G$ such that $\theta_i\chi(g)\neq\chi(g)$ for $i=1, 2, \cdots, n$. Then $\chi(g)\notin\Omega_i$, and this implies that the order of g is divisible by $p_i^{a_i}$ for $i=1, 2, \cdots, n$. This proves the corollary.

A similar method yields the slightly more general result.

COROLLARY 2. *Let χ be an irreducible character of the finite group G*

and assume that G contains elements $g_1, g_2, \cdots, g_n$ *such that* $\chi(g_i)$ *requires the* h_i*th roots of unity where the rational integers* $h_1, h_2, \cdots, h_n$ *are pairwise coprime. If* $p_i^{a_i}$ *is a prime power dividing* h_i *then G contains elements of order* $p_1^{a_1}p_2^{a_2}\cdots p_n^{a_n}$.

Another result of the same type is the following one.

COROLLARY 3. *Let* χ *be an irreducible character of the finite group G which requires the mth roots of unity. Let* m_0 *be the product of those prime power factors* $p_i^{a_i}$ *of m whose exponent* a_i *is greater than* 1. *Then G contains elements of order* m_0.

PROOF. Let Ω be the field of the mth roots of unity, and let Ω_i here be the field of the (m/p_i)th roots of unity with p_i ranging over those primes for which $p_i^2 \mid m$. Then Ω is cyclic of degree p_i over Ω_i. Choose θ_i as a generator of the Galois group of Ω over Ω_i. Suppose that a product Π of different θ_i left χ invariant. Since θ_i has order p_i and $(p_i, p_j)=1$ for $i\neq j$, each factor θ_i appearing in Π leaves χ fixed. Then χ would lie in Ω_i and this is not consistent with the choice of m.

Thus, we have Case A in Theorem 2. If $\theta_i(\chi(g))\neq\chi(g)$ for all i in question, then the order of g is divisible by $p_i^{a_i}$ for all these i and this proves the corollary.

There are still other results which can be obtained by this type of method, but it does not seem to be worthwhile to go into details. It should be mentioned that we have not used (1) fully in the proof of Theorem 2 and that more precise information is available. For instance, if $n=3$ in Theorem 2, and if we do not have Case A, then either one of the θ_i leaves χ fixed, or the four elements $\theta_1^2, \theta_2^2, \theta_3^2, \theta_1\theta_2\theta_3$ all leave χ fixed.

REFERENCES

1. H. Blichfeldt, *On the order of linear homogeneous groups.* II, Trans. Amer. Math. Soc. 5 (1904), 310–325.

2. R. Brauer and C. Nesbitt, *On the modular characters of groups*, Ann. of Math. (2) 42 (1941), 556–590.

3. W. Burnside, *Theory of groups of finite order*, 2nd ed., Cambridge Univ. Press, Cambridge, 1911.

4. I. Schur, S.-B. Preuss. Akad. Berlin (1905), 77–91.

5. R. Steinberg, *Complete sets of representations of algebras*, Proc. Amer. Math. Soc. 13 (1962), 746–747.

HARVARD UNIVERSITY

Reprinted from JOURNAL OF ALGEBRA, Vol. 1, No. 2. July 1964

Reprinted in Belgium

Some Applications of the Theory of Blocks of Characters of Finite Groups. I*

RICHARD BRAUER

Harvard University, Cambridge, Massachusetts

Received January 20, 1964

I. INTRODUCTION

If G is a finite group and p a fixed prime number, the irreducible representations of G are distributed into disjoint systems, the p-blocks, which are related closely to the arithmetic structure of the group algebra of G. These blocks have been investigated in two preceding papers, Brauer [*1*, *2*]. Different proofs of some of the results have been given by Osima [*10*], Rosenberg [*11*], Iizuka [*7*], Nagao [*9*]; see also Curtis-Reiner [*6*]. It is our aim to give some applications of the theory. In the present first part we are concerned mainly with the problem of obtaining properties of the characters of G from group theoretical information concerning G. In the later parts, we shall use the results for a study of groups of even order; the prime p will be taken as 2.

In Section II of the present paper, we list some of the known results which are used later on.

In Section III, we obtain a number of results concerning the principal p-block. As an application, we introduce certain types of groups, the groups of a given deficiency class for the prime p. These types can be defined inductively by group theoretical properties. On the other hand, they can be characterized by properties of their characters, Section IV. In Section V, the notion of a basic set for a block is introduced and it is shown how basic sets can be used for a discussion of properties of characters.

* This research was supported by the United States Air Force under contract No. AF 49 (638)-287 monitored by the AF Office of Scientific Research of the Air Research and Development Command.

II. Notation. Presupposed Results

1. In the following, G will always be a group of finite order, and p will be a fixed prime number. If M is a subset of G, we write $|M|$ for the number of elements of M. The centralizer of M in G will be denoted by $\mathfrak{C}_G(M)$ and the normalizer of M by $\mathfrak{N}_G(M)$.

By the p-regular core $\mathfrak{K}_p(G)$ of G, we mean the maximal normal subgroup of G of an order prime to p (often denoted by $\mathfrak{O}_{p'}(G)$).

2. Let $\bar{\mathbf{Q}}$ denote the algebraic closure of the field $\mathbf{Q}$ of rational numbers. Choose a fixed exponential valuation ν_p of $\bar{\mathbf{Q}}$ which extends the p-adic valuation of $\mathbf{Q}$; $\nu_p(p) = 1$. Let $\mathfrak{o}$ denote the local ring of ν_p in $\bar{\mathbf{Q}}$. Let $\mathfrak{p}$ denote the corresponding prime ideal and set $\Omega = \mathfrak{o}/\mathfrak{p}$. The residue class map of $\mathfrak{o}$ onto Ω will be denoted by an asterisk; $\alpha \to \alpha^*$.

By a representation X of G or, more clearly, by an ordinary representation of G, we shall mean a representation of G in $\bar{\mathbf{Q}}$. We use the term modular representation of G for the representations of G in Ω. An analogous terminology will be used in the case of characters. We can speak of the modular irreducible constituents of an ordinary representation. cf. Appendix of [*1*].

The group algebra of G over Ω will be denoted by $\Gamma(G)$ and its center by $Z(G)$. If M is a subset of G, we write $[M]$ for the element of $\Gamma(G)$ defined by

$$[M] = \sum_{m \in M} m. \tag{2.1}$$

In particular, if $K_1, K_2, \cdots, K_k$ are the conjugate classes of G, the elements $[K_1], [K_2], \cdots, [K_k]$ form a basis of $Z(G)$. For this reason, we refer to $Z(G)$ as the class algebra of G (over Ω). The k irreducible characters of G will be denoted by $\chi_0 = 1, \chi_1, \cdots, \chi_{k-1}$.

3. It will not be possible to list all the known results on blocks which will be used below. We shall state here some significant facts and refer for details to the papers quoted in Section I.

A p-block B of G is a set of irreducible representations of G. Here, we place two irreducible representations X and Y of G in the same block, if there exists a chain

$$X^{(0)} = X, X^{(1)}, \cdots, X^{(r)} = Y \tag{2.2}$$

of irreducible representations $X^{(i)}$ of G such that any two neighbors in (2.2) have a modular irreducible constituent in common. By the modular irreducible representations of the block B, we shall mean the modular irreducible constituents of the representations $X \in B$. Finally, we replace representations by characters and look upon a block B as a set of characters. It will be convenient to use the notation $s(B)$ for the set of indices i for which $\chi_i \in B$.

If F is a modular irreducible representation of G and hence of $\Gamma(G)$, the elements $x \in Z(G)$ are represented by scalar multiples $\psi(z) I$ of the identity; $\psi \in \Omega$. Here, ψ is a linear character of $Z(G)$, i.e. an algebra-homomorphism of $Z(G)$ onto Ω. Actually, ψ depends only on the p-block B to which F belongs. If we write ψ_B for ψ, then $B \to \psi_B$ is a one-to-one correspondence between the set of p-blocks of G and the set of linear characters of the class algebra $Z(G)$. If σ_j is a representative for the conjugate class K_j, we have $\chi_i \in B$, if and only if

$$\psi_B([K_j]) = (|K_j| \chi_i(\sigma_j)/\chi_i(1))^* \tag{2.3}$$

for all j. (As explained above, the asterisk indicates the residue class map $\mathfrak{o} \to \Omega$).

The principal p-block $B_0 = B_0(G)$ of G is the p-block containing the principal character $\chi_0 = 1$. It follows from (2.3) that

$$\psi_{B_0}([K_j]) = (|K_j|)^*. \tag{2.4}$$

4. Let H be a subgroup of G and let λ be a linear function on $Z(H)$ with values in Ω. We denote by λ^G the unique linear function on $Z(G)$ for which

$$\lambda^G([K_j]) = \lambda([K_j \cap H]) \tag{2.5}$$

for $j = 1, 2, \cdots, k$. In particular, if b is a p-block of H and ψ_b the corresponding linear character of $Z(H)$, we may take $\lambda = \psi_b$. Then $\psi_b{}^G$ may or may not be a linear character of $Z(G)$. If we have the former case and if $\psi_b{}^G = \psi_B$, we say that b^G is defined and we set $b^G = B$.

Each p-block B of G determines a class of conjugate p-subgroups of G, the defect groups of B. Their order is p^d where d is the defect of B which can be determined, if we know the degrees $\chi_i(1)$ of the characters $\chi_i \in B$. If the block b of H has the defect group D_0 (in H), then b^G is always defined when $\mathfrak{C}_G(D_0) \subseteq H$. If the block B of G has defect group D, there exist blocks b of $DC_G(D)$ such that $b^G = B$. Each such b has necessarily defect group D.

5. Let π be a fixed p-element of G, say, of order p^α. If v is a p-regular element of $C_G(\pi)$, we have formulas

$$\chi_i(\pi v) = \sum_\rho d_{i\rho}{}^\pi \varphi_\rho{}^\pi(v) \tag{2.6}$$

for $\chi_i \in B$. Here the $\varphi_\rho{}^\pi$ range over the modular irreducible characters of the blocks b of $\mathfrak{C}_G(\pi)$ for which $b^G = B$. The decomposition numbers $d_{i\rho}{}^\pi$ are algebraic integers of the field of the p^α-th roots of unity which do not depend on v. If $c_{\rho\sigma}{}^\pi$ is the Cartan invariant of $\mathfrak{C}_G(\pi)$ belonging $\varphi_\rho{}^\pi$ and $\varphi_\sigma{}^\pi$, then

$$\sum_{i \in s(B)} d_{i\rho}{}^\pi d_{i\sigma}{}^\pi = c_{\rho\sigma}{}^\pi. \tag{2.7}$$

On the other hand, if π and π' are two p-elements of G which are not conjugate in G,

$$\sum_{i \in s(B)} d_{i\rho}{}^{\pi} d_{i\sigma}{}^{\pi'} = 0. \tag{2.8}$$

By the kernel of a character, we shall always mean the kernel of the corresponding representation. Since π lies in the center of $\mathfrak{C}_G(\pi)$, the kernel of the modular irreducible characters $\varphi_i{}^{\rho}$ of $\mathfrak{C}_G(\pi)$ contains $\{\pi\}$. Hence the $\varphi_i \in b$ can be considered as modular irreducible characters of $\mathfrak{C}_G(\pi)/\{\pi\}$. They form the modular irreducible characters of a block $\tilde{b}$ of $\mathfrak{C}_G(\pi)/\{\pi\}$. If $\tilde{b}$ has defect d_0, b has defect $d_0 + \alpha$. If the Cartan invariant of $\mathfrak{C}_G(\pi)/\{\pi\}$ corresponding to $\varphi_\rho{}^{\pi}$, $\varphi_\sigma{}^{\pi}$ is $\tilde{c}_{\rho\sigma}$, then $c_{\rho\sigma}{}^{\pi}$ in (2.7) is given by $c_{\rho\sigma}{}^{\pi} = p^{\alpha}\tilde{c}_{\rho\sigma}$.

6. If Y is a representation of G and τ an automorphism of G, a representation Y^{τ} of G is defined by $g \to Y(\tau^{-1}(g))$ for all $g \in G$. An analogous remark applies to modular representations. It is clear that τ maps each p-block B on a p-block B^{τ}.

In particular, if G is a normal subgroup of a group T, we may take τ as the automorphism $g \to t^{-1}gt$ of G produced by a fixed element t of T. We write here Y^t for Y^{τ}, B^t for B^{τ}, and we say that Y and Y^t are representations of G associated in T, and that B and B^t are p-blocks of G associated in T.

III. Properties of the Principal Block

We need a simple lemma.

Lemma 1. *Let H be a normal subgroup of G. For each p-block B of G, there exists a family $\{b^g\}$ of p-blocks of H associated in G with the following property: If W is an ordinary or modular irreducible representation in B, the irreducible constituents of $W \mid H$ lie in blocks b^g.*

Proof. It follows from Clifford's theorem that for each W, the irreducible constituents of $W \mid H$ are associated in G and hence lie in a family of associated blocks of H. If we apply this to an ordinary irreducible representation $X \in B$ and to a modular irreducible constituent F of X, we obtain both times the same family of blocks of H. It then follows for all X_i in a chain (2.2) that the irreducible constituents of $X_i \mid H$ lie in a family of associated blocks of H and this implies the lemma.

As already mentioned, the principal p-block $B_0 = B_0(G)$ of G is the p-block containing $\chi_0 = 1$. We prove

Theorem 1. *The intersection of the kernels of the ordinary irreducible representations $X_i \in B_0$ is the p-regular core $\mathfrak{K}_p(G)$.*

Proof. Since $\chi_0 \mid \mathfrak{K}_p(G)$ is the principal character of $\mathfrak{K}_p(G)$, it follows from Lemma 1 with $H = \mathfrak{K}_p(G)$ that for $\chi_i \in B_0$ all irreducible constituents of $\chi_i \mid \mathfrak{K}_p(G)$ lie in the principal p-block b_0 of $\mathfrak{K}_p(G)$. Since $\mathfrak{K}_p(G)$ is p-regular, b_0 consists only of one ordinary character, the principal character. If X_i is the representation with the character χ_i, it follows that $X_i \mid \mathfrak{K}(G)$ has the principal representation of $\mathfrak{K}_p(G)$ as its only irreducible constituent. This implies that $\mathfrak{K}_p(G)$ belongs to the kernel of $X_i \in B_0$.

Let H denote the intersection of the kernels of the representations X_i. As shown, $H \supseteq \mathfrak{K}_p(G)$. We may consider the ordinary and modular representations in $B_0(G)$ as representations of G/H and then the principal blocks $B_0(G)$ and $B_0(G/H)$ coincide. In particular, both blocks have the same decomposition numbers and Cartan invariants. This implies that both have the same defect (for instance by [*1*, (6C)]). Since the defect of the principal block is the exponent of the highest power of p which divides the group order, we find $\nu_p(\mid G \mid) = \nu_p(\mid G/H \mid)$. It follows that H is p-regular and this implies $H = \mathfrak{K}_p(G)$.

Lemma 2. *If the irreducible character χ_i belongs to B_0, each algebraically conjugate character χ_i' belongs to B_0.*

This follows from (2.3), (2.4), if we use the fact that the value of χ_i' for the elements of a class K_j is equal to the value of χ_i for the elements of a class $K_{j'}$, with $\mid K_j \mid = \mid K_{j'} \mid$.

Corollary 1. *Set*

$$N = \sum_{i \in s(B_0)} \chi_i(1).$$

Then $\mid G : \mathfrak{K}_p(G) \mid$ lies between N and an upper bound depending only on N.

Proof. It follows from Theorem 1 that

$$\sum_{i \in s(B_0)} \chi_i(1)^2 \leq \mid G : \mathfrak{K}_p(G) \mid$$

and this implies $N \leqslant \mid G : \mathfrak{K}_p(G) \mid$. On the other hand, by Lemma 2,

$$\theta = \sum_{i \in s(B_0)} \chi_i$$

is a rational-valued character of G which by Theorem 1 belongs to a faithful representation of $G/\mathfrak{K}_p(G)$. Now a theorem of Schur [*11*] gives the required upper estimate of $\mid G : \mathfrak{K}_p(G) \mid$.

We can give an analogue for modular representations. We have

Theorem 2. *The intersection L of the kernels of the modular irreducible* *representations F_ρ in $B_0(G)$ is the group $\mathfrak{O}_{p'p}(G)$, i.e. the maximal normal subgroup $L_1 \supseteq \mathfrak{K}_p(G)$ of G for which $L_1/\mathfrak{K}_p(G)$ is a p-group.*

Proof. Clearly, $L \supseteq \mathfrak{K}_p(G)$. For each p-regular element v of L and for each modular irreducible character, $\varphi_\rho \in B_0$, we have $\varphi_\rho(v) = \varphi_\rho(1)$. It then follows from (2.6) with $\pi = 1$ that $\chi_i(v) = \chi_i(1)$ for each $\chi_i \in B_0$. Hence $v \in \mathfrak{K}_p(G)$. This shows that $L/\mathfrak{K}_p(G)$ is a p-group. Hence $L \subseteq \mathfrak{O}_{p'p}(G)$. On the other hand, if we consider $\varphi_\rho \in B_0$ as modular irreducible character of $G/\mathfrak{K}_p(G)$, its kernel includes the normal p-subgroup $\mathfrak{O}_{p'p}(G)/K_p(G)$ (cf. [*1*], (9D)). This implies $\mathfrak{O}_{p'p}(G) \subseteq L$ and we must have equality, q.e.d.

We can now prove the following analogue of Corollary 1.

COROLLARY 2. *Let N_0 be the sum of the degrees of the nonprincipal irreducible characters $\varphi_\rho \in B_0$. Then $G/\mathfrak{O}_{p'p}(G)$ is isomorphic with a subgroup of $GL(N_0, p)$.*

Indeed, if we form the modular representation F of degree N_0, which splits completely into the different representations with the characters $\varphi_\rho \in B_0$, $\varphi_\rho \neq 1$, then tr $F(g)$ lies in the prime field $\Lambda \subset \Omega$. It follows that F is equivalent with a representation in Λ. By Theorem 2, F has the kernel $\mathfrak{O}_{p'p}(G)$ and the corollary becomes evident.

Since we have $\mathfrak{O}_{p'p}(G) = G$, if and only if G has a normal p-complement, Theorem 2 also implies the following known result ([*5*], §29).

COROLLARY 3. *A group G has a normal p-complement, if and only if the modular principal character $\varphi_0 = 1$ is the only modular irreducible character in $B_0(G)$.*

In the case of the principal block, the formulas (2.6) take a somewhat simpler form. This is a consequence of the following result.

THEOREM 3. *Let H be a subgroup of G, let b be a block of H with the defect group D (in H), and assume that $\mathfrak{C}_G(D) \subseteq H$. Then $b^G = B_0(G)$, if and only if b is the principal block $B_0(H)$ of H.*

Proof.[1] (a) Set $B = B_0(H)^G$, $B_0 = B_0(G)$. Using the definition of the block B, we see from (2.4) for $B_0(H)$ and (2.5) that

$$\psi_B(K_j) = |K_j \cap H|^*, \qquad (j = 1, 2, \cdots, k). \tag{3.1}$$

Partition K_j into subclasses L_μ of elements conjugate with regard to D. Then $|L_\mu|$ is a power of p and $|L_\mu| = 1$, if and only if the only element of L_μ lies in $\mathfrak{C}_G(D) \subseteq H$. It follows that the complement of $K_j \cap H$ in K_j is a union

[1] A somewhat different proof will be presented elsewhere in a more general setting.

of subclasses L_μ with $|L_\mu| > 1$. This implies $|K_j| \equiv |K_j \cap H| \pmod{p}$ and (3.1) and (2.4) for $B_0 = B_0(G)$ yield

$$\psi_B(K_j) = \psi_{B_0}(K_j)$$

for all j. It follows that $B_0(H)^G = B = B_0(G)$.

(b) Conversely, assume that $b^G = B_0 = B_0(G)$. If it is not true that $b = B_0(H)$, choose a counterexample in which $\nu_p(G:D)$ is minimal and among the counterexamples which satisfy this condition, choose one in which $|G|$ is minimal.

According to Section II, there exist blocks $\tilde{b}$ of $D\mathfrak{C}_H(D) = D\mathfrak{C}_G(D)$ with the defect group D for which $\tilde{b}^H = b$. Then $\tilde{b}^G = b^G = B_0$. Here, $\tilde{b}$ cannot be the principal block of $D\mathfrak{C}_H(D)$, since otherwise by part (a), we would have $b = B_0(H)$. Changing the notation, we may as well assume

$$H = DC_G(D)$$

replacing b by $\tilde{b}$.

Then b contains a unique irreducible character θ whose kernel includes D and which as character of $D\mathfrak{C}_G(D)/D = H/D$ has defect 0.

Suppose first that there do not exist p-elements $x \in \mathfrak{N}_G(D)$, $x \notin H$ for which $\theta^x = \theta$. Then in the sense of [*1*], (12A), θ is associated with a p-block B of G with the defect group D and, by [*2*, (2D)], $B = b^G = B_0$. But in the sense of [*1*, (12A)], the principal block B_0 of G is associated only with the principal character of $D\mathfrak{C}_G(D)$. Hence θ is this principal character. This is impossible, since $\theta \in b$ and $b \neq B_0(D\mathfrak{C}_G(D))$.

It follows that there exist p-elements $x \in \mathfrak{N}_G(D)$ such that $x \notin D\mathfrak{C}_G(D)$ and that $\theta^x = \theta$. We may assume without restriction that

$$x^p \in D\mathfrak{C}_G(D) \lhd \mathfrak{N}_G(D).$$

Then $R = \{D\mathfrak{C}_G(D), x\}$ is a group containing $D\mathfrak{C}_G(D)$ as normal subgroup of index p. Set $b_1 = b^R$. Then $b_1{}^G = b^G = B_0$. Since b_1 has larger defect than b (cf. [*2*], (2F)), it follows from the minimal choice of our counterexample that $b_1 = B_0(R)$. Since then $b^R = B_0(R)$ it follows in the same manner that $R = G$. Thus,

$$D \lhd G; \qquad H = D\mathfrak{C}_G(D) \lhd G; \qquad |G:H| = p.$$

Since $\theta^x = \theta$, we can extend θ to an irreducible character χ of G; $\theta(1) = \chi(1)$. Let B be the p-block of G containing χ and let D_1 be a defect group of B. Since $D \lhd G$, we have $D \subseteq D_1$, (cf. [*1*, (9F)]) and hence $\mathfrak{C}_G(D_1) \subseteq \mathfrak{C}_G(D) \subseteq H$. We shall show that $B = B_0$ by showing that $\psi_B = \psi_{B_0}$. Since $\chi \in B$, by (2.3)

$$\psi_B(K_j) = (|K_j|\,\chi(\sigma_j)/\chi(1))^*. \tag{3.2}$$

On the other hand since $B_0 = b^G$ and $\theta \in b$, by (2.3) for b and by (2.5)

$$\psi_{B_0}(K_j) = \Big(\sum_{\mu} | k_{j\mu} | \theta(\sigma_{j\mu})/\theta(1)\Big)^* \tag{3.3}$$

where $k_{j\mu}$ ranges over all conjugate classes of H contained in K_j and where $\sigma_{j\mu} \in k_{j\mu}$. Since χ extends θ, we may replace θ by χ. We have $\chi(\sigma_{j\mu}) = \chi(\sigma_j)$. If K_j meets H, then K_j is the disjoint union of the $k_{j\mu}$, $| K_j | = \Sigma_\mu | k_{j\mu} |$ and comparison of (3.2) with (3.3) yields $\psi_{B_0}(K_j) = \psi_B(K_j)$. If $K_j \cap H = \emptyset$, then by (3.3) $\psi_{B_0}(K_j) = 0$. Suppose that $\psi_B(K_j) \neq 0$. Then ([*1*], (8A)) there exist elements $\sigma \in K_j$ such that the defect group D_1 belongs to $\mathfrak{C}_G(\sigma)$. This implies

$$\sigma \in \mathfrak{C}_G(D_1) \subseteq \mathfrak{C}_G(D) \subseteq H$$

and we have $\sigma \in K_j \cap H$, a contradiction.

We now have $\psi_B = \psi_{B_0}$ and hence $B = B_0$. Apply now Lemma 1 to B_0. Since $\chi_0 \in B_0$, $\chi_0 | H \in B_0(H)$, it follows that all irreducible constituents of $\chi | H$ lie in $B_0(H)$. Since $\chi | H = \theta \in b \neq B_0(H)$, this is a contradiction and Theorem 3 is proved.

If H is the centralizer of a p-element π as in (2.6), then π belongs to the defect group D of every p-block b of H. Hence $\mathfrak{C}_G(D) \subseteq \mathfrak{C}_G(\pi) = H$, and, by Theorem 3, $b^G = B_0(G)$, if and only if $b = B_0(H)$. Hence we have

COROLLARY 4. *If $B = B_0(G)$, then $\varphi_p{}^\pi$ in (2.6) ranges over the modular irreducible characters of $B_0(\mathfrak{C}_G(\pi))$.*

For $B \neq B_0(G)$, we may have several blocks b of H with $b^G = B$. They need not even have the same defect.

On combining Corollaries 3 and 4, we obtain

COROLLARY 5. *Suppose that π is a p-element of G for which $\mathfrak{C}_G(\pi)$ has a normal p-complement. Then for every p-regular element v of $\mathfrak{C}_G(\pi)$ and for every $\chi_i \in B_0(G)$,*

$$\chi_i(\pi v) = \chi_i(\pi).$$

The following remark of a different nature is sometimes useful.

REMARK. The characters χ_i of degree 1 in $B_0(G)$ form a group under multiplication and hence the number of such characters χ_i is a divisor of the index of the commutator subgroup G' of G.

This is an immediate consequence of (2.3) and (2.4).

IV. Groups of a Given Deficiency Class

We first prove a lemma.

Lemma 3. *Let B be a block of G of defect d with the defect group D. If π is an element of the center $\mathfrak{Z}(D)$, there exist blocks b of $\mathfrak{C}_G(\pi)$ of defect d such that $b^G = B$.*

Proof. Set $a = \nu_p(|G|)$. There exist characters $\chi_i \in B$ with

$$\nu_p(\chi_i(1)) = a - d.$$

By the definition of the defect group in [*1*], we can find a p-regular element v of G such that D is a p-Sylow group of $\mathfrak{C}_G(v)$ and that $\psi_B([K]) \neq 0$ for the conjugate class K of v. It follows from (2.3) that $\chi_i(v) \not\equiv 0 \pmod{\mathfrak{p}}$. Since $v \in \mathfrak{C}_G(\pi)$, this implies $\chi_i(\pi v) \not\equiv 0 \pmod{\mathfrak{p}}$. Now (2.6) shows that there exist a block b of $\mathfrak{C}_G(\pi)$ with $b^G = B$ and a modular irreducible character $\varphi_\rho{}^\pi$ in b such that

$$\varphi_\rho{}^\pi(v) \not\equiv 0, \qquad d_{i\rho}{}^\pi \not\equiv 0 \pmod{\mathfrak{p}}. \tag{4.1}$$

Since the modular characters of a block can be expressed by the ordinary characters of the block (restricted to p-regular elements) with integral coefficients, it follows that there exists an ordinary irreducible character $\xi \in b$ with $\xi(v) \not\equiv 0 \pmod{\mathfrak{p}}$.

Let L be the conjugate class of $\mathfrak{C}_G(\pi)$ which contains v. Since $|L| \, \xi(v)/\xi(1)$ is an algebraic integer, we find

$$\nu_p(\xi(1)) \leqslant \nu_p(|L|) = \nu_p(|\mathfrak{C}_G(\pi)|) - \nu_p(|\mathfrak{C}_G(\pi v)|).$$

Now, D is a p-Sylow group of $\mathfrak{C}_G(v)$ and also of $\mathfrak{C}_G(\pi v)$. Thus,

$$\nu_p(|\mathfrak{C}_G(\pi v)|) = d$$

and we find

$$\nu_p(\xi(1)) \leqslant \nu_p(|\mathfrak{C}_G(\pi)|) - d.$$

This shows that the block b of ξ has at least defect d. Since the defect of b cannot be larger than the defect d of $b^G = B$, the lemma is proved.

We now define types of finite groups using characters.

Definition. Let r be an integer. A finite group G will be called of *deficiency class* r for the prime p, if *there does not exist a nonprincipal block of defect* $d \geqslant r$.

The following Remarks 1 and 2 are obvious. Remark 3 follows from [*1*], (9F), and Remark 4 from [*2*], (2G).

REMARKS. 1. If p^r does not divide the order $|G|$, G is of deficiency class r.

2. Every p-group has deficiency class 0 for p.

3. If Q of order p^n is a normal p-subgroup of G and if G is of deficiency class $r \leqslant n$, then G is of deficiency class 0.

4. If Q of order p^n is a p-subgroup of the center of G, then G is of deficiency class r, if and only if G/Q is of deficiency class $r - n$. The same fact still holds if Q of order p^n is a subgroup of G for which $G = Q\mathfrak{C}_G(Q)$.

The reason for our interest in the deficiency classes lies in the fact that for $r > 0$ we can characterize them inductively by properties of the abstract group. We prove

THEOREM 4. *If G is of deficiency class r for the prime p then for every p-element π of G, $\mathfrak{C}_G(\pi)$ is of deficiency class r. Conversely, if for every element π of order p the group $\mathfrak{C}_G(\pi)$ is of deficiency class r and if r is positive, then G is of deficiency class r.*

Proof. Let π be a p-element. If $\mathfrak{C}_G(\pi)$ has a non-principal p-block b of defect d, then $b^G = B$ is defined and has defect at least d. By Theorem 3, $B \neq B_0(G)$. If G has deficiency class r, then $d < r$. It follows that $\mathfrak{C}_G(\pi)$ has deficiency class r.

Conversely, assume that G is not of deficiency class r. Let B be a non-principal p-block of defect $d \geqslant r$ and let D be the defect group. Since $r > 0$, we can choose an element of order p in the center of D. Now Lemma 3 shows that $\mathfrak{C}_G(\pi)$ contains a p-block b of defect d with $b^G = B$. Again, b cannot be $B_0(\mathfrak{C}_G(\pi))$ and hence $\mathfrak{C}_G(\pi)$ is not of deficiency class r as we had to show.

In the case of the groups $\mathfrak{C}_G(\pi)$ with $\pi \neq 1$, we can apply Remark 4 above. Hence Theorem 4 furnishes a characterization of the groups of deficiency class $r > 0$. However, we cannot distinguish the groups of deficiency class 0 in an analogous manner among the groups of deficiency class 1.

A group G has deficiency class 0, if $B_0(G)$ is the only p-block. If $\nu_p(|G|) = a$, then $B_0(G)$ contains at most p^{2a} irreducible characters, cf. Brauer-Feit [*3*]. Hence G has at most class number p^{2a}. It follows that for given a, there are only finitely many groups G, Landau [*8*]. Thus, we have

THEOREM 5. *There exist only finitely many groups G of deficiency class 0 for p for which the exact power p^a of p in $|G|$ is given.*

For $p = 2$, we can go a bit further.

THEOREM 6. *Consider groups G of even order of deficiency class 1 for $p = 2$ for which the exact power 2^a of 2 in the group order $|G|$ is given. There exist only finitely many such simple groups. Also, there exist only finitely many groups G which have more than one conjugate class consisting of involutions.*

Proof. If x is an element of order p of a group G of deficiency class 1, then by Theorem 4 and Remark 3 above, $\mathfrak{C}_G(x)$ has deficiency class 0. If $\nu_p(|G|) = a$ is given, the order of $\mathfrak{C}_G(x)$ is bounded by Theorem 5. If $p = 2$, the results of Brauer-Fowler [*4*] can be applied and yield the statements.

The C.I.T. groups of Suzuki [*13*] and in particular Suzuki's simple groups have deficiency class 1 for $p = 2$ and a number of other simple groups of this class are known. The Mathieu group M_{24} has deficiency class 0 for $p = 2$.

V. Basic Sets

In the formulas (2.6), the modular irreducible characters of the groups $\mathfrak{C}_G(\pi)$ appear. Since it may be difficult to determine these characters, we shall discuss a modification of the method.

If G is a group and p a prime, we shall denote by G^0 the set of p-regular elements of G and by $\chi_i{}^0$ the restriction of the character χ_i to G^0. If B is a fixed p-block of G, the functions $\chi_i{}^0$ with $\chi_i \in B$ generate a module M_B with regard to the ring $\mathbf{Z}$ of integers. By a basic set φ_B for B, we mean any basis $\{\varphi_\rho\}$ of M_B. Thus, we have formulas

$$\chi_i{}^0 = \sum_\rho d_{i\rho}\varphi_\rho \qquad \text{for} \qquad i \in s(B) \tag{5.1}$$

with $d_{i\rho} \in \mathbf{Z}$. Moreover, we set

$$c_{\rho\sigma} = \sum_{i\in s(B)} d_{i\rho}d_{i\sigma} \tag{5.2}$$

for $\varphi_\rho, \varphi_\sigma \in \varphi_B$.

In particular, the modular irreducible characters in B form a basic set φ_B. If this φ_B is used, the $d_{i\rho}$ in (5.1) are the decomposition numbers and the $c_{\rho\sigma}$ the Cartan invariants. We use the same terms in the case of an arbitrary basic set. It is obvious in which manner the $d_{i\rho}$ and the $c_{\rho\sigma}$ will change, if the basic set φ_B is replaced by another basic set. In particular, the matrix $(c_{\rho\sigma})$ is the matrix of a quadratic form Q. If the basic set is changed, Q is replaced by an equivalent form. We note

Theorem 7. *Let p and d be given. The quadratic form Q associated with a p-block of defect d belongs to one of a finite number of classes of positive definite quadratic forms.*

Indeed, it is clear that Q is nonnegative. If we use the set of modular irreducible characters of B for φ_B, we know that the number of members of φ_B is at most equal to p^{2d} [*3*], and that the largest elementary divisor of $(c_{\rho\sigma})$ is p^d and, in particular, that det $(c_{\rho\sigma})$ is apower of p ([*1*], (6C)). This suffices to establish Theorem 7.

COROLLARY 6. *There exist bounds $\gamma(p^d)$ depending only on p^d such that for each p-block of defect d of any finite group, a basic set can be chosen such that the Cartan invariants are at most equal to $\gamma(p^d)$.*

Using the reduction theory of quadratic forms, we can give explicit estimates for $\gamma(p^d)$. Actually, because of the form (5.2) of the $c_{\rho\sigma}$, better estimates can be obtained by direct methods. This will be shown elsewhere.

COROLLARY 7. *If $B = B_0$ is the principal p-block, a basic set φ_B can be chosen such that $1 \in \varphi_B$ and that the Cartan invariants $c_{\rho\sigma}$ lie below a bound $\gamma_0(p^n)$ depending only on the highest power p^n of p which divides $|G|$.*

Proof. We first choose a basic set φ_B in accordance with Corollary 1. Since $\chi_0 = 1$ belongs to $B = B_0$, we have

$$1 = \sum d_{0\rho}\varphi_\rho$$

where φ_ρ ranges over φ_B. It follows from (5.2) with $\rho = \sigma$ that $d_{0\rho}^2 \leqslant c_{\rho\rho} \leqslant \gamma(p^n)$. Moreover, if φ_B consists of l_B functions, the l_B numbers $d_{0\rho}$ are relatively prime. Hence we can find an $(l_B \times l_B)$ matrix T with the following properties: (1) In the first row, the l_B coefficients $d_{0\rho}$ appear. (2) All coefficients lie in $\mathbf{Z}$ and their absolute value lies below a bound depending only on p^n. (3) The determinant is ± 1.

Now the linear transformation with the matrix T changes the set φ_B into a basic set with the required properties.

Let π now be a p-element of a given finite group and let b be a p-block of $\mathfrak{C}_G(\pi)$ such that b^G is a given p-block B of G. Select a basic set φ_b of b. It is clear that we can introduce the elements of φ_b instead of the irreducible modular characters in b in (2.6). We then obtain formulas of the same form as (2.6), which we write again in the form

$$\chi_i(\pi v) = \sum_b \sum_\rho d_{i\rho}{}^\pi \varphi_\rho{}^\pi(v), \qquad (\chi_i \in B), \tag{5.3}$$

where v is a p-regular element of $\mathfrak{C}_G(\pi)$, where b ranges over the set of blocks of $\mathfrak{C}_G(\pi)$ which satisfy $b^G = B$, and where $\varphi_\rho{}^\pi$ ranges over the members of a basic set φ_b for b. If π has order p^α, the $d_{i\rho}{}^\pi$ are algebraic integers of the field of the p^α-th roots of unity. Exactly as in (2.7), we have

$$\sum_{i \in s(B)} d_{i\rho}{}^\pi \bar{d}_{i\sigma}{}^\pi = c_{\rho\sigma}{}^\pi \tag{5.4}$$

where $c_{\rho\sigma}{}^\pi$ is now the Cartan invariant belonging to φ_b; $\varphi_\rho{}^\pi, \varphi_\sigma{}^\pi \in \varphi_b$.

Let π' also be a p-element, let b' be a p-block of $\mathfrak{C}_G(\pi')$ with $(b')^G = B$ and let $\varphi_\sigma^{\pi'}$, be a member of a basic set $\varphi_{b'}$. If either $\pi = \pi'$, $b \neq b'$ or if π and π' are not conjugate in G, it follows from (2.7) and (2.8) that

$$\sum_{i \in s(B)} d_{i\rho}^{\pi} d_{i\sigma}^{\pi'} = 0 \tag{5.5}$$

We state

THEOREM 8. *Let p and d be given. For all finite groups G, there exist only finitely many "types" of p-blocks B of defect d. For each type, the defect group D of B is determined as abstract group. Furthermore, we have a fixed system of elements of D*

$$\pi_0 = 1, \pi_1, \cdots, \pi_m \tag{5.6}$$

which represent the different conjugate classes of G which meet D. Moreover for each $\pi = \pi_\mu$, $1 \leqslant \mu \leqslant m$, the coefficients $d_{i\rho}^{\pi}$ with $i \in s(B)$ are determined provided that suitable basic sets are used.

This is fairly obvious. For given p and d, we have only finitely many possibilities for D and for the choice of the system (5.6). We have already used that the number k_B of characters $\chi_i \in B$ is at most p^{2d}. For fixed $\pi = \pi_j$, let $l_B^{(j)}$ denote the total number of functions φ_ρ^{π} in basic sets φ_b belonging to blocks b of $\mathfrak{C}_G(\pi)$ with $b^G = B$. If we use for a moment again the irreducible modular characters as basic set, we have ([2], (7D))

$$\sum_{j=0}^{m} l_B^{(j)} = k_B \leqq p^{2d}. \tag{5.7}$$

For each block b with $b^G = B$, the defect d_0 is at most equal to d. Suppose that in each case a basic set φ_b is used which satisfies the condition of Corollary 6. Consider a fixed φ_ρ^{π}. The corresponding $d_{i\rho}^{\pi}$ are algebraic integers of a known algebraic number field. Now (5.4) with $\rho = \sigma$ yields

$$\sum_i | d_{i\rho}^{\pi} |^2 \leq \gamma(p^{d_0}). \tag{5.8}$$

For each algebraic conjugate $(d_{i\rho}^{\pi})'$ of $d_{i\rho}^{\pi}$, there exists an algebraically conjugate character $\chi_{i'}$ of χ_i which belongs to B and for which

$$d_{i'\rho}^{\pi} = (d_{i\rho}^{\pi})'.$$

This follows easily from (5.3). Now (5.8) shows that we have only finitely many possibilities for each $d_{i\rho}^{\pi}$. Since (5.7) shows that we have at most p^{4d} coefficients $d_{i\rho}^{\pi}$, this implies Theorem 8.

COROLLARY 8. *Let B be a p-block of a group G of given type. Let π be a p-element of G and let v be a p-regular element of $\mathfrak{C}_G(\pi)$. If π is one of the elements (5.6), and if the corresponding basic sets φ_b are known, the values of the characters $\chi_i \in B$ for the element πv are determined. If π is not conjugate in G to an element (5.6), then $\chi_i(\pi v) = 0$ for $\chi_i \in B$.*

The first statement follows from (5.3) while the second is a consequence of [*2*], (7A).

The proof of Theorem 8 does not give a practical way of finding all possible types of p-blocks, even in fairly simple cases. We shall therefore discuss some modifications.

Let B be a p-block of G and let $s(B)$ as before denote the set of indices i for which $\chi_i \in B$. It will be convenient to speak of a column $\mathfrak{z}$ belonging to B if for each $i \in s(B)$ a complex number z_i is given. These columns form a vector space with the usual operations. If $\mathfrak{z} = \{z_i\}$ and $\mathfrak{u} = \{u_i\}$ are two columns, we introduce an inner product by

$$(\mathfrak{z}, \mathfrak{u}) = \sum_{i \in s(B)} z_i \bar{u}_i .$$

If T is a ring, we say that $\mathfrak{z}$ lies in T, if all $z_i \in T$.

In discussing types of p-blocks, we may assume that the defect d is positive, since the case $d = 0$ is trivial, cf. [*1*], (6E). We take the view that the case of a fixed defect group D and a definite system (5.6) of elements of D are considered. For each $\pi = \pi_\mu$, we have to consider all blocks b of $\mathfrak{C}_G(\pi)$ with $b^G = B$. Then the defect group D_0 of b can be taken as a subgroup of D. Actually, the discussion can be reduced to that of blocks $\bar{b}$ of $\mathfrak{C}_G(\pi)/\{\pi\}$ with the defect group $\bar{D} = D_0/\{\pi\}$. If $\pi \neq 1$, i.e., if $\mu > 0$, then $|\bar{D}| < |D|$ and we may assume in the way of induction that we know the Cartan invariants for suitable basic sets of $\bar{b}$ and then of b. Now, as in the proof of Theorem 8, the formulas (5.4) can be used to discuss the possibilities for the columns $\mathfrak{d}_\rho^\pi$ with the coefficients $d_{i\rho}^\pi$, $i \in s(B)$. We do not know in general how many blocks b of $\mathfrak{C}_G(\pi)$ with $b^G = B$ exist. However, (5.7) shows that the total number of columns $\mathfrak{d}_\rho^\pi$ with $\pi \neq 1$ is less than p^{2d}. Of course, the equations (5.5) represent further necessary conditions for these columns.

Suppose that we have made a definite choice for the $d_{i\rho}^\pi$ with $\pi = \pi_1, \pi_2, \cdots, \pi_m$. It remains to discuss the columns $\mathfrak{d}_\rho^1$ with $\pi = \pi_0 = 1$. Let $\mathfrak{M}$ denote the Z-module consisting of all columns in $\mathbf{Z}$ which are orthogonal to all columns $\mathfrak{d}_\rho^\pi$ with $\pi = \pi_1, \pi_2, \cdots, \pi_m$. By (5.7), the dimension of $\mathfrak{M}$ is $l_B^{(0)}$. In fact, if we take the modular irreducible characters of B as basic set φ_B, ($\pi = 1$, $b = B$), we see that the corresponding $l_B^{(0)}$ columns $\mathfrak{d}_\rho^1$ form a $\mathbf{Z}$-basis of M. This follows from the known fact (cf. [*1*, §5]) that the greatest common divisor of all minors of degree $l_B^{(0)}$ of these columns is 1.

If we now choose an arbitrary $\mathbf{Z}$-basis $\{\mathfrak{d}_n^{(1)}\}$ of M, this means that we decide on a particular choice of the basic set φ_B. The corresponding Cartan invariants can then be obtained from (5.4).

In order to obtain further restricting conditions, we choose a generalized character θ of D and set

$$a_i(\theta) = (\chi_i \mid D, \theta) \tag{5.9}$$

for $i \in s(B)$; we have a column $\mathfrak{a}(\theta)$ in $\mathbf{Z}$. Using (5.3), we can write this in the form

$$\mid D \mid a_i(\theta) = \sum_{\pi \in D} \chi_i(\pi)\,\bar{\theta}(\pi) = \sum_{\pi} \sum_{b} \sum_{\rho} d_{i\rho}{}^{\pi} \varphi_\rho{}^{\pi}(1)\,\bar{\theta}(\pi). \tag{5.10}$$

If $\mid D \mid = p^d$ and if we know the values $\varphi_\rho{}^{\pi}(1)$, this can be considered as a congruence mod p^d for the columns $\mathfrak{d}_\rho{}^{\pi}$. We always have at least some information concerning the $\varphi_\rho{}^{\pi}(1)$. Indeed, if $p^{h(\pi)}$ divides the order of $\mathfrak{C}_G(\pi)$, it follows from [*1*], (3F), that

$$\sum_{\sigma} c_{\rho\sigma}{}^{\pi} \varphi_\sigma{}^{\pi}(1) \equiv 0 \pmod{p^{h(\pi)}} \tag{5.11}$$

where the sum extends over all $\varphi_\sigma{}^{\pi} \in \varphi_b$.

In our discussion so far, we used the estimate (5.7) for k_B. Actually, it is not difficult to show that if π is an element of the center of D, then there exists a p-block b of $\mathfrak{C}_G(\pi)$ with $b^G = B$ with the following property. For each $\chi_i \in B$, there exists a $\varphi_\rho{}^{\pi} \in b$ with $d_{i\rho}{}^{\pi} \neq 0$. This can often be used to obtain much better estimates for k_B. There are a number of further remarks which facilitate the discussion.

We note the following corollary of Theorem 8.

COROLLARY 9. *If B is a p-block of the group of a given type, if π is a p-element of G and if the group $\mathfrak{C}_G(\pi)$ is known, the values of the characters $\chi_i \in B$ for the elements πv are determined where v is a p-regular element of $\mathfrak{C}_G(\pi)$.*

As a final result, we mention

THEOREM 9. *Let G be a finite group, let p be a prime and let*

$$\pi_0 = 1, \pi_1, \cdots, \pi_m$$

be a set of representatives for the conjugate classes of G consisting of p-elements. Assume that, for $i > 0$, the group $\mathfrak{C}_G(\pi_i)$ is known and that we know which of the conjugate classes of these groups are fused in G, i.e., which of them belong to the same conjugate class of G. Then the values of the linear characters $\psi_B([K])$ for p-singular classes can be determined. (However, we are not able to say how *many blocks B of defect 0 occur). For each block b of $\mathfrak{C}_G(\pi_i)$ with $i > 0$, we can determine which ψ_B corresponds to b^G.*

If we know the $d_{i\rho}^{\pi}$ for $\pi = \pi_\mu$ with $\mu > 0$ as assumed in Theorem 8, we can then find the values of $\chi_i \in B$ for p-singular elements. Then (5.10) with $\theta = 1$ can be used to find the value of $\chi_i(1)$ (mod p^n) where p^n is the highest power of p dividing $|G|$. In a similar manner, we obtain congruences for $\chi_i(v)$ where v is a p-regular element for which $|\mathfrak{C}_G(v)|$ is divisible by p. Finally, it follows from (5.5) that

$$\sum_{i \in s(B)} \chi_i(w)\, d_{i\rho}^{\pi} = 0 \tag{5.12}$$

where w is a p-regular element of G and $\pi = \pi_\mu$ with $\mu > 0$.

References

1. Brauer, R. Zur Darstellungstheorie der Gruppen endlicher Ordnung I. *Math. Z.* **63** (1956), 406-444.
2. Brauer, R. Zur Darstellungstheorie der Gruppen endlicher Ordnung II. *Math. Z.* **72** (1959), 25-46.
3. Brauer, R., and Feit, W. On the number of irreducible characters of finite groups in a given block. *Proc. Natl. Acad. Sci.* **45** (1959), 361-365.
4. Brauer, R., and Fowler, K. A. On groups of even order. *Ann. Math.* **62** (1955), 565-583.
5. Brauer, R., and Nesbitt, C. J. On the modular characters of groups. *Ann. Math.* **42** (1941), 556-590.
6. Reiner, I., and Curtis, C. W. "Representation Theory of Finite Groups and Associative Algebras." Interscience, New York, London, 1962.
7. Iizuka, K. On Brauer's theorem on sections in the theory of group characters. *Math. Z.* **75** (1961), 299-304.
8. Landau, E. Über die Klassenanzahl der binären quadratischen Formen von negativer Diskriminante. *Math. Ann.* **56** (1903), 671-676.
9. Nagao, H. A proof of Brauer's theorem on generalized decomposition numbers. *Nagoya J. Math.* **22** (1963), 73-77.
10. Osima, M. Notes on blocks of group characters. *Math. J. Okayama Univ.* **4** (1955), 175-188.
11. Rosenberg, A. Blocks and centres of group algebras. *Math. Z.* **76** (1961), 209-216.
12. Schur, I. Über eine Klasse von endlichen Gruppen linearer Substitutionen. *Sitzber. Preuss. Akad. Wiss.* (1905), 77-91.
13. Suzuki, M. Finite groups with nilpotent centralizers. *Trans. Am. Math. Soc.* **99** (1961), 425-470.

PRINTED IN BRUGES, BELGIUM, BY THE ST CATHERINE PRESS LTD.

Reprinted from JOURNAL OF ALGEBRA, Vol. 1, No. 4, December 1964

Reprinted in Belgium

Some Applications of the Theory of Blocks of Characters of Finite Groups. II*

RICHARD BRAUER

Department of Mathematics, Harvard University, Cambridge, Massachusetts

Received April 15, 1964

I. INTRODUCTION

Since it has now been proved by W. Feit and J. G. Thompson [*19*] that finite groups of odd order are solvable, our attention will naturally turn to groups of finite even order. There are methods available here which do not have analogues in the case of an odd order.

We may pose the following question: Given a 2-group P, what can be said about the finite groups G which have P as their 2-Sylow group? Usually, it will be assumed that also information concerning the centralizers of involutions is available. Still, it seems rather surprising that nontrivial results for G can be obtained in these circumstances.

There are different ways in which the elements of P can react to G and, because of this, it is natural to subdivide our problem. For instance, for each G, we have the G-classes in P, i.e. the intersections of the conjugate classes of G with P. It is natural to assume that these G-classes of P are known. There are further conditions of this type, but it will not be necessary to go into this in this paper.

After some preliminary discussions (Sections II-IV), we discuss first two special cases. In Section VI, groups P are considered which are direct products of two cyclic groups of order $2^m > 2$. If the 2-Sylow group of P of G is a quaternion-group and if G has no normal subgroup of odd order greater than 1, it had been proved in Brauer-Suzuki [*18*] that G has a center of order 2, see also Suzuki [*21*]. The proof was complicated in the case of an ordinary quaternion group and had only been sketched. In Section VII, a simplified proof is given.

In the following sections, P may be an arbitrary 2-group. Choose arbitrarily an element π of P and two involutions (i.e., elements of order 2) y_1, y_2 of P,

* This research was supported by NSF grant G14100.

[1] We refer to the first paper of this series as I. It appeared in this Journal, vol. 1 (1964), pp. 152-167. Bracketed numbers [*1*]-[*13*] refer to the bibliography in I.

equal or distinct. There may or may not exist G-conjugates $\eta_1 \in P$ of y_1 and $\eta_2 \in P$ of y_2 such that $\eta_1\eta_2$ is G-conjugate to π. In either case, conclusions concerning G can be drawn. We shall leave the discussion of the former case to the third paper of this series and assume here that we have elements π, y_1, y_2 for which no η_1, η_2 with the stated properties exist. It will be shown that there exists a normal subgroup $N \neq G$ of G which includes the 2-regular core $\mathfrak{K}_2(G)$, (I, §2, 1), and whose index $(G : N)$ lies below a bound β depending only on the order $|P| = 2^n$ and the *core-indices* $|\mathfrak{C}_G(y_i) : \mathfrak{K}_2(\mathfrak{C}_G(y_i))|$ of the centralizers of the given involutions y_1, y_2. In particular, if G is simple, $|G| \leqslant \beta$.

If the orders of the centralizers of the involutions of G are known, and if G has more than one conjugate classes of involutions, upper bounds for $|G|$ have been given in Brauer-Fowler [*4*]. In Section IX, an analogous result is given for the case that G has only one class of involutions. However, the bounds in this case are much larger and when P is a cyclic group or a quaternion group, we can only give bounds for the core index $|G : \mathfrak{K}_2(G)|$ of G.

We shall say that a group G is *core-free* (for the prime $p = 2$), if $\mathfrak{K}_2(G) = \{1\}$. If G is arbitrary, of course $\bar{G} = G/\mathfrak{K}_2(G)$ is core-free and its 2-Sylow groups are isomorphic to P. Since $\mathfrak{K}_2(G)$ is solvable by the Feit-Thompson theorem, G and $\bar{G}$ have the same simple noncyclic constituents.

There exist 2-groups P for which there occur only finitely many non-isomorphic, core-free groups G with P as their 2-Sylow groups. Results of this nature are obtained in Section X. For instance, we may take for P the 2-Sylow groups of the simple Suzuki-groups $Su(2^{2m+1})$.

A. PREPARATIONS

II. Fusion in p-Groups

Let P be a given finite p-group. For the time being, p may be an arbitrary prime. By a *fusion* $\mathscr{F}$ in P, we mean a set of ordered pairs (x, y) of elements of P which satisfy the conditions (I), (II), and (III) below. The point is that if G is a finite group with P as its p-Sylow group, the set $\mathscr{F}(G)$ of all G-conjugate pairs of elements of P satisfies our conditions.

We require

(I) *$\mathscr{F}$ defines an equivalence relation on P.*

(II) *$\mathscr{F}(P) \subseteq \mathscr{F}$, i.e. if $x, y \in P$ are conjugate in P, then $(x, y) \in \mathscr{F}$.*

If $(x, y) \in \mathscr{F}$, we call y an $\mathscr{F}$-conjugate of x, and we can speak of the *$\mathscr{F}$-class* K of x consisting of the $\mathscr{F}$-conjugates of x. If $y \in K$ is chosen such that $|\mathfrak{C}_P(y)| \geqslant |\mathfrak{C}_P(z)|$ for $z \in K$, y will be said to be an *extreme* element.

If there exist finite groups G with P as their p-Sylow-group and if $\mathcal{F} = \mathcal{F}(G)$, then for extreme elements y, the group $\mathfrak{C}_P(y)$ is a p-Sylow group of $\mathfrak{C}_G(y)$.

Our final condition for fusions $\mathcal{F}$ is

(III) *If* $(x, y) \in \mathcal{F}$ *and if* y *is extreme, there exists an isomorphism* θ *of* $\mathfrak{C}_P(x)$ *into* $\mathfrak{C}_P(y)$ *such that* $y = \theta(x)$ *and that* $(z, \theta(z)) \in \mathcal{F}$ *for all* $z \in \mathfrak{C}_P(x)$.

It follows from the preceding remark that if groups G with P as Sylow groups exist, then $\mathcal{F} = \mathcal{F}(G)$ satisfies (III).

The conditions (I), (II), (III) imply that elements in the same $\mathcal{F}$-class have the same orders. If $(x, y) \in \mathcal{F}$, then $(x^r, y^r) \in \mathcal{F}$ for all integers r.

Given a p-group P and a fusion $\mathcal{F}$ in P, we can propose to study the groups G with P as their p-Sylow groups such that $\mathcal{F}(G) = \mathcal{F}$. No such G need exist, since there are further conditions besides (I), (II), (III) which are satisfied by all $\mathcal{F}(G)$. No necessary and sufficient conditions for the existence of G for given $\mathcal{F}$ are known.

If $\mathcal{F}$ is a fusion in the p-group P, the focal group $P^*(\mathcal{F})$ is defined as the subgroup of P generated by all quotients $x^{-1}y$ of elements x, y with $x^{-1}y \in P$. Then

$$P' \subseteq P^*(\mathcal{F}) \lhd P.$$

If there exists groups G with P as their p-Sylow group and with $\mathcal{F}(G) = \mathcal{F}$, then, as is well known, there exist normal subgroups H of G with $P^*(\mathcal{F})$ as their p-Sylow group P such that $P/P^*(\mathcal{F}) \simeq G/H$ is the maximal abelian p-quotient of G.

If Q is a normal subgroup of P and if $\bar{P} = P/Q$, we shall say that a fusion $\mathcal{F}$ in P *covers* a fusion $\bar{\mathcal{F}}$ in $\bar{P}$, if for every $(\bar{x}, \bar{y}) \in \bar{\mathcal{F}}$ there exist coset representatives $x, y \in P$ of $\bar{x}, \bar{y}$ such that $(x, y) \in \mathcal{F}$.

III. Some Lemmas

Let G be a finite group and let p be a fixed prime. Let the principal p-block $B_0 = B_0(G)$ consist of the irreducible characters χ_i of G with i ranging over the index set $s(B_0)$. If v is an element of G. we shall set

$$\lambda(G, v) = \sum_{i \in s(B_0)} |\chi_i(v)|^2 \leqslant |\mathfrak{C}_G(v)|. \tag{3.1}$$

In particular, we write $\lambda(G)$ for $\lambda(G, 1)$,

$$\lambda(G) = \lambda(G, 1) = \sum_{i \in s(B_0)} x_i^2 \tag{3.2}$$

where we wrote x_i for the degree of χ_i,

$$x_i = \chi_i(1). \tag{3.3}$$

Proposition 1. *The number* $\lambda(G, v)$ *is a positive rational integer. If* $\bar{G} = G/\mathfrak{K}_p(G)$, *then*

$$\lambda(G, v) = \lambda(\bar{G}, \bar{v}) \leqslant |\,\mathfrak{C}_{\bar{G}}(\bar{v})\,| \tag{3.4}$$

where $\bar{v} = v\mathfrak{K}_p(G)$ *is the image of* v *in* $\bar{G}$.

$$\lambda(G) = \lambda(\bar{G}) \leqslant |\, G : \mathfrak{K}_p(G)\,|\,. \tag{3.5}$$

The equality sign holds in (3.5), if and only if $\bar{G}$ *is of deficiency class* 0 *for* p *(cf. I, §4).*

Indeed, I §3, Lemma 2 shows that $\lambda(G, v)$ is a rational integer. Since $\chi_0 = 1 \in B_0$, $\lambda(G, v) > 0$. The equality in (3.4) is an immediate consequence of *I*, Theorem 1, since we can identify $B_0(G)$ and $B_0(\bar{G})$. The second part of (3.4) follows from (3.1) applied to $\bar{G}$. Taking $v = 1$, we obtain (3.5). The equality sign holds in (3.5), if and only if $B_0(\bar{G})$ if the only p-block of $\bar{G}$.

Proposition 2. ***Let*** π ***be a p-element of G and let*** v ***be a p-regular element*** *of* $\mathfrak{C}_G(\pi)$. *Then*

$$\lambda(G, \pi v) = \lambda(C_G(\pi), v) \tag{3.6}$$

and, in particular

$$\lambda(G, \pi) = \lambda(\mathfrak{C}_G(\pi)). \tag{3.7}$$

Proof. **If the notation is as in I, (2.6), we find**

$$\lambda(G, \pi v) = \sum_{i \in s(B_0)} \sum_{\rho,\sigma} d^{\pi}_{i\rho} \bar{d}^{\pi}_{i\sigma} \varphi^{\pi}_{\rho}(v) \bar{\varphi}^{\pi}_{\sigma}(v)$$

where φ^{π}_{ρ}, φ^{π}_{σ} range over the modular irreducible characters of $B_0(\mathfrak{C}_G(\pi))$, cf. I §3, Corollary 4. Now I, (2.7) yields

$$\lambda(G, \pi v) = \sum_{\rho,\sigma} c^{\pi}_{\rho\sigma} \varphi^{\pi}_{\rho}(v)\, \bar{\varphi}^{\pi}_{\sigma}(v). \tag{3.8}$$

If this formula is used for $\mathfrak{C}_G(\pi)$ instead of G and for $\pi = 1$, we obtain the same expression and this yields (3.6). For $v = 1$, we have (3.7).

Remarks. On combining (3.8) with [1], (3.10) and (3.F), we see that $\lambda(G, \pi v)$ is divisible by the highest power of p which divides $|\,\mathfrak{C}_G(\pi v)\,|$. In particular, $\lambda(G)$ is divisible by the highest power of p which divides $|\,G\,|$. Furthermore, $|\,G : \mathfrak{K}_p(G)\,|$ lies below a bound depending only on $\lambda(G)$, cf. I §3, Corollary 1.

Lemma 1. *Let H be a normal subgroup of G. Let B be a p-block of G and let* χ *be an irreducible character in B. If an irreducible constituent* ψ *of* $\chi \mid H$ *belongs to a p-block b of H, then for every character* $\psi_1 \in b$, *there exists a character* $\chi_1 \in B$ *such that* ψ_1 *is a constituent of* $\chi_1 \mid H$.

Proof. It suffices to prove this in the case that ψ and ψ_1 have a modular irreducible constituent ξ in common. Then there exists a modular irreducible constituent φ of χ such that ξ occurs in $\varphi \mid H$. Let F_φ be the modular representation of G with the character φ and let U_φ denote the corresponding indecomposable component of the regular representation. Let F_ξ and U_ξ have the analogous significance for ξ. By Clifford's theorem, $F_\varphi \mid H$ is completely reducible and F_ξ may be taken as an irreducible bottom constituent of $F_\varphi \mid H$ and hence of $U_\varphi \mid H$. By a theorem of T. Nakayama [*20*; *5*, §26], $U_\varphi \mid H$ is a direct sum of chief indecomposable representations of H. It follows that U_ξ is one of them and now Nakayama's theorem implies that the modular character ξ^* of G induced by ξ contains φ as an irreducible constituent. Consequently, the character ψ_1^* of G induced by ψ_1 contains φ as a modular constituent. It follows that there exists an irreducible character χ_1 in ψ_1^* which has φ as modular constituent and therefore belongs to B. Since by Frobenius' reciprocity theorem, ψ_1 is a constituent of $\chi_1 \mid H$, this proves the lemma.

If χ is taken as the principal character of G, Lemma 3 shows that every character ψ_1 in the principal block of H appears in the restriction of some $\chi_1 \in B_0$ to H. We now obtain easily

PROPOSITION 3. *If H is a normal subgroup of G,*

$$\lambda(G) \geqslant \lambda(H). \tag{3.9}$$

The following Lemma will be used occasionally.

LEMMA 2. *Assume that G is soluble. If the p-Sylow group of G has order p^n, there exist bounds $s(p^n)$ depending only on p^n, such that*

$$\mid G : \mathfrak{K}_p(G) \mid \leqslant s(p^n).$$

Proof. We may take $s(1) = 1$ for $n = 0$. We use induction on n. Without restriction, we may assume that $\mathfrak{K}_p(G) = \{1\}$. Since G is soluble, there exists a normal elementary abelian subgroup $S \neq \{1\}$, and as $\mathfrak{K}_p(G) = \{1\}$, the group S is a p-group, say $\mid S \mid = p^r > 1$. Now G in its action on S defines a modular representation of G of degree r in the field with p elements. The kernel is $C = \mathfrak{C}_G(S)$ and G/C is isomorphic with a subgroup of $GL(r, p)$. Hence

$$\mid G : C \mid \leqslant \mid GL(r, p) \mid.$$

Set $\mathfrak{K}_p(G/S) = L/S$. By induction,

$$\mid G : L \mid \leqslant s(p^{n-r}). \tag{3.10}$$

Since $\mid L : S \mid$ is prime to p, S is a p-Sylow subgroup of L and of $L \cap C$.

No two distinct elements of S are conjugate in $L \cap C$ and Burnside's theorem shows that $L \cap C$ has a normal p-complement V. As a characteristic subgroup of $L \cap C \lhd G$, then V is normal in G, and since V is p-regular, $V = \{1\}$. Hence $L \cap C = S$ and

$$|L : S| = |LC : C| \leqslant |GL(r, p)|. \tag{3.11}$$

Now (3.10) and (3.11) imply that $|G|$ lies below a bound depending only on p^n, and the Lemma is proved.

IV. Relations between the Values of the Characters of a Fixed p-Block

We now derive the relations which are basic for this paper. The order $|G|$ will be assumed to be even.

If $\sigma \in G$, the elements $\tau \in G$ with $\tau^{-1}\sigma\tau = \sigma^{\pm 1}$ form a subgroup $\mathfrak{C}_G^*(\sigma)$, the *extended centralizer* of σ. Either $\mathfrak{C}_G(\sigma) = \mathfrak{C}_G^*(\sigma)$, or $\mathfrak{C}_G(\sigma)$ has index 2 in $\mathfrak{C}_G^*(\sigma)$. If $\tau^{-1}\sigma\tau = \sigma^{-1}$, we say that τ *inverts* σ.

Proposition 4. *Let G be a group of even order and let p be a prime dividing $|G|$. Let π be a p-element of G and let y_1 and y_2 be involutions of G, equal or different. Assume that there do not exist G-conjugates η_1 of y_1 and η_2 of y_2 such that $\eta_1\eta_2$ has the p-factor π. Then, for each p-block B of G,*

$$\sum_{i \in s(B)} d_{i\rho}^{\pi} \chi_i(y_1)\, \chi_i(y_2)/x_i = 0 \tag{4.1}$$

where the χ_i with $i \in S(B)$ are the irreducible characters in B, $x_i = \chi_i(1)$, and where the $d_{i\rho}^{\pi}$ are the decomposition numbers belonging to fixed members φ_ρ^{π} of basic sets φ_b for the p-blocks b of $\mathfrak{C}_G(\pi)$. Also,

$$\sum_{i \in s(B)} \chi_i(\pi)\, \chi_i(y_1)\, \chi_i(y_2)/x_i = 0. \tag{4.2}$$

Proof. Let K_i denote the conjugate class of y_i, $i = 1, 2$. Let v be a p-regular element of $\mathfrak{C}_G(\pi)$ and let K be the conjugate class fo πv. Express the product $[K_1]\,[K_2]$ of the class sums of K_1, K_2 in the group algebra (I §2, 2) by class sums. Our assumption shows that πv cannot be written in the form $\eta_1\eta_2$ with $\eta_1 \in K_1$, $\eta_2 \in K_2$. Hence the coefficient γ of $[K]$ in $[K_1]\,[K_2]$ is 0. Using the well known formula for γ, we obtain

$$\sum_i \chi_i(\pi v)\, \chi_i(y_1)\, \chi_i(y_2)/x_i = 0 \tag{4.3}$$

where χ_i ranges over all irreducible characters of G. Apply now the formulas I, (2.6) to express $\chi_i(\pi v)$ by the members φ_ρ^π of basic sets φ_b for the blocks b of $\mathfrak{C}_G(\pi)$ and the decomposition numbers $d_{i\rho}^\pi$. It follows from the linear independence of the φ_ρ^π that the coefficient of each φ_ρ^π vanishes. This yields

$$\sum_i d_{i\rho}^\pi \chi_i(y_1)\, \chi_i(y_2)/x_i = 0.$$

If $\varphi_\rho^\pi \in \varphi_b$, here $d_{i\rho}^\pi = 0$ except when χ_i belongs to the block b^G of G, (I, §2, 5). This yields (4.1). The equation (4.2) is obtained on multiplying (4.1) with $\varphi_\rho^\pi(v)$ and adding over all $\varphi_\rho^\pi \in \varphi_b$ for all b with $b^G = B$.

The number of χ_i appearing in (4.2) is at most $| P |^2$. This is the advantage of (4.2) compared to (4.3).

COROLLARY 1. *If π is a p-element of G, if y_1, y_2 are involutions of G and if no G-conjugate of y_1 inverts π, then (4.1) and (4.2) hold.*

Indeed suppose that we have G-conjugates η_1 of y_1 and η_2 of y_2 such that $\eta_1\eta_2$ has the p-factor π, say $\eta_1\eta_2 = \pi v$ where v is a p-regular element of $\mathfrak{C}_G(\pi)$. Then

$$\eta_1^{-1}(\pi v)\, \eta_1 = \eta_1^{-1}\eta_1\eta_2\eta_1 = \eta_2\eta_1 = (\pi v)^{-1}.$$

Since η_1 inverts πv, it inverts the power π of πv and we have a contradiction.

We now take $p = 2$. We formulate our results in a slightly different manner.

Let P be a 2-group and let $\mathscr{F}$ be a fusion in P. Let π be an element of P and let y_1, y_2 be two involutions. As discussed in the Introduction, we are interested here in the case that the following condition (*) on (π, y_1, y_2) is satisfied:

There do not exist $\mathscr{F}$-conjugate ξ_1 of y_1 and ξ_2 of y_2 in P such that $\xi_1\xi_2$ is an extreme $\mathscr{F}$-conjugate of π. (*)

PROPOSITION 5. *Assume that P is a 2-group, $\mathscr{F}$ a fusion in P, and that elements $\pi \in P$ and the involutions y_1, $y_2 \in P$ satisfy condition (*). If G is a group with the 2-Sylow group P and fusion $\mathscr{F}(G) = \mathscr{F}$, there do not exist G-conjugates η_1 of y_1 and η_2 of y_2 such that $\eta_1\eta_2$ has the 2-factor π. In particular, (4.1) and (4.2) hold.*

Proof. Suppose that η_1, η_2 are G-conjugates of y_1, y_2 respectively and set

$$\zeta = \eta_1\eta_2 = \rho v$$

where ρ is a 2-element and where $v \in \mathfrak{C}_G(\rho)$ is 2-regular. Since η_1 and η_2 are involutions,

$$\eta_1^{-1}\zeta\eta_1 = \eta_2\eta_1 = \zeta^{-1}, \qquad \eta_2^{-1}\zeta\eta_2 = \eta_2\eta_1 = \zeta^{-1}.$$

This implies

$$\eta_1^{-1}\rho\eta_1 = \rho^{-1}, \qquad \eta_1^{-1}v\eta_1 = v^{-1}.$$

Set

$$\xi_2 = \eta_1\rho = \rho^{-1}\eta_1 = v\eta_2 .$$

Then ξ_2 is conjugate to η_2 in the dihedral group $\{\zeta, \eta_2\}$ and hence conjugate to y_2 in G. Furthermore, $\eta_1 \in \mathfrak{C}_G^*(\rho)$. Since $\{\eta_1, \rho\}$ is a 2-group, it is contained in a 2-Sylow group Q of $\mathfrak{C}_G^*(\rho)$. After replacing $\zeta = \rho v$ and Q by suitable G-conjugates, we may assume $Q \subseteq P$. Then η_1 and ρ belong to P. Now, a 2-Sylow group of $C_G(\rho) \subseteq \mathfrak{C}_G^*(\rho)$ lies in P. Hence ρ is extreme with regard to $\mathscr{F} = \mathscr{F}(G)$. Since η_1 is G-conjugate to y_1 and ξ_2 to y_2, it follows from (*) that $\rho = \eta_1\xi_2$ is not a G-conjugate of π. This proves Proposition 5.

Corollary 2. *Let $\mathscr{F}$ be a fusion in the 2-group P and let y_1, y_2 be two involutions in P. If there do not exist $\mathscr{F}$-conjugates η_1 of y_1 and η_2 of y_2 such that $\eta_1\eta_2$ is an extreme $\mathscr{F}$-conjugate of y_1y_2, then no finite group G can have the 2-Sylow group P and fusion $\mathscr{F}(G) = \mathscr{F}$.*

Remark 1. If no $\mathscr{F}$-conjugate of the involution $y_1 \in P$ inverts an extreme $\mathscr{F}$-conjugate τ of $\pi \in P$, then the condition (*) is satisfied by (π, y_1, y_2) for any choice of the involution $y_2 \in P$.

Remark 2. If the center $\mathfrak{Z}(P)$ of P contains elements π of order larger than 2, then the condition is (*) is satisfied by (π, y_1, y_2) for any choice of the involutions y_1, y_2 of P.

This follows from Remark 1, since an extreme F-conjugate τ of $\pi \in \mathfrak{Z}(P)$ lies itself in $\mathfrak{Z}(P)$.

Remark 2 shows that if no triples exist in P which satisfy condition (*), then $Z(P)$ is elementary abelian. This implies that all factor groups in the ascending central series of P are elementary abelian.

B. DISCUSSION OF SOME SPECIAL 2-GROUPS P[1]

V. Construction of Special Columns for the Principal p-Block

It happens frequently that in (4.2), the involutions y_1, y_2 are fixed while for π several (nonconjugate) p-elements can be taken. We can then obtain new equations by linear combination of the equations (4.2). Then the $\chi_i(\pi)$ will be replaced by coefficients a_i and we are interested in doing this such

[1] Part C of the paper is independent of Part B.

that the a_i are integers and that $\sum a_i^2$ is relatively small. The purpose of this section is to facilitate this work. Actually, the method can be modified and extended in a number of ways.

Let G be a finite group and let Q be a p-subgroup of G for some given prime p. If θ is a generalized character of Q and if χ_i is an irreducible character of the principal p-block B_0, set

$$a_i(\theta) = (\chi_i \mid Q, \theta) = |Q|^{-1} \sum_{\alpha \in Q} \chi_i(\alpha) \bar{\theta}(\alpha). \tag{5.1}$$

Then $a_i(\theta)$ is an integer and (5.1) defines a "column" $\mathfrak{a}(\theta)$ in $\mathbf{Z}$ for the block B_0, (I, §5).

LEMMA 3. *Let Q be a p-subgroup of G, let $N \supseteq Q$ be a subgroup of $\mathfrak{N}_G(Q)$' and let U be a subset of Q which is invariant under N. Assume that the following conditions are satisfied.*

(I) *If $\alpha \in U$ is a G-conjugate of $\beta \in Q$, there exist $\sigma \in N$, with $\beta = \alpha^\sigma$.*

(II) *For $\alpha \in U$, the number*

$$w = \lambda(\mathfrak{C}_G(\alpha))/|\mathfrak{C}_N(\alpha)| \tag{5.2}$$

is independent of α.

If θ, η are generalized characters of Q and if θ vanishes outside U, then

$$(\mathfrak{a}(\theta), \mathfrak{a}(\eta)) = (w/|Q|) \cdot \left(\theta, \sum_{\tau \in N} \eta^\tau\right). \tag{5.3}$$

Proof. If we denote the inner product on the left of (5.3) by S, we find from (5.1) that

$$|Q|^2 \cdot S = \sum_{\alpha \in Q} \sum_{\beta \in Q} (\chi(\alpha), \chi(\beta)) \bar{\theta}(\alpha) \eta(\beta)$$

where $(\chi(\alpha), \chi(\beta))$ is the inner product of the columns with the entries $\chi_i(\alpha)$, $\chi_i(\beta)$ respectively for the block B_0. By I, (2.8) and (2.6), this product vanishes, if α and β are not G-conjugate. On the other hand, if α and β are G-conjugate and hence $\chi_i(\alpha) = \chi_i(\beta)$, the formulas (3.1) and (3.7) above show that

$$(\chi(\alpha), \chi(\beta)) = (\chi(\alpha), \chi(\alpha)) = \lambda(G, \alpha) = \lambda(\mathfrak{C}_G(\alpha)).$$

Since θ vanishes outside U, it suffices to let α range over U and we obtain

$$|Q|^2 \cdot S = \sum_{\alpha \in U} \lambda(\mathfrak{C}_G(\alpha)) \bar{\theta}(\alpha) \sum_{\beta}{}' \eta(\beta)$$

where β ranges over the G-conjugates of α in Q. On account of assumption (I), β in Σ' ranges over the distinct elements of the form α^{σ} with $\sigma \in N$. Then

$$\sum_{\tau \in N} \eta(\alpha^{\tau}) = |\mathfrak{C}_N(\alpha)| \cdot \sum_{\beta}{}' \eta(\beta).$$

Using assumption (II), we find

$$|Q|^2 \cdot S = w \sum_{\alpha \in U} \bar{\theta}(\alpha) \sum_{\tau \in N} \eta^{\tau}(\alpha).$$

Since we can again let α range over Q, this yields (5.3).

Lemma 4. *Let Q be a p-subgroup of G, let U be a subset of Q and assume that there exists an involution J of G such that no $\pi \in U$ is inverted by a G-conjugate of J. If θ is a generalized character of Q which vanishes outside U, then*

$$\sum_i a_i(\theta)\, \chi_i(J)^2/x_i = 0, \tag{5.4}$$

$$\sum a_i(\theta)\, x_i = 0, \tag{5.5}$$

$$\sum a_i(\theta)\, \chi_i(J) = 0. \tag{5.6}$$

where i ranges over $s(B_0)$. If $(1, \theta) \neq 0$, then either there exist at least two positive and two negative $a_i(\theta)$ with $i \in s(B)$ or G has a proper normal subgroup $H \supseteq \{\mathfrak{K}_2(G), J\}$.

Proof. It follows from Corollary 1, Section IV that (4.2) holds for $y_1 = y_2 = J$ and all $\pi \in U$. On multiplying (4.2) with $|Q|^{-1}\bar{\theta}(\pi)$ and adding over all $\pi \in U$, we obtain (5.4), cf. (5.1). The hypothesis of Lemma 4 implies that $1 \notin U$ and that no conjugate of J belongs to U. Now I, (2.8) and (2.6) show that the columns $\chi(1)$ and $\chi(J)$ for the block B_0 are orthogonal to the column $\chi(\alpha)$ with $\alpha \in U$. Applying again (5.1), we obtain (5.5) and (5.6).

Let s be the subset of $s(B_0)$ consisting of those i for which $a_i(\theta) \neq 0$. Of course, it suffices to let i range over s in (5.4), (5.5), (5.6). Suppose now that $(1, \theta) \neq 0$. For $\chi_0 = 1$, we find from (5.1) that $a_0(\theta) = (1, \theta) \neq 0$. Thus, $0 \in s$.

If $\{u_i\}$ and $\{v_i\}$ are variables, we introduce the nonsingular symmetric bilinear form

$$T(u, v) = \sum_{i \in s} a_i(\theta)\, u_i v_i / x_i\,.$$

Interpret our variables as coordinates in a real projective space Let x be the point with the coordinates $\{x_i \mid i \in s\}$ and x' the point with the coordinates $\{\chi_i(J) \mid i \in s\}$. Now (5.5), (5.4), (5.6) read

$$T(x, x) = T(x', x') = T(x, x') = 0. \tag{5.7}$$

Hence x and x' are points of a nondegenerate quadric Γ which are conjugate with regard to Γ.

It follows already from (5.5) that not all $a_i(\theta)$ have the same sign and that s contains indices $i \neq 0$. If all but one of the $a_i(\theta)$ have the same sign, then Γ does not contain a real line and (5.7) implies that $x = x'$. For $i = 0$, we have $x_0 = \chi_0(J) = 1$. Hence $x_i = \chi_i(J)$ for all $i \in s$. Choose here $i \neq 0$ and let H denote the kernel of χ_i. Then $J \in H$, $H \lhd G$, $H \neq G$. Since I, Theorem 1 shows that $\mathfrak{K}_2(G) \subseteq H$, this completes the proof.

VI. Groups with an Abelian 2-Sylow Group of Type $(2^m, 2^m)$ with $m > 2$

THEOREM 1. *Let G be a group whose 2-Sylow group P is abelian of type $(2^m, 2^m)$ with $m \geqslant 2$. If G is core-free (i.e. if $\mathfrak{K}_2(G) = \{1\}$), then $P \lhd G$, $\mathfrak{C}_G(P) = P$, and $| G : P |$ is 1 or 3.*

Proof. Since P is abelian, any fusion of elements of P in G takes place already in $\mathfrak{N}_G(P)$. There exist exactly three involutions $J_1, J_2, J_3 = J_1 J_2$ in P. Any $\sigma \in \mathfrak{N}_G(P)$ effects a permutation $\pi(\sigma)$ of J_1, J_2, J_3. As is well known and proved easily, the mapping $\sigma \to \pi(\sigma)$ has the kernel $\mathfrak{C}_G(P)$. Since $\mathfrak{N}_G(P)/\mathfrak{C}_G(P)$ is 2-regular,

$$| \mathfrak{N}_G(P) : \mathfrak{C}_G(P) | = 1 \text{ or } 3.$$

If $\mathfrak{C}_G(P) = \mathfrak{N}_G(P)$, Burnside's theorem shows that G has a normal 2-complement W. Since G is core-free, then $W = \{1\}$, and $G = P$.

We may therefore assume that

$$\mathfrak{N}_G(P) = \{\mathfrak{C}_G(\pi), t\}.$$

where t has order 3 over $\mathfrak{C}_G(P)$ and interchanges J_1, J_2, J_3 cyclically.

Let $\alpha \neq 1$ be an element of P. If we apply the preceding argument to $\mathfrak{C}_G(\alpha)$ instead of G, then as $t \notin \mathfrak{C}_G(\alpha)$, we see that $\mathfrak{C}_G(\alpha)$ has a normal 2-complement. By (3.5) and (3.2),

$$\lambda(\mathfrak{C}_G(\alpha)) = \lambda(P) = 2^{2m} \qquad \text{for} \qquad \alpha \in P, \alpha \neq 1. \tag{6.1}$$

Let $\psi_0 = 1, \psi_1, \cdots$ denote the irreducible character of P. Set

$$r = (2^{2m} - 1)/3 \geqslant 5. \tag{6.2}$$

Since t acts without fixed points on $P - \{1\}$, the ψ_i can be arranged such that no two of

$$\psi_0, \psi_1, \cdots, \psi_r$$

are associated under t or t^{-1}, and that the remaining irreducible characters of P are $\psi_i^t, \psi_i^{t^{-1}}$ with $i = 1, 2, \cdots, r$.

We shall apply Lemma 3 with $Q = P$, $U = Q - \{1\}$ and $N = \{Q, t\}$. By (6.1) and (5.2), $w = 3/|N : P|$. It follows from (5.3) that

$$(\mathfrak{a}(\psi_i - \psi_0), \mathfrak{a}(\psi_j)) = \delta_{ij} \quad \text{for} \quad 1 \leqslant i, j \leqslant r,$$

$$(\mathfrak{a}(\psi_i - \psi_0), \mathfrak{a}(\psi_0)) = -3.$$

For

$$\mathfrak{b}_i = \mathfrak{a}(\psi_i - \psi_0) = \mathfrak{a}(\psi_i) - \mathfrak{a}(\psi_0),$$

this yields

$$(\mathfrak{b}_i, \mathfrak{b}_j) = 3 + \delta_{ij}, \qquad 1 \leqslant i \leqslant j \leqslant r. \tag{6.3}$$

If we set $\mathfrak{c}_i = \mathfrak{b}_i - \mathfrak{b}_r$ for $1 \leqslant i \leqslant r - 1$ then $(\mathfrak{c}_i, \mathfrak{c}_j) = 1 + \delta_{ij}$. An easy discussion (cf. Brauer-Leonard [*16*, p. 438]) shows that the nonzero coefficients of the $r - 1$ columns $\mathfrak{c}_i$ appear in r rows and, if the rows are taken in suitable order, the matrix in the first $r - 1$ of the rows and in the $r - 1$ columns is ϵI_{r-1} with $\epsilon = \pm 1$ while all coefficients in the r-th row are $-\epsilon$.

By (6.3), $(\mathfrak{b}_r, \mathfrak{c}_i) = -1$. It follows that $\mathfrak{b}_r$ has the same coefficient z in the first $r - 1$ of our rows and the coefficient $z + \epsilon$ in the last row. There occur further rows in which all coefficients of $\mathfrak{c}_1, \cdots, \mathfrak{c}_{r-1}$ vanish. If the coefficients of $\mathfrak{b}_r$ in these rows are $\delta_0, \delta_1, \cdots$, then (6.3) yields

$$4 = \sum \delta_j^2 + (r - 1) z^2 + (z + \epsilon)^2. \tag{6.4}$$

For the row corresponding to $\chi_0 = 1$, we have $a_0(\psi_0) = 1$, $a_0(\psi_i) = 0$ by (5.1). Hence the coefficient of $\mathfrak{b}_r$ in this row is -1, while the coefficients of all $\mathfrak{c}_j$ vanish. This shows that one δ_j in (6.4) has the value -1. It follows from (6.4) that $z = 0$ and that we have exactly three nonzero δ_j, each of which is ± 1. If we take these three rows above the r rows discussed before, we see now readily from the information obtained, that the nonzero coefficients of the columns $\mathfrak{b}_1, \mathfrak{b}_2, \cdots, \mathfrak{b}_r$ appear in rows of the following form

$$\begin{matrix} \delta_0 & \delta_0 & & \delta_0 \\ \delta_1 & \delta_1 & \cdots & \delta_1 \\ \delta_2 & \delta_2 & \cdots & \delta_2 \\ \epsilon & 0 & \cdots & 0 \\ 0 & \epsilon & \cdots & 0 \\ \vdots & \vdots & & \vdots \\ 0 & 0 & \cdots & \epsilon \end{matrix} \tag{6.5}$$

Denote the character corresponding to the $(i-1)$st row by χ_i; $\chi_0 = 1$, $\delta_0 = -1$.

In particular, (6.5) shows that $a_i(\psi_j) - a_i(\psi_0) = \delta_i$ for $i = 0, 1, 2$ and $j = 1, 2, \cdots, r$. Since $\chi_i(t^{-1}\alpha t) = \chi_i(\alpha)$ for $\alpha \in P$, it follows from (5.1) that

$$a_i(\psi_j) = a_i(\psi_j^t) = a_i(\psi_j^{t^{-1}}).$$

Hence $a_i(\psi) - a_i(\psi_0) = \delta_i$ for every nonprincipal irreducible character ψ of P and $i = 0, 1, 2$. Since (5.1) implies

$$\chi_i(\alpha) = \sum_{\psi} a_i(\psi)\,\psi(\alpha) \qquad (\text{for } \alpha \in Q = P)$$

with the sum ranging over all irreducible characters ψ of P, we obtain

$$\chi_i(\alpha) = -\delta_i \qquad (\text{for } \alpha \in P,\ \alpha \neq 1;\ 0 \leqslant i \leqslant 2). \tag{6.6}$$

As $(\chi_i \mid P, 1)$ is an integer, we see from (6.6) that

$$x_i \equiv -\delta_i \ (\mathrm{mod}\ 2^{2m}), \qquad (i = 1, 2). \tag{6.7}$$

We may choose ψ_1 as a nonprincipal character of P with $\psi_1^2 = \psi_0$. Then all elements of order less than 2^m of P belong to the kernel of ψ_1 and in

$$a_i(\psi_1 - \psi_0) = |Q|^{-1} \sum_{\alpha} \chi_i(\alpha)\,(\bar{\psi}_1(\alpha) - \psi_0(\alpha)),$$

it suffices to let α range over the elements of P of order 2^m. Since $m \geqslant 2$, and since the 2-Sylow group of G is abelian, none of these elements is inverted by a conjugate of J_1.

Now Lemma 4 can be applied to $Q = P$ with U chosen as the set of elements of order 2^m in P and $\theta = \psi_1 - \psi_0$. Taking (6.5) and (6.6) for $\alpha = J$ into account, we obtain

$$-1 + \delta_1/x_1 + \delta_2/x_2 + \epsilon\chi_3(J)^2/x_3 = 0, \tag{6.8}$$

$$-1 + \delta_1 x_1 + \delta_2 x_2 + \epsilon x_3 = 0, \qquad -3 + \epsilon\chi_3(J) = 0. \tag{6.9}$$

Suppose that $x_1 \neq 1$, $x_2 \neq 1$. By (6.7), $x_1 \geqslant 15$, $x_2 \geqslant 15$ and (6.8) yields $9\epsilon/x_3 \geqslant 13/15$ whence $\epsilon = 1$, $x_3 \leqslant 10$. Since (6.9) implies $x_3 \equiv 3 \ (\mathrm{mod}\ 16)$, we have $x_3 = 3$. Now (6.8) leads to a contradiction.

Hence $x_i = 1$ for $i = 1$ or 2. By (6.7), $\delta_i = -1$ and (6.6) shows that P belongs to the kernel H_0 of χ_i.

Then H_0 is a proper normal subgroup of G and since G is core-free so is H_0. Furthermore, H_0 still has the 2-Sylow group P and we may assume in the way of induction that the theorem holds for H_0. Hence $P \lhd H_0$ and this implies $P \lhd G$. It follows that $\mathfrak{C}_G(P) \lhd G$ and then that $\mathfrak{K}_2(\mathfrak{C}_G(P)) = \{1\}$. Since $\mathfrak{C}_G(P)$ has a normal 2-complement, we obtain $\mathfrak{C}_G(P) = P$ and

$$| G : P | = | \mathfrak{N}_G(P) : \mathfrak{C}_G(P) | = 1 \text{ or } 3, \qquad \text{q.e.d.}$$

For applications in Section VII, we add some results for the case that P is elementary abelian of type (2, 2). Theorem 1 no longer is valid. In fact, there exist infinitely many simple groups of type $LF(2, q)$ with the 2-Sylow group P.

PROPOSITION 6. *Let G be a finite group with a 2-Sylow group P which is elementary abelian of type (2, 2). Either G has three or one conjugate class of involutions. In the former case, G has a normal 2-complement and the principal 2-block B_0 of G consists of the four characters of $G/\mathfrak{K}_2(G) \cong P$ of degree 1. In the latter case, B_0 consists of four characters $\chi_0 = 1, \chi_1, \chi_2, \chi_3$ such that $\chi_i(\sigma) = \epsilon_i = \pm 1$ for 2-singular elements of G. Moreover,*

$$x_i = X_i(1) \equiv \epsilon_i \pmod 4 \qquad \textit{for} \qquad i = 1, 2, 3$$

and

$$1 + \epsilon_1\chi_1(v) + \epsilon_2\chi_2(v) + \epsilon_3\chi_3(v) = 0 \tag{6.10}$$

for 2-regular elements v of G.

Proof. Again, $| \mathfrak{N}_G(P) : \mathfrak{C}_G(P) | = 1$ or 3. In the former case, G has a normal 2-complement and the statements for this case are immediate. In the latter case, G has only one class of involutions. If J is an involution, $\mathfrak{C}_G(J)$ has a normal 2-complement. We apply the results of I, Section 2 and 3. In particular, I (2.6) for $B = B_0$, $\pi = J$ reads

$$\chi_i(Jv) = d_{i0}^J, \qquad (i \in s(B)).$$

By I (2.7), $(\mathfrak{d}_0^J, \mathfrak{d}_0^J) = 4$ for the corresponding column. We cannot have $d_{i0}^J = 0$ since then χ_i would vanish for all 2-singular elements and would have defect 0. Since $d_{i0}^J \in \mathbf{Z}$, it is now clear that B_0 consists of four irreducible characters χ_i and that if we set $\epsilon_i = d_{i0}^J$, then $\epsilon_i = \pm 1$. The relation (6.10) is a consequence of I (2.6) and (2.7) with $\pi = J$, $\pi' = 1$. The congruence for x_i is obtained by considering $(\chi_i \mid P, 1)$.

Remark. Assume that G has one class of involutions in Proposition 6. For $v = 1$, (6.10) reads

$$1 + \epsilon_1 x_1 + \epsilon_2 x_2 + \epsilon_3 x_3 = 0. \tag{6.11}$$

Since $x_i > 0$, at least one ϵ_i with $i > 0$ is positive and one is negative, say $\epsilon_1 = 1$, $\epsilon_3 = -1$. (6.10) shows that we can choose $1, \chi_1, \epsilon_2\chi_2$ as basic set for B_0 (I, Section V). If this is done, the corresponding matrices of decomposition numbers and Cartan invariants are

$$D = \begin{pmatrix} 1 & 0 & 0 \\ 0 & 1 & 0 \\ 0 & 0 & \epsilon_2 \\ 1 & 1 & 1 \end{pmatrix}, \qquad C = \begin{pmatrix} 2 & 1 & 1 \\ 1 & 2 & 1 \\ 1 & 1 & 2 \end{pmatrix} = (1 + \delta_{ij}). \tag{6.12}$$

VII. Groups Whose 2-Sylow Group is a Quaternion Group

THEOREM 2. *Let G be a finite core-free group whose 2-Sylow group P is a quaternion group. Then G has a center of order 2,* [*18, 21*].

Proof. Let P have order 2^n, $P = \{\sigma, \tau\}$ with

$$\sigma^{2^{n-1}} = 1, \qquad \tau^2 = \sigma^{2^{n-2}}, \qquad \tau^{-1}\sigma\tau = \sigma^{-1}.$$

We set $\tau^2 = J$.

We first deal with the case $n \geqslant 4$. Apply Lemma 3 with $Q = \{\sigma\}$, $N = P$, taking for U the set of elements of orders 2^{n-1} and $2^{n-2} \geqslant 4$ in Q. Since the center of P has order 2, the group $\mathfrak{C}_G(\alpha)$ for $\alpha \in U$ cannot contain any 2-Sylow group of G. Hence the cyclic group Q is a 2-Sylow group of $\mathfrak{C}_G(\alpha)$ and $\mathfrak{C}_G(\alpha)$ has a normal 2-complement for $\alpha \in U$. Thus, $\lambda(\mathfrak{C}_G(\alpha)) = 2^{n-1}$ and, by (5.2), $w = 1$. If $\alpha \in U$ is conjugate in G to an element $\beta \in Q$, then β must be a power α^r with odd exponent r. If $g^{-1}\alpha g = \alpha^r$ with $g \in G$, the 2-regular factor of g commutes with α and we may assume that g is a 2-element. Now, $\{g, \alpha\}$ belongs to a 2-Sylow group. However, the only powers of an element α of a quaternion group P_1 conjugate to α in P_1 are $\alpha^{\pm 1}$. It is now clear that the condition (I) of Lemma 3 is satisfied too.

Let ψ denote the character of degree 1 of Q for which $\psi(\sigma) = \sqrt{-1}$ and take $\theta = \psi - 1$. Then θ vanishes on $Q - U$ and (5.3) yields

$$(\mathfrak{a}(\theta), \mathfrak{a}(\theta)) = (\psi - 1, \psi - 1 + \psi^{\tau} - 1) = 3.$$

Since $\mathfrak{a}(\theta)$ is a column in $\mathbf{Z}$, it has three nonzero coefficients, each of which is ± 1. Now Lemma 4 applies and G has a proper normal subgroup H containing J. Since G is core-free, so is H. The 2-Sylow group $P_0 = P \cap H$ of H can contain only one involution and hence is either cyclic or a quaternion group. In the former case, H has a normal 2-complement and $H = P_0$. In the latter case, we may assume by induction that $\mathfrak{Z}(H) = \{J\}$. In both cases, J is the only involution in H and this implies $J \in \mathfrak{Z}(G)$. On the other hand, $\mathfrak{Z}(G)$ is a 2-group since $\mathfrak{R}_2(G) = \{1\}$. Hence $\mathfrak{Z}(G) \subseteq \mathfrak{Z}(P) = \{J\}$ and we have $\mathfrak{Z}(G) = \{J\}$.

It remains to deal with the case $n = 3$ where P is an ordinary quaternion group. The conjugate classes of P are represented by

$$1, \quad J, \quad \sigma, \quad \tau, \quad \sigma\tau.$$

If no distinct of these elements are G-conjugate, G has a normal 2-complement and the theorem is trivial. Suppose then that two of the elements are conjugate in G, say $g^{-1}\sigma g = \tau$ with $g \in G$. Since P is a 2-Sylow group of the extended centralizers of σ and τ and since τ is conjugate to τ^{-1} in P, a Sylow group argument shows that we may assume that $g \in \mathfrak{N}_G(P)$. Then g has odd order over $P\mathfrak{C}_G(P)$, and it permutes the three cyclic groups $\{\sigma\}$, $\{\tau\}$, $\{\sigma\tau\}$ cyclically;

$$g^{-1}\sigma g = \tau, \qquad g^{-1}\tau g = (\sigma\tau)^{\pm 1}. \tag{7.1}$$

This shows that G has only one class of elements of order 4. The group $\mathfrak{C}_G(\sigma)$ has the cyclic 2-Sylow group $\{\sigma\}$, and, consequently, it has a normal 2-complement. By I, Section III, Corollary 3, the modular principal character is the only modular irreducible character in the principal 2-block of $\mathfrak{C}_G(\sigma)$. The corresponding Cartan invariant is 4. Now, I (2.6) with $\pi = \sigma$, $v = 1$ yields

$$\chi(\sigma) = \mathfrak{d}_0^\sigma \tag{7.2}$$

where $\chi(\sigma)$ and $\mathfrak{d}_0^\sigma$ are the columns for $B_0 = B_0(G)$ with the entries $\chi_i(\sigma)$ and d_{i0}^σ respectively; $i \in s(B_0)$. By I (2.7), (2.8)

$$(\mathfrak{d}_0^\sigma, \mathfrak{d}_0^\sigma) = 4, \qquad (\mathfrak{d}_0^\sigma, \chi(1)) = 0. \tag{7.3}$$

We also need information concerning the columns $\mathfrak{d}_0^J$ of B_0. We first have to construct a basic set φ_b for the principal 2-block b of $\mathfrak{C}_G(J)$. As discussed in I Section V, we can replace this group by $\bar{G} = \mathfrak{C}_G(J)/\{J\}$. If bars indicate residue classes mod $\{J\}$, the 2-Sylow group $\bar{P}$ of $\bar{G}$ is elementary abelian $\bar{P} = \{\bar{\sigma}, \bar{\tau}\}$. Since (7.1) implies that g commutes with $\sigma^2 = \tau^2 = J$, we have $g \in \mathfrak{C}_G(J)$, and $\bar{g} \in \bar{G}$ interchanges $\bar{\sigma}$, $\bar{\tau}$, $\bar{\sigma}\bar{\tau}$ cyclically. Hence $\bar{G}$ has one class of involutions and Proposition 6 and the remark at the end of Section VI apply. It follows that the basic set for $B_0(\bar{G})$ consists of three functions $\psi_0 = 1, \psi_1, \psi_2$ and that we may assume that the Cartan invariants are $1 + \delta_{ij}$. In the case of $B_0(\mathfrak{C}_G(J))$, the same ψ_0, ψ_1, ψ_2 can be used. Here the Cartan invariants are twice those for the quotient group $\bar{G}$. Now the formulas I (2.6) for $\pi = J$, $v = 1$ read

$$\chi_i(J) = \sum_{j=0}^{2} d_{ij}^J \psi_j(1), \qquad (i \in s(B_0)), \tag{7.4}$$

(cf. I Section III, Corollary 4). Moreover, for the columns $\mathfrak{b}_j^J$ appearing here, we find from I (2.7) and (2.8) that

$$(\mathfrak{b}_i^J, \mathfrak{b}_j^J) = 2c_{ij}^J = 2(1 + \delta_{ij}), \tag{7.5}$$

$$(\chi(1), \mathfrak{b}_i^J) = (\mathfrak{b}_0^\sigma, \mathfrak{b}_i^J) = 0. \tag{7.6}$$

The $\psi_\rho(1)$ are odd integers. Since $\chi_i(J) \equiv \chi_i(\sigma) \pmod 2$, it follows from (7.2) and (7.4) that

$$\mathfrak{b}_0^\sigma + \mathfrak{b}_0^J + \mathfrak{b}_1^J + \mathfrak{b}_2^J \equiv 0 \pmod 2.$$

If we set

$$\begin{aligned} 2\mathfrak{u}_1 &= \mathfrak{b}_0^\sigma + \mathfrak{b}_0^J - \mathfrak{b}_1^J - \mathfrak{b}_2^J \\ 2\mathfrak{u}_2 &= \mathfrak{b}_0^\sigma + \mathfrak{b}_0^J - \mathfrak{b}_1^J + \mathfrak{b}_2^J, \\ 2\mathfrak{u}_3 &= \mathfrak{b}_0^\sigma + \mathfrak{b}_0^J + \mathfrak{b}_1^J - \mathfrak{b}_2^J \end{aligned} \tag{7.7}$$

$\mathfrak{u}_1, \mathfrak{u}_2, \mathfrak{u}_3$ are columns in $\mathbf{Z}$ for B_0. It follows easily from (7.3), (7.5), and (7.6) that

$$(\chi(1), \mathfrak{u}_j) = 0, \qquad (\mathfrak{b}_0^\sigma, \mathfrak{u}_j) = 2, \qquad (\mathfrak{u}_i, \mathfrak{u}_j) = 1 + 2\delta_{ij}.$$

In particular, this shows that each $\mathfrak{u}_i$ has three nonzero coefficients and that these coefficients are ± 1. A coefficient 1 appears in the row corresponding to $\chi_0 = 1$, which we choose as the first row. We may assume that the two other nonzero coefficients δ_1 and δ_2 of $\mathfrak{u}_1$ appear in the second and third row. Moreover, since $(\mathfrak{b}_0^\sigma, \mathfrak{u}_1) = 2$, we may assume that the coefficients of $\mathfrak{b}_0^\sigma$ in the first three rows are $1, \delta_1, 0$.

If $\mathfrak{u}_i$ with $i = 1$ or 2 has a nonzero coefficient in the second or third row, we see easily from $(\mathfrak{u}_1, \mathfrak{u}_i) = 1$ that $\mathfrak{u}_1 - \mathfrak{u}_i$ has only one nonzero coefficient. This is impossible, since $(\chi(1), \mathfrak{u}_1 - \mathfrak{u}_i) = 0$. Hence the coefficients of $\mathfrak{u}_2$ and $\mathfrak{u}_3$ in the second and third row vanish. We can now find the coefficients of $\mathfrak{b}_0^J, \mathfrak{b}_1^J$, and $\mathfrak{b}_2^J$ from (7.7) and using (7.4), we obtain

$$\chi_1(J) = -\delta_1(1 + \psi_1(1) + \psi_2(1)), \qquad \chi_2(J) = -\delta_2(\psi_1(1) + \psi_2(1)).$$

It follows that

$$1 + \delta_1\chi_1(J) - \delta_2\chi_2(J) = 0. \tag{7.8}$$

The column $\mathfrak{u}_1$ is a linear combination of columns $\chi(\pi)$ with $\pi \in P$, $\pi \neq 1$. Since J is the only involution in P, the condition (*) is satisfied for (π, J, J) and hence (4.2) holds for $y_1 = y_2 = J$. In (4.2), we may replace $\chi_i(\pi)$ by the ith coefficient of $\mathfrak{u}_1$. On account of the form of $\mathfrak{u}_1$, we then have

$$1 + \delta_1\chi_1(J)^2/x_1 + \delta_2\chi_2(J)^2/x_2 = 0. \tag{7.9}$$

Consider the quadric Γ with the equation

$$1 + \delta_1 z_1^2/x_1 + \delta_2 z_2^2/x_2 = 0.$$

By (7.9), the point $(1, \chi_1(J), -\chi_2(J))$ lies on Γ. Since $(u_1, \chi(1)) = 0$, also $(1, x_1, x_2)$ lies on Γ. Finally, (7.8) shows that the two points are conjugate. As in the proof of Lemma 4, we conclude that the two points are equal and that $\chi_1(J) = x_1$. Hence J belongs to the kernel H of χ_1. We can now finish the proof as in the case $n \geqslant 4$.

C. GENERAL RESULTS

VIII. Existence of Normal Subgroups

Proposition 7. *Let G be a group of even order with the 2-Sylow group P. Assume that condition (*) is satisfied for the element $\pi \in P$, the involutions y_1, y_2 of G, and the fusion $\mathscr{F} = \mathscr{F}(G)$. Then there exists an irreducible character $\chi_\mu \neq 1$ in the principal 2-block B_0 of G such that*

$$x_\mu^2 = \chi_\mu^2(1) = \gamma\lambda(\mathfrak{C}_G(y_1))\,\lambda(\mathfrak{C}_G(y_2)) \tag{8.1}$$

where γ lies below a bound $\gamma_0(2^\nu)$ depending only on the order 2^ν of the 2-Sylow group of $\mathfrak{C}_G(\pi)$. Moreover, for any 2-regular element v of $\mathfrak{C}_G(\pi)$, we have $\gamma \leqslant \lambda(\mathfrak{C}_G(\pi), v)$ and, in particular, $\gamma \leqslant \lambda(\mathfrak{C}_G(\pi))$.

Proof. It follows from Proposition 5 that (4.1) and (4.2) hold. On account of I, Theorem 7, Corollary 7, we may choose the basic set φ_b for the principal 2-block b of $\mathfrak{C}_G(\pi)$ such that $\varphi_0^\pi = 1 \in \varphi_b$ and that all Cartan invariants $c_{\rho\rho}^\pi$ lie below a bound $\gamma_0(2^\nu)$ depending only on 2^ν. We take $\rho = 0$ in (4.1). The term corresponding to the principal character χ_0 is 1. It follows that there exists a subset s of $s(B_0)$ with $0 \notin s$ such that

$$1 \leqslant \left| \sum_{i \in s} d_{i0}^\pi\, \chi_i(y_1)\, \chi_i(y_2)/x_i \right|$$

Let x_μ denote the minimal value of x_i with $i \in s$. Using Cauchy's inequality and I (2.7), we obtain

$$x_\mu^2 \leqslant c_{00}^\pi \sum_{i \in s} \chi_i(y_1)^2\, \chi_i(y_2)^2 \leqslant \gamma_0(2^\nu) \sum_{i \in s} \chi_i(y_1)^2 \sum_{i \in s} \chi_i(y_2)^2.$$

Summing over all $i \in s(B_0)$ on the right and using (3.1), we find

$$x_\mu^2 \leqslant \gamma_0(2^\nu)\,\lambda(G, y_1)\,\lambda(G, y_2).$$

Now, (3.7) yields (8.1) with $\gamma \leqslant \gamma_0(2^\nu)$.

An analogous argument applied to (4.2) gives (4.1) with

$$\gamma \leqslant \lambda(G, \pi v) = \lambda(C_G(\pi), v).$$

We can modify the procedure slightly by taking first the real parts in (4.2). The sum consisting of the negative terms then is less than or equal to -1. We now see that χ_μ in (8.1) can be chosen such that

$$\operatorname{Re}\,(\chi_\mu(\pi v)\,\chi_\mu(y_1)\,\chi_\mu(y_2)) < 0. \tag{8.2}$$

and that $\gamma \leqslant \lambda(\mathfrak{C}_G(\pi), v)$.

THEOREM 3. *Let G be a group of even order with the 2-Sylow group P; $|P| = 2^n$. Assume that condition (*) is satisfied for the element $\pi \in P$, the involutions y_1, y_2 of P and the fusion $\mathscr{F} = \mathscr{F}(G)$. There exists a normal subgroup H of G such that*

$$\mathfrak{K}_2(G) \subseteq H \subset G; \tag{8.3}$$

$$|\,G : H\,| < [2^{2n+1}\gamma^{1/2}\lambda(\mathfrak{C}_G(y_1))^{1/2}\,\lambda(\mathfrak{C}_G(y_2))^{1/2}]!\,. \tag{8.4}$$

Here, $\gamma \leqslant \gamma_0(2^\nu)$ where 2^ν is the order of the 2-Sylow *group of $\mathfrak{C}_G(\pi)$ and $\gamma \leqslant \lambda(\mathfrak{C}_G(\pi), v)$ for any 2-regular $v \in \mathfrak{C}_G(\pi)$.*

Remarks. On account of (3.5), we may replace $\lambda(\mathfrak{C}_G(y_j))$ in Theorem 1 by the core index $|\,\mathfrak{C}_G(y_j) : \mathfrak{K}_2(\mathfrak{C}_G(y_j))\,|$ of $\mathfrak{C}_G(y_j)$. Also, $\lambda(\mathfrak{C}_G(\pi), v)$ may be replaced by $|\,\mathfrak{C}_G(\pi v)\,|$. As shown in Brauer [*17*], explicit values for $\gamma_0(2^\nu)$ can be given. They are very large. We may conjecture that $\gamma_0(2^\nu) = 2^\nu$ suffices.

Proof of Theorem 3. Choose χ_μ as in Proposition 4. All algebraic conjugates of χ_μ lie in B_0. Let Ξ be their sum. Then Ξ has rational values. If H is the kernel of Ξ, then Ξ can be considered as the character of a faithful representation of G/H, and a theorem of I. Schur [*12*] yields $|\,G : H\,| \leqslant (2\,\Xi(1))!$. Now, B_0 contains at most 2^{2n} irreducible characters. Hence $\Xi(1) \leqslant 2^{2n}x_\mu$ and the statement is an immediate consequence of Proposition 7.

On combining Theorem 3 and Lemma 2, we have the

COROLLARY 1. *If in Theorem 3, the groups $\mathfrak{C}_G(y_1)$ and $\mathfrak{C}_G(y_2)$ are solvable, the index $|\,G : H\,|$ lies below a bound $\gamma^*(n)$ which depends only on n. In particular, if G is simple, $|\,G\,| \leqslant \gamma^*(n)$.*

COROLLARY 2. *If we replace* γ *in (8.4) by* $\lambda(\mathfrak{C}_G(\pi), v)$ *with* $v \in \mathfrak{C}_G(\pi)$, v *2-regular, and if Theorem 3 applies with* $y_1 = y_2$, *we can impose the condition* $\pi v \notin H$ *in the theorem.*

This is an immediate consequence of the remark following Proposition 7 since (8.2) shows that $\chi_\mu(\pi v) \neq x_\mu$ and hence πv does not belong to the kernel of χ_μ.

If more information was available about the Cartan invariants, a similar argument could be used in connection with the discussion of (4.2) in order to show that π, y_1, y_2 cannot all belong to H. We must use a detour and weaken (8.4) in order to obtain a result of this nature.

THEOREM 4. *Let* G, P, π, y_1, y_2 *satisfy the same assumptions as in Theorem 3. There exists a bound* β *depending only on* $|P| = 2^n$ *and* $\lambda(\mathfrak{C}_G(y_1))$ *and* $\lambda(\mathfrak{C}_G(y_2))$ *with the following property: There exists a normal subgroup* $H \supseteq \mathfrak{K}_2(G)$ *which does not contain all three elements* π, y_1, y_2 *and for which*

$$(G : H) \leqslant \beta.$$

Proof. Suppose that for some positive integer r, we have a bound $\beta^{(r)}$ depending only on r, n, $\lambda(\mathfrak{C}_G(y_1))$, and $\lambda(\mathfrak{C}_G(y_2))$ with the following property: There exist r nonprincipal characters $\chi_1, \chi_2, \cdots, \chi_r$ in B_0 such that if H_r is the intersection of their kernels, then

$$|G : H_r| \leqslant \beta^{(r)}. \tag{8.5}$$

For $r = 1$, the existence of $\beta^{(1)}$ follows from Proposition 7 and Theorem 3. Suppose that H_r still contains π, y_1, y_2. Then these elements belong to the 2-Sylow group $P_r = P \cap H_r$, and it is clear that they satisfy condition (*) in P_r for $\mathscr{F}_r = \mathscr{F}(H_r)$. Apply our result to H_r instead of G. By Proposition 3, $\lambda(\mathfrak{C}_{H_r}(y_i)) \leqslant \lambda(\mathfrak{C}_G(y_i))$. Now, Proposition 7 shows that there exists a character $\psi \neq 1$ in the principal 2-block $B_0(H_r)$ of H_r such that $\psi(1)$ lies below a bound β^* depending only on n, $\lambda(\mathfrak{C}_G(y_1))$, $\lambda(\mathfrak{C}_G(y_2))$. The restrictions $\chi_i \mid H_r$ are trivial for $i = 0, 1, \cdots, r$. On account of Lemma 1, ψ is a constituent of the restriction $\chi \mid H_r$ of a character $\chi \in B_0$. Take $\chi_{r+1} = \chi$. Since χ_{r+1} is a constituent of the character of G induced by ψ, it follows from (8.5) that

$$\chi_{r+1}(1) \leqslant \beta^* \beta^{(r)}.$$

If H^* is the kernel of χ_{r+1}, we can give an estimate for $|G : H^*|$ as in the proof of Theorem 3. It is then clear for $H_{r+1} = H^* \cap H_r$ that we have an inequality (8.5) for $r + 1$ instead of r with a suitable $\beta^{(r+1)}$ of the required kind.

Since B_0 contains at most 2^{2n} characters χ_j, this process must terminate. Hence we find an H_r with $r \leqslant 2^{2n}$ which cannot contain π, y_1, y_2. This yields the statement.

Remark. In Theorem 4, we can replace $\lambda(\mathfrak{C}_G(y_i))$ by the core index $(\mathfrak{C}_G(y_i) : \mathfrak{K}_2\mathfrak{C}_G(y_i))$ of $\mathfrak{C}_G(y_i)$.

This follows again from (3.5). Since the core index lies below a bound depending on $\lambda(\mathfrak{C}_G(y_i))$, (cf. I Section III, Corollary 1) and since β is very large anyway, it does not make much difference in which form the result is stated.

IX. Relations between the Order of G and the Orders of Centralizers of Involutions

We treat the two factors $|\mathfrak{K}_2(G)|$ and $|G : \mathfrak{K}_2(G)|$ of $|G|$ separately.

PROPOSITION 8. *If y_1 and y_1 are two distinct commuting involutions of G, and if we set $Q = \{y_1, y_2\}$, $H = \{K_2(G), Q\}$, $y_3 = y_1y_2$, then*

$$|H| \cdot |\mathfrak{C}_H(Q)|^2 = \prod_{i=1}^{3} |\mathfrak{C}_H(y_i)|. \tag{9.1}$$

Proof. Let L_i denote the conjugate class of y_i in H. Clearly, L_1, L_2, L_3 are distinct. Since the 2-Sylow group Q of H meets L_i in one element y_i, every 2-Sylow group of H meets each L_i in exactly one element.

Let η_2 be an element of L_2. Then y_1 inverts $y_1\eta_2$ and $D = \{y_1, y_1\eta_2\}$ is dihedral. If the order of $y_1\eta_2$ was odd, the two involutions $y_1 \in L_1$ and $\eta_2 \in L_2$ would be conjugate in D and hence in H, which is not true. Hence there exists an involution ζ which is a power of $y_1\eta_2$. Then ζ commutes with y_1 and η_2 and

$$A = \{y_1, \zeta\}, \qquad B = \{\eta_2, \zeta\}$$

are two 2-Sylow groups of H. It follows that $\zeta \notin L_1, L_2$ and hence $\zeta \in L_3$. Here, ζ is uniquely determined by y_1, η_2 for if an involution $\zeta' \neq y_1, \eta_2$ commutes with y_1 and η_2, then also $\zeta' \in L_3$. Also, ζ' commutes with ζ which is a power of $y_1\eta_2$. If $\zeta \neq \zeta'$, the 2-Sylow group $\{\zeta, \zeta'\}$ of H would meet L_3 in more than one element which is not true.

It follows that if we take a 2-Sylow group $A \supseteq \{y_1\}$ of H and a 2-Sylow group B of H containing $A \cap L_3$, then $B \cap L_2$ consists of one element η_2 and each $\eta_2 \in L_2$ is obtained once in this fashion. Since A is an arbitrary 2-Sylow group of $\mathfrak{C}_H(y_1)$ we have $|\mathfrak{C}_H(y_1) : \mathfrak{C}_H(Q)|$ possibilities for A. Likewise, if $\zeta \in L_3$, there exist $|C_H(y_3) : C_H(Q)|$ Sylow groups B of H containing ζ. Hence the number of elements of L_2 is

$$|\mathfrak{C}_H(y_1) : \mathfrak{C}_H(Q)| \cdot |\mathfrak{C}_H(y_3) : \mathfrak{C}_H(Q)|.$$

On the other hand, L_2 consists of $|H : C_H(y_2)|$ elements, and we obtain (9.1).

If we set $K = \mathfrak{K}_2(G)$, (9.1) can be written as

$$|K| \cdot |\mathfrak{C}_K(Q)|^2 = \prod_{i=1}^{3} |\mathfrak{C}_K(y_i)|. \tag{9.1*}$$

Remark. If K is a group of odd order and if y_1, y_2 are two distinct commuting involutory automorphisms of K, we may form the semidirect product H of K and $\{y_1, y_2\}$. Then (9.1*) can be applied and yields a relation between fixed point numbers on K. This is a special case of a general result of Wielandt [*23*]. We may also obtain (9.1*) as a special case of the results of Brauer [*15*]. It seemed worthwhile to give a direct elementary proof.

COROLLARY 1. *Let G be a group of even order with the 2-Sylow group P and assume that P is neither cyclic nor a quaternion group. Let y_1 and y_2 be two distinct commuting involutions in P. If we set $K = \mathfrak{K}_2(G)$*, then

$$|K| \leqslant |\mathfrak{C}_K(y_1)| \cdot |\mathfrak{C}_K(y_2)| \cdot |\mathfrak{C}_K(y_1y_2)|. \tag{9.2}$$

If P is neither cyclic nor a quaternion group we can choose y_1 as an involution in $\mathfrak{Z}(P)$ and y_2 as an involution of P which is different from y_1. Then a relation (9.2) holds. In the excluded cases, $|\mathfrak{K}_2(G)|$ can be arbitrarily large while the centralizer of each involution of G is core-free.

We now turn to the investigation of $G/\mathfrak{K}_2(G)$.

THEOREM 5. *Let G be a group of even order. Set*

$$\mu = \mathrm{Max}\,(|\mathfrak{C}_G(J) : \mathfrak{K}_2(\mathfrak{C}_G(J))|) \tag{9.3}$$

where J ranges over all involutions of G. There exists a bound $\beta_1(\mu)$ depending only on μ with the following property: All factor groups F_ρ of even order in a composition series of G with at most one exception have orders less than $\beta_1(\mu)$. If an exceptional F_ρ occurs, there exists a unique normal subgroup $T \supset \mathfrak{K}_2(G)$ of G for which $T/\mathfrak{K}_2(G) \simeq F_\rho$.

Proof. We first show that it suffices to prove the statement for core-free groups G, i.e. for groups for which $\mathfrak{K}_2(G) = \{1\}$. Indeed, if the theorem is true in this case, it holds for $\bar{G} = G/\mathfrak{K}_2(G)$. If $\bar{\mu}$ plays the same role for $\bar{G}$ as μ does for G and if we can show that

$$\bar{\mu} = \mu, \tag{9.4}$$

the proposition will hold for G.

Set $K = \mathfrak{K}_2(G)$. Let $\bar{J}$ be an involution of $\bar{G}$. There exist involutions J of G which correspond to $\bar{J}$ under the residue class map (mod K). Let W be the reciprocal image in G of $\mathfrak{C}_{\bar{G}}(\bar{J})$. Then W consists of all $w \in G$ for which

$$J^{-1}wJ = wk \qquad \text{with} \qquad k \in K.$$

Since $J^2 = 1$, we have $J^{-1}kJ = k^{-1}$. If we set $k = k_0^2$, k_0 a suitable power of k, we see that $wk_0 \in \mathfrak{C}_G(J)$. It follows that $W = \mathfrak{C}_G(J)\,K$ and this implies $\mathfrak{K}_2(W) = \mathfrak{K}_2(\mathfrak{C}_G(J))\,K$. Since the residue class map (mod K) maps $\mathfrak{K}_2(W)$ onto $\mathfrak{K}_2(C_{\bar{G}}(\bar{J}))$, we have

$$| \mathfrak{C}_G(J) : \mathfrak{K}_2(\mathfrak{C}_G(J)) | = | \mathfrak{C}_{\bar{G}}(\bar{J}) : \mathfrak{K}_2(\mathfrak{C}_{\bar{G}}(\bar{J})) |. \tag{9.5}$$

This implies (9.4). Note also

$$| \mathfrak{C}_{\bar{G}}(\bar{J}) | = | W : K | = | \mathfrak{C}_G(J) : \mathfrak{C}_K(J) |. \tag{9.5*}$$

Assume now $\mathfrak{K}_2(G) = \{1\}$. If P is again a 2-Sylow group of G and $| P | = 2^n$, clearly $2^n \leqslant \mu$. Let m denote the number of all ordered triples (π, y_1, y_2) where $\pi \in P$ and where y_1, y_2 are involutions of P. If (π, y_1, y_2) satisfies the condition (*) for $\mathscr{F} = \mathscr{F}(G)$, we may apply Theorem 4 and construct the corresponding normal subgroup H of G. Since

$$\lambda(\mathfrak{C}_G(y_i)) \leqslant | \mathfrak{C}_G(y_i) : \mathfrak{K}_2(\mathfrak{C}_G(y_i)) | \leqslant \mu,$$

the index $| G : H |$ lies below a bound $\beta_0(\mu)$ depending only on μ. Let S be the intersection of all normal subgroups H obtained for the different possible choices of π and y_1, y_2. Then $S \lhd G$,

$$| G : S | \leqslant \beta_0(\mu)^m. \tag{9.6}$$

Moreover for each choice of π, y_1, y_2 such that (*) is satisfied, the 2-Sylow group $Q = P \cap S$ cannot contain all three elements π, y_1, y_2.

If S has odd order, $S = \{1\}$ and Theorem 5 will be true, if $\beta_1(\mu) \geqslant \beta_0(\mu)^m$. Assume then that S has even order. Let T be a minimal normal subgroup of G included in S. Since $\mathfrak{K}_2(T)$ is characteristic in T, it follows that $\mathfrak{K}_2(T) = \{1\}$. If y is an involution in T, clearly T is the group generated by the conjugate class of y in G. We may choose $y \in Q$. For any $\pi \in Q$, the condition (*) does not hold for (π, y, y). It follows that a G-conjugate of π is a product of two G-conjugates of y. As $T \lhd G$, we see that $\pi \in T$. Hence $Q \subseteq T$ and this implies

$$| S : T | \not\equiv 0 \pmod 2 \tag{9.7}$$

In particular, this implies that T is unique.

The group T is a direct product of isomorphic simple groups, say

$$T = T_1 \times T_2 \times \cdots \times T_r. \tag{9.8}$$

We see easily that $\mathfrak{K}_2(T_1) = \{1\}$.

If $r > 1$, and if we choose $y \in T_1$, we have

$$\mathfrak{C}_T(y) = \mathfrak{C}_{T_1}(y) \times T_2 \times \cdots \times T_r,$$

$$\mathfrak{K}_2(\mathfrak{C}_T(y)) \simeq \mathfrak{K}_2(\mathfrak{C}_{T_1}(y)),$$

$$|\, \mathfrak{C}_T(y) : \mathfrak{K}_2(\mathfrak{C}_T(y)) \,| \geqslant |\, T_1 \,|^{r-1} = |\, T \,|^{(r-1)/r}.$$

Since $T \lhd G$, we have $\mathfrak{C}_T(y) \lhd \mathfrak{C}_G(y)$ and hence

$$\mathfrak{K}_2(\mathfrak{C}_T(y)) = \mathfrak{K}_2(\mathfrak{C}_G(y)) \cap \mathfrak{C}_T(y),$$

and we obtain

$$|\, T \,| \leqslant |\, \mathfrak{C}_T(y) : \mathfrak{K}_2(\mathfrak{C}_T(y)) \,|^{r/r-1} \leqslant |\, \mathfrak{C}_G(y) : \mathfrak{K}_2(\mathfrak{C}_G(y)) \,|^{r/r-1}.$$

It now follows that

$$|\, T \,| \leqslant \mu^{r/r-1} \leqslant \mu^2. \tag{9.9}$$

If we set $\beta_1(\mu) = \text{Max}\,(\beta_0(\mu)^m, \mu^2)$, then in the case $r > 1$, (9.6), (9.7), and (9.9) show that Theorem 5 holds. No exceptional F_ρ occurs. If $r = 1$ in (9.8), the group T is simple and our statement is still true. Here, we may have an exceptional F_ρ with $|\, F_\rho \,| \geqslant \beta_1(\mu)$.

Corollary 2. *If in Theorem 5, a composition factor F_ρ with $|\, F_\rho \,| \geqslant \beta_1(\mu)$ occurs, its 2-Sylow group has an elementary abelian center.*

Proof. We may again assume that $\mathfrak{K}_2(G) = \{1\}$. If the notation is as in the preceding proof, then $r = 1$ and $T \cong F_\rho$ is simple. There cannot exist elements $\pi \in Q$ and involutions $y_1, y_2 \in Q$ such that condition (*) is satisfied for (π, y_1, y_2) and $\mathscr{F}(T)$, since otherwise Theorem 4 would yield $|\, T \,| \leqslant \beta \leqslant \beta_1(\mu)$. Now, Remark 2 in Section IV shows that $\mathfrak{Z}(Q)$ is elementary abelian.

In Brauer-Fowler [*4*], it was proved that if a group G contains more than one class of involutions, there exist involutions J in G such that $|\, G \,| \leqslant |\, C_G(J) \,|^3$. The following result gives a substitute for groups with one class of involutions.

Theorem 6. *Let G be a group which contains one conjugate class of involutions. There exist a bound $\beta_2(\mu)$ depending only on μ with the following property: If J is an involution of G and $\mu = |\, \mathfrak{C}_G(J) : \mathfrak{K}_2(\mathfrak{C}_G(J)) \,|$, then*

$$|\, G : \mathfrak{K}_2(G) \,| \leqslant \beta_2(\mu) \cdot |\, \mathfrak{C}_G(J) \,|^3. \tag{9.10}$$

Proof. We use the same notation as in the proof of Theorem 5. Because of (9.7), the groups S and T have the same 2-Sylow group Q. By Sylow's theorems, this implies

$$S = \mathfrak{N}_S(Q)\, T.$$

Moreover, there exists a subgroup of odd order M of $\mathfrak{N}_S(Q)$, such that $\mathfrak{N}_S(Q) = MQ$ and hence $S = MT$. If we set $M \cap T = R$, then

$$| S : T | = | M : R |.$$

If $| S |$ is odd, $S = \mathfrak{K}_2(G)$ and the statement follows from (9.6). If $| S |$ is even, choose the involution J in $Q = P \cap S$. Since M normalizes Q, the number of M-conjugates of J is at most equal to the number ν of involutions in Q. This implies $| M | \leqslant \nu \, | \mathfrak{C}_M(J) |$. Hence

$$| S : T | = | M : R | \leqslant \nu \, | \mathfrak{C}_M(J) : \mathfrak{C}_R(J) | \, .$$

As is seen easily,

$$| \mathfrak{C}_M(J) : \mathfrak{C}_R(J) | \leqslant | \mathfrak{C}_S(J) : \mathfrak{C}_T(J) | \, .$$

Thus,

$$| S : T | \leqslant \nu \, | \mathfrak{C}_S(J) : \mathfrak{C}_T(J) | < 2^n \, | \mathfrak{C}_S(J) : \mathfrak{C}_T(J) | \, . \tag{9.11}$$

Set $\bar{T} = T/\mathfrak{K}_2(G)$, $\bar{J} = J\mathfrak{K}_2(G)$. On applying (9.5) to T instead of G, we have

$$| \mathfrak{C}_{\bar{T}}(\bar{J}) : \mathfrak{K}_2(\mathfrak{C}_{\bar{T}}(\bar{J})) | \leqslant \mu.$$

We also note (cf. (9.5*))

$$| \mathfrak{C}_{\bar{T}}(\bar{J}) | \leqslant | \mathfrak{C}_T(J) | \, .$$

Apply now Theorems 1 and 3 of Brauer [*15*] to the simple group $\bar{T}$. The element denoted by P in [*15*] can be chosen as $\bar{J}$. For $A = 2$, it follows that there exists a bound $\beta^*(\mu)$ depending only on μ such that

$$| T : \mathfrak{K}_2(G) | \leqslant \beta^*(\mu) \, | \mathfrak{C}_T(J) |^3. \tag{9.12}$$

(actually, if we have Case I in Theorem 1 of [*15*], the index in (9.12) lies below $2\gamma_0(2^n)^{1/2} \, | \mathfrak{C}_T(J) |^3$ and $2^n \leqslant \mu$. If we have Case II, the index lies below a bound depending only on μ. All this will be discussed in the third paper). Now Theorem 6 is obtained on combining (9.6), (9.11), and (9.12).

THEOREM 6*. *If the assumptions are the same as in Theorem 6 and if the 2-Sylow group P is neither cyclic nor a quaternion group, then*

$$| G | \leqslant \beta_2(\mu) \cdot | \mathfrak{C}_G(J) |^3.$$

This is seen by applying Theorem 6 to $\bar{G} = G/\mathfrak{K}_2(G)$ and using (9.5*) and (9.2).

If P is cyclic, it is well known that G has a normal 2-complement. Thus $G/\mathfrak{K}_2(G) \simeq P$. If P is a quaternion group, $G/\mathfrak{K}_2(G)$ has a center of order 2, cf. Theorem 2. Thus, for the two exceptional cases,

$$| G : \mathfrak{K}_2(G) | = | \mathfrak{C}_G(J) : \mathfrak{K}_2(\mathfrak{C}_G(J) | = \mu.$$

X. Cases in Which Only Finite Many Core-Free Groups Exist

THEOREM 7. *Let P be a 2-group which is not elementary abelian and which satisfies the following conditions:*

(a) *Every involution $y \in P$ lies in the center $\mathfrak{Z}(P)$.*

(b) *If u, v are two elements of P of the same order and if $\{u\} \cap \{v\} \neq \{1\}$, then $u^{-1}v$ belongs to the Frattini subgroup $\mathfrak{F}(P)$ of P.*

There exist only finite many nonisomorphic core-free groups G with P as their 2-Sylow group.

Proof. Take a fixed fusion $\mathscr{F}$ in P. We use a double induction, first on $| P |$ and then on the number of pairs (y_1, y_2) of involutions in $\mathscr{F}$. Suppose that the group G has the 2-Sylow group P and that $\mathscr{F}(G) = \mathscr{F}$.

1. Assume first that no two different involutions of P are $\mathscr{F}$-conjugate. Let σ and τ be two $\mathscr{F}$-conjugate elements of P, say of order 2^m. Then $\sigma^{2^{m-1}}$ and $\tau^{2^{m-1}}$ are $\mathscr{F}$-conjugate involutions and hence $\sigma^{2^{m-1}} = \tau^{2^{m-1}}$. It follows from (b) that $\sigma^{-1}\tau \in \mathfrak{F}(P)$. This implies $P^*(\mathscr{F}) \subseteq \mathfrak{F}(P)$. Now, a theorem of Tate-Thompson (cf. Tate [21]) shows that G has a normal 2-complement. Since G is core-free, $G = P$.

2. Suppose next that there exist involutions $y \in P$ which are $\mathscr{F}$-conjugate to elements $y' \neq y$, $y' \in P$. Let Ξ be the set of all such y. Since $y, y' \in \mathfrak{Z}(P)$, a Sylow group argument shows that there exist elements $\sigma \in \mathfrak{N}_G(P)$ with $y' = y^{\sigma}$. The group $\mathfrak{C}_G(y)$ also has 2-Sylow group P and in $\mathfrak{C}_G(y)$ the elements y and y' are no longer conjugate. Hence $\mathscr{F}_1 = \mathscr{F}(\mathfrak{C}_G(y))$ contains fewer pairs of involutions than F. Using induction we may assume that we have an upper bound μ^* not depending on G for all $| \mathfrak{C}_G(y) : \mathfrak{K}_2(\mathfrak{C}_G(y)) |$ with $y \in \Xi$.

Let π be an element of P of order at least 4. If $y \in \Xi$, the triple (π, y, y) satisfies condition (*), since all $\mathscr{F}$-conjugates of y lies in $\mathfrak{Z}(P)$, cf. (a). Now, Theorem 4 shows that G contains a normal subgroup H, which does not contain both π and y and for which $| G : H |$ lies below a fixed bound. Apply this argument for all possible choices of π and of y and let S be the intersection of the corresponding groups H. Then $S \lhd G$ and the index $| G : S |$ lies again below a fixed bound. Moreover, if the 2-Sylow group $Q = S \cap P$ of S meets Ξ, then Q cannot contain elements of order 4 and is therefore elementary abelian.

The group $S_1 = PS$ has the 2-Sylow group P. Set $\mathscr{F}_1 = \mathscr{F}(S_1)$. If two elements σ and τ of P are conjugate in S_1, say $\tau = \sigma^{\pi s}$ with $\pi \in P$, $s \in S$, then $\tau \equiv \sigma^{\pi} \bmod S$ and hence mod $S \cap P$. Thus, we can set

$$\tau = \sigma^{\pi s} = \sigma^{\pi} q \qquad \text{with} \qquad s \in S, \quad \pi \in P, \quad q \in Q. \tag{10.1}$$

3. Assume first that $Q \cap \Xi \neq \emptyset$. As we have seen, Q then is elementary

abelian. Thus, $q^2 = 1$. On the other hand, P is not elementary abelian and contains elements ρ of order 4. It follows from (a) that ρ and ρq have the same square and from (b) that $q \in \mathfrak{Z}(P)$. Now (10.1) implies $\sigma^{-1}\tau \in \mathfrak{Z}(P)$. Consequently, $P^*(\mathscr{F}_1) \subseteq \mathfrak{Z}(P)$. Again, the Tate-Thompson Theorem [22] shows that S_1 has a normal 2-complement. Since $\mathfrak{K}_2(G) = \{1\}$, we have $\mathfrak{K}_2(S) = \{1\}$, and since S_1/S is a 2-group, $\mathfrak{K}_2(S_1) = \{1\}$. It follows that $S_1 = P$, i.e., $S \subseteq P$. Hence the order of G is bounded.

4. We now turn to the case $Q \cap \Xi = \emptyset$. If two involutions σ, τ of P are conjugate in S_1, then (10.1) applies. Since here $\sigma, \tau \in \mathfrak{Z}(P)$, we have

$$\tau = \sigma^s = \sigma q \tag{10.2}$$

and the element $s \in S$ can be chosen in $\mathfrak{N}_S(P)$. By the Schur-Zassenhaus Theorem, $\mathfrak{N}_S(P) = PW$ with 2-regular W. It follows that s can be taken as 2-regular element. Since the involutions σ and τ commute by (a) either $q = 1$ or q is an involution of Q. In the latter case, $q \notin \Xi$ and since $q^s \in P$, the definition of Ξ shows that $q^s = q$. By (10.2), $\sigma^{s^m} = \sigma q^m$, and since s is 2-regular, we conclude $q = 1$. Hence no two distinct involutions σ, τ of P are S_1-conjugate. Taking S_1 instead of G, we have the case treated in part 1) of the proof. It follows that S_1 has a normal 2-complement; again this implies that $S \subseteq P$ and that $|G|$ is bounded. This completes the proof.

Remark. The 2-Sylow groups P of the simple Suzuki groups $Su(2^{2m+1})$, ($m \geqslant 1$), satisfy the assumptions of Theorem 7. Hence, these P occur as 2-Sylow groups of simple groups G, but there can be only finitely many such G for each P. There are other cases in which similar results can be obtained by related methods. In Brauer [14], it was shown that for abelian 2-groups P there exist only finitely many nonisomorphic core-free groups G with P as their Sylow group, if and only if P has no direct factor of type (2, 2). In generalizing the method, we can study the following situation:

Let $\mathscr{S}$ be a set of pairs $(P, \mathscr{F})$ where P is a 2-group and $\mathscr{F}$ a fusion in P. Assume that the $(P, \mathscr{F}) \in \mathscr{S}$ satisfy the following conditions:

(A) If $P^*(\mathscr{F}) \neq P, \{1\}$, there exists a group P_1 with $P \supset P_1 \supseteq P^*(\mathscr{F})$ such that for all fusions $\mathscr{F}_1$ of P_1 with $\mathscr{F}_1 \subseteq \mathscr{F}$, we have $(P_1, \mathscr{F}_1) \in \mathscr{S}$.

(B) If $P^*(\mathscr{F}) = P$, then

(i) for each fusion $\mathscr{F}_1 \subset \mathscr{F}$ of P with $P^*(\mathscr{F}_1) \neq P$, we have $(P, \mathscr{F}_1) \in \mathscr{S}$.

(ii) there exist elements $\pi \in P$ and involutions y_1, $y_2 \in P$ such that condition (*) holds for (π, y_1, y_2). Moreover, if η_i is an extreme $\mathscr{F}$-conjugate of y_i and if $Q_i = \mathfrak{C}_P(\eta_i)/\{\eta_i\}$, then $(Q_i, \mathscr{F}_i^{\#}) \in \mathscr{S}$ for any fusion $\mathscr{F}_i^{\#}$ of Q_i which is covered by a fusion $\mathscr{F}_i \subseteq \mathscr{F}$ of $\mathfrak{C}_G(\eta_i)$ $(i = 1, 2)$.

PROPOSITION 9. *If the set $\mathscr{S}$ satisfies conditions (A) and (B), then for every $(P, \mathscr{F}) \in \mathscr{S}$, there exist only ~~finite many nonisomorph icore freeroupsG~~ with 2-Sylow group P and $\mathscr{F}(G) = \mathscr{F}$.*

Since the proof is similar to that in [*14*], we shall not give details. Actually, the use of Theorem 4 leads to a simplification, since the discussion of [*14*, Case II(*b*), pp. 57-58] becomes unnecessary.

REFERENCES

Bracketed numbers [*1*]-[*13*] appear in the bibliography of I (this Journal, vol. 1 (1964), pp. 152-167).

14. BRAUER, R. On groups of even order with an Abelian 2-Sylow subgroup. *Arch. Math.* 13 (1962), 55-60.

15. BRAUER, R. Investigation of groups of even order, I. *Proc. Natl. Acad. Sci. U.S.* 47 (1961), 1891-1893.

16. BRAUER, R. AND LEONARD, Jr. S. On finite groups with an abelian Sylow group. *Can. J. Math.* 14 (1962), 436-450.

17. BRAUER, R. On certain classes of positive definite quadratic forms. *Acta Arith.*, in press.

18. BRAUER, R. AND SUZUKI, M. On finite groups of even order whose 2-Sylow group is a quaternion group. *Proc. Natl. Acad. Sci. U.S.* 45 (1959), 1757-1759.

19. FEIT, W. AND THOMPSON, J. G. Solvability of groups of odd order. *Pacific J. Math.* 13 (1963), 775-1029.

20. NAKAYAMA, T. Some remarks on regular representations, induced representations, and modular representations. *Ann. Math.* 39 (1938), 361-369.

21. SUZUKI, M. Applications of group characters. "Proceedings of Symposia in Pure Mathematics," Vol. 6, pp. 101-105. Am. Math. Soc., Providence, 1962.

22. TATE, J. Nilpotent quotient groups. *Topology*, in press.

23. WIELANDT, H. Beziehungen zwischen den Fixpunktzahlen von Automorphismen einer endlichen Gruppe. *Math. Z.* 73 (1960), 146-158.

l.2: finitely many nonisomorphic corefree groups G [R.B.]

BRAUER, R.
Math. Zeitschr. 90, 117–123 (1965)

On finite Desarguesian planes. I

Dedicated to KURT REIDEMEISTER

By
RICHARD BRAUER*

§ 1. Introduction

A Desarguesian projective plane Π determines its group G of projectivities. Conversely, if we are given the group G, we can reconstruct the plane Π. In principle, any axiomatic description of Π yields a characterization of the projective group G. Of course, these characterizations may appear somewhat artificial from a group theoretical point of view, and we may ask for other characterizations which are more closely related to group theoretical principles.

We are concerned here with the case of finite planes Π. Our question then is of interest in connection with the problem of characterizing the simple or nearly simple finite groups. If Π belongs to the Galois field $GF(q)$ with q elements, G is the group $G=PGL(q)$. Our final aim is that of giving a characterization of G by means of properties of the centralizer of an involution, as indicated in [1]. In this introductory paper, we shall deal with a set of postulates for a group G which permit us to define a projective plane Π in terms of G, such that G has a natural representation by collineations of Π. These results are still close to a direct translation of projective geometry into group theory. In a later paper, they will be used to obtain characterizations of G by means of centralizers of involutions. The number q will be odd in our work.

It seems appropriate to dedicate this work to KURT REIDEMEISTER who has contributed so much to geometry and group theory, and to our knowledge of the relationships between the two fields.

§ 2. The first three postulates

We start by considering a group G of finite order $|G|=g$ which satisfies the following conditions:

(I) G contains an elementary abelian subgroup D of order 4. If the elements of D are $1, J, J_1$ and $J_2=J\,J_1$, there exists an element $y \in G$ such that

$$y:\quad J \to J_1 \to J_2 \to J. \tag{2.1}$$

Here, we write $y: \xi \to \eta$ in order to indicate that $\eta = y^{-1}\,\xi\, y = \xi^y$.

* This research was supported by the United States Air Force under contract No. AF 49(638)—1381 monitored by the AF Office of Scientific Research of the Air Research and Development Command.

(II) There exist subgroups M and $M^* \neq 1$ of G with the following properties:

(*a*) M and M^* have the same order.

(*b*) M and M^* are normalized by the centralizer $\mathfrak{C}(J)$ of J. In symbols,

$$\mathfrak{C}(J): \quad M \to M, M^* \to M^*. \tag{2.2}$$

(*c*) $M \cap M^* = M \cap \mathfrak{C}(J) = M^* \cap \mathfrak{C}(J) = 1$.

(III) All involutions[1]) different from J in $\mathfrak{C}(J)$ are conjugate in $\mathfrak{C}(J)$.

In order to define a geometry, we shall need two further postulates in § 3, but we shall use a geometric language even before introducing them. It should be noted that our postulates are satisfied, if G is the projective group of a Desarguesian plane Π. We may choose here J as an arbitrary involution of G. Then J fixes all points of a line l and a further point p of Π. It will suffice to give the subgroups M and M^*. If l is taken as infinite line, M^* is to be taken as the group of translations of the affine plane. In other words, the non-identity elements of M^* are the elements of G which fix every point of l, but no other point of Π. Dually, M is to consist of 1 and of the elements of G which leave every line through p invariant but no other line of Π.

The involution J determines the pair (p, l) consisting of a point and a line l not through p. Here, p corresponds to the point (J) of the geometry of § 2 and l to the line $[J]$. An involution of G is interpreted as a pair (p, l) of a point p of Π and a line l not through p. Returning now to the case of an arbitrary group G which satisfies conditions (I), (II), and (III), we give the following

Definitions. A *point* p is a subset of G of the form $p = \xi^{-1} JM\xi$ with $\xi \in G$. A *line* l is a subset of G of the form $l = \eta^{-1} JM^* \eta$ with $\eta \in G$. The point p *lies on* l, if $p \cap l = \emptyset$. The *plane* Π associated with G is the set of all points and lines.

Of course, we shall have *duality* in Π. With every statement, the dual statement will hold. It will be clear what we mean by a *collineation* of Π. Let τ be a fixed element of G. If p is a point, so is $p^\tau = \tau^{-1} p \tau$. If l is a line, so is $l^\tau = \tau^{-1} l \tau$. Clearly, the mapping

$$\gamma(\tau): \quad p \to p^\tau, \quad l \to l^\tau$$

is a collineation of Π. Moreover, γ is a homomorphism of G into the group of all collineations of Π.

It follows from (IIb) that J maps $M \to M$, $M^* \to M^*$. Since only 1 remains fixed by ((IIc)), M and M^* are abelian of odd order and J "inverts" M and M^*, i.e. $J: \mu \to \mu^{-1}$ for $\mu \in M$ and for $\mu \in M^*$. It also follows from (IIb) that

$$J_1, J_2: \quad M \to M, M^* \to M^*.$$

Since M has odd order, each $\mu \in M$ has a unique square root $\mu^{\frac{1}{2}}$ in M. An analogous remark applies to M^*. If we set

$$M_i = M \cap \mathfrak{C}(J_i), \quad M_i^* = M^* \cap \mathfrak{C}(J_i), \quad (i = 1, 2), \tag{2.3}$$

[1]) We use the word *involution* to designate group elements of order 2.

we have

(2.4) $$M = M_1 \times M_2, \quad M^* = M_1^* \times M_2^*.$$

Indeed, if $\mu \in M$, we form the elements

$$\mu_1 = (\mu \, \mu^{J_1})^{\frac{1}{2}} \in M_1, \quad \mu_2 = (\mu \, \mu^{J_2})^{\frac{1}{2}} \in M_2.$$

Since $J_2 = J J_1$, $\mu^{J_2} = (\mu^{-1})^{J_1}$ and $\mu = \mu_1 \mu_2$. Since $J_2 = J J_1$ inverts M_1, we have $M_1 \cap M_2 = 1$. This yields the first equation (2.4). The proof of the second equation is dual.

It follows from (III) that there exists an element $t \in \mathfrak{C}(J)$ such that $t\colon J_1 \to J_2$. Then $t\colon J J_1 \to J J_2$, i.e. we have

(2.5) $$t\colon \quad J \to J, J_1 \to J_2 \to J_1.$$

Since $t\colon M \to M$, $M^* \to M^*$ by (IIb), it follows from (2.3) that

$$t\colon \quad M_1 \to M_2 \to M_1, \quad M_1^* \to M_2^* \to M_1^*.$$

In particular, M_1 and M_2 have the same odd order q. By (IIa),

$$|M_1| = |M_2| = |M_1^*| = |M_2^*| = q.$$

For $\mu \in M$, we have $J \mu = \mu^{-\frac{1}{2}} J \mu^{\frac{1}{2}}$. Hence each point p consists of q^2 involutions $\xi^{-1} \mu^{-\frac{1}{2}} J \mu^{\frac{1}{2}} \xi$ which all belong to the conjugate class K of J. Likewise, each line l consists of q^2 involutions in K. Actually, all involutions τ of G lie in K. We show:

(2A) *The group G contains only one class of involutions.*

Proof. Let S be a 2-Sylow subgroup of G which contains J. If J' is an involution in the center $Z(S)$ of S, then $J' \in \mathfrak{C}(J)$ and, on account of (III) and (2.1), J' is conjugate to J. After replacing S by a conjugate, we may assume $J = J'$. Now any involution τ of G is conjugate to an involution in S and hence in $\mathfrak{C}(J)$. It follows that $\tau \in K$.

(2B) *If* p and p' *are two points, either* $p = p'$ *or* $p \cap p' = \emptyset$.

Proof. After performing a collineation $\gamma(\tau)$ with $\tau \in G$, we may assume that $p' = JM$. If an element $J \mu = \mu^{-\frac{1}{2}} J \mu^{\frac{1}{2}}$ of JM belongs to $p = \xi^{-1} JM \xi$, we have an equation

$$\mu^{-\frac{1}{2}} J \mu^{\frac{1}{2}} = \xi^{-1} \mu_1^{-\frac{1}{2}} J \mu_1^{\frac{1}{2}} \xi$$

with $\mu_1 \in M$. Then $\mu_1^{\frac{1}{2}} \xi \mu^{-\frac{1}{2}} \in \mathfrak{C}(J)$. As shown by (IIb), $\mu_1^{\frac{1}{2}} \xi \mu^{-\frac{1}{2}}$ normalizes M and hence JM. Since $\mu_1^{\frac{1}{2}}$ and $\mu^{-\frac{1}{2}}$ normalize JM, it follows that ξ normalizes JM, and we have $p = \xi^{-1} JM \xi = JM = p'$. The statement is now evident.

If x is an involution of G, there exist $\xi \in G$ with $x = \xi^{-1} J \xi$. Then $x \in \xi^{-1} JM \xi$. Hence every involution x belongs to exactly one point p of G. We shall denote this point by (x). Dually, x belongs to exactly one line which will be denoted by $[x]$.

(2C) *If p is a point and if x ranges over the q^2 involutions $x \in p$, then $[x]$ ranges over the set of lines l which do not go through p. Each such line l is obtained exactly once in the form $[x]$.*

Proof. Since $x \in (x) = p$ and $x \in [x]$, we have $p \cap [x] \neq \emptyset$. This shows that p does not lie on $[x]$. Conversely, if l is a line not through p, by definition, $l \cap p \neq \emptyset$. It follows that there exist involutions $x \in l \cap p$. Then $l = [x]$. It remains to show that if x and x' belong to p and $[x] = [x']$, then $x = x'$. Using a suitable collineation $\gamma(\xi)$ with $\xi \in G$, we can assume without loss that $x = J$. Then $(x) = JM$, $[x] = JM^*$. Since $x' \in (x)$, we have $x' = J\mu$ with $\mu \in M$. Likewise, since $x' \in [x'] = [x]$, we can set $x' = J\mu^*$ with $\mu^* \in M^*$. Thus, $\mu = \mu^* \in M \cap M^* = \{1\}$, cf. (IIc). Hence $x' = J = x$ as we wished to show.

(2D) *If x and y are two distinct commuting involutions, then the point (x) lies on the line $[y]$.*

Proof. Again, we assume without loss that $x = J$. Then $y \in \mathfrak{C}(J)$, $y \neq J$, and (III) shows that y has the form $y = \xi^{-1} J_1 \xi$ with $\xi \in \mathfrak{C}(J)$. After performing $\gamma(\xi^{-1})$, we may assume that $x = J$, $y = J_1$.

Suppose that (J) does not lie on $[J_1]$. Then there exist elements $\eta \in (J) \cap [J_1]$. Here, $(J) = JM$. Since $J_1 = J^y$ by (2.1), $[J_1] = [J]^y = (JM^*)^y$. Hence there exist elements $\mu \in M$, $\mu^* \in M^*$ such that

$$\eta = J\mu = y^{-1} J \mu^* y = J_1 (\mu^*)^y.$$

It follows that

$$J_2 = J_1 J = (\mu^*)^y \mu^{-1} = (\mu^*)^y \mu_2^{-1} \mu_1^{-1}, \tag{2.6}$$

where we wrote $\mu = \mu_1 \mu_2$ with $\mu_i \in M_i$, cf. (2.4). Since J inverts μ_2 and J_2 centralizes μ_2, the element $J_1 = J J_2$ inverts μ_2. Also, $J^y = J_1$ inverts $(\mu^*)^y$. On transforming (2.6) with J_1, we obtain

$$J_2 = (\mu^{*y})^{-1} \mu_2 \mu_1^{-1}$$

Comparison with (2.6) yields $(\mu^{*y})^2 = \mu_2^2$ whence $\mu^{*y} = \mu_2$. Now (2.6) becomes $J_2 = \mu_1^{-1}$ which is absurd. Consequently, (J) lies on $[J_1]$ and the proposition is proved.

We shall set

$$r = (G : \mathfrak{C}(J)\, M) = (G : \mathfrak{C}(J)\, M^*). \tag{2.7}$$

Then $(G : \mathfrak{C}(J)) = rq^2$ and G contains exactly rq^2 involutions. Since q^2 involutions form a point and a line, (2B) and its dual show that Π consists of r points and r lines. By (2C), there exist exactly q^2 lines which do not go through a given point. We now have

(2E) *The total number of points (lines) of Π is given by (2.7). There are exactly $r - q^2$ lines through a given point and $r - q^2$ points on a given line.*

The dual of (2C) shows that the points $(J\mu_2^*)$ with $\mu_2^* \in M_2^*$ are distinct and do not lie on $[J]$. By (2D), the points $(J\mu_2^*)$ lie on $[J_2]$. Likewise, (J_1) lies on $[J_2]$ and also on $[J]$. Hence

(2F) *The* $q+1$ *points* (J_1) *and* $(J\mu_2^*)$ *with* $\mu_2^* \in M_2^*$ *are distinct and lie on* $[J_2]$.

(2G) *There are at least* $q+1$ *points on a given line. There are at least* $q+1$ *lines through a given point.*

The first statement is a consequence of (2F) since $\gamma(G)$ is transitive on the set of lines. The second statement is the dual.

§ 3. Construction of a projective plane

We now add the following two postulates

(IV) $(G\colon M\mathfrak{C}(J)) \leqq q^2+q+1$.

(V) Every class of $\mathfrak{C}(J)$-conjugate elements of M meets $M_1 = M \cap \mathfrak{C}(J_1)$.

It will not be necessary to require the dual of (V). Of course, we then cannot use duality.

We first note

(3A) *We have*

$$r = (G\colon M\mathfrak{C}(J)) = q^2+q+1. \tag{3.1}$$

There are exactly $q+1$ *points on a given line and there are exactly* $q+1$ *lines through a given point.*

Proof. On combining (2E) and (2G), we obtain

$$r - q^2 \geqq q+1.$$

Since (IV) yields the reverse inequality, we have (3.1). The last two statements of (3A) are now consequences of (2E).

Theorem (3B). *The set* Π *is a projective plane.*

Proof. (*a*) Let l_1 and l_2 be two distinct lines. There are $q+1$ points on l_1 and $q+1$ points on l_2. Since $q>1$,

$$2q+2 < q^2+q+1.$$

It follows that there exist points p neither on l_1 nor on l_2. After performing a suitable $\gamma(\xi)$ with $\xi \in G$, we may assume that $p=(J)$. Then by (2C), l_1 has the form $l_1 = [J\mu_0]$ with $\mu_0 \in M$. Perform now $\gamma(\mu_0^{-\frac{1}{2}})$ which maps

$$(J) \to (\mu_0^{\frac{1}{2}} J \mu_0^{-\frac{1}{2}}) = (J\mu_0^{-1}) = (J); \qquad [J\mu_0] \to [\mu_0^{\frac{1}{2}} J \mu_0 \mu_0^{-\frac{1}{2}}] = [J].$$

Hence we may assume that $p=(J)$, $l_1=[J]$. Since l_2 does not go through (J), we can set $l_2=[J\mu]$ with $\mu \in M$. It follows from (V) that there exist $\eta \in \mathfrak{C}(J)$ such that μ^η lies in M_1. Then $\mu^\eta \in \mathfrak{C}(J_1)$. Now $\gamma(\eta)$ maps $l_1=[J] \to l_1=[J]$; $l_2=[J\mu] \to [J\mu^\eta]$. By (2D), (J_1) lies on $[J]$ and on $[J\mu^\eta]$. Therefore, $\gamma(\eta^{-1})$ maps (J_1) on a common point of l_1 and l_2.

Thus every two lines have a common point.

(*b*) Consider the $q+1$ points on a given line l_1. Through each of these points, we have q lines $l \neq l_1$. It follows from (*a*) that each line $l_2 \neq l_1$ is obtained at least once in the form l. Since we have q^2+q lines l and since there are $r-1=q^2+q$ possibilities for l_2, each line l_2 is obtained exactly once in the form l. This shows that two lines l_1 and $l_2 \neq l_1$ have exactly one point in common.

(*c*) Let p_1 be a given point. We have $q+1$ lines through p_1 and on each of them, we have q points $z \neq p_1$. If two of the $q(q+1)$ points z obtained in this fashion were equal, then contrary to (*b*), the points p and z would lie simultaneously on two distinct lines. Hence there are q^2+q distinct points z. Since this is the total number of points $p_2 \neq p_1$ of Π, each p_2 is obtained exactly once in the form z. Hence there is exactly one line through two distinct points p_1 and p_2.

(*d*) It follows from (2D) that (J), (J_1), (J_2) are the vertices of a non-degenerate triangle with the sides $[J]$, $[J_1]$, $[J_2]$. Since there is a total of $3q$ points which lie on a side of Δ and since

$$3q < q^2+q+1,$$

there exist points p not on a side of Δ.

This completes the proof of the theorem.

(3C) *If p is a given point lying on a given line l, there exist elements $\xi \in G$ such that $\gamma(\xi) \neq 1$ and that $\gamma(\xi)$ leaves every point on l and every line through p fixed.*

Proof. Without restriction, we may assume that $l=[J_2]$. Then $p=(J\mu)$ with $\mu \in M_2^*$ or $p=(J_1)$. In the former case, $\gamma(\mu^{-\frac{1}{2}})$ maps $(J\mu) \to (J)$, $[J_2] \to [J_2]$. Since $t\,y\colon J \to J_1$, $J_2 \to J_2$ by (2.1) and (2.5), it follows that $\gamma(\mu^{-\frac{1}{2}}\,t\,y)$ maps $p \to (J_1)$, $[J_2] \to [J_2]$. Hence we may take $p=(J_1)$, $l=[J_2]$. For $\eta \in M_2^*$, $\gamma(\eta)$ maps $[J_2] \to [J_2]$, $[J] \to [J\eta^2]=[J]$. Hence $(J_1) \to (J_1)=(J)^y$. It follows that $\eta^2 \in M_1^y M_2^y$. Since $\eta \in \mathfrak{C}(J_2)$ then $\eta^2 \in M_1^y$. This implies $M_2^*=M_1^y$. But M_1 centralizes M_2 and hence M_2^* centralizes M_2^y. By (2F), $\gamma(\xi)$ with $\xi \in M_2^y$ then leaves all points of $[J_2]$ fixed. A similar argument shows that $(M_1^*)^y=M_2$, that M_2^y centralizes $(M_2^*)^{y^2}=(M_1)^{y^3}=M_1$, and that $\gamma(\xi)$ leaves all lines trough (J_1) fixed. For $\xi \neq 1$, $[J_1]$ is not fixed. Thus, ξ satisfies the conditions.

We can apply Theorem 1.8 of GLEASON [2] and obtain:

Theorem (3D). *The plane Π is Desarguesian.*

As a further application of (3C), we have

(3E) *The group $\gamma(G)$ includes the little projective group of Π* [2]).

---

[2]) I am indebted to H. LÜNEBURG for the remark that this is an immediate consequence of the results of WAGNER [3]. Actually, it is not difficult to give a direct proof. WAGNER'S results also give alternate proofs of Theorem (3D).

§ 4. The group G

(4A) *The collineation $\gamma(\xi)$ with $\xi \in G$ leaves the point (J) and the line $[J]$ fixed, if and only if $\xi \in \mathfrak{C}(J)$.*

Proof. If $\gamma(\xi)$ leaves (J) fixed, then $\xi^{-1} J \xi \in JM$. This implies $\xi^{-1} J \xi = \mu^{-1} J \mu$ for some $\mu \in M$. Hence $\xi \in \mathfrak{C}(J) M$. If in addition $\gamma(\xi)$ leaves the line $[J]$ fixed, we conclude similarly that also $\xi \in \mathfrak{C}(J) M^*$. Thus $\xi = \rho_1 \mu = \rho_2 \mu^*$ with $\rho_1, \rho_2 \in \mathfrak{C}(J)$, $\mu \in M$, $\mu^* \in M^*$. Then $\mu^* \mu^{-1}$ commutes with J. Since J inverts μ^* and μ, it follows that $\mu^{*2} = \mu^2$. On account of (IIc), we have $\mu^2 = 1$. Then $\mu = 1$, $\xi \in \mathfrak{C}(J)$.

Conversely, if $\xi \in \mathfrak{C}(J)$, it is clear that $\gamma(\xi)$ leaves (J) and $[J]$ fixed.

(4B) *The kernel K of the homomorphism γ is the intersection of the centralizers of the involutions of G.*

This is an immediate consequence of (4A). We also note

(4C) *The kernel K has odd order.*

Indeed, if this was not true, the normal subgroup K of G would contain involutions and, because of (2A), we would have $J \in K$. However, $\gamma(J)$ maps $(J\mu^*) \to (J\mu^{*-1})$ for $\mu^* \in M^*$. If $\mu^* \neq 1$, the dual of (2C) shows that $(J\mu^*) \neq (J\mu^{*-1})$. Thus $\gamma(J) \neq 1$, and we have a contradiction.

Since $\gamma(G)$ is a subgroup of the collineation group of Π, the reciprocal image $\gamma^{-1}(PGL(3, q)) = G_0$ of the projective group $PGL(3, q)$ is a normal subgroup of G with cyclic quotient group G/G_0. Then G_0/K is isomorphic with a subgroup of $PGL(3, q)$ which includes $PSL(3, q)$ by (3E). Since $PSL(3, q)$ has index 1 or 3 in $PGL(3, q)$ we have

Theorem (4D). *The group G has a chief series*

$$G \supseteq G_0 \supseteq K \supseteq 1$$

where G/G_0 is cyclic, where the group G_0/K is isomorphic with $PGL(3, q)$ or $PSL(3, q)$, and where K has odd order.

For $q \not\equiv 1 \pmod 3$, the groups $PGL(3, q)$ and $PSL(3, q)$ coincide.

References

[1] Brauer, R.: On the structure of groups of finite order. Proc. International Congress of Mathematicians, Amsterdam, 1954, vol. I, p. 209–217.

[2] Gleason, A. M.: Finite Fano planes. Amer. Journal of Mathematics **78**, 797–807 (1956).

[3] Wagner, A.: On perspectivities of finite projective planes. Math. Zeitschrift **71**, 113–123 (1959).

Dept. of Math., Harvard University, Cambridge 38, Mass. (U.S.A.)

(Received April 29, 1965)

Druck der Universitätsdruckerei H. Stürtz AG., Würzburg

On finite Desarguesian planes. II

RICHARD BRAUER

§ 1. Introduction

We are interested in characterizations of the projective group $PGL(3, q)$ of a finite Desarguesian plane of order q by means of properties of the centralizers of involutions. In this paper, we are concerned with the case $q \equiv -1$ (mod 4). The case $q \equiv 1$(mod 4) will be treated in the third paper of this series. The case of an even q has already been dealt with by SUZUKI [9]. Our principal result is the following one.

Theorem (1A). *Let $\mathfrak{G}$ be a finite group which satisfies the conditions:*

(I) *There exists an involution J of $\mathfrak{G}$ whose centralizer $\mathfrak{C}(J)$ is isomorphic with a group of the form $GL(2, q)/\mathfrak{L}$ where $\mathfrak{L}$ is a subgroup of the center $\mathfrak{Z}(GL(2, q))$ and where $q \equiv -1$(mod 4).*

(II) *The group $\mathfrak{G}$ does not have a normal subgroup of index 2.*

We then have one of the cases.

(a) $\mathfrak{G} \simeq PGL(3, q)$, $PSL(3, q)$, *or* $SL(3, q)$.

(b) $\mathfrak{G}$ *is isomorphic to a direct product of $PSL(3, q)$ with a cyclic group of order* 3; $q \equiv 1$(mod 3), $q \not\equiv 1$(mod 9).

(c) $\mathfrak{G} \simeq \mathfrak{M}_{11}$ *the Mathieu group of order* 7920, $q = 3$.

If $q \not\equiv 1$(mod 3), the three groups in (a) are isomorphic. We have $|\mathfrak{L}| = 3$ when $\mathfrak{G} \simeq PSL(3, q)$ and $q \equiv 1$(mod 3) while in all other cases $|\mathfrak{L}| = 1$.

Actually, the assumption (II) serves only the purpose to exclude cases of little interest. With an additional assumption, the part of the theorem referring to the case $\mathfrak{L} = 1$ had been stated without proof in [1].

The 2-Sylow groups of $GL(2, q)$ are of different types in the cases $q \equiv -1$ (mod 4), $q \equiv 1$(mod 4), q even. It is for this reason that the three cases have to be separated. In the case $q \equiv -1$(mod 4), the 2-Sylow subgroup of $GL(2, q)$ is quasi-dihedral. The same then is true for the 2-Sylow subgroup $\mathfrak{S}$ of $\mathfrak{G}$ in Theorem (1A). The proof of the theorem proceeds in the following steps. The results of [3] III yield information about the characters of a group $\mathfrak{G}$ with the 2-Sylow subgroup $\mathfrak{S}$. If we add the assumption that the order g of G is divisible by q^3, we can determine the degrees of the characters in the principal 2-block of $\mathfrak{G}$. From this, the order g can be obtained. Also, the results concerning the characters enable us to study the subgroup $\mathfrak{Q}$ of order q^3 of $\mathfrak{G}$.

It can now be shown that $\mathfrak{G}$ satisfies the assumptions made in [4]. This gives us then the construction of a Desarguesian plane Π and a representation of $\mathfrak{G}$ by collineations, and we can complete the proof.

In the last two sections, the assumption $q^3 \mid g$ is eliminated.

It is hoped that the notation is essentially self-explanatory. We write $\mathfrak{C}(X)$ for the centralizer of a subset X in a group $\mathfrak{G}$ and $c(X)$ for the order $|\mathfrak{C}(X)|$ of X. If necessary, the group $\mathfrak{G}$ is indicated as subscript. The normalizer of X is denoted by $\mathfrak{N}(X)$. By an involution, we mean an element of order 2. The abbreviation S_p-group stands for p-Sylow subgroup.

§ 2. Properties of $\mathfrak{C}(J)$

Let $q=p_0^r$ be a power of a prime p_0 such that $q \equiv -1 \pmod 4$. Consider a group $\mathfrak{H}$ of the form

$$\mathfrak{H} \simeq GL(2,q)/\mathfrak{L} \tag{2.1}$$

where $\mathfrak{L}$ is a subgroup of the center $\mathfrak{Z}(GL(2,q))$. We shall assume that $\mathfrak{H}$ is the centralizer $\mathfrak{C}(J)$ of an involution J of a finite group $\mathfrak{G}$. Then J belongs to the center of $\mathfrak{H}$ and this implies that $l=|\mathfrak{L}|$ has to be odd. Since $\mathfrak{Z}(GL(2,q))$ has order $q-1$, we can set

$$q-1=2ul, \quad q+1=2^{n-2}v \tag{2.2}$$

where u and v are odd integers and where $n \geqq 4$. Then

$$c(J)=|\mathfrak{H}|=(q-1)^2 q(q+1)/l=2^n u^2 l v q. \tag{2.3}$$

We collect here some results concerning the group $\mathfrak{H}$ which will be used repeatedly. The proofs are easy and well known and we shall omit most of the details. Denote by ρ a homomorphism of $GL(2,q)$ onto $\mathfrak{H}$ with the kernel $\mathfrak{L}$. Let γ denote a primitive root in the Galois field $GL(q^2)$ with q^2 elements. Then $\gamma_0=\gamma^{q+1}$ is a primitive root in $GF(q)$.

1. The center $\mathfrak{Z}(\mathfrak{H})$ is a cyclic group of order $(q-1)/l=2u$. If we set $J=\rho(-I)$, $U=\rho(\gamma_0^2 I)$, $\mathfrak{U}=\langle U\rangle$ we have

$$\mathfrak{Z}(\mathfrak{H})=\langle J\rangle \times \mathfrak{U}=\langle JU\rangle \tag{2.4}$$

with $|\mathfrak{U}|=u$.

2. We introduce the involutions

$$J_1=\rho\begin{pmatrix}1 & 0\\ 0 & -1\end{pmatrix}, \quad T=\rho\begin{pmatrix}0 & 1\\ 1 & 0\end{pmatrix}. \tag{2.5}$$

Then $\mathfrak{D}=\langle J, J_1\rangle$ is elementary abelian of order 4 and contains the three involutions J, J_1, and $J_2=JJ_1$. Since T (under conjugation) maps

$$J \to J, \quad J_1 \to J_2 \to J_1, \tag{2.6}$$

the group $\mathfrak{D}_1 = \langle \mathfrak{D}, T \rangle$ is dihedral of order 8 with the central involution J. The centralizer $\mathfrak{C}(\mathfrak{D})$ of $\mathfrak{D}$ (in $\mathfrak{H}$ or in $\mathfrak{G}$) consists of the ρ-images of the diagonal matrices. We may then set

$$\mathfrak{C}(\mathfrak{D}) = \mathfrak{D} \times \mathfrak{U} \times \mathfrak{W} \tag{2.7}$$

where $\mathfrak{W}$ is cyclic of order $u\,l$. For instance,

$$W = \rho \begin{pmatrix} 1 & 0 \\ 0 & \gamma_0^2 \end{pmatrix}$$

can be used as generator of $\mathfrak{W}$. Then

$$U^T = U, \qquad W^T = U\,W^{-1}. \tag{2.8}$$

For any $\xi \in \mathfrak{C}(\mathfrak{D})$ which does not belong to $\mathfrak{Z}(\mathfrak{H})$, we have

$$\mathfrak{C}_{\mathfrak{H}}(\xi) = \mathfrak{C}(\mathfrak{D}). \tag{2.9}$$

We also note

$$\mathfrak{C}(\mathfrak{D}_1) = \mathfrak{C}(\langle \mathfrak{D}, T \rangle) = \mathfrak{Z}(\mathfrak{H}). \tag{2.10}$$

3. There exist elements $M \in GL(2, q)$ with the characteristic roots γ and γ^q and then M has order $q+1$ over $\mathfrak{Z}(GL(2, q))$. It follows that $R = \rho(M)$ has order $q+1$ over $\mathfrak{Z}(\mathfrak{H})$. The order of R itself is $(q^2-1)/l$ and we may set

$$R = S\,V\,U^k$$

where S, V, U^k are powers of R of orders 2^{n-1}, v, u respectively. After replacing R by a conjugate, we may assume that S lies in an S_2-group $\mathfrak{S} \supset \mathfrak{D}_1$ of $\mathfrak{H}$. Here, $\mathfrak{S}$ has order 2^n and $\langle S \rangle$ is a cyclic subgroup of order 2^{n-1}. Then $\langle S \rangle \cap \mathfrak{D}_1$ has necessarily order 4. This implies that $\mathfrak{D}_1$ contains the element $F = S^{2^{n-3}}$. Without restriction, we may assume that

$$F = S^{2^{n-3}} = T\,J_1 \tag{2.11}$$

and this implies that $S^{2^{n-2}} = F^2 = J$.

The element J_1 of $\mathfrak{S}$ must transform S into a power $S' \neq S$ of S. Since the two characteristic roots of a reciprocal image $\rho^{-1}(S)$ of S are conjugate over $GF(q)$, we can only have $S' = S^q$. It follows that $\mathfrak{S} = \langle S, J_1 \rangle$ is quasi-dihedral,

$$S^{2^{n-1}} = 1, \quad J_1^2 = 1, \quad J_1^{-1}\,S\,J_1 = J\,S^{-1}$$

with $J = S^{2^{n-2}}$. We also note that

$$\mathfrak{N}_{\mathfrak{G}}(\langle S \rangle) = \mathfrak{N}_{\mathfrak{H}}(\langle S \rangle) = \langle \mathfrak{C}(S), J_1 \rangle. \tag{2.12}$$

4. The reciprocal images in $GL(2, q)$ of the elements of $\mathfrak{C}(F)$ are the matrices of the form

$$\begin{pmatrix} a & b \\ -b & a \end{pmatrix} \neq 0 \qquad (a, b \in GF(q)).$$

Because of $q \equiv -1 \pmod 4$, they are the non-zero elements of a field isomorphic to $GL(q^2)$. This implies $c(F)=(q^2-1)/l$. It follows easily that

$$\mathfrak{C}(S^\mu)=\mathfrak{C}(F)=\langle S\rangle \times \langle V\rangle \times \mathfrak{U} \qquad (\text{for } S^\mu \neq 1, J). \tag{2.13}$$

5. For $a \in GF(q)$, set

$$X_1=\rho\begin{pmatrix}1 & 0\\ a & 1\end{pmatrix}, \quad X_2=\rho\begin{pmatrix}1 & a\\ 0 & 1\end{pmatrix}. \tag{2.14}$$

The elements X_1 with arbitrary a form a subgroup $\mathfrak{X}_1$ of order q and the elements X_2 form a subgroup $\mathfrak{X}_2=\mathfrak{X}_1^T$. Clearly, J_1 inverts X_1 and X_2. It is seen without difficulty that if ξ is a p_0-element of $\mathfrak{H}$ which is inverted by J_1, then $\xi \in \mathfrak{X}_1$ or $\xi \in \mathfrak{X}_2$. We have $\mathfrak{C}_\mathfrak{H}(\mathfrak{X}_1)=\langle \mathfrak{X}_1, \mathfrak{Z}(\mathfrak{H})\rangle$.

The group $\mathfrak{C}(\mathfrak{D})$ normalizes $\mathfrak{X}_1$ acting transitively on $\mathfrak{X}_1-\{1\}$. As is well known,

$$\mathfrak{H}=\langle \mathfrak{C}(\mathfrak{D}), \mathfrak{X}_1, \mathfrak{X}_2\rangle. \tag{2.15}$$

§ 3. Summary of the results of [3], III, § 8

Assume that $\mathfrak{G}$ is a group of finite order g with quasi-dihedral S_2-group $\mathfrak{S}$, with the notation chosen as in § 2. We shall assume that F and $S J_1$ are conjugate in $\mathfrak{G}$. We summarize the results of [3], III, § 8; the integer denoted there by m will be denoted by $-q$. We do not know yet that this is the same q as above.

The principal 2-block $B_0=B_0(\mathfrak{G}; 2)$ of $\mathfrak{G}$ consists of $4+2^{n-2}$ irreducible characters. Four of these, denoted by $\chi_0=1$, χ_1, χ_2, χ_3 have odd degrees $x_i=\chi_i(1)$. There exist signs δ_1, δ_2, δ_3 such that

$$1+\delta_1 x_1+\delta_2 x_2+\delta_3 x_3=0. \tag{3.1}$$

Two of δ_1, δ_2, δ_3 are -1 and one is $+1$.

The remaining characters of B_0 can be denoted by χ_4, $\chi^{(\mu)}$ with $1 \leqq \mu \leqq 2^{n-2}-1$. For 2-regular $G \in \mathfrak{G}$,

$$\begin{cases} 1+\delta_1\chi_1(G)+\delta_2\chi_2(G)+\delta_3\chi_3(G)=0, \\ \delta_1+\chi_1(G)=\chi^{(\mu)}(G), \quad \delta_2+\chi_2(G)=\chi_4(G), \end{cases} \tag{3.2}$$

for any μ. In particular, all $\chi^{(\mu)}$ have degree δ_1+x_1 and χ_4 has degree $x_4=\delta_2+x_2$. Moreover, there exists an irreducible character φ in the principal 2-block of $\mathfrak{C}(J)$ and a sign ε such that, for 2-regular $G \in \mathfrak{C}(J)$, we have

$$\begin{cases} \chi_1(G)=-\delta_1+\delta_1\varepsilon\varphi(G), \quad \chi_2(G)=\delta_2-\delta_2\varepsilon\varphi(G), \\ \chi_3(J G)=-\delta_3, \quad \chi_4(J G)=-2\delta_2+\delta_2\varepsilon\varphi(G), \quad \chi^{(\mu)}(J G)=\pm\varphi(G). \end{cases} \tag{3.3}$$

Here, $\varphi=\varepsilon\varphi_1^J$ in the notation of [3], and φ has degree $-\varepsilon(q-1)$ with

$$q \equiv -1+2^{n-2} \pmod{2^{n-1}}. \tag{3.4}$$

We have to distinguish two cases.

Case I. J and J_1 are conjugate in $\mathfrak{G}$.

Then the group order $g=|\mathfrak{G}|$ is given by

$$g=\frac{c(J)^3}{c(\mathfrak{D})^2}\,\frac{x_1(x_1+\delta_1)}{(x_1+\delta_1 q)^2}\,\frac{q-1}{q}=\frac{c(J)^3}{c(\mathfrak{D})^2}\,\frac{x_2(x_2+\delta_2)}{(x_2-\delta_2 q)^2}\,\frac{q+1}{q}. \tag{3.5}$$

We have the additional relations

$$x_1 x_2=q^2 x_3, \tag{3.6}$$

$$\delta_1 x_1\equiv 2+q, \quad \delta_2 x_2\equiv q, \quad \delta_3 x_3\equiv -1+2^{n-1} \pmod{2^n}. \tag{3.7}$$

We note that (3.1) and (3.6) imply that

$$-q^2(1+\delta_1 x_1)=\delta_2(q^2+\delta_1 x_1)\,x_2. \tag{3.8}$$

Case II. J and J_1 are not conjugate in $\mathfrak{G}$.

Here, $\mathfrak{G}$ has a normal subgroup $\mathfrak{G}_0$ of index 2 with the S_2-group $\langle S, S J_1\rangle$. We have

$$g=\frac{c(J_1)^2\,c(J)}{c(\mathfrak{D})^2}. \tag{3.9}$$

We add some remarks concerning Case II which will not be used below. It is clear that $\mathfrak{G}$ and $\mathfrak{G}_0$ have the same maximal normal subgroup $\mathfrak{K}$ of odd order. Since $\langle S, S J_1\rangle$ is a quaternion group, a theorem of SUZUKI and the author [5] shows that $\mathfrak{G}_0/\mathfrak{K}$ has a center of order 2. This is necessarily the group $\langle J\rangle\,\mathfrak{K}/\mathfrak{K}$. Since every $\mathfrak{G}$-conjugate of J is a $\mathfrak{K}$-conjugate of J, it follows that

$$\mathfrak{G}=\mathfrak{C}(J)\,\mathfrak{K}.$$

In particular, if $\mathfrak{K}=1$, we have $\mathfrak{G}=\mathfrak{C}(J)$.

Finally, we discuss the case that $\mathfrak{G}$ is the group $\mathfrak{H}$ studied in § 2. The symbol q now appears in two different meanings. We shall soon see that they coincide, but for the moment we shall replace q in the sense of § 2, by q_0, i.e., take $\mathfrak{H}\cong GL(2, q_0)/\mathfrak{L}$. We know that $\mathfrak{H}$ has the S_2-group $\mathfrak{S}$. The center element J of $\mathfrak{H}$ is certainly not $\mathfrak{H}$-conjugate to $J_1\notin\mathfrak{Z}(\mathfrak{H})$. On the other hand, F and $S J_1$ are conjugate in $\mathfrak{H}$. This can be checked directly. It can also be seen from the results of [3], § 8, since if F and $S J_1$ were not conjugate, $\mathfrak{H}$ would have a normal 2-complement and this is not true.

It is now clear that the results stated above apply to $\mathfrak{H}$ and that we have here Case II. The irreducible characters of $GL(2, q_0)$ are well known, see for instance [7] and the literature quoted there. In particular, only the degrees 1, q_0, q_0-1, q_0+1 appear. The same is true for $\mathfrak{H}$ which is a quotient group of $GL(2, q_0)$. Since (3.4) shows that the degree of the irreducible character φ of $\mathfrak{C}_{\mathfrak{H}}(J)=\mathfrak{H}$ is even and not divisible by 4, we must have $\varphi(1)=q_0-1$. It follows that $q=q_0$ and that $\varepsilon=-1$.

§ 4. First consequences of the assumptions (I) and (II)

We assume now that $\mathfrak{G}$ is a group of finite order g which satisfies the assumptions (I) and (II) of Theorem (1A).

(4A) *An S_2-group $\mathfrak{S}$ of $\mathfrak{H}=\mathfrak{C}(J)$ is also an S_2-group of $\mathfrak{G}$.*

Indeed, if this was not so, there would exist a 2-subgroup $\mathfrak{S}_1 \supset \mathfrak{S}$ of $\mathfrak{G}$ which normalizes $\mathfrak{S}$. Then $\mathfrak{S}_1$ normalizes $\mathfrak{Z}(\mathfrak{S})=\langle J\rangle$ and hence $\mathfrak{S}_1 \subseteq \mathfrak{C}(J)$, a contradiction.

As remarked in § 3, the elements F and $S J_1$ are conjugate in $\mathfrak{H}$ and hence in $\mathfrak{G}$. The following result is now evident.

(4B) *The results of § 3 apply to $\mathfrak{G}$ and we have Case I. The irreducible character φ of $\mathfrak{C}(J)=\mathfrak{H}$ in (3.3) has degree $q-1$, and $\varepsilon=-1$. Here q is the same prime power $q=p_0^r$ which appears in $GL(2, q)$.*

As a corollary, we note

(4C) *If p is an odd prime dividing $q-1$ but not u, then the irreducible character φ of $\mathfrak{C}(J)$ has defect 0 for p.*

(4D) *There exists an element $Y \in \mathfrak{G}$ which maps*

$$Y\colon\ J \to J_1 \to J_2 \to J\,. \tag{4.1}$$

Proof. Since J_1 is conjugate to J in $\mathfrak{G}$, a Sylow group argument shows that there exist $\xi \in \mathfrak{G}$ with $J_1^\xi = J$ such that $\mathfrak{D}^\xi \subset \mathfrak{S}$. Since any elementary abelian subgroup of order 4 of $\mathfrak{S}$ is conjugate to $\mathfrak{D}$ in $\mathfrak{S}$, we may assume that $\mathfrak{D}^\xi = \mathfrak{D}$. Hence J and J_1 are conjugate in $\mathfrak{N}(\mathfrak{D})$. Since T in (2.6) also belongs to $\mathfrak{N}(\mathfrak{D})$, the group $\mathfrak{N}(\mathfrak{D})/\mathfrak{C}(\mathfrak{D})$ is isomorphic with the group of all permutations of the three symbols J, J_1, J_2. This shows the existence of an element Y which satisfies (4.1).

(4E) *If two elements Z_1 and Z_2 of $\mathfrak{Z}(\mathfrak{C}(J))$ are conjugate in $\mathfrak{G}$, they are equal.*

Indeed, since $\mathfrak{S}$ belongs to $\mathfrak{C}(Z_1)$ and $\mathfrak{C}(Z_2)$, a Sylow group argument shows that Z_1 and Z_2 are conjugate in $\mathfrak{N}(\mathfrak{S})$. As in the proof of (4A), we see that Z_1 and Z_2 are conjugate in $\mathfrak{C}(J)$, and since they lie in the center, they are equal.

We now come to the application of the results of § 3. On substituting (2.3) in (3.5) and using (2.7), we have

(4F) *The order g of $\mathfrak{G}$ has the form*

$$g=(q-1)^2\, q^3 (q+1)^3\, \mu/l \tag{4.2}$$

where the rational number μ is given by

$$\mu=\left(1-\frac{1}{q}\right) x_1(x_1+\delta_1)/(x_1+\delta_1 q)^2=\left(1+\frac{1}{q}\right) x_2(x_2+\delta_2)/(x_2-\delta_2 q)^2. \tag{4.3}$$

The remainder of this section will be occupied by the proof of the following proposition.

(4G) *In the proof of* (1A), *it suffices to assume that* $\mathfrak{Z}(\mathfrak{G})=1$. *If this assumption is satisfied, we have*

$$\mathfrak{C}(Z)=\mathfrak{C}(J) \tag{4.4}$$

for $Z\in\mathfrak{Z}(\mathfrak{C}(J))$, $Z\neq 1$.

The proof of the first statement is not very difficult. Suppose that $\mathfrak{G}$ satisfies the assumptions (I), (II) of (1A). Since $\mathfrak{Z}(\mathfrak{G})\subseteq\mathfrak{C}(J)$, we have $\mathfrak{Z}(\mathfrak{G})\subseteq\mathfrak{Z}(\mathfrak{C}(J))$. By (4D), $J\notin\mathfrak{Z}(\mathfrak{G})$ and it follows from (2.4) that $\mathfrak{Z}(\mathfrak{G})\subseteq\mathfrak{U}$. Set $\bar{\mathfrak{G}}=\mathfrak{G}/\mathfrak{Z}(\mathfrak{G})$ The centralizer of the involution $\bar{J}=J\,\mathfrak{Z}(\mathfrak{G})$ of $\bar{\mathfrak{G}}$ can be seen to be $\mathfrak{C}(J)/\mathfrak{Z}(\mathfrak{G})$. It is now clear that $\mathfrak{G}$ satisfies again the assumptions of (1A), with l replaced by $l\,|\,\mathfrak{Z}(\mathfrak{G})|=\bar{l}$. Suppose that (1A) has been proved for groups of order smaller than g. If $\mathfrak{Z}(\mathfrak{G})\neq 1$, then (1A) holds for $\bar{\mathfrak{G}}$. Since $\bar{l}\neq 1$, we must have $\bar{\mathfrak{G}}\simeq PSL(3,q)$ and $q\equiv 1 (\text{mod } 3)$. Then $\bar{l}=3$ and hence $l=1$, $|\mathfrak{Z}(\mathfrak{G})|=3$. It follows from the results of STEINBERG [8], §4, that $\mathfrak{G}\simeq SL(3,q)$ or that $\mathfrak{G}\simeq PSL(3,q)\times\mathfrak{Z}$ with $|\mathfrak{Z}|=3$. In the latter case, the assumption (I) is only satisfied when $q\not\equiv 1$ (mod 9). It is now clear that $\mathfrak{G}$ has a form given in (1A). It will therefore be sufficient to assume that $\mathfrak{Z}(\mathfrak{G})=1$.

Before coming to the proof of the second part of (4G), we prove a lemma.

(4H) *In* (4F), *we have*

$$\frac{5}{36}(5q+1)(q-1)<\mu q^2<\frac{5}{16}(5q+1)(q+1). \tag{4.5}$$

Proof. (a) Assume first that $x_1>5q$. Consider q as fixed and x_1 as an increasing variable. If $\delta_1=1$, the function $x_1(x_1+1)/(x_1+q)^2$ is increasing with the limit 1. Hence for $x_1>5q$.

$$\left(1-\frac{1}{q}\right)\frac{5}{36}(5q+1)/q<\mu<\left(1-\frac{1}{q}\right)$$

and this implies (4.5).

For $\delta_1=-1$, the function $x_1(x_1-1)/(x_1-q)^2$ decreases with the limit 1. Then for $x_1>5q$

$$1-\frac{1}{q}<\mu<\left(1-\frac{1}{q}\right)\frac{5}{16}(5q-1)/q$$

and it is seen easily that (4.5) holds.

(b) Assume next that $x_2>5q$. We use here the second form of μ in (4.3). The quotient $x_2(x_2+\delta_2)/(x_2-\delta_2 q)^2$ decreases for $\delta_2=1$ and increases for $\delta_2=-1$ when $x_2\to\infty$. An easy computation shows that (4.5) holds.

(c) Suppose then that $x_1\leqq 5q$, $x_2\leqq 5q$. Since $x_1\neq x_2$ (e.g. by (3.7)), it follows from (3.6) that $x_3<25$.

For $\delta_3=1$, we only have the following possibilities in (3.7).

$$x_3=7 \quad \text{or} \quad x_3=23 \quad \text{and} \quad n=4; \qquad x_3=15 \quad \text{and} \quad n=5.$$

Since $x_1+x_2=x_3+1$ by (3.1) and $x_1x_2=q^2x_3$ by (3.6), none of these cases is possible.

Likewise, for $\delta_3=-1$, we can only have

$$x_3=9, \qquad n=4 \quad \text{or} \quad x_3=17 \quad \text{and} \quad n=5.$$

Again (3.1) and (3.6) show that neither case can arise and this completes the proof of (4H).

We now complete the proof of (4G). Suppose that $\mathfrak{Z}(\mathfrak{G})=1$. Let $Z\neq 1$ be an element of $\mathfrak{Z}(\mathfrak{C}(J))$ and set $\mathfrak{G}_0=\mathfrak{C}(Z)$. Then $\mathfrak{G}_0\supseteq\mathfrak{C}(J)$ and $\mathfrak{G}_0$ satisfies again assumption (I).

If $\mathfrak{G}_0$ satisfies condition (II) too, then (4F) and (4H) apply to $\mathfrak{G}_0$, with the same value of q. Let μ_0 take the place of μ. Then, by (4.5),

$$|\mathfrak{G}:\mathfrak{G}_0|=\mu/\mu_0<\tfrac{9}{4}(q+1)/(q-1).$$

For $q\geqq 7$, it follows that $|\mathfrak{G}:\mathfrak{G}_0|<3$. Since $|\mathfrak{G}:\mathfrak{G}_0|$ is odd, we have $\mathfrak{G}=\mathfrak{G}_0$. However, this is impossible since $\mathfrak{Z}(\mathfrak{G})=1$. For $q<7$, we can only have $q=3$ and then (4G) is trivial since (2.4) shows that $Z=J$.

Suppose that $\mathfrak{G}_0$ does not satisfy condition (II). We then have Case II in § 3 for $\mathfrak{G}_0$ and (3.9) applies. Here

$$\mathfrak{C}_{\mathfrak{G}_0}(J_1)=\mathfrak{C}_{\mathfrak{G}}(J_1,Z)=\mathfrak{C}_{\mathfrak{G}}(J,Z^{Y^{-1}})^Y,$$

cf. (4.1). Since $Z\in\mathfrak{C}(\mathfrak{D})$, and $\mathfrak{D}^Y=\mathfrak{D}$, we have $Z^{Y^{-1}}\in\mathfrak{C}(\mathfrak{D})$. It follows from § 2, 2., that $\mathfrak{C}(J,Z^{Y^{-1}})$ is either $\mathfrak{C}(J)$ or $\mathfrak{C}(\mathfrak{D})$. In the former case, $Z^{Y^{-1}}$ belongs to $\mathfrak{Z}(\mathfrak{C}(J))$. Then by (4E), $Z^{Y^{-1}}=Z$, i.e. $Y\in\mathfrak{C}(Z)=\mathfrak{G}_0$. This is impossible since we would have Case I for $\mathfrak{G}_0$. Thus, $\mathfrak{C}(J,Z^{Y^{-1}})=\mathfrak{C}(\mathfrak{D})$ and we find $\mathfrak{C}_{\mathfrak{G}_0}(J_1)=\mathfrak{C}(\mathfrak{D})$. Since $\mathfrak{C}(J)\subseteq\mathfrak{G}_0$, the centralizers of J and $\mathfrak{D}$ respectively remain the same if $\mathfrak{G}_0$ is replaced by $\mathfrak{G}$. Now (3.9) becomes $|\mathfrak{G}_0|=c(J)$ and we have $\mathfrak{G}_0=\mathfrak{C}(J)$. This completes the proof of (4G).

§ 5. The degrees x_1, x_2, x_3 in the case $q^3|g$

We first prove a lemma.

(5A) *The number $x_1+\delta_1 q$ divides $(q-1)\,q(q+1)$.*

Proof. Let p be a prime. If a is an integer, let $\nu(a)$ denote the exact exponent with which p divides a. Since $c(J)$ divides g, it follows from (2.3) and (4F) that

(5.1) $$\nu(q-1)+\nu(q)+2\nu(q+1)+\nu(x_1)+\nu(x_1+\delta_1)\geqq 2\nu(x_1+\delta_1 q).$$

Suppose that for some p, we have

(5.2) $$\nu(x_1+\delta_1 q)>\nu(q-1)+\nu(q)+\nu(q+1).$$

On account of (5.1),

$$v(x_1)+v(x_1+\delta_1)>v(q-1)+v(q). \qquad (5.3)$$

If $v(x_1)>0$, then since $v(x_1+\delta_1 q)>v(q)$ by (5.2), we have $v(x_1)=v(q)$ and (5.3) is impossible. If $v(x_1+\delta_1)>0$, then since

$$x_1+\delta_1 q=(x_1+\delta_1)+\delta_1(q-1)$$

and since $v(x_1+\delta_1 q)>v(q-1)$ by (5.2), we have $v(q-1)=v(x_1+\delta_1)>0$. Again, (5.3) cannot hold. Hence (5.2) is false for all p, and (5A) is proved.

(5B) *If $q^3|g$, then $x_1=q^3$, $x_2=q^2+q+1$, $x_3=q(q^2+q+1)$.* Moreover, $\delta_1=\delta_2=-1$, $\delta_3=1$.

Proof. If $q=p_0^r$, and if v is defined as in the proof of (5A) with $p=p_0$, by (5A),

$$0\leqq v(x_1+\delta_1 q)\leqq v(q).$$

If we have inequality in both cases, then $v(x_1)=v(x_1+\delta_1 q)$ and (4F) would imply $v(g)<2v(q)$, contrary to the hypothesis.

If $v(x_1+\delta_1 q)=0$, we have $v(x_1)=0$. Since $q^3|g$, it follows from (4F) that $x_1\equiv-\delta_1(\operatorname{mod} q)$. On account of (5A), we can set

$$q^2-1=s(x_1+\delta_1 q) \qquad (s\in Z). \qquad (5.4)$$

Here, $s\equiv\delta_1(\operatorname{mod} q)$. Since then $s=-\delta_1=1$ is impossible, $x_1<q^2$. By (3.8), $q^3|x_2$. Now, (3.1) shows that $\delta_1=-1$. Again, by (3.8), $\delta_2=1$, $x_1-1\geqq q(q^2-x_1)$ whence $x_1\geqq q^2-q+1$. On the other hand, by (5.4) $2x_1\leqq q^2+2q-1$ (as $s\neq 1$). On combining the two inequalities for x_1, we find $x_1=7$, $q=3$. However this is inconsistent with (3.7). Thus, $v(x_1+\delta_1 q)\neq 0$.

Hence we must have $v(x_1+\delta_1 q)=v(q)$. On account of (5A), we can set $x_1+\delta_1 q=s\,q$ where s is a divisor of q^2-1. Hence $x_1\leqq q^3$ and the equality can only hold when $\delta_1=-1$, $s=q^2-1$. On the other hand, since $v(g)\geqq 3\,v(q)$ and since $v(x_1)>0$, (4F) yields

$$3\,v(q)\leqq v(g)=2\,v(q)+v(x_1)-2\,v(x_1+\delta_1 q).$$

It follows that $3\,v(q)\leqq v(x_1)$.

It is now clear that $x_1=q^3$, $\delta_1=-1$. Substituting this in (3.8), we find

$$q^3-1=(\delta_2+\delta_3\,q)\,x_2.$$

Since $\delta_2=-\delta_3$, we have $x_2=q^2+q+1$, $\delta_3=1$, $\delta_2=-1$. The value of x_3 is obtained from (3.6).

On combining (5B) and (4F), we find

(5C) *The order g is given by*

$$g=(q-1)^2(q+1)\,q^3(q^2+q+1)/l=(q^3-1)(q^2-1)\,q^3/l. \qquad (5.5)$$

On applying (3.2) for $G=1$, we have

(5D) *The degrees of the remaining degrees in* $B_0(\mathfrak{G}, 2)$ *are given by*

$$\chi_4(1)=q^2+q, \quad \chi^{(\mu)}(1)=q^3-1. \tag{5.6}$$

§ 6. The centralizers of the p_0-elements in the case $q^3 | g$

In addition to the assumptions (I), (II), we assume in §§ 6–8 that $\mathfrak{G}$ satisfies the further conditions

(III) $\qquad \mathfrak{Z}(\mathfrak{G})=1.$

(IV) $\qquad q^3 | g.$

We have seen in (4G) that (III) does not represent a restriction in the proof. Condition (IV) will be discussed in §§ 9–10.

As before, we set $q=p_0^r$ where p_0 is a prime.

(6A) *If p is a prime dividing $v=(q+1)/2^{n-2}$, then $\mathfrak{G}$ does not contain elements of order $p\,p_0$.*

Proof. Let $\mathfrak{P}$ be an S_p-group of $\langle V\rangle$ in (2.13). It follows from (5.5) that $\mathfrak{P}$ is an S_p-group of $\mathfrak{G}$. For $\xi\in\mathfrak{P}$, we have $S\in\mathfrak{C}(\xi)$. On account of (2.10), $\mathfrak{C}(\mathfrak{S})$ has order prime to p. We now see that for $\xi\neq 1$, the group $\langle S\rangle$ is an S_2-group of $\mathfrak{C}(\xi)$. This implies that if ξ is conjugate in $\mathfrak{G}$ to $\xi_1\in\mathfrak{P}$, then ξ and ξ_1 are conjugate in $\mathfrak{N}(\langle S\rangle)=\langle\mathfrak{C}(S), J_1\rangle$, c.f. (2.12). Thus $\xi_1=\xi$ or $\xi_1=\xi^{J_1}$. Since J_1 normalizes $\langle S\rangle$, it normalizes $\mathfrak{C}(S)$ and $\mathfrak{P}$. If $\xi\in\mathfrak{P}$ is centralized by J_1, it belongs to $\mathfrak{C}(\mathfrak{D})$. By (2.7), this is impossible for $\xi\neq 1$. Consequently, J_1 inverts $\mathfrak{P}$ and we see that $\xi_1=\xi^{\pm 1}$.

In particular, every element of $\mathfrak{N}(\mathfrak{P})$ centralizes or inverts ξ. Since we must have the same situation for all $\xi\in\mathfrak{P}-\{1\}$, it follows that

$$\mathfrak{N}(\mathfrak{P})=\langle\mathfrak{C}(\mathfrak{P}), J_1\rangle.$$

This shows that the assumptions of [3], III, § 9 are satisfied with $m=2$. Hence the principal p_0-block $B_0(\mathfrak{G}, p)$ of $\mathfrak{G}$ consists $\frac{1}{2}(|\mathfrak{P}|+3)$ irreducible characters $1, \psi_1, \psi^{(\mu)}$ with $1\leq\mu\leq\frac{1}{2}(|\mathfrak{P}|-1)$. These are all real. There exists a sign δ^* such that

$$\psi_1(G)=\delta^* \tag{6.1}$$

for all p-singular $G\in\mathfrak{G}$ and we have

$$\delta^*+\psi_1(1)=\psi^{(\mu)}(1). \tag{6.2}$$

It follows from the results of [2] that an irreducible real character of odd degree of a group belongs to the principal 2-block. If $\psi_1(1)$ was even, $B_0(\mathfrak{G}, 2)$ would contain all $\psi^{(\mu)}$. This is impossible for $|\mathfrak{P}|>3$, since $B_0(\mathfrak{G}, 2)$ does not contain two distinct characters of the same odd degree. Hence $\psi_1\in B_0(\mathfrak{G}, 2)$. We may assume that this is still true for $|\mathfrak{P}|=3$, since we can here interchange ψ_1 and $\psi^{(1)}$.

It is now clear that ψ_1 is one of the characters χ_1, χ_2, χ_3. Since (6.1) shows that $\psi_1(S^\lambda \xi)=\delta^*$ for all $S^\lambda \in \langle S\rangle$ and $\xi \in \mathfrak{P}-\{1\}$, Theorem (8A) of [3], III shows that we must have $\psi_1=\chi_1$, $\delta^*=\delta_1$.

If G was an element of order $p\,p_0$, we would have $\chi_1(G)=0$ since χ_1 is of defect 0 for p_0 (cf. (5B) and (5C)). This is not consistent with (6.1), and (6A) is proved.

(6B) *If $Q \neq 1$ is a p_0-element, $c(Q)$ divides $2u\,q^3$. If Q is not conjugate to an element of $\mathfrak{X}_1$ (§ 2, 5), $c(Q)\,|\,q^3$.*

Proof. (a) As is well known, the number

$$\omega^{(1)}(G)=g\,\chi^{(1)}(G)/(c(G)\,\chi^{(1)}(1))$$

is an algebraic integer for every $G\in\mathfrak{G}$. As in the proof of (6A), we have $\chi_1(Q)=0$ for $G=Q$ and then $\chi^{(1)}(Q)=\delta_1=-1$ by (3.2). If we substitute the values of g and of $\chi^{(1)}(1)$ from (5.5) and (5.6), we see that $c(Q)$ divides $(q^2-1)\,q^3/l$. By (6A), $c(Q)$ is relatively prime to $(q+1)/2^{n-2}$. If $c(Q)$ was divisible by 4, Q would centralize a group of order 4, that is, a conjugate of $\langle F\rangle$ or a conjugate of $\mathfrak{D}$. On account of (2.13) and (2.7), both possibilities are excluded. Hence $c(Q)\,|\,2u\,q^3$.

(b) Choose $Q\neq 1$ in $\mathfrak{X}_1$. Then $\mathfrak{U}\subseteq\mathfrak{C}(Q)$. Let p be a prime dividing u. It follows from (a) that an S_p-group $\mathfrak{P}$ of $\mathfrak{U}$ is an S_p-group of $\mathfrak{C}(Q)$. Moreover, the arguments in (a) show that $\omega^{(1)}(Q)$ is relatively prime to p. Hence, the S_p-group $\mathfrak{P}$ of $\mathfrak{C}(Q)$ includes a defect group of the p-block of $\chi^{(1)}$, cf. [2], I, (8A). This implies that if $\chi^{(1)}(G)\neq 0$ for some $G\in\mathfrak{G}$, the p-factor of G is conjugate to an element of $\mathfrak{U}$, [2], II, (7A).

(c) Let $Q\neq 1$ be again an arbitrary p_0-element of $\mathfrak{G}$. If $c(Q)$ is even, Q commutes with an involution and there exist conjugates of Q in $\mathfrak{C}(J)$ and then in $\mathfrak{X}_1$.

If $c(Q)$ is odd but not a power of p_0, it is divisible by some prime p dividing u. There exist elements P of order p in $\mathfrak{C}(Q)$. For $G=P\,Q$, we have again $\chi_1(G)=0$, $\chi^{(1)}(G)=\delta_1=-1$. Now the result of (b) shows that P is conjugate to an element of $\mathfrak{U}$. After replacing Q by a conjugate, we may assume $P\in\mathfrak{U}$. Now (4G) shows that $Q\in\mathfrak{C}(P)=\mathfrak{C}(J)$. Then $c(Q)$ is even, a contradiction.

(6C) *If $Q\neq 1$ is a p_0-element, the S_{p_0}-group $\mathfrak{R}$ of $\mathfrak{C}(Q)$ is characteristic in $\mathfrak{C}(Q)$.*

This is trivial if $\mathfrak{C}(Q)$ is a p_0-group. In the remaining case, by (6B), we may assume $Q\in\mathfrak{X}_1$. Then $\langle J\,U\rangle=\mathfrak{Z}(\mathfrak{C}(J))$ is a subgroup of $\mathfrak{C}(Q)$ whose index in $\mathfrak{C}(Q)$ is a power of p_0. Since (4E) shows that no two distinct elements of $\langle J\,U\rangle$ are conjugate, the theorems of BURNSIDE and FROBENIUS imply that $\mathfrak{R}$ is normal and then characteristic in $\mathfrak{C}(Q)$.

We conclude this section with some auxiliary results which will be used repeatedly. We set $J_0=J$. We also write $\mathfrak{X}_i^{(\lambda)}$ for $\mathfrak{X}_i^{Y^\lambda}$, $i=1$ or 2.

(6D) *The group* $\mathfrak{C}(\mathfrak{D})$ *normalizes* $\mathfrak{X}_i^{(\lambda)}$, *acting transitively on* $\mathfrak{X}_i^{(\lambda)}-\{1\}$. In $\mathfrak{X}_i^{(\lambda)}$, *the superscript* λ *can be taken* (mod 3). *We have*

$$(6.3)\qquad \mathfrak{X}_1^T=\mathfrak{X}_2,\quad (\mathfrak{X}_1^{(1)})^T=\mathfrak{X}_2^{(2)},\quad (\mathfrak{X}_2^{(1)})^T=\mathfrak{X}_1^{(2)}.$$

For $\lambda=0, 1, 2$, *the element* J_λ *centralizes* $\mathfrak{X}_i^{(\lambda)}$ *and it inverts* $\mathfrak{X}_i^{(\lambda-1)}$ *and* $\mathfrak{X}_i^{(\lambda+1)}$.

Proof. As shown by (4.1), Y normalizes $\mathfrak{D}$ and hence $\mathfrak{C}(\mathfrak{D})$. Since $\mathfrak{C}(\mathfrak{D})$ normalizes $\mathfrak{X}_i$, acting transitively on $\mathfrak{X}_i-\{1\}$, transformation with Y^λ shows that $\mathfrak{C}(\mathfrak{D})$ normalizes $\mathfrak{X}_i^{(\lambda)}$, acting transitively on $\mathfrak{X}_i^{(\lambda)}-\{1\}$.

Also, $Y^3\in\mathfrak{C}(\mathfrak{D})$ and hence $\mathfrak{X}_i^{(\lambda+3)}=\mathfrak{X}_i^{(\lambda)}$. By (2.6) and (4.1), $YT\equiv TY^{-1}$ mod $\mathfrak{C}(\mathfrak{D})$. For the first equation (6.3), cf. § 2, 5. Then

$$\mathfrak{X}_1^{(1)\,T}=\mathfrak{X}_1^{Y\,T}=\mathfrak{X}_1^{T\,Y^{-1}}=\mathfrak{X}_2^{Y^{-1}}=\mathfrak{X}_2^{(2)}.$$

The proof of the last equation (6.3) is analogous.

Since $\mathfrak{X}_i$ is centralized by J and inverted by J_1 and $J_2=J\,J_1$, transformation with Y^λ yields the last statement in (6D).

(6E) *Let* $\mathfrak{M}$ *be a subgroup of* $\mathfrak{G}$ *of odd order which is normalized by* $\mathfrak{D}$. *If we set* $\mathfrak{M}_\lambda=\mathfrak{M}\cap\mathfrak{C}(J_\lambda)$ for $\lambda=0, 1, 2$, then

$$(6.4)\qquad \mathfrak{M}=\mathfrak{M}_0\,\mathfrak{M}_1\,\mathfrak{M}_2.$$

If $\mathfrak{M}$ *is a* p_0*-group,* J_0 *inverts* $\mathfrak{M}_1$ *and* $\mathfrak{M}_2$, J_1 *inverts* $\mathfrak{M}_0$ *and* $\mathfrak{M}_2$, J_2 *inverts* $\mathfrak{M}_0$ *and* $\mathfrak{M}_1$. *Moreover.*

$$(6.5)\qquad 1=\mathfrak{M}_0\,\mathfrak{M}_1\cap\mathfrak{M}_2=\mathfrak{M}_0\,\mathfrak{M}_2\cap\mathfrak{M}_1=\mathfrak{M}_1\,\mathfrak{M}_2\cap\mathfrak{M}_0.$$

Proof. A slight modification of the proof of Proposition 8 of [*2*], II, yields a direct elementary proof of (6.4), see also Gorenstein and Walter [*6*], § 2.

If $\mathfrak{M}$ is a p_0-group, then (2.7) shows that $\mathfrak{M}\cap\mathfrak{C}(\mathfrak{D})=1$. Since J normalizes $\mathfrak{M}_1$ and $\mathfrak{M}_2$ centralizing only 1, it follows that J inverts $\mathfrak{M}_1$ and $\mathfrak{M}_2$. Similarly, we see that J_1 inverts $\mathfrak{M}_0$ and $\mathfrak{M}_2$ and that J_2 inverts $\mathfrak{M}_0$ and $\mathfrak{M}_1$. If $M_0\,M_1=M_2$ with $M_i\in\mathfrak{M}_i$, transformation with J_2 yields

$$M_0^{-1}\,M_1^{-1}=M_2=M_0\,M_1.$$

Hence $M_0^{-2}=M_1^2$ and, since $\mathfrak{M}_i$ has odd order, we have $M_0^{-1}=M_1$. This yields the first statement (6.5). The two others are proved in a similar fashion.

§ 7. Continuation. The p_0-Sylow subgroup of $\mathfrak{G}$

Let $Q_0\neq 1$ be an element of $\mathfrak{X}_1$. Let $\mathfrak{R}\supseteq\mathfrak{X}_1$ be an S_{p_0}-group of $\mathfrak{C}(Q_0)$ and let $\mathfrak{Q}\supseteq\mathfrak{R}$ be an S_{p_0}-group of $\mathfrak{G}$. Since J_i maps $Q_0\to Q_0^{\pm 1}$ for $i=0, 1, 2$, it follows from (6 C) that $\mathfrak{D}$ normalizes $\mathfrak{R}$. Applying (6 E) to $\mathfrak{M}=\mathfrak{R}$, we set

$$(7.1)\qquad \mathfrak{R}=\mathfrak{R}_0\,\mathfrak{R}_1\,\mathfrak{R}_2$$

with $\mathfrak{N}_i=\mathfrak{N}\cap\mathfrak{C}(J_i)$. Here, $\mathfrak{N}_0=\mathfrak{X}_1$. Since $\mathfrak{N}_1$ is centralized by J_1 and inverted by J_2, the group $\mathfrak{N}_1^{Y^{-1}}$ is centralized by $J_1^{Y^{-1}}=J$ and inverted by $J_2^{Y^{-1}}=J_1$. It follows from § 2, 5., that $\mathfrak{N}_1^{Y^{-1}}\subseteq\mathfrak{X}_\alpha$ with $\alpha=1$ or 2. A similar argument shows that $\mathfrak{N}_2^{Y}\subseteq\mathfrak{X}_\beta$ with $\beta=1$ or 2. Thus, we have

(7A) In (7.1), $\mathfrak{N}_0=\mathfrak{X}_1$, $\mathfrak{N}_1\subseteq\mathfrak{X}_\alpha^{(1)}$, $\mathfrak{N}_2\subseteq\mathfrak{X}_\beta^{(2)}$ with α, $\beta=1$ or 2.

We next prove a lemma.

(7B) *If* $c(Q_0)\not\equiv 0 \pmod{q^3}$, *either* $\mathfrak{N}_1=\mathfrak{X}_2^{(1)}\subseteq\mathfrak{C}(\mathfrak{X}_1)$ *or* $\mathfrak{N}_2=\mathfrak{X}_2^{(2)}\subseteq\mathfrak{C}(\mathfrak{X}_1)$.

Proof. Let $Z\neq 1$ be an element of $\mathfrak{Z}(\mathfrak{Q})$. Then $Z\in\mathfrak{C}(Q_0)\cap\mathfrak{Q}=\mathfrak{N}$. Set $Z=Z_0Z_1Z_2$ with $Z_i\in\mathfrak{N}_i$. Since $Z\in\mathfrak{C}(\mathfrak{X}_1)$ and $Z_0\in\mathfrak{X}_1\subseteq\mathfrak{C}(\mathfrak{X}_1)$, we have $Z_1Z_2\in\mathfrak{C}(\mathfrak{X}_1)$. Transformation with J_2 shows that $Z_1^{-1}Z_2\in\mathfrak{C}(\mathfrak{X}_1)$. It follows that $Z_1Z_2(Z_1^{-1}Z_2)^{-1}=Z_1^2\in\mathfrak{C}(\mathfrak{X}_1)$. This implies $Z_1\in\mathfrak{C}(\mathfrak{X}_1)$ and then $Z_2\in\mathfrak{C}(\mathfrak{X}_1)$.

If we had $Z_1=Z_2=1$, then $Z=Z_0\in\mathfrak{X}_1$ and Z would be conjugate to Q_0 in $\mathfrak{G}$ (even in $\mathfrak{C}(J)$). This is impossible since $c(Z)\equiv 0 \pmod{q^3}$ while we assumed $c(Q_0)\not\equiv 0 \pmod{q^3}$.

Suppose that $Z_1\neq 1$. Since $Z_1\in\mathfrak{C}(\mathfrak{X}_1)\cap\mathfrak{X}_\alpha^{(1)}$, transformation with $\mathfrak{C}(\mathfrak{D})$ shows that $\mathfrak{X}_\alpha^{(1)}\subseteq\mathfrak{C}(\mathfrak{X}_1)$, cf. (6 D). Hence $\mathfrak{X}_\alpha^{(1)}\subseteq\mathfrak{C}(Q_0)$ and by (6 C), $\mathfrak{X}_\alpha^{(1)}\subseteq\mathfrak{N}$. It is now clear that $\mathfrak{X}_\alpha^{(1)}=\mathfrak{N}_1\subseteq\mathfrak{C}(\mathfrak{X}_1)$. If here $\alpha=1$, transformation with Y and Y^{-1} shows that $\mathfrak{X}_1$, $\mathfrak{X}_1^{(1)}$, $\mathfrak{X}_1^{(2)}$ centralize each other. It follows that $\mathfrak{N}=\mathfrak{X}_1\mathfrak{X}_1^{(1)}\mathfrak{X}_1^{(2)}$ has order q^3 contrary to the assumption. Hence $\alpha=2$.

Similarly, if $Z_2\neq 1$, we have $\mathfrak{N}_2=\mathfrak{X}_2^{(2)}\subseteq\mathfrak{C}(\mathfrak{X}_1)$.

(7C) *We have* $c(Q_0)\equiv 0 \pmod{q^3}$.

Proof. (a) If this was not so, (7 B) would apply. Suppose we have the first case in (7 B). Since $\mathfrak{X}_1\cap\mathfrak{X}_2^{(1)}\subseteq\mathfrak{C}(\mathfrak{D})$ and since $\mathfrak{C}(\mathfrak{D})$ is p_0-regular (cf. 2.7), $\mathfrak{X}_1\cap\mathfrak{X}_2^{(1)}=1$. It follows that $\mathfrak{X}_1\times\mathfrak{X}_2^{(1)}$ is direct and so is the transform $\mathfrak{X}_1^{(1)}\times\mathfrak{X}_2^{(2)}$. Set

$$\mathfrak{M}=\mathfrak{X}_1^{(1)}\times\mathfrak{X}_2^{(2)}, \qquad \tilde{\mathfrak{G}}=\mathfrak{N}(\mathfrak{M}), \qquad \mathfrak{A}=\tilde{\mathfrak{G}}\cap\mathfrak{C}(J).$$

Then $|\mathfrak{M}|=q^2$. Since $q^2p_0\mid g$, we conclude that

$$|\tilde{\mathfrak{G}}|\equiv 0 \pmod{q^2p_0}. \tag{7.2}$$

It follows from (6 D) that $\mathfrak{A}\supseteq\langle\mathfrak{C}(\mathfrak{D}), T\rangle$.

(b) We show next that $\mathfrak{A}$ is p_0-regular. Suppose that this was not so. Then $\mathfrak{A}$ contains an element A of order p_0.

Since $\mathfrak{C}(J)\simeq GL(2,q)/\mathfrak{L}$ with $\mathfrak{L}\subseteq\mathfrak{Z}(GL(2,q))$, we may represent $\mathfrak{C}(J)$ by projectivities of a projective line Λ of order q. The p_0-elements of $\mathfrak{C}(J)$ leaving a point of Λ fixed form an S_{p_0}-subgroup of $\mathfrak{C}(J)$. In this fashion, the $q+1$ points of Λ correspond to the $q+1$ S_{p_0}-subgroups of $\mathfrak{C}(J)$. Let π_i be the point corresponding to $\mathfrak{X}_i$. Using coordinates, we see that $\mathfrak{C}(\mathfrak{D})$ acts transitively on the $q-1$ points of Λ different from π_1 and π_2. Furthermore, T interchanges π_1 and π_2.

Now $A \in \mathfrak{C}(J)$. Let π be the fixed point of A. It follows from the preceding remarks that some element of

$$\langle A, \mathfrak{C}(\mathfrak{D}), T \rangle \subseteq \mathfrak{A}$$

maps π on π_1. After replacing A by a conjugate, we may assume that $\pi = \pi_1$, and then $A \in \mathfrak{X}_1$. Now transformation with $\mathfrak{C}(\mathfrak{D})$ shows that $\mathfrak{X}_1 \subseteq \mathfrak{A}$. Thus, $\mathfrak{X}_1$ normalizes $\mathfrak{M}$ and

$$\mathfrak{Q}_1 = \mathfrak{X}_1 \mathfrak{X}_1^{(1)} \mathfrak{X}_2^{(2)}$$

is a p_0-group. Since $\mathfrak{D}$ normalizes $\mathfrak{Q}_1$, it normalizes $\mathfrak{Z}(\mathfrak{Q}_1)$ and we have

$$\mathfrak{Z}(\mathfrak{Q}_1) = \mathfrak{Z}_0 \mathfrak{Z}_1 \mathfrak{Z}_2; \qquad \mathfrak{Z}_i = \mathfrak{Z}(\mathfrak{Q}_1) \cap \mathfrak{C}(J_i),$$

cf. (6 E). Not all three groups $\mathfrak{Z}_i$ can be trivial. Hence some $\mathfrak{Z}_i$ contains an element Z of order p_0 with $c(Z) \equiv 0 \pmod{q^3}$. On transforming with Y^{-i}, we see that $\mathfrak{C}(J)$ contains such a Z. Then Z is conjugate to Q_0. But this is impossible since $c(Q_0) \not\equiv 0 \pmod{q^3}$. This proves that $\mathfrak{A}$ is p_0-regular.

(c) Let $\tilde{\mathfrak{S}} \supseteq \mathfrak{D}_1$ be an S_2-group of $\tilde{\mathfrak{G}}$. Then $\tilde{\mathfrak{S}} \subseteq \mathfrak{S}^\xi$ for some $\xi \in \mathfrak{G}$. If we set $\tilde{\mathfrak{S}} = \mathfrak{S}_0^\xi$, $\mathfrak{D}_1 = \mathfrak{S}_1^\xi$, we have

$$\mathfrak{S}^\xi \supseteq \mathfrak{S}_0^\xi \supseteq \mathfrak{S}_1^\xi = \mathfrak{D}_1 = \langle J_1, T \rangle.$$

Set $|\mathfrak{S}_0| = 2^m$. Then $|\mathfrak{S}_0 \cap \langle S \rangle| \geqq 2^{m-1} \geqq 4$. It follows that $S^{2^{n-m}} = S_0 \in \mathfrak{S}_0$. Likewise, $F = S^{2^{n-3}}$ belongs to the dihedral group $\mathfrak{S}_1$. Hence $F^\xi \in \mathfrak{D}_1$ and we must have $F^\xi = F^{\pm 1}$. Since ξ can be replaced by $T\xi$, we may assume $F^\xi = F$. Now (2.13) shows that $\xi \in \mathfrak{C}(S)$ and it is clear that $\tilde{\mathfrak{S}} = \mathfrak{S}_0^\xi = \langle S_0, J_1 \rangle$. In particular, $\tilde{\mathfrak{S}}$ is quasi-dihedral for $m = n$ and dihedral for $m < n$.

We discuss the various possibilities which we can have for fusion in $\tilde{\mathfrak{S}}$, cf. [3], III § 7 and § 8. Note that J and J_1 are not conjugate in $\tilde{\mathfrak{G}}$. Indeed, if $J = J_1^\rho$ with $\rho \in \tilde{\mathfrak{G}}$, then $(\mathfrak{X}_1^{(1)})^\rho \subseteq \mathfrak{C}(J) \cap \tilde{\mathfrak{G}}$ and this is excluded by the result of (b). We can then only have one of the following cases:

1. No fusion occurs in $\tilde{\mathfrak{G}}$.

Then $\tilde{\mathfrak{G}}$ has a normal 2-complement $\mathfrak{R}$. On applying (6 E), we can set

$$\mathfrak{R} = \mathfrak{R}_0 \mathfrak{R}_1 \mathfrak{R}_2, \qquad \mathfrak{R}_i = \mathfrak{R} \cap \mathfrak{C}(J_i).$$

Here, by the result in (b), $|\mathfrak{R}_0| \not\equiv 0 \pmod{p_0}$. Furthermore $|\mathfrak{R}_i|$ contains p_0 at most to the order q, since $\mathfrak{R}_i \subseteq \mathfrak{C}(J_i)$. It follows that $|\mathfrak{R}|$ and then $|\tilde{\mathfrak{G}}|$ are not divisible by $q^2 p_0$ cf. [3] II, Proposition 8. This is inconsistent with (7.2).

2. $m = n$ and F is conjugate to $S J_1$ in $\tilde{\mathfrak{G}}$. Here, we can apply the results of § 3, Case II, to $\tilde{\mathfrak{G}}$. By (3.9),

$$|\tilde{\mathfrak{G}}| = |\mathfrak{C}(J) \cap \tilde{\mathfrak{G}}| \cdot |\mathfrak{C}(J_1) \cap \tilde{\mathfrak{G}}|^2 \cdot |\mathfrak{C}(\mathfrak{D}) \cap \tilde{\mathfrak{G}}|^{-2}.$$

Part (b) shows that the first factor is prime to p_0. Since $|\mathfrak{C}(J_1) \cap \tilde{\mathfrak{G}}|$ contains p_0 at most to the power q, it follows that $|\tilde{\mathfrak{G}}|$ is not divisible by $q^2 p_0$. Again, this is a contradiction with (7.2).

10*

3. $m<n$ and J conjugate to $S_0\,J_1$ in $\tilde{\mathfrak{G}}$.

Here, $\mathfrak{D}^*=\langle J, S_0\,J_1\rangle$ is elementary abelian of order 4 and since $\mathfrak{D}^*\subseteq\tilde{\mathfrak{G}}$, the group $\mathfrak{D}^*$ normalizes $\mathfrak{M}$. Applying (6 E) to $\mathfrak{D}^*$ instead of $\mathfrak{D}$, we can set

(7.3) $$\mathfrak{M}=(\mathfrak{M}\cap\mathfrak{C}(J))(\mathfrak{M}\cap\mathfrak{C}(S_0\,J_1))(\mathfrak{M}\cap\mathfrak{C}(J\,S_0\,J_1)).$$

On account of the result of part (b), the first factor is 1. Since there exists an element $y\in\tilde{\mathfrak{G}}$ which carries $S_0\,J_1\to J$, we have

$$(\mathfrak{M}\cap\mathfrak{C}(S_0\,J_1))^y\subseteq\mathfrak{M}\cap\mathfrak{C}(J)=1.$$

It follows that the second factor of (7.2) is 1. Similarly, the third factor is 1, and we have a contradiction.

We have now shown that the first case in (7 B) is impossible. If we had the second case, we would take $\mathfrak{M}=\mathfrak{X}_1^{(2)}\times\mathfrak{X}_2^{(1)}$, $\tilde{\mathfrak{G}}=\mathfrak{N}(\mathfrak{M})$. An analogous method leads to a contradiction. This finishes the proof of (7 C).

It is now clear that the S_{p_0}-group $\mathfrak{R}$ of $\mathfrak{C}(Q_0)$ has order q^3. We then have

(7.4) $$\mathfrak{Q}=\mathfrak{R}=\mathfrak{X}_1\,\mathfrak{X}_\alpha^{(1)}\,\mathfrak{X}_\beta^{(2)},$$

cf. (7.1) and (7 A). Since Q_0 centralizes $\mathfrak{X}_\alpha^{(1)}$ and $\mathfrak{X}_\beta^{(2)}$, transformation with $\mathfrak{C}(\mathfrak{D})$ yields

(7.5) $$\mathfrak{X}_\alpha^{(1)}\subseteq\mathfrak{C}(\mathfrak{X}_1),\qquad \mathfrak{X}_\beta^{(2)}\subseteq\mathfrak{C}(\mathfrak{X}_1).$$

(7D) *In* (7.4), $\alpha=\beta=2$.

Proof. If we have $\alpha=1$ or $\beta=1$, transformation of (7.5) with powers of Y shows that $\mathfrak{X}_1$, $\mathfrak{X}_1^{(1)}$, $\mathfrak{X}_1^{(2)}$ centralize each other. It follows easily that

$$\mathfrak{R}=\mathfrak{Q}=\mathfrak{X}_1\times\mathfrak{X}_1^{(1)}\times\mathfrak{X}_1^{(2)}.$$

In particular, $\mathfrak{Q}$ is abelian.

On account of (6 B), we have the following classes of p_0-elements: 1) The class of 1. 2) The class of Q_0 with $g/(2u\,q^3)$ elements. 3) Classes with g/q^3 elements. If there are s classes of type 3), by the theorem of Frobenius, we have

$$1+g/(2u\,q^3)+s\,g/q^3\equiv 0\ (\mathrm{mod}\ q^3).$$

By (5.5), $g/(2u\,q^3)\equiv -(q+1)(\mathrm{mod}\ q^3)$ and we find

(7.6) $$q+2u\,s(q+1)\equiv 0\ (\mathrm{mod}\ q^3).$$

In particular, $s\neq 0$. It follows that $\mathfrak{C}(\mathfrak{R})\subseteq\mathfrak{R}$ and hence that $\mathfrak{C}(\mathfrak{R})=\mathfrak{R}$.

If two elements of $\mathfrak{R}$ are conjugate in $\mathfrak{G}$, they are conjugate in $\mathfrak{N}(\mathfrak{R})$. By (6 D) and (7.4), $\mathfrak{C}(\mathfrak{D})\subseteq\mathfrak{N}(\mathfrak{R})$. On account of (2.7), we can set

$$|\mathfrak{N}(\mathfrak{R})|=t(q-1)^2\,q^3/l$$

with integral t. We count the elements of $\mathfrak{R}$ according to their $\mathfrak{N}(\mathfrak{R})$-classes. This yields

(7.7) $$q^3=1+t(q-1)+s\,t(q-1)\,2u.$$

Here, $t(q-1)$ is the number of elements in the $\mathfrak{N}(\mathfrak{N})$-class of Q_0. Since the elements of $\mathfrak{X}_1^{(i)}-\{1\}$ with $i=0, 1, 2$ lie in this class, we have $t\geqq 3$. As (7.7) implies

$$q^2+q+1=t(1+2us),$$

we find $2us<(1/3)(q^2+q+1)$. But then the left side of (7.6) is less than q^3. We have a contradiction and (7 D) is proved.

(7E) *We have*

$$\mathfrak{Q}=\mathfrak{X}_1\mathfrak{X}_2^{(1)}\mathfrak{X}_2^{(2)}, \tag{7.8}$$

$$\mathfrak{X}_2^{(1)}\subseteq\mathfrak{C}(\mathfrak{X}_1), \qquad \mathfrak{X}_2^{(2)}\subseteq\mathfrak{C}(\mathfrak{X}_1), \tag{7.9}$$

$$[\mathfrak{X}_2^{(1)}, \mathfrak{X}_2^{(2)}]\subseteq\mathfrak{X}_1. \tag{7.10}$$

Proof. On account of (7.4), (7.5), and (7 D), we only have to prove (7.10). For any $\tau\in\mathfrak{G}$, we shall write τ' for τ^Y, τ'' for τ^{Y^2}. For $\xi, \eta\in\mathfrak{X}_2$, we have $\xi', \eta''\in\mathfrak{Q}$ and we can write

$$\xi'^{-1}\eta''\xi'=\zeta\rho'\sigma'' \tag{7.11}$$

with $\zeta\in\mathfrak{X}_1$; $\rho, \sigma\in\mathfrak{X}_2$, cf. (7.8). Transformation with J_1 inverts the left side while the right side is changed into $\zeta^{-1}\rho'(\sigma'')^{-1}$. Since ζ commutes with $\rho'\sigma''$ by (7.9), we obtain $(\sigma'')^{-1}\rho'\sigma''=(\rho')^{-1}$. As σ'' has odd order, it follows that $\rho'=1$. Thus,

$$\xi'^{-1}\eta''\xi'=\zeta\sigma''.$$

Transforming now with J_2, we obtain

$$\xi'\eta''\xi'^{-1}=\zeta^{-1}\sigma''=\zeta^{-2}\zeta\sigma''=\zeta^{-2}\xi'^{-1}\eta''\xi'.$$

Since ζ commutes with ξ' and η'', this can be written in the form $[\eta'',(\xi')^{-2}]=\zeta^{-2}\in\mathfrak{X}_1$, and (7.10) becomes evident.

§ 8. Proof of Theorem (1A) in the case $q^3|g$

(8A) *If* $\mathfrak{G}$ *satisfies the conditions* (I), (II), (III), (IV), *then it satisfies the postulates* (I)–(V) *of* [*4*].

Proof. The postulate (I) of [*4*] is evident since $\mathfrak{D}$ is elementary abelian of order 4 and Y satisfies (4.1). Note that, by (7.9), $\mathfrak{X}_1^{(1)}$ centralizes $\mathfrak{X}_2^{(2)}$ and $\mathfrak{X}_1^{(2)}$ centralizes $\mathfrak{X}_2^{(1)}$. It is seen easily that $\mathfrak{X}_1^{(1)}\cap\mathfrak{X}_2^{(2)}=1$ and $\mathfrak{X}_1^{(2)}\cap\mathfrak{X}_2^{(1)}=1$. We may then set

$$\mathfrak{M}=\mathfrak{X}_1^{(1)}\times\mathfrak{X}_2^{(2)}, \qquad \mathfrak{M}^*=\mathfrak{X}_2^{(1)}\times\mathfrak{X}_1^{(2)}. \tag{8.1}$$

It is clear that $\mathfrak{M}$ and $\mathfrak{M}^*$ have the same order q^2. By (6 D), both groups are inverted by J. This implies

$$\mathfrak{M}\cap\mathfrak{C}(J)=\mathfrak{M}^*\cap\mathfrak{C}(J)=1.$$

If $\xi \in \mathfrak{M} \cap \mathfrak{M}^*$, we have $\xi = \alpha_1' \beta_2'' = \alpha_2' \beta_1''$ with $\alpha_i, \beta_i \in \mathfrak{X}_i$. Transformation with J_1 yields $\xi = 1$. Hence $\mathfrak{M} \cap \mathfrak{M}^* = 1$. Using transformations by Y and T, we conclude from (8.1) and (7 E) that $\mathfrak{X}_1$ and $\mathfrak{X}_2$ normalize $\mathfrak{M}$ and $\mathfrak{M}^*$. By (6 D), $\mathfrak{C}(\mathfrak{D})$ normalizes $\mathfrak{M}$ and $\mathfrak{M}^*$ and now (2.15) shows that $\mathfrak{C}(J)$ normalizes $\mathfrak{M}$ and $\mathfrak{M}^*$. Hence $\mathfrak{G}$ satisfies the condition (II) of [4].

The condition (III) of [4] is obvious: There are two conjugate classes of involutions in $\mathfrak{C}(J)$. They are represented by J and J_1. Since $\mathfrak{C}(J)$ normalizes $\mathfrak{M}$, the group $\mathfrak{C}(J)\,\mathfrak{M}$ has order $q^2\, c(J)$ and (2.3) and (5.5) show that

$$|\mathfrak{G} : \mathfrak{C}(J)\,\mathfrak{M}| = q^2 + q + 1.$$

This is condition (IV) of [4]. Finally, the condition (V) of [4] holds. Indeed, if $\xi \neq 1$ is an element of $\mathfrak{M}$, the number of images of ξ under transformations by elements of $\mathfrak{C}(J)$ is $c(J)/c(J, \xi)$, It follows from (2.3) and (6 B) that this number is divisible by $q^2 - 1$. Hence $\mathfrak{C}(J)$ acts transitively on $\mathfrak{M} - \{1\}$, and this a stronger statement than (V).

Now [4] can be applied. We see that there exists a Desarguesian plane Π of order q and a representation γ of $\mathfrak{G}$ by collineations of Π with the following properties:

(α) The kernel $\mathfrak{K}$ is the intersection of the centralizers of the involutions of $\mathfrak{G}$.

(β) The image $\gamma(\mathfrak{G})$ of $\mathfrak{G}$ includes the group $PSL(3, q)$.

Since $\mathfrak{D}_1$ is generated by the involutions J_1 and T, we have $\mathfrak{K} \subseteq \mathfrak{C}(\mathfrak{D}_1)$ and (2.10) shows that $\mathfrak{K} \subseteq \mathfrak{Z}(\mathfrak{C}(J))$. If $\xi \neq 1$ belongs to $\mathfrak{K}$, we have $\mathfrak{C}(\xi) = \mathfrak{C}(J)$ by (4 G). Since $\xi^Y \in \mathfrak{K}$, likewise $\mathfrak{C}(\xi^Y) = \mathfrak{C}(J)$ and this implies $\mathfrak{C}(\xi) = \mathfrak{C}(J^{Y^{-1}}) = \mathfrak{C}(J_2)$. Since $\mathfrak{C}(J) \neq \mathfrak{C}(J_2)$, we have a contradiction. Thus $\mathfrak{K} = 1$. We also note that the reciprocal image $\mathfrak{G}_0 = \gamma^{-1}(PGL(3, q))$ is a normal subgroup of $\mathfrak{G}$ and that $\mathfrak{G}/\mathfrak{G}_0$ is cyclic.

If $\gamma(\mathfrak{G})$ includes $PGL(3, q)$, then as $PGL(3, q)$ has order $q^3(q^3-1)(q^2-1)$, it follows from (5.5) that $l=1$ and that γ establishes an isomorphism of $\mathfrak{G}$ onto $PGL(3, q)$. We then have Case (a) of Theorem (1 A). In particular, this is true when $q \not\equiv 1 \pmod 3$ since in this case $PGL(3, q)$ and $PSL(3, q)$ coincide.

Assume that $q \equiv 1 \pmod 3$ and that $\gamma(\mathfrak{G})$ does not include $PGL(3, q)$. By (β), $\gamma(\mathfrak{G})$ includes $PSL(3, q)$ which here has index 3 in $PGL(3, q)$. Comparison of the orders shows that $|\mathfrak{G} : \mathfrak{G}_0|\, l = 3$. We have to rule out the possibility that $|\mathfrak{G} : \mathfrak{G}_0| = 3$.

Suppose then that $|\mathfrak{G} : \mathfrak{G}_0| = 3$. The normal subgroup $\mathfrak{G}_0$ of $\mathfrak{G}$ contains all 2-elements of $\mathfrak{G}$ and hence all "real" elements of $\mathfrak{G}$ (i.e. all elements which are conjugate to their inverses). In particular, $\mathfrak{S} \subseteq \mathfrak{G}_0$ and $\mathfrak{X}_1, \mathfrak{X}_2 \subseteq \mathfrak{G}_0$ and $W^{-2}U \in \mathfrak{G}_0$. If $\xi \in \mathfrak{U}$, then in the notation of [4], $\gamma(\xi)$ leaves the point (J) and every point of the line $[J]$ in the plane Π fixed. It follows that $\gamma(\xi)$ is a projectivity. Hence $\mathfrak{U} \subseteq \mathfrak{G}_0$ and (2.15) and (2.7) show that $\mathfrak{C}(J) \subseteq \mathfrak{G}_0$.

Since $|\mathfrak{G} : \mathfrak{G}_0| = 3$, there exists a linear character λ of $\mathfrak{G}$ with the kernel $\mathfrak{G}_0$. As $\lambda(\xi) = 1$ for $\xi \in \mathfrak{C}(J)$, it is seen easily (e.g. by the results of [2], II) that λ belongs to the principal 2-block of $\mathfrak{G}$. But this is excluded by the results of § 3.

Hence $|\mathfrak{G}:\mathfrak{G}_0|=1$, $l=3$ and $\mathfrak{G}\simeq PSL(3,q)$. Thus, we have Case (a) of Theorem (1A).

Since Case (c) is excluded by our present hypothesis $q^3|g$, the proof of the theorem under this hypothesis is complete.

§ 9. The hypothesis (IV). The case $u\neq 1$

We assume now that $\mathfrak{G}$ is a group which satisfies conditions (I), (II), (III), but not (IV). Thus, from now on, we have

$$g\not\equiv 0 \pmod{q^3}. \tag{9.1}$$

We shall show in § 9 that then $u=1$.

(9A) *If* $u\neq 1$, *there exists an irreducible character* ϑ of $\mathfrak{U}\times\mathfrak{W}$ *which has six distinct associates in* $\mathfrak{N}(\mathfrak{D})$.

Proof. By (2.7), (2.6), and (4.1), we have

$$\mathfrak{C}(\mathfrak{D})=\mathfrak{D}\times\mathfrak{U}\times\mathfrak{W}, \qquad \mathfrak{N}(\mathfrak{D})=\langle\mathfrak{C}(\mathfrak{D}), T, Y\rangle. \tag{9.2}$$

The group $\mathfrak{N}(\mathfrak{D})/\mathfrak{C}(\mathfrak{D})$ is isomorphic with a symmetric group on three letters.

(α) We first deal with the case $l\neq 1$. Let η be a complex $u\,l$-th root of unity. If μ is an integer (mod u) and if ρ is an integer (mod $u\,l$), there exists a unique irreducible character $\vartheta_{\mu\rho}$ of $\mathfrak{U}\times\mathfrak{W}$ such that

$$\vartheta_{\mu\rho}(U)=\eta^{\mu l}, \qquad \vartheta_{\mu\rho}(W)=\eta^{\rho}.$$

It is seen easily from (2.8) that

$$\vartheta_{\mu\rho}^{T}=\vartheta_{\mu,\mu l-\rho}.$$

In particular, $\vartheta_{\mu\rho}$ is only invariant under T, if $\mu\, l-\rho\equiv\rho \bmod u\, l$. Then $l|\rho$ and $\vartheta_{\mu\rho}^{u}=1$. It follows that if $\vartheta_{\mu\rho}^{u}\neq 1$, $\vartheta_{\mu\rho}$ is not invariant under T nor under any conjugate of T in $\mathfrak{N}(\mathfrak{D})$.

We wish to prove that $\vartheta=\vartheta_{01}$ satisfies the condition of (9A). The preceding remark shows that if this was not so, ϑ_{01} would have to be invariant under Y. Set $U^Y=U^{\alpha}W^{\beta}$. If $\vartheta_{01}^{Y^{-1}}=\vartheta_{01}$, we have

$$1=\vartheta_{01}(U)=\vartheta_{01}(U^Y)=\vartheta_{01}(U^{\alpha}W^{\beta})=\eta^{\beta}.$$

This implies $W^{\beta}=1$ and then $U^Y=U^{\alpha}\in\mathfrak{U}\cap\mathfrak{U}^Y$. On account of (4 G), we may assume that (4.4) holds. Then

$$\mathfrak{C}(U^Y)=\mathfrak{C}(U)^Y=\mathfrak{C}(J)^Y=\mathfrak{C}(J^Y)=\mathfrak{C}(J_1).$$

Likewise, by (4.4), $\mathfrak{C}(U^{\alpha})=\mathfrak{C}(J)$. Since $\mathfrak{C}(J)\neq\mathfrak{C}(J_1)$, we have a contradiction, and (9A) is proved in the case $l\neq 1$.

(β) If $l=1$, the group $\mathfrak{U}\times\mathfrak{W}$ has order u^2. The argument in the preceding paragraph shows that (4.4) implies that $\mathfrak{U}\cap\mathfrak{U}^Y=1$. Hence we may choose U

and $U_1 = U^Y$ as a basis of $\mathfrak{U} \times \mathfrak{W}$. We then have formulas

(9.3) $$U^T = U, \quad U_1^T = U^\mu U_1^\rho; \qquad U^Y = U_1, \quad U_1^Y = U^\gamma U_1^\delta.$$

Since $Y^2 T \equiv T Y \bmod \mathfrak{C}(\mathfrak{D})$, we find

$$(U^Y)^{YT} = U^{TY} = U^Y.$$

Thus, YT leaves U_1 fixed and (9.3) yields

$$U_1 = U_1^{YT} = (U^\gamma U_1^\delta)^T = U^{\gamma + \mu\delta} U_1^{\delta\rho}.$$

It follows that

(9.4) $$\gamma + \mu\delta \equiv 0, \qquad \delta\rho \equiv 1 \pmod{u}.$$

On the other hand, since $T^2 = 1$, we see from (9.3) that

(9.5) $$\mu + \mu\rho \equiv 0, \qquad \rho^2 \equiv 1 \pmod{u}.$$

Finally, (2.6) and (4.1) imply $YTY \equiv T(\bmod \mathfrak{C}(\mathfrak{D}))$ and hence

$$U = U^{YTY} = U_1^{TY} = (U^\mu U_1^\rho)^Y = U^{\gamma\rho} U_1^{\mu + \delta\rho}.$$

Thus,

(9.6) $$\gamma\rho \equiv 1, \qquad \mu + \delta\rho \equiv 0 \pmod{u}.$$

It follows from (9.4), (9.5), and (9.6) that

$$\mu \equiv \delta \equiv \gamma \equiv \rho \equiv -1 \pmod{u}.$$

Let η here denote a complex primitive u-th root of unity. If α and β are integers (mod u), let $\vartheta_{\alpha\beta}$ now be the irreducible character of $\mathfrak{U} \times \mathfrak{W}$ for which

$$\vartheta_{\alpha\beta}(U) = \eta^\alpha, \qquad \vartheta_{\alpha\beta}(U_1) = \eta^\beta.$$

Straightforward computation shows that

$$\vartheta_{01}^T = \vartheta_{0,-1}, \quad \vartheta_{01}^{YT} = \vartheta_{-1,1}, \quad \vartheta_{01}^{TY} = \vartheta_{10}, \quad \vartheta_{01}^Y = \vartheta_{-1,0}, \quad \vartheta_{01}^{Y^{-1}} = \vartheta_{1,-1}.$$

Hence $\vartheta = \vartheta_{01}$ has six distinct associates, and (9A) holds in the case $l = 1$ too.

Each irreducible character ζ of $\mathfrak{U} \times \mathfrak{W}$ can be interpreted as an irreducible character of $\mathfrak{C}(\mathfrak{D}) = \mathfrak{D} \times \mathfrak{U} \times \mathfrak{W}$. Let $b(\zeta)$ denote the block of $\mathfrak{C}(\mathfrak{D})$ to which ζ belongs. If ϑ is chosen as in (9A), it follows from the theory of blocks developed in [2] that $b(\vartheta)^{\mathfrak{G}} = B$ is a block of $\mathfrak{G}$ with the defect group $\mathfrak{D}$. On the other hand, we have three distinct blocks

$$\tilde{B}_1 = b(\vartheta)^{\mathfrak{C}(J)}, \quad \tilde{B}_2 = b(\vartheta^Y)^{\mathfrak{C}(J)}, \quad \tilde{B}_3 = b(\vartheta^{Y^{-1}})^{\mathfrak{C}(J)}$$

of $\mathfrak{C}(J)$ with the defect group $\mathfrak{D}$. For $i = 1, 2, 3$, we have $\tilde{B}_i^{\mathfrak{G}} = B$.

The conjugate classes of $\mathfrak{C}(J)$ which meet $\mathfrak{D}$ have $1, J, J_1$ as representatives. Since there are two blocks b of $\mathfrak{C}_{\mathfrak{C}(J)}(J_1) = \mathfrak{C}(\mathfrak{D})$ with $b^{\mathfrak{C}(J)} = \tilde{B}_i$ for fixed i, there are two columns of decomposition numbers for $\tilde{B}_i$ which belong to the section of J_1. For each of these columns, the Cartan invariant is 4 and both columns are orthogonal. Moreover, we have one column for the section J, again with the Cartan invariant 4. A discussion of the decomposition numbers shows that $\tilde{B}_i$ consists of four irreducible characters $\psi_1^{(i)}, \psi_2^{(i)}, \psi_3^{(i)}, \psi_4^{(i)}$, all of the same degree f_i. There is one modular irreducible character in $\tilde{B}_i$ which can be taken as $\psi_1^{(i)}$. Moreover, since the two columns of decomposition numbers in the section J_1 belong to linear modular characters, the discussion of the decomposition numbers shows that the characters $\psi_\mu^{(i)}$ can be arranged such that

$$\psi_1^{(i)}(J_1)=2, \quad \psi_2^{(i)}(J_1)=0, \quad \psi_3^{(i)}(J_1)=0, \quad \psi_4^{(i)}(J_1)=-2;$$
$$\psi_1^{(i)}(J)=\varepsilon_i f_i, \quad \psi_2^{(i)}(J)=-\varepsilon_i f_i, \quad \psi_3^{(i)}(J)=-\varepsilon_i f_i, \quad \psi_4^{(i)}(J)=\varepsilon_i f_i.$$

Here, ε_i is a sign.

Turning now to the block B, we have three columns of decomposition numbers in the section J. These belong to the modular characters $\psi_1^{(1)}, \psi_1^{(2)}, \psi_1^{(3)}$. The columns are mutually orthogonal and each has the Cartan invariant 4. A discussion of the decomposition numbers shows that B also consists of four irreducible characters $\Xi_1, \Xi_2, \Xi_3, \Xi_4$, all of the same degree z. For suitable arrangement of the characters, we have

$$\Xi_1(J)=f_1+f_2+\varepsilon_0 f_3, \quad \Xi_2(J)=f_1-f_2-\varepsilon_0 f_3,$$
$$\Xi_3(J)=-f_1+f_2-\varepsilon_0 f_3, \quad \Xi_4(J)=-f_1-f_2+\varepsilon_0 f_3$$

where ε_0 too is a sign. The coefficients of f_1 are the entries in the column $\mathfrak{d}_1^J$ of the section J, which belongs to the modular character $\psi_1^{(1)}$.

We now have all the information needed to apply Theorem (2A) of [*2*], III. Both involutions are taken as J, the column is taken as $\mathfrak{d}_1^J$ and we use the subgroup $\mathfrak{C}(J)$. The computation yields

$$g=\pm\frac{c(J)^3}{c(\mathfrak{D})^2}\,\frac{z}{f_1 f_2 f_3}.$$

The degrees of the irreducible characters of $\mathfrak{C}(J)$ are $1, q, q+1, q-1$. Since f_i is the degree of such a character $\psi_1^{(i)}$ of defect 2 for the prime 2, we have $f_i=q+1$. It is now evident that $q^3 \mid g$. This yields

(9B) *If* $\mathfrak{G}$ *satisfies* (I), (II), (III) *and if* $u\neq 1$, *then* (IV) *holds.*

§ 10. The case $u=1$

The method of § 9 breaks down for $u=1$ and we have to use different arguments here. From now on, we assume $u=1$. We first note that then (2.7) reads

(10.1) $$\mathfrak{C}(\mathfrak{D})=\mathfrak{D}\times\mathfrak{W}$$

where $\mathfrak{W}$ is a cyclic group of order $(q-1)/2=l$.

(10A) *For $\xi \in \mathfrak{W}$, $\xi \neq 1$, we have* $\mathfrak{W} \triangleleft \mathfrak{C}(\xi)$. If we set $\mathfrak{C}_0 = \langle \mathfrak{W}, \mathfrak{D}, Y \rangle$, then $\mathfrak{C}_0 \subseteq \mathfrak{C}(\xi)$ and $\mathfrak{C}_0/\mathfrak{W} \simeq \mathfrak{A}_4$[1]. Moreover, we have one of the following cases

Case A. $\mathfrak{C}(\xi)/\mathfrak{W} \simeq \mathfrak{A}_4$, $\mathfrak{C}(\xi) = \mathfrak{C}_0$.

Case B. $\mathfrak{C}(\xi)/\mathfrak{W} \simeq \mathfrak{A}_5$, $(\mathfrak{C}(\xi):\mathfrak{C}_0) = 5$.

Proof. We note first that $c(\xi)$ is not divisible by 8. Indeed, any subgroup of order 8 of $\mathfrak{S}$ contains the element F of order 4. As shown by (2.13), $c(F)$ is relatively prime to l when $u=1$. Hence $\mathfrak{C}(\xi)$ cannot contain a subgroup of order 8. Thus, $\mathfrak{D}$ is an S_2-group of $\mathfrak{C}(\xi)$.

Since Y and T normalize $\mathfrak{D}$, they normalize $\mathfrak{C}(\mathfrak{D})$ and $\mathfrak{W}$. It follows from the preceding remark that T leaves only the element 1 of $\mathfrak{W}$ fixed. This implies that T inverts $\mathfrak{W}$. As $TY \equiv Y^{-1} T \bmod \mathfrak{C}(\mathfrak{D})$, we see that Y centralizes $\mathfrak{W}$. Hence $Y \in \mathfrak{C}(\xi)$. It follows that $\mathfrak{C}(\xi)$ has only one class of involutions. By (2.9), we have

(10.2) $$\mathfrak{C}(\xi, J) = \mathfrak{C}(\xi, \mathfrak{D}) = \mathfrak{D} \times \mathfrak{W}.$$

We now apply the results of [3], III § 5, in particular (5A) and Remark 1, § 5. This yields

(10.3) $$c(\xi) = 8\alpha_0 \, |\mathfrak{D} \times \mathfrak{W}| = 32 \alpha_0 \, l$$

where the rational number α_0 lies between 3/8 and 15/8. Moreover, if we set $\mathfrak{K} = \mathfrak{K}_2(\mathfrak{C}(\xi))$, then we have $\alpha_0 = 3/8$ only when $\mathfrak{C}(\xi)/\mathfrak{K} \simeq \mathfrak{A}_4$, and we have $\alpha_0 = 15/8$ only when $\mathfrak{C}(\xi)/\mathfrak{K} \simeq \mathfrak{A}_5$. Since $\mathfrak{C}(\xi)$ includes $\mathfrak{C}_0 = \langle \mathfrak{W}, \mathfrak{D}, Y \rangle$ of order $12\,l$, $c(\xi)$ is an odd multiple of $12\,l$. By (10.3) and the remark following it,

$$12\,l \leqq c(\xi) \leqq 60\,l.$$

Hence $c(\xi)$ has one of the values $12l$, $36l$, $60l$.

On account of (6 E), we may set

$$\mathfrak{K} = \mathfrak{K}_0 \, \mathfrak{K}_1 \, \mathfrak{K}_2$$

with $\mathfrak{K}_\lambda = \mathfrak{K} \cap \mathfrak{C}(J_\lambda)$. Then by (10.2), $\mathfrak{K}_0 \subseteqq \mathfrak{W}$, and since $\mathfrak{W}$ is centralized by $\mathfrak{D}$, we see that $\mathfrak{K}_0 \subseteq \mathfrak{K}_1$, $\mathfrak{K}_0 \subseteq \mathfrak{K}_2$. On the other hand, $\mathfrak{K}_1^{Y^{-1}} \subseteq \mathfrak{K}_0$, $\mathfrak{K}_2^{Y} \subseteq \mathfrak{K}_0$. It follows first that $\mathfrak{K}_0$, $\mathfrak{K}_1$, $\mathfrak{K}_2$ have the same order and then that they are equal. Thus, $\mathfrak{K} = \mathfrak{K}_0 \subseteq \mathfrak{W}$. If $\alpha_0 = 3/8$, as remarked above $\mathfrak{C}(\xi)/\mathfrak{K} \simeq \mathfrak{A}_4$. By (10.3), $|\mathfrak{C}(\xi):\mathfrak{W}| = 12$ and we find $\mathfrak{K} = \mathfrak{W}$. We then have Case A of (10A). Likewise, if $\alpha_0 = 15/8$ we have $\mathfrak{C}(\xi)/\mathfrak{K} \simeq \mathfrak{A}_5$. By (10.3), $|\mathfrak{C}(\xi):\mathfrak{W}| = 60 = |\mathfrak{A}_5|$. Again, $\mathfrak{K} = \mathfrak{W}$ and we are led to Case B of (10A).

Finally, assume that $c(\xi) = 36\,l$, i.e. that $\alpha_0 = 9/8$. By Remark 3 of [2], III § 5, we have $\mathfrak{C}(\xi) = \langle \mathfrak{C}(\xi, J), Y \rangle$. Since $\mathfrak{W}$ is normalized by $\mathfrak{C}(\xi, J)$ and Y, we see that $\mathfrak{W} \triangleleft \mathfrak{C}(\xi)$. This again implies $\mathfrak{K} = \mathfrak{W}$. Hence $|\mathfrak{C}(\xi)/\mathfrak{K}| = 36$. Then the degrees in the principal 2-block of $\mathfrak{C}(\xi)$ are all less than 5. It now follows easily from [3], III § 5, that the case $\alpha_0 = 9/8$ is impossible.

[1] We denote by $\mathfrak{A}_r$ the alternating group on r letters and by $\mathfrak{S}_r$ the symmetric group on r letters.

(10B) *If two elements* ξ_1 *and* ξ_2 *of* $\mathfrak{W}$ *are conjugate in* $\mathfrak{G}$, *then* $\xi_2=\xi_1^{\pm 1}$

Indeed, a Sylow group argument shows that if ξ_1 and ξ_2 are conjugate in $\mathfrak{G}$, they are conjugate in

$$\mathfrak{N}(\mathfrak{D})=\langle\mathfrak{C}(\mathfrak{D}), T, Y\rangle=\langle\mathfrak{D}, \mathfrak{W}, T, Y\rangle.$$

Since $\mathfrak{D}$, $\mathfrak{W}$, Y centralize $\mathfrak{W}$, we have $\xi_2=\xi_1$ or $\xi_2=\xi_1^T=\xi_1^{-1}$.

We shall need a number of further lemmas.

(10C) *If* $\xi\in\mathfrak{W}$, $\xi\neq 1$, *then* $c(\xi, T)=2$ *or* $c(\xi, T)=6$ *according as to whether we have Case A or Case B in* (10A).

Proof. If $\mathfrak{C}(\xi, T)$ contained an S_2-group $\mathfrak{D}_0$ of $\mathfrak{C}(\xi)$, then $\langle\mathfrak{D}_0, T\rangle$ would be an elementary abelian group of order 8 which is impossible since no such subgroup occurs in the S_2-group $\mathfrak{S}$ of $\mathfrak{G}$. Hence $\langle J\rangle$ is an S_2-group of $\mathfrak{C}(\xi, T)$ and we may set $c(\xi, T\rangle=2c$ with odd c.

Since T inverts $\mathfrak{W}$, $\mathfrak{C}(\xi, T)\cap\mathfrak{W}=1$. Consequently, $\mathfrak{W}\,\mathfrak{C}(\xi, T)/\mathfrak{W}$ is a subgroup of order $2c$ of $\mathfrak{C}(\xi)/\mathfrak{W}$. In the case A, we must have $c=1$, since $\mathfrak{A}_4$ does not have a subgroup of order 6.

Suppose then that $\mathfrak{C}(\xi)/\mathfrak{W}\simeq\mathfrak{A}_5$. The extended centralizer of ξ is given by $\mathfrak{C}^*(\xi)=\langle\mathfrak{C}(\xi), T\rangle$. Since $\mathfrak{C}^*(\xi)/\mathfrak{W}$ is an extension of order 120 of $\mathfrak{C}(\xi)/\mathfrak{W}$ and since $\mathfrak{C}^*(\xi)/\mathfrak{W}$ has a dihedral S_2-group, it is seen easily that $\mathfrak{C}^*(\xi)/\mathfrak{W}\simeq\mathfrak{S}_5$. Each involution in $\mathfrak{S}_5-\mathfrak{A}_5$ centralizes an element of order 3. It follows that there exist elements $\zeta\in\mathfrak{C}(\xi)$, of order 3 over $\mathfrak{W}$, such that $\zeta^T=\zeta\,\eta$ with $\eta\in\mathfrak{W}$. Then T centralizes $\zeta\,\eta^{\frac{1}{2}}$. Hence $|\mathfrak{C}(\xi, T)\,\mathfrak{W}/\mathfrak{W}|$ is divisible by 3. Since $\mathfrak{A}_5$ does not have subgroups of order 30, we can only have $c=3$.

(10D) *Let* p *be a prime dividing* $l=(q-1)/2$ *and let* $\xi\neq 1$ *be a p-singular of* $\mathfrak{W}$. *Then*

$$(10.4)\quad \chi_1(J)=-\delta_1 q,\quad \chi_2(J)=\delta_2 q,\quad \chi_3(J)=-\delta_3,\quad \chi_4(J)=-\delta_2(q+1).$$

$$(10.5)\quad \chi_1(J\,\xi)=-\delta_1,\quad \chi_2(J\,\xi)=\delta_2,\quad \chi_3(J\,\xi)=-\delta_3,\quad \chi_4(J\,\xi)=-2\delta_2.$$

Indeed, (10.4) is obtained by combining (3.3) for $G=1$ with (4 B). Similarly, (10.5) is obtained from (3.3) for $G=\xi$ since the irreducible character φ of $\mathfrak{C}(J)$ has defect 0 for p, cf. (4 C).

(10E) *Suppose that for* $p\,|\,l$, *the* S_p-*group* $\mathfrak{P}$ *of* $\mathfrak{G}$ *is abelian. Then we cannot have* $\mathfrak{N}(\mathfrak{P})=\langle\mathfrak{C}(\mathfrak{P}), \tau\rangle$ *where* τ *is an involution which inverts* $\mathfrak{P}$.

Proof. Suppose that this is false and that $\mathfrak{N}(\mathfrak{P})=\langle\mathfrak{C}(\mathfrak{P}), \tau\rangle$ where τ is an involution which inverts $\mathfrak{P}$. We may assume that $\mathfrak{P}\cap\mathfrak{W}\neq 1$. As in the proof of (6A), we can apply the results of [3], III, § 9. In particular, the principal p-block of $\mathfrak{G}$ contains a real irreducible character $\psi_1\neq 1$ such that, for all p-singular elements G

$$(10.6)\qquad \psi_1(G)=\delta^*$$

where δ^* is a fixed sign. Again, ψ_1 is one of the characters χ_1, χ_2, χ_3 of § 3. We may also apply [3], III, (9 C). If $\xi \neq 1$ is an element of $\mathfrak{P} \cap \mathfrak{W}$, we have

$$g = \alpha^* \frac{c(J)^2}{c(\xi)} N^2, \qquad \alpha^* = \frac{\psi_1(1)(\psi_1(1) + \delta^*)}{(\psi_1(1) - \psi_1(J))^2} \tag{10.7}$$

where N is the number of conjugates σ of J which invert ξ. These are exactly the conjugates σ of T in $\mathfrak{C}^*(\xi)$ and hence

$$N = |\mathfrak{C}^*(\xi) : \mathfrak{C}^*(\xi) \cap \mathfrak{C}(T)| = c(\xi)/c(\xi, T).$$

The number $c(\xi)$ can be obtained from (10A) and then N can be found from (10 C). Substituting the value of $c(J)$ from (2.3) in (10.7), we have

$$g = 2\alpha^* \gamma (q-1)^3 q^2 (q+1)^2 \tag{10.8}$$

where $\gamma = 3$ in the case A and $\gamma = 5/3$ in the Case B.

Let $\nu(a)$ denote the exponent with which the prime 2 divides the rational number a. By (10.8)

$$n = \nu(g) = 1 + \nu(\alpha^*) + 3 + 2n - 4.$$

Hence $\nu(\alpha^*) = -n$.

Suppose we had $\psi_1 = \chi_2$. On comparing (10.5) and (10.6), we see that $\delta^* = \delta_2$. Now (10.7) and (10.4) yield

$$\alpha^* = x_2(x_2 + \delta_2)/(x_2 - \delta_2 q)^2.$$

But then (3.7) yields $\nu(\alpha^*) \leqq n - 2 - 2n < -n$ which is impossible. Similarly, the possibility that $\psi_1 = \chi_3$ can be ruled out. Thus, $\psi_1 = \chi_1$ and then

$$\alpha^* = x_1(x_1 - \delta_1)/(x_1 + \delta_1 q)^2.$$

On comparing (10.8) with (4.2), we now find

$$\gamma(x_1 - \delta_1)(q-1) = (x_1 + \delta_1)(q+1).$$

Case A. Here $\gamma = 3$ and $q-1$ divides $x_1 + \delta_1$. Set $x_1 + \delta_1 = t\,(q-1)$. Then $3(x_1 - \delta_1) = t\,(q+1)$. It follows that $3\delta_1 = t\,(q-2)$. Hence $q=3$ or $q=5$. The latter case is impossible since $q \equiv -1 \pmod 4$. On the other hand. for $q=3$, we have $l=1$ and no prime p exists.

Case B. Since $\gamma = 5/3$, here $q-1$ divides $3(x_1 + \delta_1)$. Set $3(x_1 + \delta_1) = t\,(q-1)$. Then $5(x_1 - \delta_1) = t\,(q+1)$ and $15\,\delta_1 = t\,(q-4)$. Since $q \equiv -1 \pmod 4$ and $q \neq 3$, we can only have $q = 19$, $t = 1$ or $q = 7$, $t = 5$. Also, $\delta_1 = 1$. In the former case, $x_1 = 5$ and in the latter, $x_1 = 9$. In neither case, the conditions (3.1) and (3.6) can be satisfied.

This completes the proof.

(10F) *The number l is divisible only by the primes* 3 *and* 5. *If* $5 \mid l$, *and if* ζ *is an element of order* 5 *of* $\mathfrak{W}$, *we have Case B in* (10 A) *for* $\xi = \zeta$.

Proof. Suppose that $p \neq 3$, 5 divides l. Let $\mathfrak{P}_0 = \langle \pi \rangle$ be an S_p-group of $\mathfrak{W}$ and let $\mathfrak{P} \supseteq \mathfrak{P}_0$ be an S_p-group of $\mathfrak{G}$. It follows from (10 A) that $\mathfrak{P}_0$ is an

S_p-group of $\mathfrak{C}(\pi)$. If $\xi \neq 1$ is an element of $\mathfrak{Z}(\mathfrak{P})$, then $\xi \in \mathfrak{C}(\pi) \cap \mathfrak{P} = \mathfrak{P}_0$. Again, by (10 A), $\mathfrak{P}_0$ is an S_p-group of $\mathfrak{C}(\xi)$. Since $\mathfrak{P} \subseteq \mathfrak{C}(\xi) \cap \mathfrak{P} = \mathfrak{P}_0$, we have $\mathfrak{P} = \mathfrak{P}_0$. Thus, $\mathfrak{P}$ is cyclic. It is an easy consequence of (10 B) that $\mathfrak{N}(\mathfrak{P}) = \langle \mathfrak{C}(\mathfrak{P}), T \rangle$ and now (10 E) yields a contradiction.

If $p=5$ and if we have Case A for ξ equal to an element ζ of order 5 in $\mathfrak{W}$, the same argument leads to a contradiction.

(10 G) *If* $\xi_0 \neq 1$, $\xi_0 \in \mathfrak{W}$, *then* $\mathfrak{C}(\xi_0) = \mathfrak{C}(\mathfrak{W})$.

Proof. We have to show that $\mathfrak{C}(\xi_0)$ centralizes $\mathfrak{W}$. This is clear, if we have Case A in (10 A). Suppose that we have Case B. Since $\mathfrak{C}(\xi_0)$ normalizes $\mathfrak{W}$, we have a representation of $\mathfrak{C}(\xi_0)$ by automorphisms of $\mathfrak{W}$. The kernel $\mathfrak{C}(\mathfrak{W})$ of this representation includes $\mathfrak{W}$. Since $\mathfrak{C}(\xi_0)/\mathfrak{W} \simeq \mathfrak{A}_5$ is simple and non-abelian, we have $\mathfrak{C}(\xi_0) = \mathfrak{C}(\mathfrak{W})$.

(10 H) *If* $p=3$ *or* 5 *and if* p *divides* l *with the exact exponent* $m \geqq 2$, *then the* S_p-*group* $\mathfrak{P}$ *of* $\mathfrak{G}$ *is abelian of order* p^{m+1}.

Proof. Let $\mathfrak{P}_0$ of order p^m be an S_p-group of $\mathfrak{W}$ and let $\mathfrak{P} \supseteq \mathfrak{P}_0$ be an S_p-group of $\mathfrak{G}$. If $\mathfrak{Z}(\mathfrak{P})$ contains elements $\xi \neq 1$ of $\mathfrak{P}_0$, then $\mathfrak{P} \subseteq \mathfrak{C}(\xi)$. By (10 G), $\mathfrak{C}(\xi) = \mathfrak{C}(\mathfrak{P}_0)$. It can now be seen from (10A) and (10 G) that $|\mathfrak{P}:\mathfrak{P}_0| = p$. Hence $\mathfrak{P}$ is abelian of order p^{m+1}.

Suppose then that $\mathfrak{Z}(\mathfrak{P}) \cap \mathfrak{P}_0 = 1$. Let η be an element of order p in $\mathfrak{Z}(\mathfrak{P})$. Then $\eta \in \mathfrak{C}(\mathfrak{P}_0)$, and by (10 A), $\langle \mathfrak{P}_0, \eta \rangle = \mathfrak{P}_1$ is an S_p-group of $\mathfrak{C}(\mathfrak{P}_0)$. Moreover, $\mathfrak{P}_1$ is abelian of type (p^m, p). If $\mathfrak{P}_1$ is not an S_p-group of $\mathfrak{G}$, there exist elements $\rho \in \mathfrak{P}$ such that ρ has order p over $\mathfrak{P}_1$ and that ρ normalizes $\mathfrak{P}_1$. Set $\mathfrak{P}_0 = \langle \pi \rangle$. We have an equation of the form

$$\pi^\rho = \pi^a \eta^b.$$

It follows that $(\pi^p)^\rho = \pi^{ap}$. By (10 B), $(\pi^p)^\rho = \pi^{\pm p}$, and as ρ is a p-element, ρ centralizes $\pi^p \neq 1$. Then $\rho \in \mathfrak{C}(\pi^p) = \mathfrak{C}(\mathfrak{P}_0)$, i.e. $\rho \in \mathfrak{C}(\mathfrak{P}_0) \cap \mathfrak{P} = \mathfrak{P}_1$, a contradiction.

We continue this investigation making the same assumptions as in (10 H). The abelian group $\mathfrak{P}$ is either cyclic of order p^{m+1} or of type (p^m, p). If two elements of $\mathfrak{P}$ are conjugate in $\mathfrak{G}$, they are conjugate in $\mathfrak{N}(\mathfrak{P})$. If $\mathfrak{P}_0 = \langle \pi \rangle$, then π is conjugate to π^{-1} in $\mathfrak{G}$. We now see that $|\mathfrak{N}(\mathfrak{P}):\mathfrak{C}(\mathfrak{P})|$ is even.

Suppose first that $\mathfrak{P}$ is cyclic of order p^{m+1}. It follows easily from (10 B) that $\mathfrak{N}(\mathfrak{P})$ has the form $\mathfrak{N}(\mathfrak{P}) = \langle \mathfrak{C}(\mathfrak{P}), \tau \rangle$ where τ is a 2-element which inverts $\mathfrak{P}$. If τ had order more than 2, the involution $J^* \in \langle \tau \rangle$ would centralize $\mathfrak{P}$ of order p^{m+1}. This is impossible, since J^* is a conjugate of J. Thus, τ is an involution. But then (10 E) leads to a contradiction.

Suppose then that $\mathfrak{P} = \langle \pi, \eta \rangle$ where π is a generator of $\mathfrak{P}_0$ and where η has order p. For $\rho \in \mathfrak{N}(\mathfrak{P})$, we have formulas

$$\left\{ \begin{array}{l} \pi^\rho = \pi^a \eta^b \\ \eta^\rho = \pi^c \eta^d \end{array} \right. \tag{10.9}$$

where a, c are integers mod p^m and where b, d are integers mod p. Moreover, $p^{m-1}|c$. It follows that $\pi^{\rho p}=\pi^{a p}$ and, by (10 B), $(\pi^p)^\rho=\pi^{\pm p}$. Since π and π^{-1} are conjugate in $\mathfrak{G}$, there exists an element $\tau\in\mathfrak{N}(\mathfrak{P})$ which inverts π. Either ρ or $\rho\,\tau$ centralizes π^p. Now (10 G) shows that ρ or $\rho\,\tau$ centralizes π, i.e. we have $a\equiv\pm 1$, $b\equiv 0$ in (10.9). We see that $|\mathfrak{N}(\mathfrak{P}):\mathfrak{C}(\mathfrak{P})|$ divides $2p(p-1)$ and since $\mathfrak{N}(\mathfrak{P})/\mathfrak{C}(\mathfrak{P})$ is p-regular, its order divides $2(p-1)$. Moreover, if η is replaced by a suitable element of $\mathfrak{P}$, we may assume that in (10.9) $c\equiv 0$ for all $\rho\in\mathfrak{N}(\mathfrak{P})$[1]. We note that if ρ has even order, we cannot have $d\equiv 1$. Indeed, if this was not so, η would commute with an involution. Then η would be conjugate to an element of $\mathfrak{C}(J)$ and hence to a power of π and this is impossible since $a\equiv\pm 1$, $b\equiv 0$ for each $\rho\in\mathfrak{N}(\mathfrak{P})$.

For $p=3$, it follows that $\mathfrak{N}(\mathfrak{P})$ could only have the form excluded by (10 E). Hence $p\neq 3$, that is, $q-1\not\equiv 0 \pmod 9$.

For $p=5$, there is an additional case where $\mathfrak{N}(\mathfrak{P})=\langle\mathfrak{C}(\mathfrak{P}),\rho\rangle$ and

$$\pi^\rho=\pi^{-1},\qquad \eta^\rho=\eta^2. \tag{10.10}$$

Here, ρ has order 4 over $\mathfrak{C}(\mathfrak{P})$. We wish to show that this case leads to a contradiction.

Let $\nu(\alpha)$ now denote the exponent with which 5 divides the rational number α. Since $\nu(q-1)=m$, $\nu(g)=m+1$, it follows from (4 F) that

$$1+2\,\nu(x_2-\delta_2\,q)=\nu(x_2)+\nu(x_2+\delta_2). \tag{10.11}$$

If we had $\nu(x_2-\delta_2\,q)>0$, then $\nu(x_2)=\nu(x_2+\delta_2)=0$ and (10.11) yields a contradiction. Hence $\nu(x_2-\delta_2\,q)=0$, and we have one of the two cases:

Case (A). $\nu(x_2)=1$, $\nu(x_2+\delta_2)=0$.

Case (B). $\nu(x_2)=0$, $\nu(x_2+\delta_2)=1$.

By (10.5), $\chi_2(J\,\pi)\neq 0$, $\chi_4(J\,\pi)\neq 0$ and hence π belongs to a 5-defect group of χ_2 and χ_4. In the case (A), the same formula shows that $g\,\chi_2(J\pi)/(c(J\,\pi)\,x_2)$ is prime to 5. This implies that $\mathfrak{C}(J\,\pi)$ includes a 5-defect group of χ_2. Hence $\mathfrak{P}_0$ is a 5-defect group of χ_2. Similarly, we see that in the case (B), $\mathfrak{P}_0$ is a 5-defect group of χ_4.

We use the results of [*2*] to study the 5-blocks of $\mathfrak{G}$ with the defect group $\mathfrak{P}_0$. Let $\mathfrak{W}=\mathfrak{P}_0\times\mathfrak{W}_0$. Our results show that $|\mathfrak{W}_0|=3$ or $|\mathfrak{W}_0|=1$. Moreover, $\mathfrak{C}(\mathfrak{P}_0)/\mathfrak{P}_0$ has a normal subgroup $\overline{\mathfrak{W}}_0\simeq\mathfrak{W}_0$ such that the quotient group is isomorphic with $\mathfrak{A}_5$. It is then seen without difficulty that we may set

$$\mathfrak{C}(\mathfrak{P}_0)/\mathfrak{P}_0=\overline{\mathfrak{G}}_1\times\overline{\mathfrak{W}}_0$$

where $\overline{\mathfrak{G}}_1\simeq\mathfrak{A}_5$. Note that $\overline{\mathfrak{G}}_1$ has an irreducible character ϑ of degree 5.

If $|\mathfrak{W}_0|=3$, then $\mathfrak{C}(\mathfrak{P}_0)/\mathfrak{P}_0$ has three irreducible characters ϑ, $\vartheta\,\lambda$, $\vartheta\,\bar{\lambda}$ of 5-defect 0. Here λ is a linear character which we may interpret as a linear character of $\mathfrak{C}(\mathfrak{P}_0)$ whose kernel is the reciprocal image $\mathfrak{G}_1$ of $\overline{\mathfrak{G}}_1$ in $\mathfrak{C}(\mathfrak{P}_0)$.

[1]) For example, this can be seen by applying SYLOW's theorem to the subgroup of order $2p(p-1)$ of the automorphism group of $\mathfrak{P}$ consisting of the automorphisms which map $\pi\to\pi^{\pm 1}$.

Likewise, ϑ may be interpreted as a character of degree 5 of $\mathfrak{C}(\mathfrak{P}_0)$ whose kernel is $\mathfrak{W}$. We have $\vartheta(J)=1$. If $\mathfrak{W}_0=\langle W_0\rangle$, $\lambda(W_0)$ is a primitive third root of unity.

It follows from (10 B), that

$$\mathfrak{N}(\mathfrak{P}_0)=\langle\mathfrak{C}(\mathfrak{P}_0), T\rangle.$$

Clearly, $\vartheta^T=\vartheta$, $\lambda^T=\bar{\lambda}$. Now, the results of [2], I, show that $\mathfrak{G}$ has two 5-blocks $\mathfrak{B}_1$ and $\mathfrak{B}_2$ with the defect group $\mathfrak{P}_0$; $\mathfrak{B}_1$ corresponding to the family of characters of $\mathfrak{P}_0\,\mathfrak{C}(\mathfrak{P}_0)=\mathfrak{C}(\mathfrak{P}_0)$ consisting only of ϑ, and $\mathfrak{B}_2$ corresponding to the family $\{\vartheta\,\lambda, \vartheta\,\bar{\lambda}\}$.

Let $\xi\neq 1$ be an element of $\mathfrak{P}_0$. By (10 G), $\mathfrak{C}(\xi)=\mathfrak{C}(\mathfrak{P}_0)$. Since $\mathfrak{P}_0\lhd\mathfrak{C}(\mathfrak{P}_0)$, $\mathfrak{C}(\xi)$ has only 5-blocks whose defect group includes $\mathfrak{P}_0$. It follows that the only 5-blocks $\mathfrak{b}_i$ of $\mathfrak{C}(\xi)$ with $\mathfrak{b}_i^{\mathfrak{G}}=\mathfrak{B}_2$ are the blocks containing $\vartheta\,\lambda$ and $\vartheta\,\bar{\lambda}$ respectively as their only modular irreducible characters. Now, the results of [2], II, show that for $\chi_\mu\in\mathfrak{B}_2$ and 5-regular $\sigma\in\mathfrak{C}(\mathfrak{P}_0)$, we have formulas

$$\chi_\mu(\xi\,\sigma)=d_\mu^{(1)}(\xi)\,\vartheta(\sigma)\,\lambda(\sigma)+d_\mu^{(2)}(\xi)\,\vartheta(\sigma)\,\bar{\lambda}(\sigma),$$

Here, the decomposition numbers $d_\mu^{(1)}(\xi)$, $d_\mu^{(2)}(\xi)$ do not depend on σ. They lie in the field of the 5^m-th roots of unity. In particular, this implies that either $\chi_\mu(\xi\,J)\neq\chi_\mu(\xi\,J\,W_0)$ or that $\chi_\mu(\xi\,J)=0$. Now, (10 D) shows that neither χ_2 nor χ_4 can lie in $\mathfrak{B}_2$.

In the case $|\mathfrak{W}_0|=1$, there is only one 5-block $\mathfrak{B}_1$ of $\mathfrak{G}$ with the defect groups $\mathfrak{P}_0$. This is the block corresponding to the family $\{\vartheta\}$ of characters of $\mathfrak{P}_0\,\mathfrak{C}(\mathfrak{P}_0)$.

It is now clear that in the case (A), $\chi_2\in\mathfrak{B}_1$, and in the case (B), $\chi_4\in\mathfrak{B}_1$. This holds as well for $|\mathfrak{W}_0|=3$ as for $|\mathfrak{W}_0|=1$.

If we apply the same arguments to $\mathfrak{B}_1$ as above to $\mathfrak{B}_2$, we see that for $\chi_\mu\in\mathfrak{B}_1$, $\xi\in\mathfrak{P}_0-1$, and for 5-regular $\sigma\in\mathfrak{C}(\mathfrak{P}_0)$, we have formulas

(10.12) $$\chi_\mu(\xi\,\sigma)=d_\mu(\xi)\,\vartheta(\sigma).$$

If we replace ξ by $\xi^{-1}=\xi^T$ and use $\vartheta^T=\vartheta$, we find $d_\mu(\xi^{-1})=d_\mu(\xi)$.

In the terminology of [3], the $d_\mu(\xi)$ are the entries of a column $\mathfrak{d}(\xi)$ belonging to the block $\mathfrak{B}_1$. Let ψ be a generalized character of $\mathfrak{P}_0$ which vanishes for the element 1 and define a column for $\mathfrak{B}_1$ by the formula

$$\mathfrak{a}(\psi)=(\mathfrak{d},\psi)_{\mathfrak{P}_0}=5^{-m}\sum_{\xi\in\mathfrak{P}_0}\mathfrak{d}(\xi)\,\bar{\psi}(\xi).$$

It follows easily from the orthogonality relations for the (generalized) decomposition numbers that

$$(\mathfrak{a}(\psi), \mathfrak{a}(\psi)\,=(\psi, \psi+\psi^T)_{\mathfrak{P}_0}.$$

Since $d_\mu(\xi)=\chi_\mu(\xi\,J)$ by (10.12), we see that $\mathfrak{a}(\psi)$ is an integral rational column.

In particular, take $\psi=\omega-1$ where ω is a non-principal linear character of $\mathfrak{P}_0$. We then have $(\mathfrak{a}(\psi), \mathfrak{a}(\psi))=3$. Hence $\mathfrak{a}(\psi)$ has exactly three non-zero entries all of which are ± 1.

In the case (B), it follows from (10.5) and (10.12) that $d_4(\xi)=-2\delta_2$ for $\chi_\mu=\chi_4$ and for all $\xi\neq 1$. Then $\mathfrak{a}(\psi)=a(\omega-1)$ has the entry $2\delta_2$ for χ_4, a contradiction. Thus, the case (B) is impossible.

Assume then that we have the case (A). Here $\mathfrak{a}(\omega-1)$ has the entry $-\chi_2(J\xi)=-\delta_2$ for $\chi_\mu=\chi_2$. If τ is a 5-regular element of $\mathfrak{G}$, the column $\mathfrak{a}(\omega-1)$ is orthogonal to the column $\chi(\tau)$ with the entries $\chi_\mu(\tau)$ for the block $\mathfrak{B}_1$. For $\tau=1$, this shows that $\mathfrak{B}_1$ contains a character $\chi_\alpha\neq\chi_2$ of odd degree. In the present case $u=1$, the principal 2-block is the only 2-block of full defect, since here $\mathfrak{C}(\mathfrak{S})=\langle J\rangle$ by (2.10). Hence χ_α is one of the characters χ_0, χ_1, χ_3. Since $5|\chi_\alpha(1)$, only the possibilities $\chi_\alpha=\chi_1, \chi_3$ remain. Since $5|x_2$, it can be deduced from (3.8) that 5 divides $1+\delta_1 x_1$. Hence $x_1\not\equiv 0 (\mathrm{mod}\ 5)$. Thus, $\chi_\alpha=\chi_3$. Then $\mathfrak{a}(\omega-1)$ has the entry $-\chi_3(J\xi)=\delta_3$ for χ_3.

Let $\chi_\beta\in\mathfrak{B}_1$ be the third character for which $\mathfrak{a}(\omega-1)$ has a non-zero entry. If this entry is δ^*, then $\delta^*=\pm 1$ and

$$-\delta_2\chi_2(\tau)+\delta_3\chi_3(\tau)+\delta^*\chi_\beta(\tau)=0$$

for 5-regular τ. In particular, χ_β has degree $\delta^*(\delta_2 x_2-\delta_3 x_3)$. For $\tau=J$, we obtain

$$\chi_\beta(J)=\delta^*(\delta_2\chi_2(J)-\delta_3\chi_3(J))=\delta^*(q+1).$$

We now apply [3], III, (2A), working with the column $\mathfrak{d}(\omega-1)$ which is a linear combination of columns $\mathfrak{d}(\xi)$, $\xi\neq 1$. Both involutions y_1, y_2 are taken as J and $\tilde{G}$ is to be $\mathfrak{N}(\mathfrak{P}_0)$. The computation yields

$$\frac{g}{c(J)^2}\left(-\delta_2\frac{q^2}{x_2}+\delta_3\frac{1}{x_3}+\frac{(q+1)^2}{\delta_2 x_2-\delta_3 x_3}\right)=-\frac{1}{6}(q-1).$$

Substitute the values of g from (4F) and that of $c(J)$ from (2.3). On account of (3.6), the left side can be written in the form

$$\frac{1}{2}(q+1)\frac{x_1(x_1+\delta_1)}{(x_1+\delta_1 q)^2}\,\frac{\delta_1 x_2+\delta_1 x_1 x_3+2q x_3}{x_3(\delta_2 x_2-\delta_3 x_3)}.$$

Since

$$x_3(x_1+\delta_1 q)^2=x_1(x_1 x_3+2\delta_1 q x_3+x_2),$$

our result reads

$$3\delta_1(q+1)(x_1+\delta_1)=-(q-1)(\delta_2 x_2-\delta_3 x_3).$$

On account of (3.1), this becomes

$$x_3=x_2\frac{q+2}{2q+1}.$$

Now (3.6) yields $x_1=q^2(q+2)/(2q+1)$. Since this cannot be integral, we have a contradiction.

Thus, we have the result:

(10I) *The number l is neither divisible by 9 nor by 25.*

It follows from (10F) and (10I) that $q=2l+1$ can have only one of the following values:

$$q=3, 7, 11, 31.$$

In each of these cases, we can find the values of the degrees x_1, x_2, x_3 which satisfy the conditions of § 3 and can occur in a group $\mathfrak{G}$ of the given kind. Then g is determined by (4F).

We have the five cases:

(1) $q=3$; $x_1=11$, $x_2=45$, $x_3=55$; $g=7920$.
(2) $q=7$; $x_1=55$, $x_2=441$, $x_3=495$; $g=2^5\cdot 3^3\cdot 5\cdot 7^2\cdot 11$.
(3) $q=11$; $x_1=99$, $x_2=539$, $x_3=441$; $g=2^4\cdot 3^5\cdot 5^2\cdot 7^2\cdot 11$.
(4) $q=11$; $x_1=451$, $x_2=165$, $x_3=615$; $g=2^4\cdot 3^5\cdot 5^2\cdot 11\cdot 41$.
(5) $q=31$; $x_1=991$, $x_3=31^2\cdot 33$, $x_3=33\cdot 991$; $g=2^7\cdot 3^2\cdot 5\cdot 11\cdot 31^2\cdot 991$.

Case (2) can be ruled by considering 11-blocks, Case (3) by considering 5-blocks, Case (4) by considering 41-blocks, and Case (5) by considering 991-blocks.

In the case (1), it is not difficult to complete the table of characters and to show that the characters of $\mathfrak{G}$ are identical with the characters of $\mathfrak{M}_{11}$. This suffices to show that $\mathfrak{G}$ is isomorphic with $\mathfrak{M}_{11}$. We refer to Wong [10]. Actually, Wong's paper includes a full treatment of the case $q=3$.

Thus, if q^3 does not divide g, we have $\mathfrak{G}\simeq\mathfrak{M}_{11}$. This is Case (c) of Theorem (1A) and the proof of the theorem is complete.

References

[1] Brauer, R.: On the structure of groups of finite order. Proceedings of the International Congress of Mathematicians, Amsterdam 1954, vol. I, 209—217.
[2] — Zur Darstellungstheorie der Gruppen endlicher Ordnung. I. Math. Zeitschr. **63**, 406—444 (1956). — II. Math. Zeitschr. **72**, 25—46 (1959).
[3] — Some applications of the theory of blocks of characters of finite groups. I. J. Algebra **1**, 152—167 (1964). — II. J. Algebra **1**, 307—334 (1964). — III to appear in the J. Algebra.
[4] — On finite Desarguesian planes I. Math. Zeitschr. **90**, 117—123 (1965).
[5] —, and M. Suzuki: On finite groups of even order whose 2-Sylow group is a quaternion group. Proc. Nat. Acad. Sci. U.S.A. **45**, 1757—1759 (1959).
[6] Gorenstein, D., and J. Walter: On finite groups with dihedral Sylow 2-subgroups. Illinois J. Math. **6**, 553—593 (1962).
[7] Steinberg, R.: Representations of $GL(3, q)$, $GL(4, q)$ $PGL(3, q)$, $PGL(4, q)$. Canadian J. Math. **3**, 225—235 (1951).
[8] — Generators, relations, and coverings of algebraic groups, Colloquium on algebraic groups. Brüssel 1962.
[9] Suzuki, M.: On characterization of linear groups II. Trans. Amer. Math. Soc. **92**, 205—219 (1959).
[10] Wong, W.: On finite groups whose 2-Sylow subgroups have cyclic subgroups of index 2. J. Australian Math. Soc. **4**, 90—112 (1964).

Dept. of Math., Harvard University, Cambridge, Massachusetts, U. S. A.

(Received July 24, 1965)

Druck der Universitätsdruckerei H. Stürtz AG., Würzburg

Reprinted from the PROCEEDINGS OF THE NATIONAL ACADEMY OF SCIENCES
Vol. 55, No. 2, pp. 254–259. February, 1966.

INVESTIGATION ON GROUPS OF EVEN ORDER, II*

BY RICHARD BRAUER

HARVARD UNIVERSITY

Communicated December 1, 1965

1. *Results on Groups G with a Given 2-Sylow Subgroup.*—(*a*) If p is a prime and P a given p-group, we can propose to study the finite groups G which possess P as their p-Sylow subgroup. The theory of blocks of characters[1] can be applied. In I, results concerning the irreducible characters of G were obtained. As was shown in II, additional methods are available for $p = 2$. The nature of the results depends strongly on the structure on P. At least for certain P, an amazing amount can be said about G in the case $p = 2$. Elsewhere, the cases of quaternion dihedral, quasidihedral P have been investigated and at least partial results have been obtained for abelian P. In this section, we shall report on some further results of this nature.

(*b*) The problem of characterizing the projective groups of Desarguesian planes of order q for $q \equiv 1 \pmod 4$ requires the investigation of groups P with generators σ_1, σ_2, τ defined by the relations

$$\sigma_1^{2^m} = \sigma_2^{2^m} = 1, \qquad \tau^2 = 1; \qquad \sigma_1\sigma_2 = \sigma_2\sigma_1, \qquad \tau^{-1}\sigma_1\tau = \sigma_2.$$

Here, m is an integer; $m \geqq 2$. Thus, P has order 2^{2m+1}, and P is the wreath product of a cyclic group of order 2^m by a group of order 2. We use the following notation,

$$J_1 = \sigma_1^{2^{m-1}}, \quad J_2 = \sigma_2^{2^{m-1}}, \quad J = J_1J_2, \quad \alpha = \sigma_1\sigma_2, \quad \beta = \sigma_1\sigma_2^{-1}, \quad S = \langle\sigma_1,\sigma_2\rangle.$$

If G is a finite group with P as its 2-Sylow group, we have the following four cases:

Case Ia. $|N(S):C(S)| = 6$; τ and J_2 are conjugate in $C(\alpha)$.
Case Ib. $|N(S):C(S)| = 6$; τ and J_2 not conjugate in $C(\alpha)$.
Case IIa. $|N(S):C(S)| = 2$; τ and J_2 are conjugate in $C(\alpha)$.
Case IIb. $|N(S):C(S)| = 2$; τ and J_2 not conjugate in $C(\alpha)$.

The group G has a normal 2-complement, if and only if we have Case IIb. In the case Ib, G has a normal subgroup H of index 2 with the 2-Sylow group S. If $K = K_2(G)$ is the 2-regular core of G, then SK/K is normal in G/K of index 6. The group G/K is uniquely determined. In particular, $|G:K| = 3\cdot 2^{2m+1}$.

In the Case IIa, G has a normal subgroup H such that G/H is cyclic of order 2^m. Applying the methods of II, we find

$$|G| = |C(J_2)|^2\cdot|C(J)|\cdot|C(J,J_1)|^{-2}.$$

If again $K = K_2(G)$, we have $G = KC(J)$. In particular, if G is core-free (i.e., if $K = 1$), $G = C(J)$.

Assume then that we have Case Ia. In this case and only in this case, G does not have a normal subgroup of index 2. Our methods yield the following results.

THEOREM 1. *Let G be a group whose 2-Sylow subgroup P is a wreath product of a cyclic group of order $2^m > 2$ and a cyclic group of order 2. Let J be an involution of G. Assume that G does not have a normal subgroup of index 2. The principal 2-block of G consists of*

$$(2^{2m-1} + 3\cdot 2^{m+1} + 4)/3$$

irreducible characters. The degrees of these characters are

$$1,\ \epsilon f^3,\ \epsilon f(f+1),\ f^2+f+1,\ \epsilon f(f^2+f+1),\ \epsilon(f+1)(f^2+f+1),\ \epsilon(f-1)(f^2+f+1).$$

Here, f is an integer which is uniquely determined by the group $C(J)$ and $\epsilon = \operatorname{sign} f$. The first three degrees appear only once, while the other degrees occur with certain multiplicities.

We can also determine the linear equations which hold for these characters if the argument is restricted to the set of 2-regular elements. Moreover, we have

THEOREM 2. *If the assumptions are as in Theorem 1, there exist integers a, c, t such that*

$$|G| = a^3c\epsilon t f^3(f+1)(f-1)(f^2+f+1),$$
$$|C(J)| = ac\epsilon t(f+1)f(f-1),$$
$$|C(\beta)| = a\epsilon(f-1)c/t, \qquad |C(J,\tau)| = \epsilon ct(f-1), \qquad C(\beta,\tau) = c.$$

If the orders of $C(J)$ and of the subgroups $C(\beta)$, $C(\beta, \tau)$, $C(J, \tau)$ of $C(J)$ are known, the order of G is uniquely determined.

(c) The results of (b) can be used to prove the following theorem which completes earlier investigations on characterizations of projective groups of finite Desarguesian planes.[2]

THEOREM 3. *Let G be a finite group which contains an involution J such that the centralizer $C(J)$ of J in G has the form*

$$C(J) \simeq GL(2,q)/L, \qquad q \equiv 1 \pmod 4,$$

where L is a subgroup of odd order of $GL(2,q)$. Assume that G does not have a normal subgroup of index 2. Then G is isomorphic to one of the following groups (i) $PGL(3,q)$, (ii) $PSL(3,q)$, (iii) $SL(3,q)$, (iv) *a direct product of $PSL(3,q)$ with a cyclic group of order 3, $q \equiv 4$ or 7 (mod 9).*

If $q \not\equiv 1 \pmod 3$, the three groups (*i*), (*ii*), (*iii*) are isomorphic. We have $|L| = 3$ when $G \simeq PSL(3,q)$ and $q \equiv 1 \pmod 3$, while in all other cases $L = 1$.

In the cases treated here, $q = f$ and $\epsilon = 1$. The case $\epsilon = -1$ occurs for the groups $G = U(3,q)$, $q \equiv -1 \pmod 4$. Here, $f = -q$. It is highly probable that there exist similar characterizations of $U(3,q)$ by means of centralizers of involutions. No simple groups with the 2-Sylow group P are known except the groups $PSL(3,q)$ with $q \equiv 1 \pmod 4$ and $SU(3,q)$ with $q \equiv -1 \pmod 4$.

(*d*) As a further special case, we consider the group P of order 64 which is the 2-Sylow group of the simple Suzuki group of order 29,120. Our result is

THEOREM 4. *The Suzuki group of order* 29,120 *is the only simple group with the given 2-Sylow subgroup P.*

If P is the 2-Sylow group of any of the Suzuki groups $Sz(q)$, it is known that only finitely many simple groups with the 2-Sylow group P exist.

2. *Some Special Cases of Schreier's Conjecture.*—(*e*) Our methods can be used to obtain results concerning automorphisms of finite groups G. If H is a group of automorphisms of G, we can form the semidirect product $\mathfrak{G} = GH$ of G with H. If H normalizes the 2-Sylow group P of G and if Q is the 2-Sylow group of H, the 2-Sylow group $\mathfrak{P}$ of $\mathfrak{G}$ is the semidirect product $\mathfrak{P} = PQ$. Our methods can be applied to $\mathfrak{G}$ and, under favorable conditions, this will yield results concerning H.

(*f*) In particular, we can try to apply this to an investigation of Schreier's conjecture which states that the outer automorphism group of a simple group G is solvable. If A is the full automorphism group Aut G of G and B is the subgroup I Aut G of inner automorphisms, then following Wielandt, we term A/B the outer automorphism group of G.

The author has been aware for some time that for groups with certain types of 2-Sylow groups, Schreier's conjecture can be proved by means of the method indicated in (*e*).[3] The following elementary and rather obvious remarks are useful. Let p be an arbitrary prime. If P is a p-Sylow subgroup of G, set

$$U = N_A(P), \qquad V = C_A(P),$$

i.e., let U denote the subgroup of A which maps P onto P while V is the subgroup of U which leaves P elementwise fixed. Then $A = UB$ by Sylow's theorem and hence

$$A/B \cong U/(U \cap B).$$

It is now clear that A/B breaks up into the two constituents

$$X_1 = U/(U \cap B)V, \qquad X_2 = V/(V \cap B),$$

where X_2 is isomorphic to a normal subgroup of A/B and X_1 to the corresponding quotient group. Each $u \in U$ induces an automorphism $[u]$ of P. In this manner,

U is mapped homomorphically on a subgroup S of Aut P. If T is the image of $(U \cap B)V$, then $X_1 \simeq S/T$. Let $F(P)$ denote the Frattini subgroup of P and set $|P:F(P)| = p^r$. Now $[u]$ induces an automorphism of $P/F(P)$. If M is the subgroup of Aut P which induces the trivial mapping of $P/F(P)$, we see that S/T splits into two constituents

$$S/T(S \cap M), \qquad (S \cap M)/(T \cap M).$$

By a well-known result of P. Hall, M is a p-group and so is then the second constituent. The first constituent can be identified with a group of the form S_0/T_0, where T_0 is a normal subgroup of a subgroup S_0 of $GL(r,2)$,

$$T_0 \subseteq S_0 \subseteq GL(r,p).$$

If $T_0 = 1$ or if more generally T_0 is a p-group, we must have $N(P) = PC(P)$.

In order to prove Schreier's conjecture, we have to show that X_2 and S_0/T_0 are solvable.

(*g*) The following remarks may be helpful for the investigation of X_2. The group $V \cap B$ consists of the inner automorphisms (σ) of G induced by the elements $\sigma \in C_G(P)$. Its p-Sylow subgroup is the group $(Z(P))$ consisting of the elements (σ) such that σ belongs to the center $Z(P)$ of P. Moreover, $V \cap B = (Z(P)) \times R$, where R is a subgroup of an order prime to p. Let $Q/(V \cap B)$ be an S_p-subgroup of $V/(V \cap B)$ and let Q_0 be an S_p-subgroup of Q. Then $Q_0 \cap V \cap B = (Z(P))$.

Take now $p = 2$. If H is a subgroup of Q_0, we can apply the method of (*e*). The 2-Sylow subgroup of the semidirect product $\mathfrak{G} = GH$ is $\mathfrak{P} = P \times H$. In many cases, the following theorem of G. Glauberman[4] is useful:

THEOREM (G. Glauberman). *Let G be a finite group with the 2-Sylow group P and assume that there exists an involution J of P which is not conjugate in G to an element $J' \neq J$ of P. If $K = K_2(G)$, then the subgroup $\langle J \rangle K$ of G is normal. In particular, if $K = 1$, then J lies in the center of G.*

In dealing with the cases of Schreier's conjecture mentioned, the author used special cases of Glauberman's theorem, which were proved separately in each case. Glauberman's proof is also based on the theory of blocks of representations.

(*h*) As an example, we consider the case that P is an elementary abelian 2-group. We show that if $K_2(G) = 1$, then X_2 has odd order. If this were not so, there would exist an element α of Q_0 of order 2 over $(Z(P)) = (P)$. Then α has order 4 or 2. We take $H = \langle\alpha\rangle$ and form the direct product $\mathfrak{G} = \langle\alpha\rangle G$. If α has order 4, the element $J = \alpha^2$ satisfies the assumption of Glauberman's theorem. Since $K_2(\mathfrak{G}) = K_2(G) = 1$, we have $J \in Z(\mathfrak{G})$. Then $\alpha^2 = 1$, a contradiction. If α has order 2, Glauberman's theorem can be applied to $J = (x)\alpha$, where (x) is the inner automorphism of G induced by a suitable element x of P. Then $(x)\alpha \in Z(\mathfrak{G})$ and this implies $\alpha = (x) \in (P)$, again a contradiction. Hence $|X_2|$ is odd and, by the Feit-Thompson Theorem, X_2 is solvable. If $|P| \leq 8$, the method of (*f*) shows that S/T is solvable, except when $G = P$ is of order 8. Hence we have

THEOREM 5. *Suppose that G is a finite group whose 2-Sylow group is elementary abelian of order at most 8. Assume that $K_2(G) = 1$. If $|G| \neq 8$, the outer automorphism group of G is solvable.*

By methods of the same nature, we can prove

THEOREM 6. *Suppose that G is a finite group whose 2-Sylow group P is dihedral,* *quasi-dihedral, or a wreath product of a cyclic group of order 2^m with a cyclic group of order 2. If $K_2(G) = 1$, the outer automorphism group of G is solvable.*

The case of a dihedral P here is slightly more difficult than the other cases. We show then that Q_0 cannot contain elements of order 4 and that $|Q:(Z(P))| \leqq 2$. This implies again that X_2 and A/B are solvable.[5]

3. *Real Irreducible Characters of Odd Degree.*—(*i*) In this last section, we shall discuss a very simple remark which can be used in the investigation of special groups.[6] If G is a finite group with the 2-Sylow group P, the centralizer $C_G(P)$ has the form $C_G(P) = Z(P) \times M$ where M is a group of odd order. The 2-blocks B of G of full defect correspond to classes γ of irreducible characters θ of M associated in $N(P)$. Since $N(P)/PC(P)$ has odd order, it follows that for $\theta \neq 1$, the conjugate complex characters $\bar{\theta}$ for $\theta \in \gamma$ form a second such class $\bar{\gamma} \neq \gamma$. This implies the following result.

THEOREM 7. *The 2-blocks of full defect different from the principal block can be arranged in pairs B, $\bar{B} \neq B$ where $\bar{B}$ consists of the conjugate complexes of the characters in B.*

As an immediate consequence, we have

COROLLARY 1. *All real-valued irreducible characters of full defect for $p = 2$ lie in the principal 2-block. In particular, if $|G| = 2^n g_0$ with odd g_0, then G has at most 2^{2n} real-valued irreducible characters of odd degree.*

It is likely that in the last statement, the number 2^{2n} can be replaced by 2^n and that, if the number of these characters is 2^n, then the 2-Sylow subgroup P of G is abelian.

COROLLARY 2. *If the notation is as in Corollary 1,*

$$\sum \chi_i(1)^2 \equiv 2^n \pmod{2^{n+1}},$$

where the sum extends over all irreducible characters in the principal 2-block.

COROLLARY 3. If *χ is a real-valued irreducible character of odd degree, then for all elements ξ of the centralizer $C(P)$ of the 2-Sylow group P*

$$\chi(\xi) \equiv \chi(1) \qquad \text{mod } \mathfrak{p},$$

where $\mathfrak{p}$ is a prime ideal divisor of 2. If $C(P) = Z(P) \times M$, χ appears with odd multiplicity in the character of the permutation representation of G belonging to the subgroup M.

COROLLARY 4. *If p is an odd prime dividing the order g of a group G of even order, we have at least one of the following two cases.*

Case (a). *The principal p-block of G contains a real-valued irreducible character $\chi \neq 1$ of odd degree.*

Case (b). *No p-element $\pi \neq 1$ of G is conjugate to its inverse.*

Remarks: (1) If in Corollary 4, p^2 does not divide g and if $p \equiv 1 \pmod 4$, we must have Case (*a*).

(2) If we do not have Case (*a*), and if ξ is a p-singular element of G, there exist irreducible characters χ of odd degree in the principal p-block, for which $\chi(\xi)$is not real.

* This research was supported by the United States Air Force under contract no. AF 49(638)-1381, monitored by the Air Force Office of Scientific Research of the Air Research and Development Command.

[1] For this theory cf. Brauer, R., these PROCEEDINGS, **47,** 1888–1890 and 1891–1893 (1961). These papers are quoted as I and II. See also the literature mentioned there. We use the same notation as in I and II. In particular, $C(S)$ denotes the centralizer and $N(S)$ the normalizer of a subset S of a group G. If p is a prime, the p-regular core $K_p(G)$ is the maximal normal subgroup of G of an order prime to p. For detailed proofs, see Brauer, R., *J. Algebra,* **1,** 152–167, 307–334 (1964).

[2] Brauer, R., in *Proceedings of the International Congress of Mathematicians,* Amsterdam, 1954, vol. 1, pp. 209–217; *Math. Z.*, **91,** 119–135 (1965).

[3] *Symposium on Group Theory* (Mimeographed Notes, Harvard University, 1963).

[4] G. Glauberman's theorem will appear in a paper in *J. Algebra.* The author is indebted to him for his permission to quote the result.

[5] D. Gorenstein and J. Walter have succeeded in determining all groups G with dihedral 2-Sylow group which satisfy the conditions of Theorem 6. Of course, this makes it possible to check the Schreier conjecture directly for these G. The work of Gorenstein and Walter is now being published in a series of papers in *J. Algebra.*

[6] For an application of this method, see the second paper quoted in ref. 2.

Reprinted from JOURNAL OF ALGEBRA Vol. 3, No. 2, March 1966
Printed in Belgium

Some Applications of the Theory of Blocks of Characters of Finite Groups III*[1]

RICHARD BRAUER

Harvard University, Cambridge Massachusetts

Received March 1, 1965

I. INTRODUCTION

Let G be a finite group of even order g with the given Sylow 2-group P and let π be a fixed element of P. If y_1 and y_2 are two given involutions of P, we discussed in II the special case that π is not the product of two G-conjugates η_1 of y_1 and η_2 of y_2. In direct continuation of this work, we now take up the general case. The result is that there is an alternative. We may have essentially the same situation as in II. Then there exists a normal subgroup $N \neq G$ such that $\mathfrak{K}_2 G \subseteq N$ and that the index $| G : N |$ lies below a bound depending on the order $| P |$ and the core indices $| \mathfrak{C}_G(y_i) : \mathfrak{K}_2 \mathfrak{C}_G(y_i)|$ with $i = 1, 2$. It is also possible that we have an entirely different situation. In that case, we can give an approximate formula for the order g which depends on the orders of $\mathfrak{C}_G(y_1)$ and $\mathfrak{C}_G(y_2)$ and on the structure of the extended centralizer $\mathfrak{C}_G{}^*(\pi)$ of π.

Actually, the situation considered is slightly more general. The element π can also be a p-element of G for some odd prime p while y_1 and y_2 are taken as involutions of G.

The basic formulas are developed in Section II. The results mentioned are obtained in Section III by a somewhat crude application of these results. In Section IV, a similar idea is exploited in connection with formulas of Frobenius and Schur [4].

If the group P is known explicitly, much more precise results can be given. This is done in the later sections for certain classes of groups P. I have included the dihedral 2-groups (Sections V and VII). We have here approximate formulas for the group order g which I had given without proof in [2].

---

* This research was supported by the United States Air Force under contracts No. AF 49(638)-287 and AF 49(638)-1381 monitored by the AF Office of Scientific Research of the Air Research and Development Command.

[1] We refer to the first and second papers of this series as I and II. They appeared in this *Journal*, vol. 1, pp. 152–167 and pp. 307–334.

They have also been obtained recently by *D.* Gorenstein and J. Walter [5] by a combination of the methods used here with methods of M. Suzuki. It should be mentioned that Gorenstein and Walter have now succeeded in determining all 2-core-free groups with dihedral Sylow 2-group. This result goes, of course, far beyond what we are doing here, and depends on much more difficult and complicated methods. In Section VIII, we consider quasi-dihedral 2-groups. Finally, in Section IX abelian p-groups P are considered with a restricting condition on $\mathfrak{N}(P)$.

In all these cases, we have two types of results. The first ones are concerned with the values of the characters for the 2-singular elements of G (or, the p-singular elements in Section IX). These results may serve as illustrations for the general developments in Part I, Section V. These formulas then are used to obtain approximate formulas for the group order g. For the sake of simplicity we consider only the principal block, but the same work can be done with the other blocks. We have also concentrated our attention on the cases in which G does not contain a normal subgroup of index 2. One can give a great number of corollaries. Some of them appear as Remarks in Section V. The same ideas could have been used in other Sections.

Notation. The notation used is the same as in Part I, Section II. We shall denote by $c_H(S)$ the order of the centralizer $\mathfrak{C}_H(S)$ of a set S in a group H. If H is the full group G investigated, we often write $c(S)$ for $c_G(S)$ and $\mathfrak{C}(S)$ for $\mathfrak{C}_G(S)$. The subgroup of G generated by a subset S of G is denoted by $\langle S \rangle$. We use the abbreviation S_p-group for "Sylow p-subgroup". The conjugate class of an element x in a group G is denoted by $ccl_G(x)$.

II. The Basic Formulas

Let G be a finite group of order g and let y_1, y_2 be two fixed elements, equal or distinct. For $\sigma \in G$, let $a(\sigma)$ denote the number of ordered pairs (ξ, η) of elements of G, for which

$$\xi\eta = \sigma, \qquad \xi \in ccl_G(y_1), \qquad \eta \in ccl_G(y_2). \tag{2.1}$$

Then $a(\sigma)$ is the coefficient with which the class sum belonging to $ccl_G(\sigma)$ appears in the product of the class sums belonging $ccl_G(y_1)$ and $ccl_G(y_2)$. As is well known, $a(\sigma)$ can be expressed by the irreducible characters $\chi_0 = 1$, $\chi_1, \ldots, \chi_{k-1}$ in the form

$$a(\sigma) = gc(y_1)^{-1}c(y_2)^{-1} \sum_{i=0}^{k-1} \chi_i(y_1)\chi_i(y_2)\bar{\chi}_i(\sigma)/x_i\,, \tag{2.2}$$

where x_i is the degree of χ_i,

$$x_i = \chi_i(1).$$

Assume now that G has even order g and choose y_1 and y_2 as involutions. Then ξ and η in (2.1) are involutions too and we have

$$\xi^{-1}\sigma\xi = \eta\xi = \sigma^{-1}, \; \eta^{-1}\sigma\eta = \eta\xi = \sigma^{-1}. \tag{2.3}$$

Hence, ξ and η belong to the extended centralizer $\mathfrak{C}^*(\sigma)$ of σ, (cf. II, Section IV).

Let p be a fixed prime dividing g and let π be a p-element of G. Choose σ as an element with the p-factor π, i.e., as an element of the form $\sigma = \pi\rho$ where ρ is a p-regular element of $\mathfrak{C}(\pi)$. Then (2.3) implies that $\xi, \eta \in \mathfrak{C}^*(\pi)$.

Consider now a fixed subgroup $\tilde{G}$ of G with

$$\tilde{G} \supseteq \mathfrak{C}^*(\pi).$$

The intersections $ccl_G(y_i) \cap \tilde{G}$ are disjoint unions of conjugate classes of $\tilde{G}$, say

$$ccl_G(y_1) \cap \tilde{G} = \bigcup_\mu ccl_{\tilde{G}}(y_\mu^{(1)}); \; ccl_G(y_2) \cap \tilde{G} = \bigcup_\nu ccl_{\tilde{G}}(y_\nu^{(2)}). \tag{2.4}$$

If $\tilde{a}_{\mu\nu}(\sigma)$ has the same significance for $\tilde{G}$ and $ccl_{\tilde{G}}(y_\mu^{(1)})$, $ccl_{\tilde{G}}(y_\nu^{(2)})$ as $a(\sigma)$ had for G and $ccl_G(y_1)$, $ccl_G(y_2)$,

$$a(\pi\rho) = \sum_{\mu,\nu} \tilde{a}_{\mu\nu}(\pi\rho).$$

By (2.2), $a(\pi\rho)$ can be expressed by $\chi_0, \chi_1, \ldots$. Likewise, $\tilde{a}_{\mu\nu}(\pi\rho)$ can be expressed by the irreducible characters $\tilde{\chi}_0 = 1, \tilde{\chi}_1, \ldots$ of $\tilde{G}$. Since the values of characters for involutions are real (even rational) we find:

$$g c_G(y_1)^{-1} c_G^{-1}(y_2) \sum_i \chi_i(y_1)\chi_i(y_2)\chi_i(\pi\rho)$$

$$= \tilde{g} \sum_{\mu,\nu} c_{\tilde{G}}(y_\mu^{(1)})^{-1} c_{\tilde{G}}(y_\nu^{(2)})^{-1} \sum_j \tilde{\chi}_j(y_\mu^{(1)})\tilde{\chi}_j(y_\nu^{(2)})\tilde{\chi}_j(\pi\rho)/\tilde{x}_j$$

with $\tilde{x}_j = \tilde{\chi}_j(1)$. This holds for all p-regular $\rho \in \mathfrak{C}(\pi)$. Use I, (5.3) to express $\chi_i(\pi\rho)$ by means of the members ϕ_α^π of basic sets ϕ_b for the various p-blocks b of $\mathfrak{C}(\pi)$ and the corresponding decomposition numbers $d_{i\alpha}^\pi$ of G. Likewise, express $\tilde{\chi}_j(\pi\rho)$ by the ϕ_α^π and the decomposition numbers $\tilde{d}_{j\alpha}^\pi$ of $\tilde{G}$. Since the resulting formula holds for all p-regular $\rho \in \mathfrak{C}_G(\pi) = \mathfrak{C}_{\tilde{G}}(\pi)$ and since the ϕ_α^π are linearly independent, ϕ_α^π appears with the same coefficient on both sides. This yields

$$g \sum_i d_{i\alpha}^\pi \chi_i(y_1)\chi_i(y_2)/x_i = c_G(y_1)c_G(y_2)\tilde{g} \sum_j \tilde{d}_{j\alpha}^\pi h_j^{(1)} h_j^{(2)}/\tilde{x}_j \tag{2.5}$$

where we introduced the abbreviations

$$h_j^{(1)} = \sum_\mu \tilde{\chi}_j(y_\mu^{(1)})/c_G(y_\mu^{(1)}),\ h_j^{(2)} = \sum_\nu \tilde{\chi}_j(y_\nu^{(2)})/c_G(y_\nu) \tag{2.6}$$

with μ and ν ranging over the same values as in (2.4).

This holds for each α. If ϕ_α^π belongs to the block b of $\mathfrak{C}(\pi)$, then $d_{i\alpha}^\pi = 0$, except when χ_i belongs to $B = b^G$, cf. I, (2.6). Likewise, $\tilde{d}_{i\alpha}^\pi = 0$, except when $\tilde{\chi}_j$ belongs to the block $\tilde{B} = b^{\tilde{G}}$ of $\tilde{G}$. Hence it suffices to let χ_i range over the irreducible characters in B and $\tilde{\chi}_j$ over the irreducible characters in $\tilde{B}$. We have now shown:

THEOREM (2A). *Let G be a group of even order g. Let π be a p-element of G for some prime p dividing g and let $\tilde{G}$ be a subgroup of G with $\tilde{G} \supseteq \mathfrak{C}^*(\pi)$. If y_1 aud y_2 are two involutions of G, then*

$$g \sum_i d_{i\alpha}^\pi \chi_i(y_1)\chi_i(y_2)/x_i = c(y_1)c(y_2)\tilde{g} \sum_j \tilde{d}_{j\alpha}^\pi h_j^{(1)} h_j^{(2)}/\tilde{x}_j$$

for each member ϕ_α^π of a basic set ϕ_b of a p-block b of $\mathfrak{C}(\pi)$ and the corresponding decomposition numbers $d_{i\alpha}^\pi$, $\tilde{d}_{j\alpha}^\pi$ of G and $\tilde{G}$ respectively. On the left, i ranges over the values for which $\chi_i \in B = b^G$ and on the right, j over the values for which $\tilde{\chi}_j \in \tilde{B} = b^{\tilde{G}}$. The rational numbers $h_j^{(1)}$, $h_j^{(2)}$ are defined in (2.6).

Let B be a fixed p-block of G. Let ρ be a p-regular element of $\mathfrak{C}(\pi)$. Multiply our formula with $\phi_\alpha^\pi(\rho)$ and add over all $\phi_\alpha^\pi \in \phi_b$ for all p-blocks b of $\mathfrak{C}(\pi)$ with $b^G = B$. Then for $\tilde{B} = b^{\tilde{G}}$, we have $\tilde{B}^G = b^G = B$. Conversely, if $\tilde{B}$ is a p-block of $\tilde{G}$ with $\tilde{B}^G = B$, and if not each $\tilde{\chi}_j \in \tilde{B}$ vanishes for all elements $\pi\rho$ with p-regular $\rho \in \mathfrak{C}(\pi)$, then by I (2.6), there exist blocks b of $\mathfrak{C}(\pi)$ with $b^{\tilde{G}} = \tilde{B}$ and then $b^G = B$. We obtain:

THEOREM (2B). *Let G, $\tilde{G}$, π, y_1, y_2 have the same significance as in* (2A). *If B is a fixed p-block of G, then, for all p-regular elements ρ of $\mathfrak{C}(\pi)$,*

$$g \sum_i \chi_i(y_1)\chi_i(y_2)\chi_i(\pi\rho)/x_i = c(y_1)c(y_2)\tilde{g} \sum_j h_j^{(1)} h_j^{(2)} \tilde{\chi}_j(\pi\rho)/\tilde{x}_j,$$

where on the left, χ_i ranges over all irreducible characters in B and, on the right, $\tilde{\chi}_j$ ranges over all irreducible characters of $\tilde{G}$ which belong to p-blocks $\tilde{B}$ of $\tilde{G}$ with $\tilde{B}^G = B$.

III. Application of the Basic Formulas

The following results are obtained by a somewhat crude application of the preceding results.

Theorem (3A). *Let G be a group of finite even order g, let y_1 and y_2 be two involutions of G, let π be a p-element of G for some prime p, and let $\tilde{G} \supseteq \mathfrak{C}^*(\pi)$ be a subgroup of G; $|\tilde{G}| = \tilde{g}$. For any $A > 1$, at least one of the following two possibilities occurs:*

Case I. There exists an approximate formula for g:

$$\frac{A}{A+1} c(y_1)c(y_2)\tilde{g}w < g < \frac{A}{A-1} c(y_1)c(y_2)\tilde{g}w. \tag{3.1}$$

Here, w depends only on $\tilde{G}$ and the intersections $\tilde{G} \cap ccl_G(y_1)$, $\tilde{G} \cap ccl_G(y_2)$.

Case II. There exists a normal subgroup N of G with $\mathfrak{K}_p G \subseteq N \subset G$ such that

$$|G : N| \leqslant [2p^{2n} A \gamma_0(p^m)^{1/2} c(y_1)^{1/2} c(y_2)^{1/2}]!. \tag{3.2}$$

Here, p^n is the exact power of p dividing g and p^m is the exact power of p dividing $|\mathfrak{C}^*(\pi)|$. The expression $\gamma_0(p^m)$ depends only on p^m.

Proof. Choose b in (2A) as the principal p-block and use a basic set ϕ_b with $1 \in \phi_b$, say $\phi_0^\pi = 1$. By I, Theorem 3, the blocks $B = b^G$ and $\tilde{B} = b^{\tilde{G}}$ are the principal blocks of G and $\tilde{G}$, respectively. Apply now (2A) with $\alpha = 0$. Set

$$z = \sum_i d_{i0}^\pi \chi_i(y_1)\chi_i(y_2)/x_i\,, \quad w = \sum_j d_{j0}^\pi h_j^{(1)} h_j^{(2)}/\tilde{x}_j \tag{3.3}$$

with χ_i ranging over B and $\tilde{\chi}_j$ ranging over $\tilde{B}$. Then w depends only on $\tilde{G}$ and the intersections of $\tilde{G}$ with $ccl_G(y_1)$ and $ccl_G(y_2)$, cf. (2.6). If

$$(A-1)/A < z < (A+1)/A, \tag{3.4}$$

clearly (3.1) will hold.

Suppose then that (3.4) does not hold. The term of z with $i = 0$ is 1, since χ_0 is the principal character. Let x_τ be the minimal degree of characters $\chi_i \neq \chi_0$ in B. It follows that

$$A \sum_{i \neq 0} |\, d_{i0}^\pi \chi_i(y_1)\chi_i(y_2)\,| \geqslant x_\tau\,.$$

Using Lagrange's inequality, I(5.4), and II(3.1), we obtain

$$x_\tau^2 \leqslant A^2 c_{00}^\pi \lambda(G, y_1)\lambda(G, y_2) \leqslant A^2\gamma_0(p^m)c(y_1)c(y_2)$$

where $\gamma_0(p^m)$ has the same significance as in I, Section V. Corollary 7. Now the same method as in the proof of II, Theorem 3 leads to the result of Case II.

For $p = 2$ it follows from II (3.6), (3.2), and (3.5) that

$$\lambda(G, y_i) \leqslant |\ \mathfrak{C}(y_i) : \mathfrak{K}_2\mathfrak{C}(y_i)|.$$

Hence we have

THEOREM (3A*). *If $p = 2$, then in* (3A), *Case II*, (3.2) *can be replaced by*

$$|\ G: N\ | \leqslant [2^{2n+1}A\gamma_0(2^m)^{1/2}t_1^{1/2}t_2^{1/2}]! \qquad (3.2^*)$$

with $t_i = |\ \mathfrak{C}(y_i) : \mathfrak{K}_2\,\mathfrak{C}(y_i)|$, $(i = 1, 2)$.

We give an upper estimate for w in (3.1).

THEOREM (3B). *In Theorem* 3.1,

$$w \leqslant \gamma_0(p^m)^{1/2}l_1^{1/2}l_2^{1/2}, \qquad (3.5)$$

where l_i is the number of conjugate classes of $\tilde{G}$ contained in $ccl_G(y_i)$, $i = 1, 2$. *For $p = 2$,* (3.5) *can be replaced by*

$$w \leqslant \gamma_0(2^m)^{1/2}. \qquad (3.5^*)$$

Proof. It follows from (2.6) and Lagrange's inequality that

$$h_j^{(1)2} \leqslant \sum_\mu \tilde{\chi}_j(y_\mu^{(1)})^2/c_G(y_\mu^{(1)}) \sum_\mu 1/c_G(y_\mu^{(1)}).$$

The second sum on the right is less than 1 since

$$\sum_\mu \tilde{g}/c_G(y_\mu^{(1)}) = |\ \tilde{G} \cap ccl_G(y_1)\ | < \tilde{g}. \qquad (3.6)$$

The index μ ranges over l_1 values and we find

$$\sum_j h_j^{(1)2} \leqslant \sum_\mu \sum_j \tilde{\chi}_j(y_\mu^{(1)})^2/c_G(y_\mu^{(1)}) \leqslant l_1.$$

Similarly, we have $\sum_j h_j^{(2)2} \leqslant l_2$. Now, (3.3) yields

$$w^2 \leqslant \sum_j |\, d_{j0}^{\pi} \,|^2 \sum_j h_j^{(1)2} \sum_j h_j^{(2)2} = c_{00}^{\pi} l_1 l_2 \leqslant \gamma_0(p^m) l_1 l_2 ,$$

cf. I (5.4) and I, Section V, Corollary 7. This proves (3.5).

For $p = 2$, we can write

$$\sum_j h_j^{(1)2} = \sum_{\mu,\kappa} c_G(y_\mu^{(1)})^{-1} c_G(y_\kappa^{(1)})^{-1} \sum_j \chi_j(y_\mu^{(1)}) \chi_j(y_\kappa^{(1)})$$

where κ has the same ranges as μ. The inner sum on the right vanishes for $\kappa \neq \mu$, cf. [*1*], (7C). Hence

$$\sum_j h_j^{(1)2} = \sum_\mu c_G(y_\mu^{(1)})^{-2} \sum_j \tilde{\chi}_j(y_\mu^{(1)})^2 \leqslant \sum_\mu c_G(y_\mu^{(1)})^{-1} < 1.$$

cf. (3.6). An analogous result holds in the case of $h_j^{(2)}$. Now (3.5*) is obtained easily from (3.3).

COROLLARY (3C). *There exists a bound* $\beta(n, t)$ *depending on two parameters* n, t *with the following property: If* y *is an involution of a simple group* G *of order* $g = 2^n g_0$ *with odd* g_0 *and if* $|\,\mathfrak{C}(y) : \mathfrak{R}_2 \mathfrak{C}(y)| = t$, *then*

$$g \leqslant \beta(n, t) c_G(y)^3. \tag{3.7}$$

Proof. Apply the theorem with $\pi = y_1 = y_2 = y$, $\bar{G} = \mathfrak{C}_G(y)$, $A = 2$. In the case I, (3.1) and (3.5*) imply that (3.7) will hold if $\beta(n, t) \geqslant 2\gamma_0(2^m)^{1/2}$, where 2^m here is the exact power of 2 dividing t. If we have Case II, then $N = 1$, and (3.2*) reads

$$g \leqslant [2^{2n+2}\gamma_0(2^m)^{1/2} t]!.$$

If we choose $\beta(n, t)$ as the maximum of these expressions for $m = 1, 2, ..., n$, then (3.7) holds.

Corollary (3C) had been anticipated in the proof of II, Theorem 6 in order to obtain II (9.12). In the notation used there, $t \leqslant \mu$ and $2^n \leqslant \mu$.

COROLLARY (3D). *If in Theorem* (3A*), *the groups* $\mathfrak{C}(y_1)$, $\mathfrak{C}(y_2)$ *are solvable, then* (3.2*) *can be replaced by* $(G : N) \leqslant f(n, A)$ *where* f *is a function of* n *and* A *alone.*

This is an immediate consequence of II, Section III, Lemma 2.

Arguments quite similar to the ones used above can also be applied in connection with (2B). The decomposition numbers $d_{i\alpha}^{\pi}$ and $d_{j\bar{\alpha}}^{\pi}$ are replaced by $\chi_i(\pi\rho)$ and $\tilde{\chi}_j(\pi\rho)$, respectively. Observe that if χ ranges over the principal block of G, by II, Section III:

$$\sum_i |\chi_i(\pi\rho)|^2 = \lambda(G, \pi\rho) = \lambda(\mathfrak{C}_G(\pi), \rho).$$

In giving estimates for w, use $|\tilde{\chi}_j(\pi\rho)| \leqslant \tilde{x}_j$. We obtain

THEOREM (3E). *Let G, π, y_1, y_2, $\tilde{G}$ and A have the same significance as in* (3A). *If ρ is a p-regular element of $\tilde{G}$ we have at least one of the following two possibilities:*

Case I. The approximate formula (3.1) *holds with w depending on $\tilde{G}$, $\tilde{G} \cap ccl_G(y_1)$, $\tilde{G} \cap ccl_G(y_2)$ and ρ. Moreover, $w < l_1^{1/2} l_2^{1/2}$ where l_1, l_2 have the same significance as in* (3B). For $p = 2$, we have $w < 1$.

Case II. There exists a normal subgroup N of G with $\mathfrak{K}_p \subseteq N \subset G$ such that

$$|G : N| \leqslant [2p^{2n} A\lambda(\mathfrak{C}(\pi), \rho)^{1/2} \lambda(G, y_1)^{1/2} \lambda(G, y_2)^{1/2}]!.$$

Note that by II , Section III:

$$\lambda(\mathfrak{C}(\pi, \rho) \leqslant c(\pi\rho), \qquad \lambda(G, y_i) \leqslant c(y_i), \qquad (i = 1, 2)$$

and, for $p = 2$,

$$\lambda(G, y_i) \leqslant |\ \mathfrak{C}(y_i) : \mathfrak{K}_2 \mathfrak{C}(y_i)|.$$

COROLLARY (3F). *If y is an involution of the finite group G and if $A > 1$, then (Case I) $g \leqslant A/(A-1)c(y)^3$ or (Case II), there exists a normal subgroup N of G with $\mathfrak{K}_2 G \subseteq N \subset G$ such that*

$$|G : N| \leqslant [2^{2n+1} A\lambda(\mathfrak{C}(y))^{3/2}]!$$

where 2^n is the power of 2 dividing $|G|$.

This is immediate from (3E), if we take $p = 2$, $\pi = y_1 = y_2 = y$, $\rho = 1$, $\tilde{G} = \mathfrak{C}(\pi)$. Our method shows that in the Case II, N can be obtained as the kernel of an irreducible character $\chi_i \neq 1$ in the principal 2-block B_0 of G.

COROLLARY (3G). *Let G be a group of finite even order $g = 2^n g_0$, g_0 odd. Let $\{y_1, \ldots, y_r\}$ be a set of involutions of G. Set*

$$\lambda = \operatorname*{Max}_i \lambda\mathfrak{C}(y_i) \leqslant \operatorname*{Max}_i (\mathfrak{C}(y_i) : \mathfrak{K}_2 \mathfrak{C}(y_i)),$$
$$c = \operatorname*{Max}_i c(y_i).$$

l. 6: $\mathfrak{C}(\pi)$ [R.B.]

There exists a function $\beta(n, \lambda)$ depending only on n and λ such that at least one of the following two statements is true:

Case I.

$$g \leqslant \beta(n, \lambda)c^3. \tag{3.8}$$

Case II. There exists a normal subgroup $H \supseteq \mathfrak{K}_2 G$ of G which does not contain any y_i, $1 \leqslant i \leqslant r$, and for which

$$| G : H | \leqslant \beta(n, \lambda). \tag{3.9}$$

Proof. Apply (3F) with $y = y_1$ taking say $A = 2$. If we have Case I, $g \leqslant 2c^3$. If we have Case II, we can find an irreducible character $\chi_1 \neq 1$ in the principal 2-block B_0 whose kernel N satisfies

$$| G : N | \leqslant [2^{2n+2}\lambda^{3/2}]!.$$

If N still contains some y_j, we can again apply (3F) to N. If we have here Case I, then

$$g \leqslant [2^{2n+2}\lambda^{3/2}]!\, 2c^3.$$

If we have Case II, then as in the proof of II, Theorem 4, we can find two characters χ_1, χ_2 in B_0 such that if N_2 is the intersection of their kernels, $| G : N_2 |$ lies below a bound depending only on n and λ. If N_2 still contains some y_j we can continue, and so on. Since B_0 contains at most 2^{2n} characters, this process must terminate after $r \leqslant 2^{2n}$ steps. If we have Case I for N_r, then $| N_r | \leqslant 2c^3$ and (3.8) holds for suitable choice of $\beta(n, \lambda)$. The other possibility is that N_r does not contain any element $y_1, ..., y_r$. If we set $H = N_r$, then (3.9) holds for suitable choice of $\beta(n, \lambda)$. We have $H \supseteq K_2 G$, cf. I, Section III, Theorem 1.

Corollary (3H). *If in (3G), $y_1, y_2, ..., y_r$ are chosen as the involutions in the center $\mathfrak{Z}(P)$ of the Sylow 2-group P of G, then*

$$| G : \mathfrak{K}_2 G | \leqslant \beta(n, \lambda)c^3.$$

Indeed, if we apply (3G), the Sylow 2-subgroup $Q = H \cap P$ of H meets $\mathfrak{Z}(P)$ in $\{1\}$. Since $Q \lhd P$, then $Q = 1$. Hence $| H |$ is odd, and this implies $H = \mathfrak{K}_2 G$.

Corollary (3I). *Let G be a finite group with the Sylow 2-group P. For each normal subgroup Q of P, set $\mathfrak{S}(Q) = \mathfrak{K}_2 \mathfrak{C} Q$. There exist a function $f(n, s)$ of two arguments n, s with the following property: If $| P | = 2^n$ and if $| \mathfrak{S}(Q)| \leqslant s$ for all $Q \lhd P$, then $| G | \leqslant f(n, s)$.*

Proof. We construct $f(n, s)$ recursively with respect to n. Since

$$\mathfrak{S}(\langle 1\rangle) = \mathfrak{R}_2 G,$$

we have $|\mathfrak{R}_2 G| \leqslant s$. We may set $f(0, s) = s$. Suppose $n > 0$. If y is an involution in $\mathfrak{Z}(P)$, set $\bar{G} = \mathfrak{C}_G(y)/\langle y\rangle$. Then $\bar{G}$ has the Sylow 2-group $\bar{P} = P/\langle y\rangle$. Every normal subgroup $\bar{Q}$ of $\bar{P}$ is the image of a normal subgroup Q of P with $y \in Q$. It is easy to see that $\langle \mathfrak{R}_2\mathfrak{C}_G(Q), y\rangle/\langle y\rangle = \mathfrak{R}_2\mathfrak{C}_{\bar{G}}\bar{Q} = \mathfrak{S}(\bar{Q})$ whence $|\mathfrak{S}(\bar{Q})| \leqslant s$. If $f(n-1, s)$ has already been constructed, we find $c_G(y) \leqslant 2f(n-1, s)$. Now (3H) shows that we may take

$$f(n, s) = 8sf(n-1, s)^3\beta(n, 2f[n-1, s]).$$

(3I) represents a slight refinement of a recent result of John G. Thompson [7]. Indeed, each $\mathfrak{S}(Q)$ for $Q \lhd P$ is a 2-signalizer of G, since $\mathfrak{S}(Q)$ has odd order and $\mathfrak{S}(Q)$ is normalized by P.

Remarks

1. An argument similar to the one used in the proofs of Theorems 1 and 3 shows that if $0 < A \leqslant 1$ and if we do not have the situation described in Case II, the left half of the inequality (3.1) holds.

2. The estimates given for w can be improved. For instance, the left side of (3.6) is at most equal to the numbers of involutions of $\bar{G}$, which is usually much smaller than $\bar{g}$. If $p = 2$, we can show without difficulty that

$$\sum_j h_j^{(1)2} < 1/K,$$

where $K = \min_\mu |\mathfrak{R}_2\mathfrak{C}_G(y_\mu^{(1)})|$.

3. Corollary (3H) contains II, Theorem 6.

IV. On a Formula of Frobenius and I. Schur

If χ is an irreducible character of a finite group G of order g, set

$$\epsilon(\chi) = (1/g)\sum_{\sigma\in G}\chi(\sigma^2). \qquad (4.1)$$

As shown by Frobenius and Schur [4], we have:

(a) $\epsilon(\chi) = 1$, if the irreducible representation X corresponding to χ can be written in the field $\mathbf{R}$ of real numbers.

(b) $\epsilon(\chi) = -1$, if χ is real, but if X is not similar to a representation in $\mathbf{R}$. In this case, the degree $x = \chi(1)$ must be even.

(c) $\epsilon(\chi) = 0$, if χ is nonreal.

For $\sigma \in G$, the number $N(\sigma)$ of elements $\tau \in G$ with $\tau^2 = \sigma$ is given by

$$N(\sigma) = \sum_i \epsilon(\chi_i)\chi_i(\sigma) \tag{4.2}$$

with the sum extending over all irreducible characters χ_i of G.

THEOREM (4A). *Let G be a group of finite order, let π be a p-element of G for some prime p and let $\tilde{G} \supseteq \mathfrak{C}(\pi)$ be a subgroup of G. For each member ϕ_α^π of a basic set ϕ_b of a* p-block of $\mathfrak{C}(\pi)$ *and for the corresponding decomposition numbers $d_{i\alpha}^\pi$ and $\tilde{d}_{j\alpha}^\pi$ of G and $\tilde{G}$ respectively, we have*

$$\sum_i \epsilon(\chi_i)d_{i\alpha}^\pi = \sum_j \epsilon(\tilde{\chi}_j)\tilde{d}_{j\alpha}, \tag{4.3}$$

where $\tilde{\chi}_0, \tilde{\chi}_1, \ldots$ again denote the irreducible characters of $\tilde{G}$.

The proof is quite analogous to that of (2A). We only have to note that any solution τ of $\tau^2 = \pi\rho$ belongs to $\mathfrak{C}(\pi)$. Also, the method leading to (2B) can be applied and yields:

THEOREM (4B). *Let G, π, p and $\tilde{G}$ be as in* (4A). *If B is a p-block of $\tilde{G}$, then for every p-regular element ρ of $\mathfrak{C}(\pi)$,*

$$\sum_i \epsilon(\chi_i)\chi_i(\pi\rho) = \sum_j \epsilon(\tilde{\chi}_j)\tilde{\chi}_j(\pi\rho), \tag{4.4}$$

where on the left, χ_i ranges over B and, on the right, $\tilde{\chi}_j$ ranges over the irreducible characters $\tilde{\chi}_j$ of $\tilde{G}$ which belong to p-blocks $\tilde{B}$ with $\tilde{B}^G = B$.

REMARK (4C). *If $N(\pi) = 0$, the right hand sides of* (4.3) *and* (4.4) *can be replaced by* 0. *This can only happen for $p = 2$.*

We use (4.1) to prove a rather peculiar special result.

THEOREM (4D). *Assume that the finite group G of even order g has an irreducible character χ of a degree x which is divisible* (1) *by the full contribution 2^n of the prime* 2 *to g,* (2) *by the full contribution $p^{a(p)}$ to g of each odd prime p for which there exist elements of order $2p$ in G. Then G has only one class of involutions. For each involution J of G, $\mathfrak{C}(J)$ is a Hall subgroup of G of order x.*

In particular, x is not divisible by odd primes p for which no elements of order $2p$ exist. Finally, $\epsilon(\chi) = 1$.

Proof. Write (4.1) in the form

$$g\epsilon(\chi) = \sum{}' \chi(\sigma^2) + \sum{}'' \chi(\sigma^2),$$

where in the first sum on the right, σ ranges over the elements of even order. Since every element of odd order has a unique square root in G, σ^2 in the second sum can be replaced by σ. Since $\chi \neq \chi_0$, the equation $(\chi, \chi_0) = 0$ yields

$$g\epsilon(\chi) = \sum{}' [\chi(\sigma^2) - \chi(\sigma)]. \tag{4.5}$$

Our assumptions imply that χ has defect 0 for 2 and for each odd prime p for which elements of order $2p$ exist. Hence, for the terms of the sum in (4.5), $\chi(\sigma) = 0$. Likewise, $\chi(\sigma^2) = 0$, except when σ is an involution in which case $\chi(\sigma^2) = x$. If G contains m involutions, we obtain

$$g\epsilon(\chi) = mx.$$

This shows already that $\epsilon(\chi) = 1$. For each involution J of G, certainly $m \geqslant g/c(J)$ and hence $c(J) \geqslant x$. Since our assumptions imply that $c(J)$ divides x, we have $c(J) = x$, and $|\, ccl(J)| = m$. Now all the statements are evident.

Remarks. As examples, we may mention the groups $PSL(2, q)$. We may also remark that if G is a CIT-group in the sense of M. Suzuki [6], then G is of deficiency class 1 for 2, I, Section IV. Our result shows that if G has characters χ of 2-defect 0 (i.e., if G is not of deficiency class 0), these characters χ have all degree 2^n and G has one class of involutions.

V. Groups of Order $4g_0$, g_0 Odd

As first special case, we consider groups G of an order $g = 4g_0$ with an odd g_0. If G does not have a normal 2-complement, the S_2-group P of G is elementary abelian. Moreover (cf. II, Section VI, Proposition 6), G has only one class of involutions. The principal 2-block B_0 consists of four characters $\chi_0 = 1, \chi_1, \chi_2, \chi_3$. There exist signs $\epsilon_i = \pm 1$ such that, for all 2-singular elements ξ of G,

$$\chi_i(\xi) = \epsilon_i, \qquad 0 \leqslant i \leqslant 3, \tag{5.1}$$

while for 2-regular $\rho \in G$

$$1 + \epsilon_1\chi_1(\rho) + \epsilon_2\chi_2(\rho) + \epsilon_3\chi_3(\rho) = 0. \tag{5.2}$$

In particular, for $\rho = 1$,

$$1 + \epsilon_1 x_1 + \epsilon_2 x_2 + \epsilon_3 x_3 = 0. \tag{5.3}$$

Finally,

$$x_i \equiv \epsilon_i \pmod 4, \qquad 0 \leqslant i \leqslant 3. \tag{5.4}$$

Let J denote an involution in P and apply (2B) with $\pi = y_1 = y_2 = J$, $\tilde{G} = \mathfrak{C}(J)$. Since $\tilde{G}$ has the S_2-group P and has more than one class of involutions, it has a normal 2-complement. Hence the characters of the principal 2-block of G can be identified with the irreducible characters of $P \simeq \tilde{G}/\mathfrak{K}_2 G$, cf. I, Section III, Theorem 1. If these characters are taken in suitable order, the equations (2.6) read $h_0^{(i)} = c(J)^{-1} + 2c(P)^{-1}$, $h_1^{(i)} = c(J)^{-1} - 2c(P)^{-1}$, $h_2^{(i)} = h_3^{(i)} = -c(J)$. Using (5.1) for $\xi = J$ we obtain easily:

Theorem (5.A). *Let G be a group of order $g = 4g_0$ with odd g_0. Let P be an S_2-group of G and let J be an involution of P. If G does not have a normal 2-complement, then*

$$g = 8\alpha \frac{c(J)^3}{c(P)^2} \qquad \text{with} \qquad \alpha = \frac{x_1 x_2 x_3}{(x_1 + \epsilon_1)(x_2 + \epsilon_2)(x_3 + \epsilon_3)}. \tag{5.5}$$

Here, $1, x_1, x_2, x_3$ are the degrees of the characters of the principal 2-block of G and $x_i \equiv \epsilon_i = \pm 1 \pmod 4$.

At least one of the ϵ_i in (5.2) must be $+1$ and one must be -1, say $\epsilon_1 = 1$, $\epsilon_3 = -1$. Set $\epsilon_2 = \delta = \pm 1$.

In the case $\delta = 1$, we can apply (4B), (4C). Since J is not a square in G, we find

$$1 + \epsilon(\chi_1) + \epsilon\chi_2) - \epsilon(\chi_3) = 0. \tag{5.6}$$

With χ_3, the conjugate character $\bar{\chi}_3$ belongs to B_0, cf. I, Section III, Lemma 2. It follows from (5.4) that $\bar{\chi}_3 = \chi_3$ and, as x_3 is odd, $\epsilon(\chi_3) = 1$. Now (5.6) shows that χ_1 cannot be real and that we must have $\bar{\chi}_1 = \chi_2$. In particular, $x_1 = x_2$. If we set $x_3 = q$, $x_1 = (q - 1)/2$ by (5.3). Then (5.5) reads

$$g = 8q(q-1)(q+1)^{-2}c(J)^3 c(P)^{-2}.$$

Since $g\chi_i(J)/(c(J)x_i)$ is an algebraic integer for $i = 1, 2, 3$, it follows from (5.1) and (5.5) that $(q + 1)/4$ is an integer dividing $c(J)/c(P)$. We now have

THEOREM (5B). *Suppose that in* (5A), $\epsilon_1\epsilon_2\epsilon_3 = -1$. *If the notation is chosen such that* $\epsilon_3 = -1$ *and if we set* $q = x_3 \equiv -1 \pmod 4$, *we have* $c(J) = mc(P)(q+1)/4$ *with integral m. Then*

$$g = (m^3c(P)/4)(q^3 - q)/2.$$

If in the case $\delta = -1$ we have $x_2 = x_3$ and if we set $q = x_1$, we can rewrite (5.5) in a similar manner. It follows as a corollary of deep results of Gorenstein and Walter (cf. the Introduction) that the degrees x_2 and x_3 are always equal. However, no direct proof seems to be available.

Remarks

1. It is seen easily that in (5.5) we have $3/8 \leqslant \alpha < 1$ for $\delta = 1$ and $1 < \alpha \leqslant 15/8$. The case $\alpha = 3/8$ occurs only if $G/\mathfrak{K}_2G \simeq A_4$ and the case $\alpha = 15/8$ only if $G/\mathfrak{K}_2G \simeq A_5$. In all other cases, $55/72 \leqslant \alpha \leqslant 91/72$.

2. The three involutions $J, J_1, J_2 = JJ_1$ are conjugate in G and hence in $\mathfrak{N}(P)$. It follows that there exist elements $t \in G$ such that

$$t^{-1}Jt = J_1, \qquad t^{-1}J_1t = J_2, \qquad t^{-1}J_2t = J.$$

Here, t can be chosen as 3-element.

3. If we set $G_0 = \mathfrak{C}(J), t\rangle$ then G_0 satisfies the same assumptions as G. If we apply (5A) to G_0 and if α_0 here has the same significance as α for G, then $|G : G_0| = \alpha/\alpha_0$. Using Remark 1, we can now see that $G = \langle\mathfrak{C}(J), t\rangle$ except when $G/\mathfrak{K}_2G \simeq A_5$.

4. If $c(P) > 4$, there exist nonprincipal blocks B of defect 2. Each such B consists of four irreducible characters. Application of (2A) shows that α can be expressed by the degrees of these characters and of suitable characters of $\mathfrak{C}(J)$ and $\mathfrak{C}(P)$.

5. If P is elementary abelian of order 4, but if G has a normal 2-complement K, we can still apply (2B). We obtain

$$|K|c_K(P)^2 = c_K(J)c_K(J_1)c_K(J_2).$$

This is the special case of Wielandt's fixed point formula proved directly in II, Section IX.

VI. AUXILIARY REMARKS

If the S_p-group P is known explicitly, the decomposition numbers can be discussed by the methods of I, Section V. If g is even, for each type (I, Section V, Theorem 8) much more precise results than (3A) can be obtained. In the following sections this will be done for certain groups P. In order to avoid repetitions, we collect here a number of remarks. The p-block B will always be taken as the principal p-block B_0.

1. As in I, Section V, we speak of a *column* $\mathfrak{a}$ for B if with each $\chi_i \in B$, there is associated a complex number, the "entry" for χ_i. We shall denote this entry by $(\mathfrak{a})_i$ (or if the character is denoted by $\chi^{(j)}$, by $(\mathfrak{a})^{(j)}$). For two columns $\mathfrak{a}$ and $\mathfrak{b}$, the inner product $(\mathfrak{a}, \mathfrak{b})$ is defined as

$$(\mathfrak{a}, \mathfrak{b}) = \sum (\mathfrak{a})_i \overline{(\mathfrak{b})_i}$$

with the sum extending over all $\chi_i \in B$. A column $\mathfrak{a}$ is *integral*, if all $(\mathfrak{a})_i$ are integers.

In particular, if $\sigma \in G$, we have the column $\chi(\sigma)$ with the entries $\chi_i(\sigma)$. If $\pi \in P$, and if a fixed basic set $\phi_b{}^\pi$ has been chosen for the principal p-block b of $\mathfrak{C}(\pi)$, then for each $\phi_\mu{}^\pi \in \phi_b{}^\pi$, we have the column of decomposition numbers $\mathfrak{d}_\mu{}^\pi$ with the entries $(\mathfrak{d}_\mu{}^\pi)_i = \mathfrak{d}_{i\mu}^\pi$, cf. I (5.3). These satisfy the orthogonality relations I (5.5). If a column $\mathfrak{a}$ is a linear combination of columns $\mathfrak{d}_\mu{}^\pi$ with π belonging to a subset U of P, then for each $\xi \in G$ whose p-factor is not G-conjugate to an element of U.

$$(\mathfrak{a}, \chi(\xi)) = 0. \tag{6.1}$$

2. In particular, if $1 \notin U$, we can choose $\xi = 1$ in (6.1). It follows that $\mathfrak{a}$ cannot have one nonzero entry. In the case of integral $\mathfrak{a} \neq 0$, we can add that if $(\mathfrak{a}, \mathfrak{a}) = m \leqslant 4$, then $\mathfrak{a}$ has exactly m nonzero entries. These are ± 1 and both signs occur. If $(\mathfrak{a})_0 \neq 0$, both signs occur among the other nonzero entries when $m = 3$ or 4.

3. It is of importance to construct sets of columns for which all inner products have small values. Often, II, Section V, Lemma 3 can be used as a tool. If Q is an abelian subgroup of P and if θ is a generalized character of Q, define an integral column $\mathfrak{a}(\theta)$ by the formula

$$\mathfrak{a}(\theta) = |Q|^{-1} \sum_{\pi \in Q} \chi(\pi)\bar{\theta}(\pi). \tag{6.2}$$

If $\psi_0 = 1, \psi_1, \ldots$ are the irreducible characters of Q, then

$$\chi(\pi) = \sum_i \mathfrak{a}(\psi_i)\psi_i(\pi). \tag{6.3}$$

The entry for the principal character χ_0 is given by

$$(\mathfrak{a}(\psi_i))_0 = \delta_{i0}. \tag{6.4}$$

We assume that we have a subgroup $N \supseteq Q$ of $\mathfrak{N}_G(Q)$ and a subset U of Q such that the following conditions are satisfied.

(VIa). If $\pi \in U$ is G-conjugate to $\xi \in Q$, then $\xi \in U$ and ξ and π are N-conjugate.

(VIb) For each $\pi \in U$, $\mathfrak{C}_G(\pi)$ has a normal p-complement.

(VIc) For each $\pi \in U$, there exists an S_p-group P_0 of $\mathfrak{C}_G(\pi)$ such that $\mathfrak{C}_N(\pi) = P_0\mathfrak{C}_N(Q)$, $Q = P_0 \cap \mathfrak{C}_N(Q)$; ($P_0$ may depend on π).

It follows from II, Section III, Proposition 1 that $\lambda(\mathfrak{C}_G(\pi)) = | P_0 |$. The assumptions of II, Section V, Lemma 3 are satisfied with $1/w = | \mathfrak{C}_N(Q) : Q |$. Let M denote a residue system of N (mod $\mathfrak{C}_N(Q)$). If θ, η are generalized characters of Q with $\theta \mid (Q - U) = 0$, then II (5.3) shows that

$$(\mathfrak{a}(\theta), \mathfrak{a}(\eta)) = \left(\theta, \sum_{\mu \in M} \eta^{\mu}\right). \tag{6.5}$$

Since $\chi(\pi) = \chi(\pi^{\nu})$ for $\nu \in N$, it follows from (6.2) that

$$\mathfrak{a}(\psi_i^{\nu}) = \mathfrak{a}(\psi_i).$$

We set

$$\mathfrak{b}_i = \mathfrak{a}(\psi_0 - \psi_i) = \mathfrak{a}(\psi_0) - \mathfrak{a}(\psi_i) \tag{6.7}$$

It follows from (6.1) and (6.2) that

$$(\mathfrak{b}_i, \chi(\rho)) = 0 \qquad (\text{for } p\text{-regular } \rho \in G). \tag{6.8}$$

Choose a set X of indices $j \neq 0$ such that the characters ψ_j with $j \in X$ represent the distinct classes of irreducible characters $\psi_i \neq 1$ of Q associated in N. For each $j \in X$, let θ_j denote the sum of the distinct associates,

$$\theta_j = \sum_{\nu}{}' \psi_j^{\nu} \quad (\text{distinct summands}). \tag{6.9}$$

Since for $\pi \neq 1$ in Q, the sum of all $\psi_i(\pi)$ vanishes, we can rewrite (6.3) in the form

$$\chi(\pi) = -\sum_{j \in X} \mathfrak{b}_j \theta_j(\pi) \qquad (\text{for } \pi \in Q - \{1\}). \tag{6.10}$$

4. We add some remarks of a different nature. Suppose that π is a p-element, such that $\mathfrak{C}_G(\pi)$ has a normal p-complement. Then the function ϕ_0^{π} forms a basic set for the principal block of $\mathfrak{C}_G(\pi)$, cf. I, Section III. It follows that if ξ is an element of G with the p-factor π, then

$$\chi(\xi) = \chi(\pi) = \mathfrak{d}_0^{\pi}. \tag{6.11}$$

Moreover, by II, Section III, Propositions 1 and 2, if the S_p-group of $\mathfrak{C}(\pi)$ has order p^ν,

$$(\chi(\pi), \chi(\pi)) = p^\nu. \tag{6.12}$$

5. The orthogonality relations for group characters show that $(\chi_i \mid P, 1) \equiv 0 \pmod{|P|}$. If the values of $\chi_i(\pi)$ are known for all $\pi \neq 1$ in P, the value of $x_i \pmod{|P|}$ can be found. In particular, if $\chi_i \in B_0$ and if $P \neq 1$, x_i is not divisible by $|P|$ and hence $\chi_i(\pi)$ cannot vanish for all $\pi \neq 1$ in P. If $|P| > p$. and if S is a subgroup of index p in P, $\chi_i(\pi)$ cannot even vanish for all $\pi \neq 1$ in S.

VII. Groups with a Dihedral S_2-Group

In this section, G will be a finite group with a dihedral S_2-group $P = \langle \sigma, \tau \rangle$,

$$\sigma^{2^{n-1}} = 1, \qquad \tau^2 = 1, \qquad \tau^{-1}\sigma\tau = \sigma^{-1} \tag{7.1}$$

of order $2^n \geqslant 8$. We introduce the abbreviations

$$J = \sigma^{2^{n-2}}, \qquad \tau' = \sigma\tau; \; S = \langle \sigma \rangle. \tag{7.1*}$$

Note that $\mathfrak{Z}(P) = \langle J \rangle$, $P' = \langle \sigma^2 \rangle$. All P-classes of elements of order larger than 2 consist of two powers σ^μ and $\sigma^{-\mu}$. There are three classes of involutions represented by J, τ, and τ'.

For $\sigma^\lambda \neq 1$, J the cyclic group S is an S_2-group of $\mathfrak{C}(\sigma^\lambda)$. Hence by Burnside's theorem, $\mathfrak{C}(\sigma^\lambda)$ has a normal 2-complement. Moreover, if σ^λ and σ^μ are conjugate in G, a Sylow group argument shows that they are conjugate in $\mathfrak{N}(S)$. Since $\mathfrak{N}(S)/\mathfrak{C}(S)$ is a 2-group, it is necessarily of order 2. It follows that $\mathfrak{N}(S) = \langle \mathfrak{C}(S), \tau \rangle$ and this implies that $\sigma^\mu = \sigma^{\pm\lambda}$. Hence G-conjugate elements of S are P-conjugate. This is trivially true for $\sigma^\lambda = 1$ or J.

Thus the only fusions of P-classes in G can occur between $ccl_P(J)$, $ccl_P(\tau)$, $ccl_P(\tau')$. Set $\mathscr{F} = \mathscr{F}(G)$, $P^* = P^*(\mathscr{F})$, cf. II, Section II. If τ and τ' were G-conjugate, but not τ and J, we would have $P^* = \langle P', \tau^{-1}\tau' \rangle = S$. There would then exist a normal subgroup H of index 2 of G with the cyclic S_2-group S. But then H has a normal 2-complement $\mathfrak{K}_2 H$ and this would be a normal 2-complement of G. This is impossible as $P^* \neq P'$.

If we allow interchanging τ and τ', we have to consider the following three cases:

Case I. J, τ, τ' are G-conjugate

Case II. J and τ' are G-conjugate, but not J and τ.

Case III. J, τ, τ' lie in three distinct G-classes.

We have $P^* = P$ in the Case I; $P^* = \langle\sigma^2, \tau'\rangle$ in the Case II; $P^* = P'$ in the Case III. Hence in the Case I and only in that case, there does not exist a normal subgroup of index 2 of G. In the Case III and only in that case, there exists a normal 2-complement of G.

We first show

LEMMA (7A). *If G has a dihedral S_2-group P, then $\mathfrak{C}_G(\pi)$ has a normal 2-complement for every 2-element $\pi \neq 1$.*

Proof. It suffices to take $\pi \in P$. If π has order more than 2, then $\pi = \sigma^\lambda \neq 1$, J and the statement has already been shown. If $\pi = J$, then $\mathfrak{C}(J)$ has the S_2-group P. Since J is not "fused" in $\mathfrak{C}(J)$, we have here Case III and hence $\mathfrak{C}(J)$ has a normal 2-complement. If π is an involution different from J, we may assume that π is not G-conjugate to J. Then $\mathfrak{C}(\pi)$ has the S_2-group $\mathfrak{C}_P(\pi) = \langle\pi, J\rangle$ of order 4. Since π is not fused here, we have again a normal 2-complement, cf. Section V.

It follows from (7A) that in order to know the values of characters $\chi_\mu \in B_0$ for 2-singular elements ξ it suffices to know $\chi_\mu(\pi)$ for $\pi \neq 1$ in P, cf. Section VI, **5**. We shall determine these $\chi_\mu(\pi)$.

We work with columns for the principal block $B = B_0$ and use the same notation as in Section VI, **3**. We take $Q = S$, $N = P$, $U = S - \{1\}$. Let ζ be a primitive 2^{n-1}-st root of unity. Set $r = 2^{n-2}$. The characters ψ_i of Q are given by

$$\psi_i(\sigma^\lambda) = \zeta^{i\lambda}, \qquad (-r < i \leqslant r).$$

The set X of indices can then be chosen as

$$X = \{1, 2, ..., r\}$$

and (6.9) becomes

$$\theta_j(\sigma^\lambda) = \zeta^{j\lambda} + \zeta^{-j\lambda} \quad \text{for} \quad j \neq r, \qquad \theta_r(\sigma^\lambda) = (-1)^\lambda.$$

We determine the columns $\mathfrak{b}_1, \mathfrak{b}_2, ..., \mathfrak{b}_r$, cf. (6.7). Since $\psi_0 - \psi_j$ vanishes for 1, (6.5) yields

$$(\mathfrak{b}_i, \mathfrak{b}_j) = 2 + \delta_{ij} \quad \text{for} \quad j \neq r,$$
$$(\mathfrak{b}_r, \mathfrak{b}_r) = 4.$$

For $\chi_0 = 1$, all $\mathfrak{b}_j$ have the entry $(\mathfrak{b}_j)_0 = 1$. Take first $j \neq r$. Then each $\mathfrak{b}_j$ has two further nonzero entries and these have opposite values $+1$ and -1, cf. Section VI, **2**. It is seen easily that there exists a character say $\chi_1 \in B_0$ such that $(\mathfrak{b}_j)_1$ has the same value ± 1 for all $j \neq r$. Set $(\mathfrak{b}_j)_1 = \delta_1$. For each $j \neq r$, there exists an irreducible character $\chi^{(j)} \in B_0$ such that $(\mathfrak{b}_j)^{(j)} = -\delta_1$;

the characters $\chi_0, \chi_1, \chi^{(1)}, \ldots, \chi^{(r-1)}$ are distinct. The column $\mathfrak{b}_r$ has three nonzero entries besides $(\mathfrak{b}_r)_0$. Each of them is ± 1. If $n \geqslant 4$, then $r \geqslant 4$ and since $(\mathfrak{b}_j, \mathfrak{b}_r) = 2$ for $j < r$, we must have $(\mathfrak{b}_r)_1 = \delta_1$, $(\mathfrak{b}_r)^{(j)} = 0$. We can assume that this is true for $n = 3$ too since we may here interchange χ_1 and $\chi^{(1)}$ if necessary. For all n, there occur two further characters χ_2, χ_3 in B_0 such that $(\mathfrak{b}_r)_i = \delta_i = \pm 1$ for $i = 2, 3$.

Now all nonzero entries of $\mathfrak{b}_1, \ldots, \mathfrak{b}_r$ have been identified. If still another character χ_μ would occur in B_0, then $(\mathfrak{b}_j)_\mu = 0$ for all μ. Now (6.10) would imply $\chi_\mu(\pi) = 0$ for all $\pi \neq 1$ in S. We have seen in Section VI, **5** that this is impossible. Thus, B_0 consists of the $r + 3$ characters for which the nonzero entries of the columns $\mathfrak{b}_j$ occur. For each of these characters, the value for each $\pi = \sigma^\lambda \neq 1$ can be found from (6.10). We obtain:

THEOREM (7B). *Let G be a finite group with dihedral S_2-group of order 2^n. The principal 2-block consists of $3 + 2^{n-2}$ characters*

$$\chi_0 = 1, \qquad \chi_1, \chi_2, \chi_3; \qquad \chi^{(1)}, \chi^{(2)}, \ldots, \chi^{(2^{n-2}-1)}.$$

If σ is an element of order 2^{n-1} in P and if ζ is a primitive 2^{n-1}-st root of unity, then for $\sigma^\lambda \neq 1$, $\chi_1(\sigma^\lambda) = \delta_1$, $\chi_2(\sigma^\lambda) = \delta_2(-1)^{\lambda+1}$, $\chi_3(\sigma^\lambda) = \delta_3(-1)^{\lambda+1}$,

$$\chi^{(j)}(\sigma^\lambda) = \delta_1(\zeta^{j\lambda} + \zeta^{-j\lambda}). \tag{7.2}$$

Here $\delta_1, \delta_2, \delta_3$ are signs ± 1.

For 2-regular ρ, the equation (6.8) can be applied. This yields:

THEOREM (7C). *For 2-regular ρ and all j*

$$1 + \delta_1\chi_1(\rho) = -\delta_2\chi_2(\rho) - \delta_3\chi_3(\rho) = \delta_1\chi^{(j)}(\rho). \tag{7.3}$$

In particular, all $\chi^{(j)}$ have the same degree x and

$$1 + \delta_1 x_1 = -\delta_2 x_2 - \delta_3 x_3 = \delta_1 x. \tag{7.4}$$

We shall assume now that we have Case I. Since we know $\chi(\pi)$ here for all $\pi \neq 1$ in P, we can apply the argument in Section VI, **5**. This yields:

COROLLARY (7D). *In the Case I, (mod 2^n)*

$$x_1 \equiv \delta_1, \quad x \equiv 2\delta_1; \qquad x_i \equiv -\delta_i + 2^{n-1} \quad \text{for} \quad i = 2, 3. \tag{7.5}$$

We can also find a basic set ϕ_B for the block $B = B_0$. The method of I, Section V, p. 165 shows that ϕ_B consists of three functions. Now (7.3) shows that the three functions $\phi_0 = 1$, $\phi_1 = \delta_1\chi^{(1)}$, $\phi_2 = \delta_2 = \delta_2\chi_2$ can serve as a basic set. (More precisely, the restrictions of the functions to the set of

2-regular elements ρ should be taken.) The formulas (7.3) also give the corresponding decomposition numbers. The Cartan invariants are obtained by means of I (5.4). This leads to the result:

COROLLARY (7E). *In the Case I, the functions* $1, \delta_1\chi^{(1)}, \delta_2\chi_2$ *can serve as a basic set. The matrix of Cartan invariants then is*

$$\begin{pmatrix} 2 & -1 & 0 \\ -1 & 2^{n-2}+1 & 1 \\ 0 & 1 & 2 \end{pmatrix}. \tag{7.6}$$

We now come to the applications of (2A). We take $y_1 = y_2 = J$, $\tilde{G} = \mathfrak{C}_G(J)$. Since $\tilde{G}$ has the S_2-group P, (7A) holds for the group $\tilde{G}$. We use here the same notation as for G with tilda signs added. Since $\tilde{G}$ has a normal 2-complement, we can identify the principal block of $\tilde{G}$ with that of $P \simeq \tilde{G}/\mathfrak{K}_2\tilde{G}$. Then $\tilde{\chi}_0 = 1, \tilde{\chi}_1, \tilde{\chi}_2, \tilde{\chi}_3$ are necessarily the four linear characters of P. Since $\tilde{\chi}_1(\pi) = \tilde{\delta}_1$ for $\pi \in S - \{1\}$, we have $\tilde{\delta}_1 = 1$, $\tilde{\chi}_1(\sigma) = 1$, $\tilde{\chi}_1(\tau) = -1$. Thus $\tilde{\chi}_i(\sigma) = -1$ for $i = 2, 3$ and $\tilde{\delta}_2 = \tilde{\delta}_3 = -1$. Now (2.6) reads

$$h_0^{(i)} = c(J)^{-1} + c(J,\tau)^{-1} + c(J,\tau')^{-1}, h_1^{(i)} = c(J)^{-1} - c(J,\tau)^{-1} - c(J,\tau')^{-1}$$

while the number $((h^{(i)})^{(1)}$ corresponding to $\tilde{\chi}^{(1)}$ is $-2c(J)^{-1}$.

Since $\mathfrak{C}_G{}^*(\pi) \subseteq \tilde{G}$ for all $\pi = \sigma^\lambda \neq 1$, we can apply (2A) for each $\mathfrak{d}_0{}^\pi = \chi(\pi)$. Since $\mathfrak{b}_1$ is a linear combination of such columns, (2A) remains valid if the column $\mathfrak{d}_\alpha{}^\pi$ is replaced by $\mathfrak{b}_1$. This yields

$$g(1 + \delta_1/x_1 - 4\delta_1/(x_1 + \delta_1)) = c(J)^3(h_0^{(1)2} + h_1^{(1)2} - \tfrac{1}{2}(h^{(1)})^{(1)2}),$$

$$g(x_1 - \delta_1)^2 x_1^{-1}(x_1 + \delta_1)^{-1} = 2c(J)^3(c(J,\tau)^{-1} + c(J,\tau')^{-1})^2.$$

It follows that $x_1 \neq \delta_1$, and we have:

THEOREM (7F). *In the Case I, we have*:

$$g = \frac{2x_1(x_1 + \delta_1)}{(x_1 - \delta_1)^2} c(J)^3 \left(\frac{1}{c(J,\tau)} + \frac{1}{c(J,\sigma\tau)}\right)^2. \tag{7.7}$$

We may also use the colum $\mathfrak{b}_r - \mathfrak{b}_1$ instead of $\mathfrak{b}_1$. An analogous computation yields

$$g\delta_1\delta_2\delta_3(\delta_2 x_2 - \delta_3 x_3)^2 x_2^{-1} x_3^{-1}(x_1 + \delta_1)^{-1} = 2c(J)^3(c(J,\tau)^{-1} - c(J,\tau')^{-1})^2.$$

The following result is now evident

THEOREM (7G). *In the Case I, we have* $x_2 = x_3$, *if and only if* $c(J, \tau) = c(J, \tau')$. *If* $x_2 \neq x_3$, *then* $\delta_1\delta_2\delta_3 = 1$ *and*

$$g = \frac{2x_2x_3(x_1 + \delta_1)}{(\delta_2x_2 - \delta_3x_3)^2}\, c(J)^3 \left(\frac{1}{c(J, \tau)} - \frac{1}{c(J, \sigma\tau)}\right)^2 \tag{7.8}$$

On multiplying (7.7) and (7.8), we find

COROLLARY (7H). *If* $x_2 \neq x_3$ *in* (7F), *then* $x_1x_2x_3$ *is a square of an integer.*

Consider now Case II. We have here an additional class consisting of 2-elements, the class $ccl_G(\tau)$. It follows from (7A) and (6.12) that $(\chi^{(\tau)}, \chi^{(\tau)}) = 4$. Since $x_i \equiv \chi_i(J) \equiv \chi_i(\tau)$ (mod 2), (7B) shows that, for $0 \leqslant i \leqslant 3$, the degree x_i is odd and then, $\chi_i(\tau)$ is odd. Hence $\chi_i(\tau) = \pm 1$ for $0 \leqslant i \leqslant 3$, $\chi^{(j)}(\tau) = 0$ for all j. Since $(\mathfrak{b}_1, \chi(\tau)) = 0$, $\chi_1(\tau) = -\delta_1$. Then $(\mathfrak{b}_\tau, \chi(\tau)) = 0$ implies that $\delta_2\chi_2(\tau) = -\delta_3\chi_3(\tau)$. Here, χ_2 and χ_3 may be interchanged. We may therefore assume that

$$\chi_2(\tau) = -\delta_2, \qquad \chi_3(\tau) = \delta_3.$$

Again, we have $(\chi(\rho), \chi(\tau)) = 0$ for 2-regular ρ. Hence

$$1 - \delta_1\chi_1(\rho) - \delta_2\chi_2(\rho) + \delta_3\chi_3(\rho) = 0.$$

Now (7.3) yields $\delta_3\chi_3(\rho) = -1$, $\delta_1\chi_1(\rho) = -\delta_2\chi_2(\rho)$. We then find for $\rho = 1$ that $\delta_3 = -1$, $x_3 = 1$, $\delta_1 = -\delta_2$, $x_1 = x_2$. This gives the result

THEOREM (7I). *In the Case II, we have* $\delta_1 = -\delta_2$, $\delta_3 = -1$,

$$\chi_1(\tau) = -\delta_1, \qquad \chi_2(\tau) = \delta_1, \qquad \chi_3(\tau) = -1, \qquad \chi^{(j)}(\tau) = 0. \tag{7.9}$$

For 2-regular ρ, *we have*

$$\chi_1(\rho) = \chi_2(\rho), \qquad \chi_3(\rho) = 1 \tag{7.10}$$

and, in particular, $x_1 = x_2$, $x_3 = 1$.

Now analogous arguments can be given as in the Case I. It will suffice to state some of the results.

THEOREM (7J). *In the Case II,* $x_1 \equiv \delta_1 + 2^{n-1}$, $x \equiv 2\delta_1 + 2^{n-1} \bmod 2^n$. *The functions* $\phi_0 = 1$, $\phi_1 = \delta_1\chi^{(1)}$ *form a basic set for* B_0 *and the corresponding Cartan matrix is*

$$\begin{pmatrix} 4 & -2 \\ -2 & 2^{n-2} + 1 \end{pmatrix}. \tag{7.11}$$

THEOREM (7K). *In the Case II,*

$$g = 2\frac{x_1(x_1+\delta_1)}{(x_1-\delta_1)^2}\frac{c(J)^3}{c(J,\sigma\tau)^2}$$

Remarks

1. Methods similar to those discussed in the Remarks in Section V can be used here. We shall not go into details.

2. The character χ_1 is rational. Application of (4B), (4C) shows that either χ_2, χ_3 are real and exactly one of $\delta_1, \delta_2, \delta_3$ is $+1$ or $\bar{\chi}_2 = \chi_3$, $x_2 = x_3$, $\delta_1 = -1$, $\delta_2 = \delta_3 = 1$.

3. Suppose that we have Case I and that $c_1 = c(J, \tau)$ and $c_2 = c(J, \sigma\tau)$ are not equal, say that $c_1 < c_2$. Set $\gamma = (c_1 + c_2)^2/(c_1 - c_2)^2$. If the notation is chosen such that $x_3 < x_2$, it can be shown that $x_3 < \gamma(1 + 2^{-n})(1 + 2^{-n+1})$ when $\delta_1 = 1$ and that $x_2 < 1 + 6\gamma$ or $x_1 < 3\gamma$ when $\delta_1 = -1$. This can be used to prove that $c_2 \leqslant 3c_1$ where in the case $\mathfrak{R}_2 G = 1$ the equality sign holds only for $G \simeq A_7$.

VIII. GROUPS WITH QUASI-DIHEDRAL S_2-GROUP

The quasi-dihedral 2-group $P = \langle\sigma, \tau\rangle$ is defined by

$$\sigma^{2^{n-1}} = 1, \qquad \tau^2 = 1, \qquad \tau^{-1}\sigma\tau = \sigma^{2^{n-2}}\sigma^{-1}. \tag{8.1}$$

with $n \geqslant 4$. We use the abbrreviations

$$J = \sigma^{2^{n-2}}, \qquad F = \sigma^{2^{n-3}}, \qquad \tau' = \sigma\tau; \qquad S = \langle\sigma\rangle. \tag{8.1*}$$

As in the case of the dihedral group, $\mathfrak{Z}(P) = \langle J\rangle$, $P' = \langle\sigma^2\rangle$, but since $\tau'^2 = J$, τ' has order 4. We introduce the sets

$$Y_1 = \{\pm 1, \pm 3, ..., \pm(2^{n-3} - 1)\}, \qquad Y_2 = \{2, 4, ..., 2^{n-2} - 2\}. \tag{8.2}$$

The conjugate classes of P can be described as follows. There are 2^{n-3} classes $\{\sigma^\lambda, J\sigma^{-\lambda}\}$ with $\lambda \in Y_1$ consisting of elements of order 2^{n-1}, $2^{n-3} - 1$ classes $\{\sigma^\lambda, \sigma^{-\lambda}\}$ with $\lambda \in Y_2$, and finally the classes $ccl_P(J)$, $ccl_P(\tau)$, $ccl_P(\tau')$, $ccl_P(1)$.

Let G now be a finite group with the S_2-group P. As in the dihedral case, we see that $\mathfrak{C}(\sigma^\lambda)$ has a normal 2-complement for $\sigma^\lambda \neq 1$, J. Furthermore, two elements of S are G-conjugate only if they are P-conjugate. It follows that we have the following cases for the fusion $\mathscr{F} = \mathscr{F}(G)$.

Case I. $(J, \tau) \in \mathscr{F}, (F, \tau') \in \mathscr{F}$.
Case II. $(J, \tau) \notin \mathscr{F}. (F, \tau') \in \mathscr{F}$.
Case III. $(J, \tau) \in \mathscr{F}, (F, \tau') \notin \mathscr{F}$.
Case IV. $(J, \tau) \notin \mathscr{F}, (F, \tau') \notin \mathscr{F}$.

Set again $P^* = P^*(\mathscr{F})$. Then $P^* = P$ (Case I); $P^* = \langle \sigma^2, \tau' \rangle$ (Case II), $P^* = \langle \sigma^2, \tau \rangle$ (Case III); $P^* = P'$ (Case IV). In the Case I and only in that case, there does not exist a normal subgroup of index 2. In the Case IV and only in that case, there exists a normal 2-complement in G.

We shall assume that we have Case I or II. There exist $\xi \in G$ such that $\xi^{-1}F\xi = \tau'$. Then ξ transforms $F^2 = J$ into $\tau'^2 = J$, that is, $\xi \in \mathfrak{C}(J)$. It is clear that $\mathfrak{C}(J)$ has the S_2-group P and that we have here Case II. In constructing a basic set for the principal 2-block of $\mathfrak{C}(J)$, we can replace $\mathfrak{C}(J)$ by $G_0 = \mathfrak{C}(J)/\langle J \rangle$, cf. I, Section II, 5. The S_2-group here is $\bar{P} = P/\langle J \rangle$, which is dihedral of order 2^{n-1} with the central involution $F\langle J \rangle$. In G_0, $F\langle J \rangle$ and $\tau'\langle J \rangle$ are conjugate. On the other hand, $F\langle J \rangle$ and $\tau\langle J \rangle$ are not G_0-conjugate, since F is neither G-conjugate to τ nor to τJ. Hence in the notation of Section VII, we have Case II. We use the same notation for G_0 as in Section VII with asterisks added. Then $\phi_0{}^J = 1$ and $\phi_1{}^J = \delta_1{}^*\chi^{(1)*}$ can be used as basic set for the principal blocks of G_0 and of $\mathfrak{C}(J)$. The Cartan matrix in the case of G_0 is given by (7.11) with n replaced by $n - 1$. In the case of $\mathfrak{C}(J)$, we have to multiply the matrix with 2, (cf. I, Section II, 5). Hence we have

$$\chi(J\rho) = \mathfrak{d}_0{}^J + \mathfrak{d}_1{}^J\phi_1{}^J(\rho) \tag{8.3}$$

for 2-regular $\rho \in \mathfrak{C}_G(J)$ and

$$(\mathfrak{d}_0{}^J, \mathfrak{d}_0{}^J) = 8, \qquad (\mathfrak{d}_0{}^J, \mathfrak{d}_1{}^J) = -4, \qquad (\mathfrak{d}_1{}^J, \mathfrak{d}_1{}^J) = 2^{n-2} + 2. \tag{8.4}$$

Since (7I) shows that $\phi_1{}^J(1) = \delta_1{}^*x^* \equiv 2 + 2^{n-2} \pmod{2^{n-1}}$, it follows from (8.3) that

$$\chi(J) \equiv \mathfrak{d}_0{}^J + 2\mathfrak{d}_1{}^J \qquad (\text{mod } 2^{n-2}). \tag{8.5}$$

We apply again the considerations in Section VI, **3** with $Q = S$, $N = P$ taking here $U = S - \{1, J\}$. Set again $r = 2^{n-2}$. The set X of indices can here be chosen as

$$X = Y_1 \cup Y_2 \cup \{r\}.$$

Let ζ denote a primitive 2^{n-1}-st root of unity, and as in Section VII, define the characters ψ_i of Q by $\psi_i(\sigma^\lambda) = \zeta^{i\lambda}$. Here for $j \in X$

$$\theta_j(\sigma^\lambda) = \zeta^{j\lambda} + (-\zeta)^{-\lambda j} \quad \text{for} \quad j \neq r; \qquad \theta_r(\sigma^\lambda) = (-1)^\lambda.$$

We shall reserve the symbols α, α' for indices ranging over Y_2 and the symbols β, β' for indices ranging over Y_1. We work with the $2^{n-3}-1$ columns $\mathfrak{b}_\alpha$, the column $\mathfrak{b}_r$, and the $2^{n-3}-1$ columns

$$\mathfrak{c}_\beta = \mathfrak{b}_\beta - \mathfrak{b}_1 = \mathfrak{a}(\psi_1 - \psi_\beta) \qquad \text{with} \qquad \beta \neq 1.$$

Since $\psi_\alpha - 1$, $\psi_r - 1$ and $\psi_\beta - \psi_1$ vanish on $\{1, J\}$, (6.5) can be applied and yields

$$(\mathfrak{b}_\alpha, \mathfrak{b}_{\alpha'}) = 2 + \delta_{\alpha\alpha'}, \qquad (\mathfrak{c}_\beta, \mathfrak{c}_{\beta'}) = 1 + \delta_{\beta\beta'}, \qquad (\mathfrak{b}_\alpha, \mathfrak{c}_\beta) = 0,$$
$$(\mathfrak{b}_\alpha, \mathfrak{b}_r) = 2, \qquad (\mathfrak{b}_r, \mathfrak{b}_r) = 4, \qquad (\mathfrak{b}_r, \mathfrak{c}_\beta) = 0.$$

Moreover, $\mathfrak{b}_\alpha$, $\mathfrak{b}_r$, $\mathfrak{c}_\beta$ are linear combinations of columns $\chi(\sigma^\lambda)$ with $\sigma^\lambda \neq 1, J$. It follows that they are orthogonal to $\mathfrak{d}_0^J$, $\mathfrak{d}_1^J$ and to $\chi(\rho)$ with 2-regular ρ.

For $\sigma^\lambda \neq 1, J$, the sum of the $\theta_\beta(\sigma^\lambda)$ vanishes. Hence (6.10) can be written in the form

$$\chi(\sigma^\lambda) = -\mathfrak{b}_r\theta_r(\sigma^\lambda) - \sum_\alpha b_\alpha\theta_\alpha(\sigma^\lambda) - \sum_\beta \mathfrak{c}_\beta\theta_\beta(\sigma^\lambda) \tag{8.6}$$

for $\sigma^\lambda \neq 1, J$. For $\sigma^\lambda = J$, we have $\theta_\alpha(J) = 2$, $\theta_\beta(J) = -2$ and the sum of the $\theta_\beta(J)$ is -2^{n-2}. We obtain

$$\chi(J) \equiv -b_r - 2\sum_\alpha \mathfrak{b}_\alpha + 2\sum_\beta \mathfrak{c}_\beta \pmod{2^{n-2}}. \tag{8.7}$$

The same discussion as in Section VII shows that the nonzero entries of the columns $\mathfrak{b}_\alpha$ and $\mathfrak{b}_r$ occur for characters χ_0, χ_1, χ_2, χ_3 and $\chi^{(\alpha)}$ with $\alpha \in Y_2$ such that the nonzero entries are

$$(\mathfrak{b}_i)_0 = 1, \qquad (\mathfrak{b}_\alpha)_1 = \delta_1, \qquad (\mathfrak{b}_\alpha)^{(\alpha)} = -\delta_1;$$
$$(\mathfrak{b}_r)_\mu = \delta_\mu, \qquad (0 \leqslant \mu \leqslant 3).$$

Here, the δ_μ are ± 1; $\delta_0 = 1$. Since $\mathfrak{d}_0^J \equiv \mathfrak{b}_r \pmod 2$ by (8.5) and (8.7), $(\mathfrak{d}_0^J)_\mu$ is odd for $0 \leqslant \mu \leqslant 3$ while all other entries of $\mathfrak{d}_0^J$ are even. Since $(\mathfrak{d}_0^J, \mathfrak{d}_0^J) = 8$, we have $(\mathfrak{d}_0^J)_\mu = \pm 1$ for $0 \leqslant \mu \leqslant 3$, and there is an entry ± 2 in $\mathfrak{d}_0^J$ while all other entries vanish. Of course, $(\mathfrak{d}_0^J)_0 = 1$. We use $(\mathfrak{d}_0^J, \mathfrak{b}_\alpha) = 0$, $(\mathfrak{d}_0^J, \mathfrak{b}_r) = 0$. If $(\mathfrak{d}_0^J)^{(\alpha)} \neq 0$ for some α, we find $(\mathfrak{d}_0^J)_1 = \delta_1$, $(\mathfrak{d}_0^J)^{(\alpha)} = 2\delta_1$. If we set $2\mathfrak{h} = \mathfrak{d}_0^J + \mathfrak{b}_r$ the column $\mathfrak{h}$ is integral, $(\mathfrak{h}, \mathfrak{h}) = 3$ and $(\mathfrak{h})_0 = 1$, $(\mathfrak{h})_1 = (\mathfrak{h})^{(\alpha)} = \delta_1$. This is impossible, cf. Section VI, 2. Thus $(\delta_0^J)^{(\alpha)} = 0$ for all α.

It follows in the same manner that $(\mathfrak{d}_0^J)_1 = -\delta_1$. If we interchange χ_2 and χ_3 if necessary we can assume that $(\mathfrak{d}_0^J)_2 = \delta_2$, $(\mathfrak{d}_0^J)_3 = -\delta_3$. The entry ± 2 in d_0^J occurs for a new character χ_4. Since $\mathfrak{b}_r + \mathfrak{d}_0^J$ is orthogonal to $\chi(1)$, $(\mathfrak{d}_0^J)_4 = -2\delta_2$.

Thus,

$$(\mathfrak{d}_0{}^J)_0 = 1, \quad (\mathfrak{d}_0{}^J)_1 = -\delta_1, \quad (\mathfrak{d}_0{}^J)_2 = \delta_2, \quad (\mathfrak{d}_0{}^J)_3 = -\delta_3, \quad (\mathfrak{d}_0{}^J)_4 = -2\delta_2.$$

Each column $\mathfrak{c}_\beta$ with $\beta \neq 1$ has two nonzero entries and these are $+1$ and -1. In each column orthogonal to $\mathfrak{b}_\beta$ the corresponding entries are equal. Our preceding results imply that the nonzero entries of $\mathfrak{c}_\beta$ cannot belong to any character previously discussed.

We also see that there is a character for which all c_β have the same entry $\epsilon = \pm 1$. This character will be denoted by $\chi^{(1)}$. For each $\beta \neq 1$, there is a character $\chi^{(\beta)}$ with $(\mathfrak{c}_\beta)^{(\beta)} = -\epsilon$. We now have identified $2^{n-2} + 4$ distinct characters of B_0 :

$$\chi^0 = 1, \qquad \chi_1, \chi_2, \chi_3, \chi_4, \chi^{(\alpha)}, \qquad \chi^{(\beta)}. \tag{8.8}$$

It remains to discuss $\mathfrak{d}_1{}^\lambda$. On comparing (8.5) and (8.7) and using our previous results, we can compute $\mathfrak{d}_1{}^J \bmod 2^{n-3}$. This yields

$$(\mathfrak{d}_1{}^\pi)_1 \equiv \delta_1, \qquad (\mathfrak{d}_1{}^J)_2 \equiv -\delta_2, \qquad (\mathfrak{d}_1{}^J)_4 \equiv \delta_2,$$
$$(\mathfrak{d}_1{}^J)^{(\alpha)} \equiv \delta_1, \qquad (\mathfrak{d}_1{}^J)^{(\beta)} \equiv -\epsilon,$$

while all other entries are congruent 0. There are then $2^{n-2} + 2$ nonzero entries. Since $(\mathfrak{d}_1{}^J, \mathfrak{d}_1{}^J) = 2^{n-2} + 2$, all entries of $\mathfrak{d}_1{}^J$ are 0, 1, -1, and for $n \geqslant 5$, our congruences can be replaced by equalities. The same is true for $n = 4$ after possibly interchanging $\chi^{(1)}$ and $\chi^{(-1)}$. This is seen easily by using (8.4) and the orthogonality of $\mathfrak{d}_1{}^J$ with $\mathfrak{b}_\alpha$, $\mathfrak{b}_r$, and $\mathfrak{c}_\beta$.

For any n, the columns $\mathfrak{b}_\alpha, \mathfrak{b}_r, \mathfrak{c}_\beta, \mathfrak{d}_0{}^J, \mathfrak{d}_1{}^J$ are orthogonal to $\chi(\rho)$ for 2-regular ρ. This yields

$$1 + \delta_1\chi_1(\rho) = \delta_1\chi^{(\alpha)}(\rho) = -\delta_2\chi_2(\rho) - \delta_3\chi_3(\rho),$$
$$1 + \delta_2\chi_2(\rho) = \delta_2\chi_4(\rho); \qquad \delta_1\chi^{(\alpha)}(\rho) = \epsilon\chi^{(\beta)}(\rho).$$

In particular, for $\rho = 1$, we find $\epsilon = \delta_1$.

All nonzero entries of our columns have been identified. If B_0 contained a character χ_μ different from the characters (8.8), then (8.6) and (8.3) would show that $\chi_\mu(\pi) = 0$ for all $\pi \neq 1$ in S. This is impossible. Hence B_0 consists exactly of the characters (8.8).

Substituting in (8.6) and (8.3) the entries of the columns, we obtain

THEOREM (8A). *Let G be a group with quasi-dihedral S_2-group $P = \langle \sigma, \tau \rangle$ of order 2^n; (8.1). Assume that we have Case I or Case II. Then the principal 2-block B_0 consists of $4 + 2^{n-2}$ characters $\chi_\mu, \chi^{(j)}$ with $0 \leqslant \mu \leqslant 4$ and $j = \pm 1, \pm 3, ..., \pm(2^{n-3} - 1); j = 2, 4, ..., 2^{n-2} - 2$. Let ζ be a primitive*

2^{n-1}*-st root of unity. If* ξ *is an element of* G *with the 2-factor* $\sigma^\lambda \neq 1$, J, *then*

$$\chi_1(\xi) = \delta_1, \quad \chi_2(\xi) = \delta_2(-1)^{\lambda+1}, \quad \chi_3(\xi) = \delta_3(-1)^{\lambda+1}, \quad \chi_4(\xi) = 0,$$
$$\chi^{(j)}(\xi) = \delta_1(\zeta^{j\lambda} + (-\zeta)^{-j\lambda}).$$

If $\xi = J\rho$ *has the 2-factor* J, *then*

$$\chi_1(\xi) = -\delta_1 + \delta_1\phi_1^J(\rho), \quad \chi_2(\xi) = \delta_2 - \delta_2\phi_1^J(\rho), \quad \chi_3(\xi) = -\delta_3,$$
$$\chi_4(\xi) = -2\delta_2 + \delta_2\phi_1^J(\rho); \quad \chi^{(j)}(\xi) = (-1)^j\delta_1\phi_1^J(\rho).$$

Here, $\delta_1, \delta_2, \delta_3$ *are signs and* $\pm\phi_1^J$ *is a suitable character of the principal 2-block of* $\mathfrak{C}_G(J)/\langle J\rangle$; $\phi_1^J(1) \equiv 2 + 2^{n-2} \pmod{2^{n-1}}$.

We have also shown

THEOREM (8B). *If* ρ *is 2-regular, all* $\chi^{(j)}(\rho)$ *are equal. In particular, all* $\chi^{(j)}$ *have the same degree. Furthermore,*

$$1 + \delta_1\chi_1(\rho) = \delta_1\chi^{(j)}(\rho) = -\delta_2\chi_2(\rho) - \delta_3\chi_3(\rho), \quad 1 + \delta_2\chi_2(\rho) = \delta_2\chi_4(\rho),$$
$$1 + \delta_1 x_1 = \delta_1 x = -\delta_2 x_2 - \delta_3 x_3, \quad 1 + \delta_2 x_2 = \delta_2 x_4.$$

In the Case II, we have to determine $\chi(\tau)$. We have $(\chi(\tau), \chi(\tau)) = 4$. An argument analogous to that in Section VII yields

THEOREM (8C). *In the Case II,* $\delta_3 = -1$, $\delta_2 = -\delta_1$. We have

$$\chi_0(\tau) = 1, \quad \chi_1(\tau) = -\delta_1, \quad \chi_2(\tau) = \delta_1, \quad \chi_3(\tau) = -1.$$

All other characters of B_0 *vanish for* τ. *The same formulas hold for elements with the 2-factor* τ. *For 2-regular* ρ, $\chi_3(\rho) = 1$, $\chi_1(\rho) = \chi_2(\rho)$. *In particular,* $x_3 = 1$, $x_1 = x_2$.

Assume now that we have Case I. Then in $\tilde{G} = \mathfrak{C}(J)$, we have Case II. We use here the same notation as for G with tilda signs added. Since the principal block of $G_0 = \tilde{G}/\langle J\rangle$ may be considered as a subset of the principal block of $\tilde{G}$, we conlude that the kernels of $\tilde{\chi}_1, \tilde{\chi}_2, \tilde{\chi}_3$ contain J. If $\chi^{(1)*}$ has the same significance as above, we have $\chi^{(1)*}(\rho) = \tilde{\delta}_1 + \tilde{\chi}_1(\rho)$ for 2-regular $\rho \in \tilde{G}$; $\delta_1^* = \tilde{\delta}_1$. We can therefore take $\phi_1^J = 1 + \delta_1^*\tilde{\chi}_1$. We shall set now

$$m = \tilde{\delta}_1\tilde{\chi}_1(1) = \phi_1^J(1) - 1. \tag{8.9}$$

Then

$$m \equiv 1 + 2^{n-2} \pmod{2^{n-1}}, \quad \text{sign } m = \delta_1^*. \tag{8.10}$$

It follows from (8A) that $\chi_1(J) = \delta_1 m$, $\chi_2(J) = -\delta_2 m$, $\chi_3(J) = -\delta_3$. Application of the method in Section VI 5 now yields

COROLLARY (8D). *In the Case I, we have*

$$x_1 \equiv \delta_1(2 - m), \qquad x_2 \equiv -\delta_2 m, \qquad x_3 \equiv -\delta_3 + 2^{n-1} \pmod{2^n}.$$

This implies that $x_4 \equiv 0 \pmod{2^{n-2}}$ so that χ_4 has the maximal possible height $n - 2$.

Apply (2A) with $y_1 = y_2 = J$, $\tilde{G} = \mathfrak{C}(J)$ and with the column $\mathfrak{d}_\rho^{\,\pi}$ replaced by $\mathfrak{b}_2$ which is a linear combination of columns $\mathfrak{d}_\rho^{\,\pi}$. A straight forward computation yields

THEOREM (8E). *Let G be a finite group with the quasi-dihedral S_2-group P, (8.1). If G does not have a normal subgroup of index 2, then*

$$g = \frac{c(J)^3}{c(J, \tau)^2} \frac{x_1(x_1 + \delta_1)}{(x_1 - \delta_1 m)^2} \frac{m + 1}{m}, \tag{8.11}$$

where m is the integer in (8.9) *which is uniquely determined by* $\mathfrak{C}(J)$.

We can also use the column $\mathfrak{b}_r$ in (2A). Since $\mathfrak{b}_r$ is a linear combination of column $\chi(\sigma^\lambda)$ with odd λ, and since then σ^λ is not conjugate to $\sigma^{-\lambda}$ in G, the the right hand side of the equation is 0, (cf. II, Section IV, Proposition 4). We obtain

$$1 + \delta_1 m^2/x_1 + \delta_2 m^2/x_2 + \delta_3/x_3 = 0.$$

Using the equation connecting $1, x_1, x_2, x_3$ in (8B), we have

$$\delta_1\delta_2\delta_3 = 1, \tag{8.12}$$

$$x_1 x_2 = m^2 x_3. \tag{8.13}$$

Finally, we can use the column $\mathfrak{b}_r + \mathfrak{d}_0^{\,J}$ in (2A) and obtain

$$g = \frac{c(J)^3}{c(J, \tau)^2} \frac{x_2(x_2 + \delta_2)}{(x_2 + \delta_2 m)^2} \frac{m - 1}{m}. \tag{8.14}$$

This can also be obtained from (8.11) and (8.13).

Write the formula for g in the form $g = \mu c(J)^3/c(J, \tau)^2$. It follows from (8D) that $x_1 \geqslant 2^{n-2} - 1$, $x_2 \geqslant 2^{n-2} - 1$, $x_3 \geqslant 2^{n-1} - 1$. Likewise, by (8.10), $|m| \geqslant 2^{n-2} - 1$. By (8.13) $x_i \geqslant m x_3^{1/2}$ for $i = 1$ or for $i = 2$. If we use (8.11) or (8.14) accordingly, we obtain easily a numerical bound $\beta(n)$ which depends only on n such that $|\mu - 1| \leqslant \beta(n)$. For $n \to \infty$, $\beta(n) \to 0$. We always have $2/3 < \mu < 4/3$.

In the Case II, the corresponding method yields $x_1 = m\delta_1{}^*$, $\delta_1 = \delta_1{}^*$. We may also apply here (2B) with $\tilde{G} = \mathfrak{C}(J)$, $y_1 = J$, $y_2 = \tau$. This yields

THEOREM (8F). *If we have Case II for G, then*

$$g = c^2(\tau)c(J)/c(J, \tau)^2.$$

As a final result, we prove

COROLLARY (8*G*). *If G is as in* (8E), *then*

$$\lambda(G) \leqslant 36(4 + 2^{n-2})m^8.$$

In particular, there exist only finitely many core-free G with given m.

Proof. Since x_1, x_2, x_3 are coprime, it follows from (8.13) that if t_{ij} is the greatest common divisor of x_i and x_j, we can set

$$x_1 = t_{12}t_{13}z_1{}^2, \qquad x_2 = t_{12}t_{23}z_2{}^2. \qquad x_3 = t_{13}t_{23}$$

with positive integral z_1, z_2. Then

$$t_{12}z_1z_2 = |\, m\,|.$$

On account of the relation for the x_i, (8.13) can be written in the form

$$(m^2 - 1)x_3 = (x_1 + \delta_2 x_3)(x_2 + \delta_1 x_3).$$

Hence

$$m^2 - 1 = (t_{12}z_1{}^2 + \delta_2 t_{23})(t_{12}z_2{}^2 + \delta_1 t_{13}).$$

Since $t_{12}z_1{}^2 \leqslant m^2$, $t_{12}z_2{}^2 \leqslant m^2$, we find $t_{13} < 2m^2$, $t_{23} < 2m^2$ and then $x_3 < 4m^4$.

If $\delta_3 = 1$, we have x_1, $x_2 < x_3$. If $\delta_3 = -1$, choose $i = 1, 2$ such that $\delta_i = -1$, cf. (8.12). Then $x_i{}^2 \leqslant x_1x_2 = m^2x_3$. It follows that $x_i < 2m^3$. The remaining degree x_1, x_2 is less than $x_i + x_3 < 6m^4$. Now (8B) shows that the degree of all characters in B_0 lies below $6m^4$ and (8E) is evident.

IX. A CASE WITH ODD p

In Section IX, we consider a group G of order g with an S_p-group P such that the following conditions are satisfied (p an odd prime or $p = 2$).

(IXa) P is abelian; $P \neq 1$.

(IXb) $\mathfrak{N}(P)/\mathfrak{C}(P)$ is cyclic of order $m \neq 1$.

(IXc) If $\xi \in \mathfrak{N}(P)$, $\xi \notin \mathfrak{C}(P)$, then ξ does not commute with an element $\pi \neq 1$ of P.

We can apply the remarks in Section VI, 3 with $Q = P$, $N = \mathfrak{N}(P)$, $U = P - \{1\}$. We set

$$N = \langle \mathfrak{C}(P), \tau \rangle. \tag{9.1}$$

If $\pi \in U$, our conditions imply that $\mathfrak{C}_N(\pi) = \mathfrak{C}_G(P)$. It follows easily that the assumptions (VIa), (VIb), (VIc) are satisfied. We use the same notation as in Section VI. The class of N-associates of an irreducible character $\psi_i \neq 1$ of P has m members. If we choose a set of representatives ψ_j for these classes, the set X of indices j consists of

$$r = (|P| - 1)/m$$

members. We then have r columns $\mathfrak{b}_j$, $j \in X$, and (6.5) yields

$$(\mathfrak{b}_i, \mathfrak{b}_j) = m + \delta_{ij}, \qquad (i, j \in X).$$

If λ is a fixed index in X and if we set $\mathfrak{c}_j = \mathfrak{b}_j - \mathfrak{b}_\lambda$ for $j \in X - \{\lambda\}$, we have

$$(\mathfrak{c}_i, \mathfrak{c}_j) = 1 + \delta_{ij}, \qquad (i, j \in X - \{\lambda\}).$$

It is now easy to discuss the form of the columns $\mathfrak{c}_i$ and $\mathfrak{b}_i$; cf. also [3], Section 2. This leads to the following results:

THEOREM (9A). *Assume that G satisfies the assumptions* (IXa), (IXb), *and* (XIc) *for some prime p. The principal p-block B_0 consists of $r = (|P|-1)/m$* "exceptional" characters $\chi^{(j)}$, $(j \in X)$, and *$s \leqslant m$ "nonexceptional" characters $\chi_0 = 1, \chi_1, \ldots, \chi_{s-1}$ such that for p-singular elements ξ with the p-factor $\pi \in P$, we have*

$$\chi^{(j)}(\xi) = d + \delta \sum_{\mu \in M} \psi_j{}^{\mu}(\pi), \quad (j \in X);$$

$$\chi_i(\xi) = a_i, \qquad (i = 0, 1, \ldots, s-1).$$

Here M is a residue system of $\mathfrak{N}(P)$ mod $\mathfrak{C}(P)$; *$\delta = \pm 1$; $d, a_0, \ldots, a_{s-1} \in \mathbf{Z}$ and $a_i \neq 0$. Moreover,*

$$(d - \delta)^2 + (r - 1)d^2 + \sum_{i=0}^{s-1} a_i^2 = m + 1. \tag{9.1}$$

For p-regular ρ, all $\chi^{(j)}(\rho)$ take the same value and

$$(rd - \delta)\chi^{(j)}(\rho) + \sum_{i=0}^{s-1} a_i \chi_i(\rho) = 0. \tag{9.2}$$

In particular, all $\chi^{(j)}$ have the same degree x. It is also clear that $x_i \equiv a_i$ (mod $|P|$). It follows from (9.2) with $\rho = 1$ and from (9.1) that $s \geqslant 2$. If $r > m$, we must have $d = 0$.

We shall assume from now on that $m = 2$. In particular, g is even and p must be odd. Here, $s = 2$. For $|P| > 5$, we have $r > 2$ and $d = 0$. It is seen easily that we may assume $d = 0$ for $|P| = 3$ or 5 if the characters of B_0 are labeled suitably. If $d = 0$, it follows that $a_1 = \delta$. Since each algebraic conjugate of a character of B_0 lies in B_0 (cf. I, Section III, Lemma 2), we have

Corollary (9B). *If $m = 2$ in* (9A), *we may assume $d = 0$. Then $s = 2$, $a_1 = \delta$. Here, all $\chi^{(j)}$ have degree $x = x_1 + \delta$. All $\chi^{(j)}$ are real and χ_1 is rational.*

Application of (2B) leads to the result

Theorem (9C). *Let G be a group which, for some p, satisfies the assumptions* (IXa), (IXb), (IXc) *with $m = 2$. Let J be an involution of G and let $\pi \neq 1$ be an element of P. Denote by N the number of G-conjugates of J which transform π into π^{-1}. If $N = 0$, then $J \in \mathfrak{K}_p(G)$. If $N \neq 0$, we have*

$$|G| = \alpha \frac{c(J)^2 N^2}{c(\pi)} \quad \text{with} \quad \alpha = \frac{x_1(x_1 + \delta)}{(x_1 - \chi_1(J))^2}, \tag{9.3}$$

where χ_1 has the same significance as in (9B).

The proof of the following remark is not difficult and will be omitted.

Remark (9D). *Assume that in* (9C), *there exists an involution J of $\mathfrak{N}(P)$ which transforms π into π^{-1}. If $x_1 = 1$, $\alpha = 1/2$ and G has a normal subgroup $P\mathfrak{K}_p(G)$ of index 2. If $x_1 \neq 1$ and if we set $|P| = q$, we have*

$$q^2(q+1)^{-1}(q+2)^{-1} < \alpha \leqslant (q+1)(q+2)(q-1)^{-2} \quad \text{for} \quad \delta = 1,$$
$$q^2(q+2)^{-2} < \alpha < q^2(q-1)^{-2} \quad \text{for} \quad \delta = -1, \quad q \neq 3.$$

This remark shows that α lies very close to 1 for large q. For $q = 3$, we can assume that $\delta = 1$ by choosing the notation properly.

We shall mention only one example for (9C). This is the simple Ree group of order 10,073,444,472. For $p = 13$, we have $c(J) = 19{,}656$, $c(\pi) = N = 26$. $\alpha = 18981/18928$; $x_1 = 3^9$, $\delta = -1$.

References

1. Brauer, R. Zur Darstellungstheorie der Gruppen endlicher Ordnung II. *Math. Z.* **72** (1959), 25–46.

2. Brauer, R. Les groupes d'order fini et leur caractères, Séminaire P. Dubreil, M. L. Dubreil-Jacotin et C. Pisot, 1958/59, pp. 6–01 to 6–16.

3. Brauer, R., and Leonard, H. S., Jr. On finite groups with an abelian Sylow group. *Can. J. Math.* **14** (1962), 436–450.
4. Frobenius, G., and Schur, I. Über die reellen Darstellungen der endlichen Gruppen. *Sitzber. Preuss. Akad. Wiss.* **1906**. 186–208.
5. Gorenstein, D., and Walter, J. H. On finite groups with dihedral Sylow 2-subgroups. *Illinois J. Math.* **6** (1962). 553–593.
6. Suzuki, M. Investigations on finite groups. *Proc. Natl. Acad. Sci. U.S.A.* **46** (1960), 1611–1614.
7. Thompson, J. G. 2-signalizer of finite groups. *Pacific J. Math.* **14** (1964), 363–364.

PRINTED IN BRUGES, BELGIUM, BY THE ST CATHERINE PRESS LTD.

A CHARACTERIZATION OF THE MATHIEU GROUP $\mathfrak{M}_{12}$

BY

RICHARD BRAUER AND PAUL FONG (1)

In this paper we give a characterization of the simple Mathieu group $\mathfrak{M}_{12}$ of order 95,040. The character table of $\mathfrak{M}_{12}$ was computed by Frobenius [6], and from his results it can be immediately seen that there exists an element F in $\mathfrak{M}_{12}$ of order 8 such that (i) the cyclic subgroup $\langle F \rangle$ generated by F is self-centralizing, (ii) F is conjugate to its odd powers. Elementary arguments show that there is then a Sylow 2-subgroup $\mathfrak{P}$ of $\mathfrak{M}_{12}$, of order 64, such that (i), (ii) hold in $\mathfrak{P}$. We are thus led to a consideration of 2-groups $\mathfrak{P}$ of order 64 in which (i), (ii) hold, and groups $\mathfrak{G}$ containing $\mathfrak{P}$ as a Sylow 2-subgroup. Our main result is the following:

THEOREM (6A). *Let $\mathfrak{G}$ be a finite group of order $64g'$, where g' is odd. Suppose there is an element F of order* 8 *in $\mathfrak{G}$ such that $\langle F \rangle$ is self-centralizing in some Sylow* 2*-subgroup $\mathfrak{P}$, and F is conjugate to its odd powers in $\mathfrak{P}$. Then one of the following possibilities hold*:

(a) *$\mathfrak{G}$ has a subgroup of index* 2.

(b) *$\mathfrak{G}$ has one class of involutions.*

(c) *If $O_{2'}(\mathfrak{G})$ is the maximal normal subgroup of $\mathfrak{G}$ of odd order, then $\mathfrak{G}/O_{2'}(\mathfrak{G}) \simeq \mathfrak{G}_{1344}$ or $\mathfrak{M}_{12}$, where $\mathfrak{G}_{1344}$ is a uniquely determined nonsimple, nonsolvable group of order* 1344, *and $\mathfrak{M}_{12}$ is the Mathieu group on* 12 *symbols.*

In particular, the only simple group with more than one class of involutions satisfying the assumptions of the theorem is $\mathfrak{M}_{12}$. A recent characterization of $\mathfrak{M}_{12}$ by Wong [11], where additional assumptions on the centralizer of a center involution were made, is included in the above result.

In §1, 2-groups $\mathfrak{P}$ in which (i), (ii) hold are investigated. It will be shown in §2 that if such a group $\mathfrak{P}$ is a Sylow subgroup of a group $\mathfrak{G}$ with no subgroups of index 2, then the structure of $\mathfrak{P}$ is completely determined. The possible distributions of the involutions of $\mathfrak{P}$ into the conjugate classes of $\mathfrak{G}$ fall essentially into three cases I, II, III. In I, $\mathfrak{G}$ has one class of involutions; in II, III $\mathfrak{G}$ has two classes of involutions. The §§4–6 are largely concerned with cases II, III, which

Received by the editors February 15, 1965.

(1) This research was partially supported by the Air Force contract AF 49 (638)-1381, and by the National Science Foundation contract NSF GP-1610. We would like to thank the referee for his careful reading of the manuscript and for his valuable suggestions.

correspond to part (c) of the above theorem. Case I is incomplete as yet, but we hope to continue the work later on.

NOTATION. All groups $\mathfrak{G}$ considered are finite. If $\mathfrak{H}$ is a subgroup of $\mathfrak{G}$, we write $\mathfrak{H} \leqq \mathfrak{G}$; in case $\mathfrak{H}$ is normal in $\mathfrak{G}$, we write $\mathfrak{H} \trianglelefteq \mathfrak{G}$. If $\mathfrak{A}$ is a subset of $\mathfrak{G}$, then $\mathfrak{N}(\mathfrak{A})$ and $\mathfrak{C}(\mathfrak{A})$ are the normalizer and centralizer of $\mathfrak{A}$ in $\mathfrak{G}$; their orders are respectively $n(\mathfrak{A})$ and $c(\mathfrak{A})$. At times it will be necessary to attach a subscript $\mathfrak{G}$ to $\mathfrak{N}_{\mathfrak{G}}(\mathfrak{A})$ and $\mathfrak{C}_{\mathfrak{G}}(\mathfrak{A})$ when $\mathfrak{A}$ is contained in several groups. The subgroup of $\mathfrak{G}$ generated by $\mathfrak{A}$ will be denoted by $\langle\mathfrak{A}\rangle$. If $\mathfrak{A}$ consists of only one element A, we will write A for $\mathfrak{A}$.

Let p be a fixed rational prime; ν will then be the exponential valuation of the rational numbers determined by p, normalized by setting $\nu(p) = 1$. If $G \in \mathfrak{G}$, we will write $\nu(G)$ for $\nu(c(G))$. G is a p-element, a p-regular element, or a p-singular element if G has order a power of p, relatively prime to p, or divisible by p respectively. Each $G \in \mathfrak{G}$ is a unique product $G = G_1G_2$, where G_1 is a p-element, G_2 is p-regular, and each G_i is a power of G. G_1 is called the p-factor of G, G_2 the p-regular or p'-factor of G. If P is a p-element, then the section $S(P)$ of P is the set $\{G \in \mathfrak{G} \mid p\text{-factor of } G \text{ is conjugate to } P \text{ in } \mathfrak{G}\}$. An S_p-subgroup of $\mathfrak{G}$ is a Sylow p-subgroup of $\mathfrak{G}$.

If $\mathfrak{P}$ is an S_p-subgroup of $\mathfrak{G}$, we will denote the $\mathfrak{P}$-conjugate classes by ccl(P), where P is a representative of the class. At times all the elements of the class will be displayed between the parenthesis signs. If two classes of $\mathfrak{P}$ are conjugate in $\mathfrak{G}$, we will say they are fused in $\mathfrak{G}$. *A* class of $\mathfrak{P}$ which is not fused to any other $\mathfrak{P}$-class is said to be isolated. The symbol $\sim$ will mean conjugate to. The transform of A by G is $G^{-1}AG = A^G$; we write $G\colon A \to G^{-1}AG$.

1. Let $\mathfrak{P}$ be a 2-group of order 64 satisfying the following condition: There exists a self-centralizing element F of order 8 which is conjugate in $\mathfrak{P}$ to its odd powers, i.e.

(i) $\mathfrak{C}(F) = \langle F\rangle$,

(ii) $F \sim F^3 \sim F^5 \sim F^7$.

The determination of the structure of $\mathfrak{P}$ will proceed in several steps.

(a) Let $\mathfrak{N}$ be the normalizer of $\langle F\rangle$. By conditions (i), (ii) $\mathfrak{N}$ has order 32, and hence $\mathfrak{N} \trianglelefteq \mathfrak{P}$. Moreover, $\mathfrak{N}/\langle F\rangle$ is abelian of type (2,2), since the automorphism group of Z_8 is noncyclic.

(b) Let the class of F be ccl($F, F^3, F^5, F^7, F_1, F_1^3, F_1^5, F_1^7$). $\mathfrak{P}$ permutes by transformation the elements of ccl(F) among themselves. In particular $F_1^{-1}FF_1 \in$ ccl(F). Condition (i) implies $F_1^{-1}FF_1 = F^\alpha$ for some α. Thus $F_1 \in \mathfrak{N}$ and $F_1^2 \in \langle F\rangle$. Replacing F_1 by F_1^3 if necessary, we may assume that $F_1^2 = F^2$. Let $\mathfrak{F} = \langle F, F_1\rangle$; $\mathfrak{F}$ has order 16, and at least 8 elements of order 8. A check of the groups of order 16, which are completely known, shows that we may take $\alpha = 5$. If we set $X = F_1F$, then $\mathfrak{F}$ is the group $\langle X, F \mid X^2 = F^8 = 1, XFX = F^5\rangle$. The class ccl($F$) can then be written as ccl($F, F^3, F^5, F^7, XF, XF^3, XF^5, XF^7$).

The center of $\mathfrak{F}$ is $\langle F^2\rangle$; for notational convenience, set $J = F^4$. Since $\mathfrak{F}$ is generated by ccl(F), $\mathfrak{F} \trianglelefteq \mathfrak{P}$. It is easily seen that the conjugate classes of $\mathfrak{P}$ in $\mathfrak{F}$ are

$$\text{ccl}(F),\ \text{ccl}(F^2, F^{-2}),\ \text{ccl}(J),\ \text{ccl}(1),\ \text{ccl}(XF^2, XF^{-2}),\ \text{ccl}(X, XJ).$$

(c) For any $P \in \mathfrak{P}$, $P: F \to \text{ccl}(F)$, $X \to \text{ccl}(X)$. Since there are 8 possible images for F and 2 for X, there are in all 16 possible actions for P. If $P: F \to F$, $X \to X$, then P centralizes $\mathfrak{F}$, and hence belongs to the center $\mathfrak{z}\mathfrak{F}$ of $\mathfrak{F}$. A comparison of the orders of $\mathfrak{P}$ and $\mathfrak{z}\mathfrak{F}$ implies that all 16 possibilities occur. In particular, there exist elements N and E in $\mathfrak{P}$ such that

$$(1.1) \qquad N: \begin{array}{l} F \to F^{-1}, \\ X \to X, \end{array} \qquad E: \begin{array}{l} F \to XF, \\ X \to XJ. \end{array}$$

$N \in \mathfrak{N}$, so that $\mathfrak{N} = \langle N, X, F\rangle$. However, $E \notin \mathfrak{N}$, and so $\mathfrak{P} = \langle E, N, X, F\rangle = \langle E, N, F\rangle$.

(d) Now $N^2: F \to F$, and thus $N^2 \in \langle F\rangle$. Since $N \notin \mathfrak{C}(F^2)$, N^2 must be 1 or J. We write $N^2 = J^\rho$, where $\rho = 0$ or 1. Also $E^2: F \to F^5$, $X \to X$ so that $E^2X \in \mathfrak{C}(\mathfrak{F})$ or $E^2X \in \mathfrak{z}\mathfrak{F}$. If $E^2X = 1$ or J, then $E \in \mathfrak{C}(X)$, which is not the case. We may thus write $E^2 = XF^2J^\sigma$, where $\sigma = 0$ or 1. To complete the description of $\mathfrak{P}$, it remains to compute $E^{-1}NE$. Now $E^{-1}NE: X \to X$, $F \to F^3$, so that $E^{-1}NE(NX)^{-1} \in \mathfrak{C}(\mathfrak{F})$. Hence $E^{-1}NE = F^{2\tau}NX = NXF^{-2\tau}$. Since we may replace E by F^2E in (1.1), we may assume $\tau = 0$ or 1. We have proved the first part of

PROPOSITION (1A). *Let $\mathfrak{P}$ be a 2-group of order 64 such that there exists a self-centralizing element F of order 8 conjugate in $\mathfrak{P}$ to its odd powers. Set $F^4 = J$. Then $\mathfrak{P}$ has a normal series*

$$\langle F\rangle \trianglelefteq \langle X, F\rangle \trianglelefteq \langle N, X, F\rangle \trianglelefteq \langle E, N, X, F\rangle = \mathfrak{P}$$

where $X^2 = 1$, $XFX = FJ$; $N^2 = J^\rho$, $N^{-1}XN = X$, $N^{-1}FN = F^{-1}$; $E^2 = XF^2J^\sigma$, $E^{-1}NE = NXF^{-2\tau}$, $E^{-1}XE = XJ$, $E^{-1}FE = XF$. Here ρ, σ, τ are 0 or 1. The commutator subgroup $\mathfrak{P}'$ of $\mathfrak{P}$ is $\langle X, F^2\rangle$; $\mathfrak{P}/\mathfrak{P}'$ is of type (2,2,2). The center of $\mathfrak{P}$ is $\langle J\rangle$.

Proof. It is clear from the above relations that $\mathfrak{P}' \geqq \langle X, F^2\rangle$, and that $\langle X, F^2\rangle \trianglelefteq \mathfrak{P}$. Modulo $\langle X, F^2\rangle$ the generators E, N, F of $\mathfrak{P}$ commute and have exponent 2. The assertions about $\mathfrak{P}'$ and $\mathfrak{P}/\mathfrak{P}'$ now follow. $\mathfrak{z}\mathfrak{P}$ must lie in $\langle F\rangle$. Since $N \notin \mathfrak{C}(F^2)$, $\mathfrak{z}\mathfrak{P} = \langle J\rangle$. This completes the proof. These groups are incidentally the ones on pp. 222 and 224 of Hall-Senior [7](2).

(2) These groups were pointed out to us by Professor P. N. Burgoyne.

The following remarks are easy consequences of (1A). Each element of $\mathfrak{P}$ is uniquely expressible in the form $E^iN^jX^kF^m$, where $i,j,k=0,1$; and $0 \leqq m \leqq 7$. We have

$$
\begin{array}{lllll}
 & E \rightarrow EXJ, & E \rightarrow EJ, & E \rightarrow EXF^{2\tau}, & N \rightarrow NXF^{-2\tau}, \\
(1.2) & F\colon N \rightarrow NF^2, & X\colon N \rightarrow N, & N\colon X \rightarrow X, & E\colon X \rightarrow XJ, \\
 & X \rightarrow XJ, & F \rightarrow FJ, & F \rightarrow F^{-1}, & F \rightarrow XF.
\end{array}
$$

In particular, the subgroups $\langle E^2,EF\rangle$, $\langle F^2,EF\rangle$ for $\sigma=1$ and the subgroups $\langle NF,X\rangle=\langle NXF,X\rangle$, $\langle NF^{-1},X\rangle=\langle NXF^{-1},X\rangle$ are dihedral of order 8.

For later use it will be necessary to have a complete set of representatives of the conjugate classes of $\mathfrak{P}$, and the centralizers of these representatives. In the tables below the columns contain from left to right (i) a representative of the class, (ii) the elements of the class, (iii) the centralizer of the representative, (iv) the order of the representative, (v) the square of the representative. The calculations for the most part are straightforward; the details are omitted in such cases.

(a) Classes in $\mathfrak{F}$. The results in this case are easily derived from the steps leading up to the proof of (1A).

F	$F^{2i+1},XF^{2i+1},0\leqq i\leqq 3$	$\langle F\rangle$	8	F^2
F^2	F^2,F^{-2}	$\langle\mathfrak{P}',F,EN\rangle$	4	J
XF^2	XF^2,XF^{-2}	$\langle\mathfrak{P}',E,NF\rangle$	4	J
J	J	$\mathfrak{P}$	2	1
X	X,XJ	$\langle\mathfrak{P}',N,EF\rangle$	2	1
1	1	$\mathfrak{P}$	1	1

(b) Classes in $\mathfrak{N}-\mathfrak{F}$. Under repeated transformations by F and E, we have $N\sim NF^{2i},N\sim NXF^{2i},NF\sim NF^{2i+1},NXF\sim NXF^{2i+1}$ for $0\leqq i\leqq 3$. Since $\mathfrak{C}(N)\geqq\langle N,X,J\rangle$ has order $\geqq 8$, the class ccl(N) contains exactly 8 elements. If $\tau=0$, $\mathfrak{C}(NF)\geqq\langle E\rangle\langle NF\rangle,\mathfrak{C}(NXF)\geqq\langle EX\rangle\langle NXF\rangle$; both centralizers have order $\geqq 16$. If $\tau=1$, $\mathfrak{C}(NF)\geqq\langle E^2,EF\rangle\langle NF\rangle$, $\mathfrak{C}(NXF)\geqq\langle E^2,EF\rangle\langle NXF\rangle$ also have order $\geqq 16$. The classes of NF, NXF therefore each contain exactly 4 elements.

N	NF^{2i},NXF^{2i}, $0\leqq i\leqq 3$	$\langle N,X,J\rangle$	$2^{1+\rho}$	J^ρ
NF	NF,NF^3,NF^5,NF^7	$\langle E\rangle\langle NF\rangle$ if $\tau=0$ $\langle E^2,EF\rangle\langle NF\rangle$ if $\tau=1$	$2^{1+\rho}$	J^ρ
NXF	NXF,NXF^3,NXF^5,NXF^7	$\langle EX\rangle\langle NXF\rangle$ if $\tau=0$ $\langle E^2,EF\rangle\langle NXF\rangle$ if $\tau=1$	$2^{2-\rho}$	$J^{1+\rho}$

(c) Classes in $\mathfrak{P}-\mathfrak{N}$. Under repeated transformations by F, we have $E \sim EX \sim EJ \sim EXJ$, $EF^2 \sim EXF^2 \sim EXF^{-2} \sim EF^{-2}$, $EF \sim EXF \sim EXF^5 \sim EF^5$, $EF^3 \sim EXF^3 \sim EXF^{-1} \sim EF^{-1}$. If $\tau = 0$ we may use the results on $\mathfrak{C}(NF)$ and the fact that $EXF^2 = E^3$ or E^{-1} to conclude that $\mathfrak{C}(E) = \mathfrak{C}(EXF^2) \geqq \langle E\rangle\langle NF\rangle$, the latter being a subgroup of order 16. The classes ccl(E), ccl(EXF^2) then each have exactly 4 elements. If $\tau = 0$, $N: EF \to EXF^{-1}$; ccl(EF) then has exactly 8 elements, since $\mathfrak{C}(EF) \geqq \langle EF, X, J\rangle$. If $\tau = 1$, $N: E \to EXF^2$ and ccl(E) has exactly 8 elements. Note that E is then a self-centralizing element of order 8 conugate to its odd powers. If $\tau = 1$, then $\mathfrak{C}(EF) \geqq \langle EF\rangle\langle X, NF\rangle$, $\mathfrak{C}(EF^3) \geqq \langle EF^3\rangle\langle X, NF^{-1}\rangle$, both centralizers then having order $\geqq 16$. The classes ccl(EF), ccl(EF^3) each contain exactly 4 elements.

E	E, EX, EXJ, EJ	$\langle E\rangle\langle NF\rangle$	8	E^2	$\tau = 0$,
EXF^2	$EF^2, EXF^2, EXF^{-2}, EF^{-2}$	$\langle E\rangle\langle NF\rangle$	8	E^{-2}	
EF	$EF^{2i+1}, EXF^{2i+1}, 0 \leqq i \leqq 3$	$\langle EF\rangle\langle X, J\rangle$	$2^{2-\sigma}$	$J^{1-\sigma}$	
E	$EF^{2i}, EXF^{2i}, 0 \leqq i \leqq 3$	$\langle E\rangle$	8	E^2	$\tau = 1$.
EF	EF, EXF, EXF^5, EF^5	$\langle EF\rangle\langle X, NF\rangle$	$2^{2-\sigma}$	$J^{1-\sigma}$	
EF^3	$EF^3, EXF^3, EXF^{-1}, EF^{-1}$	$\langle EF^3\rangle\langle X, NF^{-1}\rangle$	$2^{2-\sigma}$	$J^{1-\sigma}$	

Under repeated transformations by F and X, we have $EN \sim ENJ \sim ENXF^2 \sim ENXF^{-2}$, $ENF^2 \sim ENF^{-2} \sim ENX \sim ENXJ$. If $\tau = 0$, then $N: EN \to ENX$; since EN has order 8, ccl(EN) has exactly 8 elements. If $\tau = 1$, then $\mathfrak{C}(EN) \geqq \langle EN\rangle\langle F^2, EF\rangle$, $\mathfrak{C}(ENX) \geqq \langle ENX\rangle\langle F^2, EF\rangle$; both centralizers have order $\geqq 16$, so that ccl(EN), ccl(ENX) each contain exactly 4 elements. Finally, if $\tau = 0$, ENF, ENF^{-1}, ENF^3 have order 8. Since $N: ENF \to ENXF^{-1}$, and $\mathfrak{C}(ENF) \geqq \langle E, X, NF\rangle$ has order $\geqq 32$, ccl(ENF) has just 2 elements. Multiplication by J yields another class $ENF^5 \sim ENXF^3$. Transformation of ENF^3 by E, N, F respectively yield ENF^{-1}, $ENXF^{-3}$, $ENXF$. Since $\mathfrak{C}(ENF^3) \geqq \langle ENF^3\rangle\langle X\rangle$, ccl$(ENF^3)$ has exactly 4 elements. If $\tau = 1$, transformation of ENF by E, N, F respectively give ENF^{-1}, $ENXF^{-3}$, $ENXF^{-1}$, and since $\mathfrak{C}(ENF) \geqq \langle\mathfrak{P}', ENF\rangle$, ccl (ENF) has exactly 4 elements. Multiplication by yields the remaining class.

EN	$ENF^{2i}, ENXF^{2i}, 0 \leqq i \leqq 3$	$\langle EN\rangle$	8	$F^{\pm 2}$	$\tau = 0$,
ENF	$ENF, ENXF^{-1}$	$\langle E, X, NF\rangle$	8	$XF^{\pm 2}$	
ENF^{-3}	$ENF^{-3}, ENXF^3$	$\langle E, X, NF\rangle$	8	$XF^{\pm 2}$	
ENF^3	$ENF^3, ENF^{-1}, ENXF^{-3}, ENXF$	$\langle ENF^3\rangle\langle X\rangle$	8	$XF^{\pm 2}$	
EN	$EN, ENJ, ENXF^2, ENXF^{-2}$	$\langle EN\rangle\langle F^2, EF\rangle$	$3 + (-1)^{\rho+\sigma}$	$J^{1+\rho+\sigma}$	$\tau = 1$.
ENX	$ENX, ENXJ, EN^2F, ENF^{-2}$	$\langle ENX\rangle\langle F^2, EF\rangle$	$3 - (-1)^{\rho+\sigma}$	$J^{\rho+\sigma}$	
ENF	$ENF, ENF^{-1}, ENXF^{-3}, ENXF^{-1}$	$\langle\mathfrak{P}', ENF\rangle$	4	$XJ^{1+\rho+\sigma}$	
ENF^3	$ENF^3, ENF^{-3}, ENXF, ENXF^3$	$\langle\mathfrak{P}', ENF\rangle$	4	$XJ^{\rho+\sigma}$	

2. Let $\mathfrak{P} = \mathfrak{P}(\rho, \sigma, \tau)$ be the group of order 64 considered in §1. We assume in this section that $\mathfrak{P}$ is a Sylow subgroup of a larger group $\mathfrak{G}$. Let $\mathfrak{P}^{\#}$ be the focal subgroup of $\mathfrak{P}$ in $\mathfrak{G}$, i.e. $\mathfrak{P}^{\#} \langle QP^{-1} \mid Q = P \in \mathfrak{P}$ and $Q \sim P$ in $\mathfrak{G}\rangle$. The S_2-subgroup of $\mathfrak{G}/\mathfrak{G}'$ is then isomorphic to $\mathfrak{P}/\mathfrak{P}^{\#}$ by [1], Theorem 8.

LEMMA (2A). *$E^2 \nsim F^2$ in $\mathfrak{G}$.*

Proof. Suppose $E^2 \sim F^2$ in $\mathfrak{G}$. Since $\langle E^2 \rangle$ and $\langle F^2 \rangle$ are normal in $\mathfrak{P}$, there exists by Burnside's theorem [12, Lemma, p. 139] a $G \in \mathfrak{N}(\mathfrak{P})$ such that $G\colon \langle E^2 \rangle \to \langle F^2 \rangle$; since $F^2 \sim F^{-2}$ in $\mathfrak{P}$ we may assume that $G\colon E^2 \to F^2$, and that G has odd order. G then permutes the four self-centralizing cyclic subgroups of $\mathfrak{P}$ of order 8, $\langle F \rangle, \langle XF \rangle, \langle EN \rangle, \langle ENX \rangle$ if $\tau = 0$; $\langle F \rangle, \langle XF \rangle, \langle E \rangle, \langle EF^2 \rangle$ if $\tau = 1$. Hence G fixes one of these subgroups, and then G would fix E^2 or F^2, which is impossible. Thus $E^2 \nsim F^2$ in $\mathfrak{G}$.

LEMMA (2B). *Suppose $Y \sim Z$, where Y, Z are elements of order 2 or 4 in $\mathfrak{P}$, and $vc_{\mathfrak{P}}(Y) = 3$, $vc_{\mathfrak{P}}(Z) = 5$. Then* ccl(J) *is not isolated.*

Proof. The assumptions imply that $Y \in \mathrm{ccl}(N)$ or $\mathrm{ccl}(EF)$ if $\tau = 0$, $Y \in \mathrm{ccl}(N)$ if $\tau = 1$. Suppose ccl(J) is isolated. Let $\mathfrak{C}^0(Y) = \langle T \in \mathfrak{C}(J) \mid T\colon Y \to Y$ or $YJ\rangle$, and let $\mathfrak{L} = \mathfrak{C}^0_{\mathfrak{P}}(Y) = \mathfrak{C}^0(Y) \cap \mathfrak{P}$. $\mathfrak{L} = \langle \mathfrak{P}', N \rangle$ or $\langle \mathfrak{P}', EF \rangle$ according as $Y = N$ or $Y = EF$; thus $\mathfrak{L} \trianglelefteq \mathfrak{P}$. Since ccl$(J)$ is isolated, $Y \sim Z$ in $\mathfrak{C}(J)$. In particular, there exists $G \in \mathfrak{C}(J)$ such that $G\colon Y \to Z$, and we may assume $G\colon \mathfrak{C}^0(Y) \to \mathfrak{C}^0(Z)$. Let $\mathfrak{P}_1$ be an S_2-subgroup of $\mathfrak{C}^0(Y)$ containing $\mathfrak{L}$; $\mathfrak{P}_1$ is then a Sylow subgroup of $\mathfrak{G}$. If $\mathfrak{L} \trianglelefteq \mathfrak{P}_1$, then there exists an $H \in \mathfrak{N}(\mathfrak{L})$ such that $H\colon \mathfrak{P}_1 \to \mathfrak{P}$. If Z_1 is the image of Y under H, then necessarily $vc_{\mathfrak{P}}(Z_1) = 5$ and $Z_1 \in \mathfrak{P}'$. In particular, the images of E^2, F^2 under H are elements of $\mathfrak{L}$ not commuting with Z_1; in other words, they lie in $\mathfrak{L} - \mathfrak{P}' = \mathrm{ccl}(Y)$. But then $E^2 \sim F^2$, which is impossible by (2A). Hence $\mathfrak{L}$ is not a normal subgroup of $\mathfrak{P}_1$. Choose $\mathfrak{W}$ such that $\mathfrak{L} \lhd \mathfrak{W} \lhd \mathfrak{P}_1$, and an S_2-subgroup $\mathfrak{P}_2$ of $\mathfrak{G}$ such that $\mathfrak{W} < \mathfrak{P}_2 \leqq \mathfrak{N}(\mathfrak{L})$. Then there exists an $H \in \mathfrak{N}(\mathfrak{L})$ such that $H\colon \mathfrak{P} \to \mathfrak{P}_2$. Since H must fix J, there are elements Y_1, Y_2 in ccl(Y) such that $H\colon Y_1 \to Y_2$. This implies $vc_{\mathfrak{P}_2}(Y) = 3$. However, $vc_{\mathfrak{P}_2}(Y) \geqq 4$, since $vc_{\mathfrak{P}_1}(Y) = 5$. Thus ccl$(J)$ is not isolated.

PROPOSITION (2C). *If $\tau = 0$, then $\mathfrak{P}^{\#} < \mathfrak{P}$ and $\mathfrak{G}$ has a subgroup of index 2.*

Proof. From the tables in §1, J is the only involution in $\mathfrak{P}$ which is a square in $\mathfrak{P}$. Suppose $G\colon Y \to J$ for some $Y \neq J$ in $\mathfrak{P}$ and some G in $\mathfrak{G}$. We may assume $G\colon \mathfrak{C}_{\mathfrak{P}}(Y) \to \mathfrak{P}$. Let $\mathfrak{P}_1$ be an S_2-subgroup of $\mathfrak{C}(Y)$ containing $\mathfrak{C}_{\mathfrak{P}}(Y)$. For every $T \in \mathfrak{P}_1, T^4 \in \langle Y \rangle$. If $vc_{\mathfrak{P}}(Y) \geqq 4$, we may assume $Y = X$, NF, or NXF; then $ENF \in \mathfrak{C}_{\mathfrak{P}}(Y)$. But this is impossible, since $(ENF)^4 = J$. If $vc_{\mathfrak{P}}(Y) = 3$, we may assume $Y = N$ or EF. If $Y = N$, then G must transform NX into ccl(N) or ccl(EF) by what has just been proved. If $G\colon NX \to \mathrm{ccl}(N)$, then $G\colon X = N\,(NX) \to J\,\mathrm{ccl}(N) = \mathrm{ccl}(N)$, which is impossible. If $G\colon NX \to \mathrm{ccl}(EF)$, then $G\colon X \to J\,\mathrm{ccl}(EF) = \mathrm{ccl}(EF)$, which is also impossible. If $Y = EF$, a similar argument can be applied to EFX. Thus ccl(J) is isolated.

Let $Y = N$ or EF, and suppose $Y \sim NF$ or $Y \sim NXF$. Then there exists a $G \in \mathfrak{G}$ with $G: Y \to NF$ or NXF, and $G: \mathfrak{C}_{\mathfrak{P}}(Y) \to \mathfrak{C}_{\mathfrak{P}}(NF)$ or $\mathfrak{C}_{\mathfrak{P}}(NXF)$ by (2B). Now $X \in \mathfrak{C}_{\mathfrak{P}}(Y)$; therefore $G: X \to \mathrm{ccl}(NF)$ or $\mathrm{ccl}(NXF)$. But this is impossible by (2B), since it would imply $Y \sim X$. Hence the only possible fusions among elements of order 2 and 4 are between X, NF or $NXF \mid N, EF \mid E^2, NF$ or $NXF \mid F^2, NF$ or NXF. By (2A) the only possible fusions among elements of order 8 are between $F, EN \mid E, EF^2, ENF, ENF^{-3}, ENF^3$. These fusions, if they occur, contribute at most $\langle \mathfrak{P}', E, NF \rangle$ to $\mathfrak{P}^{\#}$, and thus $\mathfrak{P}^{\#} < \mathfrak{P}$.

From now on we assume $\tau = 1$, since we want to find out which $\mathfrak{P}$ can occur as an S_2-subgroup of a group without subgroups of index 2.

LEMMA (2D). *Suppose* $\tau = 1$ *and* $\mathrm{ccl}(J)$ *is not isolated. Then* $\rho = 0, \sigma = 1$. *Moreover* $X \sim J$, *and after a suitable relabeling of the generators* E, N, F, *we may assume* $E^2 \sim ENF^3$, $F^2 \sim ENF$, $N \sim EF$.

Proof. There exists $Y \neq J$ in $\mathfrak{P}$ such that $Y \sim J$ in $\mathfrak{G}$. We first show we may take $Y = X$. Choose $G \in \mathfrak{G}$ such that $G: Y \to J$ and $G: \mathfrak{C}_{\mathfrak{P}}(Y) \to \mathfrak{P}$. If $vc_{\mathfrak{P}}(Y) = 4$, then J is a square in $\mathfrak{C}_{\mathfrak{P}}(Y)$. Its image under G is an involution $\neq J$ which is a square in $\mathfrak{P}$. Hence $G: J \to \mathrm{ccl}(X)$ and $X \sim J$. If $Y = N$, and if $G: NX \to \mathrm{ccl}(X)$ or $\mathrm{ccl}(Y_1)$ with $vc_{\mathfrak{P}}(Y_1) = 4$, then $X \sim J$. But if $G: NX \to \mathrm{ccl}(N)$, then $G: X = N(NX) \to J\,\mathrm{ccl}(N) = \mathrm{ccl}(N)$. Hence $X \sim J$ in all cases.

Let $\mathfrak{U} = \mathfrak{C}_{\mathfrak{P}}(X) = \langle \mathfrak{P}', EF, N \rangle$, and let $\mathfrak{P}_1$ be an S_2-subgroup of $\mathfrak{C}(X)$ containing $\mathfrak{U}$. Since $\mathfrak{U} \lhd \mathfrak{P}$, $\mathfrak{U} \lhd \mathfrak{P}_1$, there exists $B \in \mathfrak{N}(\mathfrak{U})$ such that $B: \mathfrak{P} \to \mathfrak{P}_1$, $B: J \to X$. Furthermore, B necessarily maps $X \to XJ$, and hence $B: XJ \to J$. We may therefore assume B has order a power of 3, if we only use the properties that $B \in \mathfrak{N}(\mathfrak{U})$, and $B: J \to X \to XJ \to J$ in the remainder of this proof. If $\rho = 1$, the 8 elements of $\mathrm{ccl}(N) \subseteq \mathfrak{U}$ are mapped by B onto 8 elements of $\mathfrak{U}$ whose squares are X. Since there are only 4 such elements in $\mathfrak{U}$, ρ must be 0. Similarly, to show $\sigma = 1$, the same argument can be applied to the 8 elements of $\mathrm{ccl}(EF) \cup \mathrm{ccl}(EF^3)$. B maps E^2, F^2 onto elements of $\mathfrak{U}$ whose squares are X. Since the map $E \to F$, $F \to E, N \to N$ defines an automorphism of $\mathfrak{P}$, we may assume $F^2 \sim ENF$, $E^2 \sim ENF^3$. Indeed, if $E^2 \sim ENF$, then in the new notation $F^2 \sim FNE \sim EFN \sim ENF$. Finally, $B: \mathrm{ccl}(N) \to \mathrm{ccl}(N) \cup \mathrm{ccl}(EF) \cup \mathrm{ccl}(EF^3)$, the latter being a characteristic set of $\mathfrak{U}$. Suppose $B: \mathrm{ccl}(N) \to \mathrm{ccl}(N)$. Since B has order a power of 3, and $\mathrm{ccl}(N)$ has 8 elements, it follows that B fixes at least two elements $N_1 \neq N_2$ of $\mathrm{ccl}(N)$. But then B would fix $N_1 N_2^{-1} \in \mathfrak{P}'$, which is impossible. Hence $N \sim EF^{\alpha}$ for some $\alpha = 1$ or 3. The map $E \to F^3, F \to E, N \to N$ defines an automorphism of $\mathfrak{P}$ which interchanges $\mathrm{ccl}(E)$, $\mathrm{ccl}(F)$, as well as $\mathrm{ccl}(ENF)$, $\mathrm{ccl}(ENF^3)$, so that we may assume $\alpha = 1$.

COROLLARY (2E). *Under the assumptions of* (2D) *there is a* $B \in \mathfrak{G}$ *of order a power of* 3 *such that*

$$B: J \to X \to XJ \to J.$$

Moreover, B either fuses ccl(N), ccl(EF), ccl(EF^3) *or B fuses* ccl(N), ccl(EF) *and* B: ccl$(EF^3) \to$ ccl(EF^3). *In either case B maps the union of these three classes onto itself.*

Lemma (2F). *Suppose* $\tau = 1$. *Let Y be one of the elements* $EN, NF, ENX, NXF, EF, EF^3$; *and Z one of the elements* X, E^2, F^2. *Let* $\mathfrak{L} = \langle \mathfrak{P}', EN, NF \rangle$. *If* $Y \sim Z$ *in* $\mathfrak{G}$, *then there exists a* $G \in \mathfrak{G}$ *such that* $G: J \to J$, $Y \to Z$, *and* $\mathfrak{C}_{\mathfrak{P}}(Y) \to \mathfrak{L} \cap \mathfrak{C}_{\mathfrak{P}}(Z)$.

Proof. We first show that $Y \sim Z$ in $\mathfrak{C}(J)$. If Y, Z have order 4, then $Y^2 = Z^2 = J$, and this is clear. Suppose then that $Z = X$. Let $\mathfrak{P}_1$ be an S_2-subgroup of $\mathfrak{C}(X)$ such that $\mathfrak{P}_1 \geqq \mathfrak{C}_{\mathfrak{P}}(X)$. Then X, XJ, J are the only involutions in $\mathfrak{P}_1$ which are squares of elements in $\mathfrak{P}_1$, since they are already squares of elements in $\mathfrak{C}_{\mathfrak{P}}(X)$. Now there exists $A \in \mathfrak{G}$ such that $A: \mathfrak{C}_{\mathfrak{P}}(Y) \to \mathfrak{P}_1$, and $A: Y \to X$. Since J is a square in $\mathfrak{C}_{\mathfrak{P}}(Y)$, A: $J \to J$ or XJ. If A: $J \to J$, then $Y \sim X$ in $\mathfrak{C}(J)$. If A: $J \to XJ$, then ccl(J) is not isolated. If B is the element of (2E), then AB: $J \to J$, $Y \to XJ$, and $Y \sim X$ in $\mathfrak{C}(J)$. Thus in all cases, $Y \sim Z$ in $\mathfrak{C}(J)$.

Let $\mathfrak{C}^0(Y) = \langle T \in \mathfrak{C}(J) \mid T: Y \to Y \text{ or } YJ \rangle$. $\mathfrak{L}$ is then $\mathfrak{C}^0(Y) \cap \mathfrak{P}$ and $\mathfrak{L} \lhd \mathfrak{P}$. Let $\mathfrak{P}_1$ be an S_2-subgroup of $\mathfrak{C}^0(Y)$ containing $\mathfrak{L}$. By the above, $(\mathfrak{P}_1 : \mathfrak{L}) = 2$ and $\mathfrak{L} \lhd \mathfrak{P}_1$. Let $G \in \mathfrak{N}(\mathfrak{L})$ such that G: $\mathfrak{P}_1 \to \mathfrak{P}$. G fixes J, since J is the only involution in $\mathfrak{L}$ which is a square in $\mathfrak{L}$. The image of Y under G has a centralizer in $\mathfrak{P}$ of order 32, and hence belongs to one of the classes ccl(X), ccl(E^2), ccl(F^2). Since no two of these classes can be fused in $\mathfrak{G}$, we may assume $G: Y \to Z$. In particular, $G: \mathfrak{C}_{\mathfrak{P}}(Y) \to \mathfrak{C}_{\mathfrak{P}}(Z) \cap \mathfrak{L}$.

In the table below the entry in the T_i-row and T_j-column is the number of elements in $\mathfrak{C}_{\mathfrak{P}}(T_i) \cap$ ccl(T_j). The union of the classes indexing the columns is $\mathfrak{L}$.

(2.1)

	1	J	X	NF	NXF	EN	ENX	EF	EF^3	E^2	F^2
NF	1	1	0	2	2	4	0	2	2	2	0
NXF	1	1	0	2	2	0	4	2	2	2	0
EN	1	1	0	4	0	2	2	2	2	0	2
ENX	1	1	0	0	4	2	2	2	2	0	2
EF	1	1	2	2	2	2	2	4	0	0	0
EF^3	1	1	2	2	2	2	2	0	4	0	0
X	1	1	2	0	0	0	0	4	4	2	2
E^2	1	1	2	4	4	0	0	0	0	2	2
F^2	1	1	2	0	0	4	4	0	0	2	2

Proposition (2G). *Suppose* $\tau = 1$. *If* ccl(J) *is isolated in* $\mathfrak{G}$, *then* $\mathfrak{P}^{\#} < \mathfrak{P}$.

Proof. (a) Suppose ccl(N) is isolated. The only possible fusions are between $ENF, ENF^3 \mid X, E^2, F^2, NF, NXF, EN, ENX, EF, EF^3$. In this case

$$\mathfrak{P}^{\#} \leqq \langle \mathfrak{P}', EN, NF \rangle < \mathfrak{P},$$

and we are done. Thus we may assume ccl(N) is not isolated.

(b) Suppose $\mathrm{ccl}(X)$ is isolated. Let Y be one of EN, NF, ENX, NXF. If $G: N \to Y$, we may assume $G: \mathfrak{C}_{\mathfrak{P}}(N) \to \mathfrak{C}_{\mathfrak{P}}(Y)$. In particular the image of X under G is not in $\mathrm{ccl}(X)$, since $\mathfrak{C}_{\mathfrak{P}}(Y) \cap \mathrm{ccl}(X) = \varnothing$, and this is impossible. By (2F), (2.1) $Y \nsim E^2$, $Y \nsim F^2$, since either fusion implies that $\mathrm{ccl}(X)$ is not isolated. Hence $vc(Y) = 4$; by (2F), (2.1) it also follows that $Y \nsim EF$, $Y \nsim EF^3$. (The results of (2F) and (2.1) are valid in case $vc(Y) = 4$.) Also $N \nsim E^2$, $N \nsim F^2$ by (2B). The only possible fusions in $\mathfrak{G}$ are between $N, EF, EF^3 \,|\, NF, NXF, EN, ENX \,|\, ENF, ENF^3$. But then $\mathfrak{P}^{\#} \leqq \langle \mathfrak{P}', ENF, N, EF \rangle < \mathfrak{P}$ and we are done. Thus we may assume $\mathrm{ccl}(X)$ is not isolated.

(c) Suppose $\sigma = 0$. Then N, NF, ENX have order $2^{\rho+1}$; EN, NXF have order $2^{2-\rho}$; EF, EF^3 have order 4. By (a) and (2B), $N \sim Y$ for some $Y \in \mathfrak{P}$ with $vc_{\mathfrak{P}}(Y) = vc(Y) = 4$. Let $G \in \mathfrak{G}$ such that $G: N \to Y$ and $G: \mathfrak{C}_{\mathfrak{P}}(N) \to \mathfrak{C}_{\mathfrak{P}}(Y)$. Now $\mathfrak{C}_{\mathfrak{P}}(N) \cap \mathrm{ccl}(N) = \{N, NJ, NX, NXJ\}$; but $\mathfrak{C}_{\mathfrak{P}}(Y) \cap (\mathrm{ccl}(NF) \cup \mathrm{ccl}(ENX))$ has only 2 elements by (2.1). Hence G necessarily fuses $\mathrm{ccl}(N)$ with $\mathrm{ccl}(EF^{\alpha})$ for some $\alpha = 1$ or 3. In other words, we may assume $Y = EF^{\alpha}$. In particular, $\rho = 1$, and N, NF, ENX have order 4. If $\mathfrak{C}^0_{\mathfrak{P}}(N) = \langle T \in \mathfrak{P} \,|\, T: N \to N \text{ or } NJ \rangle$, we may further assume $G: \mathfrak{C}^0_{\mathfrak{P}}(N) \to \mathfrak{C}^0_{\mathfrak{P}}(EF^{\alpha}) = \mathfrak{L}$, $\mathfrak{L}$ as in (2F). By (2F), (2.1) the 4 elements of $\mathrm{ccl}(N)$ in $\mathfrak{C}^0_{\mathfrak{P}}(N) - \mathfrak{C}_{\mathfrak{P}}(N)$ are mapped by G into

$$(\mathrm{ccl}(EF^{\beta}) \cup \mathrm{ccl}(NF) \cup \mathrm{ccl}(ENX)) \cap (\mathfrak{C}^0_{\mathfrak{P}}(EF^{\alpha}) - \mathfrak{C}_{\mathfrak{P}}(EF^{\alpha})),$$

where $\beta = 1$ or 3, $\beta \neq \alpha$. Thus $N \sim EF \sim EF^3$ or $N \sim EF^{\alpha} \sim NF \sim ENX$ by (2.1). Suppose we have the latter case. Then $EF^{\alpha} \sim NF$ implies by (2F), (2.1) that E^2 is fused to one of NF, ENX or EF^{α} and hence $N \sim E^2$, which is impossible by (2B). Suppose we have the first case $N \sim EF \sim EF^3$. By (b) $\mathrm{ccl}(X)$ is not isolated and hence $X \sim EN$ or $X \sim NXF$. If $X \sim EN$, then by (2F), (2.1) and (2A), $NF \sim EF \sim EF^3 \sim N$, and necessarily $ENX \sim E^2$. The fusions introduced are $X \sim EN$, $N \sim EF \sim EF^3 \sim NF$, $E^2 \sim ENX$. Similarly, if $X \sim NXF$, then $X \sim NXF$, $N \sim EF \sim EF^3 \sim ENX$, $F^2 \sim NF$. Thus $X \sim EN$ or $X \sim NXF$, but not both. The only other possible fusions are $ENF \sim ENF^3$. In all cases, $\mathfrak{P}^{\#} \leqq \langle \mathfrak{P}', EN, F \rangle$ or $\mathfrak{P}^{\#} \leqq \langle \mathfrak{P}', NF, E \rangle$, and so $\mathfrak{P}^{\#} < \mathfrak{P}$.

(d) Suppose $\sigma = 1$. Then N, EN, NF have order $2^{\rho+1}$; NXF, ENX have order $2^{2-\rho}$; EF, EF^3 have order 2. By (a) and (2B), $N \sim Y$ for some $Y \in \mathfrak{P}$ with $vc_{\mathfrak{P}}(Y) = vc(Y) = 4$. We may assume $Y = EN$, NF, or EF^{α} for some $\alpha = 1$ or 3. Let $G \in \mathfrak{G}$ such that $G: N \to Y$ and $\mathfrak{C}_{\mathfrak{P}}(N) \to \mathfrak{C}_{\mathfrak{P}}(Y)$. If $\mathfrak{C}^0_{\mathfrak{P}}(N) = \langle T \in \mathfrak{P} \,|\, T: N \to N \text{ or } NJ \rangle$, we may further assume $G: \mathfrak{C}^0_{\mathfrak{P}}(N) \to \mathfrak{C}^0_{\mathfrak{P}}(Y) = \mathfrak{L}$, $\mathfrak{L}$ as in (2F). The 4 elements of $\mathrm{ccl}(N)$ in $\mathfrak{C}^0_{\mathfrak{P}}(N) - \mathfrak{C}_{\mathfrak{P}}(N)$ are mapped into $\mathfrak{L} - \mathfrak{C}_{\mathfrak{P}}(Y)$; hence if $Y = EN$ or NF, then $N \sim EF^{\alpha}$. Therefore $N \sim EF^{\alpha}$ for some $\alpha = 1$ or 3 and $\rho = 0$. In particular, it follows that $E^2 \nsim NXF$, $F^2 \nsim ENX$, $ENX \nsim NXF$. Indeed, if any of these were not the case, then by (2F), (2.1) $E^2 \sim F^2$ which is impossible by (2A). By (b) $\mathrm{ccl}(X)$ is not isolated. If $X \sim NF$, then $F^2 \sim NXF$, $EN \sim EF \sim EF^3$ by (2F), (2.1). The only possible fusions in $\mathfrak{G}$ are $X \sim NF$, $EN \sim EF \sim EF^3 \sim N$, $ENF \sim ENF^3$, $F^2 \sim NXF$, $E^2 \sim ENX$. If $E^2 \sim ENX$

occurred, then by (2F), (2.1) NF would be fused to one of EN, EF, EF^3, and hence $X \sim N$, which is impossible by (2B). Computing $\mathfrak{P}^{\#}$, we have $\mathfrak{P}^{\#} \leqq \langle \mathfrak{P}', E, NF \rangle < \mathfrak{P}$, and we are done. If $X \sim EN$, then by a similar argument, the only possible fusions are $X \sim EN, NF \sim EF \sim EF^3 \sim N, ENF \sim ENF^3$ $E^2 \sim ENX$, and $\mathfrak{P}^{\#} \leqq \langle \mathfrak{P}', EN, F \rangle < \mathfrak{P}$. We may finally suppose $X \sim EF^{\beta}$, $\beta = 1$ or 3, $\beta \neq \alpha$. By (2F), (2.1) we then have $E^2 \sim ENX$, $F^2 \sim NXF$, $NF \sim EN \sim EF^{\alpha} \sim N, X \sim EF^{\beta}$.

(e) Since $N \sim NF^2 \sim EN$, there exist $G_1, G_2 \in \mathfrak{G}$ such that $G_1 : N \to EN$, $\mathfrak{C}_{\mathfrak{P}}(N) \to \mathfrak{C}_{\mathfrak{P}}(EN)$ and $G_2 : NF^2 \to EN$, $\mathfrak{C}_{\mathfrak{P}}(NF^2) \to \mathfrak{C}_{\mathfrak{P}}(EN)$. In particular, G_1 and G_2 both map X into $\mathrm{ccl}(EF^{\beta}) \cap \mathfrak{C}_{\mathfrak{P}}(EN)$. Conjugating if necessary by an element in $\mathfrak{C}_{\mathfrak{P}}(EN)$, we may assume both G_1 and G_2 map X onto EF^{β}. Set $G = G_1 G_2^{-1}$; then $G : J \to J, X \to X, N \to NF^2$. If $\mathfrak{A} = \langle X, J \rangle, \mathfrak{H} = \mathfrak{C}(\mathfrak{A})$, then $\langle \mathfrak{A}, F^2, EF, N \rangle$ is an S_2-subgroup of $\mathfrak{H}$. In particular, the S_2-subgroups of $\mathfrak{H}/\mathfrak{A}$ are elementary abelian of order 8. Modulo $\mathfrak{A}$ the following elements represent the elements of an S_2-subgroup of $\mathfrak{H}/\mathfrak{A}$: $1\ F^2\ EF\ N\ EF^3\ NF^2\ ENF^{-1}\ ENF$. Moreover, $N \sim NF^2$ in $\mathfrak{H}/\mathfrak{A}$ by the above. By Burnside's theorem we see that either all the nonidentity representatives above are fused in $\mathfrak{H}/\mathfrak{A}$, or 6 of them are fused in sets of 3 each. But this is impossible, since 5 of them have squares in $\langle J \rangle$ and 2 of them have squares in $\mathrm{ccl}(X)$. This completes the proof.

Remark. We note that if $\mathfrak{P} = \mathfrak{P}(\rho, \sigma, \tau)$ with $\rho = 0$, $\sigma = \tau = 1$, and if $\mathfrak{A} = \langle J, X \rangle$, then as a consequence of (e), the group $\mathfrak{C}(\mathfrak{A})$ has a normal 2-complement.

Proposition (2H). *Let $\rho = 0$, $\sigma = 1$, $\tau = 1$. If $\mathfrak{G}$ has no subgroups of index 2, then after a suitable relabeling of the generators E, N, F, there are exactly three possibilities for the fusion of involutions of $\mathfrak{P}$.*

I. $J \sim X \sim N \sim EF \sim EF^3 \sim NF \sim EN$,

II. $J \sim X \sim N \sim EF$, $EF^3 \sim NF \sim EN$,

III. $J \sim X \sim EF^3$, $N \sim EF \sim NF \sim EN$.

In all three cases, $E^2 \sim ENX \sim ENF^3$, $F^2 \sim NXF \sim ENF$.

Proof. The following are consequences of (2D) and (2F), (2.1): $E^2 \nsim NXF$, $F^2 \nsim ENX$, since either fusion implies $E^2 \sim F^2$. Secondly, if $X \sim EF^{\alpha}$ for some $\alpha = 1$ or 3, then $E^2 \sim ENX$, $F^2 \sim NXF$, $NF \sim EN \sim EF^{\beta}$, where $\beta = 1$ or 3, $\beta \neq \alpha$. By (2G) and (2E), $J \sim X$, $N \sim EF$, $E^2 \sim ENF^3$, $F^2 \sim ENF$. Suppose $N \sim X$. Then $EF \sim X$, and we have either cases I or II. Suppose that neither case I nor case II holds. If $X \sim J$ is fused to no other class, then the only possible fusions in $\mathfrak{G}$ are between

$$X \sim J \,|\, E^2 \sim ENF^3 \,|\, F^2 \sim ENF \,|\, N \sim EF, EF^3 \,|\, NF, EN$$

by (2F), (2.1). But then $\mathfrak{P}^{\#} \leqq \langle \mathfrak{P}', ENF, EF \rangle < \mathfrak{P}$, which is impossible. Thus $X \sim Y$ for some $Y \in \mathfrak{P}$ with $vc_{\mathfrak{P}}(Y) = 4$. If $Y = EF^3$, then we have case III by an above remark. If $Y = NF$, then $EN \sim EF \sim EF^3 \sim N$ by (2F), (2.1). In order that $\mathfrak{P}^{\#} = \mathfrak{P}$, E^2 must be fused to ENX. But then $EF \sim X$ by (2F), (2.1) and $N \sim X$, which has been excluded. Similarly, if $Y = EN$, a contradiction is also obtained. This completes the proof.

Summarizing the results of §§1, 2, we have:

THEOREM (2I). *Let $\mathfrak{G}$ be a finite group of order $64m$, m odd. Let $\mathfrak{P}$ be an S_2-subgroup of $\mathfrak{G}$. Suppose there is an element $F \in \mathfrak{P}$ of order 8 which is self-centralizing in $\mathfrak{P}$ and conjugate to its odd powers in $\mathfrak{P}$. If $\mathfrak{G}$ has no subgroups of index 2, then $\mathfrak{P}$ is isomorphic to $\mathfrak{P}(\rho, \sigma, \tau)$ of (1A), with $\rho = 0, \sigma = \tau = 1$. Furthermore, if notation is chosen suitably, the possibilities for the fusion of the 2-power classes of $\mathfrak{G}$ are*

I. $F|E|E^2 \sim ENX \sim ENF^3|F^2 \sim NXF \sim ENF|J \sim X \sim N \sim EF \sim EF^3 \sim NF \sim EN,$

II. $F|E|E^2 \sim ENX \sim ENF^3|F^2 \sim NXF \sim ENF|J \sim X \sim N \sim EF|EF^3 \sim NF \sim EN,$

III. $F|E|E^2 \sim ENX \sim ENF^3|F^2 \sim NXF \sim ENF|J \sim X \sim EF^3|N \sim EF \sim NF \sim EN.$

Let $\mathfrak{P}$ be the 2-group of (2I). The character table of $\mathfrak{P}$ is easily seen to be

(2.2)

E	F	F^2	NXF	ENF	E^2	ENX	ENF^3	EN	NF	EF^3	EF	N	X	J	1
1	1	1	1	1	1	1	1	1	1	1	1	1	1	1	1
1	1	1	−1	−1	1	−1	−1	−1	−1	1	1	−1	1	1	1
1	−1	1	−1	−1	1	1	−1	1	−1	−1	−1	1	1	1	1
1	−1	1	1	1	1	−1	1	−1	1	−1	−1	−1	1	1	1
−1	1	1	1	−1	1	−1	−1	−1	1	−1	−1	1	1	1	1
−1	1	1	−1	1	1	1	1	1	−1	−1	−1	−1	1	1	1
−1	−1	1	−1	1	1	−1	1	−1	−1	1	1	1	1	1	1
−1	−1	1	1	−1	1	1	−1	1	1	1	1	−1	1	1	1
0	0	2	−2	0	−2	0	0	0	2	0	0	0	−2	2	2
0	0	2	2	0	−2	0	0	0	−2	0	0	0	−2	2	2
0	0	−2	0	0	2	−2	0	2	0	0	0	0	−2	2	2
0	0	−2	0	0	2	2	0	−2	0	0	0	0	−2	2	2
0	0	−2	0	0	−2	0	0	0	0	2	−2	0	2	2	2
0	0	−2	0	0	−2	0	0	0	0	−2	2	0	2	2	2
0	0	0	0	2	0	0	−2	0	0	0	0	0	0	−4	4
0	0	0	0	−2	0	0	2	0	0	0	0	0	0	−4	4

The irreducible characters of $\mathfrak{P}$ will be denoted by ζ_i, $1 \leqq i \leqq 16$ in the above order.

3. We will require certain facts from the theory of blocks. The terminology is taken from [3]. Let $\mathfrak{G}$ be a group of order $g = p^a g'$, where p is a fixed prime

number not dividing g'. If B is a block of $\mathfrak{G}$, then the restrictions χ_μ^0 of the irreducible characters χ_μ in B to the p-regular classes generate a module M_B over Z. Any basis $\{\phi_\rho\}$ of M_B will be called a basic set ϕ_B of B. We will always assume that when B is the 1-block of $\mathfrak{G}$, then the constant function 1 is in ϕ_B. Once a basic set ϕ_B for B has been chosen, the corresponding decomposition numbers $d_{\mu\rho}$ and Cartan invariants $c_{\rho\sigma}$ of B are defined by

$$\chi_\mu^0 = \sum_\rho d_{\mu\rho}\phi_\rho \quad \text{for } \chi_\mu \in B, \ \phi_\rho \in \phi_B,$$

$$c_{\rho\sigma} = \sum_\mu d_{\mu\rho}d_{\mu\sigma} \quad \text{for } \phi_\rho, \phi_\sigma \in \phi_B.$$

$c_{\rho\sigma}$) is the matrix of a positive-definite quadratic form. If ϕ_B is replaced by another basic set, the form is replaced by an equivalent form. The irreducible modular characters of $\mathfrak{G}$ in B form a basic set, and if $(c_{\rho\sigma})$ is the corresponding Cartan matrix, then it is well known that $\sum_\sigma c_{\rho\sigma}\phi_\sigma(1) \equiv 0 \pmod{p^a}$. From this and the above remarks, it follows that the analogous congruence holds for any basic set of B.

Let P be a p-element of $\mathfrak{G}$. Let B^* run over the blocks of $\mathfrak{C}(P)$, and for each B^*, assume that a basic set ϕ_{B^*} has been selected. If V runs over p-regular elements of $\mathfrak{C}(P)$, the corresponding generalized decomposition numbers $d_{\mu\rho}^P$ are defined by

$$\chi_\mu(PV) = \sum d_{\mu\rho}^P \phi_\rho^P(V),$$

where the sum is over all $\phi_\rho^P \in \phi_{B^*}$ and all B^*. If B is the 1-block of $\mathfrak{G}$, then the above sum need only be taken over the basic set of the 1-block of $\mathfrak{C}(P)$ [3, Theorem 3]. For notational convenience, the column $(d_{\mu\rho}^P)$ will be denoted by $\mathfrak{d}_\rho^P$; its μth entry $d_{\mu\rho}^P$ will be denoted by $(\mathfrak{d}_\rho^P)_\mu$.

Assume from now on that B and B^* are the 1-blocks of $\mathfrak{G}$ and $\mathfrak{C}(P)$ respectively. Define for $\chi_\lambda, \chi_\mu \in B$,

$$S_{\lambda\mu}^P = \frac{1}{c(P)} \sum \chi_\lambda(PV)\bar{\chi}_\mu(PV), \tag{3.1}$$

where V runs over all p-regular elements of $\mathfrak{C}(P)$. Then by [4] $S_{\lambda\mu}^P = \sum_{\rho,\sigma} d_{\lambda\rho}^P d_{\mu\sigma}^P \gamma_{\rho\sigma}^P$, where $(\gamma_{\rho\sigma}^P)$ is the inverse matrix of $C^P = (c_{\rho\sigma}^P)$, the matrix of Cartan invariants of B^*. In particular, if $d = \nu(P)$, then $p^d(\gamma_{\rho\sigma}^P)$ is an integral matrix. Let Q^P denote the quadratic form determined by $p^d(\gamma_{\rho\sigma}^P)$; Q^P is positive-definite.

Let $1 = P_1, P_2, \cdots, P_r$ be a complete set of representatives for the classes of p-elements of $\mathfrak{G}$. Then from the ordinary orthogonality relation $(\chi_\lambda, \chi_\mu)_{\mathfrak{G}} = \delta_{\lambda\mu}$ grouped according to the p-sections of $\mathfrak{G}$, we have

$$\sum_{i=1}^{r} S_{\lambda\mu}^{P_i} = \delta_{\lambda\mu} = \begin{cases} 1 & \lambda = \mu, \\ 0 & \lambda \neq \mu. \end{cases}$$

In particular, for $\lambda = \mu$, $\sum_{i=1}^{r} p^a S_{\lambda\lambda}^{P_i} = p^a$. This latter equation will be referred to as the method of contribution. If $d_{\lambda\rho}^P$ is a rational integer for all ϕ_ρ^P in ϕ_{B^*}, and if $\nu(P) = d$, then $p^a S_{\lambda\lambda}^P$ is a rational integer $\geqq p^{a-d}m$, where m is the minimum value of the quadratic form Q^P for nonzero integral vectors. By [4] we also have the following congruence:

$$p^a S_{\lambda\lambda}^P \equiv p^a x_\lambda^2 S_{1,1}^P \pmod{\mathfrak{p} p^{h_\lambda}},$$

where χ_1 is the 1-character of $\mathfrak{G}$, $\chi_\lambda \in B$, $\nu(x_\lambda) = h_\lambda$ is the so-called height of χ_λ, and $\mathfrak{p}$ is a prime ideal divisor of p in a suitably large algebraic number field. We will refer to this congruence as the νth method of contribution.

If the columns $\mathfrak{d}_\rho^P$ of the 1-block B have been determined for all p-singular sections of $\mathfrak{G}$, and B has k irreducible characters χ_λ, then any set of integral, linearly independent columns of length k orthogonal to the given $\mathfrak{d}_\rho^P$ and having a maximal subdeterminant of value 1 is then the set of decomposition numbers of B relative to some basic set [3, §5]. The corresponding Cartan invariants can then be computed. This technique will be used considerably in the next two sections.

The following remarks may aid in the computations to follow.

(1) If $G \in \mathfrak{Z}\mathfrak{G}$ and G has order p^α, then the p-blocks of $\mathfrak{G}$ and $\mathfrak{G}/\langle G\rangle$ are in 1-1 correspondence. The Cartan invariants of a block of $\mathfrak{G}$ are obtained by multiplying the Cartan invariants of the corresponding block of $\mathfrak{G}/\langle G\rangle$ by p^α [3, §2].

(2) If $\mathfrak{G}$ has a normal p-complement, then each p-block consists of one modular character. In particular, the Cartan matrix of a block of defect d is (p^d) [5, §2].

(3) Let $\mathfrak{D}$ be the dihedral group $\langle A, B \mid A^4 = B^2 = ABAB = 1\rangle$ of order 8. Let $\mathfrak{G}$ be a finite group with $\mathfrak{D}$ as S_2-subgroup. The classes of involutions in $\mathfrak{D}$ are represented by A^2, B, AB. Let C be the Cartan matrix of the 1-block for the prime 2. Then one and only one of the following holds: (i) $\mathfrak{G}$ has 3 classes of involutions, $C = (8)$; (ii) $\mathfrak{G}$ has 2 classes of involutions,

$$A^2 \sim B \text{ or } A^2 \sim AB, \quad C = \begin{pmatrix} 3 & 1 \\ 1 & 3 \end{pmatrix};$$

(iii) $\mathfrak{G}$ has 1 class of involutions,

$$C = \begin{bmatrix} 2 & 0 & -1 \\ 0 & 2 & -1 \\ -1 & -1 & 3 \end{bmatrix}.$$

This can be proved by the techniques of this paragraph easily; the details will be omitted.

4. From now on $\mathfrak{G}$ will be a finite group with no subgroups of index 2 satisfying the assumptions of (2I). By that result the classes of 2-elements fall into one of three cases, which we will denote by I, II, III in the notation of (2I). We

now compute the Cartan invariants of the 1-block for the 2-singular sections of $\mathfrak{G}$ and of $\mathfrak{H} = \mathfrak{C}(J)$. Accordingly, if Y is a representative of a 2-section, C^Y and $\tilde{C}^Y$ will denote the matrix of Cartan invariants of the 1-block of $C(Y)$ and $C_{\mathfrak{H}}(Y)$ respectively. In particular, $C^Y = \tilde{C}^Y$ for $Y = E, F, E^2, F^2, J$. As we shall see, these matrices are for $\mathfrak{G}$

(4.1)

E	F	E^2	F^2	EN	
8	8	$\begin{pmatrix} 12 & 4 \\ 4 & 12 \end{pmatrix}$	$\begin{pmatrix} 12 & 4 \\ 4 & 12 \end{pmatrix}$	–	in I,
8	8	32	32	$\begin{pmatrix} 6 & 2 \\ 2 & 6 \end{pmatrix}$	in II,
8	8	32	32	16	in III

and for $\mathfrak{H}$

(4.2)

E	F	ENF	ENF^3	E^2	F^2	N	X	EN	J	
8	8	16	16	$\begin{pmatrix} 12 & 4 \\ 4 & 12 \end{pmatrix}$	$\begin{pmatrix} 12 & 4 \\ 4 & 12 \end{pmatrix}$	8	32	–	$\begin{bmatrix} 10 & 2 & 4 & 4 & 2 \\ 2 & 10 & 4 & 4 & 2 \\ 4 & 4 & 10 & 2 & 2 \\ 4 & 4 & 2 & 10 & 2 \\ 2 & 2 & 2 & 2 & 4 \end{bmatrix}$	in I,
8	8	16	16	32	32	8	32	16	$\begin{pmatrix} 6 & 2 \\ 2 & 22 \end{pmatrix}$	in II, III.

LEMMA (4A). *The only fusion in $\mathfrak{H}$ among the 2-elements are the following:*

$$E^2 \sim ENX, \quad F^2 \sim NXF, \quad EF \sim EF^3 \sim NF \sim EN \sim X \quad \textit{in } \mathrm{I},$$

$$E^2 \sim ENX, \quad F^2 \sim NXF, \quad EF^3 \sim NF \sim EN, \quad X \sim EF \quad \textit{in } \mathrm{II},$$

$$E^2 \sim ENX, \quad F^2 \sim NXF, \quad EF \sim NF \sim EN, \quad X \sim EF^3 \quad \textit{in } \mathrm{III}.$$

In particular, the labeling of the columns in (4.2) *is as indicated.*

Proof. Since $ENX \sim E^2$, $NXF \sim F^2$ in $\mathfrak{G}$, and their squares are J, these fusions occur in $\mathfrak{H}$. The elements in ccl(ENF), ccl(ENF^3) have squares X or XJ. Hence these classes are not fused in $\mathfrak{H}$ to E^2, F^2, ENX, NXF. We have thus listed the fusions of elements of order 4. If $X \sim EF$ in $\mathfrak{G}$, then $EF^3 \sim NF \sim EN$ and $X \sim EF$ in $\mathfrak{H}$ by (2F), (2.1). Similarly, if $X \sim EF^3$ in $\mathfrak{G}$, then $EF \sim NF \sim EN$ and $X \sim EF^3$ in $\mathfrak{H}$. In either case, the focal subgroup $\mathfrak{P}^\#$ of $\mathfrak{P}$ in $\mathfrak{H}$ contains $\langle \mathfrak{P}', EN, NF \rangle$ and thus $(\mathfrak{P} : \mathfrak{P}^\#) \leqq 2$. By (2G) $\mathfrak{P}^\# \neq \mathfrak{P}$, and hence ccl($N$) is isolated in $\mathfrak{H}$.

COROLLARY (4B). *$\mathfrak{H}$ has a normal subgroup $\tilde{\mathfrak{H}}$ of index 2, but none of index 4. Moreover, $\tilde{\mathfrak{H}} \cap \mathfrak{P} = \langle \mathfrak{P}', EN, NF \rangle$. Hence $\mathfrak{H}$ has a linear character λ of order 2 with $\lambda(E) = \lambda(F) = \lambda(N) = -1$.*

We now turn to the entries of (4.1) and (4.2). $\langle F \rangle, \langle E \rangle$ are S_2-subgroups of $\mathfrak{C}(F), \mathfrak{C}(E)$ respectively. Since $\mathfrak{C}(F), \mathfrak{C}(E)$ have normal 2-complements, $C^F, C^E, \tilde{C}^F, \tilde{C}^E$ are as indicated.

$\mathfrak{C}(F^2)$ has as an S_2-subgroup the subgroup $\mathfrak{Q} = \langle X, F, ENF \rangle$ of order 32. If residue classes modulo $\langle F^2 \rangle$ are denoted by a dot, then $\dot{\mathfrak{Q}}$ has order 8. If $M = ENF$, then $\dot{M}^4 = 1$, $\dot{M}^2 = \dot{X}$, $\dot{F}^2 = 1$, $\dot{F}\dot{M}\dot{F} = \dot{M}^{-1}$, and $\dot{\mathfrak{Q}}$ is dihedral. The classes of involutions are represented by $\dot{X}, \dot{F}, \dot{M}\dot{F}$. If $\dot{X} \sim \dot{M}\dot{F}$ in $\mathfrak{C}(F^2)/\langle F^2 \rangle$, then $X \sim EN$ or ENX. This is possible only in I, and indeed, occurs in that case. For since $X \sim EN$ in I in $\mathfrak{G}$, there is a $G \in \mathfrak{G}$ such that $G: EN \to X$. By (2F), (2.1), we may assume $G: F^2 \to F^{\pm 2}$, and since $F^2 \sim F^{-2}$ in $\mathfrak{C}(X)$, we may even assume G: $F^2 \to F^2$. If $\dot{X} \sim \dot{F}$ in $\mathfrak{C}(F^2)/\langle F^2 \rangle$, then $X \sim F^{2i+1}$ for some i, which is impossible. The situation for $\mathfrak{C}(E^2)$ is similar; an S_2-subgroup of $\mathfrak{C}(E^2)$ is $\langle X, E, ENF \rangle$. Thus $C^{E^2}, C^{F^2}, \tilde{C}^{E^2} \tilde{C}^{F^2}$ are as indicated.

In cases II, III, $\mathfrak{C}(EN)$ has $\langle EN \rangle \times \langle F^2, NF \rangle$ as S_2-subgroup, where $\mathfrak{Q} = \langle F^2, NF \rangle$ is dihedral of order 8. If residue classes modulo $\langle EN \rangle$ are denoted by a dot, then the classes of involutions of $\dot{\mathfrak{Q}}$ are represented by $\dot{J}$, $N\dot{F}$, $N\dot{F}^3$. Suppose we are in case II. If $\dot{J} \sim N\dot{F}^3$, then $J \sim NF^3$ or $J \sim (EN)(NF^3) = EF^3$ in $\mathfrak{C}(EN)$, which is impossible. If $\dot{J} \sim N\dot{F}$, then $J \sim NF$ or $J \sim EF$ in $\mathfrak{C}(EN)$, and only the latter can occur in case II. To see that this does occur, replace EN by the element Y in $\mathrm{ccl}(EF^3)$ which is fixed by the element B in (2E). In $\mathfrak{C}(Y)/\langle Y \rangle$, B fuses two classes of involutions, and so

$$C^{EN} = \begin{pmatrix} 6 & 2 \\ 2 & 6 \end{pmatrix}.$$

Suppose we are in III. If $\dot{J} \sim \dot{N}F$, then $J \sim NF$ or $J \sim EF$, which is impossible. If $\dot{J} \sim N\dot{F}^3$, then necessarily $J \sim EF^3$ in $\mathfrak{C}(EN)$. To see that this does not occur, take as a representative of the class of EN in $\mathfrak{G}$ the element EF. $\mathfrak{C}(EF)$ has $\langle EF \rangle \times \langle NXF, X \rangle$ as S_2-subgroup. Moreover, if $J \sim EF^3$ in $\mathfrak{C}(EN)$, then $J \sim X$ in $\mathfrak{C}(EF)$, and there is an $A \in \mathfrak{G}$ such that $A: X \to J$, $EF \to EF$. On the other hand, by (2E) there is a $B \in \mathfrak{G}$ such that $B: J \to X$, $N_1 \to \mathrm{ccl}(EF)$, where $N_1 \in \mathrm{ccl}(N)$. Since EF has 4 conjugates in $\mathfrak{C}_{\mathfrak{P}}(X)$, we may assume $B: N_1 \to EF$. Then $BA: J \to J$, $N_1 \to EF$. If $\mathfrak{P}^{\#}$ is the focal subgroup of $\mathfrak{P}$ in $\mathfrak{C}(J)$, then $ENF \in \mathfrak{P}^{\#}$, which contradicts (4B). Thus $J \nsim EF^3$ in $\mathfrak{C}(EN)$, and $C^{EN} = (16)$. This proves the validity of (4.1).

In $\mathfrak{H}/\langle J \rangle$ let small letters denote residue classes of corresponding capital letters. By (4A) the following are representatives of the classes of 2-elements of $\mathfrak{H}/\langle J \rangle$.

$$e,\ f,\ enf,\ e^2, f^2, x,\ en,\ n,\ 1$$

except that in case I, en is to be omitted. For $Y \in \mathfrak{H}$, let $\mathfrak{C}^0(Y)$ be the inverse image in $\mathfrak{H}$ of the centralizer $\dot{\mathfrak{C}}(y)$ of y in $\mathfrak{H}/\langle J \rangle$. Then $\mathfrak{C}_{\mathfrak{H}}(Y) \trianglelefteq \mathfrak{C}^0(Y)$, and the index is 1 or 2. In particular, if $\mathfrak{C}(Y)$ has a normal 2-complement, so do $\mathfrak{C}_{\mathfrak{H}}(Y)$ and $\mathfrak{C}^0(Y)$. If y is a representative of a 2-section of $\mathfrak{H}/\langle J \rangle$, let $\dot{C}^y$ be the matrix of Cartan invariants of the 1-block of $\dot{\mathfrak{C}}(y)$. We then have $\dot{C}^e = \dot{C}^f = (8)$. $\dot{\mathfrak{C}}(enf)$ has $\langle enf \rangle \times \langle f^2 \rangle$ as an S_2-subgroup, so that $\dot{\mathfrak{C}}(enf)$ has a normal 2-complement, and $\dot{C}^{enf} = (8)$. For $\dot{\mathfrak{C}}(e^2)$ and $\dot{\mathfrak{C}}(f^2)$ the above remarks apply in cases II, III, so that $\dot{C}^{e^2} = \dot{C}^{f^2} = (32)$ in these cases. In case I $\dot{\mathfrak{C}}(f^2)/\langle f^2 \rangle$ has as an S_2-subgroup $\mathfrak{Q} \times \langle N \rangle \pmod{\langle F^2 \rangle}$, where $\mathfrak{Q} = \langle X, F, ENF \rangle$. $\dot{\mathfrak{C}}(f^2)/\langle f^2 \rangle$ has a normal subgroup of index 2 with dihedral S_2-subgroups. Moreover, the center of $\mathfrak{Q} \times \langle N \rangle \pmod{\langle F^2 \rangle}$ is not contained in this subgroup From our earlier remarks and [5] (3B), it follows that

$$\dot{C}^{f^2} = \begin{pmatrix} 12 & 4 \\ 4 & 12 \end{pmatrix}.$$

Similarly,

$$\dot{C}^{e^2} = \begin{pmatrix} 12 & 4 \\ 4 & 12 \end{pmatrix}.$$

$\dot{\mathfrak{C}}(x)$ has a normal 2-complement by the remark following (2G), so that $\dot{C}^x = (32)$. $\dot{\mathfrak{C}}(n)$ has $\langle n, e^2, f^2 \rangle$ as S_2-subgroup, elementary abelian of order 8. Thus $\dot{\mathfrak{C}}(n)/\langle n \rangle$ has an S_2-subgroup of type (2,2), whose involutions modulo $\langle n \rangle$ are $e^2, f^2, x = e^2 f^2$. If these are fused in $\dot{\mathfrak{C}}(n)/\langle n \rangle$, then e^2, f^2, x are fused in $\dot{\mathfrak{C}}(n)$ which is impossible. Thus $\dot{\mathfrak{C}}(n)$ has a normal 2-complement, and $\dot{C}^n = (8)$. Finally, in cases II, III, $\mathfrak{C}_{\mathfrak{H}}(EN)$ has $\langle EN \rangle \times \langle F^2, NF \rangle$ as an S_2-subgroup, and there are no fusions of involutions in $\mathfrak{C}_{\mathfrak{H}}(EN)/\langle EN \rangle$, since J is isolated in $\mathfrak{H}$. Thus $\mathfrak{C}_{\mathfrak{H}}(EN)$ and $\dot{\mathfrak{C}}(en)$ have normal 2-complements, and $\dot{C}^{en} = (16)$. Summarizing the calculations of this paragraph, we have

(4.3)

	e	f	enf	e^2	f^2	x	n	en	
(4.3)	8	8	8	$\begin{pmatrix} 12 & 4 \\ 4 & 12 \end{pmatrix}$	$\begin{pmatrix} 12 & 4 \\ 4 & 12 \end{pmatrix}$	32	8	—	in I,
	8	8	8	32	32	32	8	16	in II, III.

The entries of (4.2) are now clear, except for those under J.

Let χ_1 be the 1-character of $\mathfrak{H}$. Then $d^e_{1,1} = d^f_{1,1} = d^{enf}_{1,1} = 1$. The existence of the character λ of (4B) shows that the nonzero entries under e, f, enf, n are four 1's and four -1's. These nonzero generalized decomposition numbers exhaust the characters of height 0, and hence the 1-block $\tilde{B}$ of $\mathfrak{H}$ has exactly 8 characters of height 0. After a rearrangement if necessary, we may assume by the orthogonality relations that χ_i for $1 \leqq i \leqq 8$ are as below:

(4.4)

e	f	enf	n
1	1	1	1
-1	-1	-1	-1
1	1	-1	-1
-1	-1	1	1
1	-1	1	-1
-1	1	-1	1
1	-1	-1	1
-1	1	1	-1

Cases II, III. For $1 \leqq \mu \leqq 8$, $d_{\mu 1}^{en}$ are odd integers occurring in pairs; hence they are ± 1, since $\sum_\mu (d_{\mu 1}^{en})^2 = 16$. The remaining nonzero entries in $\mathfrak{d}_1^{en}$ are even integers whose squares add up to 8. Hence these are ± 2, and we may assume they occur for χ_9, χ_{10}. If there is a χ_{11} in the 1-block $\dot{B}$ of $\mathfrak{H}/\langle J \rangle$, consider χ_{11} restricted to $\langle en, nf^3 \rangle$ in II, to $\langle en, nf \rangle$ in III. These subgroups are abelian of type (2,2), and their involutions are fused in $\mathfrak{H}$. Hence $3\chi_{11}(en) \equiv \chi_{11}(1) \pmod 4$. Since $\chi_{11}(en) = 0$, χ_{11} has height $\geqq 2$, and thus each of the numbers $\chi_{11}(e^2)$, $\chi_{11}(f^2)$, $\chi_{11}(x)$ is nonzero and divisible by 4 by the νth method of contribution. But then $4^2 + 4^2 + 4^2 > 32$, and this is impossible by the method of contribution applied to χ_{11}. $\dot{B}$ then has exactly 10 irreducible characters, and therefore 2 modular characters for the 1-section. For $i = 9, 10$, $\chi_i(e^2)$, $\chi_i(f^2)$, $\chi_i(x) = \pm 2$, since ± 6 is too large. The matrix of generalized decomposition numbers for $\dot{B}$ can now be essentially completed. A $\chi \in \dot{B}$ cannot assume ± 3 more than once on e^2, f^2, x by the method of contribution. Orthogonality relations are then sufficient to compute the remaining entries with the uncertainties indicated below. We omit the details.

(4.5)

e	f	enf	e^2	f^2	x	n	en	1	
1	1	1	1	1	1	1	1	1	0
-1	-1	-1	1	1	1	-1	1	1	0
1	1	-1				-1	τ_1	0	τ_1
-1	-1	1				1	τ_1	0	τ_1
1	-1	1				-1	τ_2	0	τ_2
-1	1	-1	$\mathfrak{S}_\alpha$	$\mathfrak{S}_\beta$	$\mathfrak{S}_\gamma$	1	τ_2	0	τ_2
1	-1	-1				1	τ_3	0	τ_3
-1	1	1				-1	τ_3	0	τ_3
0	0	0	$2\omega_1$	$2\omega_1$	$2\omega_1$	0	$-2\omega_1$	0	$2\omega_1$
0	0	0	$2\omega_2$	$2\omega_2$	$2\omega_2$	0	$2\omega_2$	$-\omega_2$	$-\omega_2$

Here τ_i, $\omega_i = \pm 1$, $\{\alpha, \beta, \gamma\}$ is a permutation of $\{1,2,3\}$, and

$$\mathfrak{S}_1 = \begin{bmatrix} -3\tau_1 \\ -3\tau_1 \\ \tau_2 \\ \tau_2 \\ \tau_3 \\ \tau_3 \end{bmatrix}, \quad \mathfrak{S}_2 = \begin{bmatrix} \tau_1 \\ \tau_1 \\ -3\tau_2 \\ -3\tau_2 \\ \tau_3 \\ \tau_3 \end{bmatrix}, \quad \mathfrak{S}_3 = \begin{bmatrix} \tau_1 \\ \tau_1 \\ \tau_2 \\ \tau_2 \\ -3\tau_3 \\ -3\tau_3 \end{bmatrix}. \tag{4.6}$$

The columns under 1 are determined by the methods of §3. $\tilde{C}^J$ is then

$$\begin{pmatrix} 6 & 2 \\ 2 & 22 \end{pmatrix}.$$

If f_1^J, f_2^J are the degrees of the functions in the corresponding basic set, then $f_1^J = 1$ since ϕ_1^J is the constant 1. Since $6f_1^J + 2f_2^J \equiv 0 \pmod{64}$, we see that $f_2^J \equiv -3 \pmod{32}$. This will be important later on.

The matrix of generalized decomposition numbers for the 1-block $\tilde{B}$ of $\mathfrak{H}$ in cases II, III can now be completed. A calculation of the S_2-subgroups shows that except for ENF and ENF^3, all entries under the various representatives $\neq J$ of the 2-singular sections from characters faithful on J are 0. Moreover, the nonzero entries for ENF, ENF^3 are rational even integers whose squares add up to 8. This implies the existence of χ_{11}, χ_{12} as indicated below. Since there are at least 13 columns, there is one more character χ_{13}. The entries of χ_{11}, χ_{12}, χ_{13} under J and 1 are easily determined by orthogonality relations. The arrangement of the signs is achieved by interchanging χ_{11} and χ_{12} and replacing either by its negative if necessary. The matrix of generalized decomposition numbers for $\tilde{B}$ in cases II, III is

(4.7)

E	F	E^2	F^2	ENF	ENF^3	X	N	EN	J		1	
1	1	1	1	1	1	1	1	1	1	0	1	0
−1	−1	1	1	−1	−1	1	−1	1	1	0	1	0
1	1			−1	−1		−1	τ_1	0	τ_1	0	τ_1
−1	−1			1	1		1	τ_1	0	τ_1	0	τ_1
1	−1	$\mathfrak{S}_\alpha$	$\mathfrak{S}_\beta$	1	1	$\mathfrak{S}_\gamma$	−1	τ_2	0	τ_2	0	τ_2
−1	1			−1	−1		1	τ_2	0	τ_2	0	τ_2
1	−1			−1	−1		1	τ_3	0	τ_3	0	τ_3
−1	1			1	1		−1	τ_3	0	τ_3	0	τ_3
0	0	$2\omega_1$	$2\omega_1$	0	0	$2\omega_1$	0	$-2\omega_1$	0	$2\omega_1$	0	$2\omega_1$
0	0	$2\omega_2$	$2\omega_2$	0	0	$2\omega_2$	0	$2\omega_2$	$-\omega_2$	$-\omega_2$	$-\omega_2$	$-\omega_2$
0	0	0	0	2	−2	0	0	0	−1	1	1	−1
0	0	0	0	−2	2	0	0	0	−1	1	1	−1
0	0	0	0	0	0	0	0	0	−1	−3	1	3

where ω_i, $\tau_i = \pm 1$, and $\mathfrak{S}_\alpha, \mathfrak{S}_\beta, \mathfrak{S}_\gamma$ are as in (4.6). C^J is then

$$\begin{pmatrix} 6 & 2 \\ 2 & 22 \end{pmatrix}$$

as indicated in (4.2).

Case I. Let $\dot{B}$ be the 1-block of $\mathfrak{H}/\langle J\rangle$. Then no $\chi_i \in \dot{B}$ is zero on x, since x is in the center of $\mathfrak{P}/\langle J\rangle$. If $\chi_i(x) = \pm 3$ for some $1 \leqq i \leqq 8$, then this occurs at least twice, so that $\Sigma_{i=1}^{8}\chi_i(x)^2 \geqq 24$. Since the remaining $\chi_i \in \dot{B}$ have height $\geqq 1$, this would imply that $\dot{B}$ has $\leqq 10$ characters, which is impossible. Thus we may assume $\chi_i(x) = \chi_{i+1}(x) = \tau_i$ for $i = 3, 5, 7$, $\tau_i = \pm 1$. The $\chi_i \in \dot{B}$ for $i \geqq 9$ cannot have height $\geqq 2$; otherwise the contribution from e^2, f^2, x would be too large. Hence these χ_i have height 1, and necessarily $\chi_i(x) = 2\tau_i$, $\tau_i = \pm 1$. In particular, $\dot{B}$ has 14 characters and 5 functions in the basic set from the 1-section. The quadratic forms Q^{E^2} and Q^{F^2} associated to $\mathfrak{C}(E^2)$ and $\mathfrak{C}(F^2)$ are $3u^2 - 2uv + 3v^2 = 2u^2 + 2v^2 + (u-v)^2$, which has a minimum of 3 for nonzero rational integral vectors. If a $\chi_i \in \dot{B}$ has height 0, then the sum of the contributions from $E^2, F^2, J, 1$ must be $64 - 34 = 30$. We express this as

$$2Q^{E^2} + 2Q^{F^2} + Q^J + Q^1 = 30.$$

Now $Q^J, Q^1 \geqq 1$. Since $d_{i1}^{E^2} - d_{i2}^{E^2}$ and $d_{i1}^{F^2} - d_{i2}^{F^2}$ are necessarily odd, Q^{E^2} and Q^{F^2} are $\leqq 11$. If (u, v) are the decomposition numbers of χ_i under E^2 or F^2, and if we allow for a possible change of sign $(u, v) \to (-u, -v)$ and a possible interchange $(u, v) \to (v, u)$, then the only possible values for (u, v) are $(1, 0)$ and $(2, 1)$. If $\chi_i \in \dot{B}$ has height 1, then $u + v \equiv 2 \pmod 4$. The method of contributions implies that Q^{E^2} and $Q^{F^2} \leqq 30$. With the same conventions as before, the only possible values for (u, v) are $(1, 1)$ and $(2, 0)$. Consider the entries under E^2 for all $\chi_i \in \dot{B}$. If there are a, b, c, d rows of type $(1, 0)$, $(2, 1)$, $(1, 1)$, $(2, 0)$ respectively, then from the equality $\Sigma u^2 + \Sigma v^2 = 24$, we have $a + b = 8$, $c + d = 6$, $a + 5b + 2c + 4d = 24$. Thus $b = 0$, $d = 2$. The entries can then be determined by the orthogonality relations. Similarly so for the entries under F^2. The generalized decomposition numbers for $\dot{B}$ for the 2-singular sections of $\mathfrak{H}/\langle J\rangle$ have now been determined, and are as below. There are then 6 characters of $\tilde{B}$ faithful on $\langle J\rangle$. Two of them, say χ_{15}, χ_{16}, are nonzero on ENF, ENF^3. The columns under J and 1 can then be found by the methods of §3. The matrix for $\tilde{B}$ in case I is (4.8), where $\tau_i = \pm 1$. $\tilde{C}^J$ is then as in (4.2).

5. In addition to the assumptions made on $\mathfrak{G}$ on §4, we also assume $\mathfrak{G}$ has more than one class of involutions. This excludes case I.

LEMMA (5A). *Let χ be a character of $\mathfrak{G}$. Then*
(i) $\chi(E) \equiv \chi(F) \pmod 2$,
(ii) $\chi(E^2) \equiv \chi(F^2) \equiv \chi(EN) \equiv \chi(J) \pmod 4$,
(iii) $\chi(J) \equiv 2\chi(E) - \chi(F^2) \equiv 2\chi(F) - \chi(E^2) \pmod 8$,
(iv) $-2\chi(E) - 2\chi(F) + \chi(F^2) + \chi(E^2) + 2\chi(EN) \equiv 0 \pmod 8$.

(4.8)

E	F	ENF	ENF^1	E^2		F^2		N	X	J					1				
1	1	1	1	1	0	1	0	1	1	1	0	0	0	0	1	0	0	0	0
−1	−1	−1	−1	1	0	1	0	−1	1	1	0	0	0	0	1	0	0	0	0
1	1	−1	−1	0	τ_3	0	τ_3	−1	τ_3	0	τ_3	0	0	0	0	τ_3	0	0	0
−1	−1	1	1	0	τ_3	0	τ_3	1	τ_3	0	τ_3	0	0	0	0	τ_3	0	0	0
1	−1	−1	−1	0	τ_5	τ_5	τ_5	1	τ_5	0	0	τ_5	0	0	0	0	τ_5	0	0
−1	1	1	1	0	τ_5	τ_5	0	−1	τ_5	0	0	τ_5	0	0	0	0	τ_5	0	0
1	−1	1	1	τ_7	0	0	τ^1	−1	τ_7	0	0	0	τ^2	0	0	0	0	τ^2	0
−1	1	−1	−1	τ_7	0	0	τ^2	1	τ_7	0	0	0	τ^2	0	0	0	0	τ^2	0
0	0	0	0	$-\tau_9$	$-\tau_9$	$-\tau_9$	τ_9	0	$2\tau_9$	τ_9	τ_9	τ_9	τ_9	τ_9	τ_9	τ_9	τ_9	τ_9	τ_9
0	0	0	0	$-\tau_{10}$	$-\tau_{10}$	$-\tau_{10}$	τ_{10}	0	$2\tau_{10}$	0	0	0	0	$-\tau_{10}$	0	0	0	0	$-\tau_{10}$
0	0	0	0	$-\tau_{11}$	$-\tau_{11}$	$2\tau_{11}$	0	0	$2\tau_{11}$	$-\tau_{11}$	0	$-\tau_{11}$	0	0	$-\tau_{11}$	0	$-\tau_{11}$	0	0
0	0	0	0	$-\tau_{12}$	$-\tau_{12}$	0	$2\tau_{12}$	0	$2\tau_{12}$	0	$-\tau_{12}$	0	$-\tau_{12}$	0	0	$-\tau_{12}$	0	$-\tau_{12}$	0
0	0	0	0	$2\tau_{13}$	0	$-\tau_{13}$	$-\tau_{13}$	0	$2\tau_{13}$	$-\tau_{13}$	0	0	$-\tau_{13}$	0	$-\tau_{13}$	0	0	$-\tau_{13}$	0
0	0	0	0	0	$2\tau_{14}$	$-\tau_{14}$	$-\tau_{14}$	0	$2\tau_{14}$	0	$-\tau_{14}$	$-\tau_{14}$	0	0	0	$-\tau_{14}$	$-\tau_{14}$	0	0
0	0	2	−2	0	0	0	0	0	0	1	1	1	1	0	−1	−1	−1	−1	0
0	0	−2	2	0	0	0	0	0	0	1	1	1	1	0	−1	−1	−1	−1	0
0	0	0	0	0	0	0	0	0	0	−1	1	−1	1	0	1	−1	1	−1	0
0	0	0	0	0	0	0	0	0	0	−1	1	1	−1	0	1	−1	−1	1	0
0	0	0	0	0	0	0	0	0	0	−1	−1	0	0	−1	1	1	0	0	1
0	0	0	0	0	0	0	0	0	0	0	0	−1	−1	−1	0	0	1	1	1

Proof. (i) is obvious, since $\chi(E)$ and $\chi(F)$ are rational integers congruent to $\chi(1)$ (mod 2). The others follow by restricting χ to $\mathfrak{P}$ and computing $(\chi,\phi)_{\mathfrak{P}}$, where ϕ is a suitable generalized character of $\mathfrak{P}$. For (ii) take ϕ to be respectively $\zeta_{15}-\zeta_{16}$, $\zeta_9-\zeta_{10}$, $\zeta_{13}-\zeta_{14}$ in the notation of (2.2). For (iii) take ϕ to be $\zeta=\zeta_1-\zeta_5$, which yields $2\chi(E)+\chi(F^2)+2\chi(E^2)+2\chi(EN)+\chi(J)\equiv 0 \pmod 8$. By (ii) $2\chi(F^2)\equiv 2\chi(E^2)$, $2\chi(EN)\equiv 2\chi(J) \pmod 8$. Substitution of these gives the first part of (iii). The second part is proved similarly. For (iv) take $\phi=2\zeta_1-\zeta_9$, and use the congruence $2\chi(EN)\equiv 2\chi(J) \pmod 8$.

Let B be the 1-block of $\mathfrak{G}$. If Y is a representative of a 2-singular section of $\mathfrak{G}$, we assume a basic set $\phi_B=\{\phi_\rho^Y\}$ has been chosen for the 1-block of $\mathfrak{C}(Y)$ such that the corresponding Cartan invariants are as in (4.1), and for $Y=J$, as in (4.2). Denote by f_ρ^Y the degree of ϕ_ρ^Y ; ϕ_1^Y is always the 1-character of $\mathfrak{C}(Y,)$, so that $f_1^Y=1$. In particular, $f_2^{EN}\equiv -3 \pmod 8$ in case II, and $f_2^J\equiv -3$ (mod 32). The corresponding columns of generalized decomposition numbers are then denoted by $\mathfrak{d}_\rho^Y$, except that in the case where $\mathfrak{C}(Y)$ has a normal 2-complement and there is but one such column, we will drop the index 1. For $\chi_i\in B$, $\chi_i(Y)=\sum_\rho(\mathfrak{d}_\rho^Y)_i f_\rho^Y$, where $(\mathfrak{d}_\rho^Y)_i$ is the entry $d_{i\rho}^Y$.

Define the following columns of rational integers, indexed by the $\chi_i\in B$.

$$\mathfrak{a}=\frac{\mathfrak{d}^F-\mathfrak{d}^E}{2},\qquad \mathfrak{w}=\begin{cases}\dfrac{\mathfrak{d}_1^{EN}+\mathfrak{d}_2^{EN}-\mathfrak{d}^{F^2}}{4} & \text{in II},\\[2ex] \dfrac{\mathfrak{d}^{EN}-\mathfrak{d}^{F^2}}{4} & \text{in III},\end{cases}$$

$$\mathfrak{b}=\frac{\mathfrak{d}^F+\mathfrak{d}^E}{2},$$

$$\mathfrak{S}=\frac{\mathfrak{d}_1^J-3\mathfrak{d}_2^J+\mathfrak{d}^{F^2}-2\mathfrak{d}^E}{8},\qquad \mathfrak{u}=\begin{cases}\dfrac{\mathfrak{d}^{F^2}+\mathfrak{d}^{E^2}+2(\mathfrak{d}_1^{EN}+\mathfrak{d}_2^{EN})-2\mathfrak{d}^F-2\mathfrak{d}^E}{8} & \text{in II},\\[2ex] \dfrac{\mathfrak{d}^{F^2}+\mathfrak{d}^{E^2}+2\mathfrak{d}^{EN}-2\mathfrak{d}^F-2\mathfrak{d}^E}{8} & \text{in III}.\end{cases}$$

$$\mathfrak{A}=\frac{\mathfrak{d}^{F^2}-\mathfrak{d}^{E^2}}{4},$$

That these columns are rational integral follows from (5A) and the congruences $f_2^{EN}\equiv -3 \pmod 8$, $f_2^J\equiv -3 \pmod{32}$. The inner products of these columns and the ones indicated are

(5.1)

	$\mathfrak{a}$	$\mathfrak{b}$	$\mathfrak{w}$	$\mathfrak{S}$	$\mathfrak{A}$	$\mathfrak{u}$	$\mathfrak{d}_1^J$	$\mathfrak{d}^{EN}$	$\mathfrak{d}_1^{EN}$	$\mathfrak{d}_2^{EN}$
$\mathfrak{a}$	4	0	0	1	0	0	0	0	0	0
$\mathfrak{b}$	0	4	0	−1	0	−2	0	0	0	0
$\mathfrak{w}$	0	0	3	−1	−2	0	0	4	2	2
$\mathfrak{S}$	1	−1	−1	4	2	1	0	0	0	0
$\mathfrak{A}$	0	0	−2	1	4	0	0	0	0	0
$\mathfrak{u}$	0	−2	0	1	0	3	0	4	2	2

where $\mathfrak{d}^{EN}$ occurs in III; $\mathfrak{d}_1^{EN},\mathfrak{d}_2^{EN}$ in II. By (5A),(iii) and the congruence

$\chi(E) \equiv \chi(E^2)$ (mod 2), it follows that $\mathfrak{A} \equiv \mathfrak{a}$ (mod 2). We note that $\mathfrak{a}$ and $\mathfrak{b}$ are orthogonal to the column $\chi(1)$ of degrees.

We now determine the entries of these columns. $\mathfrak{w}$ and $\mathfrak{u}$ have norm 3, and thus have 3 entries ± 1. $\mathfrak{a}, \mathfrak{b}, \mathfrak{S}, \mathfrak{A}$ have norm 4, and since $\mathfrak{a}\mathfrak{S} = 1$, $\mathfrak{b}\mathfrak{S} = -1$, each of these must have 4 entries ± 1. We can choose five characters χ such that by replacing χ by $-\chi$ if necessary, we have

$\mathfrak{w}$	$\mathfrak{A}$	$\mathfrak{a}$	$\mathfrak{b}$
1	0	0	b_1
1	−1	1	b_2
1	−1	−1	b_3
0	1	1	b_4
0	1	−1	b_5

By orthogonality $b_3 = b_4$, $b_2 = b_5$, $b_1 = -b_2 - b_3$. Suppose $b_i \neq 0$ for some $2 \leqq i \leqq 5$. Then one of $\chi_i(E), \chi_i(F)$ is ± 2, and the other is 0; say that $\chi_i(E) = \pm 2$, $\chi_i(F) = 0$. If χ_i has height 1, then by (5A),(ii) $\chi_i(F^2) \equiv \chi_i(E^2) \equiv \chi_i(EN) \equiv 2 \pmod 4$. The contribution to χ_i from E, F^2, E^2, EN is then 64, which is too large. If χ_i has height $\geqq 2$, then $\chi_i(F^2) \equiv \chi_i(E^2) \equiv 0 \pmod 4$. But (5A),(iii) implies $\chi_i(J) + \chi_i(F^2) \equiv 4 \pmod 8$ and the same congruence with the roles of E and F interchanged implies that $\chi_i(J) + \chi_i(E^2) \equiv 0 \pmod 8$. Thus one of $\chi_i(E^2), \chi_i(F^2)$ is nonzero, and the method of contribution again leads to a contradiction. The argument for the case $\chi_i(E) = 0$, $\chi_i(F) = \pm 2$ is similar. Thus $b_i = 0$ for $2 \leqq i \leqq 5$, and $b_1 = 0$ as well.

We now rearrange the characters in B so that χ_1 is the 1-character and the others are as indicated below:

(5.2)

$\mathfrak{b}$	$\mathfrak{a}$	$\mathfrak{A}$	$\mathfrak{w}$	$\mathfrak{u}$
1	0	0	0	0
1	0	0	0	−1
1	0	0	0	−1
1	0	0	0	0
0	1	−1	1	0
0	−1	−1	1	0
0	1	1	0	0
0	−1	1	0	0
0	0	0	1	0
0	0	0	0	1

The entries of $\mathfrak{u}$ are determined easily from (5.1). Since $\mathfrak{A}$ is known, the method of contributions implies that $(\mathfrak{d}^{E^2}, \mathfrak{d}^{F^2})_i = \pm(1,1)$ for $1 \leqq i \leqq 4$, and $(\mathfrak{d}^{E^2}, \mathfrak{d}^{F^2})_i = \pm(1,-3)$ or $\pm(-3,1)$ for $5 \leqq i \leqq 8$. Furthermore, knowledge of

$\mathfrak{u}$ and $\mathfrak{w}$ imply that $(\mathfrak{d}^{E^2},\mathfrak{d}^{F^2})_i = (-2,-2)$ for $i=9$, and $(2,2)$ for $i=10$. The signs of $(\mathfrak{d}^{E^2},\mathfrak{d}^{F^2})_i$ for $i=2,3,4$ can be determined by the equation $\mathfrak{u}\mathfrak{d}^{E^2}=4$. Since $\mathfrak{d}^{E^2},\mathfrak{d}^{F^2}$ have norm 32, each has exactly two ± 3's for $5 \leqq i \leqq 8$. The products $\mathfrak{w}\mathfrak{d}^{E^2}=0$, $\mathfrak{w}\mathfrak{d}^{F^2}=-8$ determine the signs completely. If the characters are arranged as in (5.2), then

F	E	E^2	F^2
1	1	1	1
1	1	−1	−1
1	1	−1	−1
1	1	1	1
1	−1	1	−3
−1	1	1	−3
1	−1	−3	1
−1	1	−3	1
0	0	−2	−2
0	0	2	2

In particular, χ_9, χ_{10} have height 1, and by (5A)(iii) χ_i has height $\geqq 3$ for $i \geqq 11$; these χ_i then actually have height 3 or 4 by [4].

The columns $\mathfrak{d}^{EN}, \mathfrak{d}_1^{EN}+\mathfrak{d}_2^{EN}$ in cases II, III respectively are now completely determined from $\mathfrak{w}$ and $\mathfrak{d}^{F^2}$. In II the entries of $\mathfrak{d}_1^{EN},\mathfrak{d}_2^{EN}$ can be determined as follows: $(\mathfrak{d}_1^{EN}+\mathfrak{d}_2^{EN})_i = \pm 1$ for $1 \leqq i \leqq 8$, 2 for $i=9,10$. Since $\mathfrak{d}_\rho^{EN}$ has norm 6, the nonzero entries are ± 1. The product $\mathfrak{u}\mathfrak{d}_1^{EN}=2$ shows that one of $(\mathfrak{d}_1^{EN})_2$, $(\mathfrak{d}_1^{EN})_3$ is -1 and the other is 0, since $(\mathfrak{d}_1^{EN})_{10} = (\mathfrak{d}_2^{EN})_{10} = 1$. Since χ_2, χ_3 may be interchanged at this point, we may assume $(\mathfrak{d}_1^{EN})_2 = -1$. From the equality $\mathfrak{b}\mathfrak{d}_1^{EN}=0$, we have $(\mathfrak{d}_1^{EN})_3 = (\mathfrak{d}_1^{EN})_4 = 0$. This completely determines the entries under EN except for the arrangement $(\mathfrak{d}_1^{EN},\mathfrak{d}_2^{EN})_i$ for $5 \leqq i \leqq 8$. Here (1,0) can occur for $i=5$ and 8, or for $i=6$ and 7; the remaining entries are $(0,1)$. This point will be settled below.

Let $s_i = (\mathfrak{S})_i$. Then $\mathfrak{S}\mathfrak{b}=-1$, $\mathfrak{S}\mathfrak{u}=1$ imply that $s_2+s_3+s_4=-1$, $-s_2-s_3+s_{10}=1$, so that $s_4=-s_{10}$. Also $\mathfrak{S}\mathfrak{a}=1$, $\mathfrak{S}\mathfrak{A}=1$ imply that $s_5=s_8$ and $s_7-s_6=1$. In particular, one of the numbers s_6, s_7 is ± 1, and the other is 0. The quadratic form Q^J is $10u^2+2v^2+(u-v)^2$. The method of contribution then implies the following: If χ_i has height 0, then $(\mathfrak{d}_1^J,\mathfrak{d}_2^J)_i$, up to a sign change, can only be (1,0), (01), (1,2), $(-1,2)$, (0,3), and for $5 \leqq i \leqq 8$, only the first three are possible. If χ_i has height 1, then only (1,1), (0,2) are possible; if χ_i has height 3, only (1,3) is possible. If χ_i has height 4, the method of contribution yields a contradiction. Since $s_i = 0$ or ± 1, the definition of $\mathfrak{S}$ forces the conclusion $(\mathfrak{d}_1^J,\mathfrak{d}_2^J)_i = (1,0)$ for $i=5$ and 8; thus $s_5=s_8=0$. Similarly, $(\mathfrak{d}_1^J,\mathfrak{d}_2^J)_i = (0,1)$ or $(-1,-2)$ for $i=6,7$, and the latter can occur only in III by the method of contribution. Thus, if $s_6=-1$, $s_7=0$, then $(\mathfrak{d}_1^J,\mathfrak{d}_2^J)_i = (0,1)$; if

$s_6 = 0$, $s_7 = 1$, then $(\mathfrak{d}_1^J, \mathfrak{d}_2^J)_i = (-1, -2)$. We call these two possibilities IIIa, IIIb respectively. Suppose $s_4 = 0$. Then one of $s_2, s_3 = \pm 1$, and the other is 0. There must then be characters χ_{11}, χ_{12} in B such that $s_{11} \neq 0$, $s_{12} \neq 0$. For $i = 11, 12$, $(\mathfrak{d}_1^J, \mathfrak{d}_2^J)_i = \pm(1,3)$, and since $\mathfrak{d}_1^J$ has norm 6, $(\mathfrak{d}_1^J)_i = 0$ for $i = 9$ or 10. But this contradicts the fact that $\mathfrak{d}_2^J$ has norm 22. Thus $s_4 \neq 0$, and we easily see that $s_4 = -1$, $s_{10} = 1$, $s_2 = s_3 = 0$. s_9 can be found from $\mathfrak{S}\mathfrak{w} = -1$. If $s_9 = 0$, B contains one more character χ_{11} for which $s_{11} = 1$. The matrix of decomposition numbers can now be completed, the uncertainty remaining in II under EN being settled from $\mathfrak{d}_1^J \mathfrak{d}_1^{EN} = 0$. We omit the details.

The complete matrices are in cases II, IIIa, IIIb respectively

(5.3)

F	E	F^2	E^2	EN		J		1		
1	1	1	1	1	0	1	0	1	0	0
1	1	−1	−1	−1	0	0	−1	0	2	1
1	1	−1	−1	0	−1	0	−1	−1	−1	−1
1	1	1	1	0	1	−1	2	0	−1	0
1	−1	−3	1	0	1	1	0	0	−1	0
−1	1	−3	1	1	0	0	1	0	0	1
1	−1	1	−3	1	0	0	1	0	0	1
−1	1	1	−3	0	1	1	0	0	−1	0
0	0	−2	−2	1	1	−1	−1	0	1	−1
0	0	2	2	1	1	0	−2	−1	1	0
0	0	0	0	0	0	−1	−3	1	−2	1

The matrix of Cartan invariants of *B* is

$$\begin{bmatrix} 4 & -2 & 2 \\ -2 & 14 & 0 \\ 2 & 0 & 6 \end{bmatrix}$$

If f_1, f_2, f_3 are the degrees of the corresponding basic set, then $f_1 = 1$, $f_2 \equiv -9$ (mod 32), $f_3 \equiv -11$ **(mod 32).**

(5.4)

F	E	F^2	E^2	EN	J		1			
1	1	1	1	1	1	0	1	0	0	0
1	1	−1	−1	−1	0	−1	0	1	2	1
1	1	−1	−1	−1	0	−1	−1	−1	−1	−1
1	1	1	1	1	−1	2	0	0	−1	0
1	−1	−3	1	1	1	0	0	0	−1	0
−1	1	−3	1	1	0	1	0	0	0	1
1	−1	1	−3	1	0	1	0	0	0	1
−1	1	1	−3	1	1	0	0	0	−1	0
0	0	−2	−2	2	−1	−1	0	0	1	−1
0	0	2	2	2	0	−2	−1	0	1	0
0	0	0	0	0	−1	−3	1	0	−2	1

.

The matrix of Cartan invariants is

$$\begin{bmatrix} 4 & 1 & -2 & 2 \\ 1 & 2 & 3 & 2 \\ -2 & 3 & 14 & 0 \\ 2 & 2 & 0 & 6 \end{bmatrix}$$

$f_1 = 1$, $f_2 \equiv 32 \pmod{64}$, $f_3 \equiv 7 \pmod{64}$, $f_4 \equiv -11 \pmod{32}$.

(5.5)

F	E	F^2	E^2	EN	J		1		
1	1	1	1	1	1	0	1	0	0
1	1	−1	−1	−1	0	−1	0	1	−1
1	1	−1	−1	−1	0	−1	0	−1	−1
1	1	1	1	1	−1	2	−1	0	2
1	−1	−3	1	1	1	0	−1	0	2
−1	1	−3	1	1	−1	−2	0	0	1
1	−1	1	−3	1	−1	−2	0	0	1
−1	1	1	−3	1	1	0	−1	0	2
0	0	−2	−2	2	0	2	1	0	−3
0	0	2	2	2	0	−2	0	0	−2

The Cartan matrix is

$$\begin{bmatrix} 5 & 0 & -9 \\ 0 & 2 & 0 \\ -9 & 0 & 29 \end{bmatrix}.$$

$f_1 = 1$, $f_2 \equiv 0 \pmod{32}$, $f_3 \equiv 29 \pmod{64}$.

6. We now use the methods of [2] to complete the investigations of cases II and III. We recall the following facts: Let $\mathfrak{G}$ be a finite group of even order g. Let p be a prime, and P a p-element of $\mathfrak{G}$. Let $\tilde{\mathfrak{G}}$ be a subgroup of order $\tilde{g}$ such that $\tilde{\mathfrak{G}} \geqq \mathfrak{C}^*(P) = \langle G \in \mathfrak{G} \mid G^{-1}PG = P^{\pm 1} \rangle$. If J_α and J_β are two involutions of $\mathfrak{G}$, and $\{\phi_\rho^P\}$ is a basic set for a p-block of $\mathfrak{C}(P)$, then

$$g \sum \chi_\mu(J_\alpha)\chi_\mu(J_\beta)d_{\mu\rho}^P/x_\mu = \tilde{g}c(J_\alpha)c(J_\beta) \sum h_{\mu\alpha}h_{\mu\beta}\tilde{d}_{\mu\rho}^P/\tilde{x}_\mu. \tag{6.1}$$

Here on the left χ_μ runs over the irreducible characters of the block B of $\mathfrak{G}$ corresponding to the block b of $\mathfrak{C}(P)$, $x_\mu = \chi_\mu(1)$, and $d_{\mu\rho}^P$ are the corresponding generalized decomposition numbers. On the right, $h_{\mu\alpha} = \sum_\lambda \tilde{\chi}_\mu(\tilde{J}_{\alpha,\lambda})/\tilde{c}(\tilde{J}_{\alpha,\lambda})$, $\tilde{\chi}_\mu$ runs over characters of the block $\tilde{B}$ of $\tilde{\mathfrak{G}}$ corresponding to b, $\tilde{x}_\mu = \tilde{\chi}_\mu(1)$, $\tilde{d}_{\mu\rho}^P$ are the corresponding generalized decomposition numbers, and $\{\tilde{J}_{\alpha,\lambda}\}$ ranges over a set of representatives of the conjugate classes of $\tilde{\mathfrak{G}}$ contained in the conjugate class of J_α in $\mathfrak{G}$. For notational convenience, we will let $L(J_\alpha, J_\beta, \mathfrak{d}_\rho^P)$, $R(J_\alpha, J_\beta, \mathfrak{d}_\rho^P)$ denote the left-hand side, the right-hand side of (6.1) respectively.

In the applications to follow, we will always take $p=2$ and $\mathfrak{b}$ the 1-block of $\mathfrak{C}(P)$. It then follows that B and $\tilde{B}$ are the 1-blocks of $\mathfrak{G}$ and $\tilde{\mathfrak{G}}$ respectively. As a second remark, we note that if $\tilde{\mathfrak{G}}$ has a normal 2-complement, then the characters $\tilde{\chi}_\mu$ in the 1-block of $\tilde{\mathfrak{G}}$ can be identified with the irreducible characters θ_μ of an S_2-subgroup $\mathfrak{Q}$ of $\tilde{\mathfrak{G}}$. Moreover, if J_α, J_β, P are in $\mathfrak{Q}$, then $c(J_\alpha)^{-1}c(J_\beta)^{-1}R(J_\alpha,J_\beta,\mathfrak{d}_1^P)$ is equal to

$$\tag{6.2}\begin{aligned}&\tilde{g}\sum_\mu\sum_\lambda\frac{\theta_\mu(\tilde{J}_{\alpha,\lambda})}{\tilde{c}(\tilde{J}_{\alpha,\lambda})}\sum_\nu\frac{\theta_\mu(\tilde{J}_{\beta,\nu})}{\tilde{c}(\tilde{J}_{\beta,\nu})}\frac{\tilde{d}^P_{\mu 1}}{\theta_\mu(1)}\\
&\qquad=\tilde{g}\sum_{\lambda,\nu}\sum_\mu\frac{\theta_\mu(\tilde{J}_{\alpha,\lambda})\theta_\mu(\tilde{J}_{\beta,\nu})\theta_\mu(P)}{\theta_\mu(1)\,c_{\mathfrak{Q}}(\tilde{J}_{\alpha,\lambda})c_{\mathfrak{Q}}(\tilde{J}_{\beta,\nu})}\frac{c_{\mathfrak{Q}}(\tilde{J}_{\alpha,\lambda})c_{\mathfrak{Q}}(\tilde{J}_{\beta,\nu})}{\tilde{c}(\tilde{J}_{\alpha,\lambda})\tilde{c}(\tilde{J}_{\beta,\nu})}\\
&\qquad=(\tilde{\mathfrak{G}}:\mathfrak{Q})\sum_{\lambda,\nu}a_{\mathfrak{Q}}(\tilde{J}_{\alpha,\lambda},\tilde{J}_{\beta,\nu},P)\frac{c_{\mathfrak{Q}}(\tilde{J}_{\alpha,\lambda})c_{\mathfrak{Q}}(\tilde{J}_{\beta,\nu})}{\tilde{c}(\tilde{J}_{\alpha,\lambda})\tilde{c}(\tilde{J}_{\beta,\nu})}\end{aligned}$$

where $a_{\mathfrak{Q}}(\tilde{J}_{\alpha,\lambda},\tilde{J}_{\beta,\nu},P)$ is the multiplicity of the class sum of P in $\mathfrak{Q}$ in the product of the class sums of $\tilde{J}_{\alpha,\lambda}$ and $\tilde{J}_{\beta,\nu}$. We assume in this that the elements $\tilde{J}_{\alpha,\lambda}$ and $\tilde{J}_{\beta,\nu}\in\mathfrak{Q}$. Finally, we note that (6.1) can be added over linear combinations of the $d^P_{\mu\rho}$ for varying $\tilde{\mathfrak{G}},P$, and ρ. If $\mathfrak{y}$ is such a combination of the $\mathfrak{d}^P_\rho$ we understand by the equality $L(J_\alpha,J_\beta,\mathfrak{y})=R(J_\alpha,J_\beta,\mathfrak{y})$ the result obtained by first applying (6.1) to the various $\mathfrak{d}^P_\rho$, and then summing appropriately. We now turn to the situation at the end of §5.

Case III. Set $\beta=f_2^J$, $\beta\equiv-3\pmod{32}$. If we set $\tilde{\mathfrak{G}}=\mathfrak{C}^*(F)$ and $\mathfrak{C}^*(E)$, and compute $L(J,J,\mathfrak{b})=R(J,J,\mathfrak{b})$, and $L(J,EN,\mathfrak{b})=R(J,EN,\mathfrak{b})$, we find that

$$\tag{6.3}\begin{aligned}1+\frac{\beta^2}{x_2}+\frac{\beta^2}{x_3}+\frac{(2\beta-1)^2}{x_4}&=0,\\
1+\frac{\beta}{x_2}+\frac{\beta}{x_3}+\frac{2\beta-1}{x_4}&=0.\end{aligned}$$

Here the right-hand sides are computed by observing that $\mathfrak{C}^*(E)$, $\mathfrak{C}^*(F)$ have normal 2-complements, and that E,F do not occur in III in the appropriate product of involutions. Subtracting the second from the first, and cancelling $\beta-1$, which is permissible since $\beta\equiv-3\pmod{32}$, we have

$$\tag{6.4}\frac{\beta}{x_2}+\frac{\beta}{x_3}+2\frac{2\beta-1}{x_4}=0.$$

Substituting (6.4) into (6.3) then yields $1-(2\beta-1)/x_4=0$, or $x_4=2\beta-1$. Thus J is in the kernel of χ_4. By (6.3) $\beta/x_2+\beta/x_3+2=0$. Since $\chi_2(J)=\chi_3(J)=-\beta$, and $|\beta|\leqq|x_2|,|x_3|$, we see that $x_2=x_3=-\beta$, and hence J belongs to the kernel of χ_i for $1\leqq i\leqq 4$.

In IIIb, $x_2=f_2-f_3$, $x_3=-f_2-f_3$; thus $f_2=0,f_3=\beta$. But $\chi_6(J)=-1-2\beta$, and $x_6=\beta$. Since $|1+2\beta|\leqq|\beta|$, we must have $\beta=-1$, which contradicts the congruence $\beta \equiv -3 \pmod{32}$. Thus IIIb is impossible. We henceforth assume we have IIIa. In this case, $\chi_4(J) = x_4$ implies that $f_3 = 1 - 2\beta$.

If $\tilde{\mathfrak{G}} = \mathfrak{C}^*(EN), \mathfrak{C}^*(F^2)$ respectively, and we compute $L(J,J,\mathfrak{w}) = R(J,J,\mathfrak{w})$, then

$$\frac{1}{x_5} + \frac{\beta^2}{x_6} + \frac{(1+\beta)^2}{x_9} = 0. \tag{6.5}$$

Here the right-hand side is computed by noting that $\mathfrak{C}^*(EN), \mathfrak{C}^*(F^2)$ have normal 2-complements, and by using (6.2). If we substitute $x_5 = -f_3, x_6 = f_4, x_9 = f_3 - f_4$ into (6.5) and simplify, we obtain the equality $f_4 + \beta f_3 = 0$, so that $f_4 = 2\beta^2 - \beta$. The equality $\chi_2(J) = x_2$ implies $f_2 = -2(\beta-1)^2$. Summarizing, we have

$$f_1 = 1,\ f_2 = -2(\beta-1)^2,\ f_3 = 1 - 2\beta,\ f_4 = \beta(2\beta - 1).$$

Let $\mathfrak{K} = \bigcap$ kernel of χ_i for $i = 1,2,3,4$. Since $J \in \mathfrak{K}, \mathfrak{K}$ must contain the subgroup $\mathfrak{A} = \langle J, X, EF^3 \rangle$ of $\mathfrak{P}$ generated by the elements of $\mathfrak{P}$ conjugate in $\mathfrak{G}$ to J. $\mathfrak{A}$ is abelian of type (2,2,2). Since $F^2, N \notin \mathfrak{K}$, $\mathfrak{A}$ is an S_2-subgroup of $\mathfrak{K}$. Assume now that $\mathfrak{G}$ has no normal 2′-subgroups $\neq 1$. Let $\mathfrak{N}$ be a minimal normal subgroup of $\mathfrak{G}$ contained in $\mathfrak{K}$. Since $|\mathfrak{N}|$ is even, $\mathfrak{A} \leqq \mathfrak{N}$.

Suppose $\mathfrak{A} \neq \mathfrak{N}$. Since $\mathfrak{N}$ is characteristically simple and $\mathfrak{A}$ is an S_2-subgroup of $\mathfrak{N}$, it follows that $\mathfrak{N}$ is simple. Hence $\mathfrak{W} = \mathfrak{N}_{\mathfrak{N}}(\mathfrak{A})/\mathfrak{C}_{\mathfrak{N}}(\mathfrak{A})$ has order $7 \cdot 3^x$ with $x = 0$ or 1. Clearly $\mathfrak{P}$ stabilizes $\mathfrak{W}$, so we may view $\mathfrak{P}/\mathfrak{A}$ as a group of operators of $\mathfrak{W}$. If $P \in \mathfrak{P}$ and $P\mathfrak{A}$ centralizes $\mathfrak{W}$, then $\mathfrak{W}$ leaves $\mathfrak{A} \cap \mathfrak{C}(P)$ invariant. Since $\mathfrak{A}$ is an irreducible $\mathfrak{W}$-group, it follows that $\mathfrak{A} \leqq \mathfrak{C}(P)$, so $P \in \mathfrak{P} \cap \mathfrak{C}(\mathfrak{A}) = \mathfrak{A}$. Thus $\mathfrak{P}/\mathfrak{A}$ is faithfully represented on $\mathfrak{W}$. This is impossible, since $|\mathfrak{P} : \mathfrak{A}| = 8$ and $|\mathfrak{W}|$ divides $3 \cdot 7$. Thus $\mathfrak{A} = \mathfrak{N}$.

Since $\mathfrak{A}$ is a self-centralizing normal subgroup of $\mathfrak{P}$, it follows that $\mathfrak{C}(\mathfrak{A}) = \mathfrak{A} \times \mathfrak{D}$, where $\mathfrak{D}$ has odd order. Hence $\mathfrak{D}$ char $\mathfrak{C}(\mathfrak{A}) \lhd \mathfrak{G}$, so $\mathfrak{D} = 1$. Thus $\mathfrak{G}/\mathfrak{A}$ is isomorphic to a subgroup of $GL(3,2)$ of order $\equiv 0 \pmod 8$ which has no subgroup of index 2. Hence $\mathfrak{G}/\mathfrak{A} \simeq GL(3,2)$. The group $\mathfrak{G}$ exists and is unique by [11].

Case II. Set $\alpha = f_2^{EN}$, $\beta = f_2^J$; $\alpha \equiv -3 \pmod 8$, $\beta \equiv -3 \pmod{32}$. If we take $\tilde{\mathfrak{G}} = \mathfrak{C}^*(E)$, $\mathfrak{C}^*(F)$ and compute $L(J,J,\mathfrak{b}) = R(J,J,\mathfrak{b})$, and $L(EN,EN,\mathfrak{b}) = R(EN,EN,\mathfrak{b})$, we find that

$$1 + \frac{\beta^2}{x_2} + \frac{\beta^2}{x_3} + \frac{(2\beta-1)^2}{x_4} = 0,$$

$$1 + \frac{1}{x_2} + \frac{\alpha^2}{x_3} + \frac{\alpha^2}{x_4} = 0.$$

The right-hand sides are 0 by observing that $\mathfrak{C}^*(F)$, $\mathfrak{C}^*(E)$ have normal 2-complements, and that in II, E, F do not occur in the appropriate products of involutions. Expressing the degrees x_i in terms of $1 = f_1, f_2, f_3$ and simplifying, we get

$$\alpha^2 - \frac{f_2(1+f_2+f_3)}{2f_2+f_3} = 0, \tag{6.6}$$

$$\beta^2(f_2-\alpha^2+4(1+f_2+f_3)) - 4\beta(1+f_2+f_3) - (1+f_2+f_3)(f_2-1) = 0, \tag{6.7}$$

(6.6) and (6.7) can be used to eliminate f_3; a little manipulation then gives the equation

$$f_2^2(\beta^2-\alpha^2) + 2\alpha^2 f_2(\beta-1)^2 + \alpha^2(\alpha^2\beta^2-(2\beta-1)^2) = 0. \tag{6.8}$$

If $\alpha^2 \neq \beta^2$, then the above equation factors into $(\beta^2-\alpha^2)(f_2-\lambda)(f_2-\mu)=0$, where

$$\lambda = \frac{\alpha-2\alpha\beta-\alpha^2\beta}{\beta+\alpha}, \quad \mu = \frac{-\alpha+2\alpha\beta-\alpha^2\beta}{\beta-\alpha} \tag{6.9}$$

If $\alpha^2 = \beta^2$, then $\alpha = \beta$ from the congruences $\alpha \equiv -3 \pmod 8$, $\beta \equiv -3 \pmod{32}$. In this case, $f_2 = -(\alpha^2+2\alpha-1)/2$.

If we take $\tilde{\mathfrak{G}} = \mathfrak{C}^*(F^2), \mathfrak{C}^*(EN)$, and compute $L(J,J,\mathfrak{w}) = R(J,J,\mathfrak{w})$, we have

$$\frac{g}{c(J)^2}\left[\frac{(1+\beta)^2}{f_2-f_3} - \frac{1}{f_2} + \frac{\beta^2}{f_3}\right] = \frac{-c(F^2)}{c(F^2,N)^2},$$

or

$$g = -c(J)^2 \frac{c(F^2)}{c(F^2,N)^2}\frac{f_2f_3(f_2-f_3)}{(\beta f_2+f_3)^2}. \tag{6.10}$$

Apply now the methods at the beginning of this section to the situation $\mathfrak{G} = \mathfrak{C}(J)$, $\tilde{\mathfrak{G}} = \mathfrak{C}^*(F^2)$, and $L(X,X,\mathfrak{d}^{F^2}) = R(X,X,\mathfrak{d}^{F^2})$. Using (4.7), we see that $L(X,X,\mathfrak{d}^{F^2}) = 2c(J)(\beta+3)^2/\beta(\beta+1)$. On the other hand, in $\mathfrak{C}(J)$ X is fused only to EF, and $(\mathrm{ccl}(X)\cup \mathrm{ccl}(EF))^2$ does not involve F^2. Thus $R(X,X,\mathfrak{d}^{F^2})=0$. and we must have $\beta=-3$. This already shows that $\alpha \neq \beta$; otherwise, $f_2=-1$, whereas $f_2 \equiv -9 \pmod{32}$. Hence $f_2=\lambda$ or μ of (6.9). If $f_2=\lambda$, then $\alpha-2\alpha\beta-\alpha^2\beta \equiv 0 \pmod{\beta+\alpha}$, and $\alpha+\beta$ divides $\beta(1-\beta)^2$. If $f_2=\mu$, then $-\alpha+2\alpha\beta-\alpha^2\beta \equiv 0 \pmod{\beta-\alpha}$, and $\alpha-\beta$ divides $\beta(1-\beta)^2$. In either case, $\alpha \pm 3$ divides 48. Only a small number of possibilities can occur for α, and since $f_2 \equiv -9, f_3 \equiv -11 \pmod{32}$, a check of these shows that only the following solution exists:

$$\alpha = 5, \quad \beta=-3, \quad f_2=55, \quad f_3=-11.$$

The degrees of the characters in B are then 1, 55, 55, 55, 11, 11, 99, 45, 66, 54, 120. In particular, $\mathfrak{G}$ has a rational character χ_6 of degree 11 faithful on $\mathfrak{P}$. Assume now that $\mathfrak{G}$ has no normal $2'$-subgroups $\neq 1$. Then by a theorem of Schur's, the order g of $\mathfrak{G}$ divides $2^6 \cdot 3^6 \cdot 5^2 \cdot 7 \cdot 11$.

The characters in the 1-block of $\mathfrak{C}(J)/\langle J\rangle$ have degrees 1, 1, 3, 3, 3, 3, 3, 3, 6, 2 by (4.7). Let $\theta = \theta_j$, for some j where $3 \leq j \leq 8$. If $\mathfrak{R}/\langle J\rangle$ is the kernel of θ, then for a suitable choice of j, the only nontrivial elements of $\mathfrak{P}/\langle J\rangle$ in $\mathfrak{R}/\langle J\rangle$ are the elements in the class of e^2. Since θ is a rational character, $(\mathfrak{C}(J):\mathfrak{R})$ is of the form 2^s3^t, so that $\mathfrak{C}(J)/\mathfrak{K}$ is solvable. If we now choose θ so that the only nontrivial elements of $\mathfrak{P}/\langle J\rangle$ in the kernel $\mathfrak{K}_1/\langle J\rangle$ of θ are the elements in the class of f^2, then a similar argument shows that $\mathfrak{C}(J)/\mathfrak{K}_1$ is solvable of order $2^{s'}3^{t'}$. But $\mathfrak{K}\cap\mathfrak{K}_1\cap\mathfrak{P}=\langle J\rangle$. It now follows that if $\mathfrak{M}$ is the maximal normal $2'$-subgroup of $\mathfrak{C}(J)$, then $\mathfrak{C}(J)/\mathfrak{M}$ is solvable, and its order is of the form 2^a3^b. Since $\mathfrak{C}(J)/\mathfrak{M}$ has only one block for the prime 2 by [5], the order of $\mathfrak{C}(J)/\mathfrak{M}$ is then 192. Set $m=|\mathfrak{M}|$.

The values of α, β, f_2, f_3, and (6.10) show that the order of $\mathfrak{G}$ is

$$g = 2^5\cdot 3^3\cdot 5\cdot 11m\frac{m}{c(F^2,N)}\,\frac{c(F^2)}{c(F^2,N)}. \tag{6.11}$$

We note that $c(F^2,N)\,|\,c(F^2)$, and $c(F^2,N)\,|\,64m$, the latter being true, since no 3-element of $\mathfrak{C}(J)/\mathfrak{M}$ commutes with N. In particular, $3^4\nmid m$ since $g\,|\,2^6\cdot 3^6\cdot 5^2\cdot 7\cdot 11$.

Suppose p divides m, where $p=7$ or 5. Since $7^2\nmid g$, $5^3\nmid g$, (6.11) implies that $p\,|\,c(F^2,N)$. Thus there exists an element Y in $\mathfrak{M}\cap\mathfrak{C}(F^2)$ of order p. In particular, by (5.3) $\chi_6(JY)=\chi_6(F^2Y)=3$ and so $\chi_6(Y)\equiv 3 \pmod 4$. Since $\chi_6(Y)$ can assume only the value 4 for $p=7$, and the values 6 or 1 for $p=5$, this is impossible. Thus $m\,|\,27$ and $g\,|\,2^6\cdot 3^6\cdot 5\cdot 11$.

Let S be an element in $\mathfrak{G}$ of order 11, and let $\mathfrak{S}=\langle S\rangle$. If $\mathfrak{N}(\mathfrak{S})=\mathfrak{C}(\mathfrak{S})$, then $\mathfrak{G}$ would have a normal 11-complement. In particular, $\mathfrak{S}$ would normalize an S_2-subgroup $\mathfrak{P}_1$ of $\mathfrak{G}$ and hence centralize the central involution of $\mathfrak{P}_1$, which is impossible. Thus $\mathfrak{N}(\mathfrak{S})\neq\mathfrak{C}(\mathfrak{S})$. The order of $\mathfrak{N}(\mathfrak{S})/\mathfrak{C}(\mathfrak{S})$ is necessarily a divisor $\neq 1$ of 10. The index $(\mathfrak{G}:\mathfrak{N}(\mathfrak{S}))$ is then of the form $32\cdot 5\cdot 3^\alpha$, $32\cdot 3^\alpha$, or $64\cdot 3^\alpha$, where $0\leq\alpha\leq 6$. By Sylow's theorem, this index is $\equiv 1 \pmod{11}$, and thus must be $64\cdot 3^3$. If $m\neq 1$, then there would exist an element T of order 3 in $\mathfrak{C}(S)$. Since $\chi_6(ST)\equiv\chi_6(T)$ (mod 11) and $\chi_6(ST)=0$, we would have $\chi_6(T)\equiv 0$ (mod 11) and this is impossible. Thus $m=1$.

We have now shown that $g=2^6\cdot 3^3\cdot 5\cdot 11=95{,}040$. Suppose $\mathfrak{G}$ has proper normal subgroups; among these choose $\mathfrak{N}$ minimal. If $\mathfrak{G}/\mathfrak{N}$ has even order, then there would exist an irreducible character $\neq 1$ in the 1-block of $\mathfrak{G}/\mathfrak{N}$ which would then belong to the 1-block B of $\mathfrak{G}$. But this is impossible, since the nontrivial characters in B are faithful. If $\mathfrak{G}/\mathfrak{N}$ has odd order, then $\mathfrak{P}$ is an S_2-subgroup of $\mathfrak{N}$. If $\mathfrak{N}$ has no subgroups of index 2, then $\mathfrak{N}$ has necessarily two classes of involutions, and the preceding work shows that $\mathfrak{N}$ has order 95,040. Hence $\mathfrak{N}$ is solvable, and $\mathfrak{N}=\mathfrak{P}$. which is impossible. Thus $\mathfrak{C}$ is simple. By a theorem of Stanton [10], $\mathfrak{G}$ must be the Mathieu group $\mathfrak{M}_{12}$. Summarizing the results of this paper, we have

THEOREM (6A). *Let $\mathfrak{G}$ be a finite group of order $64g'$, where g' is odd. Suppose there is an element F of order 8 in $\mathfrak{G}$ such that F is self-centralizing in some S_2-subgroup $\mathfrak{P}$, and F is conjugate to its odd powers in $\mathfrak{P}$. Then one of the following possibilities hold:*

(a) $\mathfrak{G}$ *has a subgroup of index* 2.

(b) $\mathfrak{G}$ *has one class of involutions.*

(c) *If* $O_{2'}(\mathfrak{G})$ *is the maximal normal subgroup of* $\mathfrak{G}$ *of odd order, then* $\mathfrak{G}/O_{2'}(\mathfrak{G}) \simeq \mathfrak{G}_{1344}$ *or* $\mathfrak{M}_{12}$, *where* $\mathfrak{G}_{1344}$ *is a uniquely specified nonsimple, nonsolvable group of order* 1344, *and* $\mathfrak{M}_{12}$ *is the Mathieu group on* 12 *symbols.*

COROLLARY (6B). *The only simple group satisfying the assumptions of* (6A) *and having more than one class of involutions is* $\mathfrak{M}_{12}$.

REFERENCES

1. R. Brauer, *A characterization of the characters of groups of finite order*, Ann. of Math. (2) **57** (1953), 357–377.
2. ———, *Investigations on groups of even order*. I, Proc. Nat. Acad. Sci. U. S. A. **47** (1961), 1891–1893.
3. ———, *Some applications of the theory of blocks of characters of finite groups*. I, J. Algebra **1** (1964), 152–167.
4. R. Brauer and W. Feit, *On the number of irreducible characters of finite groups in a given block*, Proc. Nat. Acad. Sci. U. S. A. **45** (1959), 361–365.
5. P. Fong, *On the characters of p-solvable groups*, Trans. Amer. Math. Soc. **98** (1961). 263–284.
6. G. Frobenius, *Über die Charaktere der mehrfach transitiven Gruppen*, S-B. Preuss. Akad. Wiss. (Berlin) (1904), 558–571.
7. M. Hall and J. Senior, *The groups of order* 2^n ($n \leqq 6$), Macmillan, New York, 1964.
8. P. Hall and G. Higman, *On the p-length of p-soluble groups*, Proc. London Math. Soc. **6** (1956), 1–42.
9. I. Schur, *Über eine Klasse von endlichen Gruppen linearer Substitutionen*, S-B. Preuss. Akad Wiss. (Berlin) (1905), 77–91.
10. R. Stanton, *The Mathieu groups*, Canad. J. Math. **3** (1951), 164–174.
11. W. J. Wong, *A characterization of the Mathieu group* M_{12}, Math. Z. **84** (1964), 378–388.
12. H. Zassenhaus, *The theory of groups*, Chelsea, New York, 1949.

HARVARD UNIVERSITY,
CAMBRIDGE, MASSACHUSETTS
UNIVERSITY OF CALIFORNIA,
BERKELEY, CALIFORNIA

An analogue of Jordan's theorem in characteristic p

By Richard BRAUER* and Walter FEIT**

1. Introduction

Let C be the field of complex numbers. A well-known result of C. Jordan [10] asserts the existence of an integer-valued function $J(n)$ defined on the set of positive integers such that every finite subgroup G of $\mathrm{GL}_n(C)$ contains a normal abelian subgroup A with $|G:A| < J(n)$. Several proofs of this theorem are known, see the references in [6, (36.13)]. Since a representation of a finite group in an algebraically closed field K of characteristic 0 is equivalent to a representation in an absolutely algebraic subfield K_0 of K, Jordan's theorem remains true with the same J, if C is replaced by any field of characteristic 0. On the other hand, the analogous result is false for fields of characteristic $p > 0$. For example, let K be an algebraically closed field of characteristic p and set $G_m = \mathrm{SL}_2(p^m)$. Each G_m is a subgroup of $\mathrm{GL}_2(K)$ and a normal abelian subgroup of G_m has order at most 2 while the order $|G_m|$ of G_m is arbitrarily large.

In this paper the following analogue of Jordan's theorem for fields of characteristic p is proved.

THEOREM. *Let p be a prime. There exists an integer-valued function $f(m, n) = f_p(m, n)$ such that, if K is a field of characteristic p and if G is a finite subgroup of $\mathrm{GL}_n(K)$ whose Sylow-p-subgroup has order at most p^m, then G has a normal abelian subgroup A with $|G:A| < f(m, n)$.*

This theorem had been conjectured by O. H. Kegel in a letter to us. Of course, it will suffice to prove the theorem under the additional assumption that K is algebraically closed.

Notation

Let G be a finite group, and B a non-empty subset of G. Then $\mathrm{C}_G(B)$, $\mathrm{N}_G(B)$, and $|B|$ denote respectively the centralizer of B in G, the normalizer of B in G, and the cardinality of B. If H is a subgroup of G, $|G:H|$ is the index of H in G. We write $H \lhd G$ to indicate that H is a normal subgroup of G.

* The work of the first author was partially supported by the U. S. Air Force Contract AF 49 (638)-1381 monitored by the AF Office of Scientific Research of the Air Research and Development Command.

** The work of the second author was partially supported by the U. S. Army Contract DA-31-124-ARO-D-336.

If p is a prime, a p-element is an element whose order is a power of p while a p'-element is an element of an order relatively prime to p. A group is a p-group or p'-group respectively, if all its elements are p-elements or p'-elements.

Throughout the paper, p will be a fixed prime, and K an algebraically closed field of characteristic p. All groups are assumed to be finite. The group algebra of G over K is denoted by $K[G]$ with K considered as subfield of $K[G]$. All modules are assumed to be finitely generated right unital modules. The representation of G associated with a $K[G]$-module is usually denoted by the corresponding German letter. An *invariant element* z of a $K[G]$-module L is an element $z \in L$ for which $zx = z$ for all $x \in G$. For any group G, $L_0(G)$ denotes the trivial $K[G]$-module with $\dim_K L_0(G) = 1$. If $L_1, L_2, \cdots, L_m$ are $K[G]$-modules,

$$L_1 \oplus L_2 \oplus \cdots \oplus L_m = \oplus \sum_{i=1}^{m} L_i$$

is the direct sum of the L_i. If all L_i are equal, we write mL_1 for this sum. L_1^* is the $K[G]$-module contragredient to L_1. We say that L_1 is a *constituent* of L_2, if L_1 is a quotient module of a submodule of L_2. The notation $L_1 \mid L_2$ means that L_1 is isomorphic to a direct summand of L_2. The *kernel* of L_1 is the kernel of the associated representation $\mathfrak{L}_1$. Two modules are said to be *distinct*, if they are non-isomorphic.

If H is a subgroup of G, and M a $K[H]$-module, M^G denotes the $K[G]$-module induced by M. If $x \in G$, then M^x is the $K[x^{-1}Hx]$-module consisting of the same elements as M with the operations defined in the obvious manner. Finally, if L is a $K[G]$-module, L_H is the $K[H]$-module obtained from L by restricting the operators to $K[H]$.

2. Preliminaries

For later reference, we next list a number of well-known results. The first six hold for fields of arbitrary characteristic.

(2.1) Let A and B be $K[G]$-modules, and let $\mathfrak{A}$ and $\mathfrak{B}$ respectively denote the corresponding representations. Then $A^* \otimes B$ is isomorphic to the $K[G]$-module W obtained in the following manner: W is isomorphic to $\mathrm{Hom}_K(A, B)$ as a K-space and if $w \in W$, $x \in G$, then

$$wx = \mathfrak{A}(x^{-1})w\mathfrak{B}(x) \text{ .}$$ [1]

For our purposes, we even could have defined $A^* \otimes B$ as the module W. Clearly,

(2.2) The invariant elements of $A^* \otimes B$ in (2.1) correspond to the elements of $\mathrm{Hom}_{K[G]}(A, B)$ in W.

[1] Since we work with right modules, we write $\eta\zeta$ for the transformation obtained by performing first a transformation η and then a transformation ζ; $\eta\zeta = \zeta\circ\eta$.

The next two results are due to G. W. Mackey [11], cf. [6, § 44].

(2.3) Let H be a subgroup of G and let M be a $K[H]$-module. Then

$$(M^G)_H \approx \bigoplus\sum_x ((M^x)_{H\cap x^{-1}Hx})^H$$

where x ranges over a complete system of (H, H)-double coset representatives in G.

(2.4) Let H be a subgroup of G. Let L be a $K[G]$-module and M a $K[H]$-module. Then

$$M^G \otimes L \approx (M \otimes L_H)^G .$$

(2.5) (Clifford's Theorem, cf. [6, (49.2)]). *If $H \lhd G$, and if L is an irreducible $K[G]$-module, then*

$$L_H \approx \bigoplus\sum_x M^x$$

where M is an irreducible $K[H]$-module, and x ranges over a subset of a complete system of coset representatives of H in G.

We note a corollary.

(2.6) *If $M \approx L_0(H)$ in (2.5), then H is included in the kernel of L.*

From now on, it will be essential that K has characteristic p. If L is an indecomposable $K[G]$-module, and if H is a minimal subgroup of G such that $L \mid S^G$ for some $K[H]$-module S, H is called a *vertex* of L. J. A. Green [7], using results of D. G. Higman, has shown:

(2.7) *If H is a vertex of the $K[G]$-module L, then H is a p-group which is uniquely determined up to conjugation. A subgroup H_1 of G contains a conjugate of H, if and only if $L \mid (L_{H_1})^G$.*

We define the modular character λ of a K-representation $\mathfrak{L}$ of G as the function on G whose value $\lambda(x)$ is the trace of $\mathfrak{L}(x)$ for $x \in G$.[2] We say that a module affords a character λ, if λ is the character of the associated representation.

(2.8) ([5, § 6]; [2, § 3]; or [6, § 82]). Let L_1 and L_2 be two irreducible $K[G]$-modules. Then L_1 and L_2 are isomorphic, if and only if they afford the same character.

Let Ω be a fixed algebraic number field normal of finite degree over the field Q of rational numbers and chosen such that Ω contains the $|G|^{\text{th}}$ roots of unity, and that each C-representation of a subgroup of G is equivalent to an Ω-representation.[3] We choose a fixed extension ν of the p-adic exponential valuation of Q to Ω. Throughout the paper, R will be the ring of local integers of Ω with

[2] In [2] and [5], the term modular character is used in a different sense.

[3] As is well known, our conditions are satisfied if Ω is the field of $|G|^{\text{th}}$ roots of unity over Q. This is immaterial for our purposes.

regard to ν, $\mathfrak{p}$ will be the prime ideal of R, and θ will be the residue class map of R onto $R/\mathfrak{p} = R^\theta$. Moreover, we identify R^θ with a subfield of K.

If G is a p'-group, every K-representation of G is completely reducible and every K-representation of G is equivalent to a representation obtained from an R-representation of G by the residue class map θ. Thus, Jordan's theorem implies:

(2.9) There exists an integer-valued function $J(n)$ defined on the set of positive integers such that if G is a finite p'-subgroup of $\mathrm{GL}_n(K)$, there exists a normal abelian subgroup A of G with $|G:A| < J(n)$.

(2.10) [1, p. 101] If φ is a modular irreducible character of G and if K_1 is a subfield of K such that $\varphi(x) \in K_1$ for all p'-elements x of G, then φ is afforded by some $K_1[G]$-module.

(2.11) [7, Th. 10]. Let L be an indecomposable $K[G]$-module which affords the modular character λ. If the vertex of L is properly included in a Sylow-p-subgroup P of G, then $\lambda(x) = 0$ for all p'-elements x of $C_G(P)$. In particular, $\lambda(1) = 0$, i.e., $\dim_K L$ is divisible by p.

For basic results in the theory of blocks, the reader is referred to [2] or [6, Ch. XII]. By definition, a p-block B is a set of ordinary and modular irreducible characters. We shall also speak of the irreducible modules and representations in B. A reducible module is said to belong to B, if all its irreducible constituents belong to B. Every indecomposable $K[G]$-module belongs to some block, cf. [7]. The p-block of G containing $L_0(G)$ will be denoted by $\dot{B}_0(G)$.

(2.12) [4]. Let P be a Sylow-p-subgroup of G. Each p-block contains at most $|P|^2$ modular irreducible characters.

(2.13) [5, p. 578]. Let F_i and F_j be irreducible $K[G]$-modules which lie in the same block. If $\dim_K F_i \not\equiv 0$, $\dim_K F_j \not\equiv 0 \pmod p$, then some irreducible constituent of $F_i^* \otimes F_j$ is in $B_0(G)$.

The remainer of this section is concerned with some simple lemmas which are needed for the proof of the main theorem.

(2.14) *Let L be a $K[G]$-module such that every irreducible constituent of L is equivalent to $L_0(G)$. If H is the kernel of L, then G/H is a p-group.*

PROOF. It suffices to assume that L belongs to a faithful representation $\mathfrak{L}$. By assumption, the characteristic roots of $\mathfrak{L}(x)$ are 1 for every $x \in G$. Hence $\mathfrak{L}(x) - I$ is nilpotent and so

$$\mathfrak{L}(x^{p^a}) - I = \mathfrak{L}(x)^{p^a} - I = (\mathfrak{L}(x) - I)^{p^a} = 0$$

for sufficiently large integers a. Since $\mathfrak{L}$ is faithful, each x is a p-element, and this is what we have to show.

(2.15) *Let G be a finite group of the form $P \times D$ where D is a p'-group. Let L be an indecomposable $K[G]$-module. There exists an indecomposable $K[P]$-module U and an irreducible $K[D]$-module V such that $L \approx U \otimes V$.*[4]

PROOF. Let $\mathfrak{L}$ be the representation corresponding to L. The restriction $\mathfrak{L} \mid D$ of $\mathfrak{L}$ to the p'-group D is completely reducible. For $y \in P$, $\mathfrak{L}(y)$ commutes with all $\mathfrak{L}(z)$, $z \in D$. If $\mathfrak{L} \mid D$ had two non-equivalent irreducible constituents, it would follow easily from Schur's lemma that L was decomposable. Hence all irreducible constituents of $\mathfrak{L} \mid D$ are equivalent. Introducing suitable coordinates and writing representations in matrix form we can set

$$\mathfrak{L}(z) = I_u \otimes \mathfrak{V}(z) \qquad \text{(for } z \in D) ,$$

where I_u is the unit matrix of some degree u, and where $\mathfrak{V}$ is an irreducible representation of D. Again by Schur's lemma, each $\mathfrak{L}(y)$ with $y \in P$ has the form $\mathfrak{U}(y) \otimes \mathfrak{V}(1)$ where $\mathfrak{U}(y)$ is a matrix of degree u. Clearly, $y \to \mathfrak{U}(y)$ is a representation $\mathfrak{U}$ of P. Then $\mathfrak{L}(yx) = \mathfrak{U}(y) \otimes \mathfrak{V}(z)$. If U and V are the representation modules of $\mathfrak{U}$ and $\mathfrak{V}$ respectively, we have $L \approx U \otimes V$. If U was decomposable, L would be decomposable.

(2.16) *Let N be a group which has a normal subgroup $P \times D$ such that P is a Sylow-p-subgroup of N. If L is an indecomposable $K[N]$-module, then $L_{P \times D}$ is a direct sum of the form*

$$L_{P \times D} \approx \oplus \sum_{i=1}^{e} U_i \otimes V_i ,$$

where the V_i are distinct irreducible $K[D]$-modules, the U_i are $K[P]$-modules, and $U_1 \otimes V_1, \cdots, U_e \otimes V_e$ are conjugate under the action of N. If $L_0(P) \otimes W \mid L_{P \times D}$ for some $K[D]$-module W, L is irreducible. If L is irreducible, then P is included in the kernel of L, $\dim_K L \not\equiv 0 \pmod p$, and P is a vertex of L.

PROOF. Clearly, D is a p'-group. Since $P \times D$ contains a vertex of L, there exists an indecomposable direct summand S of $L_{P \times D}$ such that $L \mid S^N$, cf. (2.7). By (2.15), S has the form $S \approx U \otimes V$ where U, V are as in (2.15). As $L_{P \times D} \mid (S^N)_{P \times D}$, it follows from (2.3) that $L_{P \times D}$ is a direct sum of modules $U^x \otimes V^x$ where U is an indecomposable $K[P]$-module, V an irreducible $K[D]$-module, and x ranges over a set of elements of N. Let $V_1, \cdots, V_e$ denote a full set of non-equivalent modules occuring among the V^x here. Let U_i denote the direct sum of the modules U^x for which $V^x \approx V_i$. Then

$$L_{P \times D} \approx \oplus \sum_{i=1}^{e} U_i \otimes V_i .$$

Since $L_{P \times D}^x \approx L_{P \times D}$ for $x \in N$, it follows easily from the Krull-Schmidt theorem

[4] Of course, U is identified with a $K[G]$-module whose kernel includes D, and V is identified with a $K[G]$-module whose kernel includes P.

that $U_1 \otimes V_1, \cdots, U_e \otimes V_e$ are conjugate under the action of N. If $L_0(P) \otimes W \mid L_{P\times D}$ for some indecomposable $K[D]$-module W, it follows from the Krull-Schmidt theorem that $L_0(P) \otimes W \mid U_i \otimes V_i$ for some i. Hence $L_0(P) \mid U_i$, and we see that $U_i \approx uL_0(P)$ for some positive integer u. This must then hold for all i. Thus, P is included in the kernel of L. Then L can be considered as an indecomposable $K[N/P]$-module. Since N/P is a p'-group, L is completely reducible and consequently, L must be irreducible.

If L is irreducible, then as $P \lhd N$, by [2, (9D)] P is included in the kernel of L, and L may be considered as an irreducible $K[N/P]$-module. Hence $\dim_K L \not\equiv 0 \pmod p$. By (2.11), P is a vertex of L.

(2.17) *Let L_1 and L_2 be two indecomposable $K[G]$-modules which afford the modular characters λ_1, λ_2, respectively. Assume that $\lambda_1(1) \neq 0$, $\lambda_2(1) \neq 0$. Let P be a Sylow-p-subgroup of G. Then L_1 and L_2 belong to the same p-block of G, if and only if*

$$\lambda_1(x)/\lambda_1(1) = \lambda_2(x)/\lambda_2(1)$$

for all p'-elements x of $\mathrm{C}_G(P)$.

PROOF. Suppose that L_i belongs to the p-block B_i, $i = 1, 2$. Since $\lambda_i(1) \neq 0$, B_i has full defect and we may choose an ordinary irreducible character $\chi_i \in B_i$ with $\chi_i(1) \not\equiv 0 \pmod p$. There exist R-representations $\mathfrak{X}_i$ with the character χ_i. Let $x \in G$, and let s be the sum of the conjugates of x in the group algebra. Then $\mathfrak{X}_i(s) = \omega_i I$ with

$$\omega_i = |G : \mathrm{C}_G(x)| \chi_i(x)/\chi_i(1) \in R .$$

If $\mathfrak{F}$ is a constituent of $\mathfrak{X}_i^\theta$, then $\mathfrak{F}_i(s) = \omega_i^\theta I$. If $\mathfrak{F}$ has the modular character φ_i we see on forming the trace that

$$|G : \mathrm{C}_G(x)|^\theta \varphi_i(x) = \omega_i^\theta \varphi_i(1) .$$

This remains true, if φ_i is replaced by any modular character in B_i, in particular, if φ_i is replaced by λ_i. Now, a known lemma on blocks ([2, (6F)] or [6 (86.25)]) applies and yields the statement.

(2.18) *If φ is a modular irreducible character in $B_0(G)$, every algebraically conjugate character φ_1 is in $B_0(G)$.*

PROOF. We may set $\varphi_1 = \varphi^\tau$ where τ is an automorphism of R^θ. There exist automorphisms σ of Ω with $R^\sigma = R$, $\mathfrak{p}^\sigma = \mathfrak{p}$ which induce τ in $R^\theta = R/\mathfrak{p}$. We can choose an absolutely irreducible R-representation $\mathfrak{X}$ of G for which $\mathfrak{X}^\theta$ has a constituent with the modular character φ. Then $(\mathfrak{X}^\sigma)^\theta$ has a constituent with the modular character $\varphi^\tau = \varphi_1$. With $\mathfrak{X}$, the algebraically conjugate representation $\mathfrak{X}^\sigma$ lies in $B_0(G)$, cf. [3, Lem. 2]. We now see that φ_1 belongs to $B_0(G)$, as we wished to show.

(2.19) *Let F be an irreducible $K[G]$-module such that* $\dim_K F \equiv 0 \pmod p$. *Then $L_0(G)$ appears at least with multiplicity 2 as irreducible constituent of $F^* \otimes F$.*

PROOF. We can interpret $F^* \otimes F$ as the module W in (2.1), (2.2) with $A = B = F$. Schur's Lemma shows that the submodule W_1 of invariant elements of W consists exactly of the scalar elements cI with $c \in K$. Let W_2 be the submodule of W consisting of the K-linear transformations of F into F of trace 0. Clearly, $W_1 \approx L_0(G) \approx W/W_2$. Since $\dim_K L \equiv 0 \pmod p$, it follows that $W_1 \subseteq W_2$. Thus, $L_0(G)$ occurs as irreducible constituent of W of multiplicity at least 2.

3. Proof of the theorem

Let P be a Sylow-p-subgroup of the finite group G. The Sylow-p-subgroup of $\mathbf{C}_G(P)$ is the center $\mathbf{Z}(P)$ of P and Burnside's theorem shows that $\mathbf{C}_G(P)$ is the direct product of $\mathbf{Z}(P)$ and a p'-group D. The p'-elements of $\mathbf{C}_G(P)$ are exactly the elements of D. Set $N = \mathbf{N}_G(P)$. Then

$$P\mathbf{C}_G(P) = P \times D \lhd N = \mathbf{N}_G(P) .$$

This notation will remain fixed throughout the rest of the paper.

(3.1) *Let F be an irreducible $K[G]$-module with* $\dim_K F \not\equiv 0 \pmod p$. *Then we have a formula*

$$F_N \approx \oplus\textstyle\sum_i M_i ,$$

where each M_i is indecomposable, $\dim_K M_1 \not\equiv 0 \pmod p$ *and* $\dim_K M_i \equiv 0 \pmod p$ *for $i > 1$. If F and M_1 afford the modular characters φ and μ_1 respectively, $\varphi(x) = \mu_1(x)$ for all $x \in D$. Also, $F \mid M_1^G$. Finally, $(M_1^G)_N$ has a unique indecomposable direct summand with vertex P.*

PROOF. Set $F_N \approx \oplus\sum M_i$ with indecomposable M_i. By (2.11), F has vertex P. As $P \subseteq N$, (2.7) shows that $F \mid M_i^G$ for some i, say $F \mid M_1^G$. If M_1 has vertex $H \subseteq P$, then $M_1 \mid ((M_1)_H)^N$ whence $M_1^G \mid ((M_1)_H)^G$. Since $F \mid M_1^G$, a vertex of F is contained in H. Thus, $H = P$, i.e., M_1 has vertex P.

Since P is the only Sylow-p-subgroup of $N = \mathbf{N}_G(P)$, it follows that $P \not\subseteq x^{-1}Nx$ for $x \in G$, $x \notin N$. Then $N \cap x^{-1}Nx$ does not contain a Sylow-p-subgroup of G, and (2.3) shows that $(M_1^G)_N$ has a unique direct summand M_1 with vertex P. The same is true for F_N since $F_N \mid (M_1^G)_N$ and $M_1 \mid F_N$. Now (2.11) can be applied to M_i with $i > 0$. If M_i affords the modular character μ_i, then $\mu_i(x) = 0$ for $x \in D$ and $i > 0$. Thus

$$\varphi(x) = \textstyle\sum_i \mu_i(x) = \mu_1(x) .$$

In particular, $\mu_1(1) = \varphi(1) \neq 0$, $\mu_i(1) = 0$, for $i > 0$. This completes the proof.

If F satisfies the assumptions of (3.1) and if M_1 is defined as in (3.1), we shall say that M_1 *corresponds* to F.

(3.2) *If F and M_1 are as in* (3.1), *then F is in $B_0(G)$, if and only if M_1 is in $B_0(N)$. We have $F \approx L_0(G)$, if and only if $M_1 \approx L_0(N)$.*

PROOF. By (2.17), F is in $B_0(G)$, if and only if $\varphi(x)/\varphi(1) = 1$ for all $x \in D$. By (3.1), then $\mu_1(x)/\mu_1(1) = 1$. Again, this is so, if and only if M_1 is in $B_0(N)$.

If $F \approx L_0(G)$, clearly $M_1 = F_N \approx L_0(N)$. Conversely, assume that $M_1 \approx L_0(N)$. By (3.1), $F \mid L_0(N)^G$; and by (2.7), $L_0(G) \mid L_0(N)^G$. If $F \not\approx L_0(G)$, this yields that $(L_0(G) \oplus F) \mid M_1^G$ and so $(L_0(N) \oplus L_0(N)) \mid (M_1^G)_N$ contrary to the fact that only one indecomposable direct summand of $(M_1^G)_N$ has vertex P. Thus, $F \approx L_0(G)$.

The crux of the proof of the theorem is contained in the next lemma.

(3.3) *Suppose that L is an irreducible $K[G]$-module with $n = \dim_K L > 1$. Then at least one of the following facts holds.*

(i) *G has a normal subgroup of index p.*

(ii) *There exists an irreducible constituent F of $L^* \otimes L \otimes L^* \otimes L$ with F in $B_0(G)$, $F \not\approx L_0(G)$.*

(iii) *There exists an irreducible constituent F of $L^* \otimes L$ with $\dim_K F \not\equiv 0 \pmod p$ and $F \not\approx L_0(G)$, such that if E is the $K[N]$-module corresponding to F, and if H is the kernel of E_D, then $|D:H| < J(n)$ where $J(n)$ is defined by* (2.9).

PROOF. Assume that neither (i) nor (ii) holds. Let $F_0 = L_0(G), F_1, \cdots, F_s$ denote the distinct irreducible constituents of $L^* \otimes L$.

Consider a $K[G]$-module Z such that each irreducible constituent of Z is isomorphic to a constituent of $L^* \otimes L \otimes L^* \otimes L$. Since (ii) is excluded, an irreducible constituent of Z is isomorphic to $L_0(G)$ or it is not in $B_0(G)$. Since (i) is excluded, (2.14) shows that an indecomposable constituent of Z lying in $B_0(G)$ is isomorphic to $L_0(G)$. This implies that if r is the multiplicity of $L_0(G)$ as irreducible constituent of Z, then r is the dimension of the K-space Z_0 of invariant elements of Z. This remark will be applied several times.

First, take $Z = F_i^* \otimes F_j$ with $0 \leqq i, j \leqq s$. By (2.2), $Z_0 \approx \mathrm{Hom}_{K[G]}(F_i, F_j)$. For $i = j$, Schur's lemma shows that $r = 1$, and it follows from (2.19) that $\dim_K F_i$ is not divisible by p. Next, take $i \neq j$. Here, $r = 0$ and now (2.13) implies that $F_0, F_1, \cdots, F_s$ lie in $s + 1$ distinct blocks of G.

Let Y be any indecomposable constituent of $L^* \otimes L$. Since all the irreducible constituents of Y lie in the same block, they must all be isomorphic to F_i for some i; $0 \leqq i \leqq s$. Let b be the length of a composition series of Y. We may then apply the remark above to $Z = F_i^* \otimes Y$. From what has already been shown, we conclude that $L_0(G)$ appears with the exact multiplicity b as irreducible constituent of Z. Thus, here $r = b$ and, by (2.2),

$$b = \dim_K \operatorname{Hom}_{K[G]}(F_i, Y) .$$

Repeat now the same argument with Y replaced by the maximal completely reducible submodule (the socle) Y_0 of Y. If b_0 is the length of a composition series of Y_0, we find

$$b_0 = \dim_K \operatorname{Hom}_{K[G]}(F_i, Y_0) .$$

Any $\eta \in \operatorname{Hom}_{K[G]}(F_i, Y)$ maps F_i either onto a simple submodule of Y or (0). Thus, $\eta(F_i) \subseteqq Y_0$ and $\operatorname{Hom}_{K[G]}(F_i, Y)$ can be identified with $\operatorname{Hom}_{K[G]}(F_i, Y_0)$. Hence $b = b_0$ and the submodule Y_0 of Y must coincide with Y. Since Y_0 is completely reducible and Y is indecomposable, we see that Y is irreducible. Since this applies to any indecomposable constituent of $L^* \otimes L$, it follows that $L^* \otimes L$ is completely reducible, say

$$L^* \otimes L \approx \oplus\textstyle\sum_{i=0}^{s} a_i F_i , \qquad (a_i > 0) . \tag{3.4}$$

Since $L_0(G)$ occurs as constituent of $L^* \otimes L$, all irreducible constituents F_i of $L^* \otimes L$ occur among the constituents of $L^* \otimes L \otimes L^* \otimes L$. We may therefore take $L^* \otimes L$ for the module Z above. Here again $r = 1$ and, by (2.19), $\dim_K L \not\equiv 0 \pmod p$. We also see that $a_0 = 1$ in (3.4).

Since $\dim_K L \not\equiv 0 \pmod p$, (3.1) applies to L and we may set

$$L_N \approx M \oplus X_1 \oplus X_2 \oplus \cdots$$

where M and the X_i are indecomposable $K[N]$-modules, the vertex of M is P while each X_i has a vertex properly contained in P. Thus,

$$(L^* \otimes L)_N \approx (M^* \otimes M) \oplus M' , \tag{3.5}$$

where M' is a direct sum of terms of the forms $M^* \otimes X_i$, $X_i^* \otimes M$, $X_i^* \otimes X_j$. It is seen easily form (2.4), that each indecomposable direct summand of M' has a vertex properly contained in P.

As we have seen, $\dim_K F_i \not\equiv 0 \pmod p$ for $i = 0, 1, \cdots, s$. Thus, (3.1) also applies to $F_0, F_1, \cdots, F_s$. Let M_i be the $K[N]$-module which corresponds to F_i; $M_0 = L_0(N)$. It follows from (3.4) and (3.5) that

$$(M^* \otimes M) \oplus M' \approx \oplus\textstyle\sum_{i=0}^{s} a_i M_i \oplus M'' , \tag{3.6}$$

where each indecomposable direct summand of M'' has a vertex properly contained in P. Clearly, $M_0 = L_0(N)$ while $M_1, M_2, \cdots, M_s$ do not lie in $B_0(N)$, cf. (3.2).

By (2.16), we can set

$$M_{P \times D} \approx \oplus\textstyle\sum_{i=1}^{e} U_i \otimes V_i ,$$

where $V_1, \cdots, V_e$ are distinct irreducible $K[D]$-modules while $U_1, \cdots, U_e$ are $K[P]$-modules which are mutually conjugate under the action of N. Thus,

$$(M^* \otimes M)_{P\times D} \approx \oplus\sum_{i,j=1}^{e} (U_i^* \otimes U_j) \otimes (V_i^* \otimes V_j)$$
$$\approx \oplus\sum_{j=1}^{e} (U_j^* \otimes U_j) \otimes L_0(D) \oplus S$$

where S is a $K[P \times D]$-module which does not contain $L_0(P \times D)$ as constituent.

Since $L_0(N)$ has vertex P, (3.6) shows that $L_0(N) \mid M^* \otimes M$. Thus $L_0(P \times D) \mid (M^* \otimes M)_{P\times D}$. It follows that $L_0(P) \mid U_j^* \otimes U_j$ for some j. Since the U_j are conjugate under the action of N, this implies that $L_0(P) \mid U_j^* \otimes U_j$ for $j = 1, \cdots, e$ and $eL_0(P \times D) \mid (M^* \otimes M)_{P\times D}$.

If $e > 1$, there exists an indecomposable direct summand $T \not\approx M_0$ of the right side of (3.6) such that $L_0(P \times D) \mid T_{P\times D}$. By (2.16) T has vertex P, and so $T = M_i$ for some $i > 0$. Also, $T = M_i$ is irreducible and by (2.6), $P \times D$ belongs to the kernel of M_i. Now, (2.17) shows that M_i lies in $B_0(N)$. This has already been seen to be false and hence we must have $e = 1$. Thus,

$$M_{P\times D} \approx U \otimes V\,, \tag{3.7}$$

where V is an irreducible $K[D]$-module and U is a $K[P]$-module such that $L_0(P) \mid U^* \otimes U$.

Set $\dim_K V = d$. If $d = 1$, then $V^* \otimes V \approx L_0(D)$ and D is in the kernel of $M^* \otimes M$. By (3.6), $M_i \mid M^* \otimes M$ for $i = 0, 1, \cdots, s$. So, D is in the kernel of each M_i. By (2.17), M_i then is in $B_0(N)$. This is possible only for $s = 0$. Then $n = 1$ by (3.4) contrary to the assumption. Thus, $d > 1$.

Let X be an irreducible constituent of M and let H_0 be the kernel of X. By (2.16), $P \subseteq H_0$ and so N/H_0 is a p'-group. Thus, by (2.9) there exists a subgroup A_0 such that $H_0 \subseteq A_0 \lhd N$, $|N : A_0| < J(n)$, and that A_0/H_0 is abelian. Set $A_1 = A_0 \cap D$, $H_1 = H_0 \cap D$. Then $A_1 \lhd N$, $|D : A_1| < J(n)$, and A_1/H_1 is abelian. By (3.7), $X_D \approx u_0 V$ for some positive integer u_0. It follows that H_1 is the kernel of V.

If A is an abelian group and V a $K[A]$-module, then $V^* \otimes V$ contains $L_0(A)$ at least with multiplicity $\dim_K V$. Applying this with $A = A_1/H_1$, we see that $(V^* \otimes V)_{A_1}$ contains $L_0(A_1)$ at least with multiplicity $d \geqq 2$. As V is irreducible, $V^* \otimes V$ contains $L_0(D)$ with multiplicity 1. It follows that there exists an irreducible constituent $W \not\approx L_0(D)$ of $V^* \otimes V$ such that $L_0(A_1) \mid W_{A_1}$. By (2.6), A_1 is in the kernel of W.

Since $L_0(P) \otimes W \mid (M^* \otimes M)_{P\times D}$ by (3.7), it follows from (3.6) that $L_0(P) \otimes W \mid T_{P\times D}$ for some direct summand T of the right hand side of (3.6). By (2.16), T is irreducible and has vertex P. Then $T \approx M_i$ for some i with $0 \leqq i \leqq s$. Clearly, $i \neq 0$. Let H be the kernel of $(M_i)_D$. Since A_1 belongs to the kernel of W, and $A_1 \lhd N$, by (2.6) A_1 belongs to H. So, $|D : H| \leqq |D : A_1| < J(n)$.

Set $F = F_i$, $E = M_i$. We have Case (iii), and (3.3) is proved.

(3.8) *Let L be an irreducible $K[G]$-module with $\dim_K L = n > 1$. Suppose that G contains no normal subgroup of index p. Then there exists an irreducible constituent F of $L^* \otimes L \otimes L^* \otimes L$ such that, if G_0 is the kernel of F, then*

$$1 < |G:G_0| < |\mathrm{GL}_{|P|^2 J(n) n^4}(p)| .$$

PROOF. Let F be an irreducible $K[G]$-module with $\dim_K F = d$, let G_0 be the kernel of F and let φ be the modular character afforded by F. Suppose that φ has e algebraic conjugates in K. The subfield generated by all $\varphi(x)$ as x ranges over the p'-elements of G has p^e elements. By (2.10), G/G_0 is isomorphic to a subgroup of $\mathrm{GL}_d(p^e)$ and so G/G_0 is isomorphic to a subgroup of $\mathrm{GL}_{de}(p)$. Hence $|G:G_0| \leqq |\mathrm{GL}_{de}(p)|$.

By assumption, (3.3) (i) does not hold.

If (3.3) (ii) holds, and if F is chosen accordingly, $G \neq G_0$ since $F \not\approx L_0(G)$. By (2.12) and (2.18), φ has at most $|P|^2$ algebraic conjugates in K. The result follows from the remarks above and the fact that $\dim_K F < n^4$.

Suppose then that (3.3) (iii) holds, and again choose F accordingly. As before, $G_0 \neq G$. Let $B_1, B_2, \cdots, B_a$ denote all the distinct p-blocks which contain algebraic conjugates of the modular character φ afforded by F. Suppose that $\varphi^{\sigma_j} \in B_j$ where σ_j is an automorphism of R^θ. Let $F^{(j)}$ denote a $K[G]$-module which affords φ^{σ_j}, let $E^{(j)}$ be the $K[N]$-module corresponding to $F^{(j)}$ and let μ_j denote the modular character afforded by $E^{(j)}$. If μ is the modular character afforded by E, clearly, $\mu_j = \mu^{\sigma_j}$ for $j = 1, 2, \cdots, a$.

We have $d = \dim_K F < n^2 < n^4$. By the remarks above, the proof will be complete when we can show that the number e of algebraic conjugates of φ is smaller than $|P|^2 J(n)$. By (2.12), it suffices to show that $a < J(n)$.

If $i \neq j$, then φ^{σ_i} and φ^{σ_j} are in distinct blocks. Thus, by (2.17), $\varphi^{\sigma_i}(x_{ij}) \neq \varphi^{\sigma_j}(x_{ij})$ for some $x_{ij} \in D$; and by (3.1), $\mu^{\sigma_i}(x_{ij}) \neq \mu^{\sigma_j}(x_{ij})$. This shows that the restriction $\mu | D$ has at least a distinct algebraic conjugates. On the other hand, the number of conjugates of $\mu | D$ is at most equal to the index $|D:H|$ of the kernel H of E_D. By (3.3) (iii), $a < J(n)$ and we are finished.

Let G be a group and let L be a $K[G]$-module. The *type* $t(G, L)$ of (G, L) is defined by $t(G, L) = (m, n, a)$ where p^m is the order of a Sylow-p-subgroup of G, $n = \dim_K L$ and a is the multiplicity with which $L_0(G)$ occurs as a constituent of $L^* \otimes L \otimes L^* \otimes L$.

Define the ordering $\prec$ by $(m_1, n_1, a_1) \prec (m, n, a)$ if one of the following is satisfied

(i) $m_1 \leqq m, \quad n_1 \leqq n, \quad a_1 \geqq a,$

(ii) $m_1 = m$, $n_1 < n$, $a_1 \geqq a$,

(iii) $m_1 = m$, $n_1 = n$, $a_1 > a$.

If H is a subgroup of G, then clearly $t(H, L_H) \prec t(G, L)$ or $t(H, L_H) = t(G, L)$. Observe that if $t(G, L) = (m, n, a)$, then $a \leqq n^4$. Thus there are only finitely many types less than a given $t(G, L)$.

(3.9) *Let P be a Sylow-p-subgroup of G, and let $|P| = p^m$. Let X be a faithful $K[G]$-module with* $\dim_K X = n$. *Let $g(m, n) = |\mathrm{GL}_{|P|^2 J(n) n^4}(p)|$ where $J(n)$ is defined by (2.9). Assume that G is not abelian. Then there exists $G_0 \triangleleft G$ with $|G: G_0| < g(m, n)$ such that $t(G_0, X_{G_0}) \prec t(G, X)$.*

PROOF. It may be assumed that G has no normal subgroup of index p; otherwise the result is trivial.

Suppose first that every irreducible constituent of X has K-dimension 1. Let X_1 be the completely reducible $K[G]$-module which has the same set of composition factors as X. By (2.14), P is the kernel of X_1 and G/P is abelian. Thus G is solvable and by a Theorem of P. Hall [9], [8, p. 141] G contains an abelian subgroup A with $|G: A| = |P|$. Let $G_0 = \bigcap_x x^{-1}Ax$. Then $G_0 \triangleleft G$, $|G: G_0| \leqq |P|! \leqq g(m, n)$ and since G is not abelian, $|P| \neq 1$ so that $t(G_0, X_{G_0}) \prec t(G, X)$.

Suppose next that some irreducible constituent L of X has K-dimension at least 2. Let F and G_0 be defined by (3.8). If $t(G, X) = (m, n, a)$ and $t(G_0, X_{G_0}) = (m_1, n_1, a_1)$, then clearly $m_1 \leqq m$ and $n_1 = n$. Since G_0 is the kernel of F and $F \not\approx L_0(G)$, $a_1 > a$. Thus $t(G_0, X_{G_0}) \prec t(G, X)$. The result follows from (3.8).

(3.10) *There exists a function $h(m, n, a)$ such that if X is a faithful $K[G]$-module, then $|G: A| \leqq h(t(G, X))$ for some normal abelian subgroup A of G.*

PROOF. Induction with respect to $\prec$. It may be assumed that the result is true for all (G_1, X_1) with $t(G_1, X_1) \prec t(G, X)$. Let $h(m, n, a) = 1$ if $(m, n, a) \neq t(G_1, X_1)$ for any (G_1, X_1). Suppose that $t(G, X) = (m, n, a)$. If G is abelian, let $G = A$. If G is non-abelian, let G_0 and $g(m, n)$ be defined by (3.9). Then by induction G_0 contains a normal abelian subgroup A_0 with $|G_0: A_0| \leqq h(t(G_0, X_{G_0}))$. Let $A = \bigcap_{x \in G} x^{-1}A_0x$. Then $|G: A| < g(m, n)h(t(G_0, X_{G_0}))^{g(m,n)}$. Define

$$h_0(t(G, X)) = \max_{t(G_1, X_1) \prec t(G, X)} h(t(G_1, X_1)) .$$

Then h can be defined by

$$h(m, n, a) = g(m, n)h_0(m, n, a)^{g(m,n)}$$

completing the proof.

The Theorem follows directly from (3.10) by defining

$$f(m, n) = \max_{0 \leqq a \leqq n^4} h(m, n, a) .$$

HARVARD UNIVERSITY
YALE UNIVERSITY

REFERENCES

1. R. BRAUER, *Über Systeme hypercomplexer Zahlen*, Math. Z., 30 (1929), 79-107.
2. ———, *Zur Darstellungstheorie der Gruppen endlicher Ordnung*: I, Math., Z. 63 (1956), 406-444.
3. ———, *Some applications of the theory of blocks of characters of finite groups*: I. J. Algebra, 1 (1964), 152-167.
4. ——— and W. FEIT, *On the number of irreducible characters of finite groups in a given block*, Proc. Nat. Acad. Sci., 45 (1959), 361-365.
5. R. BRAUER and C. NESBITT, *On the modular characters of groups*, Ann. of Math., 42 (1941), 556-590.
6. C. CURTIS and I. REINER, Representation Theory of Finite Groups and Associative Algebras, New York, 1962.
7. J. A. GREEN, *On the indecomposable representations of a finite group*, Math. Z., 70 (1959), 430-445.
8. M. HALL, The Theory of Groups, New York, 1959.
9. P. HALL, *A note on soluble groups*, J. London Math. Soc., 3 (1928), 98-105.
10. C. JORDAN, *Mémoire sur les équations differentielles linéaires à intégrale algébrique*, J. Math., 84 (1878), 89-215.
11. G. MACKEY, *On induced representations of groups*, Amer. J. Math., 73 (1951), 576-592.

(Received September 23, 1965)

SOME RESULTS ON FINITE GROUPS WHOSE ORDER CONTAINS A PRIME TO THE FIRST POWER

RICHARD BRAUER*

In memory of TADASI NAKAYAMA

The author discussed questions of the type treated here at various times with his friend Tadasi Nakayama. There had been plans of a collaboration between Nakayama and him in an effort to broaden our knowledge of the part of the character theory on which this present work is based. Nakayama's untimely death destroyed the hope of such a collaboration. I wish to dedicate this paper to his memory.

§ 1. Introduction

In a previous investigation [1], the author has studied finite groups $\mathfrak{G}$ of an order $g = pg_0$ where p is a prime and g_0 an integer not divisible by p. This work has been continued by H. F. Tuan [5]. Let t denote the number of conjugate classes of $\mathfrak{G}$ which consist of element of order p. Tuan dealt with the groups $\mathfrak{G}$ for which $t \leqq 2$ and which have a faithful representation of degree less than $p-1$. We shall assume here that $t \geqq 3$.** We shall also suppose that $\mathfrak{G}$ does not have a normal subgroup of order p. We state here two results. We shall show (Corollary, Theorem 1) that if χ is a faithful irreducible character of $\mathfrak{G}$ of degree n which has $T>1$ conjugates over the field of the g_0-th roots of unity, then

$$n > \frac{1}{3} T^{-3/4} p^{5/4}.$$

In Theorem 2, we assume that $\mathfrak{G}$ has an irreducible faithful representation of degree $n < p-1$. It is then shown that

$$p \leqq t^3 - t + 1.$$

Received June 30, 1965.

* This work has been supported in part by the United States Air Force under Contract No. AF 49(638)-1381 monitored by the AF Office of Scientific Research of the Air Research and Development Command.

** Some remarks communicated by H. F. Tuan to the author twenty years ago have been helpful.

Thus for a fixed t, only finitely many primes p are possible. In a later paper, S. Hayden and the author will study small values of t and give further extensions of the results.

§ 2. Preliminaries

Let $\mathfrak{G}$ be a finite group of order $g = pg_0$ where p is a fixed prime and where g_0 is an integer not divisible by p. Let $\mathfrak{P}$ be a p-Sylow subgroup of $\mathfrak{G}$. The centralizer $\mathfrak{C}(\mathfrak{P})$ and the normalizer $\mathfrak{N}(\mathfrak{P})$ of $\mathfrak{P}$ then have the form

$$\mathfrak{C} = \mathfrak{C}(P) = \mathfrak{P} \times \mathfrak{V}, \quad \mathfrak{N} = \mathfrak{N}(P) = \langle\ (\mathfrak{P}),\ M\rangle. \tag{2.1}$$

Here, $\mathfrak{V}$ is a group of order v prime to p and M is an element whose order m over $\mathfrak{C}(\mathfrak{P})$ divides $p-1$. If we set

$$p - 1 = mt, \tag{2.2}$$

t is the number of conjugate classes of $\mathfrak{G}$ which contain elements of order p. Distribute the irreducible characters of $\mathfrak{V}$ into classes $F_0, F_1, \ldots, F_{l-1}$ of characters associated in $\mathfrak{N}$. It is shown in [1] that $\mathfrak{G}$ has l p-blocks B_0, B_1, $\ldots, B_{l-1}$ of full defect, and we have a one-to-one correspondence $F_\lambda \to B_\lambda$. Let θ_λ be a character belonging to F_λ and suppose that F_λ consists of τ_λ characters, i.e. that the inertial group of θ_λ has index τ_λ in $\mathfrak{N}$. Then $\tau_\lambda | m$; we set

$$m = m_\lambda \tau_\lambda, \quad t_\lambda = \tau_\lambda t \tag{2.3}$$

so that

$$p - 1 = m_\lambda t_\lambda, \quad (\lambda = 0, \ldots, l-1). \tag{2.2*}$$

The degree of θ_λ will be denoted by $f_\lambda = \theta_\lambda(1)$.

As shown in [1], B_λ consists of m_λ "non-exceptional" characters $\zeta_0^{(\lambda)}, \zeta_1^{(\lambda)}, \ldots, \zeta_{m_\lambda - 1}^{(\lambda)}$ and t_λ exceptional characters $\chi_1^{(\lambda)}, \chi_2^{(\lambda)}, \ldots \chi_{t_\lambda}^{(\lambda)}$. The values of these characters for p-singular elements of $\mathfrak{G}$ can be given explictly, only certain $\pm$ signs remain undetermined. Let ρ denote a primitive p-th root of unity and let c denote a primitive root (mod p). We form the Gauss periods of length m_λ

$$\omega_k^{(\lambda)} = \sum_\nu{}' \rho^{c^\nu} \tag{2.4}$$

where ν ranges over the integers, k, $k + t_\lambda$, $k + 2t_\lambda, \ldots, k + (m_\lambda - 1)t_\lambda$. Clearly,

$\omega_k^{(\lambda)}$ remains unchanged, if k is changed (mod t_λ). If P is a fixed generator of $\mathfrak{P}$ and if $V \in \mathfrak{V}$, we have with $\theta^M(V) = \theta(M^{-1}VM)$

$$\chi_j^{(\lambda)}(PV) = \varepsilon^{(\lambda)} \sum_{\mu=0}^{\tau_\lambda - 1} \theta_\lambda^{M^\mu}(V)\,\omega_{j+\mu t}^{(\lambda)}, \tag{2.5}$$

$$\zeta_i^{(\lambda)}(PV) = \varepsilon_i^{(\lambda)} \sum_{\mu=0}^{\tau_\lambda - 1} \theta_\lambda^{M^\mu}(V) \tag{2.6}$$

with $\varepsilon^{(\lambda)}, \varepsilon_i^{(\lambda)} = \pm 1$; cf. [1] I, Theorem 4.

We take B_0 as the principal p-block of $\mathfrak{G}$. Then $\theta_0 = 1$, $\tau_0 = 1$, i.e. $t_0 = t$, $m_0 = m$. We shall choose here the notation such that $\zeta_0^{(0)} = 1$ and that $\varepsilon_i^{(0)} = 1$ for $i = 0, 1, \ldots, a-1$ and $\varepsilon_i^{(0)} = -1$ for $i = a, a+1, \ldots, m-1$.

Set $\omega_k^{(0)} = \eta_k$. The η_k are the Gauss periods of length m. It is seen easily that

$$\eta_i = \sum_{\nu=0}^{\tau_\lambda - 1} \omega_{i+\nu t}^{(\lambda)}. \tag{2.7}$$

As is well known, $\eta_1, \eta_2, \ldots, \eta_t$ form a $\mathbf{Z}$-basis for the ring of algebraic integers of the field $\mathbf{Q}(\eta_1) \subseteq \mathbf{Q}(\rho)$; *) we have

$$\sum_{i=1}^{t} \eta_i = -1. \tag{2.8}$$

With each η_i, the conjugate complex number $\bar{\eta}_i$ appears in $\{\eta_1, \eta_2, \ldots, \eta_t\}$. We shall use the notation

$$\bar{\eta}_i = \eta_{i'}. \tag{2.9}$$

We also give some formulas for the multiplication of the η_i. These can be proved easily directly, but we apply a group theoretical method.

Let $\mathfrak{M}_{p,t}$ denote the metacyclic group of order $p(p-1)/t$ defined as group with generators P, M with the relations

$$P^p = 1,\ M^m = 1,\ M^{-1}PM = P^{c^t}.$$

Then $\mathfrak{M}_{p,t}$ satisfies our conditions for $\mathfrak{G}$ and $t = (p-1)/m$ is the number of conjugate classes of elements of order p. Here, B_0 is the only p-block. The non-excepetional characters can be identified with those of $\mathfrak{M}_{p,t}/\langle P\rangle$, i.e. of a cyclic group of order m. For suitable choice of the notation, we have $\zeta_k^{(0)} = \zeta^k$, $(0 \leq k < m)$, ζ a character of degree 1. The characters $\chi_i^{(0)}$ have degree

*) We use the notation $\mathbf{Q}$ for the field of rational numbers and the notation $\mathbf{Z}$ for the ring of integers.

m, $\varepsilon^{(0)}=1$. The product $\chi_i^{(0)}\bar{\chi}_j^{(0)}$ will contain ζ^k, if and only if $\chi_j^{(0)}\zeta^k=\chi_i^{(0)}$. Taking the element P, we see that this is so, if and only if $i=j$. Hence we may set

$$\chi_i^{(0)}\bar{\chi}_j^{(0)}=\sum_{r=1}^{t}c_{ijr}\chi_r^{(0)}+\delta_{ij}\sum_{k=0}^{m-1}\zeta^k$$

with integral $c_{ijr}\geqq 0$ and Kronecker δ_{ij}. For the element P, this yields

$$\eta_i\bar{\eta}_j=\sum_{r=1}^{t}c_{ijr}\eta_r+m\delta_{ij}, \tag{2.10}$$

while for the element 1, we find

$$\sum_{r=1}^{t}c_{ijr}=m-\delta_{ij}. \tag{2.11}$$

Since c_{ijr} is the multiplicity of the principal character in $\chi_i^{(0)}\bar{\chi}_j^{(0)}\bar{\chi}_r^{(0)}$, we see that

$$c_{ijr}=c_{irj}=c_{rij}. \tag{2.12}$$

§ 3. Characters of $\mathfrak{N}$

The results stated in § 2 apply in particular to the group $\mathfrak{N}$ in (2.1). We shall write here b_λ instead of B_λ. Since $\mathfrak{P}\lhd\mathfrak{N}$, no p-block of defect 0 occurs for $\mathfrak{N}$.

The irreducible characters of $\mathfrak{V}$ have the form $\theta_\lambda^{M^\mu}$. They may be considered as characters of $\mathfrak{C}=\mathfrak{P}\times\mathfrak{V}$. Let (ρ) denote the linear character of $\mathfrak{C}$ defined by

$$(\rho)(P^iV)=\rho^i$$

for $V\in\mathfrak{V}$. If the element M is chosen suitably, we may assume that

$$(\rho)^M=(\rho)^{c^t}. \tag{3.1}$$

For any $j\in\mathbf{Z}$, the character $(\rho)^{c^j}\theta_\lambda$ of $\mathfrak{C}$ has m associates in $\mathfrak{N}$. It follows that this character induces an irreducible character

$$((\rho)^{c^j}\theta_\lambda)^*=\xi_j^{(\lambda)} \tag{3.2}$$

of $\mathfrak{N}$. Using (3.1) we see without difficulty that

$$\xi_j^{(\lambda)}|\mathfrak{C}=\sum_{\mu=0}^{\tau_\lambda-1}\theta_\lambda^{M^\mu}\sum_\nu(\rho)^{c^\nu} \tag{3.3}$$

where in the inner sum, ν ranges over all members of a residue system mod

$p-1$ for which $\nu \equiv j+\mu t \pmod{t_\lambda}$. Comparison of (3.3) with (2.5), (2.6) shows that $\xi_0^{(\lambda)}, \xi_1^{(\lambda)}, \ldots, \xi_{t_\lambda - 1}^{(\lambda)}$ are the t_λ exceptional characters in b_λ. The signs $\varepsilon^{(\lambda)}$ here are $+1$. It is now also clear that $\xi_j^{(\lambda)}$ remains unchanged, if j is changed mod t_λ. The degree of $\xi_j^{(\lambda)}$ is $m\theta_\lambda(1) = mf_\lambda$.

We next form the irreducible character θ_λ^* of $\mathfrak{N}$ induced by the character θ_λ of $\mathfrak{B}$. If ψ is an irreducible constituent of θ_λ^*, necessarily

$$\psi \mid \mathfrak{C} = e \sum_{\nu=0}^{\tau_\lambda - 1} \theta_\lambda^{M^\nu}$$

with integral $e>0$. Comparing this with (2.5), (2.6), we see that ψ is a non-exceptional character of b_λ and that $e=1$. Hence ψ has degree $\tau_\lambda f_\lambda$ and it appears only once in θ_λ^*. This shows that θ_λ^* splits into the m_λ non-exceptional characters of b_λ each appearing with multiplicity 1. In our present case, the signs $\varepsilon_i^{(\lambda)}$ are also $+1$.

The characters of the principal block b_0 have kernels including $\mathfrak{B}$. Since $\mathfrak{N}/\mathfrak{C}$ is cyclic of order m, we can find a linear character $\psi^{(0)} \in b_0$ such that $\psi^{(0)}(M)$ is a primitive m-th root of unity while $\psi^{(0)} \mid \mathfrak{C} = 1$. Then the non-exceptional characters of b_0 are

$$\psi_i^{(0)} = (\psi^{(0)})^i, \quad (i = 0, 1, 2, \ldots, m-1).$$

Let $\psi^{(\lambda)}$ now denote a non-exceptional character of b_λ. It is clear that, for any i,

$$\psi_i^{(\lambda)} = (\psi^{(0)})^i \psi^{(\lambda)} \tag{3.4}$$

is also an irreducible non-exceptional character of b_λ. Conversely, if ψ is an irreducible non-exceptional character of b_λ, then $\psi\overline{\psi}^{(\lambda)}$ must contain a constituent in b_0. This can be seen from (2.6). It also follows from the general theory. Since the kernel of $\psi\overline{\psi}^{(\lambda)}$ includes $\mathfrak{P}$, this constituent can only be a $\psi_i^{(0)}$. It then follows that ψ is the character $\psi_i^{(\lambda)}$. We see that the non-exceptional characters of b_λ are the $\psi_i^{(\lambda)}$ with $i = 0, 1, 2, \ldots, m_\lambda - 1$. Necessarily,

$$\psi_i^{(\lambda)} (\psi^{(0)})^{m_\lambda} = \psi_i^{(\lambda)}. \tag{3.5}$$

We now have

(3 A) *The block b_λ of $\mathfrak{N}$ consists of the t_λ exceptional characters $\xi_i^{(\lambda)}$ and the m_λ non-exceptional characters $\psi_i^{(\lambda)}$, $0 \leq j < t_\lambda$; $0 \leq i < m_\lambda$. The $\xi_j^{(\lambda)}$ have degree mf_λ while the $\psi_i^{(\lambda)}$ have degree $\tau_\lambda f_\lambda$. The kernel of each $\psi_i^{(\lambda)}$ includes $\mathfrak{P}$.*

Since the sign associated with each $\psi_i^{(\lambda)}$ is $+1$, the tree corresponding to b_λ is a "star" whose center corresponds to the set of exceptional characters while each free end point corresponds to a $\psi_i^{(\lambda)}$. This shows that the modular irreducible characters $\phi_i^{(\lambda)}$ of b_λ are simply the restrictions of the $\psi_i^{(\lambda)}$ to the set of p-regular elements; $i=0, 1, 2, \ldots, m_\lambda - 1$. The chief indecomposable character $\Phi_i^{(\lambda)}$ corresponding to $\phi_i^{(\lambda)}$ then is given by

$$\Phi_i^{(\lambda)} = \psi_i^{(\lambda)} + \sum_{j=1}^{t_\lambda} \xi_j^{(\lambda)}. \tag{3.6}$$

It also follows that, for p-regular elements $\sigma \in \mathfrak{N}$, we have

$$\xi_i^{(\lambda)}(\sigma) = \sum_{\nu=0}^{m_\lambda - 1} \phi_\nu^{(\lambda)}(\sigma). \tag{3.7}$$

Taking the element 1 in (3.6) and using (3 A) and (2.2), (2.3) we obtain

(3 *B*) *The character* $\Phi_i^{(\lambda)}$ *has the degree* $p\tau_\lambda f_\lambda$.

The next statement is also an immediate consequence of (3.6):

(3 *C*) *The* $\mathbf{Z}$*-module of functions spanned by the irreducible characters in* b_λ *has the basis* $\Phi_i^{(\lambda)}$, $\xi_j^{(\lambda)}$ *with* $i=0, \ldots, m_\lambda - 1$ *and* $j=1, 2, \ldots, t_\lambda$.

We conclude § 3 with the proof of three lemmas concerning the products of the characters of $\mathfrak{N}$.

(3 *D*) *For* $i \not\equiv j \pmod t$, $\xi_i^{(\lambda)}\bar{\xi}_j^{(\lambda)}$ *is a sum of exceptional characters.*

Indeed, it can be seen from (3.3) that, for $i \not\equiv j \pmod t$, $\xi_i^{(\lambda)}\xi_j^{(\lambda)}|\mathfrak{C}$ is a sum of irreducible characters $\theta(\rho)^k$ where θ is a character whose kernel includes $\mathfrak{P}$ and where $(\rho)^k \neq 1$. It follows that each constituent of $\xi_i^{(\lambda)}\bar{\xi}_j^{(\lambda)}$ is non-trivial on $\mathfrak{P}$. By (3 A), all constituents are exceptional characters.

(3 *E*) *We have formulas*

$$\Phi_i^{(\alpha)}\bar{\xi}_j^{(\beta)} = \sum_{\Upsilon=0}^{l-1} \sum_{k=1}^{m_\Upsilon - 1} q_{ik}^{\alpha\beta\Upsilon} \Phi_k^{(\Upsilon)} \tag{3.8}$$

where $q_{ik}^{\alpha\beta\Upsilon} \in \mathbf{Z}$ *and where*

$$0 \leq q_{ik}^{\alpha\beta\Upsilon} \leq \tau_\alpha f_\alpha f_\beta / f_\Upsilon. \tag{3.9}$$

Proof. We can set

$$\psi_k^{(\tau)}\psi_\nu^{(\beta)} = \sum_{\alpha=0}^{l-1}\sum_{i=0}^{m_\alpha-1} a_{i\nu k}^{\alpha\beta\tau}\psi_i^{(\alpha)}$$

with $a_{ijk}^{\alpha\beta\tau}\in \mathbf{Z}$, $a_{ijk}^{\alpha\beta\tau}\geqq 0$. This formula remains valid, if each ψ is replaced by the corresponding ϕ. Now [4], (64) and (76) show that

$$\Phi_i^{(\alpha)}\overline{\phi}_\nu^{(\beta)} = \sum_{\tau=1}^{l}\sum_{k=1}^{m_\tau-1} a_{i\nu k}^{\alpha\beta\tau}\Phi_k^{(\tau)}. \tag{3.10}$$

We claim that (3.8) holds with

$$q_{ik}^{\alpha\beta\tau} = \sum_{\nu=0}^{m_\beta-1} a_{i\nu k}^{\alpha\beta\tau}. \tag{3.11}$$

Since the chief indecomposable characters Φ vanish for p-singular elements, it suffices to check (3.8) for p-regular elements σ. Here, (3.8) is obtained by summing (3.10) for $\nu = 0, 1, \ldots, m_\beta - 1$ and using (3.7) and (3.11). It is now evident that $q_{ik}^{\alpha\beta\tau}\in \mathbf{Z}$, $q_{ik}^{\alpha\beta\tau}\geqq 0$.

We note next that $a_{i\nu k}^{\alpha\beta\tau}$ is the multiplicity of the principal character in

$$\psi_k^{(\tau)}\psi_\nu^{(\beta)}\overline{\psi}_i^{(\alpha)} = \psi^{(\tau)}\psi^{(\beta)}\overline{\psi}^{(\alpha)}(\psi^{(0)})^{k+\nu-i},$$

cf. (3.4). For fixed α, β, τ, then $a_{i\nu k}^{\alpha\beta k}$ depends only on $k+\nu-i$. Moreover, (3.5) shows that $a_{i\nu k}^{\alpha\beta\tau}$ remains unchanged if one one of the indices i, ν, or k is changed modulo the greatest common divisor d of m_α, m_β, m_τ. Let dU denote the sum of the $a_{i\nu k}^{\alpha\beta\tau}$ where one of the subscripts ranges over a residue system mod d. Then (3.11) reads

$$q_{ik}^{\alpha\beta\tau} = m_\beta U = mU/\tau_\beta. \tag{3.12}$$

On the other hand, since $a_{ijk}^{\alpha\beta\tau}$ is the multiplicity of $\psi_k^{(\tau)}$ in $\psi_i^{(\alpha)}\overline{\psi}_j^{(\beta)}$,

$$\psi_i^{(\alpha)}\overline{\psi}_j^{(\beta)} = \sum_{\tau=0}^{l-1}\sum_{k=0}^{m_\tau-1} a_{ijk}^{\alpha\beta\tau}\psi_k^{(\tau)}. \tag{3.13}$$

On comparing degrees, we have

$$\tau_\alpha\tau_\beta f_\alpha f_\beta \geqq \sum_{k=1}^{m_\tau} a_{ijk}^{\alpha\beta\tau}\tau_\tau f_\tau = m_\tau U\tau_\tau f_\tau = mUf_\tau.$$

Now (3.12) shows that (3.9) holds.

(3 *F*) *We have formulas*

$$\Phi_i^{(\alpha)}\overline{\Phi}_\nu^{(\beta)} = \sum_{\tau=0}^{l-1}\sum_{k=0}^{m_\tau-1} A_{i\nu k}^{\alpha\beta\tau}\Phi_k^{(\tau)} \tag{3.14}$$

where $A_{i\nu k}^{\alpha\beta\gamma} \in \mathbf{Z}$ *and where*

$$0 \leqq A_{i\nu k}^{\alpha\beta\gamma} \leqq (t + \tau_\gamma^{-1})\tau_\alpha\tau_\beta f_\alpha f_\beta / f_\gamma. \tag{3.15}$$

Proof. Again, it suffices to consider p-regular elements in (3.14). On adding (3.8) for $j = 1, 2, \ldots, t_\beta$ and (3.10), we see that (3.14) holds with

$$A_{i\nu k}^{\alpha\beta\gamma} = t_\beta q_{ik}^{\alpha\beta\gamma} + a_{i\nu k}^{\alpha\beta\gamma} = t\tau_\beta q_{ik}^{\alpha\beta\gamma} + a_{i\nu k}^{\alpha\beta\gamma}.$$

Now (3.15) is obtained from (3.9), since (3.13) implies $a_{i\nu k}^{\alpha\beta\gamma} \leqq \tau_\alpha\tau_\beta f_\alpha f_\beta / (\tau_\gamma f_\gamma)$.

§ 4. Some results concerning the characters of $\mathfrak{G}$

It follows from (2.5) and the results of § 3 that the difference $\chi_j^{(\lambda)} | \mathfrak{N} - \varepsilon^{(\lambda)}\xi_j^{(\lambda)}$ is a generalized character of $\mathfrak{N}$ which vanishes for all p-singular elements; $(\lambda = 0, \ldots, l-1;\ j = 1, 2, \ldots, t_\lambda)$. Consequently, the difference is a linear combination of the chief indecomposable characters $\Phi_k^{(\alpha)}$ of $\mathfrak{N}$ with coefficients u in $\mathbf{Z}$, (cf. for instance [2], Theorem 17). Moreover, since the values of the $\Phi_k^{(\alpha)}$ lie in the field Ω of the g_0-th roots of unity over $\mathbf{Q}$, application of an element of the Galois group of $\Omega(\rho)$ over Ω shows that the coefficients u do not depend on j. Hence we may set

$$\chi_j^{(\lambda)} | \mathfrak{N} = \varepsilon^{(\lambda)}\xi_j^{(\lambda)} + \sum_{\alpha, r} u_{\alpha, r}^{(\lambda)} \Phi_r^{(\alpha)} \tag{4.1}$$

where in the sum on the right, α ranges over $0, \ldots, l-1$ and, for given α, r ranges over $0, 1, \ldots, m_\alpha - 1$.

(4 *A*). *In* (4.1), *the integers* $u_{\alpha, r}^{(\lambda)}$ *are non-negative. If* $\varepsilon^{(\lambda)} = -1$, *there exists an* r *for which* $u_{\lambda, r}^{(\lambda)}$ *is positive.*

This becomes evident, if we use the formulas (3.6) to express (4.1) by means of the irreducible charcters of $\mathfrak{N}$. Since $\chi_j^{(\lambda)} | \mathfrak{N}$ is a character of $\mathfrak{N}$, the coefficients of each $\psi_k^{(\alpha)}$ and each $\xi_i^{(\lambda)}$ must be non-negative.

(4 *B*) *Choose a fixed value of* λ. *There exist coefficients* $h_{\alpha\beta r} \in \mathbf{Z}$ *such that*

$$\chi_i^{(\lambda)} \bar{\chi}_i^{(\lambda)} = \sum_{\mu=0}^{a-1} \zeta_\mu^{(0)} + \sum_{r=1}^{t} h_{iir}\chi_r^{(0)} + \Gamma + \Delta_{ii}; \tag{4.2 a}$$

$$\chi_i^{(\lambda)} \bar{\chi}_j^{(\lambda)} = \sum_{\nu=a}^{m-1} \zeta_\nu^{(0)} + \sum_{r=1}^{t} h_{ijr}\chi_r^{(0)} + \Gamma + \Delta_{ij} \tag{4.2 b}$$

for $i \not\equiv j \bmod t_\lambda$. *Here* Γ *is a character whose irreducible constituents are non-exceptional characters* $\zeta_k^{(0)} \in B_0$ *and which does not depend on* i, j, *and* $\Delta_{\alpha\beta}$ *is a character whose irreducible constituents do not lie in* B_0. (*The integers* $h_{\alpha\beta r}$ *and the characters* Γ *and* $\Delta_{\alpha\beta}$ *will depend on* λ).

Proof. Choose $i \not\equiv j \pmod{t_\lambda}$ and form the generalized character

$$\Xi = \chi_i^{(\lambda)}\bar{\chi}_j^{(\lambda)} - \chi_k^{(\lambda)}\bar{\chi}_k^{(\lambda)}.$$

Then Ξ vanishes for all p-regular elements of $\mathfrak{G}$. Express Ξ by the irreducible characters of $\mathfrak{G}$ and let Ξ_0 denote the sum of the terms which lie in B_0, say

$$\Xi_0 = \sum_{\mu=0}^{a-1} x_\mu \zeta_\mu^{(0)} + \sum_{\nu=a}^{m-1} y_\nu \zeta_\nu^{(0)} + \sum_{r=1}^{t} z_r \chi_r^{(0)}.$$

Then Ξ_0 also vanishes for p-regular elements. On the other hand, for every value of r,

$$\sum_{\mu=0}^{a-1} \zeta_\mu^{(0)} - \sum_{\nu=a}^{m-1} \zeta_\nu^{(0)} - \varepsilon^{(0)}\chi_r^{(0)}$$

vanishes for p-regular elements, cf. [1] I, Theorem 6. If we set $z = \sum z_r$, it follows that

$$\sum_{\mu=0}^{a-1} (x_\mu + \varepsilon^{(0)}z)\zeta_\mu^{(0)} + \sum_{\nu=a}^{m-1} (y_\nu - \varepsilon^{(0)}z)\zeta_\nu^{(0)}$$

vanishes for p-regular elements. Since the $\zeta_i^{(0)}$ are still linearly independent on the set of p-regular elements, we find $x_\mu = -\varepsilon^{(0)}z$, $y_\nu = \varepsilon^{(0)}z$. Now, the prinicipal character $\zeta_0^{(0)}$ appears with the multiplicity -1 in Ξ; we have $x_0 = -1$. Hence

(4.3) $$\varepsilon^{(0)} = z = \sum_r z_r, \quad x_\mu = -1, \; y_\nu = 1.$$

Thus,

(4.4) $$\chi_i^{(\lambda)}\bar{\chi}_j^{(\lambda)} = \chi_k^{(\lambda)}\bar{\chi}_k^{(\lambda)} - \sum_{\mu=0}^{a-1} \zeta_\mu^{(0)} + \sum_{\nu=a}^{m-1} \zeta_\nu^{(0)} + \sum_{r=1}^{t} z_r \chi_r^{(0)} + \Xi_{ijk}^*$$

where Ξ_{ijk}^* is a character of $\mathfrak{G}$ whose irreducible constituents lie in blocks other than B_0. In particular, this shows that $\zeta_0^{(0)}, \zeta_1^{(0)}, \ldots, \zeta_{a-1}^{(0)}$ are constituents of $\chi_k^{(\lambda)}\bar{\chi}_k^{(\lambda)}$ and that $\zeta_a^{(0)}, \ldots, \zeta_{m-1}^{(0)}$ are constituents of $\chi_i^{(\lambda)}\bar{\chi}_j^{(\lambda)}$.

We may choose our notations such that (4.2 a) holds for $i=1$ and that h_{11r}, Γ, Δ_{11} have the significance stated in (4 B). Then (4.4) with $k=1$ shows that (4.2 b) holds, if we set $h_{ijr} = h_{11r} + z_r$. It follows from (4.3) that

(4.5) $$\sum_{r=1}^{t} h_{ijr} = \sum_{r=1}^{t} h_{11r} + \varepsilon^{(0)}.$$

If we now apply (4.4) with any k, we see that (4.2 a) holds in general. In addition, we have

$$(4.6)\qquad \sum_{r=1}^{t} h_{kkr} = \sum_{r=1}^{t} h_{ijr} - \varepsilon^{(0)}.$$

This completes the proof of (4 B).

(4 C) *In (4 B), the h_{ijr} are non-negative. There exists a number H such that*

$$(4.7)\qquad \sum_{r=1}^{t} h_{ijr} = H \textit{ for } i \not\equiv j \pmod{t_\lambda},\quad \sum_{r=1}^{t} h_{iir} = H - \varepsilon^{(0)}.$$

The first statement is obvious since $h_{\alpha\beta r}$ is the multiplicity of $\chi_r^{(0)}$ in $\chi_\alpha^{(\lambda)}\bar{\chi}_\beta^{(\lambda)}$. The second statement is a consequence of (4.5) and (4.6).

Our next aim is to give an estimate for the number H. We shall now make the following assumptions:

(I). $t \geqq 3$

(II). *There does not exist a normal subgroup $\mathfrak{K}$ of $\mathfrak{G}$ such that $\mathfrak{G}/\mathfrak{K}$ is isomorphic with the metacyclic group $\mathfrak{M}_{p,t}$ of order $p(p-1)/t$* *.

Suppose that in (4.1) for $\lambda = 0$, all $u_{\alpha,k}^{(0)}$ vanish. Necessarily $\varepsilon^{(0)} = 1$ and $\chi_j^{(0)}(1) = m < (p-1)/2$. We take $\mathfrak{K}$ as the kernel of $\chi_1^{(0)}$. By (2.5), $P \not\subseteq \mathfrak{K}$ and hence $\mathfrak{K}$ has an order prime to p. Then the principal blocks of $\mathfrak{G}$ and of $\bar{\mathfrak{G}} = \mathfrak{G}/\mathfrak{K}$ can be identified, see e.g. [3], Theorem 1. In particular, we see that the number m remains the same if $\mathfrak{G}$ is replaced by $\bar{\mathfrak{G}}$. Now the results of [1] II show that $\bar{\mathfrak{G}} \simeq \mathfrak{M}_{p,t}$, a contradiction.

We may therefore assume that (4.1) for $\lambda = 0$ reads

$$(4.8)\qquad \chi_q^{(0)} \mid \mathfrak{N} = \varepsilon^{(0)}\xi_q^0 + u_{\gamma,k}^{(0)}\Phi_k^{(\gamma)} + \cdots;\quad u_{\gamma,k}^{(0)} \geqq 1.$$

Since $t \geqq 3$, we can choose $i \not\equiv j \pmod{t}$ in (4.2 b). Restrict the arguments to $\mathfrak{N}$ and express all characters of $\mathfrak{N}$ by means of the $\xi_i^{(\alpha)}$, $\Phi_j^{(\alpha)}$, cf. (3 C), (3.6). Now (4.8) shows that $\Phi_k^{(\gamma)}$ will appear at least with the coefficient H in (4.2 b).

On the other hand, on account of (4.1), we have

$$\chi_i^{(\lambda)}\bar{\chi}_j^{(\lambda)} \mid \mathfrak{N} = \xi_i^{(\lambda)}\bar{\xi}_j^{(\lambda)} + \sum_{\alpha,r}\sum_{\beta,s} u_{\alpha r}^{(\lambda)} u_{\beta s}^{(\lambda)} \Phi_r^{(\alpha)}\bar{\Phi}_s^{(\beta)} + \varepsilon^{(\lambda)}\sum_{\alpha,r} u_{\alpha,r}^{(\lambda)}[\xi_i^{(\lambda)}\bar{\Phi}_r^{(\alpha)} + \bar{\xi}_j^{(\lambda)}\Phi_r^{(\alpha)}]$$

---

* The assumption (II) is only used to eliminate the case that $\chi_j^{(0)}(1) = (p-1)/t \leqq (p-1)/3$.

where (α, r) and (β, s) range over the pairs (q, μ) with $q = 0, \ldots, l-1$ and $\mu = 0, 1, \ldots, m_q - 1$. Again express all the characters on the right in accordance with (3 C). It follows from (3 D), (3 E), (3 F) that the coefficient of $\Phi_k^{(\tau)}$ is equal to

$$\sum_{\alpha, r}\sum_{\beta, s} u_{\alpha r}^{(\lambda)} u_{\beta s}^{(\lambda)} A_{rsk}^{\alpha\beta\tau} + \varepsilon^{(\lambda)} \sum_{\alpha, r} u_{\alpha, r}^{(\lambda)} [q_{rk}^{\alpha\lambda\tau} + q_{rk^*}^{\alpha\lambda\tau^*}]$$

where we set $\overline{\Phi}_k^{(\tau)} = \Phi_{k^*}^{(\tau^*)}$. Now (3.9) and (3.15) show that in the case $\varepsilon^{(\lambda)} = 1$, this quantity and hence H is at most equal to

$$(t + \tau_\tau^{-1}) f_\tau^{-1} [\sum_{\alpha, r} u_{\alpha r}^{(\lambda)} \tau_\alpha f_\alpha]^2 + 2 f_\lambda f_\tau^{-1} \sum_{\alpha, r} u_{\alpha r}^{(\lambda)} \tau_\alpha f_\alpha.$$

In the case $\varepsilon^{(\lambda)} = -1$, the second summand can be deleted. In either case we set

(4.9) $$R = \sum_{\alpha, r} u_{\alpha r}^{(\lambda)} \tau_\alpha f_\alpha.$$

Then our result shows that

(4.10 a) $H \leqq (t+1) R^2 + 2 f_\lambda R$ for $\varepsilon^{(\lambda)} = 1$;

(4.10 b) $H \leqq (t+1) R^2$ for $\varepsilon^{(\lambda)} = -1$.

Set $n = \chi_j^{(\lambda)}(1)$. Then (4.1) in conjunction with (3 A) and (3 B) yields

(4.11) $$n = \varepsilon^{(\lambda)} m f_\lambda + p \sum_{\alpha, r} u_{\alpha r}^{(\lambda)} \tau_\alpha f_\alpha = \varepsilon^{(\lambda)} m f_\lambda + pR.$$

Suppose first that $\varepsilon^{(\lambda)} = 1$. If we put $n/p = K$, we have

$$f_\lambda = p/(K - R)/m \leqq (t+1)(K - R)$$

and (4.10) becomes

$$H \leqq (t+1)(R^2 + 2(K-R)R) < (t+1) K^2.$$

Suppose then that $\varepsilon^{(\lambda)} = -1$. As remarked in (4 A), some $u_{\lambda r}^{(\lambda)}$ is positive and hence

$$R \geqq \tau_\lambda f_\lambda,$$
$$n \geqq -m f_\lambda + p \tau_\lambda f_\lambda > -m f_\lambda + m t \tau_\lambda f_\lambda.$$

Thus, $m f_\lambda < n (t_\lambda - 1)^{-1}$. For $n/p = K$, by (4.11)

$$R < K(1 + (t_\lambda - 1)^{-1}) = K t_\lambda / (t_\lambda - 1) \leqq \frac{3}{2} K.$$

This can be substituted in (4.10 b). We have now shown.

(4 D) *If* $\mathfrak{G}$ *satisfies the assumptions* (I) *and* (II), *then*

$$H < (t+1)n^2/p^2 \qquad (\text{for } \varepsilon^{(\lambda)} = 1).$$

$$H < (t+1)(t_\lambda^2/(t_\lambda-1)^2)(n^2/p^2) \leqq \frac{9}{4}(t+1)n^2/p^2 \qquad (\text{for } \varepsilon^{(\lambda)} = -1).$$

Here, n is the degree of the exceptional characters $\chi_j^{(\lambda)}$.

§ 5. Proof of Theorem 1

THEOREM 1. *Let $\mathfrak{G}$ be a group of order $g = pg_0$, where p is a prime and where g_0 is an integer not divisible by p. Assume that the p-Sylow group $\mathfrak{P}$ of $\mathfrak{G}$ is not normal in $\mathfrak{G}$ and that $\mathfrak{G}$ contains $t \geqq 3$ conjugate classes of elements of order p. If $\chi^{(\lambda)}$ of degree n is a faithful exceptional irreducible character of $\mathfrak{G}$ (for the prime p), then*

$$(5.1) \qquad p - 1 < w(w - 2\varepsilon^{(0)})/(t-2)$$

where $w = (t+1)t_\lambda n^2/p^2$ in the case $\varepsilon^{(\lambda)} = 1$ and $w = (t+1)\cdot(t_\lambda^3/(t_\lambda-1)^2)\cdot(n^2/p^2)$ in the case $\varepsilon^{(\lambda)} = -1$. Here $\varepsilon^{(\lambda)}$ and $\varepsilon^{(0)}$ are the signs belonging to the exceptional characters of the p-blocks B_λ and B_0.

Remark. The assumption that $\chi^{(\lambda)}$ is faithful is not needed in the following two cases: 1) if $\mathfrak{M}_{p,t}$ is not a homomorphic image of $\mathfrak{G}$. 2) if $\chi^{(\lambda)}(1) > (p-1)/3$.

Proof. We shall first make the additional assumption that $\mathfrak{M}_{p,t}$ is not a homomorpic image of $\mathfrak{G}$. Then (4 D) applies. We give a lower estimate for H. Form

$$S_{\alpha\beta}^{(\lambda)} = v^{-1}\sum_{V \in \mathfrak{V}} \chi_\alpha^{(\lambda)}(PV)\bar{\chi}_\beta^{(\lambda)}(PV).$$

It follows at once from (2.5) that

$$S_{\alpha\beta}^{(\lambda)} = \sum_{\nu=0}^{t_\lambda-1} \omega_{\alpha+\nu t}^{(\alpha)}\bar{\omega}_{\beta+\nu t}^{(\alpha)}.$$

On the other hand, we can use (4.2) to find $S_{\alpha\beta}^{(\lambda)}$. The equation (2.6) implies that

$$\sum_{V \in \mathfrak{V}} \zeta_i^{(0)}(PV) = \varepsilon_i^{(0)} v.$$

Similarly, by (2.5)

$$\sum_{V \in \mathfrak{V}} \chi_j^{(0)}(PV) = \varepsilon^{(0)} v\omega_j^{(0)} = \varepsilon^{(0)} v\eta_j.$$

Since $\Delta_{\alpha\beta}$ consists of constituents not in B_0, a similar argument shows that

$$\sum_{V\in\mathfrak{B}} \Delta_{\alpha\beta}(PV) = 0.$$

Since $\varepsilon_i^{(0)} = 1$ for $0 \leq i < a$ and $\varepsilon_i^{(0)} = -1$ for $a \leq i < m$, (4.2) implies

$$S_{\alpha\beta}^{(\lambda)} = \varepsilon^{(0)} \sum_{r=1}^{t} h_{\alpha\beta r}\eta_r + x + m\delta_{\alpha\beta}^{(\lambda)}$$

where $x \in \mathrm{Z}$ does not depend on α, β and where $\delta_{\alpha\beta}^{(\lambda)} = 1$ or 0 according as to whether or not $\alpha \equiv \beta \pmod{t_\lambda}$.

Hence

$$\sum_{\nu=1}^{\tau_\lambda} \omega_{\alpha+\nu t}^{(\lambda)} \bar{\omega}_{\beta+\nu t}^{(\lambda)} = \varepsilon^{(0)} \sum_{r=1}^{t} h_{\alpha\beta r}\eta_r + x + m\delta_{\alpha\beta}^{(\lambda)}. \tag{5.2}$$

It follows from (2.7) that

$$\eta_i\bar{\eta}_j = \sum_{\nu=0}^{\tau_\lambda-1} \sum_{\mu=0}^{\tau_\lambda-1} \omega_{i+\nu t}^{(\lambda)} \bar{\omega}_{j+\mu t}^{(\lambda)} = \sum_{\nu=0}^{\tau_\lambda-1} \sum_{\mu=0}^{\tau_\lambda-1} \omega_{i+\nu t}^{(\lambda)} \bar{\omega}_{j+\mu t+\nu t}^{(\lambda)}. \tag{5.3}$$

If we set

$$C_{ijr} = \sum_{\mu=0}^{\tau_\lambda-1} h_{i,j+\mu t,r}, \tag{5.4}$$

substitution of (5.2) into (5.3) yields

$$\eta_i\bar{\eta}_j = \varepsilon^{(0)} \sum_{r=1}^{t} C_{ijr}\eta_r + \tau_\lambda x + m\delta_{ij}^{(0)} \tag{5.5}$$

where $\delta_{ij}^{(0)} = 1$ or 0 according as to whether or not $i \equiv j \pmod{t}$. On account of (4.7), we find

$$\sum_{r=1}^{t} C_{ijr} = \tau_\lambda H - \delta_{ij}^{(0)}\varepsilon^{(0)}. \tag{5.6}$$

We may now compare (5.5) and (2.10). Since $\eta_1, \eta_2, \ldots, \eta_r$ are linearly independent over Z and since their sum is -1, this yields

$$\varepsilon^{(0)} C_{ijr} - \tau_\lambda x = c_{ijr}. \tag{5.7}$$

In particular, (2.12) shows that

$$C_{\alpha\beta\gamma} = C_{\alpha\gamma\beta} = C_{\gamma\alpha\beta'}. \tag{5.8}$$

It also follows from (5.7), (5.6) and (2.11) that

$$\varepsilon^{(0)}\tau_\lambda H - t_\lambda x = m, \tag{5.9}$$

Since $t \geqq 3$, we can choose $i \not\equiv j \pmod t$. We use (5.5) to express

$$(\eta_i \bar{\eta}_i)\bar{\eta}_j = (\eta_i \bar{\eta}_j)\bar{\eta}_i$$

in two ways as linear combination of $\eta_1, \eta_2, \ldots, \eta_m$. Comparing the coefficient of $\bar{\eta}_j = \eta_{j'}$, we find

$$\sum_{r=1}^{t} C_{iir} C_{rjj'} - \varepsilon^{(0)} \sum_{r=1}^{t} C_{iir} \tau_\lambda x - \varepsilon^{(0)} m C_{iij} + \tau_\lambda x + m$$
$$= \sum_{r=1}^{t} C_{ijr} C_{rij'} - \varepsilon^{(0)} \sum_{r=1}^{t} C_{ijr} \tau_\lambda x - \varepsilon^{(0)} C_{iji} m.$$

On account of (5.8) and (5.6), this becomes

$$\sum_{r=1}^{t} C_{iir} C_{jjr} + m + 2\tau_\lambda x = \sum_{r=1}^{t} C^2_{ijr}.$$

Since all $C_{\alpha\beta\gamma}$ are non negative,

$$m + 2\tau_\lambda x \leqq \sum_{r=1}^{t} C^2_{ijr} \leqq \Big(\sum_{r=1}^{t} C_{ijr}\Big)^2 = \tau_\lambda^2 H^2.$$

On account of (5.9), this becomes

$$mt^2 + 2(\varepsilon^{(0)}\tau_\lambda H - m)t \leqq t_\lambda^2 H^2$$

which can rewritten in the form

(5.10) $$(p-1)(t-2) \leqq t_\lambda H (t_\lambda H - 2\varepsilon^{(0)}).$$

Now, (5.1) is a consequence of (4 D).

It remains to deal with the case that $\mathfrak{G}$ contains a normal subgroup $\mathfrak{K}$ such that $\bar{\mathfrak{G}} = \mathfrak{G}/\mathfrak{K} \simeq \mathfrak{M}_{p,t}$. Clearly, $\mathfrak{K}$ then is the maximal normal subgroup of $\mathfrak{G}$ of an order prime to p. This implies that the principal p-blocks of $\mathfrak{G}$ and of $\bar{\mathfrak{G}}$ coincide, cf. [3], Theorem 1. Since the p-Sylow subgroup of $\bar{\mathfrak{G}}$ is self-centralizing, the natural homorphism of $\mathfrak{G}$ onto $\bar{\mathfrak{G}}$ maps $\mathfrak{V}$ into $\mathfrak{K}/\mathfrak{K}$ and we have $\mathfrak{V} \subseteq \mathfrak{K}$. It is now clear that

(5.11) $$\mathfrak{G} = \langle \mathfrak{K}\mathfrak{P}, M \rangle$$

and that $\mathfrak{K}\mathfrak{P}$ is a normal subgroup of index m in $\mathfrak{G}$.

We assume now that χ is a faithful character of $\mathfrak{G}$.

We now use induction with regard to m to prove that

(5.12) $$(p-1)(t-2) \leqq w_0(w_0 - 2)$$

with $w_0 = (t+1)t_\lambda n^2/p^2$. This will imply (5.1). Since $t \geqq 3$, we have $p \geqq 7$. As

$\mathfrak{P}$ is not normal in $\mathfrak{G}$, it cannot be normal in $\mathfrak{K}\mathfrak{P}$.

If $m=1$, then $t=t_\lambda=p-1$. It will suffice to show $n\geqq p-1$ since then $w_0>(p-1)(p-2)$ and hence (5.12) will hold. If $n<p-1$, it follows from [1] II, Theorem 1 and Corollary 2 that $n=f_\lambda$ and that for the single non-exceptional character $\zeta_0^{(\lambda)}$ of B_λ, we have $\zeta_0^{(\lambda)}|\mathfrak{E}=\theta_\lambda$, while $\chi_j^{(\lambda)}$ is obtained from $\zeta_0^{(\lambda)}$ by multiplication with a linear character of $\mathfrak{P}\simeq\mathfrak{G}/\mathfrak{K}$. Thus, $\mathfrak{P}$ belongs to the kernel $\mathfrak{H}$ of $\zeta_0^{(\lambda)}$ and $\mathfrak{H}$ cannot contain p-regular elements $G\neq 1$ of $\mathfrak{G}$. Hence $\mathfrak{H}=\mathfrak{P}$ and we have $\mathfrak{P}\lhd\mathfrak{G}$, a contradiction.

Suppose then that $m>1$. Let s be a prime dividing m and set $M^*=M^s$,

$$\mathfrak{G}^*=\langle\mathfrak{K}\mathfrak{P},\ M^*\rangle.$$

Then $\mathfrak{G}^*$ is a normal subgroup of $\mathfrak{G}$ of index s. Moreover, $\mathfrak{G}^*$ is of the same structure as $\mathfrak{G}$ with m replaced by $m^*=m/s$, i.e. with t replaced by $t^*=ts$. In (4.2 a), the $\zeta_0^{(0)},\ldots,\zeta_{a-1}^{(0)}$ are the m linear characters of the cyclic group $\mathfrak{G}/\mathfrak{K}\mathfrak{P}$ of order m; $a=m$. Since s of them have a kernel including $\mathfrak{G}^*$, it follows from (4.2 a) that $\chi_j^{(\lambda)}|\mathfrak{G}^*$ is reducible. Consequently, $\chi_j^{(\lambda)}|\mathfrak{G}^*$ splits into s irreducible characters of degree $n^*=n/s$. The formulas (2.5) show that some of these constituents belong to the block B_λ^* of $\mathfrak{G}^*$ associated with θ_λ. If τ_λ^*, t_λ^* have the same significance for B_λ^* as τ_λ, t_λ have for B_λ, clearly, $\tau_\lambda^*\leqq\tau_\lambda$ and $t_\lambda^*\leqq st_\lambda$. It we set $w_0^*=(t^*+1)t_\lambda^*n^{*2}/p^2$, we have $w_0^*\leqq w_0$. It is now clear that the analogue of (5.12) for $\mathfrak{G}^*$ implies (5.12) and the proof of Theorem 1 is complete.

Corollary. *In Theorem 1, we have*

$$n>\frac{1}{3}t_\lambda^{-3/4}p^{5/4}.$$

Proof. If we have $w-2\varepsilon^{(0)}>2w$, then $\varepsilon^{(0)}=-1$, $w<2$ and (5.1) reads $p-1\leqq 8/(t-2)$. This is only possible for $p=7$, $t=3$. Since $w<2$ we have $12n^2<98$ and hence $n\leqq 2$. Then $n=2$. In this case, the result holds.

Assume then that $w-2\varepsilon^{(0)}\leqq 2w$. By (5.1),

$$p-1<2w^2/(t-2)\leqq 2(t+1)^2(t-2)^{-1}t_\lambda^6(t_\lambda-1)^{-4}n^4p^{-4}.$$

Here,

$$6p/7\leqq p-1,\ 3(t+1)^2\leqq 16(t-2)t,\ 2t_\lambda\leqq 3(t_\lambda-1)$$

and we obtain

$$p^5 < 63\, tt_\lambda^2 n^4 < 3^4 t_\lambda^3 n^4$$

and this yields the desired result.

Remark. If an irreducible character χ of $\mathfrak{G}$ has defect 0, it vanishes for p-singular elements. Thus, all values of χ lie in the field of the g_0-th roots of unity. This shows that the Corollary can be stated in the form given in the Introduction.

§ 6. Proof of Theorem 2

THEOREM 2. *Let $\mathfrak{G}$ be a group of order $g = pg_0$ where p is a prime and g_0 an integer not divisible by p. Assume that (1) the p-Sylow group $\mathfrak{P}$ of $\mathfrak{G}$ is not normal in $\mathfrak{G}$ and (2) that the number t of conjugate classes of elements of order p is at least 3. If $\mathfrak{G}$ has a faithful irreducible character χ of degree $n < p-1$ then $n = p - (p-1)/t$ and*

$$p \leqq t^3 - t + 1. \tag{6.1}$$

Proof. Since χ has degree $n < p-1$, it follows from [1] II, Corollary 2 that χ is an exceptional character $\chi_j^{(\lambda)}$ of a p-block of defect 1. It is also clear that we have one of the cases

Case 1. $\varepsilon^{(\lambda)} = 1$

$$\chi_j^{(\lambda)} | \mathfrak{N} = \xi_j^{(\lambda)}. \tag{6.2 a}$$

Case 2. $\varepsilon^{(\lambda)} = -1$

$$\chi_j^{(\lambda)} | \mathfrak{N} = -\xi_j^{(\lambda)} + \Phi_\mu^{(\lambda)} \tag{6.2 b}$$

where μ is one of the values $0, \ldots, m_\lambda - 1$.

Suppose that $\mathfrak{G}$ has a normal subgroup $\mathfrak{K}$ for which $\mathfrak{G}/\mathfrak{K} \simeq \mathfrak{M}_{p,t}$. As in § 5 we see that $\mathfrak{K}\mathfrak{P}$ is a normal subgroup of index m in $\mathfrak{G}$ and that $\mathfrak{V} \subseteq \mathfrak{K}$. It is also clear that the irreducible constituents of $\chi | \mathfrak{P}\mathfrak{K}$ lie in the τ_λ blocks B_μ^* of $\mathfrak{K}\mathfrak{P}$ determined by the τ_λ associates of $\theta_\lambda^{M\nu}$. All irreducible characters of B_μ^* have the same degree n^* and there is exactly one non-exceptional character $\zeta^{\mu *}$ in B_μ^*. Since $n^* < p-1$, Corollary 2 of [1] II shows that $\mathfrak{P}$ belongs to the kernel of $\zeta^{(\mu)*}$. Again, the exceptional characters of B_μ^* are obtained from the $\zeta^{(\mu)*}$ by multiplication with the linear characters of $\mathfrak{P}\mathfrak{K}/\mathfrak{K} \simeq \mathfrak{P}$. If we form the character

$$Z = \sum_{\mu} \zeta^{(\mu)*}$$

of $\mathfrak{K}\mathfrak{P}$, we see that an element $K \in \mathfrak{K}$ belongs to the kernel $\mathfrak{H}$ of Z only if it belongs to the kernel of $\chi_j^{(\lambda)}$, i.e. if $K=1$. Since $\mathfrak{P} \subseteq \mathfrak{H}$, we have $\mathfrak{P} \triangleleft \mathfrak{P}\mathfrak{K}$ and then $\mathfrak{P} \triangleleft \mathfrak{G}$, a contradiction. Hence $\mathfrak{G}$ satisfies the hypothesis (I), (II) in § 4, and all results of § 4 can be used.

We show next that the number H in (4.7) cannot vanish. Indeed, if $H=0$, (5.6) shows that, for $i, j=1, 2, \ldots, t$ with $i \neq j$, we have

$$C_{ijr} = 0, \qquad (i, j = 1, 2, \ldots, t).$$

By (5.9), $x = -m/t_\lambda$. Then (5.7) yields $c_{ijr} = m/t$ for the same i, j, r. It now follows from (2.10) that

$$\eta_i \bar{\eta}_j = -m/t \qquad (\text{for } i \neq j).$$

This is clearly impossible, since $\eta_1, \eta_2, \ldots, \eta_t$ are distinct and $t \geqq 3$. Hence $H \geqq 1$.

It is now easy to see that the first case is impossible. Indeed, in this case, $R=0$ by (4.9), while (4.10 a) shows that $R \neq 0$.

Suppose then that we have Case 2. By (6.2 b)

$$n = \chi_j^{(\lambda)}(1) = -mf_\lambda + pf_\lambda \tau_\lambda = \tau_\lambda f_\lambda (p - m_\lambda).$$

Since $m_\lambda \leqq m \leqq (p-1)/3$, the assumption $n < p-1$ implies that $\tau_\lambda = f_\lambda = 1$. Thus, $n = p - (p-1)/t$.

It now follows from the results of § 3 that

$$\xi_j^{(\lambda)} | \mathfrak{V} = m\theta_\lambda, \qquad \psi_i^{(\lambda)} | \mathfrak{V} = \theta_\lambda$$

for all i and j. Then (3.6) shows that $\Phi_\mu^{(\lambda)} | \mathfrak{V} = p\theta_\lambda$. By (6.2 b),

$$\chi_j^{(\lambda)} | \mathfrak{V} = n\theta_\lambda.$$

Since $\chi_j^{(\lambda)}$ was faithful, this implies that $\mathfrak{V}$ is cyclic and that it belongs to the center $\mathfrak{Z}(\mathfrak{G})$ of $\mathfrak{G}$. Moreover, $\chi_i^{(\lambda)} \bar{\chi}_j^{(\lambda)}$ is trivial on $\mathfrak{V}$. Then the formulas (2.5) and (2.6) show that the irreducible constituents of $\chi_i^{(\lambda)} \bar{\chi}_j^{(\lambda)} | \mathfrak{N}$ belong to b_0. We can therefore express $\chi_i^{(\lambda)} \bar{\chi}_j^{(\lambda)} | \mathfrak{N}$ as a linear combination of the $\psi_\mu^{(0)}$ and the exceptional characters in b_0. Now $\xi_\alpha^{(\lambda)} \bar{\xi}_\beta^{(\lambda)}$ contains the linear character $\psi_\mu^{(0)}$, if and only if $\xi_\alpha^{(\lambda)} = \psi_\mu^{(0)} \xi_\beta^{(\lambda)}$. Since $\psi_\mu^{(0)}$ is trivial on $\mathfrak{P} \times \mathfrak{V}$, this is so, if and only if $\alpha = \beta$. Moreover, for $\alpha = \beta$, the character $\psi_\mu^{(0)}$ appears with multiplicity 1 in $\xi_\alpha^{(\lambda)} \bar{\xi}_\alpha^{(\lambda)}$. A similar argument shows that $\psi_\mu^{(0)}$ cannot occur in $\xi_i^{(\lambda)} \bar{\psi}_\mu^{(\lambda)}$. Finally,

$\varphi_\mu^{(\lambda)}\overline{\varphi}_\mu^{(\lambda)}=1$. It now follows from (6.2 b) and (3.6) that

$$\chi_i^{(\lambda)}\overline{\chi}_j^{(\lambda)}|\mathfrak{N}=(t-1+\delta_{ij}^{(0)})\varphi_0^{(0)}+\sum_{\mu=1}^{t-1}(t-2+\delta_{ij}^{(0)})\varphi_\mu^{(0)}+\cdots \tag{6.3}$$

where the dots on the right stand for exceptional characters in b_0. On account of (3.6), this can be written in the form

$$\chi_i^{(\lambda)}\overline{\chi}_j^{(\lambda)}|\mathfrak{N}=(t-1+\delta_{ij}^{(0)})\Phi_0^{(0)}+\sum_{\mu=1}^{t-1}(t-2+\delta_{ij}^{(0)})\Phi_\mu^{(0)}+\cdots$$

where the characters not written are again exceptional characters in b_0.

We can apply the same method as in § 4 and compare our formula with (4.2) (restricted to $\mathfrak{N}$). Since $H\neq 0$, it follows at once that in (4.8), only terms $\Phi_k^{(\gamma)}$ with $\gamma=0$ can appear with coefficients $u_k^{(\gamma)}\neq 0$. Moreover, $H\leqq t-1$.

Actually, our method shows that we can have $H=t-1$ only if (4.8) has the form

$$\chi_q^{(0)}|\mathfrak{N}=\varepsilon^{(0)}\xi_q^{(0)}+\Phi_0^{(0)}. \tag{6.4}$$

We can then also compare the multiplicity of $\varphi_0^{(0)}$ in (6.3) for $i=j$ and in (4.2 a), restricted to $\mathfrak{N}$. On account of (6.4) and (4.7) this yields

$$t\geqq 1+H-\varepsilon^{(0)}.$$

Thus, if $H=t-1$, we must have $\varepsilon^{(0)}=1$. Then $\chi_q^{(0)}(1)=m+p$. Since $\varepsilon^{(0)}\neq\varepsilon^{(\lambda)}$, we have $\lambda\neq 0$. It follows that $\mathfrak{B}\neq 1$ and hence that $\mathfrak{Z}(\mathfrak{G})\neq 1$. Since $\varepsilon^{(0)}=1$, (6.1) is an immediate consequence of (5.10).

On the other hand, if $H\leqq t-2$, (5.10) yields

$$p-1\leqq t^2(t-2)+2t=t^3-2t^2+2t<t^3-t.$$

This completes the proof of Theorem 2. We also have

Theorem 2*. *If $\mathfrak{G}$ has center* 1, *the inequality* (6.1) *in Theorem* 2 *can be replaced by*

$$p\leqq t^3-2t^2+2t+1.$$

The same is true, if the degrees of the exceptional characters in the principal p-block of $\mathfrak{G}$ are different from $p+(p-1)/t$.

References

[1] R. Brauer, On groups whose order contains a prime number to the first power, I, II. American Journal of Mathematics **64** (1942), 401-420, 421-440.

[2] R. Brauer, A characterization of the characters of groups of finite order, Annals of Mathematics **57** (1953), 357-377.

[3] R. Brauer, Some applications of the theory of blocks of characters of finite groups, I. Journal of Algebra **1** (1964), 152-167.

[4] R. Brauer and C. Nesbitt, On the modular characters of groups, Annals of Mathematics **52** (1941), 556-590.

[5] H. F. Tuan, On groups whose order contains a prime to the first power, Annals of Mathematics **45** (1944), 110-140.

Harvard University

On simple groups of order $5 \cdot 3^a \cdot 2^b$

Richard Brauer*

We shall prove the following result:

__Theorem.__ If G is a simple group of an order $g = 5 \cdot 3^a \cdot 2^b > 5$, then G is isomorphic to one of the following three groups:

(i) the alternating group A_5 of order 60.

(ii) the alternating group A_6 of order 360.

(iii) the orthogonal group $O(5,3)$ of order 25,920.

In section I, some results for simple groups of an order $p^a q^b r$ with p,q,r distinct primes are given. In section II, the case $p = 3$, $q = 2$, $r = 5$ is discussed and it is shown that if G is not of type (i),(ii), its order must be 25,920. In section III, the proof is finished.

No attempt is made to develop the methods used in the most general situations in which they can be applied. We shall come back to this elsewhere.

__Notation.__ Throughout the paper, G will be a finite group; $\chi_0 = 1$, $\chi_1, \ldots$ will denote the irreducible character of G and we set $x_i = \chi_i(1)$. For any prime ℓ, $B_0(\ell)$ will denote the principal ℓ-block of G. For $\tau \in G$, $C(\tau)$ will be the centralizer of τ in G and $c(\tau) = |C(\tau)|$ the order of $C(\tau)$.

---

*This research has been supported by an NSF Grant.

I. Simple groups of order $p^a q^b r$.

Hypothesis (I). In section I, G will be a simple group of order $g = p^a q^b r$ where p,q,r are primes, $g \neq r$.

By Burnside's theorem, p,q,r are distinct and $a > 0$, $b > 0$,

(1A) The r-Sylow subgroups R of G are their own centralizers in G;

$$C_G(R) = R.$$

Proof. Suppose that this is not true. Interchanging p and q if necessary, we may assume that there exist elements π of order p in $C_G(R)$. Choose $\rho \neq 1$ in R. Since $\pi\rho$ is a p-singular element, we have

$$\sum_{\chi_i \in B_0(p)} x_i \chi_i (\pi\rho) = 0,$$

[6, Theorem 8]. Since $\chi_0 \in B_0(p)$, it follows that there exists a $\chi_i \in B_0(p)$ with $i > 0$ for which

$$x_i \not\equiv 0 \pmod q, \qquad \chi_i(\pi\rho) \neq 0.$$

Then the degree x_i cannot be divisible by r, since then χ_i would have defect 0 for the prime r and χ_i would vanish for the r-singular element $\pi\rho$.

It is now clear that x_i is a power of p. Because of the simplicity of G, $x_i > 1$. We now have a contradiction with [9, Lemma 2] and (1A) is proved.

As a corollary, we note (cf. [2]).

(**1B**) The principal r-block $B_o(r)$ of G is the only r-block of positive defect. If the normalizer N(R) of R has order rm, then m is a divisor of r-1, m ≠ 1, and N(R) is generated by R and an element μ of order m. Set $t = (r-1)/m$. Then $B_o(r)$ consists of m non-exceptional characters χ_i and t exceptional characters $\chi^{(\tau)}$, $\tau = 1,2,\cdots,t$. If ρ is r-singular, then for the non-exceptional characters, $\chi_i(\rho) = \delta_i = \pm 1$, while the values of the t exceptional characters are the t Gauss periods of length $m = (r-1)/t$ in the field of the r-th root of unity, multiplied with a fixed sign-δ. Finally, if σ is an r-regular element of G,

$$(1.1) \qquad \sum{}' \delta_i \chi_i(\sigma) + \delta\chi^{(\tau)}(\sigma) = 0$$

where χ_i in $\sum{}'$ ranges over the non-exceptional characters in $B_o(r)$.

In particular, (1.1) can be applied for $\sigma = 1$. Thus

$$(1.2) \qquad \sum{}' \delta_i x_i + \delta x = 0 .$$

Here x is the common degree of the exceptional characters $\chi^{(\tau)}$

We also note:

(**1C**) If $\chi_i \in B_o(\tau)$ is non-exceptional, then

$$(1.3) \qquad x_i \equiv \delta_i \pmod r$$

while for the degree x of the exceptional characters $\chi^{(\tau)}$

$$(1.4) \qquad tx \equiv \delta \pmod r .$$

Finally, for $\chi_i \notin B_o(r)$,

(1.5) $$x_i \equiv 0 \pmod{r}.$$

In the case $t = 1$, there is no difference between exceptional and non-exceptional characters. An arbitrary member of $B_o(r)$ can be chosen as the character $\chi^{(1)}$.

The number $m = (r-1)/t$ is the order of $\mu_.$ and hence is a divisor of g and then of g/r. We shall set

(1.6) $$m = (r-1)/t = p^u q^v.$$

Since $g/(rm)$ is the number of Sylow-r-groups, by Sylow's theorem

(1.7) $$p^a q^b \equiv p^u q^v = m \pmod{r}.$$

All the degrees x_i, x appearing in (1.2) are divisors of g/r. Since $x_o = \delta_o$, the following result is immediate.

(1D) There exist characters χ_n and χ_h with $n \neq 0$, $h \neq 0$ for which x_n is a power of p and x_h a power of q,

(1.8) $$x_n = p^\alpha > 1; \qquad x_h = q^\beta > 1.$$

Now Lemma 2 of [9] yields

(1E) The character χ_n in (1D) belongs to a p-block $B^*(p)$ of defect less than a. Likewise, χ_h belongs to a q-block $B^*(q)$ of defect less than b.

<u>Notation</u>. In what is to follow, the notation will be chosen such that $\chi_0 = 1, \chi_1, \cdots, \chi_{m-1}$ are the non-exceptional characters in $B_0(r)$ and that $\chi_m = \chi^{(1)}$. We set $\delta = \delta_m$. If $B(p)$ is any p-block, let $T(B(p))$ denote the set of indices i with $0 \leqq i \leqq m$ for which $\chi_i \in B(p)$. A corresponding notation is used for q-blocks $B(q)$.

The equation (1.1) can now be written in the form

$$\sum_{i=0}^{m} \delta_i \chi_i(\sigma) = 0$$

for r-regular elements σ. In particular, this holds for all p-singular σ. The proof of Lemma 3 of [9] yields the next result.

(<u>1F</u>) Let $B(p)$ be a p-block of G, and set

$$\psi_{B(p)} = \sum_{i \in T(B(p))} \delta_i \chi_i. \tag{1.9}$$

The generalized character $\psi_{B(p)}$ vanishes for all p-singular elements of G and

$$\psi_{B(p)}(1) = \sum_{i \in T(B(p))} \delta_i x_i \equiv 0 \pmod{p^a}. \tag{1.10}$$

(<u>1G</u>) If $T(B(p)) \neq \emptyset$, then $\psi_{B(p)}(1) \neq 0$.

<u>Proof</u>. It follows from (1E) that $T(B(p))$ does not contain all indices $0,1,\ldots,m$. If the exceptional character χ_m does not belong to $B(p)$, then by (1.3)

$$\sum_{i \in T(B(p))} \delta_i x_i \equiv \sum_{i \in T(B(p)} 1 \pmod{r}.$$

Since the number of elements in $T(B(p))$ is positive by assumption and less than $m+1 \leq r$, we have $\phi_{B(p)}(1) \not\equiv 0 \pmod r$ and this implies the statement.

If $\chi_m \in B(p)$, consider the sum $\sum \delta_i x_i$ with i ranging over the indices with $0 \leq i \leq m$ for which $\chi_i \notin B(p)$. An analogous argument shows that this sum does not vanish. Then (1.2) implies that (1.9) does not vanish either for the element 1.

(**1H**) We have

$$(1.11) \qquad p^a < (r^2-1)q^b/4 .$$

Proof. It is clear that r must be odd. If $T(B(p))$ for some p-block $B(p)$ consists of $k \neq 0$ indices, then as $|\phi_{B(p)}(1)| \geq p^a$ by (1F),(1G), Lagrange's inequality yields

$$p^{2a} \leq \Big|\sum_{i\in T(B(p))} \delta_i x_i\Big|^2 \leq k \sum_{i\in T(B)} x_i^2 .$$

In particular, this can be applied for $B_o(p)$ and for the block $B^*(p)$ defined in (1E). If $T(B_o(p))$ and $T(B^*(p))$ consist of k_1 and k_2 indices respectively, then

$$\sum_{i=0}^{m} x_i^2 \geq \sum_{i\in T(B_o(p))} x_i{}^2 + \sum_{i\in T(B^*(p))} x_i{}^2 \geq p^{2a}\left(\frac{1}{k_1} + \frac{1}{k_2}\right)$$

Here, $k_1+k_2 \leq m+1 \leq r$. As r is odd, we obtain easily

$$\sum_{i=0}^{m} x_i{}^2 \geq p^{2a}\, 4r/(r^2-1).$$

Since the sum of the squares of all degrees $x_i{}^2$ is $g = p^a q^b r$, we obtain (1.11).

(1I) Let ψ be a generalized character of G which vanishes for all p-singular elements of G, but for which $\psi(1) \neq 0$. Let T be the set of indices i for which χ_i occurs with a non-zero coefficient in ψ. Suppose that t = 1.* If B(q) is a q-block which does not contain any χ_i with $i \in T$, then all characters of B(q) have the same degree $z \not\equiv 0 \pmod{r}$. If several q-blocks B(q) of the type in question exist, the number z is the same for all of them.

Proof. Assume that B(q) contains a character χ_j with $\chi_j \equiv 0 \pmod{r}$. Then χ_j vanishes for all r-singular elements of G. Since a q-regular element $\sigma \neq 1$ is either p-singular or r-singular, $\psi(\sigma)\overline{\chi}_j(\sigma) = 0$. Hence if σ ranges over all q-regular elements of G, we have

$$\sum_{\sigma} \psi(\sigma)\overline{\chi}_j(\sigma) = \psi(1)\overline{\chi}_j(1) \neq 0 .$$

On the other hand, $\psi\overline{\chi}_j$ is a linear combination of terms $\chi_i\overline{\chi}_j$ with $i \in T$. By hypothesis, χ_i and χ_j lie in different q-blocks and the orthogonality relations for q-modular characters [6, Theorem 4] show that the sum of $\chi_i(\sigma)\overline{\chi}_j(\sigma)$ extended over all q-regular elements σ vanishes. This is a contradiction.

Thus all $\chi_j \in B(q)$ belong to $B_0(r)$. Suppose we have two such

*It is easy to remove this assumption. However, the statement of (1I) then has to be modified.

characters χ_j and χ_k of different degrees belonging both to q-blocks of the type in question. As shown by (1B), $\delta_j\chi_j-\delta_k\chi_k$ vanishes for r-singular elements. On the other hand, $\delta_j\chi_j(1)-\delta_k\chi_k(1)$ is not zero. If σ has the same range as above,

$$\sum_{\sigma} \psi(\sigma)(\delta_j\chi_j(\sigma)-\delta_k\chi_k(\sigma)) = \psi(1)(\delta_j x_j-\delta_k x_k) \neq 0.$$

On the other hand, the sum on the left again vanishes and we have a contradiction.

(1J) If all characters χ_i of a q-block B(q) of defect d have the same degree z, then B(q) consists of q^d characters χ_i.

This is a general remark which does not depend on our special assumptions. Indeed all modular irreducible characters of B(q) then have degree z. Since each character χ_i can have only one modular irreducible constituent, only one modular irreducible character occurs in B(q) and all decomposition numbers are 1. If B(q) contains k characters χ_i, the Cartan matrix is (k). Since the Cartan matrix has the elementary divisor q^d, [3,(6C)], we have $k = q^d$ as claimed. It is also clear that q^d is the exact power of q dividing $g/x_i = g/z$. In our case in II, we have

$$(1.12) \qquad z = x_i = p^{\nu}q^{b-d}$$

for some integer $\nu \geqq 0$.

We discuss some applications of (1I),(1J).

(1K) If the p-block $B(p)$ meets $B_o(r)$, it meets $B_o(q) \cap B_o(r)$.

Indeed, if this was not so, we could apply (1F),(1G) and (1I) with $\psi = \psi_{B(p)}$ and $B(q) = B_o(q)$, $T = T(B(p))$. It would follow that all characters $\chi_i \in B_o(q)$ have the same degree z. Since $\chi_o \in B_o(q)$, then $z = 1$ and $B_o(q)$ would consist only of χ_o which is absurd.

Assume $t=1$.

(1L) Let $B^*(p)$ as in (1E) be a p-block which contains a character χ_n of degree $x_n = p^\alpha > 1$ and let $B^*(q)$ be a q-block which contains a character χ_h of degree $x_h = q^\beta > 1$. If $B^*(p)$ and $B^*(q)$ can be chosen such that $B_o(r) \cap B^*(p) \cap B^*(q) = \emptyset$, each p-block of defect less than a, and in particular $B^*(p)$, consists of exactly $p^{a-\alpha}$ irreducible characters χ_i, all of degree p^α. Each q-block of defect less than b, and in particular $B^*(q)$, consists of $q^{b-\beta}$ irreducible characters, all of degree q^β.

Proof. We apply (1I) with $\psi = \psi_{B^*(p)}$ and $B(q) = B^*(q)$. It follows that $B^*(q)$ consists of characters of the same degree z. Since $\chi_h \in B^*(q)$, $z = q^\beta$ and then $B^*(q)$ has defect $b-\beta$. Now (IJ) shows that $B^*(q)$ consists of $q^{b-\beta}$ characters χ_j. Interchanging the role of p and q and of $B^*(p)$ and $B^*(q)$, we see that $B^*(p)$ consists of $p^{a-\alpha}$ characters χ_i, all of degree p^α. If $B(q)$ is a q-block of defect $< b$, all characters $\chi_j \in B(q)$ have degrees $x_j \equiv 0 \pmod q$. Then $B(q)$ does not contain a character of degree

p^{α}. Hence $B(q) \cap B^*(p) = \emptyset$. Now (1I),(1J) show that $B(q)$ consists of characters of degree $z = q^{\beta}$ and that there are $q^{b-\beta}$ such characters. Likewise, any p-block $B(p)$ of defect $< a$ consists of $p^{a-\alpha}$ irreducible characters, all of degree p^{α}.

<u>Remark.</u> In the results of Section I, we can interchange the numbers (p,a,α) and (q,b,β).

II. Simple groups of order $5 \cdot 3^a \cdot 2^b$

We now assume that $r = 5$, $p = 3$, $q = 2$. Then $t = 2$ or $t = 1$.

If $t = 2$, we have $x_o = 1$ and

$$x_1 = 3^{\alpha},\ x_2 = 2^{\beta} \qquad \text{or} \qquad x_1 = 2^{\beta},\ x_2 = 3^{\alpha} .$$

In the first case, χ_1 is the only character in $B_o(5) \cap B^*(3)$. Then (1.10) shows that $\alpha = a$. Likewise, $\beta = b$. The same conclusion can be reached in the second case. On account of (1.2) we have

$$(2.1) \qquad 2^b = 3^a \pm 1 \ .^*$$

Since $b > 1$, this equation is satisfied only in the following cases:

$$\text{(i)} \qquad 4 = 3 + 1$$

$$\text{(ii)} \qquad 8 = 9 - 1 \ .$$

*If $2^k = 3^{\ell} \pm 1$ with positive integers k,ℓ then $k = 1,2,$ or 3 and $\ell = 1$ or 2. This is seen as follows: 2 belongs to the exponent 2 (mod 3) and then to the exponent $2 \cdot 3^{\ell-1}$ (mod 3^{ℓ}). Since $2^{2k} \equiv 1$ (mod 3^{ℓ}), k is divisible by $3^{\ell-1}$. Now $2^w > 3w+1$, for $w \geq 4$. For $w = 3^{\ell-1}$ with $\ell \geq 3$, it follows that $2^k \geq 2^w > 3w+1 > 3^{\ell}+1$.

The case (i) leads to $g = 5\cdot 3\cdot 4 = 60$ and (ii) leads to $g = 5\cdot 9\cdot 8 = 360$.

The table of simple groups of small orders shows that in the case (i), we have $G \simeq A_5$ and in the case (ii), $G \simeq A_6$. If we wish to avoid these tables, we can also see this in various ways by applying known theorems. We note first that G cannot contain elements of order 6. Indeed if σ was such an element, then χ_1 and χ_2 both vanish for σ since one of the characters χ_1, χ_2 is of defect 0 for 2 and the other for 3. Now (1.1) yields a contradiction Then the result of [8] can be applied. This shows that $G \simeq PSL(2,w)$ for some prime power w. The value of g implies that $w = 4$, 5 or 9 and $PSL(2,4) \simeq PSL(2,5) \simeq A_5$; $PSL(2,9) \simeq A_6$. Instead of [8], we could also have used [12] or [13]. In any case, we have

(2A) If G is a simple group of order $5\cdot 3^a\cdot 2^b$ and if $t = 2$ for the prime $r = 5$, then $G \simeq A_5$ or $G \simeq A_6$.

From now on, we shall assume that $t = 1$. Then $m = 4$ and (1.7) becomes

$$3^a 2^b \equiv -1 \pmod 5. \tag{2.1}$$

This implies

$$a \equiv b+2 \pmod 4. \tag{2.2}$$

If $\chi_j \in B_0(5)$, then $x_j \equiv \delta_j \pmod 5$ by (1.3),(1.4). Here x_j has the form $x_j = p^{\alpha_j} q^{\beta_j}$ and then

$$(2.3)\qquad x_j = p^{\alpha_j}q^{\beta_j} \quad \text{with} \quad \begin{cases} \alpha_j \equiv \beta_j \pmod 4 & \text{for } \delta_j = 1 \\ \alpha_j \equiv \beta_j+2 \pmod 4 & \text{for } \delta_j = -1. \end{cases}$$

We next discuss the tree corresponding to the block $B_o(r)$, cf. [2]; [14]. This tree has five vertices corresponding to the five characters χ_i in $B_o(r)$. It can be arranged symmetrically with regard to its "stem" and one of the end points of the stem is the vertex corresponding to χ_o. We then have the following three possibilities:

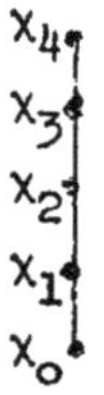

Type 1

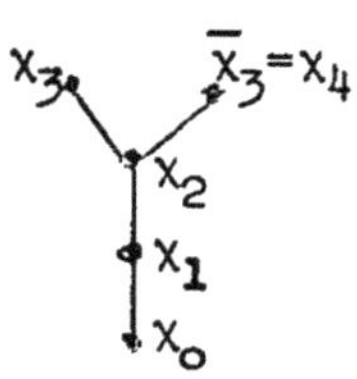

Type 2

Type 3

Here, the characters χ_3 and $\bar{\chi}_3$ in the cases 2 and 3 are conjugate complex and hence have the same degree. All the other characters χ_i are real. Since the signs δ_i corresponding to neighboring vertices are opposite, we have

$$(2.4)\qquad \begin{cases} \underline{\text{Case 1:}} & \delta_o = \delta_2 = \delta_4 = 1, \quad \delta_1 = \delta_3 = -1 \\ \underline{\text{Case 2:}} & \delta_o = \delta_2 = 1, \quad \delta_1 = \delta_3 = \delta_4 = -1 \\ \underline{\text{Case 3:}} & \delta_o = \delta_2 = \delta_3 = \delta_4 = 1, \quad \delta_1 = -1. \end{cases}$$

We shall also need some inequalities for the degrees. If χ_i is a character which does not correspond to a free end point, then as a 5-modular character, χ_i is reducible and $\chi_i\bar{\chi}_i$ contains the modular principal character φ_o at least twice. Now it can be seen from the tree that the only characters χ_j containing φ_o as modular constituent are χ_o and χ_1. Considered as an ordinary character, $\chi_i\bar{\chi}_i$ contains χ_o exactly once. Hence $\chi_i\bar{\chi}_i$ must contain χ_1. On comparing degrees, we have

(2.5) $$x_i^2 \geqq 1+x_1 \quad \left\{\begin{array}{l}\text{for } \chi_i \text{ not corresponding}\\ \text{to a free end point.}\end{array}\right.$$

Similarly, if χ_i and χ_j correspond to neighboring vertices then as 5-modular characters, they have a common irreducible constituent and then $\chi_i\bar{\chi}_j$ contains φ_o. Taken as an ordinary character, $\chi_i\bar{\chi}_j$ does not contain χ_o. Hence it must contain χ_1. This implies that $\chi_j\chi_1$ contains χ_i. Comparing degrees, we find

(2.6) $$x_i x_j \geqq x_1, \quad x_j x_1 \geqq x_i \quad \left\{\begin{array}{l}\chi_i,\chi_j \text{ corresponding}\\ \text{to neighboring vertices.}\end{array}\right.$$

We use these inequalities to prove a lemma

(2B) If the tree belonging to $B_o(5)$ is of type 1, then

(2.7) $$x_\mu^3 \geqq x_1, x_2, x_3 \qquad \text{for } \mu = 1,2,3; \quad x_4^4 > x_1, \ x_4^4 > x_3.$$

If the tree is of type 3, then

(2.8) $$x_2^3 \geqq x_1 > x_3.$$

(There are further such inequalities, but they will not be needed.)

<u>Proof</u>. Suppose the tree is of type 1. By (2.6), $x_1^2 \geqq x_2$, $x_1x_2 \geqq x_3$, $x_1x_3 \geqq x_2$. Then $x_1^3 \geqq x_1x_2 \geqq x_3$. By (2.5), $x_\mu^2 > x_1$ for $\mu = 2,3$ and hence $x_\mu^3 > x_1x_\mu \geqq x_3$. Also, $x_3^3 \geqq x_1x_3 \geqq x_2$. This proves the first part of (2.7).

Since $x_4 \neq 1$, the product $\chi_4\bar{\chi}_4$ contains at least one more character χ_μ besides χ_0 and then $\chi_4\chi_\mu$ contains χ_4. Hence $\chi_4^2\bar{\chi}_4$ contains $2\chi_4$. Since $\delta_4 = 1$, for elements ρ of order 5, $\chi_4(\rho) = 1$ by (1B). Since $\chi_4^2(\rho)\overline{\chi_4(\rho)} = 1$ and $2\chi_4(\rho) = 2$, the product $\chi_4^2\bar{\chi}_4$ contains a constituent χ_ν for which $\chi_\nu(\rho) \neq 0,1$. Then $\chi_\nu \in B_0(5)$ and $\delta_\nu = -1$. This implies that $\nu = 1$ or 3. On comparing degrees, we have $x_4^3 > x_\nu$. If $\nu = 1$, then $x_4^3 > x_1$ and as $x_1x_4 \geqq x_3$ by (2.6), then $x_4^4 > x_1x_4 \geqq x_3$. If $\nu = 3$, $x_4^3 > x_3$ and, by (2.6), $x_4^4 > x_3x_4 \geqq x_1$.

This proves (2.7). If the tree is of type 3, the same argument as used above for χ_4 shows that $\chi_2^2\bar{\chi}_2$ contains a character χ_ν with $\delta_\nu = -1$. This means here that $\nu = 1$. Hence $x_2^3 \geqq x_1$. By (1.2), $x_1 > x_3$ and we obtain (2.8).

We now show

(<u>2C</u>) If neither 3^a nor 2^b occurs as degree of an irreducible character of G, then $T(B_0(3))$ and $T(B_0(2))$ each contain at least three indices.

<u>Proof</u>. Certainly, $0 \in T(B_0(3))$. It follows from (1.10) that at least one index $i > 0$ belongs to $T(B_0(3))$. If there is only

one such i, by (1.10)

$$1 + \delta_i x_i \equiv 0 \qquad (\mathrm{mod}\ 3^a) .$$

This implies that x_i is a power of 2. We may then take $\chi_h = \chi_i$ in (1D); $x_i = 2^\beta$. By (2.3), β is even and hence $\beta \geqq 2$, $b \geqq 3$. Also, we can set

$$(2.9) \qquad 2^\beta = 3^a \xi - \delta_i$$

with integral $\xi > 0$, cf. (2.4). Clearly, ξ is odd. On account of (1H) with p and q interchanged,

$$6 \cdot 3^a > 2^b \geqq 2 \cdot 2^\beta \geqq 2(\xi \cdot 3^a - 1).$$

For $\xi \geqq 5$, this leads to a contradiction. If $\xi = 1$ or 3, (2.9) in conjunction with the footnote, p. 10 **shows** that $a \leqq 2$. If 3^α **is** chosen as in (1D), by (2.3) α is even, $\alpha = 2$. By **hypothesis,** $\alpha < a$ and this is impossible.

A similar argument applies in the case of $B_o(2)$. Again $0 \in T(B_o(2))$ and there is at least one more index i in $T(B_o(2$

If there is only one such i, then

$$1 + \delta_i x_i \equiv 0 \qquad (\mathrm{mod}\ 2^b) .$$

Then x_i is a power of 3; we may assume $x_i = 3^\alpha$ and we can set

$$(2.9^*) \qquad 3^\alpha = \xi \cdot 2^b - \delta_i$$

with integral $\xi > 0$. By (1H),

$$6\cdot 2^b > 3^a \geqq 3\cdot 3^\alpha > 3(\xi\cdot 2^b - 1).$$

For $\xi \geqq 3$, this leads to a contradiction. If $\xi = 1$ or 2, (2.9*) implies $b \leqq 3$ and then $a \leqq 3$. By (2.3), α is even and as $a > \alpha$, we have $a = 3$. Then by (2.2), $b = 1$. This is clearly impossible.

(2D) If $B^*(3)$ is a 3-block containing a character χ_n of degree $3^\alpha > 1$ and if $B^*(2)$ is a 2-block containing a character χ_h of degree $2^\beta > 1$, then $B_o(5) \cap B^*(3) \cap B^*(2) = \emptyset$.

Proof. Suppose that $\chi_j \in B_o(5) \cap B^*(3) \cap B^*(2)$. Since $B^*(3)$ has defect less than a, x_j is divisible by 3 and $\chi_j \neq \chi_h$. Similarly, we have $\chi_j \neq \chi_n$. Since $B^*(3)$ contains two distinct characters, it has positive defect and hence $0 < \alpha < a$. Similarly, $B^*(2)$ has positive defect and $0 < \beta < b$. Now (2C) shows that of the five characters in $B_o(5)$, at least three lie in $B_o(3)$ and at least three lie in $B_o(2)$. This implies that

$$T(B^*(3)) = \{n,j\}, \quad T(B^*(2)) = \{h,j\} .$$

On account of (1.10) applied to $B^*(3)$ and to $B^*(2)$, we have

$$\delta_n 3^\alpha + \delta_j x_j \equiv 0 \pmod{3^a},$$

$$\delta_h 2^\beta + \delta_j x_j \equiv 0 \pmod{2^b}.$$

It follows that x_j contains 3 to the exact power 3^α and that x_j contains 2 to the exact power 2^β. Hence $x_j = 3^\alpha 2^\beta$ and by (1.3), (1.4),

See p. 16A for insert.

$$\delta_j \equiv 3^{\alpha}\cdot 2^{\beta} \equiv \delta_n\delta_h \qquad \pmod 5$$

Hence $\delta_j = \delta_n\delta_h$ and

$$1 + \delta_n x_n + \delta_h x_h + \delta_j x_j = 1 + \delta_n 3^{\alpha} + \delta_h 2^{\beta} + \delta_n\delta_h 3^{\alpha}\cdot 2^{\beta}$$
$$= (1+\delta_n 3^{\alpha})(1+\delta_h 2^{\beta}).$$

Apart from the sign, this is the degree x_k of the fifth character χ_k ($\neq \chi_o, \chi_n, \chi_h, \chi_j$) in $B_o(5)$. Hence $3^{\alpha}+\delta_n$ must be a power 2^u of 2 and $2^{\beta}+\delta_h$ must be a power 3^v of 3; $x_k = 2^u\cdot 3^v$. The footnote on p. 10 shows that $u \leqq 3$, $\alpha \leqq 2$ and that $v \leqq 2$, $\beta \leqq 3$. By (2.3), α and β are even and $u \equiv v \pmod 2$. Hence $\alpha = \beta = 2$. It follows that $u = 3$, $\delta_n = -1$. Then $v = 1$. The five degrees in $B_o(5)$ are

$$1,\ 9,\ 4,\ 36,\ 24.$$

Since there are three degrees which are congruent to $-1 \pmod 5$, the tree associated with $B_o(5)$ must be of type 2. But then two equal degrees have to appear. This is false and (2D) is proved.

(2E) G has irreducible characters of degrees 3^a and 2^b Every irreducible character of G not in $B_o(3)$ has degree 3^a; every irreducible character of G not in $B_o(2)$ has degree 2^b

Proof. Choose χ_n, χ_h and $B^*(3)$, $B^*(2)$ as in (1D),(1E). By (2D), $B_o(5) \cap B^*(3) \cap B^*(2) = \emptyset$. Now, (1L) applies. Hence $B^*(3)$ consists of $3^{a-\alpha}$ characters of degree 3^{α} and $B^*(2)$ consists of

$2^{b-\beta}$ characters of degree 2^{β}. Since all these characters lie in $B_o(5)$ we have

$$1 + 3^{a-\alpha} + 2^{b-\beta} \leqq 5 .$$

Hence α = a or a-1 and β = b or b-1. The case α = a-1, β = b-1 is impossible. If α = a-1, β = b, by (2.3) a-1 and b are even in contradiction to (2.2). Likewise, we cannot have α = a, β = b-1. Hence α = a, β = b.

By (1K), $\chi_n \in B_o(2)$ and $\chi_h \in B_o(3)$. It follows that if $B(2) \neq B_o(2)$ is a 2-block, then $B(2) \cap B^*(3) = \emptyset$ and (1I) shows that the degrees of the irreducible characters in B(2) are the same as those in $B^*(2)$. Hence B(2) consists of one irreducible character and the degree of this character is 2^b. Likewise, every p-block $B(3) \neq B_o(3)$ consists of one irreducible character and the degree of this character is 3^a. This proves (2E).

It follows from (2E) that every $\chi_i \in B_o(5)$ and not of degree $1, 3^a, 2^b$ has a degree $x_i \equiv 0 \pmod 6$. It also follows from (1.2) that 3^a occurs an odd number of times as degree. If it appears more than once, the degrees in $B_o(5)$ are

$$1,\ 3^a,\ 3^a,\ 3^a,\ 2^b.$$

Then $3^{a+1}-2^b = \pm 1$ and this implies a+1 $\leqq$ 2, that is, a = 1. However, by (2.3), a is even, a contradiction.

Thus, 3^a appears only once as degree. Since by (2.3), b is even, $2^b \equiv 1 \pmod 3$, we have one of the following two cases:

Case (a). Exactly two characters χ_h and χ_k have degree 2^b, $\delta_h = \delta_k = 1$.

Case (b). Exactly one character χ_h has degree 2^b, $\delta_h = -1$.

(2F) Case (a) is impossible.

Proof. By (2.3) and (2.2), $b \equiv 0 \pmod 4$, $a \equiv 2 \pmod 4$ and then $\delta_n = -1$. If $x_\mu = 2^{\alpha_\mu} \cdot 3^{\beta_\mu}$ is the degree different from $1, 2^b, 3^a$ in $B_o(5)$, (1.2) yields

$$(2.10) \qquad 1 + 2^b + 2^b + \delta_\mu 2^{\beta_\mu} \cdot 3^{\alpha_\mu} = 3^a .$$

Here $2^b \equiv 0 \pmod{16}$, $3^a \equiv 9 \mod 16$. Now, (2.10) shows that $\beta_\mu = 3$. Since $\alpha_\mu < a$, it follows that 3^{α_μ} is the highest power of 3 dividing $2^{b+1} + 1$. Also $0 < \alpha_\mu$. As 2 belongs to the exponent $2 \cdot 3^{\alpha_\mu - 1} \pmod{3^{\alpha_\mu}}$, b+1 is divisible by $3^{\alpha_\mu - 1}$, $b \geq 3^{\alpha_\mu - 1} - 1$.

We deal first with the case $\delta_\mu = -1$. Here, $B_o(5)$ contains three characters with $\delta_j = 1$; we have a tree of type 1. Then $x_2 = x_4 = 2^b$ and $8 \cdot 3^{\alpha_\mu} = x_1$ or x_3. By (2.7), $x_1^3 \geq x_2$, $x_3^3 \geq x_2$ and hence

$$(2.11) \qquad (8 \cdot 3^{\alpha_\mu})^3 > 2^b > 2^{3^{\alpha_\mu - 1} - 1} .$$

This yields

$$3\alpha_\mu \log 3 > (3^{\alpha_\mu - 1} - 10)\log 2$$

where the logarithms are taken with the basis 10. Since

$$\log 3 < 0.5, \qquad \log 2 > 0.3, \qquad \text{for } \alpha_\mu \geqq 3$$

$$5\alpha_\mu > \frac{1}{3} 3^{\alpha_\mu} - 10.$$

However this is false for $\alpha_\mu \geqq 5$. Since $\delta_\mu = -1$ and $\beta_\mu = 3$, we have $\alpha_\mu \equiv 1 \bmod 4$. Hence we must have $\alpha_\mu = 1$, $x_\mu = 24$. In the case of a tree of type 1, χ_4 is a 5-modular constituent of χ_3 and hence $x_4 < x_3$. If $x_3 = 24$, $2^b < 24$ and $b \leqq 4$. If $x_1 = 24$, $x_2 = 2^b < 24^2$ by (2.6) with $j = 1$, $i = 2$. Then $b \leqq 8$.

Since $b \equiv 0 \pmod 4$ by (2.3), we have $b = 4$ or $b = 8$. If $b = 8$, the degrees in $B_0(5)$ are

$$1, 256, 256, 24, 3^a$$

and by (1.2), $3^a = 489$ which is absurd. Hence $b = 4$ and the five degrees are

$$1, 16, 16, 24, 9,$$

$a = 2$, $b = 4$. But then χ_μ of degree 24 belongs to a 3-block of defect 1. No such block can exist, cf. (2E). Hence the case $\delta_\mu = -1$ is impossible.

If $\delta_\mu = 1$, the tree is of type 3 and, by (2.10),

$$(2.12) \qquad 1 + 2^b + 2^b + 8\cdot 3^{\alpha_\mu} = 3^a$$

In the notation of p. 12, $x_1 = 3^a$, $x_2 = 8\cdot 3^{\alpha_2}$, $\mu = 2$; $x_3 = x_4 = 2^b$. Here, (2.8) implies that (2.11) again holds. As above, we see that $\alpha_\mu = \alpha_2 \leqq 4$. Here, by (2.3) $\alpha_2 \equiv 3 \pmod 4$ and then $\alpha_2 = 3$, $x_2 = 2^3\cdot 3^3 = 216$. By (2.8),

$$3^a < 2^9\cdot 3^9 < 3^6\cdot 3^9$$

and hence $a \leqq 14$. Since $a \equiv 2 \pmod 4$ and since (2.12) shows that $a > 2$, we have $a = 6$, $a = 10$, or $a = 14$.

By (2.12), $2^{b+1} = 3^a-217 \equiv 3^a+39 \pmod{256}$. For $a = 6$, we find $2^b = 256$. If $a = 10$ or $a = 14$ were possible, we would have to have $3^a \equiv 3^6 \pmod{256}$ for this a, that is, 256 would have to divide 3^4-1 or $3^8-1 = (3^4-1)(3^4+1)$, which are both false. Thus, $a = 6$, and $b = 8$; the five degrees in $B_0(5)$ are

$$1, \quad 256, \quad 256, \quad 216, \quad 729$$

and we find

$$(2.13) \qquad g = 5\cdot 3^6\cdot 2^8$$

Let J be an involution in the center of a Sylow-2-subgroup of G. Then $c(J)$ is divisible by 2^8. Since 5 does not divide $c(J)$, (cf. (1A)), we can set

(2.14) $$c(J) = 2^8 \cdot 3^s$$

with $0 \leq s \leq 6$. Set $\chi_2(J) = \xi$. We have $\chi_3(J) = \chi_4(J) = 0$, as χ_3 and χ_4 have 2-defect 0. By (1.1), $\chi_1(J) = \xi+1$. Thus, if $R = \langle\rho\rangle$,

	1	ρ	J
χ_0	1	1	1
χ_1	729	-1	$1+\xi$
χ_2	216	1	ξ
χ_3	256	1	0
χ_4	256	1	0
χ_5	x_5	0	*
χ_6	x_6	0	*
..	...	...	...

Since $N(R)$ is generated by $C(R) = R = \langle\rho\rangle$ and an element of order 4, there exist 5 involutions which invert ρ. Now the method of [4] yields

$$\frac{g}{c(J)^2}\left(1 + \frac{\xi^2}{216} - \frac{(1+\xi)^2}{729}\right) = \begin{cases} 5 \\ 0 \end{cases}$$

where the upper case applies, if J is conjugate to an involution in $N(R)$ and the lower case, if this is not so. If we have the lower case,

$$729 \cdot 8 + 27\xi^2 - 8(1+\xi)^2 = 0.$$

However, ξ is a rational integer and this equation is not satisfied for rational integers. Hence the upper case applies and on account of (2.13),(2.14), we find

$$729 \cdot 8 + 27\xi^2 - 8(1+\xi)^2 = 2^{11} \cdot 3^{2s} .$$

If $s \geq 2$, then 27 divides $(1+\xi)^2$. Then 9 divides $1+\xi$ and $(\xi,3) = 1$. But then ξ is prime to 3 and $27\xi^2$ is divisible by 81. This is absurd, and, we have $s = 0$ or $s = 1$, that is

$$729 \cdot 8 + 27\xi^2 - 8(1+\xi)^2 = \begin{cases} 2^{11} \\ 2^{11} \cdot 3^2 \end{cases},$$

and then

$$19\xi^2 - 16\xi + 64 \cdot 91 = \begin{cases} 2^{11} \\ 2^{11} \cdot 3^2 \end{cases}.$$

This shows that $8|\xi$, say $\xi = 8\eta$ and then

$$19\eta^2 - 2\eta = \begin{cases} -59 \\ 197 \end{cases}.$$

Neither equation has an integral solution η and the whole case is impossible. This completes the proof of (2F).

(2G) We have $g = 25{,}920$ and the five degrees in $B_0(5)$ are

$$1,\ 81,\ 6,\ 64,\ 24\ .$$

Proof. The five degrees in $B_0(5)$ have the form

$$1,\ x_h = 2^b,\quad x_n = 3^a,\quad x_i = 3^{\alpha_i} 2^{\beta_i},\quad x_k = 3^{\alpha_k} \cdot 2^{\beta_k}$$

and the exponents $\alpha_i, \beta_i, \alpha_j, \beta_j$ are positive. Also as we have seen $\delta_h = -1$. Then, by (2.3) $b \equiv 2 \pmod 4$ and by (2.2) $a \equiv 0 \pmod 4$. Now $\delta_n = 1$ by (2.3) and (1.2) reads

$$1 + 3^a + \delta_i 3^{\alpha_i} \cdot 2^{\beta_i} + \delta_k \cdot 3^{\alpha_k} \cdot 2^{\beta_k} = 2^b\ .$$

If $b = 2$, then $3^a < 24$ by (1H) and since $a \equiv 0 \pmod 4$, this is impossible. Hence $b \geq 6$. Since $a \equiv 0 \pmod 4$, $3^a \equiv 1 \pmod{16}$

and we find

$$(2.15) \qquad 2 + \delta_i 3^{\alpha_i} 2^{\beta_i} + \delta_k 3^{\alpha_k} \cdot 2^{\beta_k} \equiv 0 \pmod{16}.$$

After interchanging x_i, x_k, if necessary, we may assume that $\beta_i = 1$. If $\delta_i = 1$, then $\alpha_i \equiv 1 \pmod 4$ by (2.3) and $2^{\beta_i} 3^{\alpha_i} = 2 \cdot 3^{\alpha_i} \equiv 6 \pmod{16}$. Then (2.15) shows that $\beta_k = 3$. if $\delta_i = -1$, then $\alpha_i \equiv 3 \pmod 4$ by (2.3) and $x_i = 2 \cdot 3^{\alpha_i} \equiv 54 \equiv 6 \pmod{16}$. Here (2.15) yields $\beta_k = 2$. In either case, $x_i \neq x_k$ and since the five degrees in $B_o(5)$ are distinct, the tree belonging to $B_o(5)$ is of type 1.

According as to whether $\delta_i = 1$ or $\delta_i = -1$, we have two cases:

<u>Case (a)</u>:

$$(2.16a) \begin{cases} 1 + 3^a + 2 \cdot 3^{\alpha_i} = 2^b + 8 \cdot 3^{\alpha_k} \\ \delta_i = 1,\ \delta_k = -1;\ a \equiv 0,\ \alpha_i \equiv 1,\ b \equiv 2,\ \alpha_k \equiv 1 \pmod 4 \end{cases}$$

<u>Case (b)</u>:

$$(2.16b) \begin{cases} 1 + 3^a + 4 \cdot 3^{\alpha_k} = 2^b + 2 \cdot 3^{\alpha_i} \\ \delta_i = -1,\ \delta_k = 1;\ a \equiv 0,\ \alpha_k \equiv 2,\ b \equiv 2,\ \alpha_i \equiv 3 \pmod 4). \end{cases}$$

Let 3^w denote the highest power of 3 dividing $2^b - 1$. It follows from (2.16) that $\alpha_i = w < \alpha_k$ or $\alpha_k = w < \alpha_i$ or $\alpha_i = \alpha_k \leq w$. Hence $w > 0$ and at least one of the degrees x_i, x_k is not larger than $3^w \cdot 2^3$ On account of (2.7),

$$(3^w \cdot 2^3)^4 > 2^b$$

Since 2 belongs to the exponent $2\cdot 3^{w-1} \pmod{3^w}$, $b \geqq 2\cdot 3^{w-1}$ and hence

$$4w \log 3 > (2\cdot 3^{w-1} - 12)\log 2.$$

Since $\log 3 < .5$ and $\log 2 > .3$ for the base 10,

$$20w > 2\cdot 3^w - 36.$$

This is false for $w \geqq 4$. Hence $w \leqq 3$ and then $\alpha_i \leqq 3$ or $\alpha_k \leqq 3$. According to (2.16), $\alpha_k \neq 3$.

If $\alpha_i = 3$, we must have Case (b) and $x_i = 54$. Since here one of x_1, x_3 is 54 and the other 2^b by (2.7), $54^3 > 2^b$ Since $w = 3$, by the above, b is divisible by $2\cdot 3^{w-1} = 18$, say $b = 18c$ and then $54 > 2^{6c}$ which is impossible.

Again, by (2.16), $\alpha_i \neq 2$. If $\alpha_k = 2$, we must have Case (b) and $w = 2$. Then by (2.7), $36^4 > 2^b$ Here, b is divisible by $2\cdot 3^{w-1} = 6$, say $b = 6c$ and hence $36^2 > 2^{3c}$ $c \leqq 3$. Since $w = 2$, $3 \nmid c$. As $b \equiv 2 \pmod 4$, then $b = 6$. Since $3^a < 6\cdot 2^b = 384$ by (1H) and $a \equiv 0 \pmod 4$ we have $a = 4$. Now (2.16b) shows that

$$2\cdot 3^{\alpha_i} = 1 + 81 + 36 - 64 = 54.$$

Since then χ_i belongs to a 3-block of defect 1, we have a contradiction to (2E).

Hence $\alpha_i = 1$ or $\alpha_k = 1$, that is, $x_i = 6$ or $x_k = 24$. We then must have Case (a). Assume first that $b > 10$. If $\alpha_i = 1$, then $x_i = 6$ and by (2.7), $2^b < 6^4$. This is impossible. If $\alpha_k = 1$,

$x_k = 24$ and $k = 1$ or 3. By (2.7), $2^b < 24^3$. Then $2^{b-9} < 3^3 < 2^5$; $b < 14$. Since $b \equiv 2 \pmod 4$, this is impossible for $b > 10$. If $b = 10$ and $x_i = 6$, by (2.16), $3^a = 1017 + 8 \cdot 3^{\alpha_k}$. Then $\alpha_k = 2$ and this leads to a contradiction. If $b = 10$ and $x_k = 24$, then $3^a + 2 \cdot 3^{\alpha_i} = 1047$. Then $\alpha_i = 1$ which, too, is impossible.

Thus, $b < 10$ and since $b \equiv 2 \pmod 4$, we have $b = 2$ or $b = 6$. Also $a \equiv 0 \pmod 4$ and hence $3^a \geq 81$. Since $3^a < 6 \cdot 2^b$ by (1H), the case $b = 2$ is impossible. Hence $b = 6$ and $3^a < 6 \cdot 64 = 384$ whence $a = 4$. This already shows that

$$g = 5 \cdot 3^4 \cdot 2^6 = 25{,}920.$$

Also (2.16a) yields

$$x_k - x_i = 18 .$$

Since we know that $x_i = 6$ or $x_k = 24$, it follows that $x_i = 6$, $x_k = 24$ and the five degrees in $B_0(5)$ are 1, 6, 81, 64, 24 and the proof of (2G) is complete.

III. Simple groups of order 25 920

<u>Notation</u>. Let $\mathbb{Z}[G]$ be the group ring of the finite group G over the ring $\mathbb{Z}$ of rational integers. If K is a class of conjugate elements of G, denote by S(K) the <u>class sum</u>

$$S(K) = \sum_{\sigma \in K} \sigma \quad \in \mathbb{Z}[G] .$$

If $\tau_1, \tau_2, \tau_3 \in G$, if τ_i belongs to the conjugate classes K_i, we denote by $a(\tau_1, \tau_2; \tau_3)$ the coefficient with which $S(K_3)$ occurs in $S(K_1)S(K_2)$ if expressed as linear combination of the class sums.

It is clear that

$$(3.1) \qquad a(\tau_1, \tau_2; \tau_3) \in \mathbb{Z}, \qquad a(\tau_1, \tau_2; \tau_3) \geq 0 .$$

As is well known,

$$(3.2) \quad a(\tau_1, \tau_2; \tau_3) = \frac{g}{c(\tau_1)c(\tau_2)} \sum_i \frac{\chi_i(\tau_1)\chi_i(\tau_2)\overline{\chi_i(\tau_3)}}{x_i}$$

where the sum extends over all irreducible characters χ_i of G and where, as before, $x_i = \chi_i(1)$.

Let G now be a simple group of order

$$g = 25920 = 5 \cdot 3^4 \cdot 2^6 .$$

We apply the results of sections I and II. The degrees of the five characters in $B_0(5)$ are given in (2G). We change the indexing such that

(3.3) $\quad x_o = 1,\ x_1 = 6,\ x_2 = 81,\ x_3 = 24,\ x_4 = 64$.

Since these degrees are distinct, the five characters take only rational values. If, as before, $R = \langle\rho\rangle$ is a Sylow 5-group of G,

(3.4) $\quad \chi_i(\rho) = \delta_i = 1$ for $i = 0,1,2;$ $\chi_j(\rho) = \delta_j = -1$ for $j=3,4$.

For 5-regular elements σ, by (1.1),

(3.5) $\quad 1 + \chi_1(\sigma) + \chi_2(\sigma) = \chi_3(\sigma) + \chi_4(\sigma).$

(**3A**) The character χ_2 vanishes for 3-singular elements and the character χ_4 vanishes for 2-singular elements. For any $\tau_1, \tau_2 \in G$ and $\rho \in G$ of order 5,

(3.6)
$$\begin{cases} c(\tau_1)c(\tau_2)a(\tau_1,\tau_2;\rho) \\ \quad = 5\cdot 81\cdot 64\left(1 + \dfrac{\chi_1(\tau_1)\chi_1(\tau_2)}{6} + \dfrac{\chi_2(\tau_1)\chi_2(\tau_2)}{81}\right. \\ \quad\quad \left. - \dfrac{\chi_3(\tau_1)\chi_3(\tau_2)}{24} - \dfrac{\chi_4(\tau_1)\chi_4(\tau_2)}{64}\right). \end{cases}$$

Proof. The first statement follows from the fact that χ_2 of degree 81 has 3-defect 0 and the second statement follows from the fact that χ_4 of degree 64 has 2-defect 0. Finally, (3.6) is obtained from (3.2), (3.3), and (3.4), since all χ_i with $i > 4$ have 5-defect 0 and hence vanish for the element ρ.

(**3B**) There exist elements ρ and μ such that

$$\rho^5 = 1, \quad \mu^4 = 1, \quad \mu^{-1}\rho\mu = \rho^2$$

Then the Sylow-5-subgroup $R = \langle\rho\rangle$ has the normalizer $N(R) = \langle\rho,\mu\rangle$ of order 20.

This is evident since for a Sylow-5-subgroup $R = \langle\rho\rangle$, we have $C(R) = R$ and since as discussed in section II, we have $t = 1$ and hence $|N(R):R| = m = 4/t = 4$.

(3C) Set $J = \mu^2$. If τ_1 and τ_2 are involutions in G and not both are conjugate to J, then $a(\tau_1,\tau_2;\rho) = 0$. On the other hand, $a(J,J;\rho) = 5$.

Proof. If ρ is the product of two involutions, then each of them inverts ρ and hence lies in $N(R)$. The group $N(R)$ in (3B) has one class of involutions containing J and four other elements. Hence the product of a conjugate of τ_1 and of a conjugate of τ_2 can only equal ρ if τ_1 and τ_2 are conjugate to J. If this is not the case, $a(\tau_1,\tau_2;\rho) = 0$. If ξ and η are conjugates of J and if $\xi\eta = \rho$, again ξ must be one of the five involutions of $N(R)$ and $\eta = \xi\rho$. On the other hand, if ξ is one of the five involutions of $N(R)$ and $\eta = \xi\rho$, then $\xi\eta = \rho$ and $\eta^2 = \xi^{-1}\rho\xi\rho = 1$, $\eta \neq 1$, that is, η is an involution. Then ξ,η are both conjugate to J and hence $a(J,J;\rho) = 5$.

(3D) The group $N(R)$ has five irreducible characters λ_0, λ_1, λ_2, λ_3 and ψ. We have (with $i = \sqrt{-1}$)

$$\lambda_\nu(1) = \lambda_\nu(\rho) = 1; \quad \lambda_\nu(\mu) = i^\nu \quad ;$$
$$\psi(1) = 4, \quad \psi(\rho) = -1, \quad \psi(\mu) = \psi(\mu^{-1}) = \psi(J) = 0.$$

Moreover,

$$\chi_1 | N(R) = \psi + \lambda_o + \lambda_2;$$

$$\chi_3 | N(R) = 5\psi + c_o\lambda_o + c_1(\lambda_1+\lambda_3) + c_2\lambda_2$$

where (c_o, c_1, c_2) is one of the triples (1,1,1), (3,0,1), (1,0,3). Finally, if there exist irreducible characters χ_j of G of degree 5, then

$$\chi_j | N(R) = \psi + \lambda_2.$$

Proof. Since $N(R)/R$ is cyclic of degree 4 the existence of the four characters λ_ν is clear. Also, ψ is the character of the irreducible representation Y of $N(R)$ defined by

$$Y(\rho) = \begin{pmatrix} \epsilon & & & \\ & \epsilon^2 & & \\ & & \epsilon^{-1} & \\ & & & \epsilon^{-2} \end{pmatrix} , \qquad Y(\mu) = \begin{pmatrix} 0 & 0 & 0 & 1 \\ 1 & 0 & 0 & 0 \\ 0 & 1 & 0 & 0 \\ 0 & 0 & 1 & 0 \end{pmatrix}$$

where ϵ is a primitive 5th root of unity. Here, $\det Y(\mu) = -1$.

It is clear that these five characters are all the irreducible characters of $N(R)$.

Consider first an irreducible character χ_j of G of degree 5 (if one exists). Since χ_j vanishes for 5-singular elements of G and since ψ vanishes for 2-singular elements of $N(R)$, the orthogonality relations show that ψ appears with multiplicity 1 in $\chi_j | N(R)$. Hence we can set $\chi_j | N(R) = \psi + \lambda_\alpha$. Since G is simple the representation X_j has determinant 1. For the element μ, this yields $\lambda_\alpha(\mu) = -1$ and hence $\alpha = 2$.

The character χ_1 has the value $\delta_1 = 1$ for 5-singular elements. An analogous argument as before shows that ψ occurs again with multiplicity 1 in $\chi_1|N(R)$. Since χ_1 is real-valued, λ_1 and λ_3 occur with the same multiplicity. On considering the determinant of the representation for the element μ, we see that they do not appear and that in fact $\chi_1|N(R) = \psi + \lambda_o + \lambda_2$.

The character χ_3 has the value $\delta_3 = -1$ for 5-singular elements. We see that ψ occurs with multiplicity $(4\cdot 24+4)/20 = 5$ in $\chi_3|N(R)$. Again, χ_3 is real-valued and hence λ_1 and λ_3 have the same multiplicity in $\chi_3|N(R)$. If c_ν is the multiplicity of λ_ν in $\chi_3|N(R)$, we have

$$\chi_3|N(R) = 5\psi + c_o\lambda_o + c_1(\lambda_1+\lambda_3) + c_2\lambda_2.$$

The determinant argument yields

$$1 = (-1)^5(-1)^{c_2} .$$

Hence c_2 is odd and as $c_o + 2$[illegible] $= 4$, we have $c_2 = 1$ or 3. This leads to the three cases listed in (3D).

(3E) The following table gives the orders of the centralizers, the number of the elements in the conjugate classes, and the values of $\chi_o, \chi_1, \chi_2, \chi_3, \chi_4$ for the elements $1, \rho, \mu, J$:

	1	ρ	μ	J
c(τ)	25920	5	8	3·32
number of elements in conjugate class of τ	1	5184	3240	270
χ_o	1	1	1	1
χ_1	6	1	0	2
χ_2	81	1	-1	-3
χ_3	24	-1	0	0
χ_4	64	-1	0	0

Proof. The first two columns of this table are obtained from (3.3) and (3.4). Of course $c(1) = g$ and $c(\rho) = 5$ by (1A). On account of (3A), the two remaining values of χ_4 vanish while the values of χ_1 are obtained from (3D). Unless $c_o = c_1 = c_2$ in (3D), we have $\chi_3(J) = 4$ and then $\chi_2(J) = 1$ by (3.5). By (3C), $a(J,J;\rho) = 5$ and (3A) yields

$$5c(J)^2 = 5{\cdot}81{\cdot}64(1 + 2/3 + 1/81 - 16/24) .$$

Then $c(J)^2 = 64{\cdot}82$ which is absurd.

Hence we have $c_o = c_1 = c_2 = 1$ in (3D) and this yields the values $\chi_3(\mu)$, $\chi_3(J)$. The values of $\chi_2(\mu)$ and $\chi_2(J)$ are obtained from (3.5). Finally, by the argument used before,

$$5\ c(J)^2 = 5{\cdot}81{\cdot}64(1 + 2/3 + \tfrac{1}{9})$$

whence $c(J) = 96$. This shows that J does not belong to the center of a Sylow-2-group Q of G. If Q is chosen so that it contains μ, it follows that c(μ) is divisible by 8. On the other hand, by (3.6)

$$c(\mu)^2 a(\mu,\mu;\rho) = 5\cdot 81\cdot 64\ (1 + 1/81).$$

This shows that $c(\mu)^2$ divides $128\cdot 41$. Hence $c(\mu) = 8$ and the proof of (3E) is complete.

(3F) There exists an element ν of order 9 in G for which $c(\nu) = 9$ and

$$\chi_1(\nu) = \chi_2(\nu) = \chi_3(\nu) = 0, \quad \chi_4(\nu) = 1.$$

Proof. Let P be a Sylow-3-subgroup of G. Let $y \in P$, $y \neq 1$. Since χ_1 is rational valued and since G is simple, $|\chi_1(y)| < 6$. Also, $\chi_1(y) \equiv x_1 = 6 \pmod 3$ and hence

$$\chi_1(y) = 3,\ 0,\ -3.$$

Since by the orthogonality relations,

$$6 + \sum_{y\in P-\{1\}} \chi_1(y) \geqq 0$$

we cannot have $\chi_1(y) = -3$ for all y. Also, by (3A) and (3E)

$$c(y)a(J,y;\rho) = 5\cdot 27\cdot 2\ (1 + \chi_1(y)/3)\ .$$

For $\chi_1(y) \neq -3$, this implies $c(y) \not\equiv 0 \pmod{81}$. Consequently, P is not abelian. If X_1 is the representation of G with the character χ_1, then $X_1|P$ has an irreducible constituent U of degree 3. For $U(y)$ of order 3 in the center of $U(P)$, $U(P) = \epsilon_0 I$ where ϵ_0 is a primitive third root of unity. Since χ_1 is real valued, the representation $\overline{U}$ of P conjugate complex to U also appears in $X_1|P$ and is not equivalent to U. Hence $X_1|P$ is the direct sum of U and $\overline{U}$. It follows that U is a faithful irreducible representation of P.

If we write U in monomial form, we see that P has normal abelian subgroups A of order 27. For any such subgroup, $U|A$ splits into three linear constituents ζ_1, ζ_2, ζ_3 and

$$X_1|A = U|A \oplus \overline{U}|A = \zeta_1 \oplus \zeta_2 \oplus \zeta_3 \oplus \overline{\zeta}_1 \oplus \overline{\zeta}_2 \oplus \overline{\zeta}_3.$$

Since X_1 is rational-valued, with ζ_i all algebraic conjugates appear. The values of ζ_i must then lie in the field of the ninth roots of unity. Moreover, if they do not lie in the field of the third roots of unity, the constituents of $X_1|A$ are the six distinct algebraic conjugates of ζ_i. Since ζ_i maps A on a cyclic group of order 9, we can choose $y \neq 1$ in the kernel of ζ_i. Then y lies in the kernel of all the algebraic conjugates of ζ_i and hence in the kernel of X_1. But as X_1 is faithful, this is absurd. Hence $\zeta_i(y)^3 = 1$ and since this holds for an arbitrary irreducible constituent ζ_i of $U|A$, we see that A must be elementary abelian of order 27, and that $U|A$ consists of all diagonal matrices M of degree 3 with $M^3 = 1$. We can choose an element $\nu \in P$ such that the monomial matrix $U(\nu)$ has the form

$$U(\nu) = \begin{pmatrix} 0 & u_1 & 0 \\ 0 & 0 & u_2 \\ u_3 & 0 & 0 \end{pmatrix}$$

After replacing U by an equivalent representation, we may assume $u_1 = u_2 = 1$. Moreover, after multiplying ν with a suitable element of A, we may assume $u_1 = u_2 = 1$, $u_3 \neq 1$. Then $U(\nu^3) = u_3 I$ and since $\nu^3 \in A$, u_3 is a primitive third root of unity. It is now clear that ν is an element of order 9.

Since every abelian subgroup A_1 of order 27 of G belongs to a Sylow-3-subgroup P_1 and hence has been shown to be elementary abelian, it follows that $c(\nu)$ cannot be divisible by 27. Suppose that $c(\nu)$ is even. Then there exist involutions $\tau \in C(\nu)$. The group $\langle\tau\nu\rangle$ is cyclic of order 18. Hence $X_1|\langle\tau\nu\rangle$ splits into six linear constituents ω_j. Again, with each ω_j, all algebraic conjugates appear. Since some ω_j must represent $\langle\nu\rangle$ faithfully, ω_j has six conjugates, that is, $X_1|\langle\tau\nu\rangle$ splits into the six conjugates of ω_j. If $\omega_j(\tau) = \delta$, then $\delta = \pm 1$ and for each algebraic conjugate ω_i of ω_j, we have $\omega_i(\tau) = \delta$. This implies $X_1(\tau) = \delta I$. But then τ belongs to the center of G which is impossible. Hence $c(\nu)$ is odd. By (1A), $c(\nu) \not\equiv 0 \pmod 5$. It is now clear that $c(\nu) = 9$.

Since $\sum_i |\chi_i(\nu)|^2 = 9$, it follows that $|\chi_i(\nu)| < 3$ for each χ_i. The number $\chi_3(\nu)$ is rational and $\chi_3(\nu) \equiv x_3 \equiv 0 \pmod 3$. Hence $\chi_3(\nu) = 0$. Since $\chi_2(\nu) = 0$ by (3A), we have $\chi_4(\nu) = 1$ by (3.5) and (3F) is proved.

(3G) There exists an irreducible character χ_5 of G of degree 5.

Proof. The orthogonality reltations show that

$$\sum_j |\chi_j(\mu)|^2 = c(\mu); \quad \sum_j |\chi_j(\nu)|^2 = c(\nu); \quad \sum_j \chi_j(\mu)\chi_j(\nu) = 0$$

where χ_j ranges over all irreducible characters of G. Set $\chi_j(\mu) = \alpha_j$, $\chi_j(\nu) = \beta_j$. On account of (3E),(3F) we have

$$(3.7) \qquad \sum_{j>4} |\alpha_j|^2 = 6, \quad \sum_{j>4} |\beta_j|^2 = 7, \quad \sum_{j>4} \alpha_j\beta_j = -1.$$

Here α_j is an algebraic integer in the field of the fourth roots of unity and β_j is an algebraic integer in the field of the ninth roots of unity. Let $\mathfrak{p}_2$ be a prime ideal dividing 2 and let $\mathfrak{p}_3$ be a prime ideal dividing 3 in the field of the 36-th roots of unity. It will suffice to show that there exists a $j > 4$ with

$$(3.8) \qquad \alpha_j \not\equiv 0 \pmod{\mathfrak{p}_2}, \quad \beta_j \not\equiv 0 \pmod{\mathfrak{p}_3}.$$

Indeed, then

$$x_j \equiv \chi_j(\mu) \not\equiv 0 \pmod{\mathfrak{p}_2}; \quad x_j \equiv \chi_j(\nu) \not\equiv 0 \pmod{\mathfrak{p}_3}$$

and hence $x_j \not\equiv 0 \pmod 2$, $x_j \not\equiv 0 \pmod 3$. Since $x_j \equiv 0 \pmod 5$ because of $j > 4$, we have $x_j = 5$ and by changing the indexing, we may assume that $j = 5$.

If all the $\alpha_j \neq 0$ in (3.7) are units, the last equation (3.7) shows that there exists a $j > 4$ with $\beta_j \not\equiv 0 \pmod{\mathfrak{p}_3}$ and then (3.8) holds.

If not all $\alpha_j \neq 0$ in (3.7) are units, we have the following two possibilities for these α_j if the χ_j are suitably indexed:

<u>Case (a)</u>: $|\alpha_5| = 1$, $|\alpha_6| = 1$, $|\alpha_7| = 2$

<u>Case (b)</u>: $|\alpha_5| = 1$, $|\alpha_6| = \pm 1$, $\alpha_7 = \pm 1 + i$, $\alpha_8 = \bar{\alpha}_7$.

We may assume that

$$(3.9) \qquad \beta_5 \equiv \beta_6 \equiv 0 \qquad (\text{mod } \mathfrak{P}_3)$$

since otherwise (3.8) can be satisfied. If χ_j is an algebraic conjugate of χ_5 or χ_6 then $|\alpha_j| = 1$, $x_j > 4$. It is now clear that either χ_5, χ_6 are rational valued or $\{\chi_5, \chi_6\}$ is a full family of algebraic conjugates. A similar argument shows that in the case (a) χ_7 is rational valued while in the case (b), $\{\chi_7, \chi_8\}$ is a full family of algebraic conjugates, $\chi_8 = \bar{\chi}_7$. Moreover, the values of χ_7, χ_8 then lie in the field of the fourth roots of unity. In particular, $\beta_7 = \beta_8$ is rational.

If $\beta_5 = \beta_6 = 0$, by (3.7) $2\beta_7 = \pm 1$ in the case (a) and $\beta_7(\alpha_7 + \alpha_8) = -1$ in the case (b). Since in the latter case $\alpha_7 + \alpha_8 = \pm 2$, neither possibility can arise. Hence we may assume that $\beta_5 \neq 0$. If β_5 was rational, $|\beta_5| \geq 3$ by (3.9) while $|\beta_5| \leq \sqrt{7}$ by (3.7). Hence β_5 is not rational, $\{\chi_5, \chi_6\}$ form a family of algebraic conjugates. Since β_5 lies in the quadratic subfield Ω of the field of 9-th roots unity, the values of χ_5, χ_6 lie in Ω. Here Ω is the field of the third roots of unity. It is now clear that $\overline{\chi}_5 = \chi_6$. Also, we must have $\alpha_5 = \alpha_6$. On account of (3.9), the rational integers $|\beta_5|^2 = \beta_5\beta_6$ and $\beta_5+\beta_6$ are divisible by 3. Since $|\beta_5|^2 > 0$, by (3.9) and (3.7)

$$|\beta_5|^2 = \beta_5\beta_6 = |\beta_6|^2 = 3 \ .$$

Moreover, $\alpha_5\beta_5 + \alpha_6\beta_6 = \alpha_5(\beta_5+\beta_6)$ is divisible by 3 and (3.7) shows that β_7 cannot vanish. In the case (b) then $|\beta_7|^2 + |\beta_8|^2 \geq 2$ which leads to a contradiction. Thus we must have case (a) and $\beta_7 = \pm 1$.

Since $\beta_5 \neq \overline{\beta}_5 = \beta_6$, the element ν cannot be conjugate to ν^{-1} in G. Hence by the orthogonality relations,

$$0 = \sum_j \chi_j(\nu)^2 = \sum_j \beta_j^{\,2} = 3 + \beta_5^{\,2} + \beta_6^{\,2} \ .$$

But then, $(\beta_5+\beta_6)^2 = -3 + 2\beta_5\beta_6 = 3$. Since $\beta_5+\beta_6$ is rational, this is impossible and the proof of (3G) is complete.

(<u>3H</u>) We have

$$\chi_5(\mu) = -1, \qquad \chi_5(J) = 1,$$

$$\chi_5\overline{\chi}_5 = \chi_0 + \chi_3 \ . \tag{3.10}$$

<u>Proof</u>. The values $\chi_5(\mu)$ and $\chi_5(J)$ are obtained from (3D). Since χ_5 has 5-defect 0, $\chi_5\overline{\chi}_5$ is a sum of the chief indecomposable character Φ_0 belonging to the principal 5-modular character and another character of G, cf. [7, §24]. Here, Φ_0 is the sum of χ_0 and a character χ_i with $\delta_i = -1$. Then $i = 3$ or 4. Since x_4 is too large, we must have $\Phi_0 = \chi_0 + \chi_3$ and comparison of the degrees yields (3.10).

(3I) Let τ range over the 2-elements of G which are not conjugate to $1, \mu, J$. Then

$$(3.11) \qquad \sum_{\tau} \chi_2 \tau)^2 = 8505 .$$

Proof. Since χ_2 vanishes for the 3-singular elements of G, the orthogonality relations yield (cf. (3E))

$$25{,}920 = 81^2 + 5184 + 3240 + 9 \cdot 270 + \sum_{\tau} \chi_2(\tau)^2 .$$

This can be written in the form (3.11).

(3J) There exists an involution $J_1 \in G$ with $c(J_1) = 64 \cdot 9$ and

$$(3.12) \quad \chi_1(J_1) = -2,\ \chi_2(J_1) = 9,\ \chi_3(J_1) = 8,\ \chi_4(J_1) = 0,\ \chi_5(J_1) = -3.$$

Actually, (3.12) holds for any involution J_1 of G not conjugate to J in G.

Proof. Since $c(J)$ is not divisible by 64, (cf. (3E)), as already remarked J does not belong to the center of a Sylow-2-subgroup Q of G. Hence there exist involutions J_1 of G which are not conjugate to J.

Let J_1 be any such involution. Again, $\chi_4(J_1) = 0$, cf. (3A). By (3.5)

$$1 + \chi_1(J_1) + \chi_2\ (J_1) = \chi_3(J_1).$$

By (3C), $a(J, J_1; \rho) = a(J_1, J_1, \rho) = 0$ and using (3.6) and (3E), we find

$$1 + \chi_1(J_1)/3 = \chi_2(J_1)/27,$$

$$1 + \chi_1(J_1)^2/6 + \chi_2(J_1)^2/81 = \chi_3(J_1)^2/24.$$

Eliminating $\chi_3(J_1)$ and $\chi_2(J_1)$ in the last three equations, we have $\chi_2(J_1) = 9(3+\chi_1(J_1))$, $\chi_3(J_1) = 2(14+5\chi_1(J_1))$ and

$$6 + \chi_1(J_1)^2 + 6(3 + \chi_1(J_1))^2 = (14 \quad 5\chi_1(J_1))^2$$

that is,

$$18\chi_1(J_1)^2 + 104\chi_1(J_1) + 136 = 0$$

The only integral solution of this quadratic equation is -2 and hence

$$\chi_1(J_1) = -2, \quad \chi_2(J_1) = 9, \quad \chi_3(J_1) = 8.$$

Since $\chi_5(J_1)$ is rational, by (3.10), $\chi_5(J_1) = \pm 3$. Here the - sign applies since otherwise the representation X_5 belonging to χ_5 would have determinant -1 for J_1 (as only one characteristic root of $X_5(J_1)$ would be -1.) Now (3.12) is proved.

If J_1 lies in the center of Q, $c(J_1)$ is divisible by 64. By (1A), $c(J_1)$ has the form $c(J_1) = 64 \cdot 3^s$. Since $\chi_2(J_1) = 9$, by (3.11) for $\tau = J_1$, the conjugate class of J_1 has at most 8505/81 elements and this implies $s \geq 2$. By (3A), (3E) and (3.12)

$$8c(J_1)a(\mu, J_1; \rho) = 5 \cdot 81 \cdot 64(1 - 1/9)$$

This implies $s \leq 2$. Hence $s = 2$, $c(J_1) = 64 \cdot 9$ and the proof is complete.

We collect the information obtained in a table

	1	ρ	μ	J	J_1
Order of element	1	5	4	2	2
$c(\tau)$	25920	5	8	$3\cdot 32$	$9\cdot 64$
Number in class	1	5184	3240	270	45
χ_0	1	1	1	1	1
χ_1	6	1	0	2	-2
χ_2	81	1	-1	-3	9
χ_3	24	-1	0	0	8
χ_4	64	-1	0	0	0
χ_5	5	0	-1	1	-3

(3I) If T is the set of 2-elements which are not conjugate to $1, \mu, J, J_1$, then

$$(3.13)\qquad \sum_{\tau\in T} \chi_2(\tau)^2 = 4860; \qquad \sum_{\tau\in T} \chi_2(\tau) = -1620.$$

Proof. The first equation is obtained from (3.11) since we have $\chi_2(J_1) = 9$ and since the class of J_1 contains 45 elements, $4860 = 8505 - 81\cdot 45$. Again by the orthogonality relations,

$$\sum_{\sigma\in G} \chi_2(\sigma) = 0.$$

As χ_2 vanishes for 3-singular elements and as the classes of $1, \rho, \mu, J, J_1$ contribute

81, 5184, -3240, -810, 405

respectively to the sum (cf. the preceding table) this yields the second equation (3.13).

(3J) The center Z of the Sylow 2-group Q has order 2.

Proof. We may assume that $J_1 \in Z$ and that $\mu \in Q$. Then $Z \subseteq C(\mu)$. Now $\langle J_1, \mu\rangle \subseteq C(\mu)$. Since $\mu^2 = J \neq J_1$ and since $C(\mu)$ has order 8, $C(\mu) = \langle J_1, \mu\rangle$ and $Z \subseteq \langle J_1, \mu\rangle = \langle J_1\rangle \times \langle \mu\rangle$. If Z had order more than 2, then $Z \cap \langle \mu\rangle \neq 1$ and then $\mu^2 = J \in Z$ which is not true as $c(J)$ is not divisible by 64. Hence $Z = \langle J_1\rangle$.

(3K) Let K_ν be a conjugate class which belongs to the set T in (3I) and let $\tau_\nu \in K_\nu$. Then

$$c(\tau_\nu) = 2^{b_\nu} \cdot 3^{d_\nu}, \quad \chi_2(\tau_\nu) = \pm 3^{w_\nu}. \tag{3.14}$$

Here, b_ν, d_ν and w_ν are integers and

either $w_\nu = d_\nu = 1, \quad 4 \leqq b_\nu \leqq 5$

or $w_\nu = d_\nu = 0, \quad 3 \leqq b_\nu \leqq 5$.

If K_ν ranges over all classes contained in T,

$$\sum_\nu 2^{5-b_\nu} 3^{d_\nu} = 6. \tag{3.15}$$

$$\sum_\nu 2^{5-b_\nu} y_\nu = -2 \tag{3.16}$$

where $y_\nu = \pm 1$ is the sign of $\chi_2(\tau_\nu)$.

<u>Proof</u>. By (1A), 5 does not divide $c(\tau_\nu)$ and we can set $c(\tau) = 2^{b_\nu}\cdot 3^{d_\nu}$. On account of (3J), here $b_\nu \leqq 5$. Since χ_2 is the only irreducible character of G of degree 81, its values are rational integers. As τ_ν is a 2-element, $\chi_2(\tau_\nu) \equiv x_2 = 81$ (mod 2). Then $\chi_2(\tau_\nu)$ is odd and we may set $\chi_2(\tau_\nu) = 3^{w_\nu}y_\nu$ where y_ν is an odd integer not divisible by 3. The class K_ν consists of $5\cdot 2^{6-b_\nu}\cdot 3^{4-d_\nu}$ elements. As is well known, $5\cdot 2^{6-b_\nu}\cdot 3^{4-d_\nu}\chi_2(\tau_\nu)/81$ is an algebraic integer. Hence $w_\nu \geqq d_\nu$. Now the first equation (3.13) reads

$$10\cdot 81 \sum_\nu 2^{5-b_\nu}\, 3^{2w_\nu - d_\nu}\, y_\nu^{\,2} = 4860$$

where K_ν ranges over all classes contained in T. Hence

$$\sum_\nu 2^{5-b_\nu}\cdot 3^{2w_\nu - d_\nu}\, y_\nu^{\,2} = 6.$$

This shows that $y_\nu^{\,2} \leqq 6$ and as y_ν is relatively prime to 6, $y_\nu = \pm 1$. This proves (3.14). Also $2w_\nu - d_\nu \leqq 1$ and since $w_\nu \geqq d_\nu$ we either have $w_\nu = d_\nu = 1$ or $w_\nu = d_\nu = 0$. Then $2w_\nu - d_\nu = d_\nu$ and we obtain (3.15). If $d_\nu = 1$, this shows that $5-b_\nu \leqq 1$ while for $d_\nu = 0$, we have $5-b_\nu \leqq 2$.

Finally, the second equation (3.13) can be written as

$$10\cdot 81 \sum 2^{5-b_\nu}\cdot 3^{w_\nu - d_\nu} y_\nu = -1620$$

and this yields (3.16).

(3L) We have the following six possibilities for the values of $\chi_j(\tau_\nu)$ for τ_ν as in (3K) and of b_ν, d_ν:

	(1)	(2)	(3)	(4)	(5)	(6)
b_ν	5	4	4	4	3	3
d_ν	1	1	1	0	0	0
$\chi_0(\tau_\nu)$	1	1	1	1	1	1
$\chi_1(\tau_\nu)$	2	2	0	-2	-2	0
$\chi_2(\tau_\nu)$	-3	-3	3	1	1	-1
$\chi_3(\tau_\nu)$	0	0	4	0	0	0
$\chi_4(\tau_\nu)$	0	0	0	0	0	0

Proof. We consider $a(\tau_\nu,\mu;\rho)$ and $a(\tau_\nu,J;\rho)$. Since 2^{b_ν} divides $c(\tau_\nu)$, it follows from (3A) and the table on p. 41 that

(3.17) $$81 \qquad - \chi_2(\tau_\nu) \equiv 0 \pmod{2^{b_\nu-3}},$$

(3.18) $$27 + 9\chi_1(\tau_\nu) - \chi_2(\tau_\nu) \equiv 0 \pmod{2^{b_\nu-1}}$$

Since χ_4 has 2-defect 0, $\chi_4(\tau_\nu) = 0$.

Suppose first that $d_\nu = 1$. By (3K), $\chi_2(\tau_\nu) = \pm 3$ and $b_\nu \geq 4$. For $\chi_2(\tau_\nu) = -3$ by (3.18) $\chi_1(\tau_\nu) \equiv 2 \pmod 8$ and since $|\chi_1(\tau_\nu)| \leq 5$, we have $\chi_1(\tau_\nu) = 2$ and (3.5) yields $\chi_3(\tau_\nu) = 0$. This leads to the cases (1) and (2). For $\chi_2(\tau_\nu) = 3$, $\chi_1(\tau_\nu) \equiv 0 \pmod 8$ and hence $\chi_1(\tau_\nu) = 0$ and, by (3.5), $\chi_3(\tau_\nu) = 4$. By (3.17), $b_\nu - 3 \leq 1$ and we are led to Case (3).

In all remaining cases $d_\nu = 0$, $\chi_2(\tau_\nu) = \pm 1$. If $\chi_2(\tau_\nu) = 1$ and $b_\nu = 5$, by (3.18) $9\chi_1(\tau_\nu) \equiv 6 \pmod{16}$, $\chi_1(\tau_\nu) \equiv 6 \pmod{16}$ and this is impossible as $|\chi_1(\tau_\nu)| \leqq 5$. For $\chi_2(\tau_\nu) = -1$, we cannot have $b_\nu = 5$ because of (3.17). Thus $3 \leqq b_\nu \leqq 4$.

If $b_\nu = 4$ and $\chi_2(\tau_1) = 1$, by (3.18) $\chi_1(\tau_\nu) \equiv 6 \pmod 8$ and as $|\chi_1(\tau_\nu)| \leqq 5$, we have $\chi_1(\tau_\nu) = -2$. By (3.5) $\chi_3(\tau_\nu) = 0$. This is case (4). If $b_\nu = 4$ and $\chi_2(\tau_1) = -1$, by (3.18) $\chi_1(\tau_\nu) \equiv 4 \pmod 8$. This implies $\chi_1(\tau_\nu) = \pm 4$. Since $\sum_j |\chi_j(\tau_\nu)|^2 = c(\tau_\nu) = 16$, this case is impossible.

Finally, if $b_\nu = 3$ and $\chi_2(\tau_\nu) = 1$, by (3.18) $\chi_1(\tau_\nu) \equiv 6 \pmod 4$ and hence $\chi_1(\tau_\nu) = 2$ or $\chi_1(\tau_\nu) = -2$. In the former case, $\chi_3(\tau_\nu) = 4$ and this is impossible since $c(\tau_\nu) = 8$. Hence $\chi_1(\tau_\nu) = -2$ and then $\chi_3(\tau_\nu) = 0$. This is case (5). Finally, if $\chi_2(\tau_\nu) = -1$, we find from (3.18) that $\chi_1(\tau_\nu) \equiv 0 \pmod 4$. Since $c(\tau_\nu) = 8$, this leaves only the possibility $\chi_1(\tau_\nu) = 0$ and then $\chi_3(\tau_\nu) = 0$. This is Case (6).

(3M) For 3-regular elements σ, $\chi_5(\sigma)$ has rational values.

Proof. Suppose this is false. The table on p. 41 shows that then there must exist elements $\tau \in T$ for which $\chi_5(\tau)$ is not rational. There exists then a power τ^z with odd z such that $\tau_1 = \tau$ and $\tau_2 = \tau^z$ lie in different conjugate classes K_1 and $K_2 \subseteq T$. We necessarily have the same of the cases (1),(2),...

for $\nu = 1,2$ in (3L). Since (3.15) then implies $2^{5-b_\nu}\cdot 3^{d_\nu} \leqq 3$, we must have Case (1) or Case (4) for $\nu = 1,2$.

In either case, $\chi_3(\tau_1) = 0$ and hence $|\chi_5(\tau_1)|^2 = 1$ by (3.10).

Suppose first that τ_1 has order 4. If again X_5 is the representation with the character χ_5, then $X_5(\tau_1)$ has determinant 1. If $X_5(\tau_1)$ has the characteristic roots $i, -i$ with multipliers m_1, m_2 respectively, it follows that $m_1 \equiv m_2 \pmod 2$. Hence $\chi_5(\tau_1)$ has even imaginary part. Since $\chi_5(\tau_1)$ is an integer in the Gaussian field, it follows that $\chi_5(\tau_1) = \pm 1$, a contradiction.

The order of τ_1 cannot be larger than 8; cf. [5]. It remains to discuss the possibility that the order is equal to 8. Then τ_1^2 has order 4. Now, we had that $c(\mu) = 8$ and $C(\mu) = \langle \mu, J_1 \rangle$ does not contain an element of order 8. It follows that τ^2 cannot be conjugate to μ. Then τ^2 belongs to a third class $K_3 \subset T$. This rules out the Case (1) since (3.15) shows that in this case T consists only of K_1 and K_2. If we have Case (4) for K_1 and K_2, then (3.15) and (3.16) yield

$$\sum_{\nu > 2} 2^{5-b_\nu} 3^{d_\nu} = 2, \qquad 2 + 2 + \sum_{\nu > 2} 2^{5-b_\nu} y_\nu = -2,$$

since $y_1 = y_2 = 1$ when we have Case (4) for K_1 and K_2. Then $d_\nu = 0$ for $\nu \geqq 3$. Hence

$$\sum_{\nu > 2} 2^{5-b_\nu} y_\nu \geqq -\sum_{\nu > 2} 2^{5-b_\nu} = -2$$

and our equations are inconsistent. This completes the proof.

We are now ready to prove the final result of this section:

(3N) $G \simeq O_5(3)$.

Proof. Consider X_5 as a 3-modular representation X_5^o. The character χ_5^o is the restriction of χ_5 to the set of 3-regular elements of G. On account of (3M), χ_5^o has values in the ring of ordinary integers. This implies that X_5^o can be written in the field GF(3) with 3 elements; cf. [1], p. 101.

The 3-modular characters of N(R) are the same as the ordinary characters of N(R), since the order of N(R) is not divisible by 3. Since $\chi_5|N(R) = \psi + \lambda_2$ by (3D), we see that if χ_5^o is reducible, it could only split into an irreducible modular character of degree 4 and a modular character Λ of degree 1 of G such that $\Lambda|N(R) = \lambda_2$. Then Λ is not the principal character. Since G is simple, no such Λ exists.

Hence X_5^o is an absolutely irreducible modular representation of G.

Since χ_5^o is real, X_5^o is equivalent to its contragredient representation $(X_5^o)^*$ and this implies that the tensor product $X_5^o \otimes X_5^o$ contains the principal modular representation X_o^o as top constituent:

$$(3.19) \qquad X_5^o \otimes X_5^o \simeq \begin{pmatrix} X_5^o & 0 \\ * & * \end{pmatrix}.$$

Now $X_5^o \otimes X_5^o$ splits into the representation P_2 belonging to the symmetric tensors of rank 2 and the representation C_2 belonging to the skew symmetric tensors of rank 2 over the representation space of X_5^o.

The character of C_2 for a 3-regular element σ is $\frac{1}{2}(\chi_5(\sigma)^2-\chi_5(\sigma^2))$. Thus (cf. the table, p. 41) the character of C_2 has the values 10,0,0,-2 respectively for the elements $1,\rho,\mu^{\pm 1},J$. It follows that the character of $C_2|N(R)$ is $2\phi + \lambda_1 + \lambda_3$. Hence C_2 does not have a constituent X_o^o. Now (3.19) shows that P_2 has a top constituent X_o^o. This means that X_5^o has a non-trivial quadratic invariant. It follows that G is a subgroup of the full orthogonal group of degree 5 over GF(3). Since G is simple of order 25,920, this implies that $G \simeq O_5(3)$; [10, p. 159, p. 179], [11].

This proves (3N). The proof of the theorem on p. 1 is complete.

References

[1] R. Brauer, Math. Zeitschr. 30 (1929), 79-107.

[2] ________, Amer. J. Math. 64 (1942), 401-420.

[3] ________, Math. Zeitschr. 63 (1956), 406-444.

[4] ________, J. of Alg. 3 (1966), 225-255.

[5] ________, On primitive **linear** groups. To appear.

[6] R. Brauer-C.J. Nesbitt, U. of Toronto Studies, No. 4, 1937.

[7] ____________________, Annals of Math. 42 (1941), 556-590.

[8] R. Brauer-M. Suzuki-G.E. Wall, Ill. J. Math. 2 (1958), 718-745.

[9] R. Brauer-H.F. Tuan, Bull. Amer. Math. Soc. 51 (1945), 756-766.

[10] L.E. Dickson, Linear Groups, Teubner 1901.

[11] J. Dieudonné, La géométrie des groupes classiques, Springer, 1955.

[12] D. Gorenstein-J. Walter, J. of Alg. 2 (1965), 85-151, 218-270.

[13] M. Suzuki, Ann. of Math. 75 (1962), 105-145.

[14] H.F. Tuan, Ann. of Math. 45 (1944), 110-140.

BRAUER, R.
Math. Annalen 169, 73—96 (1967)

Über endliche lineare Gruppen von Primzahlgrad

CARL LUDWIG SIEGEL zum siebzigsten Geburtstag gewidmet

RICHARD BRAUER

§ 1. Einleitung

Ist L eine endliche Gruppe linearer Transformationen mit komplexen Koeffizienten eines gegebenen Grades n, so gibt es, wie C. JORDAN [19] gezeigt hat, eine nur von n abhängige Schranke $J(n)$, derart, daß L einen abelschen Normalteiler A hat, dessen Index $|L:A|$ unter $J(n)$ liegt. Explizite Werte für $J(n)$ sind von H. BLICHFELDT [3], L. BIEBERBACH [1] und G. FROBENIUS [17] gegeben worden. Für $n \leqq 4$ wurde die Gesamtheit aller Gruppen L von C. JORDAN, F. KLEIN, VALENTINER und H. BLICHFELDT und anderen gefunden. Eine Darstellung der Ergebnisse findet sich in BLICHFELDTs Buch [4]. Im Falle $n = 5$ gelangte BLICHFELDT nicht mehr zum Ziel.

Die neueren Entwicklungen der Darstellungstheorie der endlichen Gruppen machen es möglich, diese alten Untersuchungen fortzusetzen. So sollen hier die endlichen Gruppen fünften Grades bestimmt werden. Für festes n spielen bei der Behandlung der Gruppen diophantische Probleme eine Rolle. Damit hängt es zusammen, daß die verschiedenen auftretenden Gruppen ein ganz verschiedenes Aussehen haben können. Immerhin ist es möglich, einiges Gemeinsames auszusagen. Das soll hier für den Fall durchgeführt werden, daß der gegebene Grad n eine Primzahl p ist. Wir betrachten also eine endliche Gruppe G, die eine treue irreduzible Darstellung X durch lineare Transformationen von Primzahlgrad p mit komplexen Koeffizienten hat. Dabei wollen wir den Fall außer acht lassen, daß X sich als monomiale Gruppe schreiben läßt[1]. Das kommt hier darauf hinaus, daß wir X als primitiv voraussetzen. Da wir uns eigentlich weniger für X als für die zu X gehörige Gruppe projektiver Transformationen interessieren, dürfen wir annehmen, daß X unimodular ist. Dann hat das Zentrum $Z = \mathfrak{Z}(G)$ die Ordnung 1 oder p.

Es sei P von der Ordnung p^a eine zu p gehörige Sylow-Untergruppe von G. Ist $a = 1$, so erlauben es die Ergebnisse der Darstellungstheorie, die Untersuchung der Charaktere direkt in Angriff zu nehmen. Diskussion eines zahlentheoretischen Problems liefert die Grade der irreduziblen Charaktere im

[1] Man kann zwar nicht sagen, daß man alle monomialen Gruppen kennt. Man kennt ja nicht einmal alle transitiven Permutationsgruppen von Primzahlgrad. Für kleinere Werte von n können die monomialen Gruppen ohne Schwierigkeiten diskutiert werden.

p-Hauptblock von G. Die Durchführung für $p=5$ ist relativ einfach, und man kann dann weiter die Gruppen G selbst bestimmen. Für größere Werte von p wächst die Zahl der zu betrachtenden Fälle rasch. Auch wird die Diskussion schwieriger. Trotzdem dürfte es von Interesse sein, die Untersuchung fortzusetzen, da (nach (8 A)) die Gruppen G, die auftreten, im wesentlichen einfach sind.

Ist $a>1$, so ist sehr viel weniger über die Charaktere bekannt als im Fall $a=1$. Als ersten Schritt hat man die Sylowgruppen P und die Durchschnitte von P mit Konjugierten zu untersuchen. Dies wird im folgenden durchgeführt. Ist $p=5$, so genügen die Resultate, um elementar die Fälle mit $a>3$ zu diskutieren. Der Fall $a \leqq 3$ kann mit Hilfe von Charakteren behandelt werden.

Bezeichnung. Wir schreiben $|S|$ für die Anzahl der Elemente einer endlichen Menge S. Ist S eine Teilmenge einer endlichen Gruppe G, so sollen der Zentralisator und der Normalisator von S mit $\mathfrak{C}(S)$, bzw. mit $\mathfrak{N}(S)$ bezeichnet werden. Wenn notwendig, fügen wir G als Index hinzu. Wir setzen $c(S)=|\mathfrak{C}(S)|$, $n(S)=|\mathfrak{N}(S)|$. Das Zentrum $\mathfrak{C}_G(G)$ von G wird mit $\mathfrak{Z}(G)$ bezeichnet. Wir verwenden die Abkürzung „S_p-Gruppe von G" für „p-Sylow-Untergruppe von G".

§ 2. Bekannte Resultate

Es sei G eine endliche Gruppe und X eine treue irreduzible Darstellung von G durch lineare Transformationen eines gegebenen Grades n mit komplexen Koeffizienten.

Ist H ein Normalteiler von G, so zerfällt die entsprechende Darstellung $X|H$ nach Cliffords Satz [13] in eine gewisse Anzahl von irreduziblen Darstellungen $Y_1, Y_2, \ldots, Y_r$ von H, die in bezug auf G assoziiert sind und alle in $X|H$ mit derselben Vielfachheit e auftreten. Hat Y_1 den Grad m, so ist also

(2.1) $$n=erm.$$

Ist X primitiv, so ist für jede Wahl von H als Normalteiler von G stets $r=1$ ist. Dabei sprechen wir von einer primitiven Darstellung X nur, wenn X irreduzibel ist.

Der folgende Satz ist mit dem in der Einleitung genannten Jordanschen Satz gleichwertig:

(2A) *Es gibt eine nur von n abhängige Schranke $J_0(n)$ mit der folgenden Eigenschaft: Besitzt die endliche Gruppe G eine treue primitive Darstellung X vom Grad n, so ist $|G:\mathfrak{Z}(G)| \leqq J_0(n)$.*

Explizierte Werte für $J_0(n)$ sind von Blichfeldt [2] gegeben worden. Bezeichnet $(n!)_p$ die höchste in $n!$ aufgehende Potenz einer Primzahl p und enthält $|G:\mathfrak{Z}(G)|$ die Potenz p^ν, so gilt nach Blichfeldt [4], § 74 die Ungleichung

(2.2) $$p^\nu \leqq (n!)_p\, 6^{n-1}\;[2].$$

Ein wesentlicher Fortschritt in der Untersuchung der möglichen Gruppenordnungen g ist vor kurzem von W. Feit [15] erzielt worden. Seine Unter-

[2] Blichfeldt gibt an, daß die Konstante 6 durch 5 ersetzt werden kann, doch hat er den Beweis nicht veröffentlicht.

suchung ist eine Fortsetzung einer gemeinsamen Arbeit von ihm und J. G. THOMPSON [16]. Als unmittelbare Anwendung des Resultats haben wir:

(2 B) *Es sei G eine endliche Gruppe, die eine treue, primitive, unimodulare Darstellung X vom Grad n besitzt. Dann ist* $g=|G|$ *durch höchstens eine Primzahl* $p>n+1$ *teilbar. Nur die erste Potenz von p kann in g auftreten.*

Jetzt können ältere Resultate des Verfassers [5] angewendet werden. Man erhält:

(2 C) *Ist in* (2 B) *die Primzahl ein Teiler von g, so ist* $p\leqq 2n+1$. *Der Fall* $p=2n+1$ *tritt nur auf, wenn* $G/\mathfrak{Z}(G)\cong PSL(2,p)$ *ist*[3].

Setzen wir weiterhin voraus, daß in (2 B) die Ordnung g durch die Primzahl $p>n+1$ teilbar ist. Der Normalisator einer S_p-Gruppe von $G/\mathfrak{Z}(G)$ ist eine metazyklische Gruppe einer Ordnung $p(p-1)/t$, wo t ein Teiler von $p-1$ ist. Es muß dann $n=p-(p-1)/t$ sein. Der Fall $t=2$, d. h. $n=(p+1)/2$ ist von H. F. TUAN [22] vollständig behandelt worden. Ist $n>4$, so muß wieder $G/\mathfrak{Z}(G)\cong PSL(2,p)$ sein. In einer demnächst erscheinenden Arbeit [7] zeigen wir, daß für $t>2$ die Ungleichung $p\leqq t^3-t+1$ gilt. Daraus hat S. HAYDEN [18] hergeleitet, daß die Fälle $t=3$, 4, 5 nicht auftreten können. Jetzt folgt:

(2 D) *Ist in* (2 B) $5\leqq n\leqq 10$, *so ist g nur dann durch eine Primzahl* $p>n+1$ *teilbar, wenn* $G/\mathfrak{Z}(G)\cong PSL(2,p)$ *ist.*

Setzen wir voraus, daß die Gradzahl n eine Primzahl ist, so folgt aus (2.1), daß nur dann $r\neq 1$ sein kann, wenn $r=n$, $e=m=1$. Ist also die treue, irreduzible Darstellung X nicht primitiv, so kann man im Falle eines Primzahlgrades n die Darstellung X als monomiale Darstellung schreiben. Daher ist es gerechtfertigt, wenn man sich auf den Fall eines primitiven X beschränkt. Ist X primitiv, so ist für einen Normalteiler H entweder $X|H$ irreduzibel oder H liegt im Zentrum $\mathfrak{Z}(G)$.

§ 3. Einige Eigenschaften primitiver treuer Darstellungen endlicher Gruppen

Ist G eine gegebene endliche Gruppe und X eine Darstellung von G durch lineare Transformationen mit komplexen Koeffizienten, so kann man nach Ersetzen von X durch eine ähnliche Darstellung annehmen, daß die Koeffizienten von X einem algebraischen Zahlkörper K von endlichem Grad angehören. Wir wollen dabei K so wählen, daß es normal über dem Körper der rationalen Zahlen ist und die $|G|$-ten Einheitswurzeln enthält. Ist p eine feste Primzahl und $\mathfrak{p}_1$ ein Primidealteiler von p in K, so kann man sogar annehmen, daß die Koeffizienten von X dem Ring R der in bezug auf $\mathfrak{p}_1$ ganzen Zahlen von K angehören. Es sei $\mathfrak{p}$ das Primideal von R, $\mathfrak{p}=\mathfrak{p}_1 R$.

Ist X in der angegebenen Weise gewählt, so bilden die Elemente $\sigma\in G$ mit $X(\sigma)\equiv E(\operatorname{mod}\mathfrak{p})$ einen Normalteiler N von G, den wir den *modularen Kern* von X nennen[4].

[3] Wir bezeichnen mit $PSL(2,p)$ die projektive unimodulare Gruppe zweiten Grades über einem Körper mit p Elementen.

[4] Wir bezeichnen das Einheitselement einer Gruppe mit 1 und die Identitätstransformation und die Einheitsmatrix mit E.

(3 A) *Ist X eine treue Darstellung von G, so ist der modulare Kern N eine p-Gruppe.*

Ist nämlich $X(\sigma)-E\equiv 0 (\bmod \mathfrak{p})$, so folgt für positives ganzes r, daß $X(\sigma^{p^r})-E\equiv 0(\bmod \mathfrak{p}^r)$ ist. Wählt man r so groß, daß für kein $\tau\neq 1$ in G alle Koeffizienten von $X(\tau)-E$ zu $\mathfrak{p}^r$ gehören, so folgt $\sigma^{p^r}=1$. Also besteht N nur aus p-Elementen.

(3 B) *Ist X eine treue Darstellung von G vom Grad n und ist $p^r \geqq n$, so hat jedes p-Element σ von G höchstens die Ordnung p^r über dem modularen Kern N.*

Beweis. Ist λ_ν eine charakteristische Wurzel von $X(\sigma)$, so ist $\lambda_\nu\equiv 1 (\bmod \mathfrak{p})$. Anwendung der Cayleyschen Relation im Restklassenkörper ergibt die Kongruenz $(X(\sigma)-E)^n\equiv 0(\bmod \mathfrak{p})$. Dann ist $(X(\sigma)-E)^{p^r}\equiv 0(\bmod \mathfrak{p})$, also $X(\sigma^{p^r})\equiv$ $\equiv E(\bmod \mathfrak{p})$ und daher $\sigma^{p^r}\in N$.

Für $p\leqq 5$ liefert eine ältere Blichfeldtsche Methode [2] bessere Abschätzungen als (2.2). Da das Resultat für uns von Bedeutung ist, wollen wir dies hier in einer gegenüber BLICHFELDT vereinfachten Form durchführen.

Es sei p eine feste Primzahl. Es sei X eine Darstellung von G, die wie oben gewählt sei, wobei K, R und $\mathfrak{p}$ dieselbe Bedeutung wie zuvor haben. Der Charakter von X sei mit χ bezeichnet. Ist l eine positive ganze Zahl, so betrachten wir die Menge H_l der Elemente ξ von G mit der folgenden Eigenschaft: Ist $\sigma\in G$ und $\chi(\sigma)\equiv 0(\bmod \mathfrak{p}^\lambda)$ für ein λ mit $0<\lambda\leqq l$, so soll $\chi(\xi\sigma)\equiv 0(\bmod \mathfrak{p}^\lambda)$ sein. Für $l=0$ setzen wir $H_0=G$. Offenbar ist H_l ein Normalteiler von G und

$$H_0\supseteqq H_1\supseteqq H_2\supseteqq\cdots.$$

Zunächst zeigen wir:

(3 C) *Ist X eine primitive treue Darstellung von G, so ist $H_l=\mathfrak{Z}(G)$ für hinreichend großes l. Geht p nicht im Grad n von X auf, so gilt dies bereits für $l=1$.*

Beweis. Es ist klar, daß $\mathfrak{Z}(G)$ zu allen H_l gehört. Es sei $l>0$. Betrachten wir zunächst den Fall, daß χ für kein Element von H_l verschwindet. Ist ψ ein irreduzibler Bestandteil von $\chi|H_l$, so ist wegen der Primitivität von X auch $\psi(\sigma)\neq 0$ für $\sigma\in H_l$. Nach einer einfachen Bemerkung von BURNSIDE [12], S. 319 zeigt die Orthogonalitätsrelation

$$\sum_{\sigma\in H_l}|\psi(\sigma)|^2=|H_l|,$$

daß dann $\psi(1)=1$ sein muß. Da $\chi|H_l$ ein Vielfaches eines linearen Charakters ist, ist $H_l\subseteqq\mathfrak{Z}(G)$, also $H_l=\mathfrak{Z}(G)$.

Für $H_l\neq\mathfrak{Z}(G)$ gibt es also Elemente $\sigma\in H_l$ für die $\chi(\sigma)=0$ ist. Aus der Definition von H_l folgt $\chi(\sigma\sigma^{-1})=0(\bmod \mathfrak{p}^l)$, also $n\equiv 0(\bmod \mathfrak{p}^l)$. Insbesondere ist n durch p teilbar und l muß unter einer festen Schranke liegen. Daraus ergibt sich die Behauptung.

(3 D) *Es sei χ der Charakter einer treuen primitiven Darstellung X von G. Es sei A eine Abelsche Untergruppe der Primzahlpotenzordnung p^r, die nicht in $\mathfrak{Z}(G)$ liegt. Die Beschränkung $\chi|A$ enthalte genau m verschiedene lineare Charaktere $\lambda_1,\ldots,\lambda_m$ von A. Ist $m\leqq r$, so gibt es einen Normalteiler H von G, der nicht A*

enthält, aber mit A einen Durchschnitt mindestens von der Ordnung p^{r-m+1} *hat. Dabei kann man H als eine Gruppe* H_l *für geeignetes* $l>0$ *wählen.*

Beweis. Es sei hier K_0 der Körper der p^r-ten Einheitswurzeln, es sei $\mathfrak{p}_0$ der Primidealteiler von p in K_0, und es sei R_0 der Ring der für $\mathfrak{p}_0$ ganzen Zahlen von K_0. Man wähle m Unbestimmte $x_1, \ldots, x_m$ und bilde für jedes $\alpha \in A$ die Linearform

$$u_\alpha = \sum_{\mu=1}^{m} x_\mu \lambda_\mu(\alpha). \tag{3.1}$$

Diese Linearformen erzeugen einen R_0-Modul M. Da R_0 Hauptidealring ist, besitzt M eine R_0-Basis $v_1, \ldots, v_m$. Dabei darf man annehmen, daß $v_1 = u_1 = \sum_\mu x_\mu$ ist und daß $v_2, \ldots, v_m$ Linearformen in $x_2, \ldots, x_m$ sind.

Es sei π eine lokale Primzahl in $\mathfrak{p}_0$. Da $\mathfrak{p}_0$ den Restklassengrad 1 hat, kann man setzen

$$u_\alpha = \sum_{\mu=1}^{m} b_{\alpha,\mu} v_\mu + \pi \sum_{\mu=1}^{m} b'_{\alpha,\mu} v_\mu,$$

wo jedes $b_{\alpha,\mu}$ eine der Zahlen $0, 1, 2, \ldots, p-1$ ist und die $b'_{\alpha,\mu}$ in R_0 liegen. Vergleich der Koeffizienten von x_1 liefert $b_{\alpha,1} = 1$. Drückt man die v_μ in der zweiten Summe wieder durch die u_ξ mit $\xi \in A$ aus, so erhält man Formeln

$$u_\alpha = v_1 + \sum_{\mu=2}^{m} b_{\alpha,\mu} v_\mu + \pi \sum_{\xi \in A} c_{\alpha,\xi} u_\xi \tag{3.2}$$

mit $c_{\alpha,\xi} \in R_0$.

Es sei $\sigma \in G$, $\alpha \in A$. Da $X(\sigma\alpha) = X(\sigma) X(\alpha)$ ist, so sieht man durch Bildung der Spur, daß $\chi(\sigma\alpha)$ die Form

$$\chi(\sigma\alpha) = \sum_{\mu=1}^{m} t_\mu(\sigma)\, \lambda_\mu(\alpha) \tag{3.3}$$

hat, wo die komplexen Zahlen $t_\mu(\sigma)$ von α unabhängig sind. Setzt man $x_\mu = t_\mu(\sigma)$ und ist $v_\mu^*(\sigma)$ der entsprechende Wert der Linearform v_μ, so ergibt sich aus (3.3), (3.1) und (3.2) die Gleichung

$$\chi(\sigma\alpha) = v_1^*(\sigma) + \sum_{\mu=2}^{m} b_{\alpha,\mu} v_\mu^*(\sigma) + \pi \sum_{\xi \in A} c_{\alpha,\xi} \chi(\sigma\xi).$$

Man ordne jedem $\alpha \in A$ die Zahlenreihe $(b_{\alpha,2}, \ldots, b_{\alpha,m})$ zu. Im ganzen können höchstens p^{m-1} verschiedene Zahlenreihen auftreten. Da $|A| = p^r$ ist, gibt es p^{r-m+1} verschiedene Elemente $\alpha_1, \alpha_2, \ldots$ in A, denen dieselbe Zahlenreihe entspricht. Dann gilt

$$\chi(\sigma\alpha_j) - \chi(\sigma\alpha_1) = \pi \sum_{\xi \in A} (c_{\alpha_j,\xi} - c_{\alpha_1,\xi})\, \chi(\sigma\xi).$$

Ersetzt man hier σ durch $\sigma\alpha_1^{-1}$ und setzt man $\alpha'_j = \alpha_1^{-1}\sigma_j \in A$, so hat man

$$\chi(\sigma\alpha'_j) - \chi(\sigma) = \pi \sum_{\eta \in A} d_{j,\eta}\, \chi(\sigma\eta), \tag{3.4}$$

wo $d_{j,\eta} \in R$ ist und nicht von σ abhängt.

Man wähle jetzt l als die größte ganze Zahl für die $H_{l-1} \supsetneqq A$ ist. Das ist nach (3 C) möglich. Man beachte, daß $\mathfrak{p}_0 \subsetneqq \mathfrak{p}_1 \subsetneqq \mathfrak{p}$ ist. Ist $1 \leqq \lambda \leqq l$ und ist $\chi(\sigma) \equiv 0 (\bmod\, \mathfrak{p}^\lambda)$, so folgt wegen $A \subsetneqq H_{l-1}$ aus (3.4), daß $\chi(\sigma\alpha'_j) \equiv 0(\bmod\, \mathfrak{p}^\lambda)$ ist. Also gehören die p^{r-m+1} Elemente α'_j zu $A \cap H_l$. Damit ist gezeigt, daß man $H = H_l$ in (3 D) wählen kann.

(3 E) *Korollar. Hat G eine treue primitive Darstellung X vom Grad n und geht p nicht in n auf, so ist* $|G : \mathfrak{Z}(G)|$ *durch p höchstens in der Potenz* $(n!)_p p^{n-1}$ *teilbar.*

Beweis. Es sei P eine S_p-Gruppe von G. Dann kann man $X|P$ in monomialer Form schreiben, vgl. [4], S. 80. Ist A die Untergruppe der Diagonalelemente von P, so gilt $|A| = p^r$ mit $p^r \geqq |P|/(n!)_p$.

Wir nehmen an, daß $|G : \mathfrak{Z}(G)|$ durch $(n!)_p p^n$ teilbar ist. Es sei Z_p die S_p-Gruppe von $\mathfrak{Z}(G)$ und $|Z_p| = p^\mu$. Dann ist $Z_p \subsetneqq A$, und man findet $r \geqq n + \mu$. Wendet man (3 D) an, so sieht man, da $m \leqq n$ ist, daß für eine geeignete Gruppe H_l die Beziehungen

$$H_l \not\supseteq A, \quad |H_l \cap A| \geqq p^{\mu+1}$$

gelten. Dann ist $H_l \neq G$, $H_l \neq \mathfrak{Z}(G)$. Nach (3 C) ist dies nur möglich, wenn $p|n$ ist. Das ergibt einen Widerspruch.

Wir beweisen noch einen weiteren Hilfssatz.

(3 F) *Es sei X eine treue, primitive Darstellung vom Grad n von G. Es sei U eine Untergruppe derart, daß alle irreduziblen Bestandteile von X|U die Vielfachheit* 1 *haben. Dann ist* $\mathfrak{C}(U)$ *eine abelsche Gruppe und* $|\mathfrak{C}(U) : \mathfrak{Z}(G)|$ *ist nur durch Primzahlen p teilbar, die in n oder in* $|U| = u$ *aufgehen.*

Beweis. Aus dem Schurschen Lemma folgt unmittelbar, daß bei geeigneter Schreibweise von X in Matrizenform alle Elemente von $X|\mathfrak{C}(U)$ Diagonalmatrizen sind. Daher ist $\mathfrak{C}(U)$ jedenfalls eine abelsche Gruppe.

Es sei p eine nicht in u aufgehende Primzahl. Hat R dieselbe Bedeutung wie oben, so kann man nach einem Resultat von Zassenhaus [23] die Darstellung X in Matrizenform so wählen, daß alle Koeffizienten in R liegen und daß $X|U$ in irreduzible Bestandteile zerfällt. Allerdings handelt es sich dabei im allgemeinen um einen nicht vollständigen Zerfall. Geht aber p nicht in $|U|$ auf, so kann man nach einer klassischen Schurschen Schlußweise [21] durch eine weitere Ähnlichkeitstransformation vollständigen Zerfall von $X|U$ bewirken. Ist das der Fall, so besteht $X|\mathfrak{C}(U)$ nur aus Diagonalmatrizen. Ist σ ein p-Element von $\mathfrak{C}(U)$, so gehört offenbar σ zum modularen Kern N von X. Wegen der Primitivität von X ist $X|N$ ein Vielfaches einer irreduziblen Darstellung Y von N. Hier ist N nach (3 A) eine p-Gruppe, also der Grad w von Y eine Potenz von p. Gehört σ nicht zu $\mathfrak{Z}(G)$, so ist $N \not\subseteq \mathfrak{Z}(G)$ und wir haben $w \neq 1$. Daher gilt $p|n$, und daraus folgt die Behauptung.

§ 4. Die Sylow-*p*-Untergruppen der primitiven linearen Gruppen eines Primzahlgrades *p*

Wir nehmen jetzt an, daß G eine endliche Gruppe ist, die eine treue primitive unimodulare Darstellung X vom Primzahlgrad $p \geqq 5$ besitzt. Der Charakter von X wird durchgehend mit χ bezeichnet. Die Ordnung $g=|G|$ von G ist durch p teilbar. Wir setzen

$$g=p^a g_0, \quad (p, g_0)=1. \tag{4.1}$$

Das Zentrum $Z=\mathfrak{Z}(G)$ von G ist zyklisch von der Ordnung 1 oder p. Es sei P eine S_p-Gruppe von G.

(4 A) *Ist P eine abelsche Gruppe, so ist entweder $a=1$, $Z=1$ oder $a=2$, $|Z|=p$. Im zweiten Fall ist G ein direktes Produkt $G=G_1\times Z$ einer Untergruppe G_1 mit Z.*

Beweis. Jedenfalls ist $Z\subsetneqq P$. Wegen $\chi(1)=p$ ist $g\chi(\sigma)/(p\,c(\sigma))$ für alle $\sigma\in G$ eine ganze algebraische Zahl. Für $\sigma\in P$ ist also $\chi(\sigma)$ durch p teilbar. Daraus folgt bekanntlich, daß $\chi(\sigma)=0$ oder $|\chi(\sigma)|=p$ ist; der zweite Fall tritt dann und nur dann ein, wenn $\sigma\in Z$ ist. Es ist also

$$|P|\,(\chi|P, \chi|P)=\sum_{\sigma\in P}|\chi(\sigma)|^2=p^2|Z|. \tag{4.2}$$

Da $\chi|P$ reduzibel ist, ist die linke Seite größer als $|P|$, also $|P|\leqq p|Z|$. Für $Z=1$ müssen wir $|P|=p$ haben.

Nehmen wir also $Z\neq 1$ an[5]. Aus (4.2) folgt $a\leqq 2$. Im Hauptblock B von G muß es (z. B. nach [6], Theorem 1) eine Darstellung X_0 geben, deren Kern nicht Z enthält. Die Elemente $\sigma\in G$, für die

$$\operatorname{Det}(X_0(\sigma))^{g_0}=1$$

ist, bilden einen Normalteiler G_1 von G, dessen Index eine Potenz von p ist. Ist $|G_1|$ zu p teilerfremd, so ist $\chi|G_1$ reduzibel. Wegen der Primitivität von X ist dann $G_1\subsetneqq Z$, also $G_1=1$ und $g=p$ oder p^2, was auf einen Widerspruch führt.

Es ist also $|G_1|$ durch p teilbar. Es sei $\zeta\in Z$, $\zeta\neq 1$. Dann ist $X_0(\zeta)=\varepsilon E$, wo ε eine primitive p-te Einheitswurzel und E die Einheitsmatrix ist. Wegen $X_0\in B$ und $a\leqq 2$ ist der Grad von X_0 nicht durch p teilbar, vgl. [8], Theorem 2. Es folgt, daß $\zeta\notin G_1$ ist. Dann muß aber $G=G_1\times Z$ sein. Damit ist (4 A) bewiesen.

Hat man den zweiten Fall in (4 A), so ist $G/Z\cong G_1$ und für G_1 liegt der erste Fall vor. Daher kann der zweite Fall von der Betrachtung ausgeschlossen werden.

Wir betrachten jetzt die nicht-abelschen Gruppen P, die als S_p-Gruppen von G auftreten können. Hier ist $a\geqq 3$.

(4 B) *Ist P eine nicht-abelsche Gruppe, so ist $\mathfrak{Z}(P)=Z$ und $|Z|=p$.*

Beweis. Wäre $X|P$ reduzibel, so müßte es in p Bestandteile ersten Grades zerfallen. Da X eine treue Darstellung ist, wäre P entgegen der Voraussetzung eine Abelsche Gruppe. Also ist $X|P$ irreduzibel. Aus dem Schurschen Lemma folgt $\mathfrak{C}(P)=Z$. Insbesondere ist $\mathfrak{Z}(P)\subsetneqq Z$ und da $\mathfrak{Z}(P)\neq 1$ ist, müssen wir $\mathfrak{Z}(P)=Z$, $|Z|=p$ haben.

Wir untersuchen P im nichtabelschen Fall weiter. Schreibt man $X|P$ in monomialer Form (vgl. [4], S. 80) so sieht man, daß P einen abelschen Normal-

[5] Man kann den folgenden Beweis auch mit Hilfe der Verlagerung von G nach P führen.

teiler A vom Index p hat. Dann zerfällt $\chi|A$ in p lineare Charaktere

(4.3) $$\chi|A = \lambda_1 + \lambda_2 + \cdots + \lambda_p.$$

Ist τ ein nicht zu A gehöriges Element von P, so ist natürlich

(4.4) $$P = \langle \tau, A \rangle.$$

Wählt man geeignete Koordinate $x_1, \ldots, x_p$ im Darstellungsraum von X, so darf man annehmen, daß für $\alpha \in A$ die Transformation $X(\alpha)$ durch

(4.5) $$x_1 \to \lambda_1(\alpha)x_1, \; x_2 \to \lambda_2(\alpha)x_2, \; \ldots, \; x_p \to \lambda_p(\alpha)x_p$$

gegeben ist, während $X(\tau)$ die Form

(4.6) $$x_1 \to x_2, \; x_2 \to x_3, \; \ldots, \; x_{p-1} \to x_p, \; x_p \to cx_1$$

mit $c \in K$ hat. Wegen $p > 2$ folgt aus $\mathrm{Det}(X(\tau)) = 1$, daß $c = 1$ ist.

Es sei

(4.7) $$P \supset A = A_1 \supset A_2 \supset \ldots A_{a-1} = Z \supset 1$$

eine A und Z enthaltende Hauptreihe von P.

Nehmen wir an, daß man für ein ϱ mit $2 \leqq \varrho \leqq a-1$ bereits Elemente $\xi_\varrho, \xi_{\varrho+1}, \ldots, \xi_{a-1}$ hat derart, daß für $j \geqq \varrho$

(4.8) $$A_j = \langle \xi_j, \xi_{j+1}, \ldots, \xi_{a-1} \rangle,$$

(4.9) $$\tau^{-1} \xi_j \tau = \xi_j \xi_{j+1}$$

ist, wobei $\xi_a = 1$ zu setzen ist. Für $\varrho = a-1$ ist dies richtig, wenn man ξ_{a-1} als eine Erzeugende von $Z = A_{a-1}$ wählt.

Ist $\xi_{\varrho-1}$ in $A_{\varrho-1}$, aber nicht in A_ϱ, so gehört $\xi_{\varrho-1}^{-1} \tau^{-1} \xi_{\varrho-1} \tau = \gamma$ jedenfalls zu A_ϱ. Ändert man $\xi_{\varrho-1}$ nacheinander durch Multiplikation mit geeigneten Potenzen von $\xi_{a-2}, \xi_{a-3}, \ldots, \xi_\varrho$ ab, so sieht man, daß man annehmen kann, daß γ eine Potenz von ξ_ϱ ist. Wäre $\gamma = 1$, so wäre $\xi_{\varrho-1} \in \mathfrak{Z}(P) = Z$, was unmöglich ist. Ersetzt man schließlich $\xi_{\varrho-1}$ durch eine geeignete Potenz, so sieht man, daß man annehmen kann, daß $\gamma = \xi_\varrho$ ist. Jetzt gelten (4.8) und (4.9) für alle $j \geqq \varrho - 1$. Wir sehen also, daß es ein erzeugendes System $\xi_1, \xi_2, \ldots, \xi_{a-1}$ von A gibt, für das (4.8) und (4.9) allgemein gelten.

Da $Z = \langle \xi_{a-1} \rangle$ ist, ist

$$\lambda_1(\xi_{a-1}) = \lambda_2(\xi_{a-1}) = \cdots = \lambda_p(\xi_{a-1}) = \varepsilon,$$

wo ε eine primitive p-te Einheitswurzel ist. Wir behaupten, daß man annehmen kann, daß

(4.10) $$\lambda_i(\xi_\varrho) = \varepsilon^{\binom{i-1}{a-1-\varrho}} \qquad (i = 1, 2, \ldots, p)$$

ist. Dies ist jedenfalls für $\varrho = a-1$ richtig. Ist es für $\varrho = j+1 \geqq 2$ bewiesen, so folgt aus (4.9), (4.5) und (4.6), daß $\lambda_i(\xi_j) = \lambda_{i-1}(\xi_j)\lambda_{i-1}(\xi_{j+1})$ ist, wo $\lambda_0 = \lambda_p$ zu setzen ist. Aus (4.10) ergibt sich dann für $i = 1, 2, \ldots, p$, daß

(4.11) $$\lambda_i(\xi_j) = \lambda_1(\xi_j)\varepsilon^{\binom{i-1}{a-1-j}}$$

ist. Nehmen wir zunächst an, daß $a-j<p$ ist. Da $\mathrm{Det}(X(\xi_j))=1$, folgt dann $\lambda_1(\xi_j)^p=1$. Ersetzt man, was erlaubt ist, ξ_j durch ein geeignetes Element von $\xi_j Z$, so kann man annehmen, daß $\lambda_1(\xi_j)=1$ ist, und dann gilt (4.10) auch noch für $\varrho=j$.

Für $p\geqq 7$ ist nach (2.2) $a\leqq p$ und daher $a-j<p$. Wir zeigen, daß für $p=5$ ebenfalls $a\leqq 5$ ist. Wäre nämlich $a\geqq 6$, so zeigt (4.11) für $j=a-5$, daß $X(\xi_{a-5})$ vier charakteristische Wurzeln $\lambda=\lambda_1(\xi_{a-5})$ und die fünfte Wurzel $\lambda\varepsilon$ hat. Die Determinantenbedingung ergibt $\lambda^5\varepsilon=1$, so daß also λ eine primitive 25-te Einheitswurzel ist. Eine Blichfeldtsche Schlußweise [4], §§ 103—104 zeigt, daß dies unmöglich ist. Wir können also annehmen, daß (4.10) für alle $\varrho=1,2,\ldots,a-1$ gilt, und wir haben $a\leqq p$. Ist $p\geqq 7$ und wäre $a=p$, so würde (4.10) mit $\varrho=1$ einen Widerspruch mit dem Blichfeldtschen Satz [4], § 70, ergeben. Es ist hier also $a\leqq p-1$.

Wie (4.10) zeigt, hat A den Exponenten p. Da jedes Element von P die Form $\tau^r\xi$ mit $\xi\in A$ hat, folgt aus (4.6) und der Unimodularität von X, daß auch P den Exponenten p hat. Unsere Formeln zeigen, daß man die aufsteigende und absteigende Zentralreihe von P erhält, wenn man in (4.7) das Glied $A=A_1$ wegläßt. Zusammenfassend können wir sagen:

(4 C) *Hat P die Ordnung p^a, so ist $a\leqq p-1$ für $p\geqq 7$ und $a\leqq 5$ für $p=5$. Der Exponent von P ist p. Für gegebenes a sind P und $X|P$ eindeutig bestimmt. P hat eine Abelsche Untergruppe A vom Index p, und für $a\geqq 3$ hat P die maximale mögliche Klasse $a-1$.*

Ohne Schwierigkeit sieht man jetzt:

(4 D) *Die einzigen nicht-trivialen Normalteiler von P sind die Abelschen Gruppen A_i mit $i=1,2,\ldots,a-1$, $|A_i|=p^{a-i}$ und p weitere Gruppen $B_1,B_2,\ldots,B_p$ der Ordnung p^{a-1}. Jede Abelsche Untergruppe, deren Ordnung größer als p^2 ist, gehört zu A.*

Wir zeigen auch noch:

(4 E) *Gehört ξ zu A_j, aber nicht zu $A_{a-1}=Z$, so haben die charakteristischen Wurzeln von $X(\xi)$ höchstens die Vielfachheit $a-1-j$.*

Drückt man nämlich ξ durch $\xi_j,\xi_{j+1},\ldots,\xi_{a-1}$ aus, so folgt aus (4.10), daß $\lambda_i(\xi)$ die Form

$$\lambda_i(\xi)=\varepsilon^{f(i)} \tag{4.12}$$

hat, wo $f(x)$ ein Polynom mit ganzen rationalen Koeffizienten ist, dessen Grad kleiner als $a-j$ ist. Nicht alle Koeffizienten können durch p teilbar sein. Daher gibt es höchstens $a-j-1$ Werte $i\in\{1,2,\ldots,p\}$, für die $f(i)$ denselben Wert $(\mathrm{mod}\,p)$ hat. Daraus folgt die Behauptung.

Für $j=a-2$ sind die charakteristischen Wurzeln von $X(\xi)$ verschieden. Da $\mathfrak{C}_G(\xi)\supseteq A$ ist, aber $\mathfrak{C}_G(\xi)$ keine S_p-Gruppe von G enthalten kann, folgt jetzt aus (3 F):

(4 F) *Gehört ξ zu A_{a-2}, aber nicht zu Z, so ist $\mathfrak{C}_G(\xi)=A$.*

Gilt $\sigma^{-1}\xi_{a-2}\sigma=\xi_{a-2}\xi_{a-1}^n$ für ein $\sigma\in G$, so gehört $\sigma\tau^{-n}$ zu $\mathfrak{C}_G(\xi_{a-2})=A$. Daher hat man

(4 G) *Ist $\sigma\in G$ und ist $\sigma^{-1}\xi_{a-2}\sigma\in\xi_{a-2}Z$, so ist $\sigma\in P$.*

Für $p=2$ und $p=3$ folgt aus (3 B), daß P höchstens den Exponenten p^2 haben kann. Man sieht dann leicht, daß $a \leqq 3$ für $p=2$ und $a \leqq 4$ für $p=3$ ist. Da die linearen und ternären Gruppen völlig bekannt sind, wollen wir darauf nicht weiter eingehen.

§ 5. Nicht-Abelsche Sylow-Durchschnittsgruppen

Es sei weiterhin G eine endliche Gruppe, die eine treue unimodulare primitive Darstellung vom Primzahlgrad $p \geqq 5$ besitzt. Als *Sylow-Durchschnittsgruppe* oder SD-*Gruppe* bezeichnen wir eine Untergruppe D, die als Durchschnitt zweier verschiedener S_p-Gruppen von G dargestellt werden kann. Jedenfalls ist dann $Z \subseteqq D$. Wir wollen in diesem und dem folgenden Paragraphen zeigen, daß relativ wenige Untergruppen einer S_p-Gruppe als SD-Gruppen auftreten.

(5 A) *Ist D eine nicht-abelsche SD-Gruppe, so ist $|D| = p^3$.*

Beweis. Es sei D eine nicht-abelsche SD-Gruppe der Ordnung p^d. Es sei etwa $D = P \cap P^\eta$, wo P die in § 4 betrachtete S_p-Gruppe und $\eta \in G$ und $P^\eta \neq P$ ist. Da $D \subsetneqq A$ ist, hat $D \cap A$ die Ordnung p^{d-1}. Ebenso ist $|D \cap A^\eta| = p^{d-1}$. Ferner ist $D_0 = D \cap A \cap A^\eta$ ein Normalteiler von D, der mindestens die Ordnung p^{d-2} hat. Da A und D zu $\mathfrak{N}_P(D_0)$ gehören und $\langle A, D \rangle = P$ ist, ist D_0 ein Normalteiler von P. Für $d \geqq 4$ ist nach (4 D) $D_0 \supseteqq A_{a-2}$. Aus (4 F) folgt $\mathfrak{C}(D_0) = A$. Ebenso ist auch $D_0 \supseteqq A^\eta_{a-2}$, $\mathfrak{C}(D_0) = A^\eta$, also $A = A^\eta \subseteqq P \cap P^\eta$, was einen Widerspruch ergibt. Es muß also $d=3$ sein. Da auch $D \cap A$ Normalteiler von P ist, folgt $D \cap A = A_{a-2}$. Ebenso ist $D \cap A^\eta = A^\eta_{a-2}$. Notwendigerweise ist $A_{a-2} \neq A^\eta_{a-2}$, da man sonst wie im Falle $d \geqq 4$ einen Widerspruch erhielte. Also hat D_0 die Ordnung p und daher ist $D_0 = Z$. Dies ergibt

(5 B) *Ist $D = P \cap P^\eta$ in* (5 A), *so ist*

$$D \cap A = A_{a-2}, \quad D \cap A^\eta = A^\eta_{a-2}, \quad A \cap A^\eta = Z.$$

Wir setzen weiterhin voraus, daß nicht-abelsche SD-Gruppen auftreten.

(5 C) *Ist $D = P \cap P^\eta$ eine nicht-abelsche SD-Gruppe, so ist $\mathfrak{N}_G(D)/D \cong SL(2, p)$ und D gehört zu genau $p+1$ S_p-Gruppen von G. Wir haben*

$$(5.1) \qquad n_G(D) = p^4(p+1)(p-1), \quad n_G(P) = n_G(A_{a-2}) = p^a(p-1).$$

Außer für $p=5$, $a=5$ enthält P genau p^{a-4} nicht-abelsche SD-Gruppen[6].

Beweis. (a) Die Elemente von $N = \mathfrak{N}_G(D)$ transformieren D/Z in sich. Sie induzieren also lineare Transformationen eines 2-dimensionalen Vektorraumes über einem Körper mit p Elementen. Man hat daher einen Homomorphismus φ von N auf eine Untergruppe N^φ von $GL(2, p)$. Ohne wesentliche Einschränkung darf man annehmen, daß D das in § 4 für τ gewählte Element enthält. Dann ist $D = \langle \tau, A_{a-2} \rangle$. Für $\nu \in N$ hat man Formeln

$$\tau^\nu \equiv \tau^\alpha \xi^\beta_{a-2}, \quad \xi^\nu_{a-2} \equiv \tau^\gamma \xi^\delta_{a-2} \pmod{Z}.$$

Dann ist $\varphi(\nu)$ die aus den Exponenten gebildete Matrix $(\bmod p)$. Nach (4.9) ist $[\xi_{a-2}, \tau] = \xi_{a-1} \in Z$, also $[\xi^\nu_{a-2}, \tau^\nu] = \xi_{a-1}$. Daraus folgt $\alpha\delta - \beta\gamma \equiv 1 \pmod{p}$. Daher ist $N^\varphi \subseteqq SL(2, p)$.

[6] Wir werden in § 9 sehen, daß der Fall $p=5$, $a=5$ unmöglich ist.

Der Kern M von φ enthält jedenfalls D. Ist $\nu \in M$, so ist $\xi_{a-2}^{\nu} \in \xi_{a-2} Z$. Nach (4 G) ist also $\nu \in P$. Da $\tau^{\nu} \equiv \tau \pmod{Z}$ ist, folgt leicht aus § 4, daß $\nu \in D$ ist. Also ist $M = D$ und $N/D \cong N^{\varphi}$.

Ist P_1 eine D enthaltende S_p-Gruppe von G, so ist nach bekannten Eigenschaften von p-Gruppen $P_1 \cap \mathfrak{N}(D) \supset D$. Wie (5 A) zeigt, ist D nicht in einer SD-Gruppe als echte Untergruppe enthalten. Daher ist P_1 die einzige S_p-Gruppe von G, die $P_1 \cap N$ enthält. Daraus folgt, daß $P_1 \cap N$ eine S_p-Gruppe von N ist, und daß sich die D enthaltenden S_p-Gruppen von G und die S_p-Gruppen von N eineindeutig entsprechen. Insbesondere haben N und dann N^{φ} mehr als eine S_p-Gruppe. Nach dem Sylowschen Satz enthält N^{φ} die $p+1$ S_p-Gruppen von $SL(2, p)$. Da diese S_p-Gruppen $SL(2, p)$ erzeugen, ist $N^{\varphi} = SL(2, p)$. Zugleich sieht man, daß es in G genau $p+1$ die Gruppe D enthaltende S_p-Gruppen gibt. Damit ist der erste Teil von (5 C) bewiesen.

(b) Kommen r Konjugierte von D in P vor, und schreibt man die $rg/n(P)$ Konjugierten von D auf, die in einer der $g/n(P)$ Konjugierten von P auftreten, so erhält man nach (a) jede Konjugierte von D genau $(p+1)$-mal. Daraus folgt

$$rg/n(P) = (p+1)\, g/n(D). \tag{5.2}$$

Aus dem in (a) Bewiesenen ergibt sich $n(D) = p^4(p+1)(p-1)$. Da $a \geqq 4$ sein muß und dann A_{a-2} charakteristisch in P ist, hat man $\mathfrak{N}(P) \subseteq \mathfrak{N}(A_{a-2})$. Da $A_{a-2} = \langle \xi_{a-2}, Z \rangle$ ist, folgt leicht aus (4 F), daß $n(A_{a-2})$ ein Teiler von $p^a(p-1)$ ist. Jetzt zeigt (5.2), daß $r = p^{a-4}$, $n(P) = n(A_{a-2}) = p^a(p-1)$ ist. Damit ist (5.1) bewiesen.

(c) Wie in (a) gezeigt wurde, ist $R = P \cap N$ eine S_p-Gruppe von N. Die Struktur von $SL(2, p)$ zeigt, daß N ein Element ν der Ordnung $p-1$ enthält, das R normalisiert. Da nach (5 A) R nicht in einer SD-Gruppe von G enthalten ist, muß $P^{\nu} = P$ sein. Jetzt zeigt (5.1), daß $\mathfrak{N}_G(P) = \langle \nu \rangle P$ ist. Wegen $\nu \in N$ ist also jede Konjugierte von D in $\mathfrak{N}_G(P)$ bereits zu D in P konjugiert. Dies gilt für jede nicht-abelsche SD-Gruppe $D \subseteq P$.

(d) Ist $T \nsubseteq A$ eine A_{a-2} enthaltende Untergruppe der Ordnung p^3 von P, so kann man $T = \langle \tau\xi, A_{a-2} \rangle$ mit $\xi \in A$ setzen. Offenbar ist dabei $\xi \bmod A_{a-2}$ bestimmt. Es gibt also p^{a-3} derartige Untergruppen T von P. Ferner ist $\mathfrak{C}_P(\tau\xi) = \langle \tau\xi, Z \rangle$. Es hat also $\tau\xi$ in P genau p^{a-2} Konjugierte. Da die Kommutatorgruppe A_2 von P die Ordnung p^{a-2} ist, sind die Konjugierten von $\tau\xi$ in P die Elemente von $\tau\xi A_2$. Ist auch $\xi^* \in A$, so sieht man jetzt, daß T und $T^* = \langle \tau\xi^*, A_{a-2} \rangle$ dann und nur dann in P konjugiert sind, wenn $\xi \equiv \xi^* \bmod A_2$ ist. Die p^{a-3} Gruppen T verteilen sich also auf p Klassen $\mathfrak{K}_0, \mathfrak{K}_1, \ldots, \mathfrak{K}_{p-1}$ von je p^{a-4} in P konjugierten Gruppen T. Bei geeigneter Bezeichnung kann man $\langle \tau\xi_1^m, A_{a-2} \rangle$ als Vertreter der Klasse $\mathfrak{K}_m$ nehmen. Ist ein Element $T \in \mathfrak{K}_j$ Durchschnitt von P mit einer S_p-Gruppe von G, so gilt offenbar das Entsprechende für alle Elemente von $\mathfrak{K}_j$.

Aus (c) folgt, daß jedes Element $\nu \in \mathfrak{N}_G(P)$ jede Klasse in sich transformiert, die aus SD-Gruppen besteht. Ist wie oben τ so gewählt, daß $D = \langle \tau, A_{a-2} \rangle$ ist, so gilt dies jedenfalls für $\mathfrak{K}_0$. Daher gibt es eine ganze Zahl h, für die

$$\tau^{\nu} \equiv \tau^h \pmod{A_2} \tag{5.3}$$

6*

ist. Nehmen wir an, daß auch ein $\mathfrak{K}_m$ mit positivem m von allen $v \in \mathfrak{N}_G(P)$ in sich transformiert wird. Dann ist $(\tau\xi_1^m)^v \pmod{A_2}$ einer Potenz von $\tau\xi_1^m$ kongruent. Aus (5.3) folgt dann, daß

$$\xi_1^v \equiv \xi_1^h \pmod{A_2}$$

ist. Wendet man jetzt v nacheinander auf (4.9) mit $j=1, 2, \ldots$ an, so erhält man

$$\xi_j^v \equiv \xi_j^{h^j} \bmod A_{j+1},$$

wobei $A_a = 1$ zu setzen ist. Da $\xi_{a-1} \in Z$ ist, ergibt dies $h^{a-1} \equiv 1 \pmod{p}$. Für $j = a-2$ hat man dann

$$\xi_{a-2}^v \equiv \xi_{a-2}^{1/h} \pmod{Z}.$$

Man wähle jetzt v wie in (c). Wir zeigen, daß dann h eine primitive Wurzel $(\bmod p)$ ist. Andernfalls gäbe es nämlich eine Potenz $\mu = v^s$ mit $0 < s < p-1$, für die $\xi_{a-2}^\mu \equiv \xi_{a-2} \pmod{Z}$ ist. Nach (4 G) wäre aber $\mu \in P$, was unmöglich ist. Da h also eine primitive Wurzel ist, ist $p-1 \leqq a-1$, also $p=a$. Aus (4 C) folgt, daß $a=p$ höchstens für $p=5$ gilt. Außer für $p=5$, $a=5$ besteht also nur die Klasse $\mathfrak{K}_0$ aus SD-Gruppen. Damit ist (5 C) bewiesen.

§ 6. Abelsche Sylow-Durchschnittsgruppen

Wir setzen weiterhin voraus, daß G eine treue unimodulare primitive Darstellung vom Primzahlgrad $p \geqq 5$ hat. Es sei $a \geqq 4$.

(6 A) *Ist die abelsche Gruppe D der Durchschnitt zweier S_p-Gruppen P und P^η mit $\eta \in G$ und ist $D \neq A$, Z, so muß $|D| = p^2$ sein. Außer möglicherweise im Fall $p=a=5$ ist $D \subsetneqq A \cap A^\eta$.*

Beweis. (a) Es sei $D = P \cap P^\eta$ von der Ordnung p^d mit $a > d > 1$. Ist D eine abelsche Gruppe und ist $d > 2$, so ist nach (4 D)

$$D = A \cap A^\eta. \tag{6.1}$$

Im Fall $d=2$ wollen wir voraussetzen, daß (6.1) gilt.

Die Darstellung $X|D$ möge in die irreduziblen Bestandteile $Y_1, Y_2, \ldots, Y_m$ zerfallen, wobei Y_j mit der Vielfachheit e_j auftritt. Wir wollen diese Zerlegung durch die Formel

$$X|D = e_1 Y_1 \oplus \cdots \oplus e_m Y_m \tag{6.2}$$

andeuten. Hier haben die Y_j den Grad 1. Nach dem Schurschen Lemma zerfällt dann in $X|\mathfrak{C}(D)$ in der Form

$$X|\mathfrak{C}(D) = U_1 \oplus \cdots \oplus U_m, \tag{6.3}$$

wo U_j ein möglicherweise reduzibler Bestandteil vom Grad e_j ist. Es sei

$$U_j = \oplus \sum_\varrho V_j^{(\varrho)} \tag{6.4}$$

die Zerlegung von U_j in gleiche oder verschiedene irreduzible Bestandteile.

Offenbar sind A und A^η beides S_p-Gruppen von $\mathfrak{C}(D)$; es gibt also $\gamma \in \mathfrak{C}(D)$, für die

$$A^\eta = A^\gamma \tag{6.5}$$

gilt. Wir dürfen annehmen, daß die Koordinaten im Darstellungsraum so gewählt sind, daß jedes $V_j^{(\varrho)}(A)$ aus Diagonalmatrizen besteht.

(b) Ist jedes $V_j^{(\varrho)}(A)$ ein Normalteiler von $V_j^{(\varrho)}(\mathfrak{C}(D))$, so besteht $V_j^{(\varrho)}(A^\gamma)$ wegen $\gamma \in \mathfrak{C}(D)$ aus denselben Matrizen wie $V_j^{(\varrho)}(A)$. Da $X(A^\gamma)$ dann ebenso wie $X(A)$ aus Diagonalmatrizen besteht und da X eine treue Darstellung von G ist, so gilt $A^\gamma \subseteqq \mathfrak{C}(A) \subseteqq \mathfrak{C}(A_{a-2})$. Nach (4 F) ist dann $A^\gamma = A$, und jetzt zeigt (6.5), daß $D = A$ ist, was ausgeschlossen war.

(c) Wir dürfen also annehmen, daß etwa $V_1^{(1)}(A)$ nicht normal in $V_1^{(1)}(\mathfrak{C}(D))$ ist. Jetzt zeigt ein Satz von Feit-Thompson [16], daß der Grad $v_1^{(1)}$ von $V_1^{(1)}$ mindestens $(p-1)/2$ ist. Dann ist $e_1 \geqq (p-1)/2$.

Wir betrachten zunächst den Fall $p \geqq 7$. Ist $\sigma \in D$, $\sigma \notin Z$, so hat nach Blichfeldts Satz ([4], S. 96) $X(\sigma)$ mindestens drei verschiedene charakteristische Wurzeln. Daher ist $m \geqq 3$ in (6.2). Derselbe Satz zeigt, daß es unmöglich ist, daß $e_1 = e_2 = (p-1)/2$ und $e_3 = 1$ ist. Dies zeigt, daß für alle $V_j^{(\varrho)}$ mit Ausnahme von $V_1^{(1)}$ die Gruppe $V_j^{(\varrho)}(A)$ normal in $V_j^{(\varrho)}(\mathfrak{C}(D))$ ist. Ist M der Kern von $V_1^{(1)}$, so folgt aus Feits Resultaten [15], daß es eine Untergruppe A_0 von A vom Index p gibt, für die $A_0 M$ normal in $\mathfrak{C}(D)$ ist. Dann ist $V_1^{(1)}(A_0^\gamma) \subseteqq V_1^{(1)}(A_0)$ und daher $V_j^{(\varrho)}(A_0^\gamma) \subseteqq V_j^{(\varrho)}(A)$ für alle $V_j^{(\varrho)}$. Jetzt folgt ähnlich wie in (b), daß $A_0^\gamma \subseteqq \mathfrak{C}(A) = A$ und daher $A_0^\gamma \subseteqq D$ ist. Dann muß aber $d = a-2$ sein. Man sieht jetzt, daß $D \cap A_{a-3} \supset Z$ ist. Gehört σ zu $D \cap A_{a-3}$, aber nicht zu Z, so haben nach (4 E) die charakteristischen Wurzeln von $X(\sigma)$ höchstens die Vielfachheit 2. Dann ist aber $e_1 \leqq 2$ in (6.2). Wegen $e_1 \geqq (p-1)/2$ ergibt dies für $p \geqq 7$ einen Widerspruch.

(d) Wir nehmen jetzt $p = 5$. Dann ist $a = 4$ oder 5. Außer im Falle $a = 5$, $d = 2$ enthält $D \cap A_{a-3}$ ein Element $\sigma \notin Z$ und daher sind alle $e_i \leqq 2$. Wir haben den Fall auszuschließen, daß für eine oder zwei der Darstellungen $V_j^{(\varrho)}$ nicht $V_j^{(\varrho)}(A)$ normal in $V_j^{(\varrho)}(\mathfrak{C}(D))$ ist. Die binären linearen Gruppen sind bekannt. Man sieht, daß für die fraglichen $V_j^{(\varrho)}$ die Gruppe $V_j^{(\varrho)}(\mathfrak{C}(D))$ als projektive Gruppe die Ikosaedergruppe ist. Das bleibt auch richtig, wenn man $\mathfrak{C}(D)$ durch eine hohe Kommutatorgruppe H von $\mathfrak{C}(D)$ ersetzt. Ist $V_j^{(\varrho)}(A)$ normal in $V_j^{(\varrho)}(\mathfrak{C}(D))$, so ist $V_j^{(\varrho)}(\mathfrak{C}(D))$ auflösbar und daher $V_j^{(\varrho)}(H) = E$ für geeignete Wahl von H. Hat man *eine* Ikosaedergruppe $V_j^{(\varrho)}(H)$, so erhält man wie in [4], §§ 103—104 einen Widerspruch. Sind etwa $V_1^{(1)}(H)$ und $V_2^{(1)}(H)$ Ikosaedergruppen, so sieht man leicht, daß es ein $\sigma \in H$ gibt, für das $V_1^{(1)}(\sigma) = V_2^{(1)}(\sigma) = -E$ ist. Außerdem gibt es ein Element ξ in D mit $V_1^{(1)}(\xi) = e^{2\pi i/5} E = V_2^{(1)}(\xi)^{-1}$. Jetzt zeigt [4], Satz 8, S. 96, daß $X(\xi\sigma)$ charakteristische Wurzeln hat, die in einer primitiven Darstellung nicht auftreten können.

(6 B) *Es sei $a \geqq 4$. Gibt es SD-Gruppen D der Ordnung p^2, so gibt es außer im Fall $p = 5$, $a = 5$ auch nicht-abelsche SD-Gruppen.*

Beweis. Es sei wieder $D = P \cap P^\eta$ mit $\eta \in G$. Wir betrachten zunächst den Fall, daß

$$(6.6) \qquad D \nsubseteq A, \quad D \nsubseteq A^\eta$$

ist. Man kann $D = \langle \omega, Z \rangle$ mit $\omega \in \tau A$ setzen. Es folgt, daß ξ_{a-2} die Nebengruppe ωZ in sich transformiert. Ebenso wird ωZ von ξ_{a-2}^η in sich transformiert. Man kann dann eine ganze Zahl t derart wählen, daß $\xi_{a-2}^{\eta}\xi_{a-2}^{-t} \in \mathfrak{C}(D)$ ist. Wäre $\mathfrak{C}(D)=D$, so wäre $\xi_{a-2}^{\eta} \in P$, also $\xi_{a-2}^{\eta} \in D$ und $D=\langle \xi_{a-2}^{\eta}, Z\rangle \subseteqq A^{\eta}$. Also ist $\mathfrak{C}(D) \supset D$. Da $\chi(\omega)=0$ ist, folgt aus (4 F), daß $\mathfrak{C}(D)=\mathfrak{C}(\omega)$ eine p-Gruppe ist. Wegen $|\mathfrak{C}(\omega)|>p^2$ muß dann ω zu Elementen von A konjugiert sein, und $\mathfrak{C}(\omega)$ ist eine Konjugierte von A. Da $\xi_{a-2}^{-1}\omega\xi_{a-2}^{-1} \in \omega Z$ ist, muß ξ_{a-2} die Gruppe $\mathfrak{C}(\omega)$ normalisieren. Man sieht, daß $\langle \xi_{a-2}, \mathfrak{C}(\omega)\rangle$ eine S_p-Gruppe P^{γ} ist, $\gamma \in G$. Dann ist

$$P \cap P^{\gamma} \supseteqq \langle \xi_{a-2}, \omega, Z\rangle,$$

also eine nicht-abelsche Gruppe. Wäre $P^{\gamma}=P$, so wäre $\mathfrak{C}(\omega) \subseteqq P$, also $\omega \in A$, $D \subseteqq A$, was ausgeschlossen war. Also ist $P \cap P^{\gamma}$ eine nicht-abelsche SD-Gruppe.

Gilt (6.6) nicht, so dürfen wir wegen (6 A) annehmen, daß

$$D \nsubseteq A, \quad D \subseteqq A^{\eta} \tag{6.7}$$

ist; der Fall $D \subseteqq A$, $D \nsubseteq A^{\eta}$ ist analog.

Die Gruppe $\mathfrak{N}_p(D)$ ist eine nicht-abelsche Gruppe, da andernfalls wegen $n_p(D) \geqq p^3$ nach (4 D) $\mathfrak{N}_p(D) \subseteqq A$, also $D \subseteqq A$ wäre. Daher sind die S_p-Gruppen von $\mathfrak{N}(D)$ nicht-abelsche Gruppen. Insbesondere kann A^{η} nicht eine S_p-Gruppe von $\mathfrak{N}(D)$ sein. Da $A^{\eta} \subseteqq \mathfrak{N}(D)$ ist, sind die S_p-Gruppen von $\mathfrak{N}(D)$ auch S_p-Gruppen von G. Es gibt also eine S_p-Gruppe $P^{\zeta} \supset \mathfrak{N}_p(D)$ von $\mathfrak{N}(D)$; $\zeta \in G$. Da D wegen (4 D) und (6.7) nicht normal in P ist, ist $P \neq P^{\zeta}$. Wegen $P \cap P^{\zeta} \supseteqq \mathfrak{N}_p(D)$ ist $P \cap P^{\zeta}$ eine nicht-abelsche SD-Gruppe, und die Behauptung gilt auch in diesem Fall.

§ 7. Einige Ergänzungen

Es seien G und X wie in §§ 5—6 gewählt. Die Resultate in § 5 schließen die Möglichkeit nicht aus, daß A als SD-Gruppe auftritt. Neue noch nicht veröffentlichte Resultate von W. FEIT erlauben es zu beweisen, daß A für $p \geqq 13$ nicht als SD-Gruppe auftreten kann. Wir wollen darauf nicht eingehen.

Zunächst zeigen wir hier:

(7 A) *Ist $a=3$, so ist Z die einzige SD-Gruppe. Gehört ξ zu P, aber nicht zu Z, so ist $\mathfrak{C}_G(\xi)=\langle \xi, Z\rangle$.*

Beweis. Ist $a=3$, so verschwindet χ für alle Elemente $\xi \in P-Z$. Dann folgt sofort aus (3 F), daß $\mathfrak{C}(\xi)=\langle \xi, Z\rangle$ ist. Ist $D=\langle \xi, Z\rangle$ der Durchschnitt zweier S_p-Gruppen $P=\langle \sigma, D\rangle$ und $P_1=\langle \sigma_1, D\rangle$, so transformieren σ und σ_1 beide ξ in Elemente von ξZ. Wie im Beweis von (4 G) folgt $\sigma_1 \in \langle \sigma, D\rangle$. Also ist $P=P_1$.

(7 B) *Ist $a=4$, so ist A keine SD-Gruppe.*

Beweis. Nehmen wir an, daß es Elemente $\eta \in G$ gibt, derart, daß $P \cap P^{\eta}=A$ ist. Da A^{η} die einzige Abelsche Untergruppe der Ordnung p^3 von P^{η} ist, muß $A^{\eta}=A$ sein. Es sei ξ ein Element von A derart, daß $\chi(\xi)=0$ ist. In der Bezeichnung von § 4 müssen dann $\lambda_1(\xi), \lambda_2(\xi), \ldots, \lambda_p(\xi)$ die p-ten Einheitswurzeln sein. Dabei gilt wieder eine Formel von der Art von (4.12). Ist $\xi \notin A_2$, so ist dabei f ein quadratisches Polynom, und dann ist unsere Bedingung sicher nicht erfüllt. Daher gilt $\chi(\xi)=0$ dann und nur dann, wenn $\xi \in A_2-Z$ ist.

Daraus folgt, daß η die Gruppe A_2 in sich transformiert. Dann gilt $(\tau^\eta)^{-1}\xi_{a-2}\tau^\eta \in \xi_{a-2}Z$, und (4 G) ergibt $\tau^\eta \in P \cap P^\eta = A$. Das ist ein Widerspruch.

Für $p=5$, $a=5$ werden wir in § 8 direkt zeigen, daß A keine SD-Gruppe ist.

(7 C) *Es gibt eine Untergruppe G_0 von G mit den folgenden Eigenschaften:*

(a) *Ist T eine nicht-abelsche p-Untergruppe von G_0, so ist $\mathfrak{N}_G(T)$ in G_0 enthalten.*

(b) *Die Gruppe G_0 enthält S_p-Untergruppen von G. Enthält die S_p-Gruppe P_1 von G eine nicht-abelsche p-Untergruppe T von G_0, so ist $P_1 \subseteq G_0$.*

(c) *Bei geeigneter Wahl der Koordinaten gehören die Koeffizienten von $X|G_0$ zum Körper Ω der p-ten Einheitswurzeln.*

Beweis. Es sei P eine feste S_p-Gruppe von G. Man wähle die Koordinaten im Darstellungsraum von X wie in § 4. Es sei G_0 die Gruppe derjenigen Elemente $\sigma \in G$, für die $X(\sigma)$ Koeffizienten in Ω hat. Dann ist $P \subseteq G_0$.

Ist T eine nicht-abelsche p-Untergruppe von G_0, so ist $X|T$ irreduzibel. Ist $\nu \in \mathfrak{N}_G(T)$ und $\omega \in T$, so ist

$$X(\omega)\,X(\nu) = X(\nu)\,X(\omega^\nu)\,. \tag{7.1}$$

Es sei der Körper K wie in § 3 gewählt. Ist s ein festes Element der Galoisschen Gruppe von K in bezug auf Ω, so folgt aus (7.1), daß

$$X(\omega)\,X(\nu)^s = X(\nu)^s\,X(\omega^\nu)$$

ist. Wegen der Irreduzibilität von $X|T$ gibt es nach dem Schurschen Lemma Zahlen $\gamma(\nu) \in K$ derart, daß

$$X(\nu)^s = \gamma(\nu)\,X(\nu) \tag{7.2}$$

ist. Da $\mathrm{Det}(X(\nu)) = 1$ ist, ist $\gamma(\nu)$ eine p-te Einheitswurzel.

Offenbar ist $\gamma(\nu)$ ein linearer Charakter γ von $\mathfrak{N}_G(T)$. Der Kern H von γ hat dann den Index 1 oder p in $\mathfrak{N}_G(T)$. Ist $P_0 \supseteq T$ eine S_p-Gruppe von G_0, so gehört $\mathfrak{N}_G(T) \cap P_0 = M$ zu H. Ist $P_0 \supset T$, so ist $M \supset T$. Aus (5 A) folgt, daß M in keiner von P_0 verschiedenen S_p-Gruppe von G vorkommt, und daher ist M eine S_p-Gruppe von $\mathfrak{N}_G(T)$. Das gilt auch, wenn $T = P_0$ ist. Da also $|\mathfrak{N}_G(T) : H|$ zu p teilerfremd ist, gilt $H = \mathfrak{N}_G(T)$, also $\gamma(\nu) = 1$ für alle ν. Daher ist $X(\nu)^s = X(\nu)$, und da dies für alle Elemente der Galoisschen Gruppe gilt, hat $X(\nu)$ Koeffizienten in Ω. Es ist also $\mathfrak{N}_G(T) \subseteq G_0$.

Ist $P_1 \supseteq T$ eine S_p-Gruppe von G, so sei P_2 eine S_p-Gruppe von G_0, die $\mathfrak{N}_G(T) \cap P_1$ enthält. Aus (5 A) folgt jetzt, daß $P_1 = P_2$ ist. Also ist $P_1 \subseteq G_0$, und damit ist (7 C) in allen Teilen bewiesen.

(7 D) *Korollar. Ist r eine von p verschiedene Primzahl und ist $|G_0|$ durch r^k teilbar, so ist*

$$k \leqq \left[\frac{p}{r-1}\right] + \left[\frac{p}{r(r-1)}\right] + \left[\frac{p}{r^2(r-1)}\right] + \cdots.$$

Ist nämlich R eine S_r-Gruppe von G_0, so ist $\chi(\varrho)$ rational für jedes $\varrho \in R$. Jetzt folgt die Behauptung aus einem Satz von I. Schur [20].

§ 8. Charaktere

Bei der Behandlung der Fälle $a=1$ und $a=3$ spielen Charaktere von G eine wichtige Rolle. Im Prinzip könnte man auch für größere Werte von a die Resultate der vorangehenden Abschnitte dazu verwenden, um Aussagen über die Charaktere zu machen, doch wird sich das unten für die Behandlung des Falles $p=5$ als unnötig erweisen.

Setzen wir zunächst voraus, daß $a=1$ ist. Ist χ der Charakter einer treuen irreduziblen Darstellung X von G vom Primzahlgrad p, so hat χ für p den Defekt 0. Daraus folgt, daß χ für Elemente ξ der Ordnung p verschwindet. Gehört ξ zur S_p-Gruppe P, so folgt aus (4 E), daß $\mathfrak{C}(P)=P$ ist.

Für Gruppen, deren Ordnung eine Primzahl p zur ersten Potenz enthält, sind eine große Anzahl von Eigenschaften der Charaktere bekannt, vgl. [5]. Wir führen hier einige dieser Eigenschaften an. Dabei setzen wir voraus, daß $\mathfrak{C}(P)=P$ ist. Wir setzen $n(P)=ps$. Dann ist $s=(p-1)/t$ ein Teiler von $p-1$. Alle irreduziblen Charaktere von G gehören entweder zum p-Hauptblock B von G oder sie haben den p-Defekt 0. Der Block B besteht aus s „gewöhnlichen" irreduziblen Charakteren $\chi_1=1, \chi_2, \ldots, \chi_s$ und t „Ausnahme-Charakteren" $\chi_0^{(1)}, \ldots, \chi_0^{(t)}$ [8]. Die letzteren haben alle denselben Grad. Es gibt Zahlen $b_i=\pm 1$ und $b_0=\pm 1$, derart, daß für alle Elemente ξ der Ordnung p die Gleichungen gelten:

$$\chi_i(\xi)=b_i, \quad \sum_{k=1}^{t} \chi_0^{(k)}(\xi)=b_0 \tag{8.1}$$

$(i=1, 2, \ldots, s)$. Daraus folgt, daß

$$\chi_i(1)\equiv b_i, \quad t\chi_0^{(1)}(1)\equiv b_0 \pmod{p} \tag{8.2}$$

ist. Für p-reguläre Elemente $\sigma \in G$ hat man

$$\sum_{i=1}^{s} b_i\chi_i(\sigma)=-b_0\chi_0^{(1)}(\sigma)=-b_0\chi_0^{(2)}(\sigma)=\cdots=-b_0\chi_0^{(t)}(\sigma). \tag{8.3}$$

Man kann den Charakteren $\chi_1, \chi_2, \ldots, \chi_s$ und $\chi_0^{(1)}=\chi_0$ die Eckpunkte eines linearen Graphen, genauer eines „Baumes" B zuordnen. Dabei läßt sich B symmetrisch zu einer Achse, dem Stamm S, anordnen. Die Eckpunkte von S entsprechen den Charakteren $\chi_0^{(1)}$ und den reell-wertigen Charakteren χ_i, während zwei symmetrisch zum Stamm liegenden Eckpunkten konjugiert komplexe Charaktere entsprechen. Zwei Charaktere χ_i und χ_j (mit $j\geqq 0$) gehören dann und nur dann zu Eckpunkten, die durch eine Seite verbunden sind, wenn χ_i und χ_j als modulare Charaktere für p einen gemeinsamen Bestandteil haben; es ist dann $b_ib_j=-1$; vgl. [22].

(8 A). *Hat eine endliche Gruppe G eine treue primitive Darstellung X vom Primzahlgrad p und ist $\mathfrak{Z}(G)=1$, dann enthält jeder nicht-triviale Normalteiler die Kommutatorgruppe G', und G' ist einfach.*

Beweis. Wie in § 4 sieht man, daß $|G|=pg_0$ mit $(g_0, p)=1$ ist. Ist P eine S_p-Gruppe von G, so ist $\mathfrak{C}(P)=P$. Es sei $G_1 \neq 1$ ein Normalteiler von G. Da $\mathfrak{Z}(G)=1$ und X primitiv ist, ist $|G_1|$ durch p teilbar.

[8] Im Fall $t=1$ ist kein Unterschied zwischen gewöhnlichen und Ausnahmecharakteren. Ein beliebiger Charakter in B kann als Ausnahmecharakter gewählt werden.

Es sei ψ_1^* der vom Hauptcharakter ψ_1 von G_1 induzierte Charakter von G. Ist ζ ein irreduzibler Bestandteil von ψ_1^*, so ist

$$\zeta | G_1 = \zeta(1)\psi_1 .$$

Jetzt zeigt (8.1), daß ζ ein gewöhnlicher Charakter in B vom Grad 1 ist. Da $\psi^*(1) = |G : G_1|$ ist, gibt es $|G : G_1|$ lineare Charaktere von G, deren Kern G_1 enthält. Dies müssen alle Charaktere von G/G_1 sein, und daher ist G/G_1 eine abelsche Gruppe, also $G_1 \geqq G'$.

Da $\mathfrak{Z}(G')$ charakteristisch in G' ist, ist $\mathfrak{Z}(G')$ normal in G, also $\mathfrak{Z}(G') = G'$ oder $\mathfrak{Z}(G') = 1$. Ebenso ist $G'' = G'$ oder $G'' = 1$. Wäre G' eine abelsche Gruppe, so würde aus $G' \geqq P$ folgen, daß $G' = P$ ist, und dann könnte G keine irreduzible Darstellung von Grad p haben. Ist G' eine nicht-abelsche Gruppe, so muß $\mathfrak{Z}(G') = 1$, $G'' = G'$ sein. Wendet man das oben für G Bewiesene auf G' an, so sieht man, daß G' einfach ist. Damit ist (8 A) bewiesen.

Im Fall $a = 3$ setzen wir $\bar{G} = G/Z$. Einige der oben für $a = 1$ formulierten Eigenschaften der Charaktere haben Analoga im Fall von $\bar{G}$.

Die S_p-Gruppe $\bar{P} = P/Z$ von $\bar{G}$ ist eine abelsche Gruppe vom Typ (p, p). Aus (7 A) folgt, daß $\bar{P}$ der Zentralisator eines jeden seiner von 1 verschiedenen Elemente ist. Daher kann man die Ergebnisse von [9] anwenden. Der Normalisator von $\bar{P}$ in $\bar{G}$ ist $\mathfrak{N}_G(P)/Z$, seine Ordnung sei mit sp^2 bezeichnet. Dann ist $s = (p^2 - 1)/t$ ein Teiler von $p^2 - 1$. Die irreduziblen Charaktere von $\bar{G}$ liegen entweder im p-Hauptblock B von $\bar{G}$ oder sie haben den p-Defekt 2. Die Charaktere in B sollen mit

$$\chi_1 = 1, \chi_2, \ldots, \chi_l, \chi_0^{(1)}, \chi_0^{(2)}, \ldots, \chi_0^{(t)}$$

bezeichnet werden. Es gibt ganze Zahlen $b_1, \ldots, b_l$ und r und ein Vorzeichen δ, so daß für jedes Element ξ der Ordnung p in $\bar{G}$ die Gleichungen gelten

$$(8.4) \qquad \chi_j(\xi) = b_j, \quad \sum_{k=1}^{t} \chi_0^{(k)}(\xi) = rt - \delta,$$

$(j = 1, 2, \ldots, l)$. Die t „Ausnahmecharaktere" $\chi_0^{(k)}$ nehmen für jedes p-reguläre Element $\sigma \in \bar{G}$ denselben Wert an. Insbesondere haben sie denselben Grad. Aus (8.4) folgt

$$(8.5) \qquad \chi_j(1) \equiv b_j, \quad t\chi_0^{(1)}(1) \equiv rt - \delta \pmod{p^2}.$$

Daraus folgt, daß $b_1, \ldots, b_l$ und $rt - \delta$ nicht verschwinden. Für p-reguläres $\sigma \in G$ und für jedes k gilt

$$(8.6) \qquad \sum_{j=1}^{l} b_j \chi_j(\sigma) + \chi_0^{(k)}(\sigma)(rt - \delta) = 0.$$

Schließlich ist

$$(8.7) \qquad \sum_{j=1}^{l} b_j^2 + (t-1)r^2 + (r-\delta)^2 = s + 1.$$

Dies zeigt, daß $l \leqq s$ ist. Die b_j brauchen nicht ± 1 zu sein. Offenbar ist $|b_j| \leqq p$.

Wir wollen noch eine andersartige Formel herleiten. Ist wieder χ der Charakter von X, so kann $\chi\bar{\chi}$ als Charakter von $\bar{G}$ aufgefaßt werden. Dann tritt χ_1 genau einmal in $\chi\bar{\chi}$ auf. Alle andern irreduziblen Bestandteile haben höchstens den Grad p^2-1; sie gehören alle zu B. Da $\chi(\xi)\,\bar{\chi}(\xi)=0$ für $\xi\in\bar{P}-\{1\}$ ist, erscheinen alle $\chi_0^{(k)}$ mit derselben Vielfachheit u in $\chi\bar{\chi}$. Es sei u_j die Vielfachheit von χ_j, also

$$\chi\bar{\chi}=\chi_1+\sum_{j=2}^{l}u_j\chi_j+u\sum_{k=1}^{t}\chi_0^{(k)}. \tag{8.8}$$

Insbesondere findet man für das Argument $\xi\in\bar{P}-\{1\}$ die Gleichungen

$$0=1+\sum_{j=2}^{l}u_jb_j+u(rt-\delta). \tag{8.9}$$

Ist $u_j\neq 0$, so ist jedenfalls $\chi_j(1)<p^2$. Ist hier $b_j>0$, so folgt aus (8.5), daß $\chi_j(1)=b_j$ ist, und nach (8.4) gehört P zum Kern H_j von χ_j. Wegen der Primitivität von X ist $\chi|H_j$ irreduzibel. Dann tritt aber der Hauptcharakter von H_j nur einmal in $(\chi|H_j)(\bar{\chi}|H_j)$ auf. Jetzt zeigt (8.8), daß $j=1$ sein muß. Wir unterscheiden zwei Fälle.

Erster Fall: Für geeignetes j mit $1<j<l$ ist $u_j\neq 0$.

Es sei etwa $u_2\neq 0$. Dann ist $b_2<0$. Wie (8.5) zeigt, ist $|b_2|+\chi_2|\bar{P}$ ein Vielfaches $m\varrho_P$ des regulären Charakters ϱ_P von $\bar{P}$ mit ganzem $m\geqq 0$. Da der Hauptcharakter von $\bar{P}$ mindestens $|b_2|$-mal auftritt, ist $m\geqq|b_2|$, also $\chi_2(1)\geqq$ $\geqq|b_2|(p^2-1)$. Jetzt folgt aus (8.8), daß $b_2=-1$ ist. Wir haben dann notwendigerweise

$$\chi\bar{\chi}=\chi_1+\chi_2,\quad \chi_2(1)=p^2-1,\quad b_2=-1. \tag{8.10$_\mathrm{I}$}$$

Zweiter Fall: Es ist $u_1=u_2=\cdots=u_l=0$.

Nach (8.9) ist hier $u(rt-\delta)=-1$, so daß $u=1$, $rt-\delta=-1$ sein muß. Ist $\delta=1$, so ist $r=0$. Ist $\delta=-1$, so ist $rt=-2$. Ist $t=1$, so ist kein Unterschied zwischen den Charakteren erster und zweiter Art. Man kann dann also sagen, daß (8.10$_\mathrm{I}$) gilt. Ist $t=2$, so kann man durch Umbenennung der Charaktere immer erreichen, daß $\delta=1$ ist. Wir dürfen also $\delta=1$, $r=0$ annehmen. Wir haben dann

$$\chi\bar{\chi}=\chi_1+\sum_{k=1}^{t}\chi_0^{(k)};\quad \chi_0^{(k)}(1)=s,\quad \delta=1,\quad r=0. \tag{8.10$_\mathrm{II}$}$$

Dabei darf man hier $t>1$ annehmen.

Wir verfolgen den zweiten Fall noch etwas weiter.

(8 B) *Sind im zweiten Fall alle $b_j>0$, so ist P ein Normalteiler von G.*

Beweis. Aus (8.6), (8.7) und (8.10$_\mathrm{II}$) folgt

$$\sum_{j=1}^{l}b_j\chi_j(1)=s=\sum_{j=1}^{l}b_j^2. \tag{8.11}$$

Sind alle $b_j>0$, so ist dies nur möglich, wenn $\chi_j(1)=b_j$ für $j=1,2,\ldots,l$ ist. Jetzt zeigt (8.4), daß P zum Durchschnitt H der Kerne von $\chi_1,\ldots,\chi_l$ gehört.

Ist σ ein p-reguläres Element von H, so gehört σ nach (8.6) und (8.7) auch noch zum Kern der $\chi_0^{(k)}$. Nach [6] gehört dann σ zur maximalen normalen p-regulären Untergruppe $\mathfrak{R}_p(G)$ von G. Da $X|\mathfrak{R}_p(G)$ notwendig reduzibel ist, folgt aus der Primitivität von X, daß $\mathfrak{R}_p(G) \subseteqq Z$ ist. Es ist also $\mathfrak{R}_p(G)=1$, $\sigma=1$. Dies zeigt, daß H eine p-Gruppe und also $H=P$ ist. Daher ist P ein Normalteiler von G.

(8 C) *Liegt der zweite Fall vor und ist $s \leqq t$, so sind alle b_j positiv, und also ist P Normalteiler von G.*

Beweis. Es sei etwa $b_j<0$. Aus (8.4) und (8.6) folgt, daß $(\chi_j+|b_j|)(\chi_0^{(1)}-\chi_0^{(2)})$ identisch verschwindet. Daraus folgt unter Verwendung der Orthogonalitätsrelationen, daß $\overline{\chi}_0^{(1)}\chi_0^{(2)}$ den Bestandteil $|b_j|\chi_j$ haben muß. Da auch

$$\left(\sum_{k=1}^{t}\chi_0^{(k)}+1\right)(\chi_0^{(1)}-\chi_0^{(2)})=0$$

ist, muß $\overline{\chi}_0^{(1)}\chi_0^{(2)}$ einen Bestandteil der Form $\chi_0^{(k)}$ enthalten. Gradvergleich liefert dann

$$s^2 \geqq s+|b_j|\chi_j(1). \tag{8.12}$$

Aus (8.5) sieht man, daß $\chi_j(1) \geqq p^2-|b_j|$ ist. Da nach (8.7) $b_j^2 \leqq s$ ist, findet man $s^2 \geqq p^2$, also $s^2>st$, im Widerspruch zur Annahme $s \leqq t$.

Wir bemerken noch, daß für $a=3$ immer $s \neq 1$ ist. Wäre nämlich $s=1$, so wäre $\overline{P}$ sein eigener Normalisator in $\overline{G}$. Nach dem Burnsideschen Satz gibt es dann einen Normalteiler W/Z von G/Z vom Index p^2. Dann hätte W die Form $W=W_0\times Z$, wo W_0 ein Normalteiler von G von zu p teilerfremder Ordnung wäre. Wegen der Primitivität von X wäre $W_0=1$, also $G=P$ und daher X monomial. Wir hätten also einen Widerspruch.

§ 9. Der Fall $p=5$. Formulierung des Resultats

Die folgenden Abschnitte sind der Behandlung des Falls $p=5$ gewidmet. Wir beweisen den folgenden Satz.

(9 A) *Es sei G eine endliche Gruppe, die eine unimodulare treue irreduzible Darstellung X fünften Grades besitzt. Es werde vorausgesetzt, daß X sich nicht als monomiale Gruppe schreiben läßt.*

Dann haben wir einen der folgenden Fälle:

(I) $G/\mathfrak{Z}(G) \cong PSL(2,11)$, $|G:\mathfrak{Z}(G)|=660$.

(II) *$G/\mathfrak{Z}(G)$ ist eine alternierende oder symmetrische Permutationsgruppe vom Grade 5 oder 6.*

(III) $G/\mathfrak{Z}(G) \cong O_5(3)$; $|G:\mathfrak{Z}(G)|=25,920$.

(IV) *G ist eine eindeutig bestimmte Gruppe der Ordnung $24\cdot 5^4$, die einen Normalteiler D der Ordnung 125 hat. Dabei ist D eine nicht-abelsche Gruppe vom Exponenten 5.*

(V) *Wir haben eine von gewissen Untergruppen von G in* (IV), *deren S_5-Gruppe $P=D$ Normalteiler ist.*

Der Beweis ist folgendermaßen angeordnet: Schließen wir den Fall (I) aus, so ergibt sich aus (2 C), daß g nur durch die Primzahlen 2, 3 und 5 teilbar sein kann. Wir setzen

(9.1) $$g = 5^a \cdot 3^b \cdot 2^c$$

Aus (4 C) und (3 E) folgt

(9.2) $$a \leqq 5, \quad b \leqq 5, \quad c \leqq 7.$$

Es sei $T = g/n(P)$ die Anzahl der verschiedenen S_5-Gruppen. Wegen (9.1) und (9.2) kann man

(9.3) $$T = 3^\beta \cdot 2^\gamma; \quad \beta \leqq 5, \quad \gamma \leqq 7$$

setzen. Weiß man von jeder Untergruppe D einer S_5-Gruppe P, zu wievielen anderen S_5-Gruppen D gehört, so kann man Kongruenzen (mod 5^a) für T angeben. Im § 10 wird daraus geschlossen, daß $a < 5$ ist, und daß $a = 4$ auf den Fall (IV) führt. Weiter wird gezeigt, daß der Fall $a = 3$ nur für (V) eintritt; hierzu ist es notwendig, Charaktere heranzuziehen. Auf Grund von (4 A) kann der Fall $a = 2$ außer acht gelassen werden. Im Fall $a = 1$ spielen Charaktere die entscheidende Rolle. Im § 11 wird gezeigt, daß nur die Möglichkeiten (II) und (III) bestehen, wobei $\mathfrak{Z}(G) = 1$ ist.

§ 10. Die Fälle $a \geqq 3$ für $p = 5$

Es sei zunächst $a = 5$. Alle A enthaltenden S_p-Gruppen von G gehören offenbar zu $\mathfrak{N}(A)$. Da die abelsche Gruppe A Normalteiler von $\mathfrak{N}(A)$ ist, kann man $\mathfrak{X} | \mathfrak{N}(A)$ in monomialer Form schreiben. Aus (4 E) folgt, daß nur die Elemente von A Diagonalform haben. Es ist also $\mathfrak{N}(A)/A$ einer Untergruppe der symmetrischen Gruppe fünften Grades isomorph. Dies zeigt, daß $\mathfrak{N}(A)/A$ entweder sechs oder eine S_5-Gruppe enthält.

Hat man sechs S_5-Gruppen in $\mathfrak{N}(A)/A$, so gehört A zu sechs S_5-Gruppen von G. Ist P eine von ihnen, so ist $\mathfrak{N}(P) \subsetneqq \mathfrak{N}(A)$, da A charakteristisch in P ist. Hier ist $\mathfrak{N}(A)/A$ entweder die symmetrische oder die alternierende Gruppe fünften Grades. Man sieht leicht, daß in beiden Fällen $n(A)/n(P)$ und $n(P)$ gerade sind. Da nach (5 A) alle von A verschiedenen SD-Gruppen $D \subset P$ höchstens die Ordnung 5^3 haben, hat man $T \equiv 6 \pmod{25}$. Man sieht dann leicht aus (9.3), daß T einen der drei Werte 6, 3^4, $3^3 \cdot 2^7$ hat. Im letzten Fall wäre $n(P)$ ungerade, im Falle $T = 81$ wäre $g/n(P)$ und also $n(A)/n(P)$ ungerade. Ist $T = 6$, so tritt A in allen S_5-Gruppen von G auf, ist also ein abelscher Normalteiler von G. Dann wäre X monomial, was ausgeschlossen worden war.

Damit ist gezeigt, daß A keine SD-Gruppe ist. Daher ist $T \equiv 1 \pmod{25}$, und dies führt auf $T = 1$, 576 oder 7776. Wäre $T = 1$, so wären P und die charakteristische Untergruppe A von P normal in G, und dann wäre X monomial, was ausgeschlossen worden war. Da 576 und 7776 nicht kongruent 1 (mod 125) sind, muß es SD-Gruppen D der Ordnung 125 geben. Die Resultate von §§ 5—6 zeigen, daß D eine nicht-abelsche Gruppe ist. Es sei G_0 die in (7 C) diskutierte Gruppe. Die vorangehenden Überlegungen zeigen, daß die Anzahl T_0 der S_5-Gruppen von G_0 entweder 576 oder 7776 ist. Da (7 D) zeigt, daß $|G_0|$ nicht durch 27 teilbar ist, ist $T_0 = 7776$ unmöglich. Aus (5 C) und

(7 C) folgt, daß für eine S_5-Gruppe P_0 von G_0 die Ordnungen $|\mathfrak{N}(P_0)|$ und $|\mathfrak{N}(P_0)\cap G_0|$ durch 4 teilbar sind. Daher ist T_0 nicht durch 64 teilbar, also $T_0 \neq 576$. Damit ist gezeigt, daß $a \neq 5$ ist.

Ist $a=4$ und ist $T\equiv 1 \pmod{25}$, so muß wieder $T=576$ oder 7776 sein, und daher muß es SD-Gruppen der Ordnung $\geqq 25$ geben. Aus (6 B) und (7 B) folgt, daß es nicht-abelsche SD-Gruppen der Ordnung 5^3 gibt. Da nach (7 B) A keine SD-Gruppe ist, existieren nicht-abelsche SD-Gruppen D auch wenn $T \not\equiv 1 \pmod{25}$ ist. Jetzt folgt aus (5 C), daß $T\equiv 6 \pmod{25}$ ist. Wie im Falle $a=5$ ist dann $T=6$, 3^4 oder $3^3\cdot 2^7$. Aus (5 C) folgt, daß T und g/T gerade sind. Also ist $T=6$. Dann ist D Normalteiler von G. Nach (5 C) ist $|G|=24\cdot 5^4$. Ist $x\in G$ und $y\in D$, so ist $X(y)\,X(x)=X(x)\,X(y^x)$. Da X unimodular ist, ist nach dem Schurschen Lemma $X(x)$ durch $X|D$ und den Automorphismus $y\to y^x$ bis auf einen skalaren Faktor λ mit $\lambda^5=1$ bestimmt. Jetzt ist klar, daß G eindeutig bestimmt ist, und daß man Fall (IV) hat. Umgekehrt sieht man, daß dieser Fall auch wirklich möglich ist.

Schließlich sei $a=3$. Aus (7 A) folgt hier, daß $T\equiv 1 \pmod{25}$ ist. Wie früher führt dies zu einem der Fälle

$$T=1\,,\quad T=3^2\cdot 2^6\,,\quad T=3^5\cdot 2^5\,.$$

Ist $T=1$, so ist P Normalteiler von G. Da nach § 8 $n(P)=125s$ mit $s|24$ ist, hat man den Fall (V) von (9 A). Ist $T=3^2\cdot 2^6$, so folgt aus (9.1) und (9.2), daß $s|3^3\cdot 2$ ist. Für $T=3^5\cdot 2^5$ hat man $s|2^2$. Da $s|24$ und $s\neq 1$ ist, bestehen für $T\neq 1$ nur die Möglichkeiten

$$s=2,3,4 \text{ oder } 6\,.$$

Für $s\leqq 4$ ist $t=24/s\geqq s$, und (8 C) zeigt, daß der zweite Fall in § 8 unmöglich ist. Ist $s=6$, so folgt aus (8.12), daß im zweiten Fall ein $\chi_j(1)$, etwa $\chi_2(1)=24$ sein muß. Dann ist $b_2=-1$. Ebenso ist nach (8.10_1) im ersten Fall $x_2=24$. Jetzt erhält man aus (8.7) die verschiedenen Möglichkeiten, die für die b_j und r bestehen. Weiter zeigt man ohne Schwierigkeiten, daß die Bedingungen (8.5) und (8.6) für die Grade $\chi_j(1)$ und $\chi_0^{(1)}(1)$ sich nicht befriedigen lassen. Dabei ist zu beachten, daß jeder Grad ein Teiler von $3^5\cdot 2^7$ ist. Am einfachsten ist die Diskussion, wenn man auf das Auftreten von Graden achtet, die Potenzen von 2 oder von 3 sind.

Man sieht so, daß der Fall $a=3$ nur möglich ist, wenn $T=1$, also P Normalteiler von G ist.

§ 11. Der Fall $p=5$ und $a=1$

Wir verwenden dieselben Bezeichnungen wie in § 8. Ist $s=1$, so zeigt der Satz von Burnside, daß G einen Normalteiler N von Index 5 enthält. Wegen der Primitivität von X ist $N=1$, also $G=P$ und dies ist unmöglich.

Es ist also $s=2$, $t=2$ oder $s=4$, $t=1$. Im ersten Fall kann der zugeordnete Baum nur aus dem Stamm bestehen. Beachtet man (9.1), so sieht man, daß in (8.3) für $\sigma=1$ unter $\chi_2(1)$, $\chi_0^{(1)}(1)$ eine Potenz von 2 und eine Potenz von 3 vorkommen muß. Das führt zu den folgenden Möglichkeiten für $\chi_1(1)$, $\chi_2(1)$, $\chi_0^{(1)}(1)$

(a)	1,	1,	2;
(b)	1,	4,	3;
(c)	1,	9,	8.

Eine ähnliche Diskussion kann im Falle $s=4$, $t=1$ durchgeführt werden. Hier besteht kein Unterschied zwischen gewöhnlichen und Ausnahmecharakteren. Man hat drei Möglichkeiten für den Baum. Entweder der Baum besteht nur aus dem Stamm, oder man hat drei Eckpunkte auf dem Stamm und zwei Zweige, und diese können an dem mittleren Eckpunkt oder am Ende des Stammes beginnen. Im letzten Fall kann dies nicht der $\chi_1=1$ entsprechende Endpunkt sein. Die Grade der irreduziblen Charaktere sind kongruent zu $\pm 1 \pmod 5$, und sie sind Teiler von $3^5 \cdot 2^7$. Man beachte, daß wegen (9.1) und (9.2) die Grade kleiner als 400 sind. Besteht der Baum nur aus dem Stamm, so sind drei Grade kongruent zu 1 und zwei Grade kongruent zu -1. Dann sieht man ohne Schwierigkeiten, daß (8.3) nur in den folgenden Fällen gilt

(d)	1,	1,	16,	9,	9;
(e)	1,	1,	6,	4,	4;
(f)	1,	6,	6,	4,	9;
(g)	1,	16,	16,	9,	24;
(h)	1,	36,	36,	9,	64;
(i)	1,	6,	81,	24,	64;
(j)	1,	36,	81,	54,	64.

Hat man zwei Zweige an einem Endpunkt des Stammes, so sind zwei Grade kongruent 1 und drei Grade kongruent $-1 \pmod 5$. Zwei der letzteren Grade sind gleich. Das führt zu den Möglichkeiten

(k)	1,	16,	4,	4,	9;
(l)	1,	81,	9,	9,	64.

Enden schließlich die Zweige in dem mittleren Endpunkt des Stammes, so sieht man ähnlich, daß man die beiden Möglichkeiten hat:

(m)	1,	1,	1,	1,	4;
(n)	1,	1,	1,	6,	9.

Die Formeln (8.1) zeigen, daß P nur dann zum Kern der entsprechenden Darstellungen gehört, wenn der Grad 1 ist. Jetzt können [11] Lemma 2 und 3 angewendet werden. Man sieht:

(α). Tritt ein Grad auf, der eine von 1 verschiedene Potenz einer Primzahl r ist, und ist r^m die höchste in g aufgehende Potenz, so ist r^m ein Teiler einer Summe von Gliedern $b_i\chi_i(1)$ mit $0 \leqq i \leqq s$, derart, daß jedes $\chi_i(1) \equiv 0 \pmod r$ ist.

(β). r^m ist ein Teiler der Summe aller Glieder $b_i\chi_i(1)$ mit $0 \leqq i \leqq s$, für die r^{m-1} ein Teiler von $\chi_i(1)$ ist. Dies folgt aus den Eigenschaften von r-Blöcken vom

Defekt 1. Daraus ergibt sich die folgende Information bezüglich g in den verschiedenen Fällen

(a)	$g=5\cdot 3^b\cdot 2$;	(b)	$g=5\cdot 3\cdot 4$;	(c)	$g=5\cdot 9\cdot 8$;
(d)	$g=5\cdot 9\cdot 16$;	(e)	$g=5\cdot 3^b\cdot 4$ oder $5\cdot 3^b\cdot 8$;		
(f)	$g=5\cdot 9\cdot 4$;	(g)	$g=5\cdot 9\cdot 16$;	(h)	$g=5\cdot 27\cdot 64$ oder $5\cdot 9\cdot 64$;
(i)	$g=5\cdot 81\cdot 64$;	(j)	$g=5\cdot 81\cdot 64$;	(k)	$g=5\cdot 9\cdot 16$;
(l)	$g=5\cdot 81\cdot 64$;	(m)	$g=5\cdot 3^b\cdot 4$;	(n)	$g=5\cdot 9\cdot 2^c$.

Die Bemerkung (β) oben zeigt, daß (f), (j) und (n) unmöglich sind, und daß in den Fällen (e) und (h) der zweite Wert für g zu nehmen ist. In den Fällen (a) und (m) würde G' die Ordnung $5\cdot 3^b$ haben und einfach sein. Das ist nur möglich, wenn $G'=P$ ist, und dann hat G keine irreduzible Darstellung vom Grad 5. In den Fällen (b) und (c) ergibt sich aus der Liste der einfachen Gruppen kleiner Ordnung, daß G die alternierende Gruppe $\mathfrak{A}_5$ bzw. $\mathfrak{A}_6$ ist. Im Fall (d) hat G' den Index 2 und ist einfach von der Ordnung 360. Daher ist $G'\cong\mathfrak{A}_6$. Man identifiziere G' und $\mathfrak{A}_6$. Dann können in G die beiden irreduziblen Charaktere achten Grades von $\mathfrak{A}_6$ nicht assoziiert sein. Andererseits sind die beiden irreduziblen Charaktere fünften Grades assoziiert, da andernfalls G keinen irreduziblen Charakter fünften Grades haben könnte. Unter Verwendung der wohlbekannten Struktur der Automorphismengruppe von $\mathfrak{A}_6$ schließt man, daß ein geeignetes Element σ von G denselben Automorphismus induziert wie die Permutation (1, 2) der symmetrischen Gruppe $\mathfrak{S}_6$. Da dann $\sigma^2\in\mathfrak{Z}(\mathfrak{A}_6)$ ist, ist $\sigma^2=1$. Jetzt folgt, daß G ebenso wie $\mathfrak{S}_6$ das halbdirekte Produkt von $\mathfrak{S}_6$ mit $\langle\sigma\rangle$ ist. Es ist also $G\cong\mathfrak{S}_6$.

In den Fällen (e) und (k) kann man die bekannte Liste der Gruppen vierten Grades verwenden. Im Fall (e) findet man $G\cong\mathfrak{S}_5$. Man kann hier auch ähnlich wie im Fall (d) vorgehen. (g), (h), (k) stellen sich als unmöglich heraus, da G eine einfache Gruppe der Ordnung 720 bzw. 2880 sein müßte.

Im Fall (l) beachte man, daß $\chi\bar{\chi}$ als Bestandteil den Charakter des zur modularen Hauptdarstellung (für 5) gehörigen unzerfällbaren Bestandteil der regulären Darstellung enthalten muß, vgl. [10], § 24 (76). Durch Betrachtung des Baumes erkennt man, daß dieser Bestandteil den Grad 65 hat, und dies führt auf einen Widerspruch.

Es bleibt also nur der Fall (i), $g=5\cdot 81\cdot 64=25,920$ übrig. Da keine von der Hauptdarstellung verschiedenen Darstellungen ersten Grades auftreten, ist $G=G'$, also nach (8 A) G einfach. Nun kann man zeigen[9], daß $O_5(3)$ die einzige einfache Gruppe der Ordnung 25, 920 ist, und daß $O_5(3)$ wirklich eine irreduzible Darstellung des Grades 5 hat. Man hat also den Fall (III) in (9 A) und damit ist der Beweis des Satzes beendigt.

[9] Ein Beweis dieser Behauptung wird in einer Arbeit geführt, die Interessenten auf Wunsch von dem Department of Mathematics, Harvard University, Cambridge, Mass., USA, in mimeographierter Form erhalten können.

Literatur

[1] BIEBERBACH, L.: Über einen Satz des Herrn C. JORDAN in der Theorie der endlichen Gruppen linearer Substitutionen. Sitzber. Preuß. Akad. Wiss. 231—240 (1911).

[2] BLICHFELDT, H. F.: On the order of the linear homogeneous groups. I, II, III, IV. Trans. Am. Math. Soc. **4**, 387—397 (1903); **5**, 310—325 (1904); **7**, 523—529 (1906); **12**, 39—42 (1911).

[3] — On primitive homogeneous groups. Trans. Am. Math. Soc. **6**, 230—236 (1905).

[4] — Finite Collineation Groups. Chicago: University of Chicago Press 1917.

[5] BRAUER, R.: On groups whose order contains a prime number to the first power. I, II. Am. J. Math. **64**, 401—420, 421—440 (1942).

[6] — Some applications of the theory of blocks of characters of finite groups. I. J. Alg. **1**, 152—167 (1964).

[7] — Some results on finite groups whose order contains a prime to the first power. Erscheint im Nagoya J. Math.

[8] — and W. FEIT: On the number of irreducible characters of finite groups in a given block. Proc. Nat. Acad. Sci. U.S. **45**, 361—365 (1959).

[9] — and H. S. LEONARD JR.: On finite groups with an abelian Sylow group. Canad. J. Math. **14**, 436—450 (1962).

[10] — and C. NESBITT: On the modular characters of groups. Ann. Math. **42**, 556—590 (1941).

[11] — and H. F. TUAN: On simple groups of finite order. I. Bull. Am. Math. Soc. **51**, 756—766 (1945).

[12] BURNSIDE, W.: Theory of groups of finite order. Second ed. Cambridge: Cambridge University Press 1911.

[13] CLIFFORD, A. H.: Representations induced in an invariant subgroup. Ann. Math. **38**, 533—550 (1937).

[14] CURTIS, C. W., and I. REINER: Representation Theory of Finite Groups and Associative Algebras. New York, London: Interscience Publishers 1962.

[15] FEIT, W.: Groups which have a faithful representation of degree less than $p-1$. Trans. Am. Math. Soc. **112**, 287—303 (1964).

[16] — and J. G. THOMPSON: On groups which have a faithful representation of degree less than $(p-1)/2$. Pacific Math. J. **4**, 1257—1262 (1961).

[17] FROBENIUS, G.: Über den von L. BIEBERBACH gefundenen Beweis eines Satzes von C. JORDAN. Sitzber. Preuß. Akad. Wiss. 241—248 (1911).

[18] HAYDEN, S.: On finite linear groups whose order contains a prime larger than the degree. Thesis. Harvard University 1963.

[19] JORDAN, C.: Mémoire sur les équations differentielles linéaires à intégrale algébrique. J. reine angew. Math. **84**, 89—215 (1878).

[20] SCHUR, I.: Über eine Klasse von endlichen Gruppen linearer Substitutionen. Sitzber. Preuß. Akad. Wiss. 77—91 (1905).

[21] — Neue Begründung der Theorie der Gruppencharaktere. Sitzber. Preuß. Akad. Wiss. 406—432 (1905).

[22] TUAN, H. F.: On groups whose orders contain a prime to the first power. Ann. Math. **45**, 110—140 (1944).

[23] ZASSENHAUS, H.: Neuer Beweis der Endlichkeit der Klassenzahl bei unimodularer Äquivalenz endlicher ganzzahliger Substitutionsgruppen. Abhandl. math. Seminar hamburg. Univ. **12**, 276—288 (1938).

Prof. Dr. RICHARD BRAUER
Department of Mathematics, Harvard University
Cambridge, Mass.

(Eingegangen am 17. Februar 1966)

Reprinted from ILLINOIS JOURNAL OF MATHEMATICS
Vol. 11, No. 3, September 1967
Printed in U.S.A.

ON A THEOREM OF BURNSIDE

BY
RICHARD BRAUER[1]

A celebrated theorem of Burnside states that the order g of a non-cyclic finite simple group G is divisible by at least three distinct prime numbers. In other words, if we set $g = p^a q^b g_0$ where p and q are primes and a, b, g_0 are positive integers, then $g_0 > 1$. We prove a refinement.

THEOREM 1. *Let G be a simple group of finite order*

$$g = p^a q^b g_0 \tag{1}$$

where p, q are distinct primes, and where a, b, g_0 are integers, $a > 0$. If $g \neq p$, then

$$g_0 - 1 > \log p/\log 6. \tag{2}$$

Proof. We denote the irreducible characters of G by $\chi_1 = 1, \chi_2, \cdots$ and set $x_j = \chi_j(1)$. Let $\mathfrak{p}$ be a prime ideal divisor of p in the field of g-th roots of unity. If $\sigma \in G$, denote by $c(\sigma)$ the order of the centralizer $C(\sigma)$ of $\sigma \in G$. The principal p-block $B_0(p)$ of G consists of those characters χ_n for which

$$(g/c(\sigma))(\chi_n(\sigma)/x_n) \equiv g/c(\sigma) \pmod{\mathfrak{p}} \tag{3}$$

for every $\sigma \in G$. If $\tau \in G$ has an order divisible by p, then by Theorem VIII of [4],

$$\sum_{\chi_n \in B_0(p)} x_n \chi_n(\tau) = 0. \tag{4}$$

Of course, $\chi_1 \in B_0(p)$. Now (4) shows that there exist $\chi_j \in B_0(p)$ with $j \neq 1$ for which x_j is not divisible by q. Let p^h be the highest power of p dividing x_j so that

$$x_j = p^h m, \qquad m \mid g_0 . \tag{5}$$

Let X_j denote the irreducible representation of G with the character χ_j. Since G is simple and $j \neq 1$, X_j is a faithful representation. It follows from a theorem of Feit and Thompson [5] that $p \leqq 2x_j + 1$.

If $h = 0$, then $x_j = m \leqq g_0$. Hence $p \leqq 2g_0 + 1$, that is, $g_0 \geqq (p-1)/2$. This implies (2) for $p \geqq 5$. Since $g_0 \geqq 2$, (2) also holds for $p \leqq 3$.

Assume now that $h > 0$. Take σ as an element of order p in the center of a p-Sylow subgroup P of G. Then $g/c(\sigma)$ is not divisible by p. The left side of (3) represents an algebraic integer. It follows that $\chi_j(\sigma)$ is divisible by p^h in the ring of algebraic integers and that

$$\chi_j(\sigma)/p^h \equiv x_j/p^h = m \pmod{\mathfrak{p}}. \tag{6}$$

Received January 17, 1967.

[1] This work has been supported by a National Science Foundation grant.

Let ε be a primitive p-th root of unity. If a_n is the multiplicity of ε^n as characteristic root of $X_j(\sigma)$, we have

$$x_j = \sum_{n=0}^{p-1} a_n, \qquad \chi_j(\sigma) = \sum_{n=0}^{p-1} a_n \varepsilon^n = \sum_{n=1}^{p-1} (a_n - a_0)\varepsilon^n.$$

Now, $\varepsilon, \varepsilon^2, \cdots, \varepsilon^{p-1}$ form an integral basis for the integers in the field of the p-th roots of unity. Hence each $a_n - a_0$ is divisible by p^h, say

$$a_n = a_0 + p^h b_n .$$

Then

$$x_j = pa_0 + p^h \sum_{n=1}^{p-1} b_n, \qquad \chi_j(\sigma) = p^h \sum_{n=1}^{p-1} b_n \varepsilon^n.$$

It now follows from (5) that pa_0 is divisible by p^h. Since $\varepsilon \equiv 1 \pmod{\mathfrak{p}}$, (6) shows that $pa_0/p^h \equiv 0 \pmod{p}$. This means that a_0 and then all a_n are divisible by p^h.

Thus all characteristic roots of $X_j(\sigma)$ have a multiplicity divisible by p^h. In particular, $X_j(\sigma)$ has at most $x_j/p^h = m$ distinct characteristic roots.

We say that an irreducible representation X of a finite group G is *quasi-primitive*, if for every normal subgroup H of G the restriction $X \mid H$ does not have non-equivalent irreducible constituents. In particular, if G is simple, every non-principal irreducible representation is quasi-primitive. It has been proved by Blichfeldt [1, Theorem 9, p. 101] that if X is a faithful quasi-primitive representation of degree x of a group G and if for an element σ of G of order p not in the center of G the transformation $X(\sigma)$ has w distinct characteristic roots, then

$$p < 6^{w-1}.$$

Actually, Blichfeldt states his theorem only for primitive representations, but his proof remains valid for quasi-primitive representations as defined above. Applying this with $X = X_j$ we obtain

$$\log p < (w - 1) \log 6 \leqq (m - 1) \log 6 \leqq (g_0 - 1) \log 6$$

and this implies (2).

As a corollary, we note:

THEOREM 1*. *If G is a finite group of order* (1) *and if G is not p-solvable, the inequality* (2) *holds.*

Indeed, we may assume that G is not simple. Let H be a non-trivial normal subgroup of G. At least one of the groups H and G/H is not p-solvable and we can use induction.

It is likely that the lower bound (2) in these theorems can be improved substantially. We can prove this when the p-Sylow group P of G is of certain special types.

THEOREM 2. *Let G be a group of an order $g = p^a q^b g_0$ as in* (1) *and assume* *that G is not p-solvable.* ***If the p-Sylow group P of G is abelian then***

$$g_0 \geqq \sqrt{\frac{p-1}{2}}. \tag{7}$$

Proof. Again, it will suffice to prove this for simple groups G. Choose X_j as in the proof of Theorem 1. If again $x_j = DgX_j$ has the form (5) and if $h = 0$, then as we have seen above, $g_0 \geqq (p-1)/2$ and this implies (7) for $p \neq 2$. The case $p = 2$ is trivial.

Suppose that $h > 0$. Since the p-Sylow subgroups of G are abelian, the method in the proof of Theorem 1 applies for any element $\sigma \in G$ of order p. It follows that $X_j(\sigma)$ has at most g_0 distinct characteristic roots. Now a method of Speiser [6, Theorem 202] shows that if $2g_0^2 < p - 1$, the set of elements of G of order 1 or p forms a normal subgroup of G and this leads to a contradiction. Again, (7) must hold.

Remarks. 1. The preceding proof still applies, if G is a simple group with a non-abelian p-Sylow group P satisfying the following condition: If Y is a faithful (reducible or irreducible) representation of G and if σ is an element of order p of P not belonging to the center of P, then $Y(\sigma)$ has more than $(p-1)/2$ distinct characteristic roots. For instance, this condition is satisfied, when P is a non-abelian group of order p^3.

2. Other theorems of a similar type are given in [2].

If g_0 in Theorem 1 is a fixed number, then there are only finitely many possibilities for p and for q. One may conjecture that there exist only finitely many simple groups G of an order (1) with a given value of g_0. We can only prove this for groups G with abelian p-Sylow groups P of given rank.

THEOREM 3. *Let G be a simple group of order g of the form* (1) *considered in Theorem* 1. *Assume that the p-Sylow group P is abelian of rank N. Then there exists a bound* $\beta(g_0, N)$ *depending only on the integers* g_0 *and N only such that*

$$g \leqq \beta(g_0, N).$$

Proof. Again, determine X_j as in the proof of Theorem 1. If $x_j = Dg\ X_j$ is given by (3), then X_j is of height h. It follows that h lies below a bound $\beta_1(N)$ depending on p and N; (cf. [3]). Then x_j lies below a bound depending on g_0 and N. Now, Jordan's theorem (cf. [1]) implies the existence of bounds $\beta(g_0, N)$ as stated in the theorem.

REFERENCES

1. H. F. BLICHFELDT, *Finite collineation groups*, University of Chicago Press, Chicago, 1917.
2. R. BRAUER, *On some conjectures concerning finite simple groups*, Studies in mathematical analysis and related topics, Stanford University Press, Palo Alto, 1962.
3. ———, *On blocks and sections in finite groups II*, Amer. J. Math., to appear.

4. R. Brauer and C. Nesbitt, *On the modular characters of groups of finite order*, University of Toronto Studies, Mathematical Series no. 4, 1937.
5. W. Feit and J. G. Thompson, *On groups which have a faithful representation of degree less than* $(p - 1)/2$, Pacific J. Math., vol. 4 (1961), pp. 1257–1262.
6. A. Speiser, *Theorie der Gruppen von endlicher Ordnung*, third edition, Springer, Berlin, 1937.

Harvard University
Cambridge, Massachusetts

ON BLOCKS AND SECTIONS IN FINITE GROUPS. I.

By RICHARD BRAUER.*

1. Introduction. If we know the characters of a finite group G, we can determine many of the purely group theoretical properties of G. However, the full extent to which this is possible is not known. It seems important, both with a view to applications and for theoretical reasons, to obtain new connections between characters of G and structural properties of G. A number of such connections appear in the theory of blocks of representations which has been studied in two previous papers [1I, II]. In another investigation [2], we have given a number of applications to questions of a purely group theoretical nature. Other uses of the theory have been made by a number of authors. It seems therefore of some importance to continue the investigation of blocks, and this is the purpose of the present paper.

If G_0 is a subgroup of G and if p is a fixed prime number, under suitable conditions, a p-block b of G_0 defines a p-block $B = b^G$ of G. For instance, this is the case, if the centralizer $\mathfrak{C}_G(\mathfrak{d})$ of the defect group $\mathfrak{d}$ of b is contained in G_0. One of the results of this paper is to show that the relation between b and B then is reflected in relations between characters of $\mathfrak{C}_G(\mathfrak{d})$ and characters of groups $\mathfrak{C}_G(P)$ where P is a p-subgroup of G with $P \supseteq \mathfrak{d}$ As a special case, we obtain a result of [2I] which states that B is the principal block of G, if and only if b is the principal block of G_0. Given a p-block B of G and a p-element π of G, we are interested to obtain the p-blocks b of $\mathfrak{C}_G(\pi)$ for which $B = b^G$. The reason for this lies in the theorem of [1II] stated below as (2B). Here, our new results can be applied. In general, the situation is much more complicated than in the case of the principal block.

In § 2, we introduce the notation and state some of the presupposed results. In § 3, a few preliminary results are obtained. In § 4, relations below blocks of a group G and blocks of a normal subgroup are derived. This supplements results of P. Fong [6]. In § 5, it is shown that if B is a block of G with the defect group D, there exist blocks b of $\mathfrak{C}_G(D)$ such that $b^G = B$;

Received June 28, 1966.

* This research was supported by the United States Air Force under contract No. AF 49(638)-1381 monitored by the AF Office of Scientific Research of the Air Research and Development Command.

all these blocks b form a family of blocks conjugate in $\mathfrak{N}_G(D)$. Finally § 6 and § 7 deal with the main results discussed above.

2. Notation. Presupposed results.[1]

1. All groups considered will be finite groups. If M is a subset of a group G and H a subgroup of G, then $\mathfrak{C}_H(M)$ denotes the centralizer of M in H and $\mathfrak{N}_H(M)$ denotes the normalizer of M in H. The notation $\mathfrak{C}_H(M)$ is also used, if M is a single element of G. If $H = G$, the subscript G will often be suppressed. The center of H is denoted by $\mathfrak{Z}(H)$. We indicate by $H \lhd G$ that H is a normal subgroup of G. The symbol $|M|$ denotes the cardinality of the set M.

If Ξ is a field, $\Xi[G]$ is the *group algebra* of G with regard to Ξ. If M is a subset of G, then

$$\mathcal{S}M = \sum_{\sigma \in M} \sigma$$

will denote the sum of the elements of M in $\Xi[G]$. In particular, if $K_1, K_2, \cdots, K_k$ are the classes of conjugate elements of G, the *class sums* $\mathcal{S}K_1, \mathcal{S}K_2, \cdots, \mathcal{S}K_k$ form a basis of the center $Z(\Xi[G])$ of the algebra $\Xi[G]$. We call this center the *class algebra* of G.

2. Let n be an integer which is divisible by the orders of the elements of G and let Ω be the field of the n-th roots of unity over the field $\boldsymbol{Q}$ of rational numbers. As is well known, Ω is an absolute splitting field for $\Omega[G]$, i. e., the irreducible representations of G in Ω remain irreducible in all extension fields of Ω. Unless otherwise stated, the word *character* will always refer to the character of a representation in Ω. We have then k irreducible characters of G and these will be denoted throughout by $\chi_1, \chi_2, \cdots, \chi_k$. Here, $k = k(G)$ is the *class number* of G. To each irreducible character χ of G, there corresponds a homomorphism ω_χ of the class algebra $Z(\Omega[G])$ onto Ω. Here, if $\sigma_j \in K_j$,

$$\omega_\chi(\mathcal{S}K_j) = |K_j| \chi(\sigma_j)/\chi(1).$$

We shall find it convenient to set $\omega_\chi(\sigma_j) = \omega_\chi(\mathcal{S}K_j)$ for $\sigma_j \in K_j$. For typographical reasons, we shall often write ω_i for ω_{χ_i}. Actually $\omega_1, \cdots, \omega_k$ are distinct and they are the only algebra homomorphisms of $Z(\Omega[G])$ onto Ω.

3. Throughout the paper, p will denote a fixed prime number. The abbreviation S_p-group will be used for "*Sylow-p-subgroup.*" We choose a

[1] More details concerning the content of Sections 1-5 of § 2 can be found in [1 I, §§ 1-6]. Cf. also, [4].

fixed extension ν of the (exponential) p-adic valuation of $\boldsymbol{Q}$ to Ω; $\nu(p)=1$. Let $\mathfrak{o}$ denote the ring of local integers in Ω and let $\mathfrak{p}$ denote the corresponding prime ideal. For $\alpha, \beta \in \mathfrak{o}$, a congruence $\alpha \equiv \beta$ will always mean a congruence modulo $\mathfrak{p}$, unless another module is indicated. The residue class map of $\mathfrak{o}$ modulo $\mathfrak{p}$ will be denoted by θ; we set $\Omega^\theta = \mathfrak{o}/\mathfrak{p}$. Often, the argument of ν will not be enclosed in parentheses; i. e. we write νa for $\nu(a)$.

By a *modular representation* of G, we shall mean a representation of G in Ω^θ. Again, Ω^θ is an absolute splitting field of G. *Modular characters* are defined as in [1I].

We say that an element ρ of G is *p-regular,* if its order is prime to p. By a *p-element* π of G, we mean of course an element whose order is a power of p. Each element σ of G can be written uniquely in the form $\sigma = \pi\rho$ where π is a p-element and where ρ is a p-regular element of $\mathfrak{C}(\pi)$. We refer to π as to the p-factor of σ and to ρ as to the *p-regular factor* of σ. The element σ is *p-singular,* if $\pi \neq 1$. Occasionally, the set of *p-regular* elements in a group H will be denoted by H^0.

If exactly l of the conjugate classes K_i are p-regular, (i. e., consist of p-regular elements), then G possesses l modular irreducible characters $\phi_1, \phi_2, \cdots, \phi_l$.

4. The irreducible characters $\chi_1, \chi_2, \cdots, \chi_k$ of G are distributed into a number of disjoint sets, the *blocks* $B_1, B_2, \cdots$. Here χ_i and χ_j belong to the same block B, if and only if $\omega_i(\sigma) \equiv \omega_j(\sigma)$ for all $\sigma \in G$. We choose a $\chi_i \in B$ arbitrarily and set $\omega_B = \omega_i$; this expression ω_B will only appear in congruences mod $\mathfrak{p}$. If

$$\mu = \operatorname*{Min}_{\chi \in B}(\nu\chi(1)),$$

we call $d(B) = \nu\,|\,G\,| - \mu$ the *defect* of the block B. Thus, for $\chi_i \in B$, we can set

$$\nu\chi_i(1) = \nu\,|\,G\,| - d(B) + h_i$$

with $h_i \geqq 0$. The number h_i is called the *height* of χ_i. We define the *defect* $d(K)$ of a conjugate class K by $d(K) = \nu\,|\,G\,| - \nu\,|\,K\,|$. By a *defect group* of K, we mean an S_p-group of the centralizer of an element of K. Thus, the defect groups of K are determined only up to conjugacy; their order is $p^{d(K)}$.

Each block B possesses *defect groups* D, of order $p^{d(B)}$. These are also determined up to conjugacy. If $\omega_B(\mathcal{S}K) \not\equiv 0$ for some conjugate class K, then each defect group of K contains a conjugate of D. Since there exist classes K (even p-regular classes K) of defect $d(B)$ with $\omega_B(\mathcal{S}K) \not\equiv 0$, this condition characterizes D.

To each B of G, there corresponds an idempotent η_B of the class algebra $Z(\Omega[G])$. We have $\omega_i(\eta_B)=1$ or 0 according as to whether or not χ_i belongs to B.

Blocks of defect 0 consists of a single irreducible character χ_i and χ_i then vanishes for all p-singular elements.

5. Each χ_i, or rather its restriction $\chi_i^0=\chi_i \mid G^0$ to the set of p-regular elements G^0, can be considered as a modular character of G. If B_r is a block, let B_r^0 denote the set of modular irreducible characters ϕ_j which occur as constituents in characters χ_i^0 with $\chi_i \in B_r$. We call the $\phi_j \in B_r^0$ then the modular characters in B_r. Each modular irreducible character ϕ of G occurs in one and only one block.

6. We say that an (ordinary or modular) character ζ of G is trivial on a subset M of G, if $\zeta(\sigma)=\zeta(1)$ for $\sigma \in M$.

Assume that Q is a normal p-subgroup of G. As shown in [1I, § 9], every modular character of G is trivial on Q. The group Q belongs to the defect group of every block B of G and, as follows from [1I, (9E)], B contains characters χ_j of height 0 which are trivial on Q. If, in particular, G has the form $G=\mathfrak{C}(Q)Q$ and if B has the defect group Q, then B contains only one such χ_j, cf. [1II, (2G)]. Then χ_j is called the *canonical* character of B.

7. Let H be a subgroup of G and let ψ be a linear function on $Z(\Omega[H])$. We can define a linear function ψ^G on $Z(\Omega[G])$ by setting

$$\psi^G(\mathcal{S}K)=\psi(\mathcal{S}(K\cap H))$$

for each conjugate class K of G. In particular, let b be a block of H and take $\psi=\omega_b$. If there exists a block B of G such that $\omega_B \equiv \omega_b{}^G$, we say that b^G is defined and set $b^G=B$; [1II, § 2]. If the centralizer $\mathfrak{C}_G(\mathfrak{d})$ of the defect group $\mathfrak{d}$ of b belongs to H, then b^G is defined.

The following result is proved in [1I, § 10], cf. also [11], [12], [4].

(2A) *Let D be a fixed p-subgroup of G. There exists a one-to-one correspondence between the blocks B of G and the blocks b of $\mathfrak{N}_G(D)$ both with the defect group D. Here, b is associated with $B=b^G$.*

We also quote the main result of [1II], cf. also [5], [8], [9].

(2B) *Let B be a block of G and let π be a fixed p-element of G. For $\chi_i \in B$ and for p-regular $\rho \in \mathfrak{C}_G(\pi)$, we have formulas*

$$\chi_i(\pi\rho)=\sum_j d_{ij}{}^{\pi}\phi_j{}^{\pi}(\rho).$$

Here $\phi_j{}^{\pi}$ ranges over the modular irreducible characters of $\mathfrak{C}_G(\pi)$ belonging *to blocks b with $b^G = B$. The "decomposition numbers" $d_{ij}{}^{\pi}$ are algebraic integers which do not depend on ρ.*

8. The following remarks are obvious. Let λ be an automorphism of G. If ψ is a character of a subgroup H, a character ψ^λ of H^λ is defined by

$$\psi^\lambda(\sigma^\lambda) = \psi(\sigma) \text{ for } \sigma \in H.$$

A corresponding remark applies to modular characters. In this manner, the automorphisms λ of G act on the characters of subgroups. Then every block b of H is mapped on a block b^λ of H^λ. Of course, each conjugate class L of H is mapped by λ on a conjugate class L^λ of H^λ.

In particular, if $\tau \in G$, we may apply this to the inner automorphism $\lambda = (\tau)$ which maps $\xi \in G$ on $\xi^\tau = \tau^{-1}\xi\tau$. If $H \lhd G$ and if ψ is a character of H, the *inertial group* $T(\psi)$ of ψ in G is the group of all $\tau \in G$, which map ψ on itself. Similarly, we define the *inertial group* $T(b)$ *of a block* b of $H \lhd G$.

3. Some preliminary results.

(3A) *Let b be a block of a subgroup H of G for which the block b^G of G is defined. Let ψ be an irreducible character of b and suppose that the character ψ^* of G induced by ψ is given by*

$$\psi^* = \sum_{i=1}^{k} u_i\chi_i \tag{3.1}$$

$(u_i \in \mathbf{Z}, u_i \geqq 0)$. If B is a block of G, we have

$$\nu(\sum_{\chi_i \in B} u_i\chi_i(1)) > \nu(\psi^*(1)), \text{ if } B \neq b^G, \tag{3.2a}$$

$$\nu(\sum_{\chi_i \in B} u_i\chi_i(1)) = \nu(\psi^*(1)), \text{ if } B = b^G. \tag{3.2b}$$

Proof. Let σ be an element of the conjugate class K of G. By the definition of induced characters,

$$\sum_{i=0}^{k-1} u_i\chi_i(\sigma) = |\mathfrak{C}_G(\sigma)| \sum_j |\mathfrak{C}_H(\sigma_j)|^{-1}\psi(\sigma_j)$$

where σ_j ranges over a set of representatives for the conjugate classes L_j of H contained in K. This can be written in the form

$$\sum_i u_i\omega_i(\mathcal{S}K)\chi_i(1) = |G:H|\,\psi(1) \sum_j \omega_\psi(\mathcal{S}L_j) = \psi^*(1)\omega_\psi{}^G(\mathcal{S}K).$$

Since the class sums $\mathcal{S}K$ form a basis for the class algebra $Z(\Omega[G])$, we have

$$\sum_i u_i\chi_i(1)\omega_i = \psi^*(1)\omega_\psi{}^G.$$

Substitute here the idempotent η_B belonging to the block B of G. Now $\omega_i(\eta_B) = 1$ or 0 according as to whether or not χ_i belongs to B. Since $\omega_\psi{}^G \equiv \omega_\chi$ for $\chi \in b^G$, the statement becomes evident.

Clearly, (3A) implies that if we know the block distribution of the irreducible characters χ_i of G and if we know the decomposition of ψ^* into irreducible characters for some $\psi \in b$, we can identify b^G among the blocks B of G (provided, of course, that b^G is defined). Another immediate consequence of (3A) is the corollary:

(3B) *If in* (3A) *b and b^G have the same defect and if ψ has height* 0 *in b, there exists a character χ_i of height* 0 *in B which appears in ψ^* with a multiplicity $u_i \not\equiv 0 \pmod p$.*

We next derive a result concerning the defect groups $\mathfrak{d}$ of b and D of b^G, assuming again that b^G is defined. We know that D can be chosen such that $D \supseteq \mathfrak{d}$, cf. [1II, (2B)]. We need a lemma.

LEMMA. *Let B be a block of G with the defect group D. Let σ be an element of G such that $\omega_i(\sigma) \not\equiv 0$ for $\chi_i \in B$. There exist conjugates τ of σ such that $\tau \in \mathfrak{C}(D)$ and that the p-factor π of τ lies in $\mathfrak{Z}(D)$.*

Proof. Set $\mathfrak{g} = D\mathfrak{C}(D)$. By [1II, 2D)], there exist blocks $\mathfrak{b}$ of $\mathfrak{g}$ with $\mathfrak{b}^G = B$.[2] We can find an irreducible character ψ of height 0 in $\mathfrak{b}$ which is trivial on $D \lhd \mathfrak{g}$, (§ 2, 6). If σ belongs to the conjugate class K of G, then

$$\omega_\psi{}^G(\mathcal{S}K) \equiv \omega_B(\mathcal{S}K) \equiv \omega_i(\sigma) \not\equiv 0.$$

The definition of $\omega_\psi{}^G$ shows that there exist G-conjugates $\tau \in \mathfrak{g}$ of σ for which $\omega_\psi(\tau) \not\equiv 0$. If the p-factor π of τ did not belong to D, τD would be a p-singular element of $\mathfrak{g}/D$. Considered as character of $\mathfrak{g}/D$, ψ has defect 0. Then $\psi(\tau) = 0$, cf. § 2, **4**. This implies $\omega_\psi(\tau) = 0$, a contradiction. Thus, $\pi \in D$.

Since $D \lhd \mathfrak{g}$, the (unique) defect group of $\mathfrak{b}$ is D. Since $\omega_\psi(\tau) \not\equiv 0$, we have $D \subseteq \mathfrak{C}(\tau) \cap \mathfrak{g}$, (§ 2, **4**). Hence $\tau \in \mathfrak{C}(D)$ and then $\pi \in D \cap \mathfrak{C}(D) = \mathfrak{Z}(D)$ as stated.

(3C) *Let b be a block of a subgroup H of G such that b^G is defined. Let $\mathfrak{d}$ be a defect group of b and let D be a defect group of b^G. Every element of $\mathfrak{Z}(D)$ is conjugate in G to an element of $\mathfrak{Z}(\mathfrak{d})$.*

Proof. Choose $\chi_i \in b^G$ of height 0. It follows from the definition of defect groups that there exist p-regular elements ρ for which D is an S_p-group

[2] Actually, this is also proved in § 5 below, Of course, no use is made there on results based on the Lemma.

of $\mathfrak{C}(\rho)$ and for which $\omega_i(\rho) \not\equiv 0$. Then $\chi_i(\rho) \not\equiv 0$. If π is an element of $\mathfrak{Z}(D)$, we have $\chi_i(\pi\rho) \not\equiv 0$ and, as D is an S_p-group of $\mathfrak{C}(\pi\rho)$, $\omega_i(\pi\rho) \not\equiv 0$.

If ψ is an irreducible character of b, $\omega_i \equiv \omega_\psi{}^G$. It follows that there exists G-conjugates $\sigma \in H$ of $\pi\rho$ for which $\omega_\psi(\sigma) \not\equiv 0$. Now the lemma applied to H shows that we may choose σ such that its p-factor π_1 lies in $\mathfrak{Z}(\mathfrak{d})$. Since σ is conjugate to $\pi\rho$ and hence π to π_1, this proves (3C).

We mention some corollaries.

(3D) *If* $\mathfrak{d}$ *and* D *are as in* (3C), *the exponent of* $\mathfrak{Z}(D)$ *is at most equal to the exponent of* $\mathfrak{Z}(\mathfrak{d})$.

(3E) *If* b *in* (3C) *has defect* 0, *then* b^G *has defect* 0.

(3F) *Take* $H = 1$ *and let* b *denote the unique block of* H. *If* $|G| \equiv 0 \pmod p$, b^G *is not defined.*

Indeed, if b^G was defined, it would have defect 0. It would then consist of a single irreducible character χ. Now (3A) leads to a contradiction since ψ^* here is the regular character of G.

We conclude this section with a discussion of a special case of a theorem of J. A. Green [7]. We give here another proof, with the aim of making our work self-contained.

(3G) *Let* B *be a block of* G *with the defect group* D. *Let* χ *be an irreducible character in* B *and let* σ *be an element of* G. *If* $\nu\chi(\sigma) = \alpha$, *there exists a conjugate* D^τ *of* D *which intersects* $\mathfrak{C}(\sigma)$ *in a group of order* p^μ *with* $\mu \geqq \nu|\mathfrak{C}(\sigma)| - \alpha$.

Proof. We use induction on $|G|$. If $\nu|\mathfrak{C}(\sigma)| \leqq \alpha$, the result is obvious. Assume $\nu|\mathfrak{C}(\sigma)| > \alpha$.

(α). Assume first that σ is p-regular and that $|\mathfrak{Z}(G)| \equiv 0 \pmod p$. Since $\nu\chi(\sigma) = \alpha$, there exists a modular irreducible character $\phi \in B$ with $\nu\phi(\sigma) \leqq \alpha$. Let π be an element of order p in $\mathfrak{Z}(G)$. Then ϕ can be considered as a character of $\bar{G} = G/\langle\pi\rangle$ belonging to a block $\bar{B}$ with the defect group $D/\langle\pi\rangle$, cf. [1II, (2G)]. Set $\bar{\sigma} = \sigma\langle\pi\rangle \in \bar{G}$. There must exist ordinary irreducible characters $\chi_1 \in \bar{B}$ such that $\nu\chi_1(\bar{\sigma}) \leqq \alpha$. Since σ is p-regular, the centralizer of $\bar{\sigma}$ in $\bar{G}$ is $\mathfrak{C}_G(\sigma)/\langle\pi\rangle$. Using induction and applying (3G) for $\bar{G}$, χ_1, and $\bar{\sigma}$, we obtain easily (3G) for G, χ, and σ.

(β). Assume next that σ is p-singular. Let π be the p-factor of σ and set $\sigma = \pi\rho$ with p-regular $\rho \in \mathfrak{C}_G(\pi)$. It follows from (2B) that there exists a modular irreducible character ϕ^π of $\mathfrak{C}(\pi)$ with $\nu\phi^\pi(\rho) \leqq \alpha$ such that ϕ^π belongs to a block b with $b^G = B$. Then b also contains an ordinary irreducible character ψ with $\nu\psi(\rho) \leqq \alpha$. The defect group of b is conjugate to a subgroup of D. Applying the result of (α) to $\mathfrak{C}(\pi)$, ψ and ρ, we obtain (3G).

(γ). Suppose then that σ is p-regular. Let P of order p^λ be an S_p-group of $\mathfrak{C}(\sigma)$; $\lambda = \nu\,|\,\mathfrak{C}(\sigma)\,| > \alpha$. The orthogonality relations for the characters of P show that, in the ring of algebraic integers,

$$\sum_\pi \chi(\sigma\pi) \equiv 0 \pmod{p^\lambda}$$

where π ranges over P. If $\{\pi_i\}$ is a set of representatives for the classes of P, this can be written as

$$\sum_i |\,P: \mathfrak{C}_P(\sigma\pi_i)\,|\,\chi(\sigma\pi_i) \equiv 0 \pmod{p^\lambda}.$$

It follows from the hypothesis of (3G) that, for some $\pi_i \neq 1$,

$$\nu\chi(\sigma\pi_i) \leqq \alpha - \lambda + \nu\,|\,\mathfrak{C}_P(\sigma\pi_i)\,|.$$

As shown in (β) there exists a conjugate D^τ of D which meets $\mathfrak{C}_G(\sigma\pi_i)$ in a subgroup of order p^μ with

$$\mu \geqq \nu\,|\,\mathfrak{C}_G(\sigma\pi_i)\,| - \nu\chi(\sigma\pi_i) \geqq \lambda - \alpha.$$

Since $\mathfrak{C}_G(\sigma) \supseteq \mathfrak{C}_G(\sigma\pi_i)$ and since $\lambda = \nu\,|\,\mathfrak{C}_G(\sigma)\,|$, this implies (3G).

4. Groups with a normal subgroup. We assume in § 4 that H is a normal subgroup of G.

Definition 1. A block B of G covers a block b of H, if there exist irreducible characters $\chi \in B$ and $\psi \in b$ such that ψ is a constituent of $\chi\,|\,H$.

The following statement is a consequence of Clifford's theorem, [2I, Lemma 1], and [2II, Lemma 1].

(4A) *The blocks of H covered by the block B of G form a family F of blocks conjugate in G. If $b \in F$, for every $\chi \in B$, some constituent of $\chi\,|\,H$ belongs to b. If $\psi \in b$, there exists a $\chi \in B$ such that ψ is a constituent of $\chi\,|\,H$.*

Definition 2. A block B of G is *regular* with regard to H, if $\omega_B(\mathcal{S}K) \equiv 0$ for all conjugate classes K of G which do not meet H. The block B is *weakly regular*, if there exists a conjugate class K of G such that $K \subseteq H$ and that

$$\omega_B(\mathcal{S}K) \not\equiv 0, \qquad d(K) = d(B). \tag{4.1}$$

Clearly, regularity implies weak regularity. We note

(4B) *Let B be a block of G with the defect group D. If $\mathfrak{C}_G(D) \subseteq H$, then B is regular with regard to H.*

Indeed, if $\nu\omega_B(\mathcal{S}K) = 0$ for a conjugate class K of G, by § 2, 4 then $D \subseteq \mathfrak{C}_G(\sigma)$ for some $\sigma \in K$ and this implies $\sigma \in \mathfrak{C}_G(D) \subseteq H$ and hence $K \subseteq H$.

Weak regularity has been studied by P. Fong [6, § 1].

(4C) (*Fong*). *Let b be a block of H. There exist blocks B of G which are weakly regular with regard to H and which cover b. All these blocks B have the same defect groups. The defect group D of B can be chosen such that $\mathfrak{d} = D \cap H$ is a defect group of b. Finally, if B_1 is any block of G which covers b and is not weakly regular, the defect group of B_1 can be chosen as proper subgroup of D.*

We shall derive some further results. We begin with some obvious remarks. If χ is an irreducible character of G, by Clifford's theorem,

$$\chi \mid H = e(\psi_1 + \psi_2 + \cdots + \psi_r) \tag{4.2}$$

where $\{\psi_1, \cdots, \psi_r)$ is a family of irreducible characters of H conjugate in G, and where e is a positive integer. In particular,

$$\chi(1) = er\psi_1(1).$$

The number r is the index

$$r = \mid G : T(\psi) \mid \tag{4.3}$$

of the inertial group $T(\psi)$ of $\psi = \psi_1$ in G.

On the other hand, if K is a conjugate class of G which meets H, then K is a disjoint union

$$K = L_1 \cup L_2 \cup \cdots \cup L_s$$

of conjugate classes L_j of H. Here, s is the index $\mid G : \mathfrak{N}_G(L_1) \mid$ of the inertial group $\mathfrak{N}_G(L_1)$ of L_1 and each L_j is a G-conjugate $L_1{}^\tau$ of L_1, $\tau \in G$. Since all $\mid L_j \mid$ are equal, we have

$$\mid K \mid = s \mid L_j \mid ; \ \mid G : H \mid = s \mid \mathfrak{C}_G(\sigma_j) : \mathfrak{C}_H(\sigma_j) \mid \tag{4.4}$$

where σ_j is an element of L_j, $\sigma_1 = \sigma$. Then

$$\sum_{x \in G} \psi(x\sigma x^{-1}) = (g/r) \sum_{i=1}^{r} \psi_i(\sigma) = (g/s) \sum_{j=1}^{s} \psi(\sigma_j). \tag{4.5}$$

It follows easily from the preceding equations that

$$\omega_\chi(\mathcal{S}K) = (s/r) \sum_{i=1}^{r} \omega_{\psi_i}(\sigma) = \sum_{j=1}^{s} \omega_\psi(\sigma_j) = \omega_\psi{}^G(\mathcal{S}K). \tag{4.6}$$

We now show

(4D) *If B is a block of G, there exist blocks b of H with $b^G = B$, if and only if B is regular with regard to H. If this is so, B covers b.*

Proof. If $B = b^G$, then $\omega_B \equiv \omega_b{}^G$ and hence $\omega_B(\mathcal{S}K) \equiv 0$ for conjugate classes K with $K \cap H = \emptyset$. Conversely, assume that B is regular with regard to H. Let b be a block of H covered by B. If χ is an irreducible character in B and ψ an irreducible constituent of $\chi \mid H$, it follows from (4.6) that

$$(4.7) \qquad \omega_B(\mathcal{S}K) \equiv \omega_b{}^G(\mathcal{S}K)$$

for all conjugate classes K with $K \cap H = \emptyset$. If $K \cap H = \emptyset$, the right side of (4.7) is 0, while the left side is congruent 0 because of the regularity. Hence (4.7) holds for all conjugate classes K of G and we have $B = b^G$. The last part follows from (3A). We mention two corollaries.

(4E) *For each block b of H there exists at most one block B of G which is regular with regard to H and covers b.*

(4F) *Suppose that $H \lhd G$ and that $\mathfrak{D}$ is a normal p-subgroup of G such that $\mathfrak{C}_G(\mathfrak{D}) \subseteq H$. Then every block B of G is regular with regard to H. For every block b of H, the block b^G is defined and b^G is the only block of G which covers b.*

Indeed, the defect group D of any block B of G contains $\mathfrak{D}$, (§2,**6**). Then $\mathfrak{C}_G(D) \subseteq \mathfrak{C}_G(\mathfrak{D}) \subseteq H$ and, by (4B), B is regular with regard to H. If B is a block of G covering b, by (4D) there exist a block b_0 of H for which $B = b_0{}^G$. By (4A), b and b_0 are conjugate in G and then $B = b^G$.

Our formulas can also be used to supplement (4C).

(4G) *Let B be a block of G which is weakly regular with regard to H. Let b be a block of H covered by B and let ψ be an irreducible character of b. If D is a defect group of B, then*

$$(4.8) \qquad \nu \mid T(\psi) : H \mid \leqq \nu \mid HD : H \mid \leqq d(B) - d(b).^{8}$$

Proof. Choose the conjugate class K of G such that $K \subseteq H$ and that (4.1) holds. Then (4.6) shows that

$$(4.9) \qquad \nu(s) \leqq \nu(r) = \nu \mid G : T(\psi) \mid .$$

Moreover there exist elements $\sigma_j \in H \cap K$ with $\omega_\psi(\sigma_j) \not\equiv 0$. By §2,**4**, $\mathfrak{C}_H(\sigma_j)$ contains a defect group $\mathfrak{d}$ of b. Let $P \supseteq \mathfrak{d}$ be an S_p-group of $\mathfrak{C}_H(\sigma_j)$ and let $D_1 \supseteq P$ be an S_p-group of $\mathfrak{C}_G(\sigma_j)$. Since $\sigma_j \in K$, it follows from (4.1) that D_1 is a defect group of B. By (4.4)

$$\nu \mid G : D \mid = \nu(s) + \nu \mid H : P \mid .$$

Since $P = H \cap D_1$ and since D_1 and D are conjugate, then

[8] As shown by Fong [6], if $\chi \in B$ has height 0 and if $\psi \in b$ is an irreducible constituent of $\chi \mid H$, then $T(\psi)$ contains a defect group D_1 of B.

$$\nu \mid G \mid = \nu(s) + \nu \mid D_1 H \mid = \nu(s) + \nu \mid DH \mid.$$

Now (4.9) yields the left part of (4.8) while the right part is obtained from $\nu \mid D \mid = d(B)$, $\nu \mid P \mid \geqq d(b)$.

(4H) *Let b be a block of H with the defect group $\mathfrak{d}$, let ψ be an irreducible character of b of height* 0 *and assume that the inertial group $T(\psi)$ of ψ in G contains an S_p-group of the inertial group $T(b)$ of b in G. Then there exist p-regular elements $\sigma \in H$ such that*

$$\nu \mid \mathfrak{C}_G(\sigma) \mid \leqq \nu \mid T(\psi) : H \mid + d(b). \tag{4.10}$$

If the equality sign holds in (4.10) *for some p-regular $\sigma \in H$, then $\mathfrak{C}_G(\sigma)$ contains a defect group D of each block B covering b, and if B is weakly regular the equality sign holds in* (4.8).

Remark. The element σ in (4H) can be chosen such that, in addition, $\mathfrak{d}$ contains an S_p-group of $\mathfrak{C}_H(\sigma)$.

Proof. Let $\psi_1 = \psi, \psi_2, \cdots, \psi_r$ denote the characters of H conjugate to ψ in G. We first show that there exist p-regular elements $\sigma \in H$ for which

$$\sum_{i=1}^{r} \psi_i(\sigma) \not\equiv 0. \tag{4.11}$$

Suppose that this is not true. If we express the ψ_i by the modular irreducible characters ζ_j of H, we see that, for all p-regular $\sigma \in H$

$$\sum_i \sum_j d_{ij}\zeta_j(\sigma) \equiv 0$$

where the d_{ij} are the decomposition numbers of H. This implies that

$$\sum_i d_{ij} \equiv 0 \pmod p \tag{4.12}$$

for each j. This remains true, if we let i range over the values for which ψ_i belongs to b. The number m of such ψ_i is the index of $T(\psi)$ in the inertial group of b. If we multiply (4.12) with $\zeta_j(1)$ and add over all modular irreducible characters ζ_j in b, then as $\nu\zeta_j(1) \geqq \nu \mid H \mid - d(b)$, we find

$$\nu(m) + \nu\psi(1) > \nu \mid H \mid - d(b).$$

Now, by assumption, $\nu(m) = 0$ and ψ has height 0; we have a contradiction.

This shows that there exist p-regular $\sigma \in H$ for which (4.11) is true. Replacing σ by a conjugate, we may assume that $\psi(\sigma) \not\equiv 0$ and that, by (3G), $\mathfrak{d}$ contains an S_p-group of $\mathfrak{C}_H(\sigma)$.

Let B be a block of G covering b and choose an irreducible character $\chi \in B$ for which $\chi \mid H$ has ψ as a constituent, cf. (4A). It follows from (4.2), (4.3), and (4.11) that

$$\nu\omega_\chi(\sigma) = \nu \mid G : \mathfrak{C}_G(\sigma) \mid - \nu \mid G : T(\psi) \mid - \nu\psi(1).$$

Since ψ has height 0, $\nu\psi(1) = \nu \mid H \mid - d(b)$,

$$\nu\omega_\chi(\sigma) = \nu \mid T(\psi) : H \mid - \nu \mid \mathfrak{C}_G(\sigma) \mid + d(b).$$

As $\nu\omega_\chi(\sigma) \geqq 0$, we obtain (4.10). If the equality sign holds, $\mathfrak{C}_G(\sigma)$ contains a defect group of B *and* $d(B) \leqq \nu \mid \mathfrak{C}_G(\sigma) \mid$. If B is weakly regular with regard to H, the equality sign holds in (4.8).

5. Construction of blocks of G by means of blocks of centralizers of defect groups. We consider again an arbitrary finite group G. In this section, D will be a p-subgroup of G. Let H be a normal subgroup of $\mathfrak{N}(D)$ with $H \supseteq \mathfrak{C}(D)$. In particular, we are interested in the cases $H = \mathfrak{C}(D)$ and $H = D\mathfrak{C}(D)$.

We begin our discussion with some preliminary remarks. Since $D \lhd H$, the defect group $\tilde{D}$ of each block $\tilde{B}$ of $N = \mathfrak{N}(D)$ contains D, (§2, **6**) and $\tilde{B}$ contains irreducible characters $\tilde{\chi}$ of height 0 which are trivial on D. Likewise, the defect group $\mathfrak{d}$ of each block b of H contains $H \cap D$ and b contains irreducible characters ψ of height 0 which are trivial on $\mathfrak{d}$. By (4F), b^N is defined for each block b of H.

We show

(5A) *Let B be a block of G with the defect group D. Let H be a subgroup such that*

$$\mathfrak{C}(D) \subseteq H \lhd \mathfrak{N}(D). \tag{5.1}$$

There exist blocks b of H for which $b^G = B$ and all these b form a family of blocks of H conjugate in $\mathfrak{N}(D)$. Each such b has the defect group $\mathfrak{d} = D \cap H$.

Proof. It follows from (2A) that it will suffice to prove (5A) in the case $G = \mathfrak{N}(D)$. By (4F), then B is regular with regard to H. Now, (4D) and (4A) show that there exist blocks b of H with $b^G = B$ and that all these b are conjugate in $G = \mathfrak{N}(D)$. Since $D \cap H \lhd H$, the p-group $D \cap H$ belongs to the defect group $\mathfrak{d}$ of b. Since b^G has the unique defect group D, we have $\mathfrak{d} \subseteq D$ and as $\mathfrak{d} \subseteq H$, we have $\mathfrak{d} = D \cap H$.

(5B) *If B, D, H are as in* (5A) *and if b is again a block of H with $b^G = B$, then b contains an irreducible character ψ with the following* *properties* (i) ψ *is trivial on* $\mathfrak{d} = H \cap D$. (ii) ψ *has height* 0. (iii) *If* $T = T(\psi)$ *is the inertial group of* ψ *in* $\mathfrak{N}(D)$, *then* $T \supseteq D$ *and*

$$\nu \,|\, T : DH \,| = 0.$$

Proof. With H, the group HD satisfies the conditions (5.1). Then $B^* = b^{HD}$ is defined and we have $(B^*)^G = B$. Since B^* has the defect group D, B^* contains an irreducible character χ^* of height 0 which is trivial on D, (§ 2, **6**). There exist irreducible constituents ψ of $\chi^* \,|\, H$ which lie in b. Then ψ is trivial on $H \cap D = \mathfrak{d}$. By Clifford's theorem $\chi^*(1)$ is divisible by $|\, HD : (HD \cap T) \,|\, \psi(1)$. Now,

$$\nu\chi^*(1) = \nu \,|\, HD \,| - \nu \,|\, D \,|, \qquad \nu\psi(1) \geqq \nu \,|\, H \,| - \nu \,|\, \mathfrak{d} \,|.$$

Since $HD/D \cong H/\mathfrak{d}$, it follows that ψ has height 0 and that $\nu \,|\, HD : (HD \cap T) | = 0$. Since $T \supseteq H$ and since HD/H is a p-group, $HD = HD \cap T$, that is, $D \subseteq T$. Now, (4G) yields

$$\nu \,|\, HD : H \,| \geqq \nu \,|\, T : H \,| = \nu \,|\, T : HD \,| + \nu \,|\, HD : H \,|$$

whence $\nu \,|\, T : HD \,| = 0$. This proves (5B).

We assume that H satisfies (5.1) and the additional condition $H \subseteq D\mathfrak{C}(D)$. If $\mathfrak{d} = H \cap D$ as before, clearly $H = \mathfrak{d}\mathfrak{C}_H(\mathfrak{d})$. Each block b of H with the defect group $\mathfrak{d}$ contains a unique irreducible character ψ which is trivial on $\mathfrak{d}$, the canonical character of b, § 2, **6**. The inertial group $T(\psi)$ of ψ in $\mathfrak{N}(\mathfrak{d})$ must coincide with the inertial group $T(b)$ of b. We show:

(5C) *Let D be a p-subgroup of G. Let H be a normal subgroup of $\mathfrak{N}(D)$ such that $\mathfrak{C}(D) \subseteq H \subseteq \mathfrak{C}(D)D$. There is a one-to-one correspondence between blocks B of G with the defect D and families $\{b\}$ of blocks of H conjugate in $N = \mathfrak{N}(D)$ which satisfy the following conditions*:

(i) *b has the defect group $\mathfrak{d} = D \cap H$.*

(ii) *If $T(b)$ is the inertial group of b in N, then $\nu \,|\, T(b) \,| \leqq \nu \,|\, HD \,|$. This correspondence is defined by $b \to B = b^G$.*

Proof. On account of our previous results, we only have to show that if b satisfies the conditions (i) and (ii), then $\bar{B} = b^N$ has the defect group D. We determine the p-regular element σ as in (4H). Since $D\mathfrak{C}(D)/\mathfrak{C}(D)$ is a p-group, $\sigma \in \mathfrak{C}(D)$. Now in (4.10), we have

$$\nu \,|\, D \,| \leqq \nu \,|\, \mathfrak{C}_N(\sigma) \,| \leqq \nu(T(b) : H) + d(b).$$

On account of (ii),

$$\nu(T(b):H) \leqq \nu(HD:H) = \nu(D:\mathfrak{d}) = \nu|D| - d(b).$$

Thus, the equality sign holds in (4.10) and this implies that $d(\tilde{B}) = \nu|D|$. Since the defect group of $\tilde{B}$ contains D, it is equal to D as we had to show.

If $H = D\mathfrak{C}(D)$, we have $\mathfrak{d} = D$. This case had already been treated in [11, §11]. If $H = \mathfrak{C}(D)$, then $\mathfrak{d} = D \cap \mathfrak{C}(D) = \mathfrak{Z}(D)$.

If H is as in (5c) and if ψ is an irreducible character of H which is trivial on $\mathfrak{d} = D \cap H$, then $\psi \mid \mathfrak{C}(D)$ is an irreducible character of $\mathfrak{C}(D)$ which is trivial on $\mathfrak{Z}(D)$. Conversely, every irreducible character ξ of $\mathfrak{C}(D)$ which is trivial on $\mathfrak{Z}(D)$ can be extended to a unique irreducible character ψ of H which is trivial on $\mathfrak{d}$. We now see that if in (5c), H is replaced by the subgroup $\mathfrak{C}(D)$, a block b of H with $b^G = B$ can be replaced by the block $b^{\#}$ of $\mathfrak{C}(D)$ covered by b. The canonical character of $b^{\#}$ is simply the restriction $\psi^{\#} = \psi \mid \mathfrak{C}(D)$ to $\mathfrak{C}(D)$ of the canonical character ψ of b.

6. Relation between blocks of groups and blocks of subgroups. Let b be a block of a subgroup G_0 of G such that the defect group $\mathfrak{d}$ of b satisfies the condition $\mathfrak{C}_G(\mathfrak{d}) \subseteq G_0$. Then $b^G = B$ is defined, (§2, **7**). We wish to obtain a construction of the defect group D of B. Moreover, if $\mathfrak{b}$ is a block of $\mathfrak{g} = \mathfrak{C}_G(\mathfrak{d})\mathfrak{d}$ with $\mathfrak{b}^{G_0} = b$ and if ψ is the canonical character of b (§2, **6**), we wish to obtain a construction of the canonical character of a block $\mathfrak{B}$ of $D\mathfrak{C}(D)$ with $\mathfrak{B}^G = B$.

(6A) *Let b with the defect group $\mathfrak{d}$ be a block of a subgroup G_0 of G such that $\mathfrak{C}_G(\mathfrak{d}) \subseteq G_0$. Let $\mathfrak{b}$ be a block of $\mathfrak{g} = \mathfrak{C}_G(\mathfrak{d})\mathfrak{d}$ with $\mathfrak{b}^{G_0} = b$. Let $M \supseteq \mathfrak{g}$ be a subgroup of $\mathfrak{N}_G(\mathfrak{d})$ such that $\mathfrak{b}$ is invariant in M. The block $B^* = \mathfrak{b}^M$ of M is defined. Then B^* is the only block of M which covers $\mathfrak{b}$, and $\mathfrak{b}$ is the only block of $\mathfrak{g}$ covered by B^*. If D^* is the defect group of B^*, then $D^* \supseteq \mathfrak{d}$ and*

$$\nu|M:\mathfrak{g}| = \nu|\mathfrak{g}D^*:\mathfrak{g}| = d(B^*) - d(b); \tag{6.1}$$

$$\mathfrak{d} = \mathfrak{g} \cap D^* = G_0 \cap D^*. \tag{6.2}$$

Proof. Since $\mathfrak{d} \lhd M$, we have $\mathfrak{g} \lhd M$. By (4F), $B^* = \mathfrak{b}^M$ is defined and B^* is the only block of M which covers $\mathfrak{b}$. Since $\mathfrak{b}$ is invariant in M, it is the only block of $\mathfrak{g}$ covered by B^*. It is clear that $\mathfrak{d} \subseteq D^*$.

If ψ is the canonical character of $\mathfrak{b}$, then M belongs to the inertial group $T(\psi)$ of ψ in $\mathfrak{N}_G(d)$ and as $d(\mathfrak{b}) = d(b)$, (4G) applied to M and $\mathfrak{g}$ yields

$$\nu|M:\mathfrak{g}| \leqq \nu|\mathfrak{g}D^*:\mathfrak{g}| \leqq d(B^*) - d(b). \tag{6.3}$$

There exist irreducible characters ξ of height 0 in B^* such that ξ is trivial on the normal subgroup $\mathfrak{d}$ of M. Since all irreducible constituents of $\xi \mid \mathfrak{g}$ lie in $\mathfrak{b}$ and are trivial on $\mathfrak{d}$, the only such constituent is the canonical character ψ of $\mathfrak{b}$. Hence $\xi \mid \mathfrak{g} = e\psi$ with integral e and

$$\nu \mid M \mid - d(B^*) = \nu\xi(1) = \nu(e) + \nu \mid \mathfrak{g} \mid - d(b).$$

Now (6.1) is obtained by combining this with (6.3); we must have $\nu(e) = 0$.

We may apply (6.1) with M replaced by $G_0 \cap M$. Then B^* is replaced by $\mathfrak{b}^{G_0 \cap M}$ and $d(B^*)$ by $d(b)$. Hence $\nu \mid G_0 \cap M : \mathfrak{g} \mid = 0$. Since $G_0 \cap D^*$ is a p-subgroup of $G_0 \cap M$, we have $G_0 \cap D^* \subseteq \mathfrak{g}$ and then $G_0 \cap D^* = \mathfrak{g} \cap D^*$. Since $\mathfrak{g}D^*/\mathfrak{g} \cong D^*/\mathfrak{g} \cap D^*$, by (6.1) $\nu \mid \mathfrak{g} \cap D^* \mid = d(b)$. This implies $\mathfrak{g} \cap D^* = \mathfrak{d}$ and hence (6.2).

We note that $\mathfrak{C}(D^*) \subseteq \mathfrak{C}(\mathfrak{d}) \subseteq M$ and that

$$(6.4) \qquad (B^*)^G = (\mathfrak{b}^M)^G = \mathfrak{b}^G = (\mathfrak{b}^{G_0})^G = b^G.$$

Thus, in solving our problem, we may replace G_0 and b by M and B^*. We next remark that without essential restriction, we may take $M = \mathfrak{C}(\mathfrak{d})D^*$.

(6B) *Notation and assumptions as in* (6A). *Set*

$$H = \mathfrak{C}(\mathfrak{d})D^*, \quad \mathfrak{g}^* = \mathfrak{C}(D^*)D^*.$$

If $\mathfrak{b}^*$ is a block of $\mathfrak{g}^*$ with $(\mathfrak{b}^*)^M = B^*$, then $(\mathfrak{b}^*)^H = \mathfrak{b}^H$ and this block has defect group D^*. Moreover,

$$(\mathfrak{b}^*)^G = \mathfrak{b}^G = b^G.$$

Proof. It is clear that $(\mathfrak{b}^*)^H = \tilde{B}$ is defined and has the same defect group D^* as $\mathfrak{b}^*$ and B^*. Let $\mathfrak{b}_1$ be a block of $\mathfrak{g} \lhd H$ which is covered by $\tilde{B}$. As $\tilde{B}$ is regular with regard to $\mathfrak{g}$, then $\tilde{B} = \mathfrak{b}_1{}^H$ and

$$\mathfrak{b}_1{}^M = \tilde{B}^M = (\mathfrak{b}^{*H})^M = \mathfrak{b}^{*M} = B^* = \mathfrak{b}^M.$$

Since $\mathfrak{b}$ is conjugate only to itself in M, by (4A) and (4D) $\mathfrak{b} = \mathfrak{b}_1$ and $\tilde{B} = \mathfrak{b}^H$, as we wished to show.

(6C) *Let b with the defect group $\mathfrak{d}$ be a block of a subgroup G_0 of G such that $\mathfrak{C}_G(\mathfrak{d}) \subseteq G_0$. The following condition is necessary and sufficient in order that $d(b^G) > d(b)$: If $\mathfrak{b}$ is a block of $\mathfrak{g} = \mathfrak{C}(\mathfrak{d})\mathfrak{d}$ with $\mathfrak{b}^{G_0} = b$, there exists a p-subgroup $D^* \neq \mathfrak{d}$ with $\mathfrak{d} \lhd D^*$, $D^* \cap \mathfrak{C}(\mathfrak{d}) \subseteq \mathfrak{d}$ such that $\mathfrak{b}^\sigma = \mathfrak{b}$ for $\sigma \in D^*$. If this is so, we can choose D^* such that there exist blocks $\mathfrak{b}^*$ of $\mathfrak{C}(D^*)D^*$ with a defect group D^* for which $(\mathfrak{b}^*)^H = \mathfrak{b}^H$ with $H = \mathfrak{C}(\mathfrak{d})D^*$.*

Proof. Choose M in (6A) as the inertial group of $\mathfrak{b}$ in $\mathfrak{N}_G(\mathfrak{d})$. If

$\nu\,|\,M:\mathfrak{g}\,|=0$, it follows from (5c) that $\mathfrak{b}^G=b^G$ has the defect group $\mathfrak{d}$. No group D^* with the properties stated in (6c) can exist in this case, since we would have $D^*\subseteq M$ and then $D^*\subseteq\mathfrak{g}$ which is impossible when $D^*\cap\mathfrak{C}(\mathfrak{d})\subseteq\mathfrak{d}\subseteq D^*$.

If $\nu\,|\,M:\mathfrak{g}\,|>0$, by (6.1) $d(B^*)>d(b)$ and hence $d(b^G)>d(b)$. Now (6A) and (6B) show the existence of groups D^* with the required properties.

We now derive a criterion which permits us to decide whether or not two blocks $\mathfrak{b}$ and $\mathfrak{b}^*$ satisfy the conditions of (6c).

(6D) *Let $\mathfrak{d}_1$ and $\mathfrak{d}_2$ be two distinct p-subgroups of G such that*

$$\mathfrak{d}_1\lhd\mathfrak{d}_2,\ \mathfrak{d}_2\cap\mathfrak{C}(\mathfrak{d}_1)\subseteq\mathfrak{d}_1.$$

Let $\mathfrak{b}_i$ be a block of $\mathfrak{g}_i=\mathfrak{C}(\mathfrak{d}_i)\mathfrak{d}_i$ with the defect group $\mathfrak{d}_i$ and with the canonical character ψ_i, $i=1,2$. Assume that $\psi_1{}^\sigma=\psi_1$ for $\sigma\in\mathfrak{d}_2$. Set $H=\mathfrak{C}(\mathfrak{d}_1)\mathfrak{d}_2$. Then $\mathfrak{b}_1{}^H=\mathfrak{b}_2{}^H$, if and only if $\psi_2\,|\,\mathfrak{C}_G(\mathfrak{d}_2)$ occurs in $\psi_1\,|\,\mathfrak{C}_G(\mathfrak{d}_2)$ with a multiplicity not divisible by p.

Proof. (α). It follows from the assumption that $\mathfrak{d}_2\cap\mathfrak{g}_1=\mathfrak{d}_1$. Hence

$$H/\mathfrak{g}_1=\mathfrak{C}(\mathfrak{d}_1)\mathfrak{d}_2/\mathfrak{g}_1\cong\mathfrak{d}_2/\mathfrak{d}_2\cap\mathfrak{g}_1=\mathfrak{d}_2/\mathfrak{d}_1. \tag{6.5}$$

Clearly, $\mathfrak{d}_1\lhd H$. By (6A), applied with $M=H$, the block $B^*=\mathfrak{b}_1{}^H$ is defined and

$$d(B^*)-d(\mathfrak{b}_1)=\nu\,|\,H/\mathfrak{g}_1\,|=\nu\,|\,\mathfrak{d}_2\,|-\nu\,|\,\mathfrak{d}_1\,|. \tag{6.6}$$

Thus, $d(B^*)=\nu\,|\,\mathfrak{d}_2\,|$.

Let $\xi_1,\xi_2,\cdots,\xi_n$ denote the irreducible characters in B^* which are trivial on $\mathfrak{d}_1$. Since $\mathfrak{b}_1$ is the only block of $\mathfrak{g}_1$ covered by B^*, each irreducible constituent of $\xi_j\,|\,\mathfrak{g}_1$ lies in $\mathfrak{b}_1$. Since $\xi_j\,|\,\mathfrak{g}_1$ is trivial on $\mathfrak{d}_1$, the only irreducible constituent of $\xi_j\,|\,\mathfrak{d}_1$ is ψ_1 and we can set

$$\xi_j\,|\,\mathfrak{g}_1=e_j\psi_1 \tag{6.7}$$

with integral e_j. If $\psi_1{}^*$ is the character of H induced by ψ_1, each irreducible constituent ξ of $\psi_1{}^*$ lies in B^* and is trivial on $\mathfrak{d}_1$. Hence ξ is one of the ξ_j and, by Frobenius' reciprocity theorem

$$\psi_1{}^*=\sum_{j=1}^{n}e_j\xi_j. \tag{6.8}$$

Since ψ_1 has defect 0 when considered as character of $\mathfrak{g}_1/\mathfrak{d}_1$, it defines an irreducible modular character $\psi_1{}^0$ of $\mathfrak{g}_1/\mathfrak{d}_1$ and hence of $\mathfrak{g}_1$; $\psi_1(\rho)=\psi_1{}^0(\rho)$

for p-regular $\rho \in \mathfrak{g}_1$. Then ψ_1^0 is the only modular irreducible character in $\mathfrak{b}_1$, (§2, 4).

If ϕ is a modular irreducible character in B^*, each irreducible constituent of $\phi \mid \mathfrak{g}_1$ lies in $\mathfrak{b}_1$. Thus, we can set

$$\phi \mid \mathfrak{g}_1 = m\psi_1^0 \tag{6.9}$$

with integral m. Since (6.5) implies that every p-regular element ρ of H lies in $\mathfrak{g}_1$, it follows that ϕ is the only modular irreducible character in B^*. In particular, there exists a unique block $\bar{B}$ of $H/\mathfrak{d}_1$ whose characters considered as characters of H lie in B^*. In this sense, $\bar{B}$ consists exactly of $\xi_1, \xi_2, \cdots, \xi_n$. Let t_j denote the multiplicity with which ϕ occurs as a modular constituent of ξ_j. Since $t_1, t_2, \cdots, t_n$ then are the decomposition numbers of $\bar{B}$, $\sum t_j^2$ is the unique Cartan invariant c of $\bar{B}$. Here, c is the power of p with the exponent $d(\bar{B})$, [1I, (6c)]. Since $d(\bar{B}) = d(B^*) - \nu \mid \mathfrak{d}_1 \mid$, by (6.6)

$$\sum_{j=1}^{n} t_j^2 = \mid H : \mathfrak{g}_1 \mid.$$

By (6.9), ψ_1^0 appears with the multiplicity mt_j as modular constituent of $\xi_j \mid \mathfrak{g}_1$. By (6.7), $e_j = mt_j$. On comparing degrees in (6.7) and (6.8), we find

$$\sum_{j=1}^{n} e_j^2 = \mid H : \mathfrak{g}_1 \mid.$$

Hence $m = 1$, $t_i = e_i$. This shows that ψ_1^* considered as character of $H/\mathfrak{d}_1$ is the unique chief indecomposable character Φ belonging to $\bar{B}$.

(β). It is seen easily that $\mathfrak{g}_1 \cap \mathfrak{g}_2 = \mathfrak{C}(\mathfrak{d}_2)\mathfrak{d}_1$. Set

$$L = \mathfrak{g}_1 \cap \mathfrak{g}_2 = \mathfrak{C}(\mathfrak{d}_2)\mathfrak{d}_1. \tag{6.10}$$

It follows that $L \lhd \mathfrak{g}_2$ and that

$$H/\mathfrak{g}_1 = \mathfrak{g}_1\mathfrak{g}_2/\mathfrak{g}_1 \cong \mathfrak{g}_2/L. \tag{6.11}$$

Since $\mathfrak{C}(\mathfrak{d}_2) \subseteq L$, the block $\mathfrak{b}_2$ of $\mathfrak{g}_2$ with the defect group $\mathfrak{d}_2$ is regular with regard to L. Set $\psi_2 \mid L = \lambda$. Then λ is trivial on $\mathfrak{d}_1$ and it follows from (6.10) that λ is irreducible. Hence λ belongs to a block b^* of L covered by $\mathfrak{b}_2$. Then

$$(b^*)^H = \mathfrak{b}_2^H.$$

Since $\psi_2 \in \mathfrak{b}_2$ has height 0, we have

$$\nu\lambda(1) = \nu\psi_2(1) = \nu\,|\,\mathfrak{g}_2\,| - \nu\,|\,\mathfrak{d}_2\,|$$

and, by (6.11) and (6.5),

$$(6.12)\qquad \nu\lambda(1) = \nu\,|\,L\,| + \nu\,|\,H\,| - \nu\,|\,\mathfrak{g}_1\,| - \nu\,|\,\mathfrak{d}_2\,| = \nu|\,L\,| - \nu\,|\,\mathfrak{d}_1\,|.$$

Thus, if we consider λ as irreducible character $\dot{\lambda}$ of $L/\mathfrak{d}_1$, then $\dot{\lambda}$ has defect 0. If $\dot{\lambda}^*$ is the character of $H/\mathfrak{d}_1$ induced by $\dot{\lambda}$, by Nakayama's theorem $\dot{\lambda}^*$ is a sum of chief indecomposable characters of $H/\mathfrak{d}_1$, cf. [3, § 26], [10].[4] If Φ occurs with the multiplicity k, we can set

$$\dot{\lambda}^* = k\Phi + \dot{\mu}$$

where all irreducible constituents of $\dot{\mu}$ belong to blocks of $H/\mathfrak{d}_1$ different from $\tilde{B}$. If λ^* is the character of H induced by the character λ of L, then since $\psi_1{}^*$ considered as character of $H/\mathfrak{d}_1$ coincided with Φ,

$$(6.13)\qquad \lambda^* = k\psi_1{}^* + \mu = \sum_{j=1}^{n} ke_j\xi_j + \mu$$

where no irreducible constituent of μ lies in B^*. It now follows from (3A) that

$$\nu\lambda^*(1) \leqq \nu(k) + \nu(\sum_{j=1}^{n} e_j\xi_j(1)) = \nu(k) + \nu\psi_1{}^*(1)$$

where the equality sign holds, if and only if $(b^*)^H \doteq B^*$, that is, if and only if $\mathfrak{b}_2{}^H = B^* = \mathfrak{b}_1{}^H$. Here, on account of (6.12),

$$\nu\lambda^*(1) = \nu\,|\,H : L\,| + \nu\lambda(1) = \nu\,|\,H\,| - \nu\,|\,\mathfrak{d}_1\,|.$$

Since $\psi_1 \in \mathfrak{b}_1$ has height 0, $\nu\psi_1(1) = \nu\,|\,\mathfrak{g}_1\,| - \nu\,|\,\mathfrak{d}_1\,|$ and

$$\nu\psi_1{}^*(1) = \nu\,|\,H\,| - \nu\,|\,\mathfrak{d}_1\,| = \nu\lambda^*(1).$$

Thus, $\nu(k) = 0$ if and only if $\mathfrak{b}_2{}^H = \mathfrak{b}_1{}^H$.

It follows from (6.13) that $\xi_j\,|\,L$ contains λ with the multiplicity ke_j. Since (6.7) yields $\xi_j\,|\,L = e_j\psi_1\,|\,L$, clearly $\psi_1\,|\,L$ contains λ with the multiplicity k. All the irreducible constituents of $\psi_1\,|\,L$ are trivial on $\mathfrak{d}_1$. As shown by (6.10), each of these constitutents remains irreducible when restricted to $\mathfrak{C}(\mathfrak{d}_2)$. Moreover, distinct irreducible constituents of $\psi_1\,|\,L$ remain distinct when restricted to $\mathfrak{C}(\mathfrak{d}_2)$. It is now clear that $\lambda\,|\,\mathfrak{C}(\mathfrak{d}_2) = \psi_2\,|\,\mathfrak{C}(\mathfrak{d}_2)$ appears with the multiplicity k in $\psi_1\,|\,\mathfrak{C}(\mathfrak{d}_2)$. Since $\nu(k) = 0$, if and only if $\mathfrak{b}_1{}^H = \mathfrak{b}_2{}^H$, this completes the proof of (6D).

[4] The chief indecomposable characters can be considered as special ordinary characters which vanish for p-singular elements, [11, (3F)].

Remark. If we set $\psi_i \mid \mathfrak{C}(\mathfrak{d}_i) = \psi_i{}^\#$ for $i = 1, 2$, then $\psi_i{}^\#$ is the canonical character of the unique block $b_i{}^\#$ of $\mathfrak{C}(\mathfrak{d}_i)$ covered by $\mathfrak{b}_i$ in (6D). On the other hand ψ_i is the unique extension of $\psi_i{}^\#$ to a character of $\mathfrak{g}_i$ trivial on $\mathfrak{d}_i$. We have $\mathfrak{b}_1{}^H = \mathfrak{b}_2{}^H$, if and only if $\psi_2{}^\#$ occurs with a multiplicity prime to p in $\psi_1{}^\# \mid \mathfrak{C}(\mathfrak{d}_2)$.

Our results yield a solution of the problem formulated at the beginning of § 6. We require that we know the system Σ of the p-subgroups P of G which contain the defect group $\mathfrak{d}$ of the given block b of G_0. Moreover, we assume that the centralizers $\mathfrak{C}(P)$ of the groups $P \in \Sigma$ are known and also the automorphisms of $\mathfrak{C}(P)$ induced by elements σ of groups $P_1 \in \Sigma$ which normalize P. Indeed, set $\mathfrak{d} = \mathfrak{d}_1$, and let ψ_1 be the canonical character of a block $\mathfrak{b}_1$ of $\mathfrak{g}_1 = \mathfrak{d}_1\mathfrak{C}(\mathfrak{d}_1)$ with $\mathfrak{b}_1{}^{G_0} = b$. Then $\psi_1{}^\# = \psi_1 \mid \mathfrak{C}(\mathfrak{d}_1)$ is the canonical character of the corresponding block $b_1{}^\#$ of $\mathfrak{C}(\mathfrak{d}_1)$. If there does not exist a group $\mathfrak{d}_2 \neq \mathfrak{d}_1$ satisfying the conditions

$$(6.14) \qquad \mathfrak{d}_1 \lhd \mathfrak{d}_2;\ \mathfrak{d}_2 \cap \mathfrak{C}_G(\mathfrak{d}_1) \subseteq \mathfrak{d}_1;\ \psi_1{}^\sigma = \psi_1 \text{ for } \sigma \in \mathfrak{d}_2,$$

then by (6C), $\mathfrak{d}_1$ is the defect group of $\mathfrak{b}_1{}^G = B$ and ψ_1 is the canonical character of the block $\mathfrak{b}_1$ of $\mathfrak{g}_1 = \mathfrak{C}(\mathfrak{d}_1)\mathfrak{d}_1$.

On the other hand, if there exists $\mathfrak{d}_2 \neq \mathfrak{d}_1$ in Σ for which (6.14) holds, then $\mathfrak{d}_2$ can be chosen such that there exist blocks $\mathfrak{b}_2$ of $\mathfrak{g}_2 = \mathfrak{C}(\mathfrak{d}_2)\mathfrak{d}_2$ with the defect group $\mathfrak{d}_2$ for which $\mathfrak{b}_1{}^G = \mathfrak{b}_2{}^G$. By means of (6D), we can decide for which $\mathfrak{d}_2$ this is so, and we can find the canonical character $\psi_2{}^\#$ of a block $b_2{}^\#$ of $\mathfrak{C}(\mathfrak{d}_2)$ covered by a block $\mathfrak{b}_2$ of the kind in question. The canonical character ψ_2 of $\mathfrak{b}_2$ is the unique extension of $\psi_2{}^\#$ to a character of $\mathfrak{g}_2$ trivial on $\mathfrak{d}_2$.

We now repeat the same construction with $\mathfrak{g}_2$ and ψ_2 as above with $\mathfrak{g}_1$ and ψ_1. In this manner we obtain an increasing sequence of distinct p-subgroups.

$$(6.15) \qquad \mathfrak{d}_1 \lhd \mathfrak{d}_2 \lhd \cdots \lhd \mathfrak{d}_m$$

with $\mathfrak{d}_{j+1} \cap \mathfrak{C}(\mathfrak{d}_j) \subseteq \mathfrak{d}_j$ and a sequence of characters

$$(6.16) \qquad \psi_1, \psi_2, \cdots, \psi_m$$

such that ψ_j is the canonical character of a block $\mathfrak{b}_j$ of $\mathfrak{C}(\mathfrak{d}_j)\mathfrak{d}_j$ with the defect group $\mathfrak{d}_j$ for which $\mathfrak{b}_j{}^G = b^G$. Each $\psi_{j+1} \mid \mathfrak{C}(\mathfrak{d}_{j+1})$ occurs with a multiplicity not divisible by p in $\psi_j \mid \mathfrak{C}(\mathfrak{d}_{j+1})$. If (16.14) is chosen such that the sequence cannot be extended further in the same manner, then $D = \mathfrak{d}_m$ is a defect group of $B = b^G$ and $\psi = \psi_m$ is the canonical character of a block $\mathfrak{B} = \mathfrak{b}_m$ of $\mathfrak{C}(D)D$ with $\mathfrak{B}^G = B$.

It is clear that we may choose (16.14) such that $|\mathfrak{d}_{j+1}:\mathfrak{d}_j| = p$ for $j = 1, 2, \cdots$.

7. Corollaries. As a first application of the results of § 7, we have

(7A) *Let b be a block of a subgroup G_0 of G. Let $\mathfrak{d}$ be the defect group and assume that $\mathfrak{C}(\mathfrak{d}) \subseteq G_0$. Then there exist defect groups D of $B = b^G$ with*

$$\mathfrak{Z}(D) \subseteq \mathfrak{Z}(\mathfrak{d}) \subseteq \mathfrak{d} \subseteq D.$$

Proof. If we use the method of § 6, we may assume that $\mathfrak{d} = \mathfrak{d}_1$, $D = \mathfrak{d}_m$ and that $\mathfrak{d}_{j+1} \cap \mathfrak{C}(\mathfrak{d}_j) \subseteq \mathfrak{d}_j$ for $j = 1, 2, \cdots, m-1$. If $m = 1$, the statement is trivial. If $m > 1$, we have

$$\mathfrak{Z}(\mathfrak{d}_{j+1}) = \mathfrak{d}_{j+1} \cap \mathfrak{C}(\mathfrak{d}_{j+1}) = \mathfrak{d}_{j+1} \cap \mathfrak{C}(\mathfrak{d}_j) \cap \mathfrak{C}(\mathfrak{d}_{j+1}) \subseteq \mathfrak{d}_j \cap \mathfrak{C}(\mathfrak{d}_j) = \mathfrak{Z}(\mathfrak{d}_j)$$

and we have

$$(7.1) \qquad \mathfrak{Z}(D) = \mathfrak{Z}(\mathfrak{d}_m) \subseteq \mathfrak{Z}(\mathfrak{d}_{m-1}) \subseteq \cdots \subseteq \mathfrak{Z}(\mathfrak{d}_2) \subseteq \mathfrak{Z}(\mathfrak{d}_1) = \mathfrak{Z}(\mathfrak{d}).$$

Actually, our method shows that there exist normal series

$$\mathfrak{d} = \mathfrak{d}_1 \subset \mathfrak{d}_2 \subset \cdots \subset \mathfrak{d}_m = D$$

for which (7.1) holds and for which $|\mathfrak{d}_{j+1}:\mathfrak{d}_j| = p$.

(7B) *If $D \neq \mathfrak{d}$ in (7A), then $\mathfrak{d}_2$ in (7.1) is non-abelian. Moreover, $\mathfrak{Z}(D) \subset \mathfrak{d}$, $D \cap \mathfrak{N}(\mathfrak{d}) \not\subseteq G_0$.*

Indeed, $\mathfrak{Z}(\mathfrak{d}_2) \subseteq \mathfrak{d}_1 \subset \mathfrak{d}_2$. If $\mathfrak{Z}(\mathfrak{d}_2) = \mathfrak{d}_1$, then $\mathfrak{d}_2 \subseteq \mathfrak{C}(\mathfrak{d}_1) \cap \mathfrak{d}_2 \subseteq \mathfrak{d}_1$, a contradiction. Thus $\mathfrak{Z}(\mathfrak{d}_2) \subset \mathfrak{d}$ and hence $\mathfrak{Z}(D) \subset \mathfrak{d}$. Since $\mathfrak{d}_2 \subseteq D \cap \mathfrak{N}(\mathfrak{d}_1)$ and $\mathfrak{d}_2 \not\subseteq G_0$ by (6A), $D \cap \mathfrak{N}(\mathfrak{d}_1) \not\subseteq G_0$.

We also have the corollary:

(7C) *If D in (7A) is abelian, then b and B have the same defect group D.*

If B is a block of G with the defect group D, if $\mathfrak{b}$ is a block of $\mathfrak{C}(D)D$ with $\mathfrak{b}^G = B$, and if ψ is the canonical character of $\mathfrak{b}$, we can associate the following numerical invariants with B:

$$\gamma(B) = |\mathfrak{C}(D)|, \quad \zeta(B) = |\mathfrak{Z}(D)|, \quad t(B) = \psi(1).$$

(7D) *If B and b are as in (7A), then $\gamma(B) \mid \gamma(b)$; $\zeta(B) \mid \zeta(b)$; $t(B) \leqq t(b)$;*

$$(7.2) \qquad d(B)/\zeta(B) \mid d(b)/\zeta(b).$$

Proof. The first two statements are immediate from (7A). The third statement is also clear, since in (6.16), $\psi_{j+1}(1) \leqq \psi_j(1)$. Finally, in the notation of § 6,

$$\mathfrak{C}(\mathfrak{d}_1)\mathfrak{d}_2/\mathfrak{C}(\mathfrak{d}_1) \cong \mathfrak{g}_2/\mathfrak{g}_2 \cap \mathfrak{C}(\mathfrak{d}_1).$$

The order of the group on the right divides $|\mathfrak{g}_2/\mathfrak{C}(\mathfrak{d}_2)| = |\mathfrak{d}_2|/\mathfrak{Z}(\mathfrak{d}_2)|$ while the order of the group on the left is divisible by $|\mathfrak{g}_1/\mathfrak{C}(\mathfrak{d}_1)| = |\mathfrak{d}_1/\mathfrak{Z}(\mathfrak{d}_1)|$. Hence $|\mathfrak{d}_1 : \mathfrak{Z}(\mathfrak{d}_1)|$ divides $|\mathfrak{d}_2 : \mathfrak{Z}(\mathfrak{d}_2)|$. Similarly, $|\mathfrak{d}_j : \mathfrak{Z}(\mathfrak{d}_j)|$ divides $|\mathfrak{d}_{j+1} : \mathfrak{Z}(\mathfrak{d}_{j+1})|$ and this yields (7.2).

As another application, we obtain Theorem 3 of [2I]:

(7E) *In* (7A), *the block B of G is the principal block B_0 of G, if and only if the block b of G_0 is the principal block b_0 of G_0.*

Proof. If $b = b_0$, then in the notation used above, $\psi_1 = 1$. Since $\psi_m{}^{\#} = \psi_m \mid \mathfrak{C}(\mathfrak{d}_m)$ is a constituent of $\psi_1 \mid \mathfrak{C}(\mathfrak{d}_m)$, then $\psi_m = 1$. This implies $B = b_m{}^G = B_0$.

Conversely, if $B = B_0$, we have $\psi_m = 1$. We have to show that if $\psi_{j+1} = 1$, then $\psi_j = 1$. It will suffice to prove this for $j = 1$. Then in (6D), B^* is the principal block of H and hence it covers the principal block $\mathfrak{b}_0$ of $\mathfrak{g}_1$. Thus, $\mathfrak{b}_1 = \mathfrak{b}_0$ and then $\psi_1 = 1$ as we had to show.

In general, if B is a block of G, there may exist several blocks b of G_0 for which $b^G = B$, and which satisfy the condition of (7A). These b need not even all have the same defect.

HARVARD UNIVERSITY.

REFERENCES.

[1] R. Brauer, "Zur Darstellungstheorie der Gruppen endlicher Ordnung, *Mathematische Zeitschrift*, I, vol. 63 (1956), pp. 406-444; II, vol. 72 (1959), pp. 25-46.

[2] ———, "Some applications of the theory of blocks of characters of finite groups," *Journal of Algebra*, I, vol. 1 (1964), pp. 152-167; II, vol. 1 (1964), pp. 307-334; III, vol. 3 (1966), pp. 225-255.

[3] R. Brauer and C. J. Nesbitt, "On the modular character of groups," *Annals of Mathematics*, vol. 2 (1941), pp. 556-590.

[4] C. W. Curtis and I. Reiner, *Representation theory of finite groups and associative algebras*, Interscience, New York, London, 1962.

[5] E. C. Dade, "On Brauer's second main theorem," *Journal of Algebra*, vol. 2 (1965), pp. 299-311.
[6] P. Fong, "On the character of p-solvable groups," *Transactions of the American Mathematical Society*, vol. 98 (1961), pp. 263-284.
[7] J. A. Green, "On the indecomposable representations of a finite group," *Mathematische Zeitschrift*, vol. 70 (1959), pp. 430-445.
[8] K. Iizuka, "On Brauer's theorem on sections in the theory of group characters," *Mathematische Zeitschrift*, vol. 75 (1961), pp. 299-304.
[9] H. Nagao, "A proof of Brauer's theorem on generalized decomposition numbers," *Nagoya Mathematical Journal*, vol. 22 (1963), pp. 73-77.
[10] T. Nakayama, "Some remarks on regular representations, induced representations, and modular representations," *Annals of Mathematics*, vol. 39 (1938), pp. 361-369.
[11] M. Osima, "Notes on blocks of group characters," *Mathematical Journal of Okayama University*, vol. 4 (1955), pp. 175-188.
[12] A. Rosenberg, "Blocks and centres of group algebras," *Mathematische Zeitschrift*, vol. 76 (1961), pp. 209-216.

J. Math. Soc. Japan
Vol. 20, Nos. 1-2, 1968

On pseudo groups

Dedicated to Professor Shôkichi Iyanaga on his 60th Birthday

By Richard BRAUER

(Received June 14, 1967)

Introduction

An investigation of finite groups by means of group characters frequently leads to an array of numbers which looks like a perfectly reasonable table of characters. Actually there may exist groups G with these characters. On the other hand, it may turn out that no such groups exist.

It seems natural to study such arrays. Since they may be considered as generalizations of the system of group characters of finite groups, we will use the term *pseudo groups* for them. We shall see that they share some of the properties of groups. We require relatively little in our definition of pseudo groups in section 1. A stronger axiom is added in section 3. There are further conditions which one might impose. This is discussed briefly in the last section.

1. Definition of pseudo groups

We consider collections

$$G=\{K, \Xi, v, E\} \tag{1.1}$$

where K is a finite set

$$K=\{k_1, k_2, \cdots, k_r\}, \tag{1.2}$$

Ξ is a set of complex-valued functions

$$\Xi=\{\chi_1, \chi_2, \cdots . \chi_s\} \tag{1.3}$$

defined on K, v is a complex-valued function defined on K, and E is a set of mappings e_n of K into K, one for each rational integer n.

Each finite group G gives rise to a system (1.1), if the following interpretations are used:

(a) K is the set of conjugate classes of G.

(b) Ξ is the set of irreducible characters of G.

(c) If g is the order of G, then for each conjugate class k_j, $gv(k_j)$ is the number of elements in k_j.

(d) If σ is an element of the conjugate class k then $e_n(k)$ is the well defined conjugate class which contains σ^n.

We shall denote by $G(\mathfrak{G})$ the particular system (1.1) obtained in this manner from the given group $\mathfrak{G}$.

In the case of an arbitrary system (1), we shall use an analogous terminology. We call the element k_j the *classes*, even though they need not be sets of elements. The functions χ_i will be termed the *irreducible characters* of G. By a *character* χ of G, we mean a linear combination

$$\chi=\sum_j a_j\chi_j \tag{1.4}$$

with non-negative rational integral a_j.

It will be convenient to write k^n for $e_n(k)$.

If $\tilde{G}=\{\tilde{K}, \tilde{\Xi}, \tilde{v}, \tilde{E}\}$ is a second system, a mapping ζ of $\tilde{K}$ into K will be called an *imbedding* of $\tilde{G}$ into G, if the following conditions are satisfied:

(α) If χ is an irreducible character of G, the composite function $\chi\circ\zeta$ is a character of $\tilde{G}$.

(β) $\zeta\circ\tilde{e}_n=e_n\circ\zeta$.

(γ) If $\zeta(\tilde{k}^n)=\zeta(\tilde{k}^0)$ for some $\tilde{k}\in\tilde{K}$, then $\tilde{k}^n=\tilde{k}^0$.

If $\tilde{\mathfrak{G}}$ is a subgroup of a group $\mathfrak{G}$, we have a natural imbedding of $G(\tilde{\mathfrak{G}})$ into $G(\mathfrak{G})$.

We shall call the system G in (1) a *pseudo-group*, if the following axioms are satisfied:

(I) The number s of irreducible characters χ_i is equal to the number r of classes k_j[1]. The functions $\chi_1, \cdots, \chi_r$ satisfy the "orthogonality relations"

$$\sum_{k\in K} v(k)\chi_i(k)\overline{\chi}_j(k)=\delta_{ij} \tag{1.5}$$

for $i, j=1, 2, \cdots, r$.

(II) The product of two irreducible characters χ_i, χ_j of G is a character of G.

(III) The constant 1 is an irreducible character of G.

(IV) There is a fixed class denoted by 1 such that $k^0=1$ for each $k\in K$.

(V) For each class k, there exists a positive integer m such that the system $G(\mathfrak{Z}_m)$ belonging to a cyclic group $\mathfrak{Z}_m=\langle\sigma\rangle$ of order m have imbeddings ζ into G with $\zeta(\sigma)=k$.

Obviously, m is uniquely determined by k. We call m the *order* of k.

(VI) If k is a class of order m and if m and the integer n are coprime, then $v(k)=v(k^n)$.

It is clear that if $\mathfrak{G}$ is a group, $G(\mathfrak{G})$ is a pseudo-group.

If k is a class of order m, it follows from (V) that m is the first positive

1) It suffices to require $s\geqq r$.

integer for which $k^m=1$. We have $k^1=k$, $(k^q)^n=k^{qn}$. We may have $k^n=k^q$ for $0<n<q<m$.

By $A=A(G)$, we denote the algebra of all complex-valued functions defined on K. For two such *class functions* $\varphi, \psi \in A$, we define an inner product

$$(\varphi, \psi)=\sum_{k \in K} v(k)\varphi(k)\overline{\psi(k)}.$$

Then $\{\chi_1, \cdots, \chi_r\}$ is an orthonormal basis of A, cf. (1.5).

By a *generalized character* of G, we mean a difference of two characters. It follows from (II) that the generalized characters form a ring. An element $\varphi \in A$ is a generalized character, if and only if the product (φ, χ_j) belongs to the ring $\boldsymbol{Z}$ of integers for $j=1, 2, \cdots, r$. The element φ is a character, if and only if all (φ, χ_j) are non-negative elements of $\boldsymbol{Z}$.

Let (χ) denote the $(r\times r)$-matrix $(\chi_i(k_j))$ with i as row-index and j as column index. If V is the diagonal matrix with the entry $v(k_i)$ in the i-th row, then (1.5) can be written in the form

(1.6) $$(\chi)V(\overline{\chi})^T=I\ ^{2)}.$$

It follows that $v(k_i)\neq 0$ for each i. We set

(1.7) $$c(k_i)=v(k_i)^{-1}.$$

In the following, we shall work with the function c rather than with the function v. If several pseudo-groups occur, we shall write c_G instead of c. In particular, $c(1)$ will be called the *order* of G. Since (6) implies

$$(\overline{\chi})^T(\chi)=V^{-1},$$

we have the orthogonality relation of second kind

(1.8) $$\sum_{\chi \in \Xi} \overline{\chi}(k_i)\chi(k_j)=c(k_i)\delta_{ij},$$

$(1\leqq i, j\leqq r)$. As indicated, χ here ranges over all irreducible characters of G.

Consider a fixed class k of order m and let ζ denote the imbedding of the system $G(\mathfrak{Z}_m)$ given by Axiom (V): Clearly, ζ is unique. Let $\{\varepsilon_1, \varepsilon_2, \cdots, \varepsilon_m\}$ denote the m-th roots of unity. If χ is a character of G, the condition (α) for imbeddings implies that $\chi\circ\zeta$ is a character of the cyclic group $\mathfrak{Z}_m$. It follows that

(1.9) $$\chi(k^i)=\sum_{j=0}^{m-1} b_j\varepsilon_j^i$$

where the b_j are non-negative integers which depend on k but not on i. This shows that all $\chi(k^i)$ lie in the field Ω_m of the m-th roots of unity. If σ is an element of the Galois group of Ω_m over the field $\boldsymbol{Q}$ of rational numbers and

2) The transpose of a matrix M is denoted by M^T and $\overline{M}^T$ is the conjugate complex of M^T.

if σ maps a primitive m-th root of unity ε on ε^n with $n \in \boldsymbol{Z}$, $(m, n)=1$, then

(1.10) $$\chi(k)^\sigma = \chi(k^n).$$

In particular,

(1.11) $$\overline{\chi(k)} = \chi(k^{-1}).$$

It is also clear from (1.9) that the *degree* $\chi(1)$ of the character χ is a non-negative rational integer

(1.12) $$\chi(1) = \sum_{j=0}^{m-1} b_j\ ;$$

we have $\chi(1)=0$ if and only if $\chi=0$.

By (1.8), $c(k)$ is an algebraic integer and, if k has order m, then $c(k) \in \Omega_m$. It now follows from (1.10), (1.7) and Axiom VI that $c(k)$ is a rational integer. Clearly, $c(k)>0$. In particular, we have

(1A) *The order of a pseudo-group is a positive rational integer g.*

It follows from (1.9) and (1.12) that, for each class k, we have $|\chi(k)| \leq \chi(1)$. Then (1.8) implies that $c(k) \leq g$.

2. Sub-pseudo-groups

Consider two pseudo-groups

$$G = \{K, \Xi, 1/c, E\}\,, \qquad \tilde{G} = \{\tilde{K}, \tilde{\Xi}, 1/\tilde{c}, \tilde{E}\}$$

where we now expressed the weight-function v by its reciprocal c, cf. (1.7). If ζ is an imbedding of $\tilde{G}$ into G, we say that the pair $(\tilde{G}, \zeta)$ is a *sub-pseudo-group* of G. There may exist distinct imbeddings ζ, ζ_1 of $\tilde{G}$ into G. Then $(\tilde{G}, \zeta)$ and $(\tilde{G}, \zeta_1)$ will be considered as distinct sub-pseudo-groups.

If $(\tilde{G}, \zeta)$ is a sub-pseudo-group of G, and (H, η) a sub-pseudo-group of $\tilde{G}$, then $(H, \zeta\circ\eta)$ is a sub-pseudo-group of G.

If $\mathfrak{G}$ is a group and $\tilde{\mathfrak{G}}$ a subgroup, we have a natural imbedding of $G(\tilde{\mathfrak{G}})$ into $G(\mathfrak{G})$. Obviously, conjugate subgroups define the same sub-pseudo-group.

Let G and $\tilde{G}$ be pseudo-groups and let $(\tilde{G}, \zeta)$ be a sub-pseudo-group of G. If $\theta \in A(G)$ is a class function of G, we can define a class function θ_ζ on H by setting

(2.1) $$\theta_\zeta = \theta \circ \zeta\,.$$

Clearly, the mapping $\theta \rightarrow \theta_\zeta$ is a linear mapping of $A(G)$ into $A(\tilde{G})$. In the case of groups, this is the restriction map and the conjugate map of $A(\tilde{G}) \rightarrow A(G)$ maps each class function $\varphi \in A(\tilde{G})$ on the function $\varphi^\zeta \in A(G)$ "induced" by φ. This carries over to the case of a pseudo-group G and a sub-pseudo-group $(\tilde{G}, \zeta)$. If k is a class of G, we set

(2.2)
$$\varphi^{\zeta}(k)=c(k)\sum_{l\in\zeta^{-1}(k)}\tilde{c}(l)^{-1}\varphi(l)$$

where the sum on the right extends over all classes $l\in\tilde{K}$ of $\tilde{G}$ for which $\zeta(l)=k$. Just as in the group case, one sees that, for $\theta\in A(G)$, $\varphi\in A(\tilde{G})$, we have

(2.3)
$$\varphi^{\zeta}\cdot\theta=(\varphi\cdot\theta_{\zeta})^{\zeta},$$

and the "Frobenius reciprocity"

(2.4)
$$(\varphi^{\zeta},\theta)=(\varphi,\theta_{\zeta}).$$

The usual proof then yields.

(2A) Let G and $\tilde{G}$ be pseudo-groups and let $(\tilde{G},\zeta)$ be a sub-pseudo-group of G. If ψ is a character of $\tilde{G}$, then ψ^{ζ} is a character of G.

If G and $\tilde{G}$ have orders g and $\tilde{g}$ respectively,

(2.5)
$$\tilde{g}\cdot\psi^{\zeta}(1)=g\psi(1).$$

As a corollary, we note

(2B) *The order $\tilde{g}$ of a sub-pseudo-group $(\tilde{G},\zeta)$ of the pseudo group G divides the order g of G.*

Indeed, we see this when we take ψ as the constant 1 in (2.5). We use the ordinary notation $|G:\tilde{G}|$ for $g/\tilde{g}$ and speak of the *index* of $\tilde{G}$ in G. We see in the same manner that for each class k of G,

(2.6)
$$c(k)\sum_{l\in\zeta^{-1}(k)}\tilde{c}(l)^{-1}\in \boldsymbol{Z},$$

where $\boldsymbol{Z}$ denotes the ring of rational integers.

It follows from (2B) and Axiom V that the order m of each class k of a pseudo-group G divides the order g of G. The *exponent* e of G can be defined as the least common multiple of the orders of the classes k of G. Then e divides the order g of G.

The following statement is obvious on account of the results of Section 1.

(2C) *Let G be a pseudo-group of exponent e. The values of the characters of G lie in the field of the e-th roots of unity.*

It will be clear what we mean by an isomorphism of pseudo-groups.

(2D) *Assume that $(\tilde{G},\zeta)$ is a sub-pseudo-group of the pseudo-group G and that G and $\tilde{G}$ have the same order $g=\tilde{g}$. Then ζ defines an isomorphism of $\tilde{G}$ onto G.*

PROOF. If $\tilde{\chi}_i$ is an irreducible character of $\tilde{G}$, then by (2A), $\tilde{\chi}_i^{\zeta}$ is a character of G of the same degree as $\tilde{\chi}_i$. If χ_j is an irreducible constituent of $\tilde{\chi}_i^{\zeta}$, by the Frobenius reciprocity law (2.4), $\tilde{\chi}_i$ is an irreducible constituent of $(\chi_j)_{\zeta}$. It follows that $\tilde{\chi}_i(1)=\chi_j(1)$ and hence that $\tilde{\chi}_i^{\zeta}=\chi_j$, $(\chi_j)_{\zeta}=\chi_i$. Each irreducible character χ_j of G can be obtained in the form $\tilde{\chi}_i^{\zeta}$ and ζ establishes a one-to-one mapping of the set $\tilde{\Xi}$ of irreducible characters of $\tilde{G}$ onto the corresponding set Ξ for G.

The mapping ζ maps the set $\tilde{K}$ of classes of G into the set K of classes of G. If $\zeta(\tilde{K}) \neq K$, choose $k \in K$, $k \notin \zeta(\tilde{K})$. Then

$$\tilde{\chi}_i^{\zeta}(k) = 0$$

for each $\tilde{\chi}_i \in \tilde{\Xi}$. But then $\chi_j(k) = 0$ for each $\chi_j \in \Xi$ and this is certainly false. Since G and G have the same number of classes, ζ is a one-to-one mapping of $\tilde{K}$ onto K. By (1.8), $\tilde{c}(\tilde{k}) = c(\zeta(\tilde{k}))$ for all classes $\tilde{k}$ of $\tilde{G}$. Finally,

$$\zeta \circ \tilde{e}_n = e_n \circ \zeta$$

by the condition (β) in the definition of imbedding. This proves (2D).

(2E) Let $(\tilde{G}, \zeta)$ be a sub-pseudo-group of the pseudo-group G. Assume that ζ maps the set $\tilde{K}$ of classes of $\tilde{G}$ onto the set of classes K of G. Then G and $\tilde{G}$ have the same order and ζ establishes an isomorphism of $\tilde{G}$ onto G.

PROOF. If we use the same notation as before, then (2.6) can be written in the form

$$\sum_{l \in \zeta^{-1}(k)} \tilde{v}(l) = M(k)v(k) \tag{2.7}$$

with $M(k) \in Z$, cf. (1.7). Moreover, $M(k) > 0$.

Applying (1.5) to $\chi_i = \chi_j = 1$, we have

$$\sum_{k \in K} v(k) = 1, \qquad \sum_{l \in \tilde{K}} \tilde{v}(l) = 1,$$

and (2.7) yields

$$M(k) = \sum_{k \in K} M(k)v(k) \leqq \sum_{l \in \tilde{K}} \tilde{v}(l) = 1.$$

Thus, $M(k) = 1$ for all classes. In particular, for $k = 1$, we have $v(1) = \tilde{v}(1)$ and hence $g = \tilde{g}$. Now, (2C) applies and yields the statement.

In the group case, (2E) yields the well known result that if a subgroup $\tilde{G}$ of a group G meets every conjugate class of G, then $G = \tilde{G}$.

3. Artin's Theorem

By a *cyclic subgroup* of a pseudo-group G, we mean a sub-pseudo-group $(\tilde{G}, \zeta)$ with $\tilde{G}$ of the form $G(Z)$, Z a cyclic group.

We can state Artin's theorem in the form

(3A) *Let G be a pseudo-group. Let n be the least common multiple of the numbers $c(k)$ for the classes k of G. If χ is a character of G, then χ can be written in the form*

$$\chi = \frac{1}{n} \sum_j c_j \phi_j \tag{3.1}$$

where each ϕ_j is a character of G induced by a character of a cyclic subgroup *and where the c_j are rational integers.*

The proof does not differ substantially from a proof in the group case and we sketch it only briefly. Let e be the exponent of G and let ε be a primitive e-th root of unity. Set $R=Z[\varepsilon]$. It will suffice to prove that χ can be written in the form (3.1) with $c_j \in R$. If we then express each c_j by a Z-basis $\omega_1=1, \omega_2, \cdots$ of R, the method in [1] yields a representation of χ with coefficients c_j in Z.

Let V_R denote the set of all linear combinations

$$\sum_j b_j \psi_j$$

where each ψ_j is a character of G induced by a character of a cyclic subgroup of G and where the coefficients b_j belong to R. We have to show that $n\chi$ belongs to V_R.

Let h be a fixed class, say of order m, set $\tilde{G}=G(\mathfrak{Z}_m)$ and let ζ denote the imbedding of $\tilde{G}$ into G given by Axiom V which maps the generator σ of $\mathfrak{Z}_m$ on h. Define a class function ψ on $G(\mathfrak{Z}_m)$ by setting

$$\psi(\sigma)=m\,, \qquad \psi(\sigma^j)=0 \qquad \text{for} \quad \sigma^j \neq \sigma\,.$$

Let η_h denote the induced function on G. Then

$$\eta_h(k)=c(k)\sum_{l\in\zeta^{-1}(k)} m^{-1}\psi(l)=\begin{cases} c(h) & \text{for} \quad k=h\,, \\ 0 & \text{for} \quad k\neq h\,. \end{cases}$$

As ψ is a linear combination of characters of $G(\mathfrak{Z}_m)$ with coefficients in R, we have $\eta_h \in V_R$ for each class h of G.

On the other hand, we can set

$$\chi=\sum_{h\in K}\chi(h)c(h)^{-1}\eta_h\,.$$

Since all $nc(h)^{-1}\chi(h)$ lie in R, we have $n\chi \in V_R$ and the proof is complete.

4. Characterization of characters

Our next aim is to establish the results of [1] for pseudo-groups. This seems to require an additional axiom. We begin with some definitions. In them, p will be a fixed prime number. If k is a class of order m of a pseudo-group and if $m=p^a b$ with $(p,b)=1$, we determine u and v in Z such that

$$1=u+v\,, \qquad u\equiv 0 \pmod{p^a}, \qquad v\equiv 0 \pmod{b}.$$

Then

$$\mathfrak{R}(k)=k^u\,, \qquad \mathfrak{S}(k)=k^v$$

are uniquely determined by k. We call $\mathfrak{R}(k)$ the *p-regular factor* and $\mathfrak{S}(k)$ the *p-singular* factor of k. The class k is *p-regular,* if its order is prime to p.

Then $\mathfrak{R}(k)=k$. For each k, $\mathfrak{R}(k)$ is p-regular.

DEFINITION. Let H be a pseudo-group, let p be a prime number. We call H a *p-elementary* pseudo-group with the *base factor* h, if the following conditions are satisfied.

(*a*) The class h is p-regular. If its order is m, the classes

$$1,\, h,\, h^2,\, \cdots,\, h^{m-1}$$

are distinct and they are the only p-regular classes of H.

(*b*) Let $(G(\mathfrak{Z}_m), \zeta)$ be a cyclic subgroup of G such that ζ maps the generator σ of $\mathfrak{Z}_m$ on h, (cf. Axiom V). If ω is an irreducible character of $\mathfrak{Z}_m$ then

$$\psi(k)=\omega\zeta^{-1}(\mathfrak{R}(k)), \qquad (k\in K)$$

is a character of G.

A pseudo-group will be said to be *elementary*, if it is p-elementary for a suitable prime p.

If $\mathfrak{G}$ is a direct product of a p-group and a group of order prime to p, then $G(\mathfrak{G})$ is p-elementary.

Our additional axiom is:

AXIOM (A). Let G be a pseudo-group. Let p be an arbitrary prime. If k is a p-regular class of G, there exists a sub-pseudo group (H, ζ) such that H is p-elementary with the base factor h, that $\zeta(h)=k$, and that the following condition holds: The prime p divides $c_H(h)$ at least with the same exponent with which p divides $c_G(k)$.

In the case of group systems $G(\mathfrak{G})$, $c_G(k)$ is the order of the centralizers $\mathfrak{C}_{\mathfrak{G}}(\tau)$ for elements τ in the conjugate class k. If $\mathfrak{P}$ is a p-Sylow sub-group, if we set $\mathfrak{H}=\langle\tau\rangle\times\mathfrak{P}$ and $H=G(\mathfrak{H})$, we see immediately that the Axiom (A) holds in $G(\mathfrak{G})$.

Now the methods of [1] can be applied. They yield the following two equivalent statements.

(4A) *Let G be a pseudo-group for which Axiom* (A) *is satisfied. The following condition is necessary and sufficient in order that a class function θ on G be a generalized character: If (H, ζ) is an elementary sub-pseudo-group of G, then θ_ζ is a generalized character of H.*

(4B) *Each character χ of G in* (4A) *can be written in the form*

$$\chi=\sum c_i\psi_i$$

where each ψ_i is a character of G induced by a character of an elementary sub-pseudo-group (H, ζ) of G and where $c_i\in \mathbf{Z}$.

A class function θ on G is an irreducible character of G, if and only if in addition to the condition in (4A), the following conditions hold

$$(\theta, \theta)=1\,, \qquad \theta(1)>0\,.$$

5. Additional remarks

We are mainly interested in pseudo-groups $G(\mathfrak{G})$ belonging to groups $\mathfrak{G}$. In this case, a number of further properties of classes and characters are known. In principle, each of them could be introduced as a new axiom. We mention only a few possibilities.

AXIOM B. Let k, h, l be classes of a pseudo-group G of order g. Then

$$a(k, h, l) = gc(k)^{-1}c(h)^{-1}\sum_i \chi_i(k)\chi_i(h)\overline{\chi_i(l)}\chi_i(1)^{-1}$$

(the sum extended over all irreducible characters of G) is a non-negative rational integer.

If this is so, we can define a commutative and associative algebra T whose elements are formal linear combinations

$$\sum_{k \in K} u_k k$$

with coefficients $u_k \in \boldsymbol{Z}$. Multiplication is defined by means of

$$k \cdot h = \sum_{l \in K} a(k, h, l)\, l\,.$$

In the case of group systems $G(\mathfrak{G})$, T will be the class algebra of G.

If the Axiom (B) holds, the usual proof shows that, for each irreducible character χ_i of G and each class k, the number

$$g\, c(k)^{-1}\chi_i(k)\chi_i(1)^{-1}$$

is an algebraic integer. It follows that the degrees of the irreducible characters divide g. It is also clear that $c(k)$ must divide g. The number n in (3A) then is equal to g (as in the usual form of Artin's theorem).

AXIOM (C). Let χ_i be a character of degree x of a pseudo-group G. Let Y be one of the Schur representations of the general linear group $GL(x, \boldsymbol{C})$, and η its character, cf. [2], [3]. Then η applied to χ in the obvious manner yields a character of G.

This can be modified in various ways.

Other possible axioms might describe other known group theoretical statements. For instance, we may require that $c(k)$ is the order of a sub-pseudo-group $(C(k), \zeta)$ of a particular type. Our aim then would be to replace axiom (A) in Section IV. However, it seems that further postulates are needed.

We may also incorporate results of the theory of modular characters as axioms.

As is well known if $\mathfrak{G}$ is a group, it can be seen from the table of characters whether or not $\mathfrak{G}$ is simple. An analogous condition is to define *simplicity* of pseudo-groups.

It will be clear that a great number of questions arise. Many of them seem to be very difficult. Of course, if one could describe all simple pseudo-groups, this would have far reaching effects on the theory of simple groups. In some ways, the algebraic conditions characterizing simple pseudo-groups may seem simpler than the conditions for simple groups. (At least, they might look simpler to a computer.) On the other hand, in discussing pseudo-groups, we cannot use many of the powerful methods of group theory.

It might be of some interest to study simple pseudo-groups in which all the quantities depend in a well-specified manner on a parameter q.

Harvard University

References

[1] R. Brauer and J. Tate, On the characters of finite groups, Ann. of Math., 62 (1955), 1-7.
[2] I. Schur, Ueber eine Klasse von Matrizen, die sich einer gegebenen Matrix zuordnen lassen, Dissertation, Berlin, 1901.
[3] H. Weyl, The classical groups, Princeton University Press, 1946.

ON BLOCKS AND SECTIONS IN FINITE GROUPS, II.

By Richard Brauer.*

1. Introduction. Definitions. Throughout this paper, G will be a finite group and p will be a fixed prime number which divides the order of G. We use the same notation as in the first part [1] of this investigation. In particular, $\chi_1, \chi_2, \cdots, \chi_k$ will denote the irreducible characters of G.

By a *p-section* S of G, we mean the set of elements $\sigma \in G$ whose p-factor σ_p lies in a given conjugate class K of p-elements of G. Thus, G is a disjoint union

$$G = S_1 \cup S_2 \cup \cdots \tag{1.1}$$

of sections. We shall study the values of the characters χ_i on a fixed section S.

If π is a p-element of S, each element σ of S is conjugate to an element $\pi\rho$ where ρ is a p-regular element of the centralizer $\mathfrak{C}(\pi)$ of π. If χ_i belongs to a fixed p-block B of G, we have formulas

$$\chi_i(\sigma) = \chi_i(\pi\rho) = \sum_b \sum_j d_{ij}{}^{(\pi)} \phi_j{}^{(\pi)}(\rho) \tag{1.2}$$

cf. [1, (2B)]. Here, b ranges over the blocks of $\mathfrak{C}(\pi)$ with $b^G = B$, and for each b, $\phi_j{}^{(\pi)}$ ranges over the irreducible modular characters in b. The $d_{ij}{}^{(\pi)}$ are the corresponding decomposition numbers. For fixed π and j, we look upon the $d_{ij}{}^{(\pi)}$ as the entries of a column $d_j{}^{(\pi)}$ where the indices i with $\chi_i \in B$ serve as row indices. If these columns $d_j{}^{(\pi)}$ are known and if we know the characters of $\mathfrak{C}(\pi)$, the values of χ_i on the section S are determined.

We say that the columns $d_j{}^{(\pi)}$ appearing here *belong* to the section S and are *associated* with the block B. Specifically, we say that the columns $d_j{}^{(\pi)}$ with $\phi_j{}^{(\pi)} \in b$ form a *subsection* s of the section S. Since s is determined by the representative $\pi \in S$ and the block b of $\mathfrak{C}(\pi)$, we write $s = (\pi, b)$. This subsection is *associated* with the block $B = b^G$ of G.

In our definition of s, a fixed representative π of S was used. It will be necessary to discuss briefly what happens if instead another p-element

Received April 6, 1967.

* This research was supported by the United States Air Force under contract No. AF 49(638)-1381 monitored by the A. F. Office of Scientific Research of the Air Research and Development Command.

$\pi' \in S$ is used. Then π' is a conjugate $\pi' = \pi^\mu$ with $\mu \in G$. Of course, $\mathfrak{C}(\pi') = \mathfrak{C}(\pi)^\mu$ and if we set

$$(\phi_j^{(\pi)})^\mu(\rho^\mu) = \phi_j^{(\pi)}(\rho)$$

for p-regular $\rho \in \mathfrak{C}(\pi)$, then $(\phi_j^{(\pi)})^\mu$ is a modular irreducible character of $\mathfrak{C}(\pi^\mu)$. Moreover, for the $\phi_j^{(\pi)} \in b$, the corresponding $(\phi_j^{(\pi)})^\mu$ will be the modular irreducible characters of a block b^μ of $\mathfrak{C}(\pi')$. Actually, b^μ depends on π, b and π', but not on the particular choice of μ. It then is natural to identify the subsections (π, b) and (π', b'):

$$s = (\pi, b) = (\pi^\mu, b^\mu) \tag{1.3}$$

for $\mu \in G$. It is clear from (1.2) that $d_{ij}^{(\pi)}$ remains the same if π is replaced by π^μ and $\phi_j^{(\pi)}$ by $(\phi_j^{(\pi)})^\mu$.

Let again π be a p-element in the section S and let $\sigma \in S$ be conjugate to $\pi\rho$ with p-regular $\rho \in \mathfrak{C}(\pi)$. If b is a fixed block of $\mathfrak{C}(\pi)$ and if $s = (\pi, b)$, set

$$\chi_i^{(s)}(\sigma) = \chi_i^{(\pi,b)}(\sigma) = \sum_j d_{ij}^{(\pi)}\phi_j^{(\pi)}(\rho) \tag{1.4}$$

with $\phi_j^{(\pi)}$ ranging over the modular irreducible characters in b. For $\chi_i \in b^G = B$, $\chi_i^{(s)}(\sigma)$ is the inner sum in (1.2) while for $\chi_i \notin b^G$, $\chi_i^{(s)}(\sigma) = 0$. Hence, for $\sigma \in S$ and $i = 1, 2, \cdots, k$

$$\chi_i(\sigma) = \sum_s \chi_i^{(s)}(\sigma) \tag{1.5}$$

where s ranges over all subsections of S. If we set

$$\chi_i^{(s)}(\sigma) = 0 \tag{1.4*}$$

when s is a subsection of S and $\sigma \notin S$, then (1.5) holds for all $\sigma \in G$, if we let s range over all subsections of $S_1, S_2, \cdots$ in (1.1), that is, over all subsections of G.

An investigation of the class functions $\chi_i^{(s)}$ yields a deeper insight into the nature of the irreducible characters χ_i of G. This is the principal theme of the present investigation. In particular, the results are useful in connection with the application of the methods of [3] which requires a detailed discussion of the columns $d_j^{(\pi)}$.

If the p-element π does not belong to a defect group of the block B of G, no block b of $\mathfrak{C}(\pi)$ with $b^G = B$ exists. Accordingly, every $\chi_i \in B$ vanishes on the section S of π. On the other hand, if π belongs to defect groups of B, it is shown in § 3 that there always exist subsections s of S associated with B. If $\chi_i \in B$ has height 0, s can be chosen such that $\chi_i^{(s)}$ does not vanish identically and then χ_i itself does not vanish identically on S.

If $s = (\pi, b)$ is a subsection, we call the defect $d(b)$ of b also the *defect* of s. If s is associated with the block B of G, it is clear that $d(b) \leqq d(B)$. We say that s is a *major subsection*, if $d(b) = d(B)$. In § 4, it is shown that for a given section S and a given block B of G, there exist major subsections s of S associated with B if and only if the p-elements π of S belong to the centers of defect groups of B. We can then characterize the number of these major subsections. While we may also have 'minor' subsections besides the major subsections, in a certain sense, the contribution of the major subsections in (1.5) represents the principal part of $\chi_i(\sigma)$. If s is a major subsection associated with B, then $\chi_i^{(s)}$ does not vanish identically for $\chi_i \in B$. Actually, $\chi_i^{(s)}(\pi) \neq 0$ for p-elements π in S.

If B is a p-block of G, we are interested in the number $k(B)$ of ordinary irreducible characters $\chi_i \in B$, in the number $l(B)$ of modular irreducible characters in B, and in the values possible for the heights h_i of the characters $\chi_i \in B$. Results concerning these quantities are given in §§ 5 and 6. In § 7, the case of an abelian defect group D is discussed. Unfortunately our knowledge of the situation is still far from satisfactory.

Notation. As already remarked, the notation used is the same as in [1]. In particular, ν will be a fixed exponential valuation of a suitable cyclotomic field Ω with $\nu(p) = 1$, [1, § 2, 3]; $\mathfrak{o}$ is the ring of local integers and $\mathfrak{p}$ the prime ideal of $\mathfrak{o}$. For $\alpha, \beta \in \mathfrak{o}$, a congruence $\alpha \equiv \beta$ will mean a congruence modulo $\mathfrak{p}$, unless another module is indicated.

If B is a block the defect of B is denoted by $d(B)$. We shall denote by $k(B)$ the number of ordinary irreducible characters in B and by $l(B)$ the number of modular irreducible characters in B. The height [1, § 2, 4] of an irreducible character $\chi_i \in B$ is given by

(1.6) $$\nu\chi_i(1) = \nu \mid G \mid - d(B) + h_i.$$

2. Lemmas. We prove some simple lemmas which will be used.

(2A) *If B is a p-block of defect $d(B)$ and if χ_i is an irreducible character in B of height h_i, then for each $\sigma \in G$,*

(2.1) $$\nu\chi_i(\sigma) \geqq \nu \mid \mathfrak{C}(\sigma) \mid - d(B) + h_i.$$

If ϕ_j is a modular irreducible character in B, for each p-regular $\rho \in G$

(2.2) $$\nu\phi_j(\rho) \geqq \nu \mid \mathfrak{C}(\rho) \mid - d(B).$$

16

Proof. If ω_i has the usual significance (cf. [1, § 2, 2]) then (2.1) is equivalence with $\nu\omega_i(\sigma) \geqq 0$. This is obvious since $\omega_i(\sigma)$ is an algebraic integer. Moreover, $\phi_j(\rho)$ for p-regular ρ is a linear combination with integral coefficients of the $\chi_i(\rho)$, $\chi_i \in B$. Since $h_i \geqq 0$, (2.2) is a consequence of (2.1).

Define ω_B as in [1, § 2, 4].

(2B) *If B is a p-block and $\sigma \in G$, set*

$$\xi_B(\sigma) = | G \colon \mathfrak{C}(\sigma) |^{-1} \omega_B(\sigma).$$

If $\chi_i \in B$ has height h_i and if $\nu | \mathfrak{C}(\sigma) | \geqq d(B) - h_i$,

$$\chi_i(\sigma) \equiv \xi_B(\sigma)\chi_i(1). \tag{2.3}$$

If ϕ_j is a modular irreducible character in B and if ρ is p-regular with $\nu | \mathfrak{C}(\rho) | \geqq d(B)$,

$$\phi_j(\rho) \equiv \xi_B(\rho)\phi_j(1). \tag{2.4}$$

Indeed, (2.3) is a consequence of the congruence $\omega_i(\sigma) \equiv \omega_B(\sigma)$. Then (2.4) follows in the same way from (2.3) as (2.2) followed from (2.1).

(2C) *Let f be a complex valued function defined on G of the form*

$$f = \sum_{i=1}^{k} u_i \omega_i$$

with $u_i \in \Omega$. Assume that $f(\sigma)$ is a local integer for every p-regular $\sigma \in G$. If B is a p-block and if f_B is the sum of the terms $u_i\omega_i$ with $\chi_i \in B$, then $f_B(\sigma)$ is a local integer for p-regular σ.

Proof. We may consider f and f_B as linear functions defined on the class algebra; if σ belongs to the conjugate class K of G, we set $f(\mathcal{S}K) = f(\sigma)$ in the notation introduced in [1, § 2, 1] and $f_B(\mathcal{S}K) = f_B(\sigma)$. Let η_B denote the idempotent of the class algebra associated with B; [1, § 2, 4]. Take σ as a p-regular element and set

$$\mathcal{S}K \cdot \eta_B = \sum_{j=1}^{k} v_j \cdot \mathcal{S}K_j \tag{2.5}$$

with $v_i \in \Omega$, the v_j are local integers. Choose a representative $\sigma_j \in K_j$, $j = 1, 2, \cdots, k$. Taking ω_i in (2.5), we have

$$\omega_i(\sigma)\omega_i(\eta_B) = \sum_j v_j \omega_i(\sigma_j) = \sum_j v_j | K_j | \chi_i(\sigma_j)/\chi_i(1).$$

Then the orthogonality relations yield

$$|G|v_j = \sum_{i=1}^{k} \bar{\chi}_i(\sigma_j)\chi_i(1)\omega_i(\sigma)\omega_i(\eta_B).$$

Since $\omega_i(\eta_B) = 1$ for $\chi_i \in B$ and $\omega_i(\eta_B) = 0$ for $\chi_i \notin B$, this can be written in the form

$$(2.6) \qquad |G|v_j = |G:\mathfrak{C}(\sigma)| \sum_{\chi_i \in B} \chi_i(\sigma_j)\chi_i(\sigma).$$

Express here the $\chi_i(\sigma)$ by means of the modular irreducible characters $\phi_j \in B$. It follows from the results of [2, § 3] and in particular from [2, (3F)] that v_j in (2.6) vanishes if σ_j is p-singular. Hence it suffices to let K_j in (2.5) range over the p-regular classes. Now the hypothesis of (2C) shows that $f(\mathcal{S}K \cdot \eta_B)$ is a local integer.

On the other hand,

$$f(\mathcal{S}K \cdot \eta_B) = \sum_{i=1}^{k} u_i\omega_i(\mathcal{S}K \cdot \eta_B) = \sum_{i=1}^{k} u_i\omega_i(\mathcal{S}K)\omega_i(\eta_B) = \sum_{\chi_i \in B} u_i\omega_i(\mathcal{S}K) = f_B(\mathcal{S}K)$$
$$= f_B(\sigma)$$

and $f_B(\sigma)$ is a local integer as stated.

(2D) *Let B be a block of G with the defect group D. Let $\mathfrak{b}$ be a block of $\mathfrak{g} = D\mathfrak{C}(D)$ with $\mathfrak{b}^G = B$ and let ψ denote the canonical character in $\mathfrak{b}$; [1, § 2, 6]. If $\rho \in \mathfrak{C}(D)$, then for $\chi_i \in B$ of height h_i*

$$(2.7) \qquad \chi_i(\rho) \equiv A_i \sum_{\tau \in R} \psi(\tau\rho\tau^{-1})$$

where R is a residue system of $N = \mathfrak{N}(D)$ modulo $\mathfrak{g}$ and where A_i is defined by

$$(2.8) \qquad A_i = |G:\mathfrak{g}|^{-1}\chi_i(1)\psi(1)^{-1}.$$

We have $\nu(A_i) = h_i$.

Proof. Since $\psi \in \mathfrak{b}$ has height 0, $\nu(\psi(1)) = \nu|\mathfrak{g}| - d(B)$ and the equation $\nu(A_i) = h_i$ is immediate from (2.8).

Suppose that the conjugate class $\tilde{K}$ of ρ in N consists of the t distinct conjugate classes $k_1, k_2, \cdots, k_t$ of $\mathfrak{g}$. Since each k_j consists of the same number of elements,

$$t = |\tilde{K}|/|k_j| = |N : N \cap \mathfrak{C}(\rho)| \cdot |k_j|^{-1}.$$

Set $\tilde{B} = \mathfrak{b}^N$ and choose an irreducible character $\zeta \in \tilde{B}$. Since $\psi \in \mathfrak{b}$, then for $\sigma_j \in k_j$,

$$\omega_\zeta(\rho) \equiv \omega_\psi{}^N(\rho) \equiv \sum_{j=1}^{t} \omega_\psi(\sigma_j) = |k_j|\,\psi(1)^{-1} \sum_{j=1}^{t} \psi(\sigma_j).$$

If τ ranges over the $|N:\mathfrak{g}|$ elements of the residue system R, the element $\tau\rho\tau^{-1}$ will be in each of the t classes k_j the same number of times. Hence

$$\sum_{j=1}^{t}\psi(\sigma_j)=t\,|N:\mathfrak{g}|^{-1}\sum_{\tau\in R}\psi(\tau\rho\tau^{-1}).$$

On combining these formulas we have

$$\omega_\zeta(\rho)\equiv|\mathfrak{g}|\cdot|N\cap\mathfrak{C}(\rho)|^{-1}\psi(1)^{-1}\sum_{\tau\in R}\psi(\tau\rho\tau^{-1}). \tag{2.9}$$

As $\chi_i\in B=b^G=\tilde{B}^G$ and $\zeta\in\tilde{B}$, $\omega_i(\rho)\equiv\omega_\zeta{}^G(\rho)$.

Assume first that D is an S_p-group of $\mathfrak{C}(\rho)$. If $\omega_\zeta(\rho')\not\equiv 0$ for some $\rho'\in N$, then $\mathfrak{C}_N(\rho')$ contains the unique defect group D of B. If ρ' here is conjugate to ρ in G, then D is also an S_p-group of $\mathfrak{C}(\rho')$. It follows that ρ' is conjugate to ρ in N. Now the definition of $\omega_\zeta{}^G$ (cf. [2II, § 2]) shows that $\omega_\zeta{}^G(\rho)\equiv\omega_\zeta(\rho)$. Expressing $\omega_i(\rho)$ by $\chi_i(\rho)$, we have

$$\chi_i(\rho)\,|G:\mathfrak{C}(\rho)|\,\chi_i(1)^{-1}=\omega_i(\rho)\equiv\omega_\zeta(\rho). \tag{2.10}$$

Since D is an S_p-group of $\mathfrak{C}(\rho)$, $\nu|\mathfrak{C}(\rho)|=d(B)$ and by (1.6)

$$\nu(\chi_i(1))=\nu|G:\mathfrak{C}(\rho)|+h_i.$$

Hence $\chi_i(1)|G:\mathfrak{C}(\rho)|^{-1}$ is a local integer and (2.9) and (2.10) yield

$$\chi_i(\rho)\equiv|G:\mathfrak{g}|^{-1}\chi_i(1)\psi(1)^{-1}|\mathfrak{C}(\rho):N\cap\mathfrak{C}(\rho)|\sum_{\tau\in R}\psi(\tau\rho\tau^{-1}).$$

The product of the first three factors on the right is A_i and hence an algebraic integer. Since D is assumed to be an S_p-group of $\mathfrak{C}(\rho)$, the factor $|\mathfrak{C}(\rho):N\cap\mathfrak{C}(\rho)|$ is the number of S_p-groups of $\mathfrak{C}(\rho)$. Henc this factor can be replaced by 1. We obtain (2.7).

It remains to deal with the case that D is not an S_p-group of $\mathfrak{C}(\rho)$. Since $D\subseteq\mathfrak{C}(\rho)$, there exist S_p-groups P of $\mathfrak{C}(\rho)$ which contain D properly. Then (2.1) shows that $\chi_i(\rho)\equiv 0$. It follows from the properties of p-groups that $P\cap N\supset D$. Then

$$\nu|N\cap\mathfrak{C}(\rho)|>\nu|D|$$

and hence

$$\nu|N\cap\mathfrak{C}(\rho)|>d(B)=\nu|\mathfrak{g}|-\nu(\psi(1)).$$

Now (2.9) shows that

$$\sum_{\tau}\psi(\tau\rho\tau^{-1})\equiv 0$$

and (2.7) holds again.

3. Sections and subsections. Let π be a p-element of G and let $s = (\pi, b)$ be a subsection. If $\mathfrak{d}$ is a defect group of the block b of $\mathfrak{C}(\pi)$, we call $\mathfrak{d}$ a defect group of the subsection s. Since $\langle\pi\rangle$ is normal in $\mathfrak{C}(\pi)$ then $\pi \in \mathfrak{d}$. Let D be the defect group of the block $B = b^G$ of G with which s is associated. Then $\mathfrak{d}$ belongs to conjugates D^τ of D, [2, II (2B)] and we have

$$(3.1) \qquad \pi \in \mathfrak{d} \subseteq D^\tau \cap \mathfrak{C}(\pi).$$

We show a partial converse.

(3A) *Let B be a block of G with the defect group D. Let π be a p-element of G. Consider the family $\mathfrak{D}$ of groups $D^\tau \cap \mathfrak{C}(\pi)$ with $\tau \in G$ which contain π. If $\mathfrak{D}$ is not empty and if $\mathfrak{d}^* = D^\tau \cap \mathfrak{C}(\pi)$ is a maximal member of $\mathfrak{D}$, there exist subsections $s = (\pi, b)$ associated with B with the defect group $\mathfrak{d}^*$.*

Proof. It follows from the definition of defect groups that we can find a p-regular element ρ of G such that D^τ is an S_p-group of $\mathfrak{C}(\rho)$ and that $\omega_B(\rho) \not\equiv 0$. If $\chi_i \in B$ has height 0, then

$$\nu(\chi_i(1)) = \nu \mid G \mid - d(B) = \nu \mid G : \mathfrak{C}(\rho) \mid$$

and hence

$$\chi_i(\rho) = \mid G : \mathfrak{C}(\rho) \mid^{-1} \chi_i(1) \omega_i(\rho) \not\equiv 0.$$

Since $\pi \in D^\tau \subseteq \mathfrak{C}(\rho)$, the element $\pi\rho$ has the p-factor π and

$$\chi_i(\pi\rho) \equiv \chi_i(\rho) \not\equiv 0.$$

By (1.5), there exists a subsection $s = (\pi, b)$ associated with B such that

$$(3.2) \qquad \chi_i^{(s)}(\pi\rho) \not\equiv 0.$$

It follows from (1.4) that we can choose a modular irreducible character $\phi_j^{(\pi)} \in b$ for which

$$d_{ij}^{(\pi)} \not\equiv 0, \qquad \phi_j^{(\pi)}(\rho) \not\equiv 0.$$

Since $\phi_j^{(\pi)}$ can be expressed as a linear combination of the ordinary irreducible characters $\zeta_m \in b$ (restricted to p-regular elements), there exists $\zeta_m \in b$ with $\zeta_m(\rho) \not\equiv 0$. Now [1, (3G)] applied to $\mathfrak{C}(\pi)$ shows that there exist defect groups $\mathfrak{d}$ of b which include an S_p-group P_0 of $\mathfrak{C}(\pi) \cap \mathfrak{C}(\rho) = \mathfrak{C}(\pi\rho)$. Since $\mathfrak{d}^* = D^\tau \cap \mathfrak{C}(\pi)$ is a p-subgroup of $\mathfrak{C}(\pi\rho)$, and since P_0 can be replaced by $\mathfrak{C}(\pi\rho)$-conjugates, we may assume that

$$(3.3) \qquad \mathfrak{d}^* \subseteq P_0 \subseteq \mathfrak{d}.$$

On account of (3.1), $\mathfrak{d}$ is contained in members of $\mathfrak{D}$. On account of the maximality of $\mathfrak{d}^*$ then $\mathfrak{d}^* = \mathfrak{d}$. Thus, the subsection $s = (\pi, b)$ has the defect group $\mathfrak{d}$ and this proves (3A).

We use the same notation and derive a supplementary result. Since $\mathfrak{d}^* = \mathfrak{d}$, it follows from (3.3) that $\mathfrak{d}^*$ is an S_p-group P_0 of $\mathfrak{C}(\pi\rho)$;

$$\nu \mid \mathfrak{C}(\pi\rho) \mid = \nu \mid \mathfrak{d}^* \mid = d(b). \tag{3.4}$$

Apply (2B) to the block b of $\mathfrak{C}(\pi)$. By (2.4)

$$\phi_j^{(\pi)}(\rho) \equiv \xi_b(\rho)\phi_j^{(\pi)}(1) \tag{3.5}$$

for modular irreducible characters $\phi_j^{(\pi)}$ of b with $\xi_b(\rho)$ defined by

$$\xi_b(\rho) = \mid \mathfrak{C}(\pi) : \mathfrak{C}(\pi\rho) \mid^{-1} \omega_b(\rho). \tag{3.6}$$

On substituting (3.5) in (1.4), we have

$$\chi_i^{(s)}(\pi\rho) \equiv \xi_b(\rho) \sum_j d_{ij}^{(\pi)} \phi_j^{(\pi)}(1) \tag{3.7}$$

with $\phi_j^{(\pi)}$ ranging over all modular irreducible characters in b. Since $\nu(\phi_j^{(\pi)}(1)) \geqq \nu \mid \mathfrak{C}(\pi) \mid - d(b)$, it follows from (3.2) and (3.4) that

$$\nu(\xi_b(\rho)) \leqq d(b) - \nu \mid \mathfrak{C}(\pi) \mid = -\nu \mid \mathfrak{C}(\pi) : \mathfrak{C}(\pi\rho) \mid .$$

Now (3.6) shows that the equality sign holds. Since the sum in (3.7) is equal to $\chi_i^{(s)}(\pi)$, (cf. (1.4)), we obtain

(3B) *Let χ_i have height 0 in B. The subsection $s = (\pi, b)$ in* (3A) *can be chosen such that*

$$\nu(\chi_i^{(s)}(\pi)) = \nu \mid \mathfrak{C}(\pi) \mid - d(b). \tag{3.8}$$

In particular, there exist $d_{ij}^{(\pi)}$ with $\phi_j^{(\pi)} \in b$ for which $d_{ij}^{(\pi)} \not\equiv 0$.

The last statement follows from (3.2).

As a corollary, we note

(3C) *If $\chi_i \in B$ has height 0, a p-element π of G belongs to defect groups of B, if and only if χ_i does not vanish identically on the section S of π.*

Indeed, if π belongs to some defect group of B, the family $\mathfrak{D}$ in (3A) is not empty and there exist subsections $s = (\pi, b)$ for which $\chi_i^{(s)}$ does not vanish identically. Since the modular irreducible characters $\phi_j^{(\pi)}$ of $\mathfrak{C}(\pi)$ are linearly independent, (1.4) and (1.5) show that then χ_i cannot vanish identically on S.

On the other hand, if π does not belong to any defect group of B, by

(3.1) no blocks b of $\mathfrak{C}(\pi)$ with $b^G = B$ exist and every $\chi_i \in B$ vanishes on the section S of π.

The class functions $\chi_i^{(s)}$ share some of the properties of the irreducible characters of G. In analogy to the definition of the ω_i, we set

$$\omega_i^{(s)}(\sigma) = |G:\mathfrak{C}(\sigma)|\,\chi_i^{(s)}(\sigma)/\chi_i(1) \tag{3.9}$$

for $\sigma \in G$.

(3D) *If $s = (\pi, b)$ is a subsection of G, then $\omega_i^{(s)}(\sigma)$ is a local integer for all $\sigma \in G$. If χ_i and χ_j belong to the same block B of G, we have*

$$\omega_i^{(s)}(\sigma) \equiv \omega_j^{(s)}(\sigma). \tag{3.10}$$

Proof. It suffices to prove this for elements of the form $\sigma = \pi\rho$ with p-regular $\rho \in \mathfrak{C}(\pi)$. By (1.5) and (3.9),

$$\omega_i(\pi\rho) = \sum_s \omega_i^{(s)}(\pi\rho).$$

where s ranges over the subsections (π, b) of the section S of π. By (1.2) $\omega_i(\pi\rho)$ is a linear combination of the expressions $\omega_\zeta(\rho)$ belonging to the irreducible characters ζ of $\mathfrak{C}(\pi)$. Apply now (2C) to the group $\mathfrak{C}(\pi)$. Since $\omega_i(\pi\rho)$ is a local integer for all p-regular $\rho \in \mathfrak{C}(\pi)$, it follows at once that each $\omega_i^{(s)}(\pi\rho)$ is a local integer.

If χ_i and χ_j lie in the same block of G and if w is a local prime, we can apply a similar argument to the expression

$$(1/w)(\omega_i(\pi\rho) - \omega_j(\pi\rho))$$

which is a local integer for each p-regular $\rho \in \mathfrak{C}(\pi)$. This expression too can be written as a linear combination of the $\omega_\zeta(\rho)$. Applying (2C), we see that

$$(1/w)(\omega_i^{(s)}(\pi\rho) - \omega_j^{(s)}(\pi\rho))$$

is a local integer and this yields the second part of (3D).

The next proposition is essentially a restatement of (3D).

(3E) *Let $s = (\pi, b)$ be a subsection associated with the block B of G. If $\chi_i \in B$ has height h_i then for $\sigma \in G$*

(i) $$\nu(\chi_i^{(s)}(\sigma)) \geqq \nu\,|\mathfrak{C}(\sigma)| - d(B) + h_i.$$

If here the inequality sign holds for one $\chi_i \in B$, it holds for all $\chi_i \in B$.

(ii) $$\nu(\chi_i^{(s)}(\sigma)) \geqq \nu\,|\mathfrak{C}(\sigma)| - d(b).$$

(iii) *If $d(B) \neq d(b)$, the inequality sign holds in* (i).

Proof. The statement (i) is immediate from (3.9) and the first statement of (3D). We have the inequality sign, if and only if $\omega_i^{(s)}(\sigma) \equiv 0$. The second statement of (3D) shows that if this is so for one $\chi_i \in B$, it is so for all $\chi_i \in B$.

In proving (ii), we may assume that $\sigma = \pi\rho$ with p-regular $\rho \in \mathfrak{C}(\pi)$. Then (ii) is obtained from (1.4) in combination with (2.2) applied to $\mathfrak{C}(\pi)$ and b.

If $d(B) \neq d(b)$ and if we choose $\chi_i \in B$ of height 0, (ii) shows that we have inequality in (i). But then we have inequality in (i) for all $\chi_i \in B$.

(3F) *Let $s = (\pi, b)$ be a subsection associated with the block B. If $\nu(\omega_i^{(s)}(\sigma)) = 0$ for some $\sigma \in G$, then $d(B) = d(b)$ and $\mathfrak{C}(\sigma)$ contains a group D which is a defect group of both b and B.*

Proof. The relation $\nu(\omega_i^{(s)}(\sigma)) = 0$ implies that $\chi_i \in B$ and that σ belongs to the same section as π. Moreover, the equality sign holds in (3E)(i). It follows that $d(B) = d(b)$.

We may assume that χ_i has height 0. Then

$$\nu(\chi_i^{(s)}(\sigma)) = \nu\,|\,\mathfrak{C}(\sigma)\,| - d(b).$$

If σ is conjugate to $\pi\rho$ where π is a p-element and $\rho \in \mathfrak{C}(\pi)$ is p-regular, it follows from (1.4) that there exists a modular irreducible character $\phi_j^{(\pi)} \in b$ for which

$$\nu(\phi_j^{(\pi)}(\rho)) \leqq \nu\,|\,\mathfrak{C}(\pi\rho)\,| - d(b).$$

Again, this implies that there exists an ordinary irreducible character $\zeta \in b$ with

$$\nu(\zeta(\rho)) \leqq \nu\,|\,\mathfrak{C}(\pi\rho)\,| - d(b).$$

But then $\nu(\omega_\zeta(\rho)) = 0$. By [2I, (8A)] applied to $\mathfrak{C}(\pi)$, the centralizer $\mathfrak{C}(\pi\rho)$ of ρ in $\mathfrak{C}(\pi)$ contains a defect group D of b. Since $d(B) = d(b)$, D then is a defect group of $B = b^G$.

We can also prove certain orthogonality relations for the functions $\chi_i^{(s)}$. For instance, if s and s' are two subsections, if α and β are two elements of G, we have

$$\sum_{\chi_i \in B} \chi_i^{(s)}(\alpha)\chi_i^{(s')}(\beta^{-1}) = 0$$

unless the following conditions are satisfied

(i) $s = s'$; (ii) s is associated with B; (iii) α, β belong to the section S to which s belongs.

This is an immediate consequence of the orthogonality relations for the decomposition numbers, [2II, (7B)].

We conclude this section with a proposition which describes the relation between subsections with the same defect group $\mathfrak{d}$ and associated with a given block B of G.

(3G) *Let $s_1 = (\pi, b_1)$ be a subsection with the* same *defect group* $\mathfrak{d}$ as $B = b_1^G$. *Set* [R.B.]

$$\mathfrak{g} = \mathfrak{d}\mathfrak{C}(\mathfrak{d}), \qquad C = \mathfrak{C}(\pi), \qquad N = \mathfrak{N}(\mathfrak{d}).$$

There exists a block $\mathfrak{b}_1$ of $\mathfrak{g}$ such that $\mathfrak{b}_1{}^C = b_1$. Consider the class $\mathfrak{B}$ of conjugates of $\mathfrak{b}_1$ in N and let $\mathfrak{b}_1, \mathfrak{b}_2, \cdots, \mathfrak{b}_n$ be a set of representatives for the subclasses of $\mathfrak{B}$ of blocks conjugate in $N \cap C$. Set $\mathfrak{b}_j{}^C = b_j$, $s_j = (\pi, b_j)$ for $j = 1, 2, \cdots, n$. Then $s_1, s_2, \cdots, s_n$ are the distinct subsections of the section of π with the defect group $\mathfrak{d}$ associated with $B = b_1{}^G$.

Proof. By (3.1) $\pi \in \mathfrak{d} \subseteq C$ and then $\mathfrak{g} \subseteq C$. If b_j is a block of C with the defect group $\mathfrak{d}$, by [1, (5A)] there exists a block $\mathfrak{b}_j$ of $\mathfrak{g}$ with $\mathfrak{b}_j{}^C = b_j$. In particular, we have a block $\mathfrak{b}_1$ of $\mathfrak{g}$ with $\mathfrak{b}_1{}^C = b_1$.

Suppose that $(\pi, b_1) \cdots, (\pi, b_n)$ are the distinct subsections with the defect group $\mathfrak{d}$ associated with $B = b_1{}^G$. Then

$$B = b_j{}^G = (\mathfrak{b}_j{}^C)^G = \mathfrak{b}_j{}^G, \qquad (j = 1, 2, \cdots, n).$$

By [1, (5A)], $\mathfrak{b}_j$ and $\mathfrak{b}_1$ are conjugate in N. Conversely, for any conjugate $\mathfrak{b}^*$ of $\mathfrak{b}_1$ in N, we have $(\mathfrak{b}^*)^G = B$ and if we set $b^* = (\mathfrak{b}^*)^C$ and $s^* = (\pi, b^*)$, s^* is a subsection with the defect group $\mathfrak{d}$ associated with B. Thus, s^* is one of $s_1, \cdots, s_n$. We have $s^* = s_j$, if and only if $(\mathfrak{b}^*)^C = (\mathfrak{b}_j)^C$ and by [1, (5A)] applied to C, this is so, if and only if $\mathfrak{b}_j$ and $\mathfrak{b}^*$ are conjugate in $C \cap N$. Now (3G) is evident.

Remark. In (3G), we can set $\mathfrak{b}_j = \mathfrak{b}_1{}^{\sigma_j}$ with $\sigma_j \in N$; $j = 1, 2, \cdots, n$. If $T = T(\mathfrak{b}_1)$ is the inertial group of $\mathfrak{b}_1$ in N, then $\{\sigma_1, \cdots, \sigma_n\}$ can be constructed as a set of double coset representatives in N (modd $T, N \cap C$).

4. Major subsections. Let $s = (\pi, b)$ a subsection of G associated with the block B of G. We say that s is a *major subsection*, if b and B have the same defect.

(4A) *Let B be a block of G. Let π be a p-element. There exist major subsection $s = (\pi, b)$ associated with B, if and only if π belongs to the center of some defect group of B.*

Indeed, if (π, b) is a major subsection associated with B, a defect group $\mathfrak{d}$ of b is a defect group of B. Since $\mathfrak{d} \subseteq \mathfrak{C}(\pi)$ and $\pi \in \mathfrak{d}$ by (3.1), π belongs to the center $\mathfrak{Z}(\mathfrak{d})$ of $\mathfrak{d}$. Conversely, if π belongs to the center of a defect group D of B, (3A) shows the existence of subsections $s = (\pi, b)$ associated with B with the defect group $D = D \cap \mathfrak{C}(\pi)$. Then s is a major subsection.

(4B) *Let $s = (\pi, b)$ be a major subsection associated with the block B of G and let D be a defect group of b. Set*

$$C = \mathfrak{C}(\pi), \qquad N = \mathfrak{N}(D), \qquad \mathfrak{g} = D\mathfrak{C}(D). \tag{4.1}$$

There exist blocks $\mathfrak{b}$ of $\mathfrak{g}$ for which $\mathfrak{b}^C = b$. If $T = T(\mathfrak{b})$ is the inertial group of $\mathfrak{b}$ in N, then for $\chi_i \in B$, we have

$$\omega_i^{(s)}(\pi) \equiv |T : T \cap C| \not\equiv 0. \tag{4.2}$$

Proof. Let ρ be a p-regular element of $\mathfrak{C}(D)$. Since $\pi \in D$, $\rho \in C$. Suppose that $s^* = (\pi, b^*)$ is any subsection associated with B. If we express the $\phi_j^{(\pi)}$ in (1.4) for p-regular elements by the ordinary irreducible characters ζ_m in b^*, we obtain formulas

$$\chi_i^{(s^*)}(\pi\rho) = \sum_{\zeta_m \in b^*} r_{im}\zeta_m(\rho). \tag{4.3}$$

The centralizer of ρ in C is $\mathfrak{C}(\pi\rho)$. If $\zeta_m(\rho) \not\equiv 0$, by [1, (3G)] applied to C, there exists a defect group D^* of b^* which contains an S_p-group P^* of $\mathfrak{C}(\pi\rho)$. On the other hand, $d(b^*) \leqq d(b^{*G}) = d(B)$. Since $D \subseteq \mathfrak{C}(\pi\rho)$ and $\nu |D| = d(B)$, we have $|D^*| = |D|$. The S_p-groups D^* and D of $\mathfrak{C}(\pi\rho)$ are conjugate in $\mathfrak{C}(\pi\rho) \subseteq \mathfrak{C}(\pi)$. Hence D is also a defect group of b^*.

Let $s_j = (\pi, b_j)$ with $j = 1, 2, \cdots, n$ denote the subsections with the defect group D which are associated with B. We have then shown that $\chi_i^{(s)}(\pi\rho) \equiv 0$ unless s is one of these s_j. Now (1.5) yields

$$\chi_i(\pi\rho) \equiv \sum_{j=1}^{n} \chi_i^{(s_j)}(\pi\rho).$$

Combining this with (4.3) and using $\chi_i(\pi\rho) \equiv \chi_i(\rho)$, we have

$$\chi_i(\rho) \equiv \sum_{j=1}^{n} \sum_{\zeta_m \in b_j} r_{im}\zeta_m(\rho). \tag{4.4}$$

We can use (3G) to construct the blocks $b_1, \cdots, b_n$. In particular, there exists a block $\mathfrak{b}$ of $\mathfrak{g}$ with $\mathfrak{b}^C = b$ and elements $\sigma_1, \cdots, \sigma_n$ in N such that if we set $\mathfrak{b}_j = \mathfrak{b}^{\sigma_j}$, we have $b_j = \mathfrak{b}_j^C$; $j = 1, 2, \cdots, n$. We may take $\mathfrak{b}_1 = \mathfrak{b}$, $\sigma_1 = 1$, $b_1 = b$. Then for $j > 1$, the block $\mathfrak{b}_j$ is not conjugate to $\mathfrak{b}$ in $N \cap C$.

Let ψ be the canonical character in $\mathfrak{b}$. The canonical character in $\mathfrak{b}_j$ is $\psi_j = \psi^{\sigma_j}$; $\psi_j(1) = \psi(1)$.

Apply now (2D) to $\chi_i \in B$ in G and to $\zeta_m \in \mathfrak{b}_j$ in C. If we set

$$(4.5) \qquad A_i = |G:\mathfrak{g}|^{-1}\chi_i(1)\psi(1)^{-1}; \quad A_m{}^* = |C:\mathfrak{g}|^{-1}\zeta_m(1)\psi(1)^{-1}$$

and if R and R^* are residue systems of N and $N \cap C$ respectively modulo $\mathfrak{g}$, we have

$$(4.6) \qquad \chi_i(\rho) \equiv A_i \sum_{\tau \in R} \psi^\tau(\rho); \qquad \zeta_m(\rho) \equiv A_m{}^* \sum_{\lambda \in R^*} \psi_j{}^\lambda(\rho).$$

Substituting this in (4.4), we find

$$(4.7) \qquad A_i \sum_{\tau \in R} \psi^\tau(\rho) \equiv \sum_{j=1}^{n} \sum_{\zeta_m \in \mathfrak{b}_j} r_{im} A_m{}^* \sum_{\lambda \in R^*} \psi_j{}^\lambda(\rho).$$

Each character ψ^τ with $\tau \in N$ can be considered as an irreducible character of defect 0 of $\mathfrak{g}/D$ and hence as an irreducible modular character of $\mathfrak{g}/D$ and of $\mathfrak{g}$. Since (4.7) holds for all p-regular elements ρ of $\mathfrak{C}(D)$, that is, for all p-regular elements of $\mathfrak{g} = D\mathfrak{C}(D)$, the coefficients with which ψ appears on the left and on the right are congruent. The inertial group T of $\mathfrak{b}$ in N is also the inertial group of ψ. Since R contains $|T:\mathfrak{g}|$ elements of T, the coefficient of ψ on the left is $A_i|T:\mathfrak{g}|$. If $\psi_j{}^\lambda = \psi$, then $\mathfrak{b}_j{}^\lambda = \mathfrak{b}$. Hence $\mathfrak{b}_j$ and $\mathfrak{b}$ are conjugate in $N \cap C$. This implies $j = 1$. There are $|T \cap C:\mathfrak{g}|$ elements $\lambda \in R^*$ with $\psi^\lambda = \psi$. We now find

$$(4.8) \qquad A_i|T:\mathfrak{g}| \equiv \sum_{\zeta_m \in \mathfrak{b}} r_{im} A_m{}^* |T \cap C:\mathfrak{g}|.$$

Choose χ_i of height 0. As shown by (2D) then A_i is a local unit. On the other hand, by [1, (5B)], $|T:\mathfrak{g}|$ and then $|T \cap C:\mathfrak{g}|$ are local units. Hence

$$\sum_{\zeta_m \in \mathfrak{b}} r_{im} A_i{}^{-1} A_m{}^* \equiv |T:T \cap C| \not\equiv 0.$$

If we substitute the values of A_i and $A_m{}^*$ from (4.5), this becomes

$$|G:C|\,\chi_i(1)^{-1} \sum_{\zeta_m \in \mathfrak{b}} r_{im}\zeta_m(1) \equiv |T:T \cap C| \not\equiv 0.$$

By (4.3) the sum here is $\chi_i{}^{(s)}(\pi)$. By (3.9) the left side is $\omega_i{}^{(s)}(\pi)$ and we have (4.2) for χ_i of height 0. By (3.10), (4.2) holds for all $\chi_i \in B$.

As a corollary, we have

(4C) *Let $s = (\pi, \mathfrak{b})$ be a major subsection associated with the block B of G. If $\chi_i \in B$ has height h_i, then*

$$(4.9) \qquad \nu(\chi_i{}^{(s)}(\pi)) = \nu|\mathfrak{C}(\pi)| - d(B) + h_i.$$

In particular, $\chi_i^{(s)}(\pi) \neq 0$. *Moreover, there exist modular characters* $\phi_j^{(\pi)} \in B$ *such that* $\nu(d_{ij}^{(\pi)}) \leqq h_i$ *for the corresponding decomposition numbers.*

Proof. (4.9) is a consequence of (4.2), cf. (3.9). Since

$$\nu(\phi_j^{(\pi)}(1)) \geqq \nu\,|\mathfrak{C}(\pi)| - d(b) = \nu\,|\mathfrak{C}(\pi)| - d(B),$$

(4.9) in conjunction with (1.4) yields the last part of (4C).

If $s = (\pi, b)$ is a major subsection associated with B and if we know the Cartan matrix C_b for the block b of $\mathfrak{C}(\pi)$, we have only finitely many possibilities for columns $d_j^{(\pi)}$ which satisfy the orthogonality relations [2II, (7B)] since by (4C) no row consisting of zeros can occur. Actually, (4C) may help to eliminate some possibilities.

If we know which of the possibilities applies and if we know the degrees of the modular characters $\phi_j^{(\pi)}$ in b, then (4.9) determines the height h_i of χ_i. In this manner, we obtain information concerning the irreducible characters of G from a knowledge of the group $\mathfrak{C}(\pi)$. Actually, in this discussion, the modular irreducible characters $\phi_j^{(\pi)}$ in b can be replaced by the members of a basic set for b, cf. [3I, Section V] and these can be determined more easily. Our results are of importance in connection with the methods developed in [3].

For a later application, we state the following formula.

(4D) *Let s be a major subsection and let the notation be as in* (4B). *Let ψ be the canonical character in* $\mathfrak{b}$. *For $\chi_i \in B$ and for p-regular $\rho \in \mathfrak{C}(\pi)$, we have*

$$\chi_i^{(s)}(\pi\rho) \equiv |C:\mathfrak{g}|^{-1}\psi(1)^{-1}\chi_i^{(s)}(\pi)\sum_{\lambda}\psi^{\lambda}(\rho) \tag{4.10}$$

where λ ranges over a residue system R^* of $\mathfrak{N}(D) \cap C \pmod{\mathfrak{g}}$.

Proof. By (4.3) and (4.6) for $j = 1$,

$$\chi_i^{(s)}(\pi\rho) \equiv \sum_{\zeta_m \in b} r_{im}A_m^* \sum_{\lambda \in R^*}\psi^{\lambda}(\rho)$$

On account of (4.8) and (4.2), this can be written in the form

$$\chi_i^{(s)}(\pi\rho) \equiv A_i\,|T: T \cap C|\sum_{\lambda}\psi^{(\lambda)}(\rho) \equiv A_i\omega_i^{(s)}(\pi)\sum_{\lambda}\psi^{\lambda}(\rho).$$

If $\omega_i^{(s)}(\pi)$ is expressed by $\chi_i^{(s)}(\pi)$ by (3.9) and if A_i is substituted from (4.5), we obtain (4.10).

We next discuss the construction of the major subsections s of a given section which are associated with a given block of G. We first determine the defect groups which have to be considered.

(4E) *Let B be a block of G with the defect group D and let π be an element of the center $\mathfrak{Z}(D)$ of D. Let $\pi_1, \pi_2, \cdots, \pi_r$ be a system of representatives for the conjugate classes of $\mathfrak{N}(D)$ which lie in the intersection of $\mathfrak{Z}(D)$ with the conjugate class of π in G. Set $\pi_j = \sigma_j \pi \sigma_j^{-1}$ with $\sigma_j \in G$. Each of the group $D_j = \sigma_j^{-1} D \sigma_j$ with $j = 1, 2, \cdots, r$ occurs as defect group of a major subsection $s_j = (\pi, b)$ associated with B. Conversely, each major subsection (π, b) associated with B has exactly one of the groups $D_1, D_2, \cdots, D_r$ as defect group.*

Proof. The defect group D_j of a major subsection $s = (\pi, b)$ associated with B is a conjugate D^σ of D such that $\pi \in \mathfrak{Z}(D^\sigma)$. Conversely, each such D^σ is a maximal member of $\mathfrak{D}$ in (3A) and hence occurs as defect group of subsections $s = (\pi, b)$ associated with B. The most general defect group of b is a conjugate of D^σ in $\mathfrak{C}(\pi)$. In order to obtain a system $D_1, D_2, \cdots, D_r$ of groups such that each major subsection associated with B has exactly one D_j as defect group, we have to choose a maximal set of groups D^σ such that no two are conjugate in $\mathfrak{C}(\pi)$. This means that we have to let σ range over a system $\{\sigma_1, \cdots, \sigma_r\}$ of double coset representatives of $G(\text{modd}\, \mathfrak{N}(D), \mathfrak{C}(\pi))$. Only double cosets $\mathfrak{N}(D)\sigma_0\mathfrak{C}(\pi)$ have to be taken whose elements σ satisfy the condition $\pi \in \mathfrak{Z}(D^\sigma)$; if this condition is satisfied for one element of one double coset, it holds for all its elements.

The elements of these double cosets are the elements σ for which $\sigma\pi\sigma^{-1} \in \mathfrak{Z}(D)$. In order to obtain a set of representatives $\{\pi_1, \pi_2, \cdots, \pi_r\}$ for conjugate classes of $\mathfrak{N}(D)$ among these elements $\sigma\pi\sigma^{-1}$, we have to let σ range again over a set of double coset representatives $(\text{modd}\, \mathfrak{N}(D), \mathfrak{C}(\pi))$. Now (4E) is evident.

(4F) *Continuation. Let B and s be as in* (4B) *and use the same notation as in* (4B) *and* (4E). *If j is a fixed value with $1 \leqq j \leqq r$, set $\mathfrak{g}^* = \sigma_j^{-1} \mathfrak{g} \sigma_j$, $\mathfrak{b}^* = \mathfrak{b}^{\sigma_j}$, $b^* = (\mathfrak{b}^*)^C$. Then $s^* = (\pi, b^*)$ is a major subsection associated with B. The defect group of s^* is D_j and application of* (3G) *gives all subsections associated with B with the defect group D_j.*

Proof. Since $\pi_j \in \mathfrak{Z}(D)$, we have $\mathfrak{g} \subseteq \mathfrak{C}(\pi_j)$ and hence $\mathfrak{g}^* \subseteq \mathfrak{C}(\pi) = C$. As $\mathfrak{b}^* = \mathfrak{b}^{\sigma_j}$ is a block of $\mathfrak{g}^*$, it is now clear that $b^* = (\mathfrak{b}^*)^C$ is defined as a block of C. We have

$$b^{*G} = (\mathfrak{b}^{*C})^G = \mathfrak{b}^{*G} = \mathfrak{b}^G = B.$$

Hence $s^* = (\pi, b^*)$ is a subsection associated with B. Since $\mathfrak{b}^*$ has the defect group $\sigma_j^{-1} D \sigma_j = D_j$, there exist defect groups of s^* containing D_j. But

$d(b^*) \leqq d(B)$ and hence D_j itself is a defect group of s^*. This is all we had to show.

We conclude this section with some additional remarks. The first one is an immediate consequence of (1.5) and (3E), (ii).

(4G) *Let B be a block of G. Let π be a p-element and let d^* denote the maximal defect of a non-major subsection (π, b) associated with B. If ρ is a p-regular element of $\mathfrak{C}(\pi)$ and if we set $c = \nu \mid \mathfrak{C}(\pi\rho) \mid$, for $\chi_i \in B$,*

$$\chi_i(\pi\rho) \equiv \sum_s \chi_i^{(s)}(\pi\rho) \pmod{p^{c-d^*}} \tag{4.11}$$

where s ranges over all major subsections (π, b) associated with B.

The next remark can be obtained without much difficulty from (1.5), (3F) and (4B).

(4H) *Let B be a block of G with the defect group D and let π be a p-element of G. Let $D_1, D_2, \cdots, D_r$ have the same significance as in* (4E).[1] *Then*

$$\omega_B(\pi) \equiv \sum_{j=1}^{r} \mid \mathfrak{N}(D_j) : \mathfrak{N}(D_j) \cap C \mid. \tag{4.12}$$

Actually, this can be seen more directly without using subsections by expressing ω_B by ω_ψ where ψ has the same significance as before.

5. The contribution of a subsection. Let $s = (\pi, b)$ be a subsection of G. If χ_i, χ_j are two irreducible characters of G, we set

$$m_{ij}^{(s)} = \mid \mathfrak{C}(\pi) \mid^{-1} \sum_\rho \chi_i^{(s)}(\pi\rho) \bar{\chi}_j^{(s)}(\pi\rho) \tag{5.1}$$

where ρ ranges over the p-regular elements of $\mathfrak{C}(\pi)$.[2] We call the expression $m_{ij}^{(s)}$ the *contribution* of s to the inner product (χ_i, χ_j). If we express χ_i and χ_j by (1.4) and use the orthogonality relations for modular characters, we find

$$m_{ij}^{(s)} = \sum_{\alpha,\beta} d_{i\alpha}^{(\pi)} \gamma_{\alpha\beta} \bar{d}_{j\beta}(\pi) \tag{5.2}$$

where $\phi_\alpha^{(\pi)}$, $\phi_\beta^{(\pi)}$ range over the modular irreducible characters of b and where $(\gamma_{\alpha\beta})$ is the inverse of the Cartan matrix C_b of the block b. The following result is now evident.

---

[1] If no conjugate of D contains π in its center, we have to set $r = 0$.

[2] In the case of the p-regular subsection $s = (1, B)$, the expression $m_{ij}^{(s)}$ has already been used in earlier work of C. J. Nesbitt and the author and in [5].

(5A) *The contribution* $m_{ij}^{(s)}$ *of a subsection* $s=(\pi,b)$ *vanishes unless* χ_i, χ_j *both belong to the block* $B=b^G$. *If* π *has order* p^e, *then* $m_{ij}^{(s)}$ *belongs to the field* Ω_e *of the* p^e*-th roots of unity and* $p^{d(b)}m_{ij}^{(s)}$ *is an algebraic integer.*

A similar argument shows that if $s=(\pi,b)$ and $s'=(\pi,b')$ are two different subsections of the same section, then

$$\sum_\rho \chi_i^{(s)}(\pi\rho)\bar{\chi}_j^{(s')}(\pi\rho)=0 \tag{5.3}$$

where ρ again ranges over the p-regular elements of $\mathfrak{C}(\pi)$.

(5B) *Let* B *be a block of* G. *Let* π *range over a set* $\{\pi\}$ *of representatives for the conjugate classes of* p*-elements of* G *and, for each* π, *let* $s=(\pi,b)$ *range over the subsections of the section of* π. *Then*

$$\sum_\pi\sum_s m_{ij}^{(s)}=\delta_{ij}. \tag{5.4}$$

Proof. The orthogonality relation $(\chi_i,\chi_j)=\delta_{ij}$ for group characters can be written in the form

$$\sum_\pi |\mathfrak{C}(\pi)|^{-1}\sum_\rho \chi_i(\pi\rho)\bar{\chi}_j(\pi\rho)=\delta_{ij}$$

where for each $\pi\in\{\pi\}$, ρ ranges over the p-regular elements of $\mathfrak{C}(\pi)$. If (1.5), (5.1) and (5.3) are used, the statement becomes obvious.

For a given subsection $s=(\pi,b)$, we form the matrix

$$M^{(s)}=(m_{ij}^{(s)}) \tag{5.5}$$

where i and j range over the indices for which χ_i and χ_j respectively belong to the blocks $B=b^G$.

(5C) *If* $s=(\pi,b)$ *is a subsection, the matrix* $M^{(s)}$ *of the contributions* (5.5) *is the matrix of a non-negative Hermitian form. Its degree is the number* $k(B)$ *of irreducible characters* $\chi_i\in b^G=B$ *and its rank is the number* $l(b)$ *of modular irreducible characters in* b. *Moreover,* $M^{(s)}$ *is idempotent and its trace is* $l(b)$.

Proof. The first statement is obvious from (5.2). Form the matrix $V=(d_{i\alpha}^{(\pi)})$ with the row index i ranging over the indices for which $\chi_i\in B$ and with the column index α ranging over the indices for which the modular irreducible character $\phi_\alpha^{(\pi)}$ belongs to b. Then (5.2) can be written as

$$M^{(s)}=VC_b^{-1}\bar{V}'$$

where the prime denotes the transpose. The orthogonality relations for the decomposition numbers [2II] yield $\bar{V}'V=C_b$. Hence

$$M^{(s)}V = VC_b^{-1}\bar{V}'V = VC_b^{-1}C_b = V$$

and it is now clear that $M^{(s)2} = M^{(s)}$. This implies that the characteristic roots of $M^{(s)}$ are 0 or 1. Since $M^{(s)}$ is Hermitian, we see that its rank is the number of characteristic roots 1 and hence it is equal to the trace $\operatorname{tr}(M^{(s)})$. Then

$$\operatorname{tr}(M^{(s)}) = \operatorname{tr}(VC_b^{-1}\bar{V}') = \operatorname{tr}(C_b^{-1}\bar{V}'V) = \operatorname{tr}(C_b^{-1}C_b).$$

Since C_b has degree $l(b)$, $C_b^{-1}C_b$ is the unit matrix of degree $l(b)$ and $\operatorname{tr}(M^{(s)}) = l(b)$. The definition of $M^{(s)}$ shows that the degree of $M^{(s)}$ is the number $k(B)$ of characters $\chi_i \in B$.

The following result can be used in order to obtain upper bounds for the number $k(B)$ of irreducible characters χ_i in a block B.

(5D) *Let $s = (\pi, b)$ be a major subsection associated with the block $B = b^G$. Let Q_b be the positive definite quadratic form with integral rational coefficients with the matrix $p^{d(b)}C_b^{-1}$ where C_b is the Cartan matrix of b. If $q(b) \geqq 1$ is the minimal value of Q_b for integral rational values of the variables not all zero, then*

$$q(b)k(B) \leqq p^{d(b)}l(b). \tag{5.6}$$

Proof. On account of (5C), we have

$$\sum_i p^{d(b)}m_{ii}^{(s)} = p^{d(b)}\operatorname{tr}(M^{(s)}) = p^{d(b)}l(b) \tag{5.7}$$

where i ranges over the values for which $\chi_i \in B$. Set $|G| = p^a g_0$ with $\nu|G| = a$ and let Ξ denote the field of the g_0-th roots of unity. With any $\chi_i \in B$, all the algebraic conjugates with regard to Ξ belong again to B. We call these conjugates the *p-conjugates* of χ_i.

Let z be a row of length $l(b)$ in which the entries are the coefficients $d_{i\alpha}^{(\pi)}$ with fixed i and with the column index α ranging over the $l(b)$ values for which $\phi_\alpha^{(\pi)}$ is a modular irreducible character of b. If ϵ is a primitive p^a-th root of unity and if $v = p^{a-1}(p-1)$, we can set

$$z = \sum_{\mu=0}^{v-1} a_\mu \epsilon^\mu$$

where the a_μ are rows with integral rational coefficients. It follows from (5.2) that

$$p^{d(b)}m_{ii}^{(s)} = \sum_{\lambda,\mu=0}^{v-1} \epsilon^{\lambda-\mu} a_\lambda p^{d(b)} \gamma_{\lambda\mu} a_\mu'.$$

Here, $p^{d(b)}(\gamma_{\alpha\beta})$ is the matrix of the quadratic form Q_b. Since the determinant of the symmetric matrix C_b is a power of p and since $p^{d(b)}$ is its largest elementary divisor [2I, § 5], the coefficients of Q_b are rational integers. As C_b belongs to a positive definite quadratic form, so does C_b^{-1}. Hence Q_b is positive definite. If $Q(x,y)$ denotes the corresponding bilinear form, our formula takes the form

$$p^{d(b)}m_{ii}{}^{(s)}=\sum_{\lambda,\mu=0}^{v-1}\epsilon^{\lambda-\mu}Q(a_\lambda,a_\mu).$$

This number is an element of the field $\Xi(\epsilon)$. In the following, $\mathrm{tr}(\xi)$ denotes the trace of an element ξ of $\Xi(\epsilon)$ with regard to the field Ξ. Then $\mathrm{tr}(\epsilon^j)=p^{a-1}(p-1)$ if $j\equiv 0 \pmod{p^a}$; $\mathrm{tr}(\epsilon^j)=-p^{a-1}$ if $j\equiv 0 \pmod{p^{a-1}}$ but $j\not\equiv 0 \pmod{p^a}$; and $\mathrm{tr}(\epsilon^j)=0$ in all other cases. We find

$$\mathrm{tr}(p^{d(b)}m_{ii}{}^{(s)})=p^{a-1}\sum_{j=0}^{p^{a-1}-1}T_j$$

where

$$T_j=-\sum_{\lambda,\mu}Q(a_\lambda,a_\mu)+p\sum_{\lambda}Q(a_\lambda,a_\lambda)$$

with λ and μ ranging over the values congruent to $j \pmod{p^{a-1}}$ with $0\leqq\lambda,\mu<p^{a-1}(p-1)$. This can be written in the form

$$T_j=\sum_{\lambda<\mu}Q(a_\lambda-a_\mu,a_\lambda-a_\mu)+\sum_{\lambda}Q(a_\lambda,a_\lambda). \tag{5.8}$$

Suppose that among the $p-1$ rows a_λ with $\lambda\equiv j \pmod{p^{a-1}}$ and $0\leqq\lambda<p^{a-1}(p-1)$, there appear m distinct rows different from zero with multiplicities $t_1,t_2,\cdots,t_m$ respectively. Suppose further that t_0 of the rows are zero. Then

$$t_0+t_1+\cdots+t_m=p-1.$$

If $a_\lambda\neq 0$, then $Q(a_\lambda,a_\lambda)\geqq q(b)$, since $q(b)$ is the minimum of the quadratic form Q_b. The second sum in (5.8) is at least equal to

$$(t_1+t_2+\cdots+t_m)q(b).$$

If $t_0=0$, this is $(p-1)q(b)$. If $0<t_0<p-1$, the first sum in (5.8) is at least equal to $t_0t_1q(b)\geqq t_0q(b)$ and again $T_j\geqq(p-1)q(b)$. If $t_0=p-1$, all a_λ are zero and $T_j=0$.

By (4c), $z\neq 0$, and there exists at least one j for which not all a_λ vanish. It now follows that

$$\mathrm{tr}(p^{d(b)}m_{ii}{}^{(s)})\geqq p^{a-1}(p-1)q(b).$$

17

Distribute the characters $\chi_i \in B$ into classes of p-conjugates. If χ_i belongs to a class of w p-conjugates, w is a divisor of $p^{a-1}(p-1)$ and each p-conjugate of χ_i appears $p^{a-1}(p-1)/w$ times amongst the $p^{a-1}(p-1)$ conjugates of χ_i with regard to Ξ. Thus,

$$\operatorname{tr}(p^{d(b)} m_{ii}^{(s)}) = p^{a-1}(p-1) w^{-1} \sum_{\lambda} p^{d(b)} m_{\lambda\lambda}^{(s)},$$

where λ ranges over the w values for which χ_λ is a p-conjugate of χ_i. Hence

$$p^{d(b)} \sum_{\lambda} m_{\lambda\lambda}^{(s)} \geqq wq(b).$$

If we add this now over all all families of p-conjugates in B and compare the result with (5.7), we obtain (5.6).

We note a corollary.

(5E) *If π is an element of order $p^e > 1$ in the center of a defect group of the block B and if n_δ is the maximal value of the quotient $l(\bar{b})/q(\bar{b})$ for p-blocks $\bar{b}$ of defect δ of finite groups, then $k(B) \leqq p^{d(B)} n_{d-e}$.*

Indeed, we can find major subsections (π, b) associated with B, cf. (4A). There exists blocks $\bar{b}$ of defect $d-e$ of $\mathfrak{C}(\pi)/\langle\pi\rangle$ such that the characters of $\bar{b}$ (when interpreted as characters of $\mathfrak{C}(\pi)$) lie in b, cf. [2II, (2G)]. In the same sense, b and $\bar{b}$ contain the same modular irreducible characters; $l(b) = l(\bar{b})$. Also, $C_b = p^e C_{\bar{b}}$ and the quadratic form Q_b remains the same, if b is replaced by $\bar{b}$; $q(b) = q(\bar{b})$.

(5F) *Let $s = (\pi, b)$ be a subsection associated with the block B. If $\chi_i, \chi_j \in B$ have heights h_i, h_j respectively and if $\chi_l \in B$ has height $h_l = 0$, then in the ring $\mathfrak{o}$,*

$$(5.9) \qquad p^{d(B)} m_{ij}^{(s)} \equiv p^{d(B)} \frac{\chi_i(1)\chi_j(1)}{\chi_l(1)^2} m_{ll}^{(s)} \pmod{\mathfrak{p} p^{\max(h_i, h_j)}}.$$

Proof. On account of (3.9), the formula (5.1) can be written in the form

$$(5.10) \qquad m_{ij}^{(s)} = |G|^{-1} \chi_i(1) \sum_{\rho^*} \omega_i^{(s)}(\pi\rho^*) \bar{\chi}_j^{(s)}(\pi\rho^*)$$

where ρ^* ranges over a set of representatives for the p-regular classes of $\mathfrak{C}(\pi)$. It now follows from (3D) that

$$|G| \chi_i(1)^{-1} m_{ij}^{(s)} \equiv |G| \chi_l(1)^{-1} m_{lj}^{(s)}$$

and this yields

$$p^{d(B)} m_{ij}^{(s)} \equiv \chi_i(1) \chi_l(1)^{-1} p^{d(B)} m_{lj}^{(s)} \pmod{\mathfrak{p} p^{h_i}}.$$

Similarly, we have

$$p^{d(B)}m_{lj}^{(s)} \equiv \chi_j(1)\chi_l(1)^{-1}p^{d(B)}m_{ll}^{(s)} \pmod{\mathfrak{p}p^{h_j}}.$$

Combination of these two congruences shows that the two sides of (5.9) are congruent modulo $\mathfrak{p}p^{h_i}$. An analogous argument with the roles of i and j interchanged shows that the two sides are congruent modulo $\mathfrak{p}p^{h_j}$. This proves (5.9).

As a corollary, we note

(5G) ***If $h_i \geqq h_j$ in (5F), then $\nu(m_{ij}^{(s)}) \geqq h_i - d(B)$ and the equality sign can only hold if s is a major subsection and $h_j = 0$.***

Proof. By (5A), $\nu(m_{ll}^{(s)}) \geqq -d(b)$. Thus

$$\nu(\chi_i(1)\chi_j(1)\chi_l(1)^{-2}p^{d(B)}m_{ll}^{(s)}) \geqq h_i + h_j + d(B) - d(b).$$

Since $d(B) \geqq d(b)$, the statement follows from (5F).

We next prove the converse.

(5H) ***If $s = (\pi, b)$ is a major subsection and if $\chi_j \in B = b^G$ has height 0, then***

(5.11) $$\nu(m_{ij}^{(s)}) = h_i - d(B).$$

Proof. It will suffice to prove (5.11) for $i = j$ since then the general case of (5.11) can be obtained from (5.9) with $l = j$. By (5.10)

$$p^{d(B)}m_{jj}^{(s)} = |G|^{-1}p^{d(B)}\chi_j(1)\sum_{\rho^*}\omega_j^{(s)}(\pi\rho^*)\bar{\chi}_j^{(s)}(\pi\rho^*)$$

where ρ^* ranges over a set of representatives for the p-regular classes of $\mathfrak{C}(\pi)$. Here, $|G|^{-1}p^{d(B)}\chi_j(1)$ is a local unit. If we work $(\operatorname{mod}\mathfrak{p})$, it suffices to take only the terms for which

(5.12) $$\omega_j^{(s)}(\pi\rho^*) \not\equiv 0, \qquad \chi_j^{(s)}(\pi\rho^*) \not\equiv 0.$$ [3]

Let D be a defect group of b. By (3F) the first relation (5.12) implies that $\mathfrak{C}(\pi\rho^*)$ contains a defect group of b. If the residue system $\{\rho^*\}$ is chosen suitably, we may assume that $D \subseteq \mathfrak{C}(\pi\rho^*)$ and then $\rho^* \in \mathfrak{C}(D)$. It follows from (5.12) and (3.9) that

$$\nu|G:\mathfrak{C}(\pi\rho^*)| = \nu(\chi_j(1)) = \nu|G| - d(B).$$

Hence

$$\nu|\mathfrak{C}(\pi\rho^*)| = d(B)$$

[3] Note that

$$\omega_j^{(s)}(\pi\rho^*)\bar{\chi}_j^{(s)}(\pi\rho^*) = \omega_j^{(s)}(\pi\rho^*)\chi_j^{(s)}(\pi\rho^*)$$

and that $\omega_j^{(s)}(\pi\rho^*)$ is a local integer by (3D) applied to the character χ_j.

and

$$p^{d(B)}m_{jj}^{(s)} \equiv |G|^{-1}p^{d(B)}\chi_j(1)\sum_{\rho^*}{}' \omega_j^{(s)}(\pi\rho^*)\bar{\chi}_j^{(s)}(\pi\rho^*)$$

where the prime at the summation sign indicates that only terms are taken for which D is an S_p-group of $\mathfrak{C}(\pi\rho^*)$. This yields

$$p^{d(B)}m_{jj}^{(s)} \equiv \sum_{\rho^*}{}' |\mathfrak{C}(\pi\rho^*):D|^{-1}|\chi_j^{(s)}(\pi\rho^*)|^2.$$

Since $\rho^* \in \mathfrak{C}(D)$, we can use here (4D). Here, $|\mathfrak{C}(\pi\rho^*):D|$ is a local unit. With the notation as in (4B) and (4D), we find

$$(5.13) \qquad p^{d(B)}m_{jj}^{(s)} \equiv \sum_{\rho^*}{}' |C:\mathfrak{g}|^{-2}|\mathfrak{C}(\pi\rho^*):D|^{-1}\psi(1)^{-2}|\chi_j^{(s)}(\pi)|^2|\Sigma|^2$$

where Σ is the sum on the right in (4.10).

In particular, this formula can be applied, if we replace G by $\mathfrak{C}(\pi)$, B by b, $s=(\pi,b)$ by $(1,b)$, and χ_j by a character ζ_l of height 0 in b. If m^* is the contribution of the subsection $(1,b)$ to the inner product (ζ_l,ζ_l) and if we apply (5.13) to $p^{d(b)}m^*$, we have the same expression on the right hand side as in (5.13) except that $\chi_j^{(s)}(\pi)$ must be replaced by $\zeta_l^{(1,b)}(1)=\zeta_l(1)$. If we set $\chi_j^{(s)}(\pi)=\zeta_l(1)\eta$, then η is a local unit by (4C). Since $d(B)=d(b)$, we have

$$p^{d(B)}m_{jj}^{(s)} \equiv p^{d(b)}m^*|\eta|^2.$$

As shown in [5], we have $\nu(m_{jj}^{(1,b)}) = -d(B)$ for the subsection $(1,b)$ and χ_j of height 0, cf. [5, (8)]. Applying this to C and $(1,b)$ we see that $\nu(m^*) = -d(b) = -d(B)$. This yields $\nu(m_{jj}^{(s)}) = -d(B)$ and then (5H) holds.

(5I) *Let χ_i be a character of height h in a block B of G. Then*

$$(5.14) \qquad \sum_{(\pi,b)} \operatorname{Max}\left(\frac{q(b)}{p^{d(B)}}, \frac{p^h}{p^{d(B)}}\right) \leqq 1$$

where the sum extends over all distinct subsections $s=(\pi,b)$ for which $\chi_i^{(s)}$ does not vanish identically and where $q(b)$ has the same meaning as in (5D).

Proof. We apply (5B) for $i=j$. If $\chi_i^{(s)}$ does not vanish identically, then $m_{ii}^{(s)}>0$. With each $m_{ii}^{(s)}\neq 0$, all algebraically conjugates appear; the values of $d(b)$ and $q(b)$ are the same for the subsections in question; cf. § 7 below. If there are w distinct such conjugates and if $s=(\pi,b)$, the argument in the proof of (5D) shows that the sum S of the conjugates of $m_{ii}^{(s)}$ is at least $wq(b)p^{-d(b)}$. On the other hand, by the theorem on the arithmetical and geometrical mean, $S \geqq wP^{1/w}$ where P is the product of the w conjugates of $m_{ii}^{(s)}$. By (5A), P is a rational number whose denominator is a power of p. By (5G) for each conjugate m of $m_{ii}^{(s)}$, we have $\nu(m) \geqq p^{h-d(B)}$. Hence $\nu(P) \geqq w(h-d(B))$. It follows that $P \geqq p^{w(h-d(B))}$, $S \geqq wp^{h-d(B)}$. Now, (5.14) is obtained as a consequence of (5.4).

6. Algebraically conjugate subsections. As before, we set $\nu\,|\,G\,| = a$

$$g = |\,G\,| = p^a g_0, \qquad (g_0, p) = 1.$$

Let Ω be the field of the g-th roots of unity and let Ξ be the subfield of the g_0-th roots of unity. Denote by Γ the Galois group of Ω with regard to Ξ. If ϵ_g is a primitive g-th roots of unity, then, for $\gamma \in \Gamma$, we can set

$$\epsilon_g{}^\gamma = \epsilon_g{}^{(\gamma)} \tag{6.1}$$

where (γ) is a rational integer, determined $(\operatorname{mod} g)$ such that

$$(\gamma) \equiv 1 \ (\operatorname{mod} g_0), \qquad ((\gamma), p) = 1.$$

Application of γ permutes the irreducible characters $\chi_1, \chi_2, \cdots$ of G. If $\chi_i \rightarrow \chi_i{}^\gamma$, we have

$$\chi_i{}^\gamma(\sigma) = \chi_i(\sigma^{(\gamma)})$$

for $\sigma \in G$. In particular, this shows that $\chi_i{}^\gamma$ and χ_i coincide for p-regular elements σ and hence belong to the same p-block. In the terminology used in § 5, χ_i and $\chi_i{}^\gamma$ are p-conjugate characters. Analogous remarks apply to characters and blocks of subgroups of G.

Let $s = (\pi, b)$ be a subsection of G. We define

$$s^\gamma = (\pi, b)^\gamma = (\pi^{(\gamma)}, b). \tag{6.2}$$

It should be observed that this definition is compatible with the identification of subsections introduced in (1.3). Hence each $\gamma \in \Gamma$ defines a permutation of the distinct subsections of G. We say that two subsections s and s^γ with $\gamma \in \Gamma$ are *algebraically conjugate*. It is clear that s and s^γ are associated with the same block B of G. If $d_j{}^{(\pi)}$ is a column of the subsection s, the *algebraically conjugate* column $(d_j{}^{(\pi)})^\gamma$ appears in the subsection s^γ. Indeed, application of γ to (1.2) shows that

$$\chi_i(\pi^{(\gamma)}\rho) = \sum_b \sum_j (d_{ij}{}^{(\pi)})^\gamma \phi_j{}^{(\pi)}(\rho) \tag{6.3}$$

for p-regular $\rho \in \mathfrak{C}(\pi)$.

By the *stability group* $\Gamma_0(s)$ of a subsection s, we mean of course the group of those $\gamma \in \Gamma$ for which $s^\gamma = s$. We show

(6A) *Let* $s = (\pi, b)$ *be a subsection. Let* $\mathfrak{d}$ *be a defect group of* b. *Set* $C = \mathfrak{C}(\pi)$ *and let* $\mathfrak{b}$ *denote a block of* $\mathfrak{d}\mathfrak{C}(\mathfrak{d})$ *for which* $\mathfrak{b}^C = b$. *Finally, let* $T = T(\mathfrak{b})$ *denote the inertial group of* $\mathfrak{b}$ *in* $\mathfrak{N}(\mathfrak{d})$. *The stability group* $\Gamma_0(s)$ *consists of those* $\gamma \in \Gamma$ *for which there exist elements* $\mu \in T \cap \mathfrak{N}(\langle\pi\rangle)$ *such that* $\pi^\mu = \pi^{(\gamma)}$.

Proof. If $\gamma \in \Gamma_0(s)$, then

$$(\pi, b) = (\pi, b)^\gamma = (\pi^{(\gamma)}, b)$$

and by (1.3), this is so if and only if there exist elements $\mu \in G$ for which

$$\pi^\mu = \pi^{(\gamma)}, \qquad b^\mu = b. \tag{6.4}$$

The first equations shows that $\mu \in \mathfrak{N}(\langle\pi\rangle)$ and then μ normalizes C. It follows from the second equation that $\mathfrak{d}^\mu$ is a defect group of b. Hence $\mathfrak{d}^\mu$ and $\mathfrak{d}$ are conjugate in C. Replacing μ by an element of μC, we may assume that $\mu \in \mathfrak{N}(\mathfrak{d})$. Now the second equation (6.4) shows that $\mu \in T$ and then $\mu \in T \cap \mathfrak{N}(\langle\pi\rangle)$.

Conversely, assume that there exist $\mu \in T \cap \mathfrak{N}(\langle\pi\rangle)$ for which $\pi^\mu = \pi^{(\gamma)}$ for some $\gamma \in \Gamma$. Then, by (6.2) and (1.3)

$$(\pi, b)^\gamma = (\pi^{(\gamma)}, b) = (\pi^\mu, b) = (\pi^\mu, b^\mu) = (\pi, b)$$

and $\gamma \in \Gamma_0(s)$. This proves (6A).

As a corollary we note

(6B) *Let the notation be as in* (6A). *If* π *has order* $p^e > 1$ *and if we set* $C^* = \mathfrak{N}(\langle\pi\rangle)$, $C = \mathfrak{C}(\pi)$, *we have*

$$|\Gamma_0(s)| = p^{a-e}\,|T \cap C^* : T \cap C|. \tag{6.5}$$

The number $w = w(s)$ *of distinct subsections algebraically conjugate to* s *is*

$$w = p^{e-1}(p-1)\,|T \cap C^* : T \cap C|^{-1}. \tag{6.6}$$

If s *is associated with the block* B *of* G, *we have* $\nu(w) = e - 1 - r$ *with*

$$r = d(\mathfrak{b}^{C^* \cap T}) - d(b) \leqq d(b^G) - d(b). \tag{6.7}$$

Proof. If μ ranges over $T \cap C^*$, then π^μ ranges over $|T \cap C^* : T \cap C|$ distinct elements. For each integer m there exist p^{a-e} integers c in a complete residue system $(\operatorname{mod} p^a)$ for which

$$c \equiv m \pmod{p^e}.$$

This implies that for a fixed $\mu \in T \cap C^*$, there exist exactly p^{a-e} distinct elements $\gamma \in \Gamma$ with $\pi^{(\gamma)} = \pi^{\mu}$. Thus (6.5) is a consequence of (6A). Since $w = |\Gamma : \Gamma_0(s)|$ and $|\Gamma| = p^{a-1}(p-1)$, we have (6.6). Moreover, $\nu(w) = e - 1 - r$ with

$$r = \nu |T \cap C^* : T \cap C|. \tag{6.8}$$

Apply the results of [1, § 4] with G replaced by $T \cap C^*$ and $H = \mathfrak{d}\mathfrak{C}(\mathfrak{d})$. Since $T \subseteq \mathfrak{N}(\mathfrak{d})$, then $H \lhd T \cap C^*$ as required. By [1, (4D)], $\mathfrak{b}^{C^* \cap T} = \tilde{b}$ is regular with regard to H. Then Fong's theorem [1, (4C)] shows that if $\tilde{D}$ is a defect group of $\tilde{b}$, $\tilde{D} \cap H = \mathfrak{d}$. If ψ is the canonical character in $\mathfrak{b}$, its inertial group is $T(\mathfrak{b}) = T$ and [1, (4G)] implies

$$\nu |T \cap C^* : H| = \nu |H\tilde{D} : H|.$$

It follows that

$$\nu |T \cap C^* : H| = \nu |\tilde{D} : \tilde{D} \cap H| = \nu |\tilde{D} : \mathfrak{d}| = d(\mathfrak{b}^{C^* \cap T}) - d(\mathfrak{b}).$$

If we apply an analogous argument to $T \cap C$, we have

$$\nu |T \cap C : H| = d(\mathfrak{b}^{C \cap T}) - d(\mathfrak{b}).$$

Since $d(\mathfrak{b}^{C \cap T}) \leqq d(\mathfrak{b}^{C}) = d(\mathfrak{b})$, the right hand side vanishes and

$$r = \nu |T \cap C^* : T \cap C| = d(\mathfrak{b}^{C^* \cap T}) - d(\mathfrak{b}).$$

Here, $d(\mathfrak{b}^{C^* \cap T}) \leqq d(B)$ because

$$(\mathfrak{b}^{C^* \cap T})^G = \mathfrak{b}^G = b^G = B$$

and the prof of (6.7) is complete.

Remark 1. In (6B), we have

$$r \leqq \nu |C^* : C|. \tag{6.9}$$

This is clear from (6.8) as $C \lhd C^*$.

As an application, we prove

(6C) *Let B be a block of G with the defect group D and let p^e be the exponent of $\mathfrak{Z}(D)$. If $\chi_i \in B$ has height h, then $h \leqq d(B) - e$. If $p = 2$ and $h > 0$, we have $h < d(B) - e$.*

Proof. Choose $\pi \in \mathfrak{Z}(D)$ of order p^e. By (4A), there exist major subsections $s_0 = (\pi, b)$ associated with B. As shown by (4C), $\chi_i^{(s_0)}$ does not vanish identically and hence $m_{ii}^{(s_0)} \neq 0$. Since the statement is trivial for

$h=0$, we may assume $h>0$ and then $D\neq 1$, $e>0$. If s ranges over the w sections algebraically conjugate to s_0, it follows from (5B) that

$$(6.10)\qquad \sum_s m_{ii}^{(s)} < 1.$$

As $h>0$, (5G) shows that

$$\nu(m_{ii}^{(s)}) > h-d(B).$$

Now the argument used in the proof of (5I) yields

$$wp^{h-d(B)} < 1.$$

On the other hand, since $d(B)=d(b)$, (6.7) implies $r=0$, that is, $\nu(w)=e-1$. Hence $w\geqq p^{e-1}$ whence $h\leqq d(B)-e$.

Suppose now that $p=2$. If $e=1$, then $m_{ii}^{(s_0)}$ is rational, $m_{ii}^{(s_0)} \geqq 2^{h-d(B)+1}$ and (6.10) shows that $h<d(B)-1$. Assume then that $e>1$. Let ϵ be a primitive 2^e-th root of unity. Let Q denote the rational field. The numbers ϵ^μ with $-2^{e-2}\leqq \mu < 2^{e-2}$ form an integral basis of the field $Q(\epsilon)$. By (5A) and (5G), $2^{d(B)-h}m_{ii}^{(s)}$ is an integer of $Q(\epsilon)$ and we can write

$$2^{d(B)-h}m_{ii}^{(s)} = \sum_\mu a_\mu \epsilon^\mu$$

with rational integral a_μ. Since $m_{ii}^{(s)}$ is real, then $a_\mu=0$ for $\mu=-2^{e-2}$ and $a_\mu=a_{-\mu}$ for the other μ. By (5G), $2^{d(B)-h}m_{ii}^{(s)}\equiv 0$ and as $\epsilon\equiv 1$, we find $a_0\equiv 0$. Thus, a_0 is even. It follows here from (6B) that $w=2^{e-1}$. Thus the left side of (6.10) is the trace of $m_{ii}^{(s)}\in Q(\epsilon)$ with regard to Q. Then

$$2^{d(B)-h}\sum_s m_{ii}^{(s)} = 2^{e-1}a_0.$$

Since this expression is positive and a_0 is even, (6.10) implies $h<d(B)-e$ as stated.

Remark 2. In many cases, stronger results can be obtained by using more terms of (5.4) than appear in (6.10). If s is a non-major subsection and if it is known that $\chi_i^{(s)}$ does not vanish identically, the corresponding terms of (5.4) can also be used. It may here be advantageous to use Remark 1, rather than (6.7).

(6D) *Let B be a block. Choose an integer n such that $p^n > k(B)-l(B)$. If $s=(\pi,b)$ is a subsection associated with B and if π has order $p^e>1$, then*

$$(6.11)\qquad d(B)-d(b)\geqq e-n.$$

If D is a defect group of B and if $\pi\in D$ has order p^e, we have

(6.12) $$d(B) - \nu\,|\,\mathfrak{C}_D(\pi)\,| \geqq e - n.$$

The center $\mathfrak{Z}(D)$ has at most exponent p^n.

Proof. There exist $\chi_i \in B$ for which $\chi_i^{(s)}$ does not vanish identically. In each of the w subsections algebraically conjugate with s, we have at least one column of decomposition numbers (§ 1). Since $\pi \neq 1$, it follows that

$$k(B) - l(B) \geqq w$$

cf. [2II, (7D)]. Now (6B) yields

$$n > \nu(w) \geqq e - 1 - (d(B) - d(b))$$

and this is equivalent with (6.11).

If $\pi \in D$, by (3A), there exist subsections (π, b) associated with B of defect $d(b) \geqq \nu\,|\,\mathfrak{C}_D(\pi)\,|$ and (6.11) implies (6.12). For $\pi \in \mathfrak{Z}(D)$ this yields $e \leqq n$.

It will again be clear that in many cases (6D) can be improved.

Let $s = (\pi, b)$ be a subsection associated with B. If $l(b)$ is the number of modular irreducible characters in b, we have $l(b)$ columns of decomposition numbers $d_j^{(\pi)}$ in the subsection. The algebraically conjugate columns occur in the $w(s)$ algebraically conjugate subsections. If F_μ is such a family of algebraically conjugate columns, the number $|F_\mu|$ of members of F_μ is a multiple $u_\mu w(s)$ of $w(s)$. If we let F_μ range over the distinct families which are represented by columns in s, we have $\sum_\mu u_\mu = l(b)$. Now let s range over a system $\{s\}$ of subsections associated with B such that each subsection associated with B is algebraically conjugate to one and only one $s \in \{s\}$. We then have information concerning the numbers $|F_\mu|$ of members of the families F_μ of conjugate columns of decomposition numbers for B. Now the principle used in [4, §§ 5-6] yields information concerning the number of members of the families of p-conjugate characters in B, at least for odd p. For $p = 2$, it is necessary to adjoin $i = \sqrt{-1}$ to the ground field in order to apply this method.

This method often gives valuable information concerning the distribution of the $\chi_i \in B$ into family of p-conjugates. We state a corollary.

(6E) *Let $\pi_1, \pi_2, \cdots, \pi_m$ be a set of elements of the center $\mathfrak{Z}(D)$ of the defect group D of the block B such that $\langle\pi_i\rangle$ is not conjugate in $\mathfrak{N}(D)$ to $\langle\pi_j\rangle$ for $i \neq j$ and that each π_i has order at least p^n where $n \geqq 2$ is a fixed integer. Then B contains at least m families F_μ^* of p-conjugate characters, in which the number $|F_\mu^*|$ of members of F_μ^* is divisible by p^{n-1}. In* *particular, if* $\mathfrak{Z}(D)$ *has exponent* p^e, *there is at least one family with* $|F_\mu{}^*|$ *divisible by* p^{e-1}.

7. Blocks with abelian defect groups. The whole situation is somewhat simpler for blocks B whose defect groups are abelian. As an immediate consequence of [1, (7C)], we have

(7A) *If the block* B *has an abelian defect group, each subsection* s *associated with* B *is major.*

It follows that application of (4E), (4F), and (3G) yields all the subsections s associated with B. We mention a consequence. If for all $\pi \neq 1$ in D, we can find the number $l(b)$ of modular irreducible characters in blocks b of $\mathfrak{C}(\pi)/\langle\pi\rangle$ with the defect group $D/\langle\pi\rangle$, we can determine the difference $k(B) - l(B)$.

Somewhat different results can be obtained by using the methods of § 5. For instance, we have

(7B) *Let* B *be a block with an abelian defect group* D. *The total number* $N(B)$ *of subsections associated with* B *is at most equal to* $p^{d(B)}$ *and we can have* $N(B) = p^{d(B)}$ *only if* $k(B) = p^{d(B)}$, $l(B) = 1$. *If characters* χ_i *of height* $h > 0$ *occur in* B, *then* $N(B) < p^{d(B)-h}$.

Proof. If $\chi_i \in B$ and if s is a subsection associated with B, then (4C) shows that the contribution $m_{ii}{}^{(s)}$ does not vanish. Now (5B) implies that $N(B) \leqq p^{d(B)}$. If χ_i has positive height h, by (5G)

$$\nu(p^{d(B)}m_{ii}{}^{(s)}) > h.$$

Then (5B) yields $N(B) < p^{d(B)-h}$. This can be improved easily.

Finally, assume that $N(B) = p^{d(B)}$. Each $\chi_i \in B$ then has height 0. Our method shows that for each subsection s associated with B, we have $|p^{d(B)}m_{ii}{}^{(s)}| = 1$. In particular, we can take $s = (1, B)$. Since $m_{ii}{}^{(s)}$ here is rational and positive, then

$$m_{ii}{}^{(1,b)} = p^{-d(B)}.$$

If we set $Z = p^{d(B)}M^{(s)}$ for $s = (1, B)$ in (5C), the main diagonal coefficients of Z are 1. Since Z belongs to a non-negative quadratic form with integral coefficients and since no coefficient vanishes, cf. [5, (8)], all coefficients are ± 1. By (5C), $Z^2 = p^{d(B)}Z$ and hence $k(B) = p^{d(B)}$. Finally, by (5B), $\operatorname{tr}(Z) = p^{d(B)}l(B)$ and hence $l(B) = 1$.

One can apply our results in the case of a cyclic defect group $D = \langle\pi_0\rangle$ of order p^d. However, they are not strong enough to give the theorems obtained by E. C. Dade [6]. Still, we mention some of the facts, since they throw a different light on some of the results. Let again $\mathfrak{b}$ be a block of $D\mathfrak{C}(D) = \mathfrak{C}(\pi_0)$ with $\mathfrak{b}^G = B$. The modular irreducible characters in $\mathfrak{b}$ are the same as those of the corresponding block $\mathfrak{b}^*$ of $\mathfrak{C}(D)/D$. Since $\mathfrak{b}^*$ has defect 0, $l(\mathfrak{b}) = 1$. Now (5D) yields $k(B) \leqq p^d$.[4] By (6C), only characters χ_i of height 0 occur in B. Let $T = T(\mathfrak{b})$ be the inertial group of $\mathfrak{b}$ in $\mathfrak{N}(D)$ and set $|T: \mathfrak{C}(D)| = m$. By [1, (5B)], $m \not\equiv 0 \pmod{p}$. If $\pi \in D$ has order $p^e > 1$, clearly $\mathfrak{N}(D) \subseteq \mathfrak{N}(\langle\pi\rangle)$. It is seen easily that $T \cap \mathfrak{C}(\pi) = \mathfrak{C}(D)$. A subsection $s = (\pi, b)$ associated with B is obtained for $b = \mathfrak{b}^C$ with $\mathfrak{C} = \mathfrak{C}(\pi)$. Then (6B) shows that s has $p^{e-1}(p-1)/m$ algebraic conjugates. Taking for π generators of the subgroups of D of order p^e with $e = 1, 2, \cdots, d$ and counting the number of columns, we find

$$(7.1) \qquad k(B) - l(B) \geqq \sum_{e=1}^{d} p^{e-1}(p-1)/m = (p^d - 1)/m.$$

If $m = 1$, the inequality $k(B) \leqq p^d$ implies $l(B) = 1$. In particular, this applies, if G is replaced by $\mathfrak{C}(\pi)$ with $\pi \neq 1$ in D. Hence we have $l(b) = 1$ for blocks b of $\mathfrak{C}(\pi)$ with defect group D. Now our counting process shows that for $m \geqq 1$ the equality sign holds in (7.1). Thus

$$(7.2) \qquad k(B) = (p^e - 1)/m \quad + l(B).$$

[R.B.]

We can also discuss the families of p-conjugate characters in B.

As already mentioned, Dade [6] obtained much more complete information on blocks B with cyclic defect group. In particular, he showed that $l(B) = m$. His results enable us to prove the lemma:

(7C) *If B is a block with cyclic defect group, the minimum $q(B)$ of the quadratic form Q_B in* (5D) *(for $b = B$) has the value $l(B)$.*

Proof. We use the same notation as before. It follows from Dade's results that the non-exceptional characters form a basic set for B; [31, Section 5]. If we set $t = (p^d - 1)/m$, the corresponding Cartan matrix has coefficients $t + 1$ in the main diagonal and coefficients t outside the main diagonal; the degree is $m = l(b)$. The matrix of Q_B then has coefficients $p^d - t$ in the main diagonal and coefficients $-t$ elsewhere. If we write Q_B in the variables $x_1, \cdots, x_m$, we find

$$Q_B = \sum_{i=1}^{m} x_i^2 + t \sum_{i<j} (x_i - x_j)^2.$$

An argument similar to the one used in connection with (5.8) yields $q_B = l(B)$.

[4] This had already been stated in [5, Theorem 4].

(7D) *Let B be a block with an abelian group D of order p^d. If D has rank $r=2$, then $k(B) \leqq p^d$. If D has rank $r=3$, $k(B) \leqq p^{5d/3}$.*

Proof. Let $\langle\pi\rangle$ of order p^e be a cyclic factor of maximal order of the abelian group D. Choose a subsection (π, b) associated with B and let $\bar{b}$ denote the block of $\mathfrak{C}(\pi)/\langle\pi\rangle$ corresponding to b. Then $l(b)=l(\bar{b})$, $q(b)=q(\bar{b})$. If $r=2$, $\bar{b}$ has cyclic defect group $D/\langle\pi\rangle$. Now (5D) and (7C) show that $k(B) \leqq p^d$. If $r=3$, by the first part of (7D), $l(\bar{b}) < p^{d-e}$ and by (5D)

$$k(B) \leqq p^d l(\bar{b}) < p^{2d-e}.$$

Since $e \geqq d/3$, this yields the second statement.

Remark. There are further inequalities for $k(B)$ which can be obtained by similar methods.

(7E) *Let B be a block with an abelian group D of rank r. Assume that $|\mathfrak{N}(D):\mathfrak{C}(D)|$ is not divisible by p. The height h of a character $\chi_i \in B$ lies below a bound $\gamma(r)$ depending only on r. Moreover, there exists a bound $J(r)$ depending on r only such that for $p > J(r)$ we have $h \leqq r$.*

Proof. We may assume $D \neq 1$. Set $W=\mathfrak{N}(D)/\mathfrak{C}(D)$. Then W acts on D by conjugation, and if $D_0=\Omega_1(D)$ is the subgroup of elements of order at most p in D, W acts on D_0. This yields a representation Y of W by linear transformations of an r-dimensional vector space over the field $GF(p)$. Under the hypothesis of (7E), W is p-regular. Then $\sigma \in W$ belongs to the kernel of Y only if σ acts trivially on D, that is, if $\sigma=1$. Hence Y is faithful.

Since W is p-regular, Jordan's theorem applies. It follows that there exists a bound $J(r)$ depending only r such that W has a normal abelian subgroup A of index less than $J(r)$.

Consider $Y \mid A$ in the algebraic closure of $GF(p)$. All its irreducible constituents are linear. Suppose that they belong to families $F_1, F_2 \cdots$ of algebraically conjugate linear characters of A. If F_j consists of z_j linear characters $\lambda_j, \lambda_j', \cdots$,

$$z_1+z_2+\cdots=r.$$

Since $\lambda_j(A)$ lies in a field with p^{z_j} elements we find

$$|A| < p^{z_1+z_2+\cdots}=p^r.$$

It follows that

$$|W| < p^r J(r). \tag{7.3}$$

For each $\pi \in D$, the conjugate class of π in $\mathfrak{N}(D)$ intersects D in a set of at most $|W|$ elements. If k_0 is the total number of conjugate classes of $\mathfrak{N}(D)$ which meet D, then $k_0 \geqq p^d/|W|$. On the other hand, it follows from (4E) that the total number of subsections s associated with B is at least k_0. If $\chi_i \in B$, $\chi_i^{(s)}$ does not vanish identically for these s (cf. (4C)) and (5I) yields $k_0 p^h \leqq p^d$. Hence $p^h \leqq |W|$. By (7.3)

$$p^h < p^r J(r).$$

For $p \geqq J(r)$, then $h \leqq r$. For $p < J(r)$, it is clear that h lies below a bound $\gamma(r)$ depending only on r.

Remark. If D lies in the center of an S_p-group P of G, the assumption $|\mathfrak{N}(D) : \mathfrak{C}(D)| \not\equiv 0 \pmod{p}$ is satisfied. In particular, if P itself is abelian, (7E) applies to all blocks of G.

REFERENCES.

[1] R. Brauer, "On blocks and sections in finite groups I," *American Journal of Mathematics*, vol. 89 (1967), pp. 1115-1136.

[2] ———, "Zur Darstellungstheorie der Gruppen endlicher Ordnung," *Mathematische Zeitschrift I*, vol. 63 (1956), pp. 406-444; II, vol. 72 (1959), pp. 25-46.

[3] ———, "Some applications of the theory of blocks of characters of finite groups," *Journal of Algebra*, I, vol. 1 (1964), pp. 157-167; II, vol. 1 (1964), pp. 307-334; III, vol. 3 (1966), pp. 225-255.

[4] ———, "On the connection between the ordinary and the modular characters of groups of finite order," *Annals of Mathematics*, vol. 42 (1941), pp. 926-935.

[5] R. Brauer and W. Feit, "On the number of irreducible characters of finite groups in a given block," *Proceedings of the National Academy of Sciences, U.S.A.*, vol. 45 (1959), pp. 361-365.

[6] E. C. Dade, "Blocks with cyclic defect group," *Annals of Mathematics*, vol. 84 (1966), pp. 20-48.

Reprinted from the AMERICAN MATHEMATICAL MONTHLY
Vol. 76, No. 1, January, 1969

ON A THEOREM OF FROBENIUS

RICHARD BRAUER, Harvard University

1. A well-known theorem of Frobenius states that if G is a finite group of order g and if n is a divisor of g, the number of solutions β of the equation $\beta^n = 1$ in G is a multiple of n. This can be generalized in various ways. A number of proofs have been given [1–7]. I shall give here still another proof. I found this proof a very long time ago when I was still a student. I did not publish it, but I have used it for many years as material for problems in courses, giving a number of hints. I present the proof here in the hope that other teachers may find it interesting and useful for the same purpose.

2. We start with a lemma.

LEMMA. *Let H be a group. Let A be a normal subgroup of finite order a. If $\sigma \in H$ and if $\alpha \in A$, then σ^a and $(\sigma\alpha)^a$ are conjugate in H.*

Proof. Set $\tau = \sigma\alpha$. For each $\beta \in A$, consider the set S_β of distinct elements $\sigma^{-j}\beta\tau^j$ with $j = 0, \pm 1, \pm 2, \cdots$. Since

$$\sigma^{-j}\beta\tau^j \in \sigma^{-j}\beta(\sigma\alpha)^j A = \sigma^{-j}\sigma^j A = A,$$

S_β is a subset of A. In particular, S_β is a finite set. For two integers i and j, we have

(1) $$\sigma^{-i}\beta\tau^i = \sigma^{-j}\beta\tau^j$$

Professor Brauer did his doctoral work under I. Schur at the University of Berlin. He has been on the faculties of the Univ. of Koenigsberg, Univ. of Kentucky, Institute for Advanced Study, Univ. of Toronto, and Univ. of Michigan. Presently he holds the Perkins Professorship at Harvard. On leaves of absence he held positions at the Tata Institute, Nagoya Univ., the Akademie der Wissenschaften Göttingen (Gauss Professor), and the E. T. H. Zürich. He is best known for his contributions to the representation theory of groups and algebras and was awarded the AMS Cole Prize in Algebra in 1949. *Editor*

if and only if $\beta^{-1}\sigma^{j-i}\beta=\tau^{j-i}$. It follows that the number N_β of distinct elements in S_β is the least positive exponent N for which

$$\beta^{-1}\sigma^N\beta = \tau^N \tag{2}$$

and that (1) holds if and only if N divides $i-j$. It is also clear that if β and γ are elements of A, the sets S_β and S_γ are equal or disjoint. Hence A is a disjoint union of sets S_β, say

$$A = S_{\beta_1} \cup S_{\beta_2} \cup \cdots \cup S_{\beta_r} \tag{3}$$

with $\beta_i \in A$. Let m denote the minimal value of N_{β_i} with $1 \leq i \leq r$. Then by (2), m is the least positive exponent for which conjugation by an element β of A carries σ^m into τ^m. Moreover, if exactly k of the r numbers N_{β_i} are equal to m, the km elements of the corresponding sets S_{β_i} are exactly the elements β of A for which (2) holds with $N=m$. On the other hand, the number of these elements is the order of the centralizer of σ^m in A and hence divides a. Thus km divides a. Since σ^m and τ^m are conjugate and since m divides a, we see that σ^a and τ^a are conjugate as stated.

REMARK. The proof shows that there exist elements $\beta \in A$ which carry σ^a into τ^a.

3. Let G be a group and let H be a subgroup. We shall say that two elements β, γ of G are *equivalent with regard to H*, if $\beta^{-r}\gamma^r \in H$ for all integers r. Clearly this is an equivalence relation. If β is an element of G, denote by F_β the set of all elements $\sigma \in G$ for which $\beta^{-r}\sigma\beta^r \in H$ for all integers r.

PROPOSITION 1. *Let G be a group, H a subgroup, and β an element of G. Then F_β is a subgroup of H and $\beta^{-1}F_\beta\beta = F_\beta$. An element γ of G is equivalent to β with regard to H if and only if $\gamma \in \beta F_\beta$.*

Proof. The first statement is obvious. If $\gamma = \beta\phi$ with $\phi \in F_\beta$, then since F_β is normal in the group generated by β and F_β, for any integer r

$$\beta^{-r}\gamma^r = \beta^{-r}(\beta\phi)^r \in \beta^{-r}\beta^r F_\beta = F_\beta \subseteq H$$

and β and γ are equivalent.

Conversely, if β and γ are equivalent, set

$$\delta_r = \beta^{-r}\gamma^r \in H.$$

Then, for integral s

$$\beta^{-s}\delta_1\beta^s = \beta^{-s}\beta^{-1}\gamma\beta^s = \beta^{-s-1}\gamma^{s+1}\gamma^{-s}\beta^s = \delta_{s+1}\delta_s^{-1} \in H.$$

This shows that $\delta_1 \in F_\beta$. Hence $\gamma = \beta\delta_1 \in \beta F_\beta$ as stated.

PROPOSITION 2. *If H in Proposition 1 has finite order h and if β and γ are equivalent with regard to H, then β^h and γ^h are conjugate.*

Proof. By Proposition 1, $\gamma = \beta\phi$ with $\phi \in F_\beta$. Apply the lemma to the group $H = \langle \beta, F_\beta \rangle$ and its normal subgroup $A = F_\beta$, noting that the order of F_β divides h.

4. As before, let G be a group and H a subgroup. We shall say that two elements β, γ of G are *weakly equivalent with regard to* H, if there exists an element $\tau \in H$ such that $\beta^{-r}\tau\gamma^r \in H$ for all integers r. Again this is an equivalence relation. Since our condition can be written in the form

$$\beta^{-r}(\tau\gamma\tau^{-1})^r \in H,$$

we have

PROPOSITION 3. *Two elements* β, γ *of* G *are weakly equivalent with regard to* H *if and only if* β *is equivalent with regard to* H *to an* H*-conjugate* $\tau\gamma\tau^{-1}$ *of* γ, $(\tau \in H)$.

It follows from Propositions 3 and 1 that the class of β with regard to weak equivalence consists of the elements of the form

$$\gamma = \tau^{-1}\beta\delta\tau, \qquad \delta \in F_\beta, \qquad \tau \in H. \tag{4}$$

It is clear that it suffices to let τ range over a set T of representatives for right cosets in H modulo F_β. We claim that the corresponding representation (4) of γ is unique. This will be shown, if we can prove the following statement:

PROPOSITION 4. *Let* G, H, β, F_β *be as before. If* $\sigma \in H$ *and if for some* δ *and* δ' *in* F_β

$$\sigma^{-1}\beta\delta\sigma = \beta\delta' \tag{5}$$

then $\sigma \in F_\beta$.

Proof. It follows from (5) that

$$\beta^{-1}\sigma\beta \in F_\beta\sigma F_\beta.$$

Since β normalizes F_β, then for integral r,

$$\beta^{-r-1}\sigma\beta^{r+1} \in F_\beta\beta^{-r}\sigma\beta^r F_\beta.$$

Since $\sigma \in H$, it follows by induction on r, (both for $r>0$ and for $r<0$) that $\beta^{-r}\sigma\beta^r \in H$. Hence $\sigma \in F_\beta$ as claimed.

The next proposition is a consequence of the remark following (4), of Proposition 4, and of Proposition 2.

PROPOSITION 5. *Let* G *be a group and* H *a subgroup of finite order* h. *If* $\beta \in G$, *the class of* β *with regard to weak equivalence consists of exactly* h *elements* γ *and for all these elements* γ, *the elements* β^h *and* γ^h *are conjugate in* G.

5. One form of the Frobenius' Theorem reads

THEOREM. *Let* G *be a group of finite order* g *and let* K *be a conjugate class of* G. *If* n *is a positive integer and* $(g, n) = d$, *the number* N *of elements* $\beta \in G$ *with* $\beta^n \in K$ *is divisible by* d.

(It is well known that a slightly better result is obtained by applying the theorem to the centralizer in G of an element of K. If K consists of k elements, it follows that N is divisible by (g, nk).)

Using Sylow's Theorem, we can obtain Frobenius's Theorem as an immediate consequence of Proposition 5. Indeed, if p^r is any prime power dividing $d=(g, n)$, we choose H as a subgroup of order $h=p^r$ of G. Since the set of elements β with $\beta^n \in K$ consists of full weak equivalence classes with regard to H, the number N is divisible by p^r and hence by d.

It is clear that the same method still applies to infinite groups G, provided that G possesses suitable finite subgroups. We have

THEOREM. *Let G be a group and let K be a conjugate class of G. Let n be an integer and assume that the number N of elements β of G with $\beta^n \in K$ is finite. Then N is divisible by every prime power p^r dividing n for which there exist subgroups H of order p^r of G.*

Far-reaching generalizations of Frobenius's Theorem have been given by P. Hall [5]. These cannot be obtained here.

6. We mention another consequence of the lemma in Section 2.

PROPOSITION 6. *Let G be a finite group of order g which has a normal subgroup H of order h. If $\beta \in G$ has order relatively prime to h and if C is the centralizer of β in G, then the centralizer $\bar{C}$ of βH in G/H is CH/H.*

Proof. The reciprocal image C^* of $\bar{C}$ in G consists of the elements $\gamma \in G$ for which

$$\gamma^{-1}\beta\gamma \in \beta H. \tag{6}$$

It is clear that $CH \subseteq C^*$. Conversely, if $\gamma \in C^*$ then by the lemma and the remark following it, there exists an element $\sigma \in H$ such that

$$\sigma^{-1}\beta^h\sigma = \gamma^{-1}\beta^h\gamma.$$

Since h and the order of β are coprime, then

$$\sigma^{-1}\beta\sigma = \gamma^{-1}\beta\gamma.$$

Hence $\gamma\sigma^{-1} \in C$, $\gamma \in CH$.

References

1. W. Burnside, The Theory of Groups of Finite Order, 2nd ed., Cambridge University Press, 1907, page 49.

2. G. Frobenius, Verallgemeinerung des Sylowschen Satzes, Sitzungsberichte der Preussischen Akademie, Berlin, 1895, pp. 437–449.

3. ———, Über einen Fundamentalsatz der Gruppentheorie I, II, Sitzungsberichte der Preussischen Akademie, Berlin, 1903, 987–991, 1907, 428–437.

4. M. Hall, Jr., The Theory of Groups, Macmillan, New York, 1959, p. 139.

5. P. Hall, On a theorem of Frobenius, Proc. London Math. Soc., 40 (1936) 468–501.

6. B. Huppert, Endliche Gruppen, vol. 1, Springer, Berlin, Heidelberg, New York, 1967, p. 44.

7. H. Zassenhaus, The Theory of Groups, 2nd ed., Vandenhoeck and Ruprecht, Göttingen, 1956.

Reprinted from ILLINOIS JOURNAL OF MATHEMATICS
Vol. 13, No. 1, March 1969
Printed in U.S.A.

DEFECT GROUPS IN THE THEORY OF REPRESENTATIONS OF FINITE GROUPS

BY
RICHARD BRAUER

Dedicated to Oscar Zariski

1. Introduction[1]

Let G be a finite group. Let Ξ be an algebraically closed field. As is well known, the study of the characters of G is closely related to that of the *group algebra* $\Xi[G]$ and of its center $Z = Z(\Xi[G])$. We call Z the *class algebra* of G. We are concerned here with a further investigation of Z continuing the work in [1].

The dimension of Z as a Ξ-space is the class number $k(G)$ of G. Since we are interested in characters and related functions, we also consider the dual space $\hat{Z}$ consisting of all linear functions defined on Z with values in Ξ.

Write Z as a direct sum

$$Z = \oplus \sum B \tag{1.1}$$

of *block ideals* of Z, i.e. of indecomposable ideals of Z. This decomposition (1.1) corresponds to the decomposition

$$\hat{Z} = \oplus \sum F_B \tag{1.2}$$

where F_B is the subspace of $\hat{Z}$ consisting of those $f \in \hat{Z}$ which vanish on all block ideals $B_1 \neq B$ in (1.1). Then B and F_B are themselves dual vector spaces and they have the same dimension k_B.

Each B is a commutative ring with a unit element η_B. If 1 is the unit element of Z, we have

$$1 = \sum_B \eta_B \tag{1.3}$$

and (1.3) is the decomposition of 1 into primitive orthogonal idempotents. It follows that

$$\Xi[G] = \oplus \sum_B \eta_B \, \Xi[G]$$

is the decomposition of the group algebra into (two-sided) block ideals.

Since B is indecomposable, the residue class ring $\bar{B}$ of B modulo its radical is simple and hence an extension field of finite degree of Ξ. Since Ξ was algebraically closed, $\bar{B}$ is isomorphic to Ξ. We then have an algebra homomorphism ω of B onto Ξ. Clearly, ω can be extended to an algebra homomorphism ω_B of Z onto Ξ such that ω_B vanishes for all block ideals $B_1 \neq B$ in (1.1). Thus $\omega_B \in F_B$. Conversely, it is seen at once that each non-zero algebra homomorphism of Z into Ξ coincides with ω_B for some B.

---

[1] This work has been supported in part by a National Science Foundation grant.

The case that Ξ has characteristic 0 is well known and fairly trivial. Let $\chi_1, \chi_2, \cdots, \chi_{k(G)}$ denote the irreducible characters of G. Each χ_j defines an algebra-homomorphism ω_j onto Ξ given by Frobenius' formula

$$(1.4) \qquad \omega_j(\mathrm{S}K) = |K| \chi_j(\sigma_K)/\chi_j(1)$$

where K is a class of conjugate elements of G, where $\mathrm{S}K \in \Xi[G]$ is the sum of the $|K|$ elements of K, and where $\sigma_K \in K$. Since the $k(G)$ homomorphism ω_j are distinct, we have $k(G)$ block ideals $B \cong \Xi$ in (1.1) and Z is semisimple.

We now turn to fields of prime characteristic. Throughout this paper, p will be a fixed prime number and we shall reserve the letter Ω for an algebraically closed field of characteristic p. Take then $\Xi = \Omega$ above and set

$$Z = Z(G) = Z(\Omega[G]).$$

It is clear in principle that if we know the irreducible characters $\chi_1, \chi_2, \cdots, \chi_{k(G)}$, we can construct the block ideals B, or as we shall simply say, the *blocks* B of G. Actually, this can be done in an explicit fashion (§2, 2). In particular, the dimension k_B turns out to be the number of irreducible characters χ_i in B in the sense of [1].

In a way, our aim lies in the opposite direction. This is part of our effort to find new links between characters of G and group theoretical properties of G. The main result of [1, I] is already of this type. With each block B of G, we associate a p-subgroup D of G, the *defect group* of B. If we know[2] the normalizer $N_G(D)$ of D, we can construct the algebra homomorphism ω_B for the blocks B of G with the defect group D. This gives us the values (1.4) for the characters $\chi_j \in B$ modulo a prime ideal divisor of p in an appropriate algebraic number field.

The defect group D of B is determined up to conjugacy. We shall associate with B a system of p-subgroups of G which we shall call the lower defect groups of B. Again, they are really only determined up to conjugacy. In order to fix ideas, it will be convenient to choose a set $\mathcal{P}(G)$ of representatives for the classes of conjugate p-subgroups of G. We then take defect groups and lower defect groups in $\mathcal{P}(G)$.

Let K be a conjugate class of G. There is a unique element $P \in \mathcal{P}(G)$ such that P is a p-Sylow subgroup of the centralizer $C_G(\sigma)$ for suitable $\sigma \in K$. We then call P the *defect group* D_K *of the class* K.

Let B now be a block. A member P of $\mathcal{P}(G)$ will be called a *lower defect group* of B, if there exist elements f of the space F_B in (1.2) with the following properties:

(i) There exist conjugate classes K with the defect group P such that $f(\mathrm{S}K) \neq 0$ with $\mathrm{S}K$ defined as in (1.4).

[2] When we say that a subgroup H of G is known, we usually assume that we know H not only as an abstract group but also the imbedding of H in G, i.e. the manner in which the conjugate classes of H lie in the conjugate classes of G.

(ii) We have $f(\mathcal{S}K) = 0$ for all conjugate classes K for which the order $|D_K|$ of the defect group D_K is smaller than the order $|P|$ of P.

More generally, we consider subspaces V of F_B such that all $f \neq 0$ in V have properties (i) and (ii). Let $m_B(P)$ denote the maximal dimension of such a space V. We count P exactly $m_B(P)$ times as lower defect group of B. Let $\mathfrak{D}_B$ denote the system consisting of the groups $P \in \mathcal{P}(G)$, each P taken with the multiplicity $m_B(P) \geqq 0$. This is the *system* $\mathfrak{D}_B$ *of lower defect groups* of B. We shall show (§4) that $\mathfrak{D}_B$ consists of exactly k_B groups. In other words,

$$k_B = \textstyle\sum_P m_B(P); \qquad (P \in \mathcal{P}(G)). \tag{1.5}$$

If P is a lower defect group of B, i.e. if $m_B(P) > 0$, then P is conjugate to a subgroup of the defect group D of B, and D itself is a lower defect group of B. If we know the normalizer $N_G(P)$ of $P \in \mathcal{P}(G)$, we are able to construct a subspace V_P of dimension $m_B(P)$ of F_B with the properties (i), (ii) above such that F_B is the direct sum of the V_P for the various $P \in \mathcal{P}(G)$. If $P \neq 1$, $N_G(P)$ is a 'local subgroup' of G. However, since $P = 1$ occurs in $\mathcal{P}(G)$, our construction falls short of a full construction of F_B based on a knowledge of the local subgroups of G. In particular, in (1.5) the term $m_B(1)$ cannot be determined, and we can only give a lower estimate for k_B.

By a *p-section* $\mathfrak{S}(\tau)$ of an element τ of G, we mean the set of all elements $\xi \in G$ such that the p-factor ξ_p of ξ is conjugate to the p-factor τ_p of τ, cf. [1, II, §3]. Each p-section is a union of conjugate classes. We shall denote by Π a set of representatives for the conjugate classes of p-elements of G. Each p-section has the form $\mathfrak{S}(\pi)$ with $\pi \in \Pi$ and G is the disjoint union of these $\mathfrak{S}(\pi)$. In §6, we shall associate each lower defect group of B with one of the sections. Let $m_B^{(\pi)}(P)$ of the $m_B(P)$ members P of $\mathfrak{D}_B$ be associated with $\mathfrak{S}(\pi)$ so that

$$\textstyle\sum_\pi m_B^{(\pi)}(P) = m_B(P); \qquad (\pi \in \Pi). \tag{1.6}$$

We shall show that $m_B^{(\pi)}(P)$ can be determined when we know the centralizer $C_G(\pi)$ of π and the blocks b of $C_G(\pi)$ with $b^G = B$ (in the sense of [1, II, §2]. It suffices to know the lower defect groups of b associated with the section of the unit element in $C_G(\pi)$.

The numbers $m_B^{(1)}(P)$ have some remarkable properties. If l_B is the number of modular irreducible characters in B, then

$$l_B = \textstyle\sum_P m_B^{(1)}(P); \qquad P \in \mathcal{P}(G). \tag{1.7}$$

This is a kind of analogue of (1.5). If in (1.7) we sum only over the $P \in \mathcal{P}(G)$ of a fixed order p^r, the partial sum represents the multiplicity of p^r as elementary divisor of the Cartan matrix C_B of B. This refines a result announced without proof in [2].

Notation. Most of the notation used has been explained above. The letter G will always stand for a finite group and p will be a fixed prime number.

We shall denote by Ω an algebraically closed field of characteristic p. The class algebra $Z(\Omega[G])$ of G over Ω will be denoted by Z or $Z(G)$. Occasionally in §2, a particular field Ω will be used, but it is clear that the results concerning Z will not depend on the choice of Ω. If M is a subset of G we denote by $\mathcal{S}M$ the sum of the elements of M in the group algebra of G.

The set of conjugate classes of G will be denoted by $\mathcal{C}\ell(G)$. For $K \in \mathcal{C}\ell(G)$, we shall denote by σ_K a representative element in K. If f is a function defined on Z, we shall usually write $f(K)$ instead of $f(\mathcal{S}K)$. The set of blocks of G (for given p) will be denoted by $\mathcal{B}\ell(G)$.

We choose a set $\mathcal{P}(G)$ of representatives for the classes of conjugate p-subgroups of G. If $P, Q \in \mathcal{P}(G)$, we write $P \preceq Q$ when P is conjugate in G to a subgroup of Q. Then $\mathcal{P}(G)$ is partially ordered. A set of representatives for the conjugate classes of p-elements of G will be denoted by Π.

If M is a subset of G, the centralizer of M in G is denoted by $C_G(M)$ and the normalizer of M is denoted by $N_G(M)$. We write $|M|$ for the cardinality of M.

In summations, the range of the summation is often indicated in parentheses at the end of the line, e.g. see (1.5). We frequently have to use determinants Δ of the following kind. We have a set F of n functions f and a set X of n arguments. Each row of Δ correspond to one $f \in F$ and each column of Δ corresponds to one $x \in X$. We then write[3]

$$\Delta = \det(f(x)); \qquad (f \in F, \quad x \in X).$$

2. Preliminaries

1. In the following, a simple method developed in [1, I, §7] will play an important role. We discuss it briefly. We shall say that a pair of subgroups (T, H) of G is an *admissible pair*, if there exists a p-subgroup Q of G such that

$$(2.1) \qquad T = C_G(Q), \quad QT \subseteq H \subseteq N_G(Q).$$

(Actually, these conditions could be replaced by weaker ones.)

As shown in [1, I, §7], there exists a unique algebra homomorphism μ of $Z(G) = Z(\Omega[G])$ into $Z(H) = Z(\Omega[H])$ such that

$$(2.2) \qquad \mu : \mathcal{S}K \longrightarrow \mathcal{S}(K \cap T) \quad \text{for} \quad K \in \mathcal{C}\ell(G).$$

The dual mapping λ then maps the dual space $\hat{Z}(H)$ of $Z(H)$ into $\hat{Z}(G)$. For $\varphi \in \hat{Z}(H)$, we have

$$(2.3) \qquad \lambda : \varphi \longrightarrow \varphi^{\lambda} = \varphi \circ \mu.$$

In particular, if b is a block of H and if φ is the corresponding algebra-homomorphism ω_b of $Z(H)$ onto Ω then ω_b^{λ} is an algebra homomorphism of $Z(G)$ onto Ω. Hence $\omega_b^{\lambda} = \omega_B$ for some block B. We then write $B = b^G$;

[3] The order in which the elements of G and of X are taken will always be immaterial.

cf. [1, II, §2]. We show:

(2A) *Let (T, H) be an admissible pair of subgroups of G. Let b_0 be a block of H and let F_{b_0} denote the subspace of $\hat{Z}(H)$ corresponding to b_0. If $\varphi \in F_{b_0}$ and if λ is the mapping (2.3), then $\varphi^\lambda \in F_{B_0}$ with $B_0 = b_0^G$.*

Proof. Since μ is an algebra homomorphism, it maps the idempotent η_B of $B \in \mathcal{B}\ell(G)$ on an idempotent of $Z(H)$ or on 0. Hence we can set

$$\eta_B^\mu = \sum\nolimits_b \eta_b \tag{2.4}$$

where b ranges over a set Γ_B of blocks of H. If $b_0 \in \mathcal{B}\ell(H)$ and if $B_0 = b_0^G$, by (2.3) and (2.4),

$$\omega_{B_0}(\eta_B) = \omega_{b_0}(\eta_B^\mu) = \sum\nolimits_b \omega_{b_0}(\eta_b); \qquad (b \in \Gamma_B).$$

This shows that $\omega_{B_0}(\eta_B) = 1$, if and only if $b_0 \in \Gamma_B$. Hence Γ_B consists of exactly those $b \in \mathcal{B}\ell(H)$ for which $b^G = B$.

Suppose now that $\varphi \in F_{b_0}$. Then, for $\zeta \in Z(G)$,

$$\varphi^\lambda(\eta_B\zeta) = \varphi(\eta_B^\mu \zeta^\mu) = \sum\nolimits_b \varphi(\eta_b \zeta^\mu); \qquad (b \in \Gamma_B).$$

If $B \neq b_0^G$, then $b_0 \notin \Gamma_B$ and it follows that our expression vanishes. This shows that $\varphi^\lambda \in F_{B_0}$ with $B_0 = b_0^G$.

(2B) *Let (T, H) form an admissible pair of subgroups of G with $T = C_G(Q)$, $Q \in \mathcal{P}(G)$. Let $\varphi \in \hat{Z}(H)$ and $f = \varphi^\lambda$, cf. (2.3). If $f(K) \neq 0$ for some conjugate class, then the defect group D_K of K satisfies $D_K \geq Q$ in the partial ordering of $\mathcal{P}(G)$.*

Indeed, by (2.2) and (2.3)

$$f(K) = \varphi(\mathcal{S}(K \cap C_G(Q))).$$

If $f(K) \neq 0$, the class K meets $C_G(Q)$ and this implies $D_K \geq Q$.

2. We next discuss the connection between the algebras $Z(\Xi[G])$ and $Z(\Omega[G])$ where Ξ is an algebraically closed field of characteristic 0 and Ω (as always) an algebraically closed field of characteristip p. As we have seen in §1, the class algebra $Z(\Xi[G])$ is semi-simple and, if $k(G)$ is the class number of G, we have exactly $k(G)$ distinct algebra homomorphisms ω_i of $Z(\Xi[G])$ onto Ξ, cf. (1.4). These formulas show that this result remains valid, if Ξ is replaced by the field Ξ_0 of the $|G|$-th roots of unity over the field Q of rational numbers. Indeed, all $\chi_i(\sigma_K)$ in (1.4) lie in Ξ_0.

Let p be a fixed rational prime. Let ν denote a fixed extension of the p-adic (exponential) valuation of Q to a valuation of Ξ_0. If $\mathfrak{o}$ is the ring of local integers for ν in Ξ_0 and $\mathfrak{p}$ the corresponding prime ideal, we set

$$\mathfrak{o}/\mathfrak{p} = \Omega_0 \tag{2.5}$$

and form the subring

(2.6) $$J = \sum_K \mathfrak{o}(\mathbb{S}K); \qquad (K \in \mathcal{C}\ell(G))$$

of "integral" elements of $Z(\Xi_0[G])$. If θ_0 is the natural homomorphism of $\mathfrak{o}$ onto Ω_0 in (2.5), clearly θ_0 can be extended to a homomorphism θ of J onto the class algebra $Z(\Omega_0[G])$. If φ is a linear function defined on $Z(\Xi_0[G])$ with values in Ξ_0, and if $\varphi(\alpha) \in \mathfrak{o}$ for all $\alpha \in J$, then the map θ defines a linear function φ^θ, defined on $Z(\Omega_0[G])$ with values in Ω_0. Let Ω denote the algebraic closure of Ω_0. By linearity, φ^θ can be considered as a linear function on the class algebra $Z = Z(\Omega[G])$ with values in Ω, i.e. φ^θ can be considered as an element of the dual space $\hat{Z}$.

Since as is well known the right sides in (1.4) are algebraic integers in Ξ_0, we can apply this to the function $\varphi = \omega_j$. It is clear that ω_j^θ is an algebra homomorphism of Z onto Ω. Hence ω_j^θ must be an ω_B for some block B of G. In [1], the irreducible character χ_j of G was said to *belong* to the block B of G, if $\omega_j^\theta = \omega_B$. We shall also say now that then ω_j is *associated with* B. If this is so for k_B^* values of j, clearly

(2.7) $$k(G) = \sum_B k_B^*; \qquad (B \in \mathcal{B}\ell(G)).$$

Consider the Ξ_0-space W spanned by the ω_j associated with B,

(2.8) $$W = \sum_j \Xi_0\, \omega_j; \qquad (\chi_j \in B),$$

and take the subset M_B consisting of those $\varphi \in W$ for which $\varphi(\alpha) \in \mathfrak{o}$ for all $\alpha \in J$. Then M_B is an $\mathfrak{o}$-module of rank k_B^*. Since $\mathfrak{o}$ is a principal ideal domain, M_B has an $\mathfrak{o}$-basis. It follows that the module $(M_B)^\theta$ of all φ^θ with $\varphi \in M_B$ has again rank k_B^*. On the other hand, the method in [1, II, §4] shows that $(M_B)^\theta \subseteq F_B$. Hence

(2.9) $$\dim_\Omega B = \dim_\Omega F_B \geqq k_B^*.$$

If we add over all B, both sides have the same sum $k(G)$, cf. (1.1) and (2.7). Hence we must have equality in (2.9). Thus

(2C) *Let Ω be an algebraically closed field of characters p. Let B be a block of G. Then* $\dim_\Omega B$ *is equal to the number of ordinary irreducible characters of G in B in the sense of* [1].

With the notation introduced above, we also have

(2D) *If φ ranges over the elements of the $\mathfrak{o}$-module M_B, then φ^θ ranges over F_B.*

3. We add some remarks which will only be used in §6 and §7.

(2E) *Let B be a block of G. Suppose we have coefficients $a_K \in \Xi$ such that*

(2.10) $$\sum_K a_K\, \omega_j(K) = 0; \qquad (K \in \mathcal{C}\ell(G))$$

for every ω_j associated with B. Then (2.10) *remains valid if we let K range only over the conjugate classes which belong to a fixed p-section.*

Proof. Expressing ω_j by χ_j by means of (1.4), we have

$$\sum \mid K \mid a_K \, \chi_j(\sigma_K) = 0.$$

We may assume that the p-factor of σ_K is an element $\pi_K \in \Pi$. If $\sigma_K = \pi_K \, \rho_K$, we can express $\chi_j(\sigma_K)$ by the decomposition numbers belonging to B and the section $\mathfrak{S}(\pi_K)$ and the values of modular irreducible characters of $C_G(\pi_K)$ for the element ρ_K, cf. [1, II (3.2), (6A)]. Since [I1, II (7B)] implies that the matrix of decomposition numbers belonging to B is non-singular the statement is immediate.

(2F) *Let B be a block; $k_B = \dim_\Omega B$. Suppose we have a set F of k_B elements of F_B and a set $\mathfrak{K}$ of k_B conjugate classes such that*

$$\det f(K) \neq 0; \qquad (f \in F, \quad K \in \mathfrak{K}).$$

Let K_0 be a fixed conjugate class. There exist coefficients $c_K \in \mathfrak{o}$ such that

$$\omega_j(K_0) = \sum\nolimits_K c_K \, \omega_j(K); \qquad (K \in \mathfrak{K}) \tag{2.11}$$

for each ω_j associated with B. Here, c_K vanishes when K and K_0 belong to different p-sections. For each $f \in F_B$, then

$$f(K_0) = \sum\nolimits_K c_K^\theta f(K); \qquad (K \in \mathfrak{K}). \tag{2.12}$$

Proof. For each $f \in F_B$, there exists a $\varphi \in M_B$ with $\varphi^\theta = f$. If Φ is the system of k_B functions φ obtained from F in this manner,

$$\det\,(\varphi(K)) \not\equiv 0 \pmod{\mathfrak{p}}; \qquad (\varphi \in \Phi, \quad K \in \mathfrak{K}).$$

It follows that we can find coefficients $c_K \in \mathfrak{o}$ such that

$$\varphi(K_0) = \sum\nolimits_K c_K \, \varphi(K); \qquad (K \in \mathfrak{K})$$

for each $\varphi \in \Phi$. Since the k_B functions φ are certainly linearly independent and belong to W in (2.8), they form a Ξ_0-basis of W and hence

$$\omega_j(K_0) = \sum\nolimits_K c_K \, \omega_j(K); \qquad (K \in \mathfrak{K})$$

for each ω_j associated with B. Now (2E) shows that this result remains valid, if we replace c_K by 0 for all $K \in \mathfrak{K}$ which do not belong to the section of K_0.

The relation (2.11) remains valid if ω_j is replaced by an arbitrary element φ of W. In particular, we may take $\varphi \in M_B$. Now (2.12) is immediate from (2D).

The following result has been observed by M. Osima and K. Iizuka

(2G) *Let B be a block of G. There exists a unique idempotent $\varepsilon_B \in Z(\mathfrak{o}[G])$ such that $\omega_i(\varepsilon_B) = 1$ or 0 according as to whether or not ω_j is associated with B. If K_0 is a fixed conjugate class, we have formulas*

$$(\mathbb{S}K_0)\varepsilon_B = \sum\nolimits_K a_K(\mathbb{S}K); \qquad (K \in \mathcal{C}\ell(G)) \tag{2.13}$$

with $a_K \in \mathfrak{o}$. If K_0 belongs to the p-section $\mathfrak{S}(\pi)$, here $a_K = 0$ for all K not contained in $\mathfrak{S}(\pi)$.

Proof. As shown in [1, II, §4] there exists an idempotent $\varepsilon_B \in Z(\mathfrak{o}[G])$ for which $\omega_j(\varepsilon_B)$ has the values 1 or 0 as indicated. It is clear that ε_B is unique. Then for each $K_0 \in \mathcal{C}\ell(G)$, we have an equation (2.13) with $a_K \in \mathfrak{o}$. This implies that

$$\sum\nolimits_K a_K\, \omega_j(K) = \omega_j(K_0)$$

if ω_j is associated with B while in the other case the sum is 0. In either case, (2E) shows that

$$\sum\nolimits_K a_K\, \omega_j(K) = 0; \qquad (K \in \mathcal{C}\ell(G), \quad K \nsubseteq \mathfrak{S}(\pi)).$$

Since this holds for $j = 1, 2, \cdots, k(G)$, we have $a_K = 0$ for all K not contained in $\mathfrak{S}(\pi)$, Q.E.D.

The map θ of $Z(\mathfrak{o}[G])$ onto $Z(\Omega_0[G])$ clearly maps ε_B onto the idempotent $\eta_B \in B$. Hence

(2H) *Let B be a block of G. Let K_0 be a fixed conjugate class. There exist elements $c_K \in \Omega$ such that*

$$(\mathcal{S}K_0)\eta_B = \sum\nolimits_K c_K\, (\mathcal{S}K).$$

where K ranges over those conjugate classes which are contained in the section of K_0.

3. Selection of sets of conjugate classes for the blocks

(3A) *For each block B of G, we can select a set $\mathfrak{K}_B$ of k_B conjugate classes of G and a set X_B of k_B elements of F_B, denoted by h_K with $K \in \mathfrak{K}_B$, such that:*

(i) *The set $\mathcal{C}\ell(G)$ is the disjoint union of the sets $\mathfrak{K}_B$ with $B \in \mathcal{B}\ell(G)$.*

(ii) *The set X_B is a basis of F_B.*

(iii) *If $Q \in \mathcal{P}(G)$ and if $\mathfrak{K}_B(Q)$ is the subset of $\mathfrak{K}_B$ consisting of those classes with defect group Q, each h_K with $K \in \mathfrak{K}_B(Q)$ has the form $h_K = \varphi^\lambda$ where $\varphi \in \hat{Z}(N_G(Q))$ and where λ is the operator in (2.3) with $T = C_G(Q)$, $H = N_G(Q)$.*

(iv) *$h_K(K) = 1$; $h_K(K') = 0$ for $K, K' \in \mathfrak{K}_B(Q)$ and $K \neq K'$.*

Proof. Consider a fixed $Q \in \mathcal{P}(G)$ and set $H = N_G(Q)$. For each $b \in \mathcal{B}\ell(H)$, let F_b be the subspace of $\hat{Z}(H)$ defined in a manner analogous to the definition of F_B in $\hat{Z}(G)$. Let Y_b denote a basis of F_b.

If $B \in \mathcal{B}\ell(G)$, denote by B_H the set of blocks b of H with $b^G = B$ and let Y_B be the union of the Y_b for these b. Since

$$\hat{Z}(H) = \oplus \sum\nolimits_b F_b\,; \qquad (b \in \mathcal{B}\ell(H)),$$

the union Y of the sets Y_B for all $B \in \mathcal{B}\ell(G)$ is a basis of $\hat{Z}(H)$. Hence

(3.1) $$\det\,(\varphi(L)) \neq 0; \qquad (\varphi \in Y, \quad L \in \mathcal{C}\ell(H)).$$

It follows from (3.1) that, for each $B \in \mathcal{B}\ell(G)$, we can select a subset $\mathfrak{L}_B$ of $\mathcal{C}\ell(H)$ such that

(3.2) $$\mathcal{C}\ell(H) = \bigcup\nolimits_B \mathfrak{L}_B \text{ (disjoint)}; \qquad (B \in \mathcal{B}\ell(G))$$

and that $|\mathfrak{L}_B| = |Y_B|$ and

(3.3) $$\det(\varphi(L)) \neq 0; \qquad (\varphi \in Y_B, \quad L \in \mathfrak{L}_B).$$

For $|Y_B| = |\mathfrak{L}_B| = 0$, the determinant in (3.3) is 1 by definition and (3.3) is always satisfied.

Let $\mathfrak{L}_B(Q)$ denote the subset of $\mathfrak{L}_B$ consisting of the classes in $\mathfrak{L}_B$ with the defect group Q in H. If follows from (3.3) that we can find a subset $Y_B(Q)$ of Y_B with $|Y_B(Q)| = |\mathfrak{L}_B(Q)|$ such that

(3.4) $$\det(\varphi(L)) \neq 0; \qquad (\varphi \in Y_B(Q), \quad L \in \mathfrak{L}_B(Q)).$$

It is an immediate consequence of Sylow's theorems that if L is a conjugate class of $H = N_G(Q)$ with the defect group Q in H, then the conjugate class L^G of G which contains L has defect group Q in G. Conversely, every conjugate class K of G with defect group Q is obtained in this fashion; the corresponding class L of H is uniquely determined; $L = K \cap C_G(Q)$. Let $Y_B(Q)^\lambda$ denote the set of functions φ^λ with $\varphi \in Y_B(L)$ and with λ defined in (2.3), with $T = C_G(Q), H = N_G(Q)$. On account of (2A), $Y_B(Q)^\lambda$ is a subset of F_B. Let $\mathfrak{K}_B(Q)$ denote the set of classes L^G with $L \in \mathfrak{L}_B(Q)$. Then each class in $\mathfrak{K}_B(Q)$ has defect group Q. Moreover, for $\varphi \in Y_B(Q)$ and $K = L^G$ with $L \in \mathfrak{L}_B(Q)$, by (2.3)

$$\varphi^\lambda(K) = \varphi(\mathcal{S}(K \cap C_G(Q))) = \varphi(L).$$

Hence (3.4) implies

$$\det(f(K)) \neq 0; \qquad (f \in Y_B(Q)^\lambda, \quad K \in \mathfrak{K}_B(Q)).$$

It is now clear that we can find linear combinations h_K of the elements of $Y_B(Q)^\lambda$ which satisfy the conditions (iv) in (3A). If $\mathfrak{K}_B$ is the union of the sets $\mathfrak{K}_B(Q)$ for all $Q \in \mathcal{P}(G)$, then condition (iii) is likewise satisfied. For each $K \in \mathfrak{K}_B$, the function h_K belongs to F_B.

If K is any class of G and if Q is the defect group, then by (3.2), $L = K \cap C_K(Q)$ belongs to $\mathfrak{L}_B$ for a unique block B. It follows that K belongs to $\mathfrak{K}_B$ for a unique B. Hence condition (i) of (3A) holds.

We show that the set X_B of functions h_K with $K \in \mathfrak{K}_B$ is linearly independent. Suppose we have a non-trivial relation

(3.5) $$\textstyle\sum_K c_K h_K = 0; \qquad (K \in \mathfrak{K}_B)$$

with coefficients $c_K \in \Omega$. Since not all c_K vanish, we can choose a group $P \in \mathcal{P}(G)$ such that $c_K \neq 0$ for some $K \in \mathfrak{K}_B(P)$ while we have $c_K = 0$ for all $K \in \mathfrak{K}_B$ whose defect group D_K has smaller order than $|P|$.

Take $K' \in \mathfrak{K}_B(P)$. Then K' has defect group P. Consider a term $c_K h_K$ in (3.5). If here $K \in \mathfrak{K}_B(Q)$ with $Q \in \mathcal{P}(G)$, by (iii) and (2B), we have $h_K(K') = 0$ except when $P \geq Q$. If $P > Q$, by construction $c_K = 0$. It follows from (3.5) that, for $K' \in K_B(P)$, we have

$$\textstyle\sum_K c_K h_K(K') = 0$$

where K ranges over the classes in $\mathfrak{K}_B$ with the defect group $Q = P$. These are the $K \in \mathfrak{K}_B(P)$. It now follows from (iv) that $c_K = 0$ for all $K \in \mathfrak{K}_B(P)$, a contradiction.

Hence the set $X_B = \{h_K\}$ is linearly independent. This implies

$$| \mathfrak{K}_B | = | X_B | \leqq \dim_\Omega F_B = k_B .$$

If we add here over all $B \in \mathcal{B}\ell(G)$, the sum on the left is $k(G)$ by (i). Since the sum on the right is also $k(G)$ by (1.2), we must have equality for each B. Hence X_B is a basis of F_B. This proves (ii) and the proof of (3A) is complete.

(3B) *Let* $\mathfrak{K}_B$ *be chosen as in* (3A). *There exists a basis* $\{f_K\}$ *of* F_B *with* K *ranging over* $\mathfrak{K}_B$ *with the following properties*

$$f_K(K) = 1; f_K(K') = 0 \quad \text{for} \quad K, K' \in \mathfrak{K}_B , K \neq K'.$$

Moreover, if $f_K(K^*) \neq 0$ for some $K^* \in \mathcal{C}\ell(G)$, then $D_{K*} \geq D_K$.

Proof. Let $Q \in \mathcal{P}(G)$. Suppose that f_K has already been obtained for all $K \in \mathfrak{K}_B(P)$ with $P \in \mathcal{P}(G)$ and $P > Q$. Suppose now that $K \in \mathfrak{K}_B(Q)$ and set

$$(3.6) \qquad f_K = h_K - \textstyle\sum_{K_1} h_K(K_1)f_{K_1} ; \qquad (K_1 \in \mathfrak{K}_B , D_{K_1} > Q).$$

Here, f_{K_1} is assumed to be defined. If $f_K(K^*) \neq 0$ for $K^* \in \mathcal{C}\ell(G)$, then $h_K(K^*) \neq 0$ or $f_{K_1}(K^*) \neq 0$ for some $K_1 \in \mathfrak{K}_B$ with $D_{K_1} > Q$. In the latter case, by assumption $D_{K*} \geq D_{K_1}$ and hence $D_{K*} \geq Q$. In the former case, by (3A) (iii) and (2B), $D_{K*} \geq Q$. This shows that f_K has the last property in (3B).

Suppose now that $K' \in \mathfrak{K}_B$. If $D_{K'} > Q$ then K' is one of the K_1 in (3.6) and we see that $f_K(K') = 0$. If $D_{K'} = Q$, then K' is not one of the K_1 and (3.6) yields

$$f_K(K') = h_K(K').$$

Now (3A) (iv) shows that $f_K(K') = 0$ for $K' \neq K$ and that $f_K(K) = 1$. Finally, for the remaining $K' \in \mathfrak{K}_B$, we have $f_K(K') = 0$ since otherwise as shown above $D_{K'} \geq Q$.

Applying this successively for all $Q \in \mathcal{P}(G)$ we obtain the required system $\{f_K\}$. Since $\{h_K\}$ was a basis of F_B, so is $\{f_K\}$.

If for the local subgroups $H = N_G(P)$ with $P \in \mathcal{P}(G)$, $P \neq 1$, we know a basis of F_b with $b \in \mathcal{B}\ell(H)$, we can construct the functions f_K except for the $K \in \mathfrak{K}_B$ with $D_K = 1$.

(3C) *Let* B *be a block ideal of* $Z(G)$ *and set*

$$B^* = \oplus \textstyle\sum_{B_1} B_1 , \qquad (B_1 \in \mathcal{B}\ell(G), \quad B_1 \neq B).$$

For each $K^* \in \mathcal{C}\ell(G)$, $K^* \notin \mathfrak{K}_B$ *form the element*

$$\zeta_{K*} = \mathcal{S}K^* - \textstyle\sum_K f_K(K^*)\mathcal{S}K; \qquad (K \in \mathfrak{K}_B).$$

These elements form a basis of B^*.

Proof. It is clear that all f_K with $K \in \mathfrak{K}_B$ vanish for the elements ζ_{K^*} and this implies $\zeta_{K^*} \in B^*$. It is clear that the elements ζ_{K^*} are linearly independent and since the number of these elements is equal to $\dim_\Omega B^*$, they form an Ω-basis of B^*.

Remark. The construction in (3A), (3B) can be performed in the case when we have a partition

$$\mathcal{B}\ell(G) = \bigcup B \quad \text{(disjoint)}$$

where each B is a union of blocks. In particular, if we take

$$\mathcal{B}\ell(G) = B \cup B^*$$

with B and B^* as in (3C) and interchange the roles of B and B^*, we obtain an Ω-basis of B.

It should be mentioned that the selection $\mathfrak{K}_B$ of sets of classes for the blocks in (3A) is not uniquely determined.

4. The lower defect groups of a block

The system $\mathfrak{D}_B$ of lower defect groups of a block has been defined in the introduction. We show

(4A) *If $\mathfrak{K}_B$ is as in* (3A), *the system $\mathfrak{D}_B$ of lower defect groups of the block B coincides exactly with the system of defect groups of the k_B classes $K \in \mathfrak{K}_B$.*

Proof. We have to show that for $P \in \mathcal{P}(G)$, the multiplicity $m_B(P)$ of P in $\mathfrak{D}_B$ (cf. §1) is equal to $|\mathfrak{K}_B(P)| = k_B(P)$. Let V_0 denote the subspace of F_B spanned by the $k_B(P)$ functions f_K with $K \in \mathfrak{K}_B(P)$. It is clear from (3B) that V_0 has dimension $k_B(P)$ and that for $v \neq 0$ in V_0, there exist classes K with $D_K = P$ such that $v(K) \neq 0$. We may even choose $K \in \mathfrak{K}_B(P)$. Moreover, if $K^* \in \mathcal{C}\ell(G)$ and if $v(K^*) \neq 0$, then $f_K(K^*) \neq 0$ for some $K \in \mathfrak{K}_B(P)$ and then, by (3B), $D_{K^*} \geq P$. In particular, $|D_{K^*}| \geqq |P|$. This shows that V_0 has the properties (i) and (ii) required in the definition of $m_B(P)$ in §1 of subspaces V of F_B and hence $k_B(P) \leqq m_B(P)$.

Conversely, let V be any subspace of F_B with these properties (i), (ii), §1. Express $v \in V$ by the basis $\{f_K\}$ of F_B in (3B),

$$v = \textstyle\sum_K a_K f_K\,; \qquad (K \in \mathfrak{K}_B), \quad a_K \in \Omega.$$

Here $a_K = v(K)$ for $K \in \mathfrak{K}_B$. For any $K^* \in \mathcal{C}\ell(G)$, then

$$v(K^*) = \textstyle\sum_K v(K) f_K(K^*); \qquad (K \in \mathfrak{K}_B). \tag{4.1}$$

Because of the property §1, (ii) of V, it suffices to let K range over the classes for which $|D_K| \geqq |P|$.

If $v \neq 0$, then by §1, (i), we can choose K^* with the defect group P such that $v(K^*) \neq 0$. By (3B), $f_K(K^*) = 0$ in (4.1) except when $P \geq D_K$. It follows that there exist $K \in \mathfrak{K}_B$ with the defect group P for which $v(K) \neq 0$. Since $K \in \mathfrak{K}_B(P)$ and $|\mathfrak{K}_B(P)| = k_B(P)$, this implies that the dimension of

V is at most equal to $k_B(P)$. Hence $m_B(P) \leqq k_B(P)$. We then have equality and the proof is complete.

In particular, the numbers $| \mathfrak{K}_B(P) |$ in (3A) do not depend on the choice of $\mathfrak{K}_B$. As a corollary of (4A), we mention

(4B) *The number k_B of irreducible characters χ_i of G in the block B is given by*

$$k_B = \sum_P m_B(P); \qquad P \in \mathcal{P}(G). \tag{4.2}$$

For each P, the sum

$$\sum_B m_B(P); \qquad (B \in \mathcal{B}\ell(G)) \tag{4.3}$$

represents the number of conjugate classes of G with defect group P.

A re-examination of the proof of (3A) yields

(4C) *For any $B \in \mathcal{B}\ell(G)$ and any $Q \in \mathcal{P}(G)$*

$$m_B(Q) = \sum_b m_b(Q) \tag{4.4}$$

where b ranges over the blocks of $H = N_G(Q)$ with $b^G = B$.

Proof. It follows from (3.3) that, for each $B \in \mathcal{B}\ell(G)$ and each $b \in B_H$, we can find subsets Y_b of Y_B and $\mathfrak{L}_b$ of $\mathfrak{L}_B$ with $| Y_b | = | \mathfrak{L}_b |$ such that Y_B is the disjoint union of the Y_b, that $\mathfrak{L}_B$ is the disjoint union of the $\mathfrak{L}_b$ with b ranging over B_H and that for each b

$$\det (\varphi(L)) \neq 0; \qquad (\varphi \in Y_b, \quad L \in \mathfrak{L}_b).$$

We apply (3A) to the group $H = N_G(Q)$ instead of G. Let $\mathfrak{L}_b(Q)$ denote the set of those $L \in \mathfrak{L}_b$ which have defect group Q in H. Since $L^H = L$, we see that $\mathfrak{L}_b(Q)$ has the same significance for H and b as $\mathfrak{K}_B(Q)$ has for G and B. Hence by (3A)

$$| \mathfrak{L}_b(Q) | = m_b(Q).$$

Since $\mathfrak{L}_B(Q)$ in §3 is the disjoint union of the sets $\mathfrak{L}_b(Q)$ with $b \in B_H$ and since

$$| \mathfrak{K}_B(Q) | = | \mathfrak{L}_B(Q) |$$

cf. §3, (4.4) now is evident.

(4D) *The defect group D of B (in the sense of* [1]*) occurs in $\mathfrak{D}_B$. It is the unique maximal element of $\mathfrak{D}_B$ in the partial ordering of $\mathcal{P}(G)$.*

Proof. The algebra homomorphism ω_B in F_B (cf. §1) vanishes for all $K \in \mathcal{C}\ell(G)$ with $| D_K | < | D |$, but not for all K with $D_K = D$, [1, I, §8]. Hence $D \in \mathfrak{D}_B$.

On the other hand, if $P \in \mathfrak{D}_B$, there exist blocks b of $H = N_G(P)$ with $b^G = B$. Let d be a defect group of b in the sense of [1]. Since $P \lhd H$, then $P \subseteq d$, [1, I, (9F)] and d is conjugate in G to a subgroup of D, [1, II (2B)]. Hence $D \geq P$ as stated.

If $d = P$, then $D = P$ by [1]. If $P \subset d$, there exist blocks b_0 of $N_H(d)$ with

$b_0^H = b$ and then $b_0^G = B$. Hence we have

(4E) *If B and D are as in* (4D) *and if P is a lower defect group of B with $P \neq D$, there exists a p-subgroup d of G with*

$$P \subset d \subseteq N_G(P)$$

and a block b_0 of $N_G(P) \cap N_G(d)$ with $b_0^G = B$.

We finally prove an extension of (4A).

(4F) *Suppose that for each block B of G we have a subset $\mathfrak{K}_B^*$ of $\mathcal{C}\ell(G)$ such that*

(i) *each $K \in \mathcal{C}\ell(G)$ belongs to at least one $\mathfrak{K}_B^*$.*

(ii)) *If $|\mathfrak{K}_B^*| = k_B^*$, there exists a subset U_B of F_B with $|U_B| = |k_B^*|$* and

$$\det(h(K)) \neq 0; \qquad (h \in U_B, \quad K \in \mathfrak{K}_B^*). \tag{4.5}$$

Then $k_B^* = k_B$ and exactly $m_B(Q)$ classes of $\mathfrak{K}_B^*$ have defect group Q; $(Q \in \mathcal{P}(G))$.

Proof. It follows from (i) that

$$\textstyle\sum_B k_B^* \geqq k(G) = \sum_B k_B; \qquad (B \in \mathcal{B}\ell(G)).$$

On the other hand, (ii) implies that

$$k_B^* \leqq \dim F_B = k_B.$$

If we add over B, we conclude that $k_B^* = k_B$. Each $K \in \mathcal{C}\ell(G)$ belongs to exactly one $\mathfrak{K}_B^*$.

For any $Q \in \mathcal{P}(G)$, let $r_B(Q)$ denote the number of $K \in \mathfrak{K}_B^*$ with the defect group Q. Then

$$\textstyle\sum_B r_B(Q) = \sum_B m_B(Q); \qquad B \in \mathcal{B}\ell(G), \tag{4.6}$$

since on both sides, we have the number of conjugate classes of G with defect group Q.

If $r_B(Q) \neq m_B(Q)$ for some B and Q, choose a Q of maximal order for which this happens. On account of (4.6), we can then choose B such that

$$r_B(Q) < m_B(Q). \tag{4.7}$$

If $\{f_K\}$ has the same significance as in (3B), it follows from the assumption (ii) and $k_B^* = k_B$ that

$$\det(f_K(K^*)) \neq 0; \qquad (K \in \mathfrak{K}_B, K^* \in \mathfrak{K}_B^*). \tag{4.8}$$

Consider here the rows for which $D_K \succeq Q$. By (3B) then $f_K(K^*) = 0$ except when

$$D_{K^*} \succeq D_K \succeq Q. \tag{4.9}$$

The number of rows in question is

$$R = \textstyle\sum_P m_B(P); \qquad (P \in \mathcal{P}(G), \quad P \succeq Q).$$

As shown by (4.9), the non-zero coefficients in the rows occur in

$$C = \sum_P r_B(P); \qquad (P \in \mathcal{P}(G), \quad P \geq Q)$$

columns. Our choice of Q implies that $m_B(P) = r_B(P)$ for $|P| > |Q|$. By (4.7), $R > C$. But this is inconsistent with (4.8) and (4F) is proved.

5. The ideals I_Q of $Z(G)$

We shall give another characterization of the multiplicity $m_B(Q)$ of $Q \in \mathcal{P}(G)$ as lower defect groups of the block B. We first note

(5A) *Let $Q \in \mathcal{P}(G)$. Let K range over the conjugate classes of G which do not meet $T = C(Q)$. The corresponding class sums $\mathcal{S}K$ form the basis of an ideal I_Q of $Z(G)$.*

This is immediate since I_Q is the kernel of the homomorphism μ in (2.2) of $Z(G)$ into $Z(H)$; $H = N_G(Q)$, $T = C_G(Q)$.

(5B) *Let $B \in \mathcal{B}\ell(G)$. If I_Q is as in* (5A),

(5.1) $$\dim_\Omega (B \cap I_Q) = \sum{}' m_B(P)$$

where P in the sum ranges over the members of $\mathcal{P}(G)$ which do not contain a conjugate of Q.

Proof. Consider B as an algebra over Ω. Then $I = B \cap I_Q$ is an ideal of B. Let R denote the representation of B belonging to the B-module B/I. We then have $R(\zeta) = 0$ for $\zeta \in I$. Conversely, if $\zeta \in B$ and $R(\zeta) = 0$, then $B\zeta \subseteq I$. Since B has a unit element η_B, this implies $\zeta \in I$. Hence R has the kernel I.

Choose an Ω-basis of B/I and write R in matrix form. Each coefficient of R considered as a function of a variable element of B can be viewed as an element of the dual space $\hat{B}$ of B. Let W denote the subspace of $\hat{B}$ spanned by the different coefficients of R. Since R has the kernel I, we have

(5.2) $$\dim_\Omega W = \dim_\Omega(B/I) = k_B - \dim_\Omega I.$$

If $w \in W \subseteq \hat{B}$, we can consider w as an element of F_B. Then

$$w(\zeta) = w(\eta_B \zeta)$$

for $\zeta \in Z$. Express w by the basis $\{f_K\}$ in (3B),

(5.3) $$w = \sum_K w(K) f_K; \qquad (K \in \mathfrak{K}_B).$$

If here $\mathcal{S}K \in I_Q$, then $\eta_B(\mathcal{S}K) \in I$ and

$$w(K) = w(\eta_B(\mathcal{S}K)) = 0.$$

Therefore, it suffices to let K in (5.3) range over those elements K of $\mathfrak{K}_B$ which meet T. These are the K for which $D_K \geq Q$. Then

$$\dim_\Omega W \leqq \sum_P m_B(P); \qquad (P \in \mathcal{P}(G), \quad P \geq Q)$$

since the sum of the right represents the number of K in (5.3). By (5.2) and (5.3),

$$k_B - \sum_P m_B(P) \leqq \dim_\Omega I; \qquad (P \in \mathcal{P}(G), \quad P \geq Q).$$

Here the left side is equal to the sum in (5.1), cf. (4.2). Thus

$$(5.4) \qquad \sum\nolimits_P' m_B(P) \leqq \dim_\Omega I = \dim_\Omega(I_Q \cap B).$$

Add here over all $B \in \mathcal{B}\ell(G)$. On the left, we obtain the number of $K \in \mathcal{C}\ell(G)$ whose defect group does not contain a conjugate of Q, cf. (4.3). By (5A), this is the dimension of I_Q. Since I_Q is the direct sum of the $I_Q \cap B$ for the different blocks, we have equality after adding (5.4) and hence equality in (5.4), Q.E.D.

It is clear from (5.1) that, for each $P \in \mathcal{P}(G)$, $m_B(P)$ can be expressed by the dimensions of the ideals $B \cap I_Q$ for suitable $Q \in \mathcal{P}(G)$ if k_B is known.

6. The p-sections of G

The p-sections of a group have been defined in §1. Each $\zeta \in Z = Z(G)$ has a unique representation

$$(6.1) \qquad \zeta = \sum\nolimits_K a_K(\mathcal{S}K); \qquad (K \in \mathcal{C}\ell(G)).$$

If π is a p-element of G, let $\zeta^{(\pi)}$ denote the sum of the terms in (6.1) for which K belongs to the section $\mathfrak{S}(\pi)$ of π. Then

$$(6.2) \qquad \zeta = \sum\nolimits_\pi \zeta^{(\pi)}; \qquad (\pi \in \Pi).$$

We note

(6A) *If ζ belongs to the block B of G, each $\zeta^{(\pi)}$ in* (6.2) *does.*

Indeed, since $\zeta = \eta_B \zeta$, we have

$$\zeta = \sum\nolimits_\pi \eta_B \zeta^{(\pi)}; \qquad (\pi \in \Pi).$$

On account of (2H), each $\eta_B \zeta^{(\pi)}$ is a linear combination of class sums $\mathcal{S}K$ with $K \subseteq \mathfrak{S}(\pi)$. On comparing this with (6.2), we find

$$\zeta^{(\pi)} = \eta_B \zeta^{(\pi)} \in B.$$

We shall say that an element f of $\hat{Z}$ is *sectional* and is *associated with the section* $\mathfrak{S}(\pi)$, if $f(K) = 0$ for all $K \subseteq \mathcal{C}\ell(G)$ which are not contained in $\mathfrak{S}(\pi)$.

(6B) *The functions f_K in* (3B) *are sectional.*

Proof. Let K_0 be a fixed conjugate class, say $K_0 \subseteq \mathfrak{S}(\pi_0)$. Apply (2F) with $\mathfrak{K} = \mathfrak{K}_B$ and with F consisting of the f_K with $K \in \mathfrak{K}_B$. It follows that for $K \in \mathfrak{K}_B$ there exists elements $c_K^\theta \in \Omega$ such that, for every $f \in F_B$,

$$f(K_0) = \sum\nolimits_K c_K^\theta f(K); \qquad (K \in \mathfrak{K}_B)$$

and that $c_K^\theta = 0$ if K is contained in a section $\mathfrak{S}(\pi) \neq \mathfrak{S}(\pi_0)$. Taking $f = f_K$ with $K \subseteq \mathfrak{S}(\pi)$, we see that $f_K(K_0) = 0$ as we wished to show.

We shall denote by $F_B^{(\pi)}$ the subset of F_B consisting of the functions $f \in F_B$, which are sectional and are associated with $\mathfrak{S}(\pi)$. As a corollary of (6B), we have

(6C) *The space F_B is the direct sum of the subspaces $F_B^{(\pi)}$ with π ranging over* Π.

Indeed, the functions f_K wiht $K \in \mathfrak{K}_B$ form a basis of F_B. Here f_K is sectional and, if $K \subseteq \mathfrak{S}(\pi)$, then as $f_K(K) = 1, f_K$ is associated with $\mathfrak{S}(\pi)$ and hence $f_K \in F_B^{(\pi)}$.

We now replace F_B by $F_B^{(\pi)}$ in the definition of $m_B(P)$ in §1. For given $B \in \mathcal{B}\ell(G)$, $P \in \mathcal{P}(G)$ and $\pi \in \Pi$, we consider subspaces V of $F_B^{(\pi)}$ such that every $f \neq 0$ in V has the following properties:

(i) There exist conjugate classes K with the defect group P such that $f(K) \neq 0$.

(ii) We have $f(K) = 0$ for every $K \in \mathcal{C}\ell(G)$ for which $|D_K| < |P|$.

Of course, it will suffice here to consider only classes $K \subseteq \mathfrak{S}(\pi)$.

We denote by $m_B^{(\pi)}(P)$ the maximal dimension of a space V with the properties (i), (ii). Let $\mathfrak{D}_B^{(\pi)}$ be the system of groups consisting of the groups $P \in \mathcal{P}(G)$ with each P taken with the multiplicity $m_B^{(\pi)}(P)$. Now a proof quite analogous to that of (4A) yields

(6D) *Let $\mathfrak{K}_B$ be as in* (3A) *and let $\mathfrak{K}_B^{(\pi)}$ be the subset consisting of the $K \in \mathfrak{K}_B$ which are contained in the section $\mathfrak{S}(\pi)$, $\pi \in \Pi$. Then $\mathfrak{D}_B^{(\pi)}$ consists exactly of the defect group D_K of the $K \in \mathfrak{K}_B^{(\pi)}$.*

We shall say that $\mathfrak{D}_B^{(\pi)}$ is the system of lower defect groups of B *associated with the section* $\mathfrak{S}(\pi)$. Now (6D) yields the following:

(6E) *The system $\mathfrak{D}_B$ is the union of the systems $\mathfrak{D}_B^{(\pi)}$ for all $\pi \in \Pi$. In other words*

$$m_B(P) = \textstyle\sum_\pi m_B^{(\pi)}(P); \qquad (\pi \in \Pi) \tag{6.3}$$

for each $P \in \mathcal{P}(G)$. If we set $|\mathfrak{K}_B^{(\pi)}| = l_B^{(\pi)}$, we have

$$l_B^{(\pi)} = \textstyle\sum_P m_B^{(\pi)}(P); \qquad (P \in \mathcal{P}(G)). \tag{6.4}$$

The number of conjugate classes $K \subseteq \mathfrak{S}(\pi)$ with a given defect group P is given by

$$\textstyle\sum_B m_B^{(\pi)}(P); \qquad (B \in \mathcal{B}\ell(G)). \tag{6.5}$$

We can also extend (4F).

(6F) *Suppose that for each $B \in \mathcal{B}\ell(G)$ and each $\pi \in \Pi$, we have a set $\mathfrak{K}_B^*$ of conjugate classes $K \subseteq \mathfrak{S}(\pi)$ such that each $K \in \mathcal{C}\ell(G)$ contained in the section $\mathfrak{S}(\pi)$ belongs to $\mathfrak{K}_B^*$ for some B. Suppose further that for each B, we have* *a subset* U_B *of* $F_B^{(\pi)}$ with $|U_B| = |\mathfrak{K}_B^*|$ such that

$$\det(h(K)) \neq 0; \qquad (h \in U_B, \quad K \in \mathfrak{K}_B^*).$$

Then $|\mathfrak{K}_B^*| = l_B^{(\pi)}$ and exactly $m_B^{(\pi)}(Q)$ classes $K \in \mathfrak{K}_B^*$ have defect group Q; $(Q \in \mathcal{P}(G))$.

This is shown by the same method as (4F) considering only classes contained in the section $\mathfrak{S}(\pi)$ and replacing $m_B(Q)$ by $m_B^{(\pi)}(Q)$ and F_B by $F_B^{(\pi)}$.

Our next result shows that the numbers $m_B^{(\pi)}(P)$ can be expressed by the analogous numbers with G replaced by $C_G(\pi)$ and B replaced by blocks of $C_G(\pi)$.

(6G) *Let* $\pi \in \Pi$ *and set* $C = C_G(\pi)$. *For each* $B \in \mathcal{B}\ell(G)$ *and each* $P \in \mathcal{P}(G)$, *we have*

$$m_B^{(\pi)}(P) = \textstyle\sum_Q \sum_b m_b^{(\pi)}(Q) \tag{6.6}$$

where Q *ranges over those groups in* $\mathcal{P}(C)$ *which are conjugate to* P *in* G *and where* b *ranges over the set* B_C *of blocks* b *of* C *with* $b^G = B$.

Proof. We apply here the method of §2, 1 with $Q = \langle \pi \rangle$, $T = H = C$. Let $\mathfrak{S}_C(\pi)$ denote the p-section of π in C. If $K \in \mathcal{C}\ell(G)$ and $K \subseteq \mathfrak{S}(\pi)$, set

$$K \cap \mathfrak{S}_C(\pi) = L. \tag{6.7}$$

Then L is not empty and L consists of elements of the form $\pi\rho$ with p-regular $\rho \in C$. Any two elements of L are conjugate in G. It follows from their form that they are conjugate in C. It is now evident that L is a conjugate class of C; $L \subseteq \mathfrak{S}_C(\pi)$. Conversely, if $L \in \mathcal{C}\ell(C)$ and $L \subseteq \mathfrak{S}_C(\pi)$, the conjugate class $K = L^G$ of G containing L belongs to $\mathfrak{S}(\pi)$ and satisfies (6.7). Hence we have a one-to-one correspondence between the set of $K \in \mathcal{C}\ell(G)$ with $K \subseteq \mathfrak{S}(\pi)$ and the set of $L \in \mathcal{C}\ell(C)$ with $L \subseteq \mathfrak{S}_C(\pi)$. Moreover, if $Q \in \mathcal{P}(C)$ is the defect group of L in C, the defect group P of $K = L^G$ in G is conjugate to Q.

We now use a method similar to that used in the proof of (3A). If $b \in \mathcal{B}\ell(C)$, let F_b be the subspace of the dual space $\hat{Z}(C)$ of $Z(C)$ defined in a manner analogous to the definition of F_B in $\hat{Z}(G)$. Let Y_b denote a basis of F_b consisting of sectional functions, (cf. (6C)). Let Y_B be the union of all Y_b with $b^G = B$ and let Y be the union of Y_B for all $B \in \mathcal{B}\ell(G)$. Then Y is a basis of $\hat{Z}(C)$ and hence

$$\det(\varphi(L)) \neq 0; \qquad (\varphi \in Y, \quad L \in \mathcal{C}\ell(C)).$$

It follows that, for each B, we can select a subset $\mathfrak{L}_B$ of $\mathcal{C}\ell(C)$, such that $\mathcal{C}\ell(C)$ is the disjoint union of the sets $\mathfrak{L}_B$, that $|Y_B| = |\mathfrak{L}_B|$, and that

$$\det(\varphi(L)) \neq 0; \qquad (\varphi \in Y_B, \quad L \in \mathfrak{L}_B).$$

Let $\mathfrak{L}_B^{(\pi)}$ denote the set of all $L \in \mathfrak{L}_B$ such that $L \subseteq \mathfrak{S}_C(\pi)$. We can then find a subset $Y_B^{(\pi)}$ with $| Y_B^{(\pi)} | = | \mathfrak{L}_B^{(\pi)} |$ such that

$$(6.8) \qquad \det(\varphi(L)) \neq 0; \qquad (\varphi \in Y_B^{(\pi)}, \quad L \in \mathfrak{L}_B^{(\pi)}).$$

Since all $\varphi \in Y_B^{(\pi)}$ are sectional, it follows from (6.8) that they are associated with the section $\mathfrak{S}_C(\pi)$. Let $X_B^{(\pi)}$ denote the system of functions φ^λ with $\varphi \in Y_B^{(\pi)}$ and λ defined by (2.3) with $T = H = C$. Each $f = \varphi^\lambda$ belongs to F_B, cf. (2A). If $K \in \mathcal{C}\ell(G)$, by (2.2) and (2.3)

$$f(K) = \varphi(\mathcal{S}(K \cap C)).$$

If K is not contained in $\mathfrak{S}(\pi)$, then K does not meet $\mathfrak{S}_C(\pi)$ and hence $f(K) = 0$. If $K \subseteq \mathfrak{S}(\pi)$, $K \cap C$ is the union of L in (6.7) and of conjugate classes $L^* \in \mathcal{C}\ell(C)$ contained in sections of C different from $\mathfrak{S}_C(\pi)$. Since $\varphi(L^*) = 0$ for these L^*, we find $f(K) = \varphi(L)$.

If $\mathfrak{K}_B^*$ is the set of classes L^G with $L \in \mathfrak{L}_B^{(\pi)}$, we then have

$$\det(f(K)) \neq 0; \qquad (f \in X_B^{(\pi)} \quad K \in \mathfrak{K}_B^*).$$

Since every $L \in \mathcal{C}\ell(C)$ with $L \subseteq \mathfrak{S}_C(\pi)$ belongs to $\mathfrak{L}_B^{(\pi)}$ for some $B \in \mathcal{B}\ell(G)$, every $K \in \mathcal{C}\ell(G)$ with $K \subseteq \mathfrak{S}(\pi)$ belongs to some $\mathfrak{K}_B^*$. We can now apply (6F) and see that exactly $m_B^{(\pi)}(P)$ classes $K \in \mathfrak{K}_B^*$ have defect group P. It follows that exactly $m_B^{(\pi)}(P)$ of the classes $L \in \mathfrak{L}_B^{(\pi)}$ have defect groups Q in C with Q conjugate to P. On the other hand, if we set

$$Y_b^{(\pi)} = Y_B^{(\pi)} \cap Y_b$$

for $b \in B_C$, it follows from (6.8) that we can break up $\mathfrak{L}_B^{(\pi)}$ into subsets $\mathfrak{L}_b^{(\pi)}$ with $b \in B_C$ such that $| Y_b^{(\pi)} | = | L_b^{(\pi)} |$ and

$$\det(\varphi(L)) \neq 0; \qquad (\varphi \in Y_b^{(\pi)}, \quad L \in \mathfrak{L}_b^{(\pi)}).$$

Applying (6F) to C and b, we see that $m_b^{(\pi)}(Q)$ of the classes $L \in \mathfrak{L}_b^{(\pi)}$ have defect group $Q \in \mathcal{P}(C)$. It is now clear that (6.6) holds.

7. The section of the unit element

(7A) *Assume that π is a p-element in the center $Z(G)$ of G. Let B be a block of G. Let $\mathfrak{K}_B$ have the same significance as in* (3A). *As before, let $\mathfrak{K}_B^{(\pi)}$ denote the set of classes $K \in \mathfrak{K}_B$ with $K \subseteq \mathfrak{S}(\pi)$ and set $| \mathfrak{K}_B^{(\pi)} | = l_B^{(\pi)}$. We can find a set X_B of $l_B^{(\pi)}$ irreducible characters χ_i in B such that*

$$(7.1) \qquad \det(\chi_i(\sigma_K)) \not\equiv 0 \pmod{\mathfrak{p}}; \qquad (\chi_i \in X_B, \quad K \in \mathfrak{K}_B^{(\pi)}).$$

Here, $\mathfrak{p}$ has the same meaning as in §2, 2. *Moreover, $l_B^{(\pi)}$ coincides with the number l_B of modular irreducible characters in B. Finally, $\mathfrak{D}_B^{(\pi)} = \mathfrak{D}_B^{(1)}$.*

Proof. As shown in [1, I (5A)] there exists a set X_B of l_B irreducible characters $\chi_i \in B$ and a set $\mathfrak{L}_B^{(1)}$ of l_B conjugate classes $K \subseteq \mathfrak{S}(1)$ such that each conjugate class in $\mathfrak{S}(1)$ belongs to $\mathfrak{L}_B^{(1)}$ for exactly one B, and that for every B

$$\det(\chi(\sigma_K)) \not\equiv 0 \pmod{\mathfrak{p}}; \qquad (\chi \in X_B, \quad K \in \mathfrak{L}_B^{(1)}).$$

The section $\mathfrak{S}(1)$ consists of the p-regular elements of G. If K ranges over the classes in $\mathfrak{S}(1)$, πK ranges over the classes in $\mathfrak{S}(\pi)$. Let $\mathfrak{L}_B^{(\pi)}$ denote the set of classes πK with $K \in \mathfrak{L}_B^{(1)}$. Since

$$\chi(\sigma_K) \equiv \chi(\sigma_{\pi K}) \pmod{\mathfrak{p}}$$

we have

$$\text{(7.2)} \qquad \det(\chi(\sigma_K)) \not\equiv 0 \pmod{\mathfrak{p}}; \qquad (\chi \in X_B, \quad K \in \mathfrak{L}_B^{(\pi)})$$

Let $r_B^{(\pi)}(P)$ denote the number of classes in $\mathfrak{L}_B^{(\pi)}$ with the given defect group $P \in \mathcal{P}(G)$. Then

$$\text{(7.3)} \qquad \sum_B r_B^{(\pi)}(P) = \sum_B m_B^{(\pi)}(B); \qquad (B \in \mathcal{B}\ell(G))$$

since both sides represent the number of classes $K \subseteq \mathfrak{S}(\pi)$ with defect group P, cf. (6.5).

If some class $K \in \mathfrak{L}_B^{(\pi)}$ does not belong to $\mathfrak{R}_B^{(\pi)}$, we try to replace it by a class in $\mathfrak{R}_B^{(\pi)}$ with the same defect group such that the condition (7.2) is preserved after the replacement. We continue in this manner as long as possible.

Assume first that, for every $\pi \in Z(G)$ and for every $B \in \mathcal{B}\ell(G)$, this process only comes to an end when all classes in $\mathfrak{L}_B^{(\pi)}$ have been replaced by classes in $\mathfrak{R}_B^{(\pi)}$. Then, obviously,

$$r_B^{(\pi)}(P) \leqq m_B^{(\pi)}(P)$$

and (7.3) implies that we have equality. This means that we have replaced $\mathfrak{L}_B^{(\pi)}$ by $\mathfrak{R}_B^{(\pi)}$ and hence (7.1) holds. Also,

$$l_B^{(\pi)} = |\,\mathfrak{R}_B^{(\pi)}\,| = |\,\mathfrak{L}_B^{(\pi)}\,| = l_B\,.$$

Since the classes K and πK have the same defect group, we have $r_B^{(\pi)}(P) = r_B^{(1)}(P)$. Since our result above can be applied for $\pi = 1$, we find

$$m_B^{(\pi)}(P) = r_B^{(\pi)}(P) = r_B^{(1)}(P) = m_B^{(1)}(P)$$

Hence $\mathfrak{D}_B^{(\pi)} = \mathfrak{D}_B^{(1)}$, and (7A) holds in the case under discussion.

Assume then that for some $\pi \in Z(G)$ our exchange comes to an end before all classes in $\mathfrak{L}_B^{(\pi)}$ have been replaced. Let H_B denote the set obtained from $\mathfrak{L}_B^{(\pi)}$ when the process terminates. All classes in H_B lie in $\mathfrak{S}(\pi)$. Exactly $r_B^{(\pi)}(P)$ classes in H_B have defect group P, we have $|\,H_B\,| = l_B$ and

$$\text{(7.4)} \qquad \Delta = \det(\chi(\sigma_K)) \not\equiv 0 \pmod{\mathfrak{p}}; \qquad (\chi \in X_B, \quad K \in H_B).$$

Finally, for some B, there exist classes $K_0 \in H_B$ which do not belong to $\mathfrak{R}_B^{(\pi)}$ and which cannot be exchanged with a class in $\mathfrak{R}_B^{(\pi)}$ with the same defect group such that (7.4) is preserved. Choose here B and K_0 such that the defect group Q of K_0 has maximal order.

If $P \in \mathcal{P}(G)$ and $|\,P\,| > |\,Q\,|$, our choice implies that, for every block B_1 and every $K \in H_{B_1}$ with $D_K = P$, we have $K \in \mathfrak{R}_{B_1}^{(\pi)}$. This implies that $r_{B_1}^{(\pi)}(P) \leqq m_{B_1}^{(\pi)}(P)$. On account of (7.3), we have equality. This shows that, for $|P| > |Q|$, the same classes of defect group P occur in H_{B_1} and in $\mathfrak{K}_{B_1}^{(\pi)}$.

We shall now derive a contradiction. It follows from (2F) that there exist elements $c_K \in \mathfrak{o}$ for $K \in \mathfrak{K}_B^{(\pi)}$ such that

$$(7.5) \qquad \omega_j(K_0) = \textstyle\sum_K c_K\, \omega_j(K); \qquad (K \in \mathfrak{K}_B^{(\pi)})$$

for each ω_j associated with B. Then, by (1.4)

$$(7.6) \quad \chi_j(\sigma_{K_0}) = \textstyle\sum_K c_K(\,|K|/|K_0|)\chi_j(\sigma_K); \qquad (K \in \mathfrak{K}_B^{(\pi)}).$$

Let Δ_{K_1} denote the determinant obtained form Δ in (7.4) by replacing the column $\chi(\sigma_{K_0})$ by $\chi(\sigma_{K_1})$ with $K_1 \in \mathfrak{K}_B^{(\pi)}$. On account of (7.6), then

$$(7.7) \qquad \Delta = \textstyle\sum_K c_K(\,|K|/|K_0|\,)\Delta_K\,; \qquad (K \in \mathfrak{K}_B^{(\pi)}).$$

If here K has a defect group D_K with $|D_K| > |Q|$, then as remarked, $K \in H_B$ and hence $\Delta_K = 0$ since two columns are equal. It will therefore suffice to let K range over the classes with $|D_K| \leqq |Q|$. Since $D_{K_0} = Q$, then $|K|/|K_0| \in \mathfrak{o}$. It then follows from (7.7) that there exist classes $K_1 \in \mathfrak{K}_B^{(\pi)}$ for which

$$(7.8) \qquad c_{K_1} \not\equiv 0, \quad |\Delta_{K_1}| \not\equiv 0 \pmod{\mathfrak{p}}, \quad |D_{K_1}| = |Q|.$$

In the notation of §2, 2, the equation (7.5) remains valid if we replace ω_j by an element of M_B. Then (2F) shows that

$$f(K_0) = \textstyle\sum_K c_K^\theta f(K); \qquad (K \in \mathfrak{K}_B^{(\pi)})$$

for any $f \in F_B$. For $f = f_{K_1}$, this yields

$$c_{K_1}^\theta = f_{K_1}(K_0)$$

By (7.8), then $f_{K_1}(K_0) \neq 0$ and by (3B) $Q \succeq D_{K_1}$. Now (7.8) shows that $D_{K_1} = Q$ and that we could have exchanged K_0 with $K_1 \in \mathfrak{K}_B^{(\pi)}$ since both have the same defect group and since (7.4) would be preserved. This is a contradiction and the proof is complete.

If π is an arbitrary p-element of the group G, we can apply (7A) to the group $C = C_G(\pi)$. If $b \in \mathcal{B}\ell(C)$ and if $Q \in \mathcal{P}(C)$, then $\mathfrak{D}_b^{(1)} = \mathfrak{D}_b^{(\pi)}$ and hence $m_b^{(1)}(Q) = m_b^{(\pi)}(Q)$. Now (6G) becomes

(7B) *Let π be a p-element of G and set $C = C_G(\pi)$. For each $B \in \mathcal{B}\ell(G)$ and each $P \in \mathcal{P}(G)$,*

$$(7.9) \qquad m_B^{(\pi)}(P) = \textstyle\sum_Q \sum_b m_b^{(1)}(Q)$$

where Q ranges over the groups in $\mathcal{P}(C)$ which are conjugate to P and where b ranges over the blocks of C with $b^G = B$.

By (7A), $l_b^{(1)} = l_b$ is the number of modular irreducible characters in b. If we add (7.9) over all $P \in \mathcal{P}(G)$, (6E) yields the corollary:

(7C) *If the notation is as in (7B), we have*

(7.10) $$l_B^{(\pi)} = \sum_b l_b$$

where b ranges over the blocks of C with $b^G = B$.

On comparing (7.10) with [1, II, §7] we have

(7D) *The number $l_B^{(\pi)}$ of lower defect groups of B associated with the section $\mathfrak{S}(\pi)$ is equal to the number of modular irreducible characters of $C_G(\pi)$ which belong to B in the sense of* [1, II, §7].

In particular, there are l_B lower defect groups of B associated with the section $S(1)$. On account of (6E), we have

(7E) *The number l_B of modular irreducible characters in the block B is given by*

(7.11) $$l_B = l_B^{(1)} = \sum_P m_B^{(1)}(P); \qquad (P \in \mathcal{P}(G)).$$

The proposition (7A) can be applied for $\pi = 1$ for any G. Hence

(7F) *For every $B \in \mathcal{B}\ell(G)$, we have*

$$\det(\chi(\sigma_K)) \not\equiv 0 \pmod{\mathfrak{p}}; \qquad (\chi \in X_B\,; \;\; K \in \mathfrak{K}_B^{(1)}).$$

Since every p-regular class of G belongs to $\mathfrak{K}_B^{(1)}$ for exactly one B, we can apply the method in (1, I, §5) and obtain

(7G) *Let B be a block. Let $r \geqq 0$ be a rational integer. The multiplicity of p^r as an elementary divisor of the Cartan matrix of B is given by*

$$\sum_P' m_B^{(1)}(P)$$

where P ranges over all groups in $\mathcal{P}(G)$ of order p^r.

This is a refinement of a result stated without proof in [2].

As a consequence of (7G), we have

(7H) *If B has defect group D (in the sense of* [1]) *with D chosen in $\mathcal{P}(G)$), then D occurs exactly once in $\mathfrak{D}_B^{(1)}$.*

The results of [3] show that D occurs in $\mathfrak{D}_B^{(\pi)}$ for $\pi \in \Pi$, if and only if π is conjugate to an element of $Z(D)$; they also allow us to characterize the multiplicity $m_B^{(\pi)}(D)$. For arbitrary $\pi \in \Pi$, we can determine the maximal elements of $\mathfrak{D}_B^{(\pi)}$.

References

1. R. Brauer, *Zur Darstellungstheorie der Gruppen endlicher Ordnung, I, II*, Math. Zeitschrift, vol. 63 (1956), pp. 406–444; vol. 72 (1959), pp. 25–46.
2. ———, *On blocks of characters of groups of finite order, I*, Proc. Nat. Acad. Sci., vol. 32 (1946), pp. 182–186.
3. ———, *On blocks and sections in finite groups, I, II*, Amer. J. Math., vol. 89 (1967) pp. 1115–1135; II to appear.

Harvard University
Cambridge, Massachusetts

www.ingramcontent.com/pod-product-compliance
Lightning Source LLC
Chambersburg PA
CBHW060944210326
41598CB00031B/4714
9780262523837